An Introduction
to Differential Equations
and Linear Algebra

An Introduction to Differential Equations and Linear Algebra

STEPHEN W. GOODE
California State University, Fullerton

Prentice Hall, Englewood Cliffs, New Jersey 07632

Library of Congress Cataloging-in-Publication Data

GOODE, STEPHEN W., (date)
    An introduction to differential equations and linear algebra/
Stephen W. Goode.
        p.   cm.
    Includes index.
    ISBN 0-13-485657-0
    1. Algebras, Linear. 2. Differential equations. I. Title.
QA371.G644   1991   90-21804
512'.5—dc20   CIP

Editorial/production supervision
    and interior design: *Kathleen M. Lafferty*
Cover design: *Wanda Lubelska*
Prepress buyer: *Paula Massenaro*
Manufacturing buyer: *Lori Bulwin*
Acquisitions editor: *Steven R. Comny*

© 1991 by Prentice-Hall, Inc.
A Division of Simon & Schuster
Englewood Cliffs, New Jersey 07632

Printed in the United States of America

10 9 8 7 6 5 4 3 2

ISBN 0-13-485657-0

Prentice-Hall International (UK) Limited, *London*
Prentice-Hall of Australia Pty. Limited, *Sydney*
Prentice-Hall Canada Inc., *Toronto*
Prentice-Hall Hispanoamericana, S.A., *Mexico*
Prentice-Hall of India Private Limited, *New Delhi*
Prentice-Hall of Japan, Inc., *Tokyo*
Simon & Schuster Asia Pte. Ltd., *Singapore*
Editora Prentice-Hall do Brasil, Ltda., *Rio de Janeiro*

To Christina,
for lost time

Contents

Preface

In *An Introduction to Differential Equations and Linear Algebra* the standard material on linear algebra and linear differential equations required in many sophomore courses for mathematics, science, and engineering majors is introduced. I have endeavored to develop an appreciation for the power of the general vector space framework in formulating and solving linear problems. In particular, the theory underlying the solution of linear differential equations is derived very simply as an application of the vector space results. My aim has been to present the material in a manner that is accessible to the student who has successfully completed three semesters of calculus. It is definitely the intention that the student read the text, not just the examples. Almost all the results are proved in detail, and therefore the level of rigor is reasonably high.

The text begins with two chapters on the classical techniques for solving first-order differential equations and some of their applications. It is here that the student gains familiarity with the terminology and notation used in differential equation theory and an appreciation for the types of problems whose mathematical formulation gives rise to differential equations. The advantage of beginning with these chapters is that differential equations can then be used to motivate and to illustrate the more abstract theorems and definitions that form the basis of the linear algebra developed in Chapters 3 to 8. For example, the problem of finding the set of all solutions to the differential equation $y'' + y = 0$ is used to motivate the more abstract idea of the kernel of a linear transformation, T. Then, having proved that the kernel of T is a vector space, we return to differential equations to conclude that the set of all solutions to $y'' + ay' + by = 0$ is a vector space. The question then arises as to the dimension of the solution space. This question is answered in Chapter 9, and the remainder of the text is concerned with introducing techniques for solving linear differential equations and linear systems of differential equations.

For many students linear algebra is their first exposure to abstract mathematics, and they invariably have a very hard time of it. However, the use of differential equations in developing the fundamental ideas can give the applied-oriented reader both the motivation and direction for persevering with the abstractness of the vector space framework.

There is certainly too much material to finish the whole text in one semester, and so the chapters have been structured for maximum flexibility. At Fullerton our fourth-semester linear algebra and differential equations course is structured around Chapters 1 to 9 and 11. This material can be completed fairly easily in one semester. In teaching this course I often cover Chapter 9 (Linear Ordinary Differential Equations) directly after Chapter 7 (Linear Transformations). This enables the student to see a concrete application of the vector space framework to differential equations and also provides a short break in the abstract development. Having completed Chapter 9, I then return to Chapter 8 (Eigenvalues and Eigenvectors) before finishing the course with a discussion of linear systems of differential equations.

There are several chapters that can be covered rather rapidly if the instructor feels there is too much detail. Particular instances are Chapters 1 and 2 (really all that is needed for the remainder of the text are the introductory ideas in Section 1.1 and familiarity with linear differential equations). Also, Chapter 5 contains a fairly detailed account of determinants for this level text. This reflects my own personal feeling that the student obtains a firmer understanding of the idea of a determinant by mastering the classical definition as opposed to the simpler inductive definition. In Chapter 7 the inverse transformation is not needed unless the Laplace transform is to be studied later, and so Section 7.3 may be omitted. Section 8.4 is not required in the remainder of the text, and Section 8.3 is required only if the matrix exponential function is going to be discussed in Chapter 11.

Most of the exercises have been checked using the symbolic computer algebra system Maple that is being developed at the University of Waterloo, Canada. In fact I have used Maple quite extensively in constructing many of the exercise sets. Some of the graphics in Chapter 10 and Section 13.6 were generated by Mathematica. All the other figures were drawn using the graphics software Cricket Draw, and the whole manuscript was produced on an Apple Macintosh computer.

Acknowledgments

The text has been extensively class tested over the passed three years. I would like to thank my colleagues Harriet Edwards, Ted Hromadka, Vyron Klassen, John Mathews, Ron Miller, and Edsel Stiel, who used various versions of the manuscript at Fullerton, and especially Ernie Solheid, who checked the galley proofs thoroughly for mathematical accuracy. Their comments, criticisms, and suggestions have contributed significantly to the final product. Indeed, to a large extent it was the encouragement of Dr. Mathews and Dr. Stiel that provided the initial motivation for the development of this project from a set of supplementary class notes to a full-blown textbook.

I would also like to acknowledge the thoughtful comments of the many people involved in reviewing the several drafts of this text, in particular William L. Briggs, University of Colorado, Denver; Paul W. Britt, Louisiana State University, Baton Rouge; John E. Brown, Purdue University; David Lesley, San Diego State University; and David B. Surowski, Kansas State University. All these comments were considered carefully in the final preparation of the text, and they have been of invaluable help in reinforcing my own feelings as to how the material in this text should be presented.

The person who has made the largest contribution to the accuracy of this text is Walfred Lester. As Visiting Lecturer in the mathematics department at Fullerton during the academic year 1988–1989, Walfred taught from the manuscript and worked all the exercises. While doing so, he uncovered and corrected many of the errors that were present at the time and made several suggestions regarding the presentation of the material. I wish to express my thanks and appreciation for a helpful, lively, and enjoyable interaction.

Thanks are also due to the production editor, Kathleen Lafferty, who has done a superb job in overseeing all aspects of the production of this text.

I owe the greatest debt of gratitude to my wife, Christina Goode. Her continued support and encouragement throughout the development of this project has been a constant source of inspiration, particularly during those times when it seemed as though the manuscript would never be completed. In addition, her careful proofreading has helped to minimize errors and to clarify the explanations in several places. I dedicate this book to her.

Finally I would like to acknowledge the indirect influence of my mentor and friend Professor John Wainwright.

Stephen W. Goode

1

First-Order Differential Equations

1.1   INTRODUCTION TO DIFFERENTIAL EQUATIONS

Differential equations are among the most important equations in applied mathematics. This importance occurs because any problem involving quantities that change continuously in time (or space) can usually be described mathematically by a differential equation. In physics, for example, differential equations arise naturally from Newton's second law of motion, which states that the mass times the acceleration of an object is equal to the applied external force. If we assume that the object has mass m and moves along the y-axis of a Cartesian coordinate system, then the mathematical expression of Newton's second law is the differential equation

$$m\,\frac{d^2y}{dt^2} = F,$$

where F denotes the external force acting on the object. This equation is called a differential equation because it involves *derivatives* of the "unknown" function $y(t)$. In general we have the following:

> **Definition 1.1.1:**   A **differential equation** is an equation involving one or more derivatives of an unknown function.

Example 1.1.1   The following are examples of differential equations.

(a) $\dfrac{dy}{dx} + y = x^2$

(b) $\dfrac{d^2y}{dx^2} = -k^2 y$

(c) $\dfrac{d^4y}{dx^4} + \left(\dfrac{d^2y}{dx^2}\right)^3 + \cos x = 0$

(d) $\sin\left(\dfrac{dy}{dx}\right) + \tan^{-1}\left(\dfrac{y}{x}\right) = 1$

(e) $\phi_{xx} + \phi_{yy} - \phi_x = e^x + x \sin y$

The differential equations occurring in (a) through (d) are called **ordinary** differential equations since the unknown function $y(x)$ depends only on one variable x. In (e) the unknown function $\phi(x, y)$ depends on more than one variable, and hence the equation involves partial derivatives. Such a differential equation is called a **partial** differential equation. In this text we consider only ordinary differential equations. Our aim is to develop methods for determining all unknown functions that satisfy a given ordinary differential equation. Before doing so, we need to introduce some definitions and terminology.

> **Definition 1.1.2:**   The order of the highest derivative occurring in a differential equation is called the **order** of the differential equation.

In the preceding examples, (a) has order 1, (b) has order 2, (c) has order 4, and (d) has order 1.

Any ordinary differential equation of order n can be written in the form

$$F(x, y, y', y'', \ldots, y^{(n)}) = 0, \tag{1.1.1}$$

where we have introduced the prime notation to denote derivatives and $y^{(n)}$ denotes the nth derivative of y with respect to x (not y to the power of n). Of particular interest to us throughout the text will be linear differential equations. These arise as the special case of (1.1.1) when F is a linear function of $y, y', \ldots, y^{(n)}$. The standard form for such a differential equation is given in the next definition.

> **Definition 1.1.3:**   A differential equation that can be written in the form
>
> $$a_0(x)y^{(n)} + a_1(x)y^{(n-1)} + \cdots + a_n(x)y = F(x),$$
>
> where $a_0, a_1, \ldots, a_n$ and F are functions of x only, is called a **linear differential equation of order** n. Such an equation is linear in $y, y', y'', \ldots, y^{(n)}$.

A differential equation that does not satisfy this definition is said to be a **nonlinear** differential equation.

Example 1.1.2

$$y'' + x^2y' + (\sin x)y = e^x \quad \text{and} \quad xy''' + 4x^2y' - \frac{2}{1+x^2}\, y = 0$$

are linear differential equations of order 2 and order 3, respectively, whereas the differential equations

$$y'' + x \sin(y') - xy = x^2 \quad \text{and} \quad y'' - x^2y' + y^2 = 0$$

are nonlinear.   (Why?)

Example 1.1.3   The general first- and second-order linear equations are of the form

$$a_0(x)\frac{dy}{dx} + a_1(x)y = F(x)$$

and

$$a_0(x)\frac{d^2y}{dx^2} + a_1(x)\frac{dy}{dx} + a_2(x)y = F(x),$$

respectively.

SOLUTIONS OF ORDINARY DIFFERENTIAL EQUATIONS

We now define precisely what is meant by a solution of a differential equation.

> ***Definition 1.1.4:***   A **solution** of an nth-order differential equation on an interval I is any function $y = y(x)$ that is (at least) n times differentiable on I and that satisfies the differential equation identically for all x in I.

Example 1.1.4   Show that $y = c_1\sin x + c_2\cos x$, where c_1, c_2 are constants, is a solution of the linear differential equation $y'' + y = 0$ for x in the interval $(-\infty, \infty)$.

Solution   The function $y(x)$ is certainly twice differentiable for all real x. Furthermore,

$$y' = c_1\cos x - c_2\sin x,$$

and

$$y'' = -(c_1\sin x + c_2\cos x).$$

Consequently $y'' = -y$, so that $y'' + y = 0$.   It follows from the preceding definition that the given function is a solution of the differential equation on $(-\infty, \infty)$.

In the previous example x could assume all real values. Often, however, the independent variable will be restricted in some manner. For example, the differential equation

$$\frac{dy}{dx} = \frac{1}{2\sqrt{x}}(y - 1)$$

is undefined when $x \leq 0$, and so any solution would be defined only for $x > 0$. In fact this linear differential equation has solution

$$y = ce^{\sqrt{x}} + 1, \qquad x > 0,$$

where c is a constant. (You can check this by "plugging in" to the given differential equation, as in Example 1.1.4. In Section 1.3 we will introduce a technique that will enable you to derive this solution.)

We now distinguish two different ways in which solutions to a differential equation can be expressed. Usually, as in Example 1.1.4, we will be able to obtain a solution of a differential equation in the **explicit** form

$$y = \phi(x)$$

for some function ϕ. However, we sometimes have to be content with a solution written in the **implicit** form

$$\Phi(x, y) = 0,$$

where the function Φ defines the solution, $y(x)$, implicitly as a function of x. This is illustrated in the following example.

Example 1.1.5   Show that the relation $x^2 + y^2 = 4$ defines an implicit solution to the nonlinear differential equation $dy/dx = -x/y$.

Solution   We regard the given relation as defining y as a function of x. Differentiating with respect to x yields[1]

$$2x + 2y \frac{dy}{dx} = 0,$$

that is,

$$\frac{dy}{dx} = -\frac{x}{y},$$

as required. In this example, we can obtain y explicitly in terms of x, since $x^2 + y^2 = 4$ implies that

$$y = \pm\sqrt{4 - x^2}.$$

The implicit relation therefore contains the *two* explicit solutions

$$y = \sqrt{4 - x^2}, \qquad y = -\sqrt{4 - x^2},$$

---

[1] Note that we have used implicit differentiation in obtaining $\frac{d}{dx}(y^2) = 2y\frac{dy}{dx}$.

which correspond graphically to the two semicircles sketched in Figure 1.1.1. Since $x = \pm 2$ correspond to $y = 0$ in both of these equations, whereas the differential equation is defined only for $y \neq 0$, we must omit $x = \pm 2$ from the domains of the solutions. Thus both of these solutions of the differential equation are valid for $-2 < x < 2$.

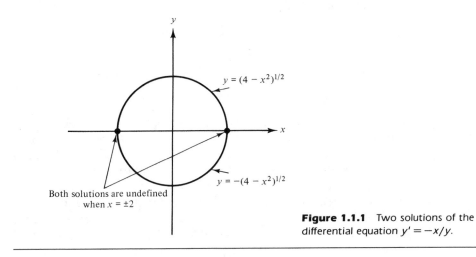

Figure 1.1.1 Two solutions of the differential equation $y' = -x/y$.

In the previous example the solutions of the differential equation are more simply expressed in implicit form, although—as we have shown—it is quite easy to obtain the corresponding explicit solutions. In the following example the solution must be expressed in implicit form because it is impossible to solve the implicit relation (analytically) for y as a function of x.

Example 1.1.6 Show that the relation $\sin(xy) + y^2 - x = 0$ defines a solution to

$$\frac{dy}{dx} = \frac{1 - y\cos(xy)}{x\cos(xy) + 2y}.$$

Solution   Differentiating the given expression with respect to x yields

$$\cos(xy)\left(y + x\frac{dy}{dx}\right) + 2y\frac{dy}{dx} - 1 = 0;$$

that is,

$$\frac{dy}{dx}[x\cos(xy) + 2y] = 1 - y\cos(xy),$$

which implies that

$$\frac{dy}{dx} = \frac{1 - y\cos(xy)}{x\cos(xy) + 2y}.$$

Now consider the simple differential equation

$$\frac{d^2y}{dx^2} = 12x.$$

We can find all solutions of this differential equation by performing two successive integrations. Indeed, integrating once yields

$$\frac{dy}{dx} = 6x^2 + c_1,$$

where c_1 is an arbitrary constant. Integrating again we obtain

$$y = 2x^3 + c_1x + c_2, \tag{1.1.2}$$

where c_2 is another arbitrary constant. The point to notice about this solution is that it contains *two* arbitrary constants. Further, by assigning appropriate values to these constants, we can determine all solutions of the differential equation. We call (1.1.2) the *general solution* of the differential equation. In this example the given differential equation was of second order and the general solution contained two arbitrary constants, which arose due to the fact that two integrations were required to solve the differential equation. In the case of an nth-order differential equation we might suspect that the most general *form* of solution that can arise would contain n arbitrary constants. This is indeed the case and motivates the following definition.

Definition 1.1.5:   A solution of an nth-order ordinary differential equation on an interval I is called the **general solution** on I if it satisfies the following conditions:

1. The solution contains n constants $c_1, c_2, \ldots, c_n$.

2. All solutions of the differential equation can be obtained by assigning appropriate values to the constants.

REMARK   Not all differential equations have a general solution. For example, consider

$$\left(\frac{dy}{dx}\right)^2 + (y - 1)^2 = 0.$$

The only solution to this differential equation is $y(x) = 1$, and hence the differential equation does not have a solution containing an arbitrary constant.

Example 1.1.7   We have shown in Example 1.1.4 that a solution of the differential equation $y'' + y = 0$ is $y = c_1\sin x + c_2\cos x$. Are all solutions of $y'' + y = 0$ of this form?   The answer is yes, although this is by no means obvious.

Example 1.1.8   Find the general solution of the differential equation $y'' = e^{-x}$.

Solution   Writing the equation as $d^2y/dx^2 = e^{-x}$ and integrating with respect to x yields

$$\frac{dy}{dx} = -e^{-x} + c_1, \tag{1.1.3}$$

where c_1 is an integration constant. Integrating (1.1.3) we obtain

$$y = e^{-x} + c_1 x + c_2, \tag{1.1.4}$$

where c_2 is another integration constant. All solutions of $y'' = e^{-x}$ are of the form (1.1.4) so that, according to Definition 1.1.5, $y = e^{-x} + c_1 x + c_2$ is the general solution of $y'' = e^{-x}$ on $(-\infty, \infty)$.

A solution of a differential equation is called a **particular solution** if it does not contain any arbitrary constants. One way in which particular solutions arise is by assigning specific values to the arbitrary constants occuring in the general solution of a differential equation. For example, from (1.1.4)

$$y = e^{-x} + x$$

is a particular solution of the differential equation $d^2y/dx^2 = e^{-x}$ (the solution corresponding to $c_1 = 1$, $c_2 = 0$).

INITIAL VALUE PROBLEMS

The unique specification of an applied problem requires more than just a differential equation. We must also specify appropriate auxiliary conditions that characterize the problem under investigation. As an example, consider the motion of an object of mass m that is falling freely under the influence of gravity. The only force acting on the object is $F = mg$, where the constant g denotes the acceleration due to gravity. If we let $y(t)$ denote the position of the object at time t and measure the positive y-direction *downward* (see Figure 1.1.2), then, according to Newton's second law, the motion of the object is determined by the differential equation

$$m\frac{d^2y}{dt^2} = mg,$$

so that

$$\frac{d^2y}{dt^2} = g.$$

The general solution of this equation is

$$y = \frac{1}{2}gt^2 + c_1 t + c_2, \tag{1.1.5}$$

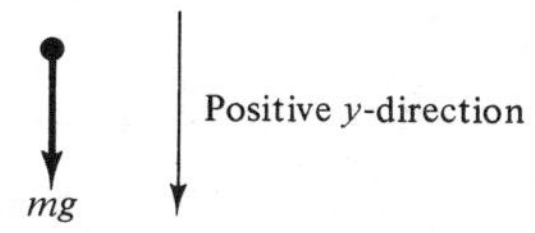

Positive y-direction

mg

Figure 1.1.2 The motion of an object falling under gravity.

where c_1 and c_2 are arbitrary constants. In order to characterize the motion of a particular object it is usual to specify its initial position, $y(t_0) = y_0$, and initial velocity,

$$\frac{dy}{dt}(t_0) = v_0,$$

at some time t_0. These auxiliary conditions are called **initial conditions** since they are imposed at the same value, t_0, of the independent variable. The corresponding problem,

solve

$$\frac{d^2y}{dt^2} = g$$

subject to

$$y(t_0) = y_0, \qquad \frac{dy}{dt}(t_0) = v_0,$$

is referred to as an *initial value problem*. Taking $t_0 = 0$ for simplicity, the first initial condition reduces to $y(0) = y_0$. Setting $t = 0$ and $y = y_0$ in (1.1.5) yields $c_2 = y_0$. Differentiating (1.1.5) we obtain

$$\frac{dy}{dt} = gt + c_1,$$

so that the initial condition $(dy/dt)(0) = v_0$ requires $c_1 = v_0$. Consequently the initial value problem has the *unique* solution

$$y = \frac{1}{2}gt^2 + v_0 t + y_0.$$

More generally, the initial value problem for an nth-order differential equation is defined as follows:

Definition 1.1.6: An nth-order differential equation together with n auxiliary conditions imposed at the *same* value of the independent variable is called an **initial value problem**.

Example 1.1.9   Solve the initial value problem

$$y'' = e^{-x}, \tag{1.1.6}$$

$$y(0) = 1, \qquad y'(0) = 4. \tag{1.1.7}$$

Solution   From Example 1.1.8 the general solution of (1.1.6) is

$$y(x) = e^{-x} + c_1 x + c_2. \tag{1.1.8}$$

We now impose the auxiliary conditions (1.1.7). $y(0) = 1$ if and only if $1 = 1 + c_2$, so that $c_2 = 0$. Setting $c_2 = 0$ in (1.1.8) and differentiating the result yields

$$y'(x) = -e^{-x} + c_1.$$

Consequently $y'(0) = 4$ if and only if $4 = -1 + c_1$, and hence $c_1 = 5$. Thus the given auxiliary conditions pick out the solution of the differential equation with $c_1 = 5$ and $c_2 = 0$, and so the initial value problem has the unique solution

$$y(x) = e^{-x} + 5x.$$

Initial value problems play a fundamental role in the theory and applications of differential equations. In the previous examples each initial value problem had a unique solution. More generally, suppose we have a differential equation that can be written in the form

$$y^{(n)} = f(x, y, y', \ldots, y^{(n-1)}).$$

The initial value problem for such an nth-order differential equation is, according to Definition 1.1.6,

$$
\begin{cases}
\text{Solve} \\[4pt]
\qquad \dfrac{d^n y}{dx^n} = f(x, y, y', \ldots, y^{(n-1)}) \\[8pt]
\text{subject to} \\[4pt]
\quad y(x_0) = y_0,\; y'(x_0) = y_1,\; \ldots,\; y^{(n-1)}(x_0) = y_{n-1},
\end{cases}
\qquad (1.1.9)
$$

where $y_0, y_1, \ldots, y_{n-1}$ are constants. It can be shown that this initial value problem always has a unique solution *provided f is a sufficiently smooth function* (all that we require is that f and its partial derivatives with respect to $y, y', \ldots, y^{(n-1)}$ are continuous in an appropriate region). This is a fundamental result in differential equation theory whose power will be illustrated in Chapter 9. The special case when $n = 1$ is stated precisely in the next section.

In the remainder of this chapter and the next we will focus our attention primarily on *first-order* differential equations and some of their elementary applications. We will show how to solve a linear first-order differential equation and also develop some special techniques for solving certain nonlinear differential equations. The theory underlying the solution of higher-order *linear* differential equations will follow as an application of the linear algebra that we introduce in Chapters 3 to 8.

EXERCISES 1.1

In problems 1–5 determine the order of the given differential equation. In each case state whether the differential equation is linear or nonlinear.

1. $\dfrac{d^2 y}{dx^2} + e^{xy} \dfrac{dy}{dx} = x^2$.

2. $\dfrac{d^3y}{dx^3} + 4\dfrac{d^2y}{dx^2} + (\sin x)\dfrac{dy}{dx} = xy + \tan x.$

3. $\dfrac{d^2y}{dx^2} + 3x\left(\dfrac{dy}{dx}\right)^3 - y = 1 + 3x.$

4. $\dfrac{dy}{dx} + (\sin x)e^{d^2y/dx^2} - \tan y = \cos x.$

5. $\dfrac{d^4y}{dx^4} + 3\dfrac{d^2y}{dx^2} = x.$

In problems 6–16 verify that the given function is a solution of the given differential equation (c_1, c_2 are arbitrary constants). In each case state the maximum interval over which the solution is valid.

6. $y = c_1e^x + c_2e^{-2x},\ y'' + y' - 2y = 0.$

7. $y = \dfrac{1}{x+4},\ y' = -y^2.$

8. $y = c_1x^{1/2},\ y' = \dfrac{y}{2x}.$

9. $y = e^{-x}\sin 2x,\ y'' + 2y' + 5y = 0.$

10. $y = c_1\cosh 3x + c_2\sinh 3x,\ y'' - 9y = 0.$

11. $y = c_1x^{-3} + c_2x^{-1},\ x^2y'' + 5xy' + 3y = 0.$

12. $y = c_1x^{1/2} + 3x^2,\ 2x^2y'' - xy' + y = 9x^2.$

13. $y = c_1x^2 + c_2x^3 - x^2\sin x,\ x^2y'' - 4xy' + 6y = x^4\sin x.$

14. $y = c_1e^{ax} + c_2e^{bx},\ \dfrac{d^2y}{dx^2} - (a+b)\dfrac{dy}{dx} + aby = 0,$ where a and b are constants and $a \neq b$.

15. $y = e^{ax}(c_1 + c_2x),\ \dfrac{d^2y}{dx^2} - 2a\dfrac{dy}{dx} + a^2y = 0,$ where a is a constant.

16. $y = e^{ax}(c_1\cos bx + c_2\sin bx),\ \dfrac{d^2y}{dx^2} - 2a\dfrac{dy}{dx} + (a^2 + b^2)y = 0,$ where a and b are constants.

In problems 17–21 show that the given relation defines an implicit solution to the given differential equation (c is an arbitrary constant).

17. $x\sin y - e^x = c,\ \dfrac{dy}{dx} = \dfrac{e^x - \sin y}{x\cos y}.$

18. $xy^2 + 2y - x = c,\ y' = \dfrac{1 - y^2}{2(1 + xy)}.$

19. $e^{xy} - x = c,\ \dfrac{dy}{dx} = \dfrac{1 - ye^{xy}}{xe^{xy}}.$ Determine the solution satisfying $y(1) = 0.$

20. $e^{y/x} + xy^2 - x = c,\ \dfrac{dy}{dx} = \dfrac{x^2(1 - y^2) + ye^{y/x}}{x(e^{y/x} + 2x^2y)}.$

21. $x^2y^2 - \sin x = c,\ \dfrac{dy}{dx} = \dfrac{\cos x - 2xy^2}{2x^2y}.$ Determine the corresponding explicit solution that satisfies $y(\pi) = 1/\pi.$

In problems 22–25 find the general solution of the given differential equation and the maximum interval on which the solution is valid.

22. $\dfrac{dy}{dx} = \sin x.$

23. $\dfrac{dy}{dx} = x^{-1/2}.$

24. $\dfrac{d^2y}{dx^2} = xe^x.$

25. $\dfrac{d^2y}{dx^2} = x^n,\ n$ an integer.

In problems 26–28 solve the given initial value problem.

26. $\dfrac{dy}{dx} = \ln x,\ y(1) = 2.$

27. $\dfrac{d^2y}{dx^2} = \cos x,\ y(0) = 2,\ y'(0) = 1.$

28. $\dfrac{d^3y}{dx^3} = 6x,\ y(0) = 1,\ y'(0) = -1,\ y''(0) = 4.$

A differential equation of order n together with n auxiliary conditions imposed at more than one value of the independent variable is called a **boundary value problem**. In problems 29 and 30 solve the given boundary value problem.

29. $\dfrac{d^2y}{dx^2} = e^{-x},\ y(0) = 1,\ y(1) = 0.$

30. $\dfrac{d^2y}{dx^2} = -2(3 + 2\ln x),\ y(1) = y(e) = 0.$

31. The differential equation $y'' + y = 0$ has general solution $y = c_1\cos x + c_2\sin x.$
(a) Show that the boundary value problem $y'' + y = 0,\ y(0) = 0,\ y(\pi) = 1$ has no solutions.
(b) Show that the boundary value problem $y'' + y = 0,\ y(0) = 0,\ y(\pi) = 0$ has an infinite number of solutions.

1.2 FIRST-ORDER DIFFERENTIAL EQUATIONS

Consider the differential equation

$$\frac{dy}{dx} = f(x, y), \qquad (1.2.1)$$

where $f(x, y)$ is a given function. The problem of determining all solutions to (1.2.1) for an arbitrary function f is too difficult to solve analytically. In the following sections we will introduce some special techniques that will enable us to solve (1.2.1) when f has certain specific forms. The basic idea behind these techniques is to rewrite (1.2.1) in the integrable form

$$\frac{d}{dx}[g(x, y) - F(x)] = 0$$

for appropriate functions g and F. Integration with respect to x will then yield the general solution of the corresponding differential equation, namely,[1]

$$g(x, y) = F(x) + c.$$

Example 1.2.1 Consider the differential equation

$$x\frac{dy}{dx} + y = 2x, \qquad x > 0.$$

Applying the product rule for differentiation we see that the left-hand side of this differential equation is just the expanded form of $(d/dx)(xy)$, so that we can write the differential equation as

$$\frac{d}{dx}(xy) - 2x = 0,$$

that is,

$$\frac{d}{dx}(xy - x^2) = 0.$$

Integrating with respect to x yields the general solution

$$xy - x^2 = c,$$

that is,

$$y = x^{-1}(c + x^2).$$

Before proceeding with the development, we give a general discussion of certain aspects of the differential equation (1.2.1) and its solutions.

---

[1] Recall from elementary calculus that the only functions satisfying $dG/dx = 0$ on an interval I are constant functions $G(x) = c$.

Often it will be useful to write (1.2.1) in "differential" form as

$$M(x, y)\, dx + N(x, y)\, dy = 0. \tag{1.2.2}$$

The differential equation (1.2.1) can always be written in differential form, although not in a unique way. Conversely, the differential equation (1.2.2) can be written uniquely in the form (1.2.1), since (1.2.2) implies that

$$\frac{dy}{dx} = -\frac{M(x, y)}{N(x, y)}.$$

Example 1.2.2   Write the following equation in differential form in two different ways:

$$\frac{dy}{dx} = \frac{x + y^2}{1 + x}.$$

Solution   The given differential equation can be written

$$(x + y^2)\, dx - (1 + x)\, dy = 0 \quad \text{or} \quad \frac{(x + y^2)}{1 + x}\, dx - dy = 0.$$

Obviously there are other possibilities.

THE INITIAL VALUE PROBLEM FOR A FIRST-ORDER ORDINARY DIFFERENTIAL EQUATION

The initial value problem corresponding to (1.2.1) is

$$\begin{cases} \dfrac{dy}{dx} = f(x, y), \\[2mm] y(x_0) = y_0, \end{cases} \tag{1.2.3}$$

where y_0 is a constant.   The key theoretical result that places differential equations on a firm mathematical basis is the following existence-uniqueness theorem.

Theorem 1.2.1:   Let $f(x, y)$ be a function that is defined and continuous on the rectangle

$$R = \{(x, y) : a \le x \le b, c \le y \le d\}.$$

Suppose further that $\partial f/\partial y$ is also continuous in R.   Then for any interior point (x_0, y_0) in the rectangle R, there exists an interval I containing x_0 such that the initial value problem

$$\begin{cases} \dfrac{dy}{dx} = f(x, y), \\[2mm] y(x_0) = y_0, \end{cases}$$

has a unique solution for x in I.

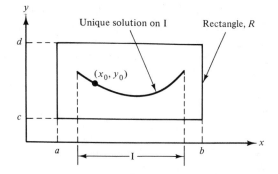

Figure 1.2.1   *Illustration of the existence-uniqueness theorem for first-order equations.*

PROOF   A full discussion of this theorem is given in Appendix 4.   For a complete proof see, for example, G. F. Simmons, *Differential Equations,* McGraw-Hill, 1972. Figure 1.2.1 gives a geometric illustration of the result.   ■

According to Theorem 1.2.1, provided $f(x, y)$ is sufficiently smooth, the initial value problem

$$\begin{cases} \dfrac{dy}{dx} = f(x, y), \\[2mm] y(x_0) = y_0, \end{cases}$$

will indeed have a unique solution for x in some interval containing x_0.   All the first-order differential equations that we consider will satisfy the assumptions of the existence-uniqueness theorem on an appropriate rectangle.

Example 1.2.3   Prove that the initial value problem

$$\begin{cases} \dfrac{dy}{dx} = x^2\cos y, \\[2mm] y(x_0) = y_0, \end{cases}$$

has a unique solution for all values of x_0 and y_0.

Solution   If we let $f(x, y) = x^2\cos y$, then $\partial f/\partial y = -x^2\sin y$.   Thus both f and $\partial f/\partial y$ are continuous for all values of x and y.   Hence the hypotheses of the existence-uniqueness theorem are satisfied in the whole xy-plane, and the initial value problem therefore has a unique solution for all values of x_0 and y_0.

GEOMETRIC INTERPRETATION OF THE SOLUTION OF A FIRST-ORDER ORDINARY DIFFERENTIAL EQUATION

The graph of any solution to the differential equation

$$\frac{dy}{dx} = f(x, y) \tag{1.2.4}$$

is called a **solution curve** or **integral curve** of the differential equation.   Usually we will be able to determine solutions of (1.2.4) that contain a single arbitrary constant c.   Such a solution can be written in the form $F(x, y, c) = 0$, for some function F.   For each value of c there is a corresponding solution curve and we say that there is a *one-parameter* family of solution curves.

Example 1.2.4   Show that

$$y = ce^{2x} \qquad\qquad (1.2.5)$$

is a solution of $y' = 2y$, and sketch some of the solution curves.

Solution   Differentiating (1.2.5) with respect to x yields

$$y' = 2ce^{2x}.$$

Substituting for $c = ye^{-2x}$ from (1.2.5), we obtain

$$y' = 2y,$$

so that (1.2.5) is a solution of the differential equation.   The solution curves defined by (1.2.5) are the family of curves shown in Figure 1.2.2.   The auxiliary condition $y(0) = 2$, for example, picks out that member of the family that passes through the point $(0, 2)$.

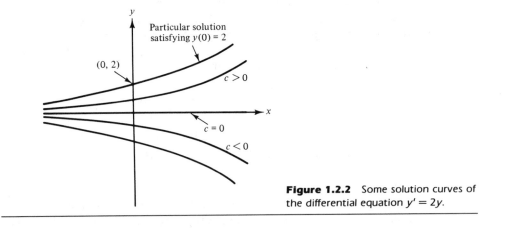

Figure 1.2.2   Some solution curves of the differential equation $y' = 2y$.

Conversely, if we have a relation of the form

$$F(x, y, c) = 0 \qquad\qquad (1.2.6)$$

that defines a family of curves in the xy-plane, then, by differentiating (1.2.6) and eliminating the constant c, we can derive a first-order differential equation that has this family as solution curves.   The resulting differential equation is called the **differential equation of the family.**

Example 1.2.5   Determine the differential equation of the family of curves

$$x^2 + y^2 = 2cy. \qquad\qquad (1.2.7)$$

Solution   We first differentiate (1.2.7) implicitly with respect to x to obtain

$$2x + 2y \frac{dy}{dx} = 2c \frac{dy}{dx}. \qquad (1.2.8)$$

We must now eliminate c from (1.2.8) in order to obtain a differential equation that applies to all curves in the family. From (1.2.7) we have

$$c = \frac{x^2 + y^2}{2y},$$

which, when substituted into (1.2.8), yields the differential equation

$$2x + 2y \frac{dy}{dx} = \left(\frac{x^2 + y^2}{y} \right) \frac{dy}{dx};$$

that is,

$$\left(\frac{x^2 + y^2}{y} - 2y \right) \frac{dy}{dx} = 2x.$$

Solving algebraically for dy/dx, we obtain the required differential equation, namely,

$$\frac{dy}{dx} = \frac{2xy}{x^2 - y^2}.$$

Geometrically, this differential equation gives the slope of the tangent line to the family of curves at any point. Completing the square in y in (1.2.7) yields

$$x^2 + (y - c)^2 = c^2,$$

which is the equation of the family of circles centered at $(0, c)$ with radius c. This family is sketched in Figure 1.2.3.

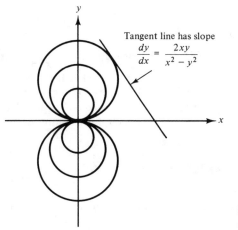

Tangent line has slope
$$\frac{dy}{dx} = \frac{2xy}{x^2 - y^2}$$

Figure 1.2.3   The family of curves $x^2 + y^2 = 2cy$.

EXERCISES 1.2

In problems 1 and 2 rewrite the given differential equation in differential form.

1. $\dfrac{dy}{dx} - x^2y^2 = x^2e^y$.

2. $(x^2 - 1)y' + \dfrac{(y^2 - x)}{x}y = x^2 + y^2$.

In problems 3–5 find the general solution of the given differential equation and sketch some of the solution curves.

3. $\dfrac{dy}{dx} = 4x$.

4. $dy - e^x\, dx = 0$.

5. $x^2\dfrac{dy}{dx} = -1$.

6. Prove that the initial value problem

$$\frac{dy}{dx} = x\sin(x + y), \qquad y(0) = 1$$

has a unique solution.

7. Use the existence-uniqueness theorem to prove that $y(x) = 3$ is the only solution to the initial value problem

$$\frac{dy}{dx} = \frac{x}{x^2 + 1}\,(y^2 - 9), \qquad y(0) = 3.$$

8. Do you think that the initial value problem

$$\frac{dy}{dx} = xy^{1/2}, \qquad y(0) = 0$$

has a unique solution? Justify your answer.

9. Consider the differential equation $dy/dx = f(x, y)$, where f is a function that is defined and continuous for all values of x and y and whose first partial derivative, $\partial f/\partial y$, is also continuous for all x and y. Do any of the solution curves of the differential equation intersect? Justify your answer.

In problems 10–13 show that the given function (or relation) defines a solution of the given differential equation and sketch some of the solution curves. If an initial condition is given, label the solution curve corresponding to the resulting unique solution. (In these problems c denotes an arbitrary constant.)

10. $x^2 + y^2 = c,\ \dfrac{dy}{dx} = -\dfrac{x}{y}$.

11. $y = cx^3,\ \dfrac{dy}{dx} = \dfrac{3y}{x},\ y(2) = 8$.

12. $y^2 = cx,\ 2x\,dy - y\,dx = 0,\ y(1) = 2$.

13. $(x - c)^2 + y^2 = c^2,\ \dfrac{dy}{dx} = \dfrac{y^2 - x^2}{2xy},\ y(2) = 2$.

In problems 14–20 determine the differential equation of the given family of curves and sketch some curves from each family.

14. $y = cx^2$.

15. $y = \dfrac{c}{x}$.

16. $y^2 = cx$.

17. $x^2 + y^2 = 2cx$.

18. $y^2 - x^2 = c$.

19. $2cy = x^2 - c^2$.

20. $(x - c)^2 + (y - c)^2 = 2c^2$.

21. The differential equation

$$\frac{dy}{dx} = -2xy^2$$

has solution $y = \dfrac{1}{x^2 + c}$ for all values of the constant c. Sketch the solution curves corresponding to $c = 0$, $c = \alpha^2$, and $c = -\alpha^2$, where α is a positive constant, and determine the intervals on which these solutions are valid.

1.3   SEPARABLE EQUATIONS

We now develop some special techniques for solving the first-order differential equation

$$\frac{dy}{dx} = f(x, y)$$

when the function $f(x, y)$ has certain specified forms.   We begin with the simplest type of equation.

Definition 1.3.1:   A first-order ordinary differential equation is said to be **separable** if it can be written in the form

$$p(y) \frac{dy}{dx} = q(x). \qquad (1.3.1)$$

The solution technique for a separable differential equation is given in the following theorem.

Theorem 1.3.1:   If $p(y)$ and $q(x)$ are continuous, then (1.3.1) has general solution

$$\int p(y) \, dy = \int q(x) \, dx + c, \qquad (1.3.2)$$

where c is an arbitrary constant.

PROOF   Integrating both sides of (1.3.2) with respect to x yields

$$\int p(y) \frac{dy}{dx} \, dx = \int q(x) \, dx + c;$$

that is,

$$\int p(y) \, dy = \int q(x) \, dx + c. \qquad \blacksquare$$

REMARK   In "differential form" (1.3.1) can be written as

$$p(y) \, dy = q(x) \, dx,$$

and the general solution (1.3.2) is obtained by integrating the left-hand side with respect to y and the right-hand side with respect to x.   This is the general procedure for solving separable equations.

Example 1.3.1   Solve $e^y \dfrac{dy}{dx} = x \cos x$.

Solution   By inspection the differential equation is separable.   Integrating both sides (using integration by parts on the right-hand side) we obtain

$$e^y = x \sin x + \cos x + c.$$

Solving for y yields the explicit solution

$$y = \ln(x \sin x + \cos x + c).$$

Notice that the differential equation $dy/dx = f(x)g(y)$ is separable because it can be written as

$$\frac{1}{g(y)} \, dy = f(x) \, dx,$$

which is certainly of the form (1.3.1). It is important to note, however, that in writing the given differential equation in this form we have assumed that $g(y) \neq 0$. Thus the general solution of the resulting differential equation may not include solutions of the original equation corresponding to any values of y for which $g(y) = 0$. We illustrate with an example.

Example 1.3.2   Solve

$$y' = -2y^2x. \tag{1.3.3}$$

Solution   Separating the variables yields

$$y^{-2} \, dy = -2x \, dx. \tag{1.3.4}$$

Integrating both sides we obtain

$$-y^{-1} = -x^2 + c,$$

which implies that

$$y = \frac{1}{x^2 - c}. \tag{1.3.5}$$

This is the general solution of (1.3.4). It is not the general solution of (1.3.3), since there are no values of the constant c for which $y = 0$, whereas by inspection $y = 0$ is clearly a solution of (1.3.3). This solution is not contained in (1.3.5), since in separating the variables we divided by y and hence assumed implicitly that $y \neq 0$. Thus the solutions to (1.3.3) are

$$y = \frac{1}{x^2 - c} \quad \text{and} \quad y = 0.$$

Example 1.3.3   Find the equation of the curve that passes through the point $(0, \frac{1}{2})$ and whose slope at each point (x, y) is $-x/4y$.

Solution   Let $y = f(x)$ be the equation of the required curve. Then we are given that

$$\frac{dy}{dx} = -\frac{x}{4y},$$

so that y is obtained by solving the initial value problem

$$\begin{cases} \dfrac{dy}{dx} = -\dfrac{x}{4y}, \\ y(0) = \dfrac{1}{2}. \end{cases}$$

Separating the variables in the differential equation yields

$$4y\, dy = -x\, dx,$$

which has general solution (integrating) $x^2 + 4y^2 = c$. Imposing the initial condition we find that $c = 1$, and hence the required curve has equation $x^2 + 4y^2 = 1$. Geometrically, the general solution of the differential equation gives a family of ellipses, and the required curve is that member of the family that passes through the point $(0, \frac{1}{2})$. (See Figure 1.3.1.)

Example 1.3.4 An object of mass m falls from rest, starting at a point near the earth's surface. Assuming that the air resistance varies as the velocity of the object, determine the subsequent motion. (See Figure 1.3.2.)

Solution Let $y(t)$ be the distance traveled by the object at time t from the point it was released and let the positive y-direction be downward. Then $y(0) = 0$, and the velocity of the object, $v(t)$, is $v = dy/dt$. The forces acting on the object are those due to gravity, $F_g = mg$, and the force due to air resistance, $F_r = -kv$, where k is a positive constant. According to Newton's second law, the differential equation describing the motion of the object is

$$m\,\frac{dv}{dt} = F_g + F_r = mg - kv.$$

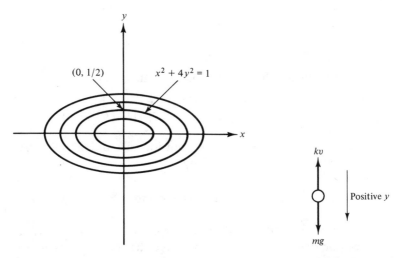

Figure 1.3.1 The solution curves of the differential equation $y' = -x/4y$.

Figure 1.3.2 Particle falling under the influence of gravity and air resistance.

We are also given the auxiliary condition $v(0) = 0$. Thus the initial value problem governing the behavior of v is

$$\begin{cases} m\dfrac{dv}{dt} = mg - kv, \\[2mm] v(0) = 0. \end{cases} \qquad (1.3.6)$$

Separating the variables in (1.3.6) yields

$$\frac{m}{mg - kv}\, dv = dt.$$

To facilitate the integration we write this as

$$\frac{1}{(mg/k) - v}\, dv = \frac{k}{m}\, dt,$$

which can be integrated directly to obtain

$$-\ln\left(\frac{mg}{k} - v\right) = \frac{k}{m}t + c.$$

Multiplying both sides of this equation by -1 and exponentiating the result yields

$$\frac{mg}{k} - v = c_1 e^{-(k/m)t},$$

where $c_1 = e^{-c}$. Consequently,

$$v = \frac{mg}{k} - c_1 e^{-(k/m)t}.$$

Imposing the initial condition $v(0) = 0$ yields

$$c_1 = \frac{mg}{k},$$

so that the solution to the initial value problem (1.3.6) is

$$v = \frac{mg}{k}[1 - e^{-(k/m)t}]. \qquad (1.3.7)$$

Notice that the velocity does not increase indefinitely but approaches a so-called limiting velocity v_L defined by:

$$v_L = \lim_{t \to \infty} v(t) = \lim_{t \to \infty} \frac{mg}{k}[1 - e^{-(k/m)t}] = \frac{mg}{k}.$$

Since $dy/dt = v$, it follows from (1.3.7) that the position of the object at time t can be determined by solving the initial value problem

$$\frac{dy}{dt} = \frac{mg}{k}[1 - e^{-(k/m)t}], \quad y(0) = 0.$$

The differential equation can be integrated directly to obtain

$$y = \frac{mg}{k}\left[t + \frac{m}{k}\,e^{-(k/m)t}\right] + c.$$

Imposing the initial condition $y(0) = 0$ yields

$$c = \frac{-m^2 g}{k^2},$$

so that

$$y = \frac{mg}{k}\left\{t + \frac{m}{k}\,[e^{-(k/m)t} - 1]\right\}.$$

EXERCISES 1.3

In problems 1–11 solve the given differential equation.

1. $\dfrac{dy}{dx} = 2xy.$

2. $\dfrac{dy}{dx} = \dfrac{y}{x \ln x}.$

3. $e^{x+y}\,dy - dx = 0.$

4. $\dfrac{dy}{dx} = \dfrac{y^2}{x^2 + 1}.$

5. $y\,dx - (x - 2)\,dy = 0.$

6. $\dfrac{dy}{dx} = \dfrac{2x(y - 1)}{x^2 + 3}.$

7. $y - x\dfrac{dy}{dx} = 3 - 2x^2\dfrac{dy}{dx}.$

8. $\dfrac{dy}{dx} = \dfrac{\cos(x - y)}{\sin x \sin y} - 1.$

9. $\dfrac{dy}{dx} = \dfrac{x(y^2 - 1)}{2(x - 2)(x - 1)}.$

10. $\dfrac{dy}{dx} = \dfrac{x^2 y - 32}{16 - x^2} + 2.$

11. $(x - a)(x - b)y' - (y - c) = 0$, where a, b, and c are constants.

In problems 12–15 solve the given initial value problem.

12. $(x^2 + 1)\dfrac{dy}{dx} + y^2 = -1$, $y(0) = 1$. [To obtain a "nice" form for the solution, the formula $\tan(a + b) = \dfrac{\tan a + \tan b}{1 - \tan a \tan b}$ will be useful.]

13. $(1 - x^2)y' + xy = ax$, $y(0) = 2a$, where a is a constant.

14. $\dfrac{dy}{dx} = 1 - \dfrac{\sin(x + y)}{\sin y \cos x}$, $y\left(\dfrac{\pi}{4}\right) = \dfrac{\pi}{4}.$

15. $\dfrac{dy}{dx} = y^3\sin x$, $y(0) = 0.$

16. One solution to the initial value problem

$$\frac{dy}{dx} = \frac{2}{3}(y - 1)^{1/2}, \qquad y(1) = 1$$

is $y(x) = 1$. Determine another solution to this initial value problem. Does this contradict the existence-uniqueness theorem (1.2.1)? Explain.

17. An object of mass m falls from rest, starting at a point near the earth's surface. Assuming that the air resistance varies as the square of the velocity of the object, a simple application of Newton's second law yields the following initial value problem for the velocity, $v(t)$, of the object at time t:

$$m\frac{dv}{dt} = mg - kv^2, \qquad v(0) = 0,$$

where k, m, g are positive constants.

(a) Solve this initial value problem for v in terms of t.

(b) Does the velocity of the object increase indefinitely? Justify.

(c) Determine the position of the object at time t.

18. The differential equation governing the velocity of an object is

$$\frac{dv}{dt} = -kv^n,$$

where $k \, (> 0)$ and n are constants. At $t = 0$ the object is set in motion with velocity v_0.

(a) Show that the object comes to rest in a finite time if and only if $n < 1$, and determine the maximum distance traveled by the object in this case.

(b) If $1 \leq n < 2$, show that the maximum distance traveled by the object in a finite time is less than $v_0^{2-n}/(2 - n)k$.

(c) If $n \geq 2$, show that there is no limit to the distance that the object can travel.

1.4   FIRST-ORDER HOMOGENEOUS EQUATIONS

Most first-order differential equations that we have to solve are not separable. However some differential equations can be reduced to separable form by making an appropriate change of variables. We now consider a particular type of differential equation where this is indeed the case. We first require a preliminary definition.

> *Definition 1.4.1:*   A function $f(x, y)$ is said to be **homogeneous of degree zero** if
>
> $$f(tx, ty) = f(x, y),$$
>
> for all positive values of t for which (tx, ty) is in the domain of f.[1]

REMARK   Equivalently, we can say that f is homogeneous of degree zero if it is invariant under a rescaling of the variables x and y.

Example 1.4.1   The simplest nonconstant functions that are homogeneous of degree zero are $f(x, y) = y/x$ and $f(x, y) = x/y$.

Example 1.4.2   If $f(x, y) = \dfrac{x^2 - y^2}{2xy + y^2}$, then $f(tx, ty) = \dfrac{t^2(x^2 - y^2)}{t^2(2xy + y^2)} = f(x, y)$, so that f is homogeneous of degree zero.

In the previous example, if we factor an x^2-term from the numerator and denominator, then the function f can be written in the form

$$f(x, y) = \frac{x^2[1 - (y/x)^2]}{x^2[2y/x + (y/x)^2]},$$

that is,

$$f(x, y) = \frac{1 - (y/x)^2}{2y/x + (y/x)^2}.$$

---

[1] More generally, $f(x, y)$ is said to be homogeneous of degree m if $f(tx, ty) = t^m f(x, y)$.

Thus f can be considered as a function of the single variable $V = y/x$.   The follow-ing theorem establishes that this is a basic property of all functions that are homo-geneous of degree zero.

Theorem 1.4.1:   A function $f(x, y)$ is homogeneous of degree zero if and only if it depends only on y/x.

PROOF   Suppose that f is homogeneous of degree zero.   We must consider two cases separately.

1. If $x > 0$, we can take $t = 1/x$ in Definition 1.4.1 to obtain

$$f(x, y) = f(1, y/x),$$

   a function only of $V = y/x$.
2. If $x < 0$, then we can take $t = -1/x$ in Definition 1.4.1.   In this case we obtain

$$f(x, y) = f(-1, -y/x),$$

   which once more depends only on y/x.

Conversely, suppose that $f(x, y)$ depends only on y/x.   If we replace x by tx and y by ty, then f is unaltered and hence is homogeneous of degree zero.   ■

REMARK   Do not memorize the formulas arising in Theorem 1.4.1.   Just remember that a function $f(x, y)$ that is homogeneous of degree zero depends only on the com-bination y/x and hence can be considered as a function of a single variable, say $F(V)$, where $V = y/x$.

Example 1.4.3   Show that the function $f(x, y) = \dfrac{\sqrt{x^2 + y^2}}{x}$ is homogeneous of degree zero and express it as a function only of y/x.

Solution   For $t > 0$,

$$f(tx, ty) = \frac{\sqrt{t^2 x^2 + t^2 y^2}}{tx} = \frac{\sqrt{x^2 + y^2}}{x} = f(x, y).$$

Thus f is homogeneous of degree zero.   We now express f as a function only of y/x.   We have

$$f(x, y) = \frac{\sqrt{x^2[1 + (y/x)^2]}}{x} = \frac{|x|\sqrt{1 + (y/x)^2}}{x}.$$

Thus

$$f(x, y) = \begin{cases} \sqrt{1 + \left(\dfrac{y}{x}\right)^2}, & \text{if } x > 0, \\[3mm] -\sqrt{1 + \left(\dfrac{y}{x}\right)^2}, & \text{if } x < 0. \end{cases}$$

> **Definition 1.4.2:**   If $f(x, y)$ is homogeneous of degree zero, then the differential equation
>
> $$\frac{dy}{dx} = f(x, y)$$
>
> is called a **homogeneous first-order ordinary differential equation**.

In general, if

$$\frac{dy}{dx} = f(x, y) \tag{1.4.1}$$

is a homogeneous first-order differential equation, then we cannot solve it directly. However, our discussion suggests that instead of using the variables x and y, we should use the variables x and V, where $V = y/x$; that is,

$$y = xV(x). \tag{1.4.2}$$

We have already seen that substitution of (1.4.2) into the right-hand side of (1.4.1) has the effect of reducing $f(x, y)$ to a function of V only, say $F(V)$. We must also determine how the derivative term dy/dx transforms. Differentiating (1.4.2) with respect to x using the product rule yields the following relationship between dy/dx and dV/dx:

$$\frac{dy}{dx} = x \frac{dV}{dx} + V. \tag{1.4.3}$$

Thus, substituting the new variables into (1.4.1), we obtain

$$x \frac{dV}{dx} + V = F(V),$$

or, equivalently,

$$x \frac{dV}{dx} = F(V) - V.$$

The variables can now be separated to yield

$$\frac{1}{F(V) - V} \, dV = \frac{1}{x} \, dx,$$

which can be solved directly by integration. We have therefore proved the following theorem.

    Theorem 1.4.2:   The change of variables $y = xV(x)$ reduces a homogeneous first-order differential equation $dy/dx = f(x, y)$ to the separable equation

$$\frac{1}{F(V) - V} \, dV = \frac{1}{x} \, dx.$$

REMARK   The separable equation that results in the preceding technique can be integrated to obtain a relationship between V and x. We then obtain the solution of the given differential equation by substituting y/x for V in this relationship.

Example 1.4.4   Solve the initial value problem $y' = \dfrac{x^2 + 5y^2}{4xy}$, $y(1) = 1$.

Solution   By inspection $f(x, y)$ is homogeneous of degree zero. We therefore let $y = xV$, which implies that $y' = xV' + V$. Substitution into the given differential equation yields

$$xV' + V = \frac{1 + 5V^2}{4V},$$

which can be written as

$$xV' = \frac{1 + V^2}{4V}.$$

Separating the variables we must solve

$$\frac{4V}{1 + V^2}\, dV = \frac{dx}{x},$$

which can be integrated directly to yield[1]

$$2\ln(1 + V^2) = \ln|x| + \ln c, \tag{1.4.4}$$

where c is an arbitrary positive constant. Combining the log terms and exponentiating both sides of this equation, we obtain

$$(1 + V^2)^2 = c|x|. \tag{1.4.5}$$

Since x cannot change sign [see (1.4.4)] we can dispose of the absolute value sign by introducing a new constant c_1 defined by

$$c_1 = \begin{cases} c, & \text{if } x > 0, \\ -c, & \text{if } x < 0. \end{cases}$$

The solution (1.4.5) can then be written as

$$(1 + V^2)^2 = c_1 x. \tag{1.4.6}$$

Substituting for $V = y/x$ into (1.4.6), we finally obtain

$$(x^2 + y^2)^2 = c_1 x^5.$$

Imposing the given initial condition $y(1) = 1$, we find that $c_1 = 4$, so that the solution to the initial value problem is

$$(x^2 + y^2)^2 = 4x^5.$$

---

[1] When there are logarithm terms arising in the solution of a differential equation, it is often useful to write the integration constant as $\ln c$. There is no loss in generality in doing so, since $\ln c$ assumes all real values as c assumes all positive values.

Notice that we have not solved explicitly for y (although we could easily do so in this problem).

Example 1.4.5   Find the general solution of

$$x\frac{dy}{dx} - y = \sqrt{9x^2 + y^2}, \quad x > 0.$$

Solution   We first rewrite the given differential equation in the equivalent form

$$\frac{dy}{dx} = \frac{\sqrt{9x^2 + y^2} + y}{x}.$$

Factoring out an x^2 from the square root yields

$$\frac{dy}{dx} = \frac{|x|\sqrt{9 + \left(\frac{y}{x}\right)^2} + y}{x}.$$

Since we are told to solve the differential equation on the interval $x > 0$, we have $|x| = x$, so that

$$\frac{dy}{dx} = \sqrt{9 + \left(\frac{y}{x}\right)^2} + \frac{y}{x}, \tag{1.4.7}$$

which we recognize as being homogeneous.   We therefore let $y = xV$, so that $y' = xV' + V$.   Substitution into (1.4.7) yields

$$xV' + V = \sqrt{9 + V^2} + V;$$

that is

$$xV' = \sqrt{9 + V^2}.$$

Separating the variables in this equation we obtain

$$\frac{1}{\sqrt{9 + V^2}} dV = \frac{1}{x} dx. \tag{1.4.8}$$

If we recall the standard integral[1]

$$\int \frac{1}{\sqrt{a^2 + x^2}} dx = \ln(x + \sqrt{a^2 + x^2}) + c,$$

then (1.4.8) can be integrated directly to yield

$$\ln(V + \sqrt{9 + V^2}) = \ln c_1 x,$$

---

[1] This integral can be evaluated by using the trigonometric substitution $x = a\tan\theta$.   Alternatively it can be obtained from an integration table (see front endpaper and Appendix 3).

where c_1 is a positive constant.   Exponentiating both sides we obtain

$$V + \sqrt{9 + V^2} = c_1 x.$$

Substituting $y/x = V$ and multiplying through by x yields the general solution

$$y + \sqrt{9x^2 + y^2} = c_1 x^2.$$

EXERCISES 1.4

In problems 1–8 determine whether the given function is homogeneous of degree zero. Rewrite those that are as functions of the single variable $V = y/x$.

1. $f(x, y) = \dfrac{x^2 - y^2}{xy}$.

2. $f(x, y) = x - y$.

3. $f(x, y) = \dfrac{x \sin(x/y) - y \cos(y/x)}{y}$.

4. $f(x, y) = \dfrac{\sqrt{x^2 + y^2}}{x - y}$, $x > 0$.

5. $f(x, y) = \dfrac{y}{x - 1}$.

6. $f(x, y) = \dfrac{x - 3}{y} + \dfrac{5y + 9}{3y}$.

7. $f(x, y) = \dfrac{\sqrt{x^2 + y^2}}{x}$, $x < 0$.

8. $f(x, y) = \dfrac{\sqrt{x^2 + 4y^2} - x + y}{x + 3y}$, $x \neq 0, y \neq 0$.

In problems 9–22 solve the given differential equation.

9. $(3x - 2y)\dfrac{dy}{dx} = 3y$.

10. $\dfrac{dy}{dx} = \dfrac{(x + y)^2}{2x^2}$.

11. $\sin\left(\dfrac{y}{x}\right)\left(x\dfrac{dy}{dx} - y\right) = x \cos\left(\dfrac{y}{x}\right)$.

12. $\dfrac{dy}{dx} = \dfrac{\sqrt{16x^2 - y^2} + y}{x}$, $x > 0$.

13. $(2x - y)\,dy - (x + 2y)\,dx = 0$.

14. $y(x^2 - y^2)\,dx - x(x^2 + y^2)\,dy = 0$.

15. $x\dfrac{dy}{dx} + y \ln x = y \ln y$.

16. $\dfrac{dy}{dx} = \dfrac{y^2 + 2xy - 2x^2}{x^2 - xy + y^2}$.

17. $2xy\,dy - (x^2 e^{-y^2/x^2} + 2y^2)\,dx = 0$.

18. $x^2\dfrac{dy}{dx} = y^2 + 3xy + x^2$.

19. $\dfrac{dy}{dx} = \dfrac{\sqrt{x^2 + y^2} - x}{y}$, $x > 0$.

20. $2x(y + 2x)\dfrac{dy}{dx} = y(4x - y)$.

21. $x\dfrac{dy}{dx} = x \tan\left(\dfrac{y}{x}\right) + y$.

22. $\dfrac{dy}{dx} = \dfrac{x\sqrt{x^2 + y^2} + y^2}{xy}$, $x > 0$.

In problems 23–25 solve the given initial value problem.

23. $\dfrac{dy}{dx} = \dfrac{2x - y}{x + 4y}$, $y(1) = 1$.

24. $\dfrac{dy}{dx} = \dfrac{2(2y - x)}{x + y}$, $y(0) = 2$.

25. $\dfrac{dy}{dx} = \dfrac{y - \sqrt{x^2 + y^2}}{x}$, $y(3) = 4$.

26. Find *all* solutions to $x\dfrac{dy}{dx} - y = \sqrt{4x^2 - y^2}$, $x > 0$.

27. Consider the differential equation

$$y' = F(ax + by + c), \qquad \text{(i)}$$

where a, b ($\neq 0$), and c are constants. Show that the change of variables from x and y to x and V defined by

$$V = ax + by + c$$

reduces (i) to the separable form

$$\frac{1}{bF(V) + a}\, dV = dx.$$

In problems 28–30 use the result from the previous problem to solve the given differential equation (in 29 impose the given initial condition).

28. $y' = (4x + y + 2)^2$.

29. $y' = (9x - y)^2$, $y(0) = 0$.

30. $\dfrac{dy}{dx} = \sin^2(3x - 3y + 1)$.

31. Consider the differential equation

$$\frac{dy}{dx} = \frac{x + 2y - 1}{2x - y + 3}. \tag{ii}$$

(a) Show that the change of variables defined by

$$x = r - 1, \quad y = s + 1,$$

transforms (ii) into the homogeneous equation

$$\frac{ds}{dr} = \frac{r + 2s}{2r - s}. \tag{iii}$$

(b) Find the general solution of (iii) and hence solve (ii).

32. Show that the differential equation

$$M(x, y)\, dx + N(x, y)\, dy = 0$$

is a first-order homogeneous equation if and only if M and N are homogeneous functions of the same degree.

1.5   FIRST-ORDER LINEAR DIFFERENTIAL EQUATIONS

In this section we derive a technique for determining the general solution of any first-order *linear* differential equation.

Definition 1.5.1:   A differential equation that can be written in the form

$$a(x)\frac{dy}{dx} + b(x)y = r(x), \tag{1.5.1}$$

where $a(x)$, $b(x)$, and $r(x)$ are functions defined on an interval (α, β), is called a **first-order linear** differential equation.

We assume that $a(x) \neq 0$ on (α, β) and divide both sides of (1.5.1) by $a(x)$ to obtain the **standard form**

$$\frac{dy}{dx} + p(x)y = q(x), \tag{1.5.2}$$

where $p(x) = b(x)/a(x)$ and $q(x) = r(x)/a(x)$. The idea behind the solution technique for (1.5.2) is to rewrite the differential equation in the form

$$\frac{d}{dx}[g(x, y)] = F(x) \tag{1.5.3}$$

for an appropriate function $g(x, y)$. The general solution of the differential equation can then be obtained by an integration. First consider an example.

Example 1.5.1   Solve the differential equation

$$\frac{dy}{dx} + \frac{1}{x}y = e^x, \qquad x > 0. \tag{1.5.4}$$

Solution   If we multiply (1.5.4) by x, it becomes

$$x\frac{dy}{dx} + y = xe^x.$$

But, from the product rule for differentiation, the left-hand side of this equation is just the expanded form of $\frac{d}{dx}(xy)$. Thus (1.5.4) can be written in the equivalent form

$$\frac{d}{dx}(xy) = xe^x.$$

Integrating both sides of this equation with respect to x yields

$$xy = xe^x - e^x + c,$$

and hence the general solution (1.5.4) is

$$y = \frac{1}{x}[e^x(x-1) + c],$$

where c is an arbitrary constant.

We now generalize the technique used in the previous example to the case of the first-order linear differential equation

$$\frac{dy}{dx} + p(x)y = q(x), \tag{1.5.5}$$

where we assume that the functions p and q are continuous on the interval (α, β). We begin by multiplying (1.5.5) by an as yet undetermined (nonzero) function $I(x)$. The result is

$$I\frac{dy}{dx} + p(x)Iy = Iq(x), \tag{1.5.6}$$

where we wish to choose I so that (1.5.6) reduces to the form (1.5.3). The left-hand side of (1.5.6) bears a striking resemblance to the derivative of the product Iy, namely,

$$\frac{d}{dx}(Iy) = I\frac{dy}{dx} + \frac{dI}{dx}y. \tag{1.5.7}$$

Indeed, comparing (1.5.6) and (1.5.7), we see that (1.5.6) can be written in the integrable form

$$\frac{d}{dx}(Iy) = Iq(x),$$

provided the function I is a solution of[1]

$$I \frac{dy}{dx} + p(x)Iy = I \frac{dy}{dx} + \frac{dI}{dx} y.$$

This will hold whenever I satisfies the separable differential equation

$$\frac{dI}{dx} = p(x)I. \qquad (1.5.8)$$

Separating the variables and integrating yields

$$\ln|I| = \int p(x) \, dx + \ln c,$$

so that

$$I = c_1 e^{\int p(x) \, dx},$$

where c_1 is an arbitrary constant. Since we require only one solution to (1.5.8), we set $c_1 = 1$, in which case

$$I = e^{\int p(x) \, dx}. \qquad (1.5.9)$$

We can therefore draw the following conclusion.

Multiplying the linear differential equation

$$\frac{dy}{dx} + p(x)y = q(x) \qquad (1.5.10)$$

by $e^{\int p(x) \, dx}$ reduces it to the integrable form

$$\frac{d}{dx} [e^{\int p(x) \, dx} y] = q(x)e^{\int p(x) \, dx}. \qquad (1.5.11)$$

The general solution to (1.5.10) can now be obtained from (1.5.11) by integration. Formally we have

$$y = e^{-\int p(x) \, dx} \left[\int q(x)e^{\int p(x) \, dx} \, dx + c \right]. \qquad (1.5.12)$$

REMARKS

1. The function $I = e^{\int p(x) \, dx}$ is called an **integrating factor** for the differential equation (1.5.10) because it enables us to reduce the differential equation to a form that is directly integrable.

2. You should not memorize (1.5.12). In a specific problem we first evaluate the integrating factor $e^{\int p(x) \, dx}$ and then use (1.5.11).

Example 1.5.2   Solve $\dfrac{dy}{dx} - \dfrac{3}{x} y = 2x^4$, $x > 0$.

---

[1] This is obtained by equating the left-hand side of (1.5.6) to the right-hand side of (1.5.7).

Solution   The given equation is first-order linear. Since it is already written in standard form, we can compute the integrating factor directly. In this case $p(x) = -3/x$; from (1.5.9) an appropriate integrating factor is

$$I = e^{-\int (3/x)\, dx} = e^{-3 \ln x} = x^{-3}.$$

It follows from (1.5.11) that the given differential equation can be written as

$$\frac{d}{dx}(x^{-3}y) = 2x,$$

which, upon integrating with respect to x, yields

$$x^{-3}y = x^2 + c.$$

Consequently, the general solution of the given differential equation is

$$y = x^3(x^2 + c).$$

Example 1.5.3   Solve the initial value problem

$$\frac{dy}{dx} + xy = xe^{x^2/2}, \quad y(0) = 1.$$

Solution   An appropriate integrating factor in this case is

$$I = e^{\int x\, dx} = e^{x^2/2}.$$

Multiplying the given differential equation by I and using (1.5.11) yields

$$\frac{d}{dx}(e^{x^2/2}y) = xe^{x^2}.$$

Integrating both sides with respect to x we obtain

$$e^{x^2/2}y = \int xe^{x^2}\, dx + c,$$

which implies that

$$y = e^{-x^2/2}\left(\frac{1}{2}e^{x^2} + c\right).$$

Imposing the initial condition $y(0) = 1$ yields

$$1 = \frac{1}{2} + c,$$

so that $c = \frac{1}{2}$.   Thus the required particular solution is

$$y = \frac{1}{2}e^{-x^2/2}(e^{x^2} + 1) = \frac{1}{2}(e^{x^2/2} + e^{-x^2/2}) = \cosh\left(\frac{x^2}{2}\right).$$

Example 1.5.4   Solve $xy' + 2y = \cos x$, $x > 0$.

Solution   We first write the given differential equation in standard form. Dividing by x we obtain

$$y' + \frac{2}{x}y = \frac{\cos x}{x}. \tag{1.5.13}$$

An integrating factor is thus

$$I = e^{\int(2/x)\,dx} = e^{2\ln x} = x^2,$$

so that the differential equation (1.5.13) can be written in the form

$$\frac{d}{dx}(x^2 y) = x\cos x.$$

Integrating both sides yields

$$x^2 y = \int x\cos x \, dx + c;$$

that is,

$$y = x^{-2}(x\sin x + \cos x + c).$$

EXERCISES 1.5

In problems 1–14 solve the given differential equation.

1. $\dfrac{dy}{dx} - y = e^{2x}$.

2. $x^2 \dfrac{dy}{dx} - 4xy = x^7\sin x$, $x > 0$.

3. $\dfrac{dy}{dx} + 2xy = 2x^3$.

4. $y' + \dfrac{2x}{(1 - x^2)}y = 4x$, $-1 < x < 1$.

5. $\dfrac{dy}{dx} + \dfrac{2x}{(1 + x^2)}y = \dfrac{4}{(1 + x^2)^2}$.

6. $2\cos^2 x\,\dfrac{dy}{dx} + y\sin 2x = 4\cos^4 x$, $0 \le x < \dfrac{\pi}{2}$.

7. $y' + \dfrac{1}{x\ln x}y = 9x^2$.

8. $y' - y\tan x = 8\sin^3 x$.

9. $t\dfrac{dx}{dt} + 2x = 4e^t$, $t > 0$.

10. $\dfrac{dy}{dx} = (y\sec x - 2)\sin x$.

11. $(1 - y\sin x)\,dx - \cos x\,dy = 0$.

12. $\dfrac{dy}{dx} - \dfrac{3}{x}y = 2x^2\ln x$.

13. $\dfrac{dy}{dx} + \alpha y = e^{\beta x}$, α, β constants.

14. $y' + \dfrac{m}{x}y = \ln x$, m constant.

In problems 15–18 solve the given initial value problem.

15. $\dfrac{dy}{dx} + \dfrac{2}{x}y = 4x$, $y(1) = 2$.

16. $\sin x\,\dfrac{dy}{dx} - y\cos x = \sin 2x$, $y\!\left(\dfrac{\pi}{2}\right) = 2$.

17. $\dfrac{dx}{dt} + \dfrac{2}{(4 - t)}x = 5$, $x(0) = 4$.

18. $(y - e^x)\,dx + dy = 0$, $y(0) = 1$.

19. Find the general solution of the second-order differential equation

$$\frac{d^2y}{dx^2} + \frac{1}{x}\frac{dy}{dx} = 9x, \ x > 0.$$

(*Hint:* Let $u = dy/dx$.)

20. (Variation-of-parameters method for first-order linear differential equations) Consider the first-order linear ordinary differential equation

$$y' + p(x)y = q(x). \tag{i}$$

(a) Show that the general solution of the associated *homogeneous* equation

$$y' + p(x)y = 0$$

is

$$y_H = c_1 e^{-\int p(x)\,dx}.$$

(b) Determine the function $u(x)$ such that

$$y = u(x)e^{-\int p(x)\,dx}$$

is a solution of (i), and hence derive the general solution of (i).

In problems 21–24 use the technique derived in the previous problem to solve the given differential equation.

21. $\dfrac{dy}{dx} + y = e^{-2x}.$

22. $\dfrac{dy}{dx} + \dfrac{1}{x}y = \cos x, \ x > 0.$

23. $x\dfrac{dy}{dx} - y = x^2 \ln x.$

24. $\dfrac{dy}{dx} + y \cot x = 2 \cos x, \ 0 < x < \pi.$

25. The current $i(t)$ in the *RL* circuit shown in Figure 1.5.1 is governed by the differential equation

$$L\frac{di}{dt} + Ri = E(t), \tag{ii}$$

where R and L are constants and $E(t)$ is a known function of t.

(a) If $E(t) = E_0 \cos \omega t, \ t > 0$, where E_0 and ω are constants, solve (ii) for $i(t)$.

(b) Show that the solution in (a) can be written in the form

$$i(t) = Ae^{-at} + \frac{E_0}{L\sqrt{a^2 + \omega^2}} \cos (\omega t - \phi),$$

where $\phi = \tan^{-1}(\omega/a)$, $a = R/L$, and A is a constant.

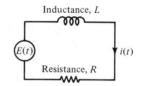

Figure 1.5.1 A simple *RL* circuit.

1.6   BERNOULLI EQUATIONS

Sometimes an ordinary differential equation that is not linear can be reduced to a linear equation by a change of variables. In this section we consider one such type of differential equation.

> ***Definition 1.6.1:*** A differential equation that can be written in the form
>
> $$\frac{dy}{dx} + p(x)y = q(x)y^n, \tag{1.6.1}$$
>
> where n is a constant, is called a **Bernoulli equation**.

If $n \neq 0$ or 1, then a Bernoulli equation is nonlinear. We can reduce such an equation to a linear equation as follows.

We first divide (1.6.1) by y^n to obtain

$$y^{-n}\frac{dy}{dx} + y^{1-n}p(x) = q(x). \tag{1.6.2}$$

We now make the change of variables

$$u = y^{1-n}, \tag{1.6.3}$$

which implies that

$$\frac{du}{dx} = (1-n)y^{-n}\frac{dy}{dx};$$

that is,

$$y^{-n}\frac{dy}{dx} = \frac{1}{(1-n)}\frac{du}{dx}.$$

Substituting into (1.6.2) for y^{1-n} and $y^{-n}\frac{dy}{dx}$ yields the *linear* differential equation

$$\frac{1}{(1-n)}\frac{du}{dx} + p(x)u = q(x),$$

or, in standard form,

$$\frac{du}{dx} + (1-n)p(x)u = (1-n)q(x). \tag{1.6.4}$$

The *linear* equation (1.6.4) can now be solved for u as a function of x. The solution of the original equation is then obtained from (1.6.3).

Example 1.6.1   Solve $\dfrac{dy}{dx} + \dfrac{3}{x}y = \dfrac{12y^{2/3}}{(1+x^2)^{1/2}}$, $x > 0$.

Solution   The differential equation is a Bernoulli equation. Thus we divide both sides by $y^{2/3}$ to obtain

$$y^{-2/3}\frac{dy}{dx} + \frac{3}{x}y^{1/3} = \frac{12}{(1+x^2)^{1/2}}. \tag{1.6.5}$$

We now let

$$u = y^{1/3}, \tag{1.6.6}$$

which implies that

$$\frac{du}{dx} = \frac{1}{3}y^{-2/3}\frac{dy}{dx}.$$

Substituting into (1.6.5) yields

$$3\frac{du}{dx} + \frac{3}{x}u = \frac{12}{(1+x^2)^{1/2}},$$

or, in standard form,

$$\frac{du}{dx} + \frac{1}{x} u = \frac{4}{(1+x^2)^{1/2}}. \tag{1.6.7}$$

An integrating factor for this linear equation is

$$I = e^{\int (1/x)\, dx} = e^{\ln x} = x,$$

so that (1.6.7) can be written as

$$\frac{d}{dx} (ux) = \frac{4x}{(1+x^2)^{1/2}}.$$

Integrating, we obtain

$$u = \frac{1}{x} [4(1+x^2)^{1/2} + c],$$

and so, from (1.6.6), the solution to the original differential equation is

$$y^{1/3} = \frac{1}{x} [4(1+x^2)^{1/2} + c].$$

Example 1.6.2 Solve the initial value problem $y' + y \tan x = 4y^2 \sin x$, $y(0) = 1$.

Solution The given differential equation is once more a Bernoulli equation with $n = 2$, so that we first divide the equation by y^2 to obtain

$$y^{-2} y' + y^{-1} \tan x = 4 \sin x. \tag{1.6.8}$$

We now let

$$u = y^{-1}, \tag{1.6.9}$$

in which case $u' = -y^{-2} y'$. Substituting into (1.6.8) for y^{-1} and $y^{-2} y'$ yields

$$-u' + u \tan x = 4 \sin x;$$

that is,

$$u' - u \tan x = -4 \sin x. \tag{1.6.10}$$

An integrating factor for this linear equation is

$$I = e^{-\int \tan x\, dx} = \cos x,$$

so that we can rewrite (1.6.10) in the equivalent form

$$\frac{d}{dx} (u \cos x) = -4 \sin x \cos x.$$

Integrating this equation yields

$$u = \frac{2 \cos^2 x + c}{\cos x},$$

so that, from (1.6.9),

$$y = \frac{\cos x}{2 \cos^2 x + c}.$$

Imposing the given initial condition, $y(0) = 1$ if and only if

$$\frac{1}{2 + c} = 1,$$

so that $c = -1$. It follows that the unique solution to the given initial value problem is

$$y = \frac{\cos x}{2 \cos^2 x - 1},$$

which we can write as

$$y = \frac{\cos x}{\cos 2x}.$$

EXERCISES 1.6

In problems 1–13 solve the given differential equation.

1. $\dfrac{dy}{dx} - \dfrac{1}{x}y = 4x^2 y^{-1}\cos x,\ x > 0.$

2. $\dfrac{dy}{dx} + \dfrac{1}{2}(\tan x)y = 2y^3\sin x.$

3. $\dfrac{dy}{dx} - \dfrac{3}{2x}y = 6y^{1/3}x^2\ln x.$

4. $\dfrac{dy}{dx} + \dfrac{2}{x}y = 6(1 + x^2)^{1/2}\sqrt{y},\ x > 0.$

5. $\dfrac{dy}{dx} + \dfrac{2}{x}y = 6y^2 x^4.$

6. $2x(y' + y^3 x^2) + y = 0.$

7. $(x - a)(x - b)\left(\dfrac{dy}{dx} - y^{1/2}\right) = 2(b - a)y,$ where a, b are constants.

8. $y' + \dfrac{6}{x}y = 3y^{2/3}\,\dfrac{\cos x}{x},\ x > 0.$

9. $y' + 4xy = 4x^3 y^{1/2}.$

10. $\dfrac{dy}{dx} - \dfrac{1}{2x\ln x}y = 2xy^3.$

11. $\dfrac{dy}{dx} - \dfrac{1}{(\pi - 1)x}y = \dfrac{3}{(1 - \pi)}xy^\pi.$

12. $2y' + y\cot x = 8y^{-1}\cos^3 x.$

13. $(1 - \sqrt{3})y' + y\sec x = y^{\sqrt{3}}\sec x.$

In problems 14 and 15 solve the given initial value problem.

14. $\dfrac{dy}{dx} + \dfrac{2x}{1 + x^2}y = xy^2,\ y(0) = 1.$

15. $y' + y\cot x = y^3\sin^3 x,\ y\!\left(\dfrac{\pi}{2}\right) = 1.$

16. Consider the differential equation

$$y^{-1}\frac{dy}{dx} + p(x)\ln y = q(x), \qquad \text{(i)}$$

where $p(x)$ and $q(x)$ are continuous functions on some interval (a, b). Show that the change of variables $u = \ln y$ reduces (i) to the linear differential equation

$$\frac{du}{dx} + p(x)u = q(x),$$

and hence show that the general solution to (i) is

$$y = \exp\left\{I^{-1}\left[\int I(x)q(x)\, dx + c\right]\right\},$$

where

$$I = e^{\int p(x)\, dx} \tag{ii}$$

and c is an arbitrary constant.

17. Use the technique derived in the previous problem to solve the following initial value problem:

$$y^{-1}\frac{dy}{dx} - \frac{2}{x}\ln y = \frac{1 - 2\ln x}{x}, \quad y(1) = e.$$

18. Consider the differential equation

$$f'(y)\frac{dy}{dx} + p(x)f(y) = q(x), \tag{iii}$$

where p and q are continuous functions on some interval (a, b), f is an invertible function, and a prime denotes differentiation with respect to y. Show that (iii) can be written as

$$\frac{du}{dx} + p(x)u = q(x),$$

where $u = f(y)$, and hence show that the general solution to (iii) is

$$y = f^{-1}\left\{I^{-1}\left[\int I(x)q(x)\, dx + c\right]\right\},$$

where I is given in (ii), f^{-1} is the inverse of f, and c is an arbitrary constant.

19. Solve $\sec^2 y \dfrac{dy}{dx} + \dfrac{1}{2\sqrt{1+x}}\tan y = \dfrac{1}{2\sqrt{1+x}}.$

1.7   EXACT ORDINARY DIFFERENTIAL EQUATIONS

For our next technique it is best to consider first-order differential equations written in differential form

$$M(x, y)\, dx + N(x, y)\, dy = 0, \tag{1.7.1}$$

where M and N are given functions. The method that we will consider is based on the idea of a *differential*.

You should recall from a previous calculus course that if $\phi = \phi(x, y)$ then the differential of ϕ, denoted $d\phi$, is defined by

$$d\phi = \frac{\partial\phi}{\partial x}\, dx + \frac{\partial\phi}{\partial y}\, dy. \tag{1.7.2}$$

The right-hand side of this expression is similar to the expression arising in (1.7.1). This observation can sometimes be used to solve a given differential equation.

Example 1.7.1   Solve

$$2x\sin y\, dx + x^2\cos y\, dy = 0. \tag{1.7.3}$$

Solution   This equation is separable; however, we will use a different technique to solve it. By inspection we notice that $2x\sin y\, dx + x^2\cos y\, dy = d(x^2\sin y)$. Consequently, (1.7.3) can be written as

$$d(x^2\sin y) = 0.$$

This implies that

$$x^2\sin y = \text{constant},$$

and hence the general solution of (1.7.3) is

$$\sin y = \frac{c}{x^2},$$

where c is an arbitrary constant.

In Example 1.7.1 we were able to write the given differential equation in the form $d\phi(x, y) = 0$, and hence obtain its solution. We cannot always do this, however. Indeed we see by comparing (1.7.1) to (1.7.2) that the differential equation $M(x, y)\, dx + N(x, y)\, dy = 0$ can be written as $d\phi = 0$ if and only if

$$M = \frac{\partial \phi}{\partial x} \quad \text{and} \quad N = \frac{\partial \phi}{\partial y}$$

for some function ϕ. This motivates the following definition.

Definition 1.7.1:   The differential equation $M(x, y)\, dx + N(x, y)\, dy = 0$ is said to be **exact** in a region R of the xy-plane if there exists a function $\phi(x, y)$ such that

$$\frac{\partial \phi}{\partial x} = M, \qquad \frac{\partial \phi}{\partial y} = N \tag{1.7.4}$$

for all (x, y) in R.

Any function ϕ satisfying (1.7.4) is called a **potential function** for the differential equation $M(x, y)\, dx + N(x, y)\, dy = 0$. We emphasize that if such a function exists, then the differential equation $M(x, y)\, dx + N(x, y)\, dy = 0$ can be written as

$$d\phi = 0.$$

This is why such a differential equation is called an *exact* differential equation. From the previous example, a potential function for the differential equation

$$2x \sin y \, dx + x^2 \cos y \, dy = 0$$

is

$$\phi = x^2 \sin y.$$

We now show that if a differential equation is exact, then—*provided we can find a potential function ϕ*—its solution can be written down immediately.

Theorem 1.7.1:   The general solution of an exact equation

$$M(x, y)\, dx + N(x, y)\, dy = 0$$

is defined implicitly by

$$\phi(x, y) = c,$$

where ϕ satisfies (1.7.4) and c is an arbitrary constant.

PROOF   We rewrite the differential equation in the form

$$M(x, y) + N(x, y)\frac{dy}{dx} = 0;$$

that is, from (1.7.4) (assuming exactness),

$$\frac{\partial \phi}{\partial x} + \frac{\partial \phi}{\partial y}\frac{dy}{dx} = 0.$$

But this is just $d\phi/dx = 0$, which implies that $\phi(x, y) = c$, where c is a constant.   ∎

REMARKS

1. The potential function ϕ is a function of two variables x and y, and we interpret the relationship $\phi(x, y) = c$ as defining y implicitly as a function of x.   The preceding theorem states that this relationship defines the general solution of the differential equation for which ϕ is a potential function.   For example, we have shown that a potential function for the differential equation (1.7.3) is $\phi(x, y) = x^2\sin y$.   It follows from the theorem that the relationship

$$x^2\sin y = c$$

defines an implicit solution to (1.7.3).

2. Geometrically Theorem 1.7.1 says that the solution curves of an exact differential equation are just the family of *level curves* of the function $\phi(x, y)$—that is, the family of curves $\phi(x, y) =$ constant.

The following two questions now arise:

1. How can we tell whether a given differential equation is exact?
2. If we have an exact equation, how do we find a potential function?

The answers are given in the following theorem.

Theorem 1.7.2 (Test for Exactness):   Let M, N, and their first partial derivatives M_y, N_x, be continuous in a (simply connected) region R of the xy-plane.[1] Then the ordinary differential equation

$$M(x, y)\, dx + N(x, y)\, dy = 0$$

is exact for all x, y in R if and only if

$$\frac{\partial M}{\partial y} = \frac{\partial N}{\partial x}. \tag{1.7.5}$$

PROOF   (Outline) We first prove that exactness implies the validity of (1.7.5).   If the differential equation is exact, then by definition there exists a potential function $\phi(x, y)$ such that $\phi_x = M$ and $\phi_y = N$.   Thus, taking partial derivatives, $\phi_{xy} = M_y$,

---

[1] Roughly speaking, *simply connected* means that the interior of any closed curve drawn in the region also lies in the region.   For example, the interior of a circle is a simply connected region, although the region between two concentric circles is not.

$\phi_{yx} = N_x$. Since M_y and N_x are continuous in R, it follows that ϕ_{xy} and ϕ_{yx} are continuous in R.   But this implies that $\phi_{xy} = \phi_{yx}$ and hence that $M_y = N_x$.

We now prove the converse.   Thus we assume that (1.7.5) holds and must prove that there exists a potential function ϕ such that

$$\frac{\partial \phi}{\partial x} = M, \tag{1.7.6}$$

and

$$\frac{\partial \phi}{\partial y} = N. \tag{1.7.7}$$

The proof is constructional; that is, we actually find a potential function ϕ. We begin by integrating (1.7.6) with respect to x, holding y fixed (this is a partial integration), to obtain[1]

$$\phi(x, y) = \int^x M(s, y) \, ds + h(y), \tag{1.7.8}$$

where $h(y)$ is an arbitrary function of y (this is our integration "constant" that we must allow to depend on y, since we held y fixed in performing our integration). We now show how to determine $h(y)$ so that the function ϕ defined in (1.7.8) also satisfies (1.7.7).   Differentiating (1.7.8) partially with respect to y yields

$$\frac{\partial \phi}{\partial y} = \frac{\partial}{\partial y} \int^x M(s, y) \, ds + \frac{dh}{dy}.$$

In order that ϕ satisfy (1.7.7), we must therefore choose $h(y)$ to satisfy

$$\frac{\partial}{\partial y} \int^x M(s, y) \, ds + \frac{dh}{dy} = N(x, y);$$

that is,

$$\frac{dh}{dy} = N(x, y) - \frac{\partial}{\partial y} \int^x M(s, y) \, ds. \tag{1.7.9}$$

Since the left-hand side of this expression is a function of y only, we must show, for consistency, that the right-hand side also depends only on y.   Taking the derivative of the right-hand side with respect to x yields

$$\frac{\partial}{\partial x} \left[N - \frac{\partial}{\partial y} \int^x M(s, y) \, ds \right] = \frac{\partial N}{\partial x} - \frac{\partial^2}{\partial x \partial y} \int^x M(s, y) \, ds$$

$$= \frac{\partial N}{\partial x} - \frac{\partial}{\partial y} \left[\frac{\partial}{\partial x} \int^x M(s, y) \, ds \right]$$

$$= \frac{\partial N}{\partial x} - \frac{\partial M}{\partial y}.$$

---

[1] Throughout the text $\int^x f(t) \, dt$ means to evaluate the indefinite integral $\int f(t) \, dt$ and replace t by x in the result.

Thus, using (1.7.5) we have

$$\frac{\partial}{\partial x}\left[N - \frac{\partial}{\partial y}\int^x M(s,\, y)\, ds\right] = 0,$$

so that the right-hand side of (1.7.9) does just depend on y. It follows that (1.7.9) is a consistent equation, and hence we can integrate both sides with respect to y to obtain

$$h(y) = \int^y N(x,\, t)\, dt - \int^y \frac{\partial}{\partial t}\left[\int^x M(s,\, t)\, ds\right] dt.$$

Finally, substitution into (1.7.8) yields the potential function

$$\phi(x,\, y) = \int^x M(s,\, y)\, ds + \int^y N(x,\, t)\, dt - \int^y \frac{\partial}{\partial t}\left[\int^x M(s,\, t)\, ds\right] dt. \qquad \blacksquare$$

REMARK You should *not* attempt to learn the final result for ϕ. In a particular example we will construct an appropriate potential function from first principles. This is illustrated in Examples 1.7.3 and 1.7.4.

Example 1.7.2 Determine whether the given differential equation is exact.
  (a) $[1 + \ln(xy)]\, dx + (x/y)\, dy = 0$.
  (b) $x^2 y\, dx - (xy^2 + y^3)\, dy = 0$.

Solution
  (a) In this case $M = 1 + \ln(xy)$ and $N = x/y$, so that $M_y = 1/y = N_x$. It follows from the above theorem that the differential equation is exact.
  (b) In this case we have $M = x^2 y$, $N = -(xy^2 + y^3)$, so that $M_y = x^2$, whereas $N_x = -y^2$. It follows that the differential equation is not exact.

Example 1.7.3 Find the general solution of

$$2xe^y\, dx + (x^2 e^y + \cos y)\, dy = 0.$$

Solution We have

$$M = 2xe^y, \qquad N = x^2 e^y + \cos y,$$

so that

$$\frac{\partial M}{\partial y} = 2xe^y = \frac{\partial N}{\partial x}.$$

It follows from Theorem 1.7.2 that the given differential equation is exact and so there exists a potential function ϕ such that (see Definition 1.7.1)

$$\frac{\partial \phi}{\partial x} = 2xe^y, \tag{1.7.10}$$

$$\frac{\partial \phi}{\partial y} = x^2 e^y + \cos y. \tag{1.7.11}$$

Integrating (1.7.10) with respect to x, holding y fixed, yields

$$\phi = x^2 e^y + h(y), \tag{1.7.12}$$

where h is an arbitrary function of y. We now determine $h(y)$ such that (1.7.12) also satisfies (1.7.11). Taking the derivative of (1.7.12) with respect to y yields

$$\frac{\partial \phi}{\partial y} = x^2 e^y + \frac{dh}{dy}. \tag{1.7.13}$$

Equations (1.7.11) and (1.7.13) are consistent if and only if

$$\frac{dh}{dy} = \cos y,$$

which implies that

$$h = \sin y,$$

where we have set the integration constant equal to zero without loss of generality, since we only require one potential function. Substitution into (1.7.12) yields the potential function

$$\phi(x, y) = x^2 e^y + \sin y.$$

It follows that the given differential equation can be written as

$$d(x^2 e^y + \sin y) = 0,$$

and so, from Theorem 1.7.1, the general solution is

$$x^2 e^y + \sin y = c.$$

Notice that the solution obtained in the preceding example is an implicit solution. Due to the nature of the way in which the potential function for an exact equation is obtained, this is usually the case.

Example 1.7.4   Find the general solution of

$$[\sin(xy) + xy \cos(xy) + 2x]\, dx + [x^2 \cos(xy) + 2y]\, dy = 0.$$

Solution   In this case $M = \sin(xy) + xy \cos(xy) + 2x$ and $N = x^2 \cos(xy) + 2y$. Thus,

$$M_y = 2x \cos(xy) - x^2 y \sin(xy) = N_x,$$

and so the differential equation is exact. It follows that there exists a potential function $\phi(x, y)$ such that

$$\frac{\partial \phi}{\partial x} = \sin(xy) + xy \cos(xy) + 2x, \tag{1.7.14}$$

$$\frac{\partial \phi}{\partial y} = x^2 \cos(xy) + 2y. \tag{1.7.15}$$

In this case (1.7.15) is the simpler equation and so we integrate it with respect to y, holding x fixed, to obtain

$$\phi = x \sin(xy) + y^2 + g(x), \qquad (1.7.16)$$

where $g(x)$ is an arbitrary function of x. We now determine $g(x)$, and hence ϕ, from (1.7.14) and (1.7.16). Differentiating (1.7.16) partially with respect to x yields

$$\frac{\partial \phi}{\partial x} = \sin(xy) + xy \cos(xy) + \frac{dg}{dx}. \qquad (1.7.17)$$

Equations (1.7.14) and (1.7.17) give two expressions for $\partial \phi / \partial x$. This allows us to determine g. Subtracting (1.7.14) from (1.7.17) yields the consistency requirement

$$\frac{dg}{dx} = 2x,$$

and hence, upon integrating,

$$g(x) = x^2,$$

where we have once more set the integration constant to zero without loss of generality, since we only require one potential function. Substituting into (1.7.16) gives the potential function

$$\phi(x, y) = x \sin(xy) + x^2 + y^2.$$

The original differential equation can therefore be written as

$$d[x \sin(xy) + x^2 + y^2] = 0,$$

and hence the general solution is

$$x \sin(xy) + x^2 + y^2 = c.$$

REMARK   At first sight this procedure appears to be quite complicated. However, with a little bit of practice you will see that the steps are, in fact, fairly straightforward. As we have shown in Theorem 1.7.2, the method works in general, *provided you start with an exact differential equation.*

EXERCISES 1.7

In problems 1–3 determine whether the given differential equation is exact.

1. $(y + 3x^2) \, dx + x \, dy = 0$.

2. $[\cos(xy) - xy \sin(xy)] \, dx - x^2 \sin(xy) \, dy = 0$.

3. $ye^{xy} \, dx + (2y - xe^{xy}) \, dy = 0$.

In problems 4–12 solve the given differential equation.

4. $2xy \, dx + (x^2 + 1) \, dy = 0$.

5. $(y^2 + \cos x) \, dx + (2xy + \sin y) \, dy = 0$.

6. $\dfrac{(xy - 1)}{x} \, dx + \dfrac{(xy + 1)}{y} \, dy = 0$.

7. $(4e^{2x} + 2xy - y^2) \, dx + (x - y)^2 \, dy = 0$.

8. $(y^2 - 2x) \, dx + 2xy \, dy = 0$.

9. $\left(\dfrac{1}{x} - \dfrac{y}{x^2 + y^2} \right) dx + \dfrac{x}{x^2 + y^2} \, dy = 0$.

10. $[1 + \ln(xy)]\, dx + \dfrac{x}{y}\, dy = 0.$

11. $[y \cos(xy) - \sin x]\, dx + x \cos(xy)\, dy = 0.$

12. $(2xy + \cos y)\, dx + (x^2 - x \sin y - 2y)\, dy = 0.$

In problems 13-15 solve the given initial value problem.

13. $(3x^2 \ln x + x^2 - y)\, dx - x\, dy = 0,\ y(1) = 5.$

14. $2x^2 \dfrac{dy}{dx} + 4xy = 3 \sin x,\ y(2\pi) = 0.$

15. $(ye^{xy} + \cos x)\, dx + xe^{xy}\, dy = 0,\ y\left(\dfrac{\pi}{2}\right) = 0.$

16. Show that if $\phi(x, y)$ is a potential function for $M(x, y)\, dx + N(x, y)\, dy = 0$, then so is $\phi(x, y) + c$, where c is an arbitrary constant. This shows that potential functions are defined only up to an additive constant.

17. Determine the exact differential equation that has potential function

$$\phi(x, y) = x^2y - e^x.$$

1.8   INTEGRATING FACTORS

Usually a given differential equation will not be exact. However, *sometimes* it is possible to multiply the differential equation by a nonzero function to obtain an exact equation that can then be solved using the technique of the previous section. (Notice that the solution of the resulting exact equation will be the same as that of the original equation, since we multiply by a *nonzero* function.)

> **Definition 1.8.1:**   A nonzero function $I(x, y)$ is called an **integrating factor** for $M(x, y)\, dx + N(x, y)\, dy = 0$ if the differential equation
>
> $$I(x, y)M(x, y)\, dx + I(x, y)N(x, y)\, dy = 0$$
>
> is exact.

Example 1.8.1   Show that $I = x^2y$ is an integrating factor for the differential equation

$$(3y^2 + 5x^2y)\, dx + (3xy + 2x^3)\, dy = 0. \tag{1.8.1}$$

Solution   Multiplying the given differential equation (which is not exact) by x^2y yields

$$(3x^2y^3 + 5x^4y^2)\, dx + (3x^3y^2 + 2x^5y)\, dy = 0. \tag{1.8.2}$$

Thus

$$M_y = 9x^2y^2 + 10x^4y = N_x,$$

so that the differential equation (1.8.2) is exact, and hence $I = x^2y$ is an integrating factor for (1.8.1). We see by inspection that (1.8.2) can be written as $d(x^3y^3 + x^5y^2) = 0$, so that its general solution [and hence the general solution of (1.8.1)] is defined implicitly by $x^3y^3 + x^5y^2 = c$, that is, $x^3y^2(y + x^2) = c$.

Using the test for exactness it is straightforward to determine the conditions that a function $I(x, y)$ must satisfy in order to be an integrating factor for the differential equation $M(x, y)\, dx + N(x, y)\, dy = 0$.

Theorem 1.8.1: $I(x, y)$ is an integrating factor for

$$M(x, y)\, dx + N(x, y)\, dy = 0 \qquad (1.8.3)$$

if and only if it is a solution of the partial differential equation

$$N\frac{\partial I}{\partial x} - M\frac{\partial I}{\partial y} = \left(\frac{\partial M}{\partial y} - \frac{\partial N}{\partial x}\right)I. \qquad (1.8.4)$$

PROOF   Multiplying (1.8.3) by I yields

$$IM\, dx + IN\, dy = 0.$$

This equation is exact if and only if

$$\frac{\partial}{\partial y}(IM) = \frac{\partial}{\partial x}(IN),$$

that is, if and only if

$$\frac{\partial I}{\partial y}M + I\frac{\partial M}{\partial y} = \frac{\partial I}{\partial x}N + I\frac{\partial N}{\partial x}.$$

Rearranging these terms yields (1.8.4).   ∎

The previous theorem is not too useful in general, since it is usually no easier to solve the partial differential equation (1.8.4) to find I than it is to solve the original equation (1.8.3). However, it sometimes happens that an integrating factor exists that depends only on one variable. We now show that Theorem 1.8.1 can be used to determine when such an integrating factor exists and also to actually find a corresponding integrating factor.

Theorem 1.8.2:   Consider the differential equation

$$M(x, y)\, dx + N(x, y)\, dy = 0.$$

1. There exists an integrating factor that depends only on x if and only if $\dfrac{M_y - N_x}{N} = f(x)$, a function of x only. In such a case an integrating factor is

$$I(x) = e^{\int f(x)\, dx}.$$

2. There exists an integrating factor that depends only on y if and only if $\dfrac{M_y - N_x}{M} = g(y)$, a function of y only. In such a case an integrating factor is

$$I(y) = e^{-\int g(y)\, dy}.$$

PROOF   We will prove only (1). Suppose first that $I = I(x)$ is an integrating factor for $M(x, y)\, dx + N(x, y)\, dy = 0$. Then $\partial I/\partial y = 0$ and so, from (1.8.4), I is a solution of

$$\frac{dI}{dx}N = (M_y - N_x)I,$$

that is,

$$\frac{1}{I}\frac{dI}{dx} = \frac{M_y - N_x}{N}.$$

Since, by assumption, I is a function of x only, it follows that the left-hand side of this expression (and hence also the right-hand side) depends only on x.

Conversely, suppose that $\dfrac{M_y - N_x}{N} = f(x)$ is a function of x only.   Then, dividing (1.8.4) by N, it follows that I is an integrating factor for

$$M(x, y)\, dx + N(x, y)\, dy = 0$$

if and only if it is a solution of

$$\frac{\partial I}{\partial x} - \frac{M}{N}\frac{\partial I}{\partial y} = If(x). \tag{1.8.5}$$

We must show that this differential equation has a solution $I = I(x)$.   However, if $I = I(x)$, then (1.8.5) reduces to

$$\frac{dI}{dx} = If(x),$$

which is a separable equation with solution

$$I(x) = e^{\int f(x)\, dx}.$$

The proof of (2) is similar and so we leave it as an exercise.   ∎

Example 1.8.2   Solve

$$(2x - y^2)\, dx + xy\, dy = 0. \tag{1.8.6}$$

Solution   The equation is not exact ($M_y \neq N_x$).   However,

$$\frac{(M_y - N_x)}{N} = \frac{(-2y - y)}{xy} = -\frac{3}{x},$$

a function only of x.   It follows from (1) of Theorem 1.8.2 that an integrating factor for (1.8.6) is

$$I(x) = e^{-\int (3/x)\, dx} = e^{-3\ln|x|} = |x|^{-3} = \begin{cases} x^{-3}, & \text{if } x > 0, \\ -x^{-3}, & \text{if } x < 0. \end{cases}$$

Multiplying (1.8.6) by I in both cases ($x > 0$ or $x < 0$) yields the exact equation

$$(2x^{-2} - x^{-3}y^2)\, dx + x^{-2}y\, dy = 0.$$

[You should check that this *is* exact ($M_y = N_x$), although it must be by the previous theorem.] We can now solve the equation using the method given in the previous section or by inspection. We find that

$$\phi(x, y) = \frac{x^{-2}y^2}{2} - 2x^{-1},$$

and hence the general solution of (1.8.6) is given implicitly by

$$\frac{x^{-2}y^2}{2} - 2x^{-1} = c,$$

that is, after multiplying through by $2x^2$ and redefining the constant,

$$y^2 - 4x = c_1 x^2.$$

Theorem 1.8.2 is useful only when there exists an integrating factor that depends on just one variable. In general this is not the case, and it is usually impossible to obtain an integrating factor. However, if $M(x, y)$ and $N(x, y)$ involve only powers of x and y, there sometimes exists an integrating factor of the form $I = x^r y^s$ for appropriate constants r and s. We illustrate how such an integrating factor may be obtained, assuming one exists, via an example.

Example 1.8.3 Determine an integrating factor of the form $I = x^r y^s$ for the differential equation

$$y(3x - 2y^2) \, dx + x(3x - 5y^2) \, dy = 0. \qquad (1.8.7)$$

Solution Multiplying (1.8.7) by $x^r y^s$ we obtain

$$(3x^{r+1}y^{s+1} - 2x^r y^{s+3}) \, dx + (3x^{r+2}y^s - 5x^{r+1}y^{s+2}) \, dy = 0. \qquad (1.8.8)$$

We now try to choose r and s such that this equation is exact. Setting $M_y = N_x$, the equation is exact if and only if

$$3(s + 1)x^{r+1}y^s - 2(s + 3)x^r y^{s+2} = 3(r + 2)x^{r+1}y^s - 5(r + 1)x^r y^{s+2},$$

that is, upon dividing by $x^r y^s$, if and only if

$$3(s + 1)x - 2(s + 3)y^2 = 3(r + 2)x - 5(r + 1)y^2.$$

Equating coefficients of x and y^2 on both sides of this equation, we therefore require r and s to satisfy

$$3(s + 1) = 3(r + 2)$$

$$5(r + 1) = 2(s + 3)$$

that is,

$$r - s = -1,$$

$$5r - 2s = 1.$$

This system of equations has solution $r = 1$, $s = 2$. Thus we have shown that an integrating factor for (1.8.7) is

$$I(x, y) = xy^2.$$

Multiplying (1.8.7) by I yields the exact equation [this can be obtained directly by setting $r = 1$ and $s = 2$ in (1.8.8)]

$$xy^3(3x - 2y^2)\,dx + x^2y^2(3x - 5y^2)\,dy = 0.$$

We leave it as an exercise to show that a potential function for this equation is

$$\phi = x^2y^3(x - y^2),$$

so that the general solution of (1.8.7) is

$$x^2y^3(x - y^2) = c.$$

EXERCISES 1.8

In problems 1–3 determine whether the given function is an integrating factor for the given differential equation.

1. $I(x, y) = \cos(xy)$, $[\tan(xy) + xy]\,dx + x^2\,dy = 0$.
2. $I(x) = \sec x$,
$[2x - (x^2 + y^2)\tan x]\,dx + 2y\,dy = 0$.
3. $I(x, y) = y^{-2}e^{-x/y}$, $y\,[x^2 - 2xy]\,dx - x^3\,dy = 0$.

In problems 4–10 determine an integrating factor for the given differential equation and hence find the general solution.

4. $(xy - 1)\,dx + x^2\,dy = 0$.
5. $y\,dx - (2x + y^4)\,dy = 0$.
6. $x^2y\,dx + y(x^3 + e^{-3y}\sin y)\,dy = 0$.
7. $(y - x^2)\,dx + 2x\,dy = 0$, $x > 0$.
8. $xy\,[2\ln(xy) + 1]\,dx + x^2\,dy = 0$, $x > 0$.
9. $\dfrac{dy}{dx} + \dfrac{2x}{1 + x^2}\,y = \dfrac{1}{(1 + x^2)^2}$.
10. $(3xy - 2y^{-1})\,dx + x(x + y^{-2})\,dy = 0$.

In problems 11–13 determine the values of the constants r and s such that $I(x, y) = x^r y^s$ is an integrating factor for the given differential equation.

11. $(y^{-1} - x^{-1})\,dx + (xy^{-2} - 2y^{-1})\,dy = 0$.

12. $y(5xy^2 + 4)\,dx + x(xy^2 - 1)\,dy = 0$.
13. $2y(y + 2x^2)\,dx + x(4y + 3x^2)\,dy = 0$.

14. Prove that if $\dfrac{M_y - N_x}{M} = g(y)$, a function of y only, then an integrating factor for

$$M(x, y)\,dx + N(x, y)\,dy = 0 \text{ is } I(y) = e^{-\int g(y)\,dy}.$$

15. Consider the general first-order *linear* ordinary differential equation

$$\frac{dy}{dx} + p(x)y = q(x), \qquad \text{(i)}$$

where $p(x)$ and $q(x)$ are continuous functions on some interval (α, β).
(a) Rewrite (i) in differential form and derive the following integrating factor for the resulting equation

$$I(x) = e^{\int p(x)\,dx}. \qquad \text{(ii)}$$

(b) Show that the general solution of (i) can be written in the form

$$y = I^{-1}\left[\int I(x)q(x)\,dx + c\right],$$

where I is given in (ii), and c is an arbitrary constant.

1.9   SUMMARY OF TECHNIQUES FOR SOLVING FIRST-ORDER ORDINARY DIFFERENTIAL EQUATIONS

The techniques that we have learned in the previous sections will always work, provided they are applied to the correct type of equation. The difficulty that is usually encountered at this stage is not in actually applying the techniques, but rather in determining which technique is appropriate. In Table 1.9.1 we summarize the five basic types of differential equations that we have developed techniques for solving.

TABLE 1.9.1 A SUMMARY OF THE BASIC SOLUTION TECHNIQUES FOR $y' = f(x, y)$

Type	Standard form	Technique
Separable	$p(y)\dfrac{dy}{dx} = q(x)$.	Separate the variables and integrate directly.
First-order homogeneous	$\dfrac{dy}{dx} = f(x, y)$, with f homogeneous of degree zero $[f(tx, ty) = f(x, y)]$.	Change variables: $y = xV(x)$ and reduce to a separable equation.
First-order linear	$\dfrac{dy}{dx} + p(x)y = q(x)$.	Rewrite as $\dfrac{d}{dx}[y\, e^{\int p(x)\,dx}] = q(x)e^{\int p(x)\,dx}$, and integrate with respect to x.
Bernoulli equation	$\dfrac{dy}{dx} + p(x)y = q(x)y^n$	Divide by y^n and make the change of variables $u = y^{1-n}$. This reduces the equation to a linear equation.
Exact	$M(x, y)\,dx + N(x, y)\,dy = 0$, with $M_y = N_x$.	The solution is $\phi(x, y) = c$, where ϕ is determined by integrating $\phi_x = M$, $\phi_y = N$.

If a given differential equation cannot be written in one of these forms, then the next step is to try to determine an integrating factor. Unless the given differential equation is written directly in differential form, it is better to leave checking exactness until last, since whether or not a differential equation is exact can depend on the particular differential form that is used.

Example 1.9.1 Determine into which of the above types, if any, the following differential equations fall.

(a) $\dfrac{dy}{dx} = -\dfrac{(8x^5 + 3y^4)}{4xy^3}$.

(b) $(x^5 + y)\,dx - x^2y^3\,dy = 0$.

Solution

(a) Since the given differential equation is written in the form $dy/dx = f(x, y)$, we first check whether it is separable or homogeneous. By inspection we see that it is neither of these. We next check to see if it is a linear or a Bernoulli equation. We rewrite the equation in the equivalent form

$$\frac{dy}{dx} + \frac{3}{4x}y = -2x^4y^{-3}, \tag{1.9.1}$$

which we recognize as a Bernoulli equation with $n = -3$. We could therefore solve the equation using the appropriate technique. Due to the y^{-3} term in (1.9.1), it follows that the equation is *not* a linear equation. Finally we check for exactness. The natural differential form to try for the given differential equation is

$$(8x^5 + 3y^4)\,dx + 4xy^3\,dy = 0. \tag{1.9.2}$$

In this form we have

$$M_y = 12y^3, \qquad N_x = 4y^3,$$

so that the equation is *not* exact. However, we see that

$$\frac{M_y - N_x}{N} = \frac{2}{x},$$

so that, according to Theorem 1.8.2, $I = x^2$ is an integrating factor. Thus we could multiply (1.9.2) by x^2 and then solve it as an exact equation.

(b) In this case the form of the differential equation suggests we first test for exactness. We have

$$M_y = 1, \qquad N_x = -2xy^3.$$

It follows that the differential equation is not exact. Further, $(M_y - N_x)/N$ and $(M_y - N_x)/M$ each depend on both x and y, so that Theorem 1.8.2 is of no help here. We leave it as an exercise to check that the given differential equation does not fit into any of the types that we have discussed and so we cannot find its solution.

EXERCISES 1.9

In problems 1–20, determine into which of the five types of differential equations that we have studied the given equation falls and use an appropriate technique to find the general solution.

1. $\dfrac{dy}{dx} = \dfrac{2\ln x}{xy}$.

2. $xy' - 2y = 2x^2\ln x$.

3. $\dfrac{dy}{dx} = -\dfrac{2xy}{x^2 + 2y}$.

4. $(y^2 + 3xy - x^2)\,dx - x^2\,dy = 0$.

5. $\dfrac{dy}{dx} + y(\tan x + y \sin x) = 0$.

6. $\dfrac{dy}{dx} + \dfrac{2e^{2x}}{1 + e^{2x}}\,y = \dfrac{1}{e^{2x} - 1}$.

7. $\dfrac{dy}{dx} - \dfrac{1}{x}\,y = \dfrac{\sqrt{x^2 - y^2}}{x}$.

8. $\dfrac{dy}{dx} = \dfrac{\sin y + y \cos x + 1}{1 - x \cos y - \sin x}$.

9. $\dfrac{dy}{dx} + \dfrac{1}{x}\,y = \dfrac{25x^2\ln x}{2y}$.

10. $e^{2x+y}\,dy - e^{x-y}\,dx = 0$.

11. $\dfrac{dy}{dx} + y \cot x = \sec x$.

12. $\dfrac{dy}{dx} + \dfrac{2e^x}{1 + e^x}\,y = 2\sqrt{y}e^{-x}$.

13. $y\left[\ln\left(\dfrac{y}{x}\right) + 1\right]dx - x\,dy = 0$.

14. $(1 + 2xe^y)\,dx - (e^y + x)\,dy = 0$.

15. $\dfrac{dy}{dx} + y \sin x = \sin x.$

16. $(3y^2 + x^2)\, dx - 2xy\, dy = 0.$

17. $2x \ln x \dfrac{dy}{dx} - y = -9x^3 y^3 \ln x.$

18. $(1 + x) \dfrac{dy}{dx} = y(2 + x).$

19. $(x^2 - 1) \left(\dfrac{dy}{dx} - 1 \right) + 2y = 0.$

20. $x \sec^2(xy)\, dy = -[y \sec^2(xy) + 2x]\, dx.$

21. Determine all values of the constants m and n, if any, for which the differential equation

$$(x^5 + y^m)\, dx - x^n y^3 \, dy = 0,$$

is each of the following.
(a) Exact.   **(b)** Separable.   **(c)** Homogeneous.
(d) Linear.   **(e)** Bernoulli equation.

1.10   SOME HIGHER-ORDER DIFFERENTIAL EQUATIONS

So far we have developed techniques for solving only special types of first-order differential equations. The methods that we have discussed do not apply directly to higher-order differential equations, and so the solution of such equations usually requires the derivation of new techniques. We note, however, that most higher-order differential equations can be replaced by an equivalent *system* of first-order equations (this will be developed further in Chapter 11). For example, any second-order differential equation that can be written in the form

$$\frac{d^2y}{dx^2} = F\left(x,\, y,\, \frac{dy}{dx}\right), \tag{1.10.1}$$

where F is a known function, can be replaced by an equivalent pair of first-order differential equations as follows.

If we let $u = dy/dx$, then $d^2y/dx^2 = du/dx$, and so solving (1.10.1) is equivalent to solving the following *two* first-order differential equations

$$\frac{dy}{dx} = u, \tag{1.10.2}$$

$$\frac{du}{dx} = F(x,\, y,\, u). \tag{1.10.3}$$

In general the differential equation (1.10.3) cannot be solved directly, since it involves *three* variables, x, y, and u. However, for certain forms of the function F, (1.10.3) will involve only two variables and then can sometimes be solved for u using one of our previous techniques. Having obtained u we can then substitute into (1.10.2) to obtain a first-order differential equation for y. We now discuss two forms of F when this is certainly the case.

CASE 1: SECOND-ORDER EQUATIONS WITH THE DEPENDENT VARIABLE MISSING
If y does not occur explicitly in the function F, then (1.10.1) assumes the form

$$\frac{d^2y}{dx^2} = F\left(x,\, \frac{dy}{dx}\right). \tag{1.10.4}$$

This can be reduced to an equivalent system of two first-order equations using the preceding technique as follows.

Let $u = dy/dx$, so that $du/dx = d^2y/dx^2$. Substituting for dy/dx and d^2y/dx^2 into (1.10.4) allows us to replace the second-order differential equation with the two first-order equations

$$\frac{dy}{dx} = u, \tag{1.10.5}$$

$$\frac{du}{dx} = F(x, u). \tag{1.10.6}$$

Thus to solve (1.10.4), we first solve (1.10.6) for u in terms of x and then solve (1.10.5) for y as a function of x. The method will work only if F in (1.10.4) does not depend explicitly on y.

Example 1.10.1   Find the general solution of

$$\frac{d^2y}{dx^2} = \frac{1}{x}\left(\frac{dy}{dx} + x^2\cos x\right), \qquad x > 0. \tag{1.10.7}$$

Solution   In (1.10.7) the dependent variable is missing and so we let

$$u = \frac{dy}{dx},$$

which implies that

$$\frac{d^2y}{dx^2} = \frac{du}{dx}.$$

Substituting into (1.10.7) yields the following equivalent first-order system

$$\frac{du}{dx} = \frac{1}{x}(u + x^2\cos x), \tag{1.10.8}$$

$$\frac{dy}{dx} = u. \tag{1.10.9}$$

Rearranging the terms in (1.10.8) yields

$$\frac{du}{dx} - \frac{1}{x}u = x \cos x, \tag{1.10.10}$$

which we recognize as being first-order linear. An appropriate integrating factor is

$$I = e^{-\int(1/x)\,dx} = \frac{1}{x}.$$

It follows that the differential equation (1.10.10) can be written in the equivalent form

$$\frac{d}{dx}\left(\frac{1}{x}u\right) = \cos x,$$

which can be integrated directly to obtain

$$\frac{1}{x}u = \sin x + c_1.$$

Thus

$$u = x \sin x + c_1 x. \qquad (1.10.11)$$

Substituting for u into (1.10.9), we must solve

$$\frac{dy}{dx} = x \sin x + c_1 x,$$

that is,

$$dy = (x \sin x + c_1 x)\, dx.$$

This can be integrated to yield

$$y = -x \cos x + \sin x + c_1 x^2 + c_2,$$

where we have absorbed a factor of $\frac{1}{2}$ into c_1.

CASE 2: SECOND-ORDER EQUATIONS WITH THE INDEPENDENT VARIABLE MISSING
If x does not occur explicitly in the function F in (1.10.1), then we must solve a differential equation of the form

$$\frac{d^2y}{dx^2} = F\left(y, \frac{dy}{dx}\right). \qquad (1.10.12)$$

If we try the change of variables suggested previously, then (1.10.12) is replaced by the first-order differential equation

$$\frac{du}{dx} = F(y, u).$$

This first-order equation cannot be solved directly, since it involves three variables. We can, however, replace (1.10.12) by a solvable first-order system by making a different change of variables. We let

$$u = \frac{dy}{dx}$$

as before, but in this case we use the chain rule to express d^2y/dx^2 in terms of du/dy (rather than du/dx) as follows:

$$\frac{d^2y}{dx^2} = \frac{du}{dx} = \frac{du}{dy}\frac{dy}{dx} = u\frac{du}{dy}.$$

Via chain rule

Substituting for dy/dx and d^2y/dx^2 into (1.10.12) reduces the second-order equation to the equivalent first-order system

$$u\frac{du}{dy} = F(y, u), \tag{1.10.13}$$

$$\frac{dy}{dx} = u. \tag{1.10.14}$$

In this case we first solve (1.10.13) for u as a function of y and then solve (1.10.14) for y as a function of x.

Example 1.10.2   Find the general solution of

$$\frac{d^2y}{dx^2} = -\frac{2}{(1-y)}\left(\frac{dy}{dx}\right)^2. \tag{1.10.15}$$

Solution   In this differential equation the independent variable (x) does not occur explicitly.   Thus we let $u = dy/dx$, so that

$$\frac{d^2y}{dx^2} = \frac{du}{dx} = \frac{du}{dy}\frac{dy}{dx} = u\frac{du}{dy}.$$

Substituting into (1.10.15) we must solve

$$u\frac{du}{dy} = -\frac{2}{(1-y)}u^2, \tag{1.10.16}$$

$$\frac{dy}{dx} = u. \tag{1.10.17}$$

Separating the variables in (1.10.16) yields

$$\frac{1}{u}\,du = -\frac{2}{(1-y)}\,dy, \tag{1.10.18}$$

which can be integrated to obtain

$$\ln|u| = 2\ln|(1-y)| + \ln c;$$

that is, combining the log terms,

$$u = c(1-y)^2, \tag{1.10.19}$$

where we have redefined c to take account of the absolute value signs. Notice that in solving (1.10.16) we implicitly assumed that $u \neq 0$, since we divided by it to obtain (1.10.18). However the general form (1.10.19) does include the solution $u = 0$, provided we allow c to equal zero. Substituting for u into (1.10.17) yields

$$\frac{dy}{dx} = c(1-y)^2.$$

Separating the variables, we must solve

$$\frac{1}{(1-y)^2}\,dy = c\,dx,$$

which has general solution

$$(1 - y)^{-1} = cx + d,$$

that is,

$$1 - y = \frac{1}{cx + d}.$$

Solving for y yields

$$y = \frac{cx + (d - 1)}{cx + d}. \tag{1.10.20}$$

If we divide the numerator and denominator by c and redefine the constants by[1] $a := (d - 1)/c$, $b := d/c$, then our solution can be written as

$$y = \frac{x + a}{x + b}. \tag{1.10.21}$$

Notice that the form (1.10.21) does not include the solution $y = $ constant, which *is* contained in (1.10.20) (set $c = 0$). This is because in dividing by c, we implicitly assumed that $c \neq 0$. Thus in specifying the solution in the form (1.10.21), we should also include the statement that $y = $ constant is a solution.

Example 1.10.3   An object is attached to the lower end of a vertical elastic spring, the other end being held fixed (see Figure 1.10.1). If the object is displaced a distance a units from its equilibrium position and released from rest, the subsequent motion of the object is described by the initial value problem

$$\frac{d^2x}{dt^2} = -\omega^2 x, \tag{1.10.22}$$

$$x(0) = a, \qquad \frac{dx}{dt}(0) = 0, \tag{1.10.23}$$

where ω is a positive constant. Solve this initial value problem to find the displacement, $x(t)$, of the object from equilibrium at time t.

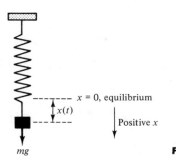

$x = 0$, equilibrium

$x(t)$

Positive x

mg

Figure 1.10.1   A spring-mass system.

---

[1] Here and throughout the text the notation := means "is defined by."

Solution The differential equation (1.10.22) has the independent variable (in this case, t) missing. We therefore let $u = dx/dt$, so that

$$\frac{d^2x}{dt^2} = u \frac{du}{dx}.$$

It then follows that (1.10.22) can be replaced by the equivalent first-order system

$$u \frac{du}{dx} = -\omega^2 x, \qquad (1.10.24)$$

$$\frac{dx}{dt} = u. \qquad (1.10.25)$$

Separating the variables and integrating (1.10.24) yields

$$\frac{u^2}{2} = -\frac{\omega^2 x^2}{2} + c,$$

which implies that

$$u = \pm\sqrt{c_1 - \omega^2 x^2},$$

where $c_1 = 2c$. Substituting for u into (1.10.25) yields

$$\frac{dx}{dt} = \pm\sqrt{c_1 - \omega^2 x^2}. \qquad (1.10.26)$$

Setting $t = 0$ in this equation and using the initial conditions (1.10.23), we find that $c_1 = \omega^2 a^2$. It follows from (1.10.26) that we must solve

$$\frac{dx}{dt} = \pm\omega \sqrt{a^2 - x^2}.$$

Once more the variables are separable:

$$\frac{1}{\sqrt{a^2 - x^2}}\, dx = \pm\omega\, dt,$$

which, upon integrating, yields

$$\sin^{-1}\left(\frac{x}{a}\right) = \pm\omega t + b,$$

where b is a constant. Thus,

$$x = a \sin(b \pm \omega t).$$

The initial condition $x(0) = a$ implies that $\sin(b) = 1$, and so we can choose $b = \pi/2$. We thus have

$$x = a \sin\left(\frac{\pi}{2} \pm \omega t\right);$$

that is,

$$x = a \cos \omega t.$$

REMARK   When we study higher-order linear differential equations in Chapter 9, you will learn how to solve the initial value problem (1.10.22), (1.10.23) in about three lines of work!

EXERCISES 1.10

In problems 1–13 solve the given differential equation.

1. $\dfrac{d^2y}{dx^2} = \dfrac{2}{x}\dfrac{dy}{dx} + 4x^2.$

2. $\dfrac{d^2y}{dx^2} = \dfrac{1}{(x-1)(x-2)}\left(\dfrac{dy}{dx} - 1\right).$

3. $\dfrac{d^2y}{dx^2} + \dfrac{2}{y}\left(\dfrac{dy}{dx}\right)^2 = \dfrac{dy}{dx}.$

4. $\dfrac{d^2y}{dx^2} = \left(\dfrac{dy}{dx}\right)^2 \tan y.$

5. $\dfrac{d^2y}{dx^2} + \tan x\,\dfrac{dy}{dx} = \left(\dfrac{dy}{dx}\right)^2.$

6. $\dfrac{d^2x}{dt^2} = \left(\dfrac{dx}{dt}\right)^2 + 2\,\dfrac{dx}{dt}.$

7. $\dfrac{d^2y}{dx^2} - \dfrac{2}{x}\dfrac{dy}{dx} = 6x^4.$

8. $t\,\dfrac{d^2x}{dt^2} = 2\left(t + \dfrac{dx}{dt}\right).$

9. $\dfrac{d^2y}{dx^2} - \alpha\left(\dfrac{dy}{dx}\right)^2 - \beta\,\dfrac{dy}{dx} = 0,\ \alpha,\ \beta$ constants $(\neq 0).$

10. $\dfrac{d^2y}{dx^2} - \dfrac{2}{x}\dfrac{dy}{dx} = 18x^4.$

11. $\dfrac{d^2y}{dx^2} = -\dfrac{2x}{(1+x^2)}\dfrac{dy}{dx}.$

12. $\dfrac{d^2y}{dx^2} + \dfrac{1}{y}\left(\dfrac{dy}{dx}\right)^2 = ye^{-y}\left(\dfrac{dy}{dx}\right)^3.$

13. $\dfrac{d^2y}{dx^2} - \tan x\,\dfrac{dy}{dx} = 1,\ 0 \le x < \dfrac{\pi}{2}.$

In problems 14 and 15 solve the given initial value problem.

14. $yy'' = 2y'^2 + y^2,\ y(0) = 1,\ y'(0) = 0.$

15. $\dfrac{d^2y}{dx^2} = \omega^2 y,\ y(0) = a,\ y'(0) = 0,\ \omega,\ a$ positive constants.

16. Consider the following general second-order linear differential equation with dependent variable missing:

$$\frac{d^2y}{dx^2} + p(x)\,\frac{dy}{dx} = q(x).$$

Replace this differential equation with an equivalent pair of first-order equations and express the solution in terms of integrals.

17. Consider the general third-order differential equation of the form

$$\frac{d^3y}{dx^3} = F\left(x, \frac{d^2y}{dx^2}\right). \tag{i}$$

(a) Show that (i) can be replaced by the equivalent first-order system

$$\frac{du_1}{dx} = u_2, \qquad \frac{du_2}{dx} = u_3, \qquad \frac{du_3}{dx} = F(x, u_3),$$

where the variables $u_1,\ u_2,\ u_3$ are defined by

$$u_1 = y, \qquad u_2 = \frac{dy}{dx}, \qquad u_3 = \frac{d^2y}{dx^2}.$$

(b) Solve:

$$\frac{d^3y}{dx^3} = \frac{1}{x}\left(\frac{d^2y}{dx^2} - 1\right).$$

18. A simple pendulum consists of a particle of mass m supported by a piece of string of length L. Assuming that the pendulum is displaced through an angle θ_0 radians from the vertical and then released from rest, the resulting motion is described by the initial value problem

$$\frac{d^2\theta}{dt^2} + \frac{g}{L}\sin\theta = 0, \quad \theta(0) = \theta_0, \quad \frac{d\theta}{dt}(0) = 0. \tag{ii}$$

(a) For small oscillations, $\theta \ll 1$, we can use the approximation $\sin \theta \approx \theta$ in (ii) to obtain the linear equation

$$\frac{d^2\theta}{dt^2} + \frac{g}{L}\theta = 0, \qquad \theta(0) = \theta_0, \qquad \frac{d\theta}{dt}(0) = 0.$$

Solve this initial value problem for θ as a function of t. Is the predicted motion reasonable?

(b) Obtain the following first integral of (ii):

$$\frac{d\theta}{dt} = \pm\sqrt{\frac{2g}{L}}\,(\cos \theta - \cos \theta_0)^{1/2}. \qquad \text{(iii)}$$

(c) Show from (iii) that the time $T\,[=(\text{period of}$

motion)$/4]$ required for θ to change from 0 to θ_0 is given by the *elliptic integral of the first kind,*

$$T = \sqrt{\frac{L}{2g}}\int_0^{\theta_0} \frac{1}{(\cos \theta - \cos \theta_0)^{1/2}}\,d\theta. \qquad \text{(iv)}$$

(d) (Harder) Show that (iv) can be written as

$$T = \sqrt{\frac{L}{g}}\int_0^{\pi/2} \frac{1}{(1 - k^2\sin^2 u)^{1/2}}\,du,$$

where $k = \sin(\theta_0/2)$.

[*Hint:* First express $\cos \theta$ and $\cos \theta_0$ in terms of $\sin^2(\theta/2)$ and $\sin^2(\theta_0/2)$.]

2

Applications of First-Order Ordinary Differential Equations

2.1 ORTHOGONAL TRAJECTORIES

In this chapter we consider several applied problems that can be formulated mathematically in terms of first-order differential equations. We begin with an important geometric problem.

Suppose

$$F(x, y, c) = 0 \tag{2.1.1}$$

defines a family of curves in the xy-plane, where the constant c labels the different curves. We assume that every curve has a well-defined tangent at each point. Associated with this family is a second family of curves, say

$$G(x, y, k) = 0, \tag{2.1.2}$$

with the property that whenever a curve from the family (2.1.2) intersects a curve from the family (2.1.1), it does so at right angles.[1] We say that the curves in the family (2.1.2) are **orthogonal trajectories** of the family (2.1.1), and conversely.

---

[1] That is, the tangent lines to each curve are perpendicular at any point of intersection.

Example 2.1.1   From elementary geometry it follows that the straight lines $y = kx$ are orthogonal trajectories of the family of concentric circles $x^2 + y^2 = c^2$ (see Figure 2.1.1).

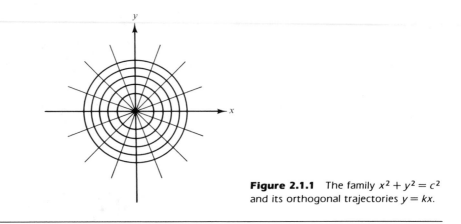

Figure 2.1.1   The family $x^2 + y^2 = c^2$ and its orthogonal trajectories $y = kx$.

Orthogonal trajectories arise in various applications. For example, a family of curves and its orthogonal trajectories can be used to define an orthogonal coordinate system in the xy-plane. In Figure 2.1.1 the families $x^2 + y^2 = c^2$ and $y = kx$ are the coordinate curves of a polar coordinate system (that is, the curves $r = $ constant and $\theta = $ constant, respectively.) In physics, the lines of electric force of a static configuration are the orthogonal trajectories of the family of equipotential curves. As a final example, if we consider a two-dimensional heated plate, then the direction of heat flow is along the orthogonal trajectories to the constant temperature curves (isotherms).

Statement of the problem: Given the equation of a family of curves, find the equation of the family of orthogonal trajectories.

Mathematical formulation: We recall that curves that intersect at right angles satisfy

product of the slopes[1] at the point of intersection $= -1$.

Thus if the given family has slope $m_1 = f(x, y)$ at the point (x, y), then the slope of the family of orthogonal trajectories is $m_2 = -1/f(x, y)$, and hence the differential equation that determines the orthogonal trajectories is

$$\frac{dy}{dx} = -\frac{1}{f(x, y)}.$$

We illustrate the procedure for determining orthogonal trajectories with an example.

---

[1] By the slope of a curve at a given point we mean the slope of the tangent line to the curve at that point.

Example 2.1.2 Find the equation of the orthogonal trajectories to the family

$$x^2 + y^2 - 2cx = 0. \qquad (2.1.3)$$

[Completing the square in x we obtain $(x - c)^2 + y^2 = c^2$, which represents the family of circles centered at $(c, 0)$ with radius c.]

Solution First we need an expression for the slope of the given family at the point (x, y). Differentiating (2.1.3) implicitly with respect to x yields

$$2x + 2y\frac{dy}{dx} - 2c = 0. \qquad (2.1.4)$$

This is not the differential equation of the given family, since it still contains the constant c, and hence is dependent on the individual curves in the family. There-fore, *we must eliminate c to obtain an expression for the slope of the family that is independent of any particular curve in the family.* From (2.1.3) we have

$$c = \frac{x^2 + y^2}{2x},$$

and so, substituting into (2.1.4), we obtain

$$2x + 2y\frac{dy}{dx} - \frac{(x^2 + y^2)}{x} = 0;$$

that is, solving algebraically for dy/dx,

$$\frac{dy}{dx} = \frac{y^2 - x^2}{2xy}.$$

Thus the slope of the given family at the point (x, y) is

$$m_1 = \frac{y^2 - x^2}{2xy},$$

so that the differential equation of the orthogonal trajectories is

$$\frac{dy}{dx} = -\frac{2xy}{y^2 - x^2}. \qquad (2.1.5)$$

This differential equation is first-order homogeneous. We therefore let $y = xV(x)$, in which case

$$\frac{dy}{dx} = x\frac{dV}{dx} + V.$$

Substituting into (2.1.5) for y and dy/dx yields

$$x\frac{dV}{dx} + V = \frac{2V}{1 - V^2};$$

that is,

$$x \frac{dV}{dx} = \frac{V + V^3}{1 - V^2}.$$

Separating the variables we obtain

$$\frac{(1 - V^2)}{V(1 + V^2)} \, dV = \frac{1}{x} \, dx.$$

Decomposing the left-hand side into partial fractions yields

$$\left(\frac{1}{V} - \frac{2V}{1 + V^2} \right) dV = \frac{1}{x} \, dx,$$

which can be integrated directly to obtain

$$\ln|V| - \ln(1 + V^2) = \ln|x| + \ln c,$$

or, equivalently,

$$\ln\left(\frac{|V|}{1 + V^2} \right) = \ln|cx|.$$

Exponentiating both sides and redefining the constant yields

$$\frac{V}{1 + V^2} = c_1 x.$$

Substituting back for $V = y/x$, we obtain

$$\frac{xy}{x^2 + y^2} = c_1 x,$$

that is,

$$x^2 + y^2 = c_2 y,$$

where $c_2 = 1/c_1$.   Completing the square in y yields

$$x^2 + (y - k)^2 = k^2, \tag{2.1.6}$$

where $k = c_2/2$.   Equation (2.1.6) is the equation of the family of orthogonal trajectories.   This is the family of circles centered at $(0, k)$ with radius k (circles along the y-axis) (see Figure 2.1.2).

WARNING   The two most common mistakes that arise in solving for orthogonal trajectories are

1. To forget to eliminate the constant in obtaining an expression for the slope of the given family.
2. To forget to replace $f(x, y)$ by $-1/f(x, y)$ (in this case the given family is obtained rather than the orthogonal trajectories upon integrating).

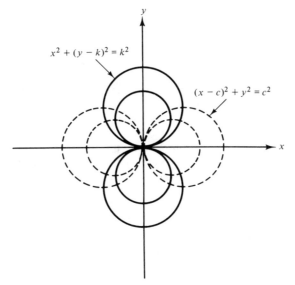

$x^2 + (y - k)^2 = k^2$

$(x - c)^2 + y^2 = c^2$

Figure 2.1.2 The family $(x - c)^2 + y^2 = c^2$ and its orthogonal trajectories $x^2 + (y - k)^2 = k^2$.

EXERCISES 2.1

In problems 1–8 find the equation of the orthogonal trajectories to the given family of curves. In each case sketch some curves from each family.

1. $x^2 + 4y^2 = c$.

2. $y = \dfrac{c}{x}$.

3. $y = cx^2$.

4. $x^2 + y^2 = 2cy$.

5. $y^2 = cx$.

6. $y^2 = 2x + c$.

7. $y = ce^x$.

8. $(x - c)^2 + (y - c)^2 = 2c^2$.

In problems 9–12, m ($\neq 0$) is a fixed constant and c is the constant labeling the different curves in the given family. In each case find the equation of the orthogonal trajectories.

9. $y = mx + c$ (guess the answer before solving).

10. $y = cx^m$.

11. $y^2 + mx^2 = c$.

12. $y^2 = mx + c$.

13. Let S_1 denote the family of circles, centered on the line $y = mx$, each member of which passes through the origin.

(a) Show that the equation of S_1 can be written in the form

$$(x - a)^2 + (y - ma)^2 = a^2(m^2 + 1),$$

where a is a constant that labels particular members of the family.
(b) Determine the equation of the family of orthogonal trajectories to S_1, and show that it consists of the family of circles centered on the line $x = -my$ that pass through the origin.

14. A family of curves whose orthogonal trajectories belong to the same family is said to be **self-orthogonal**. Show that the family

$$2cy = x^2 - c^2,$$

where c is a constant, is self-orthogonal. (*Hint:* This problem can be solved *without* integration.)

15. A coordinate system (u, v) is called orthogonal if its coordinate curves (the two families of curves u = constant, v = constant) are orthogonal trajectories (for example, a Cartesian coordinate system or a polar coordinate system). Let (u, v) be orthogonal coordinates where $u = x^2 + 2y^2$, x and y being Cartesian coordinates. Find the Cartesian equation of the v-coordinate curves, and sketch the (u, v) coordinate system.

Any curve with the property that whenever it intersects a curve of a given family it does so at an angle $\alpha \neq \pi/2$ is called an **oblique trajectory** of the given family. The remaining problems deal with determining all oblique trajectories of a given family of curves.

16. Let m_1 ($= \tan \theta_1$) denote the slope of the required family at the point (x, y), and let m_2 ($= \tan \theta_2$) denote the slope of the given family (see Figure 2.1.3). Show that

$$m_1 = \frac{m_2 - \tan \alpha}{1 + m_2 \tan \alpha}.$$

[*Hint:* From Figure 2.1.3, $\tan \theta_1 = \tan(\theta_2 - \alpha)$.] Thus the equation of the family of oblique trajectories is obtained by solving

$$\frac{dy}{dx} = \frac{m_2 - \tan \alpha}{1 + m_2 \tan \alpha}. \quad (i)$$

In problems 17–22 use (i) to determine the equation of the family of curves that cuts the given family at the given angle.

17. $y = \frac{1}{2}x + c, \ \alpha = \frac{\pi}{4}$.

18. $y = 2x^2 + c, \ \alpha = \frac{\pi}{6}$.

19. $x^2 + y^2 = c, \ \alpha = \frac{\pi}{4}$.

20. $y = cx^6, \ \alpha = \frac{\pi}{4}$.

21. $x^2 + y^2 = 2cx, \ \alpha = \frac{\pi}{4}$.

22. $(x + \sqrt{3}c)^2 + (y - c)^2 = 4c^2, \ \alpha = \frac{\pi}{6}$.

23. Let S_1 denote the family of circles with equation

$$(x - c)^2 + y^2 = c^2.$$

(a) Show that the equation of the family of curves that cuts S_1 at an angle $\alpha = \tan^{-1}(m)$, where m is a constant, is determined from

$$\frac{mV^2 + 2V - m}{(V^2 + 1)(mV + 1)} dV = -\frac{1}{x} dx, \quad V = y/x. \quad (ii)$$

(b) Integrate (ii) and hence show that the equation of the required family is

$$(x - k)^2 + (y - mk)^2 = k^2(1 + m^2). \quad (iii)$$

Describe the family of curves (iii).

24. Find the equation of the family of curves that intersects the family of hyperbolas $y = cx^{-1}$ at an angle of $\alpha = \alpha_0 \ (\neq \pi/2)$.

25. Show that the family of curves that cuts the family of concentric circles $x^2 + y^2 = c$ at an angle $\alpha = \tan^{-1}(m)$ is

$$\ln(\sqrt{x^2 + y^2}) - m \tan^{-1}\left(\frac{y}{x}\right) = k.$$

Express this result in polar coordinates, and describe the curves.

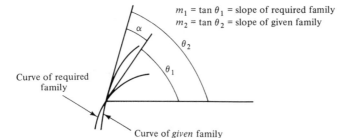

$m_1 = \tan \theta_1 =$ slope of required family
$m_2 = \tan \theta_2 =$ slope of given family

Curve of required family

Curve of *given* family

Figure 2.1.3 Oblique trajectories intersecting an angle α.

2.2 EXPONENTIAL GROWTH AND DECAY

In many applied problems we are interested in the behavior of some quantity, say $x(t)$, whose rate of change in time is proportional to the instantaneous value of x. The differential equation that describes this behavior is

$$\frac{dx}{dt} = kx, \tag{2.2.1}$$

where k is a constant and t denotes time. This separable equation has general solution

$$x = x_0 e^{kt}, \tag{2.2.2}$$

where the integration constant x_0 is the initial value $x(0)$. The resulting behavior is called **exponential growth** or **exponential decay**, depending on whether the constant k is positive or negative, respectively. Representative solutions are sketched in Figure 2.2.1 for the case when $x_0 > 0$. In this section we discuss examples of each of these phenomena.

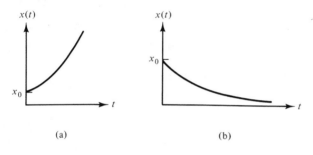

(a)   (b)

Figure 2.2.1 (a) Exponential growth: $x = x_0 e^{kt}$, $k > 0$. (b) Exponential decay: $x = x_0 e^{kt}$, $k < 0$.

A SIMPLE MODEL OF POPULATION GROWTH

The simplest mathematical model of population growth is obtained by assuming that the rate of increase of the population at any time is proportional to the size of the population at that time. If we let $P(t)$ denote the population at time t, then

$$\frac{dP}{dt} = kP, \tag{2.2.3}$$

where k is a *positive* constant. Separating the variables and integrating yields

$$P = P_0 e^{kt}, \tag{2.2.4}$$

where P_0 denotes the population at $t = 0$. This law predicts an exponential increase in the population with time that gives a reasonably accurate description of the growth of certain algae, bacteria, and cell cultures.[1] The time taken for such a culture to double in size is called the **doubling time.** This is the time, t_d, when $P = 2P_0$. Substituting into (2.2.4) yields

$$2P_0 = P_0 e^{kt_d}.$$

---

[1] The exponential law (2.2.4) does not give a good approximation for the growth of the population in a city or country. This should not be surprising, since we have not taken into account important factors such as disease, war, famine, immigration and emigration, all of which affect the change in a population.

Dividing both sides by P_0 and taking logarithms, we find

$$kt_d = \ln 2,$$

so that the doubling time is

$$t_d = \frac{1}{k} \ln 2. \tag{2.2.5}$$

Notice that t_d is independent of P_0 and hence is a property of the system.

Example 2.2.1   The number of bacteria in a certain culture grows at a rate that is proportional to the number present. If the number increased from 500 to 2000 in 2 h, determine the number present after 12 h, and also find the doubling time.

Solution   The behavior of the system is governed by the differential equation

$$\frac{dP}{dt} = kP,$$

so that

$$P = P_0 e^{kt}.$$

Taking $t = 0$ as the time when the population was 500, we have $P_0 = 500$. Thus

$$P = 500 e^{kt}.$$

Further, $P(2) = 2000$ implies that

$$2000 = 500 e^{2k},$$

so that

$$k = \tfrac{1}{2} \ln 4 = \ln 2.$$

Consequently,

$$P = 500 e^{t \ln 2}.$$

The number of bacteria present after 12 h is therefore

$$P(12) = 500 e^{12 \ln 2} = 500(2^{12}) = 2{,}048{,}000.$$

The doubling time of the system is

$$t_d = \frac{1}{k} \ln 2 = 1 \text{ h}.$$

RADIOACTIVE DECAY

Let $N(t)$ denote the number of radioactive atoms present in a sample of radioactive material. Then it is found experimentally that N decays at a rate that is proportional to the number of radioactive atoms present in the material at time t. This law of radioactive decay can be expressed mathematically as the differential equation

$$\frac{dN}{dt} = kN, \tag{2.2.6}$$

where k is a *negative* constant.   The general solution to (2.2.6) is

$$N = N_0 e^{kt}, \tag{2.2.7}$$

where N_0 denotes the number of radioactive atoms initially present in the sample. Since k is negative in (2.2.7) we see that $N(t)$ *decays* exponentially with time.   The time taken for exactly half of the radioactive atoms initially present in the sample to decay is called the **half-life** of the material.   Taking $N = \frac{1}{2}N_0$ in (2.2.7), it follows that the half-life, denoted by $t_{1/2}$, is determined from

$$\tfrac{1}{2}N_0 = N_0 e^{kt_{1/2}}.$$

Solving for $t_{1/2}$ yields

$$t_{1/2} = -\frac{1}{k}\ln 2. \tag{2.2.8}$$

Notice that $t_{1/2}$ is independent of N_0 and hence is a property of the radioactive material.   This should be compared with the doubling time for the population model considered previously.

One application of the radioactive decay law is in carbon dating.   During the lifetime of an organism it is found that the ratio of radioactive ^{14}C (carbon-14) to ordinary carbon present in the organism is approximately constant and is equal to the ratio in the surrounding medium.   However, once the organism dies, the amount of ^{14}C present in its remains decreases due to radioactive decay.   Since it is known that the half-life of ^{14}C is approximately 5600 years, by measuring the amount of ^{14}C in the remains of organisms it is possible to estimate their age.

Example 2.2.2   A fossilized bone is found to have 70% of the ^{14}C that is found in the bone of a living human.   Taking the half-life of ^{14}C to be 5600 y, determine the age of the fossil.

Solution   The decay of ^{14}C is governed by the differential equation

$$\frac{dN}{dt} = kN,$$

so that

$$N = N_0 e^{kt}. \tag{2.2.9}$$

Since the half-life of ^{14}C is 5600 y, we have $N(5600) = \frac{1}{2}N_0$, which implies that

$$\frac{1}{2} = e^{5600k},$$

so that

$$k = -\frac{1}{5600}\ln 2.$$

Substitution into (2.2.9) yields

$$N = N_0 e^{-(t/5600)\ln 2}.$$ (2.2.10)

We are given that at the present time the fossil contains 70% of the original ^{14}C, that is, $N = \frac{7}{10}N_0$. Substituting this value of N into (2.2.10) yields

$$\frac{7}{10} = e^{-(t/5600)\ln 2},$$

so that

$$t = -\frac{5600}{\ln 2}\ln\left(\frac{7}{10}\right) \approx 2881.61 \text{ y}.$$

Thus the age of the fossil is approximately 2882 y.

EXERCISES 2.2

1. The number of bacteria in a culture grows at a rate proportional to the number present. Initially there were 10 bacteria in the culture. If the doubling time of the system is 3 h, find the number of bacteria present after 24 h.

2. The number of bacteria in a culture grows at a rate proportional to the number present. After 10 h there were 5000 bacteria present, and after 12 h there were 6000 bacteria present. Determine the initial size of the culture and the doubling time of the system.

3. A certain cell culture has a doubling time of 4 h. Initially there were 2000 cells present. Assuming an exponential growth law, determine the time it takes for the culture to contain 10^6 cells.

4. At any time t, the population, $P(t)$, of a certain city is increasing at a rate that is proportional to the number of residents in the city at that time. In January 1980 ($t = 0$), the population of the city was 10,000, and 5 y later it was 20,000. Determine:
(a) What the population of the city will be at the beginning of the year 2000.
(b) The year in which the population will reach 1 million.

5. Consider the population model in which the birth rate per individual, α, and the death rate per individual, β, are constants. The differen-

tial equation governing this model is

$$\frac{dP}{dt} = (\alpha - \beta)P.$$

Solve this differential equation and determine the behavior it predicts for the population as $t \to \infty$. If $(\alpha - \beta) > 0$, determine the doubling time.

6. Let $B(t)$ and $D(t)$ denote the birth rate per individual and death rate per individual in a certain population. Then, neglecting factors such as immigration and emigration, the population, $P(t)$, at any time, t, is governed by the differential equation

$$\frac{dP}{dt} = (B - D)P.$$ (i)

(a) Suppose that $B = \alpha P$ and $D = \beta$, where α and β are positive constants. Solve (i) and show that if $\beta/\alpha < P_0$, where P_0 denotes the initial population, then there is a finite time, $t = t_0$, such that $\lim_{t \to t_0} P(t) = \infty$. This gives a mathematical model of population explosion.
(b) Now consider the case when $B = \alpha$ and $D = \beta P$, where α and β are positive constants. Solve (i) and show that $\lim_{t \to \infty} P(t) = \alpha/\beta$.

7. If it takes 20 y for 30% of a sample of radioactive material to decay, determine the half-life of the material.

8. The radioactive isotope ^{67}Cu has a half-life of 12.4 h. How long will it take for 99% of a sample to decay?

9. The radioactive decay of ^{125}I (radioactive iodine) is governed by the differential equation

$$\frac{dN}{dt} = kN,$$

where $k = -0.011552$ and t is measured in days. Determine the half-life.

10. Show that the half-life of a radioactive material is given by the formula

$$t_{1/2} = \frac{t_2 - t_1}{\ln(N_1/N_2)} \ln 2,$$

where N_1 and N_2 denote the amounts of material present at times t_1 and t_2, respectively.

11. The remains of a murder victim are found to contain 90% of the normal amount of ^{14}C. Taking the half-life of ^{14}C to be 5600 y, determine how long ago the murder took place.

Suppose A_0 dollars is invested in an account that has an annual interest rate of $r\%$. If the interest is compounded k times per year, then the amount of money in the account after t years is

$$A(t) = A_0\left(1 + \frac{r}{k}\right)^{kt}, \qquad \text{(ii)}$$

whereas if the interest is compounded continuously, then $A(t)$ is governed by the initial value problem

$$\frac{dA}{dt} = rA, \qquad A(0) = A_0. \qquad \text{(iii)}$$

Problems 12–14 deal with compound interest.

12. Show that the solution to (iii) is

$$A(t) = A_0 e^{rt}. \qquad \text{(iv)}$$

13. Show that (iv) can be obtained by letting k tend to infinity in (ii). This shows that the definition of continuously compounded interest is reasonable.

14. If $5000 is invested in an account at an annual interest rate of 5%, use (ii) and (iv) to determine the amount of money in the account after 2 y if the interest is:
(a) compounded annually;
(b) compounded quarterly;
(c) compounded continuously.

In a first-order chemical reaction, the rate at which a chemical A is converted into a chemical B is proportional to the amount of A that is unconverted. The following two problems deal with first-order reactions.

15. Determine the differential equation that governs a first-order reaction.

16. In a first-order chemical reaction, half of the original amount of substance A remains after 1 min. Determine the time taken for 90% of the sample to have been converted.

2.3   NEWTON'S LAW OF COOLING

We now consider a physical law whose mathematical formulation gives rise to a first-order differential equation. Suppose that we bring an object into a room. If the temperature of the object is hotter than that of the room, then the object will begin to cool. Further, we might expect that the major factor affecting the rate at which the object cools is the temperature difference between it and the room. Indeed, according to Newton's law of cooling:

> the rate of change of temperature of an object is proportional to the temperature difference between the object and its surrounding medium.

To formulate this law mathematically, we let T denote the temperature of the object at time t, and let T_m denote the temperature of the surrounding medium. Newton's law of cooling can then be expressed as the differential equation

$$\frac{dT}{dt} = -k(T - T_m), \tag{2.3.1}$$

where k is a constant. The minus sign in front of the constant k is traditional. It ensures that k will always be positive.[1] We will assume that T_m is a constant, in which case the differential equation (2.3.1) is separable with general solution

$$T = T_m + ce^{-kt}. \tag{2.3.2}$$

We see from (2.3.2) that Newton's law of cooling predicts that as $t \to \infty$, the temperature of the object approaches that of the surrounding medium ($T \to T_m$). This is certainly consistent with our everyday experience (see Figure 2.3.1).

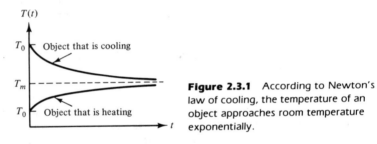

Figure 2.3.1   According to Newton's law of cooling, the temperature of an object approaches room temperature exponentially.

REMARK   You should *not* memorize (2.3.2); in a given problem you will be expected to derive it from (2.3.1).

Example 2.3.1   A hot metal bar whose temperature is 350°F is placed in a room whose temperature is 70°F. After 2 min the temperature of the bar is 210°F. Determine the temperature of the bar after 4 min and the time required for the bar to cool to 100°F.

Solution   If $T(t)$ denotes the temperature of the bar at time t, then in Newton's law of cooling we are given: $T_m = 70°F$; $T(0) = 350°F$; $T(2) = 210°F$.
     We wish to find: (a) $T(4)$ and (b) the time when $T = 100°F$. Substituting for T_m in (2.3.1) we have

$$\frac{dT}{dt} = -k(T - 70). \tag{2.3.3}$$

Separating the variables yields

$$\frac{1}{T - 70} \, dT = -k \, dt,$$

_____

[1] If $T > T_m$, then the object will cool so that $dT/dt < 0$; hence from (2.3.1) k must be positive. Similarly, if $T < T_m$, then $dT/dt > 0$, and once more (2.3.1) implies that k must be positive.

which we can integrate immediately to obtain

$$\ln|T - 70| = -kt + \ln c.$$

This can be written as

$$\ln\left(\frac{|T - 70|}{c}\right) = -kt,$$

which, upon exponentiating both sides and solving for T, yields

$$T = 70 + c_1 e^{-kt}. \tag{2.3.4}$$

The two unknown constants c_1 and k can be determined from the given auxiliary conditions as follows.

$T(0) = 350°F$ requires that $350 = 70 + c_1$. Hence

$$c_1 = 280.$$

Substituting into (2.3.4) yields

$$T = 70(1 + 4e^{-kt}). \tag{2.3.5}$$

Further,

$$T(2) = 210°F$$

if and only if

$$210 = 70(1 + 4e^{-2k})$$

so that

$$e^{-2k} = \tfrac{1}{2}.$$

Consequently,

$$k = \tfrac{1}{2}\ln(2),$$

and so, from (2.3.5),

$$T = 70[1 + 4e^{-(t/2)\ln(2)}]. \tag{2.3.6}$$

We can now answer (a) and (b).

(a) $T(4) = 70[1 + 4e^{-2\ln(2)}] = 70\left[1 + 4\left(\dfrac{1}{2^2}\right)\right] = 140°F.$

(b) From (2.3.6), $T = 100°F$ when

$$100 = 70[1 + 4e^{-(t/2)\ln(2)}];$$

that is, when

$$e^{-(t/2)\ln(2)} = \tfrac{3}{28}.$$

Taking the natural logarithm of both sides and solving for t yields

$$t = \frac{2\ln\left(\frac{28}{3}\right)}{\ln 2} \approx 6.4 \text{ min.}$$

EXERCISES 2.3

1. An object whose temperature is 615°F is placed in a room whose temperature is 75°F. At 4 P.M. ($t = 0$) the temperature of the object is 135°F, whereas an hour later its temperature is 95°F. At what time was the object placed in the room?

2. An inflammable substance whose initial temperature is 50°F is inadvertently placed in a hot oven whose temperature is 450°F. The temperature of the substance is 150°F after 20 min. Find the temperature of the substance after 40 min. If the substance ignites when its temperature reaches 350°F, find the time of combustion.

3. At 4 P.M. a hot coal was pulled out of a furnace and allowed to cool at room temperature (75°F). If after 10 min the temperature of the coal was 415°F, and after 20 min its temperature was 347°F, find:
(a) The temperature of the furnace.
(b) The time when the temperature of the coal was 100°F.

4. At 2 P.M. on a cool (34°F) afternoon in March, Sherlock Holmes measured the temperature of a dead body to be 38°F. One hour later the temperature was 36°F. After a quick calculation using Newton's law of cooling and taking normal living body temperature as 98°F, Holmes concluded that the time of death was 10:00 A.M. Was Holmes right?

5. A chicken roast is cooked when its internal temperature reaches 200°F. An uncooked chicken, initial temperature 70°F, is placed in an oven whose temperature is 450°F. After 1 h the temperature of the chicken is 150°F. Given that an average-sized chicken requires 20 min per pound to cook, determine the size of the roast.

6. A hot object is placed in a room whose temperature is 72°F. After 1 min the temperature of the object is 150°F and its rate of change of temperature is 20°F per minute. Find the initial temperature of the object and the rate at which its temperature is changing after 10 min.

7. Suppose that an object is placed in a medium whose temperature is increasing at a constant rate of α°F per minute. Show that, according to Newton's law of cooling, the temperature of the object at time t is given by

$$T(t) = \alpha\left(t - \frac{1}{k}\right) + C_1 + C_2 e^{-kt}.$$

where C_1 and C_2 are constants.

8. Between 8:00 A.M. and 12 noon on a hot summer day the temperature rose at a rate of 10°F per hour from an initial temperature of 65°F. At 9:00 A.M. the temperature of an object was measured to be 35°F and was, at that time, increasing at a rate of 5°F per hour. Show that the temperature of the object at time t was

$$T = 10t - 15 + 40\, e^{(1-t)/8}, \qquad 0 \le t \le 4.$$

9. Consider the situation depicted in Figure 2.3.2, in which an object is placed in a room whose temperature T_1 is itself changing according to Newton's law of cooling. In this case we must solve the system of differential equations

$$\frac{dT_1}{dt} = -k_1(T_1 - T_2),$$

$$\frac{dT}{dt} = -k(T - T_1),$$

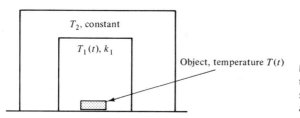

Figure 2.3.2 A situation in which the temperature of the medium surrounding an object is itself changing according to Newton's law of cooling.

where k, k_1, and T_2 are constants. Show that for $k \neq k_1$, the temperature of the object at time t is given by

$$T(t) = T_2 + \frac{c_1 k}{(k - k_1)} e^{-k_1 t} + c_2 e^{-kt},$$

where c_1 and c_2 are constants.

10. An object is placed in an ice chest whose temperature is 35°F. The external temperature is constant at 80°F. After 5 min the temperature of the object is 60°F and is decreasing at a rate of 1°F per minute. After 10 min the temperature in the ice chest is 40°F. Determine:

(a) The initial temperature of the object.

(b) The time when the temperature of the object begins to increase.

2.4 CHEMICAL REACTIONS

Consider a chemical reaction in which two chemicals A and B combine to form a third chemical C. Let $x(t)$ denote the amount of C that has been formed at time t. Then, according to the law of mass action,

> the rate of change of x at time t is proportional to the product of the amounts of A and B that are untransformed at that time.

To give a mathematical description of this law, suppose that we begin with A_0 grams of A and B_0 grams of B and that the reaction is such that A and B combine in the ratio $a : b$ (that is, any sample of C is a made up of a parts of A and b parts of B). Then, when x grams of C are formed, they consist of

$$\frac{a}{a + b} x \text{ grams}$$

of A and

$$\frac{b}{a + b} x \text{ grams}$$

of B. Consequently the amount of A that is unconverted at time t is

$$A_0 - \frac{a}{a + b} x, \tag{2.4.1}$$

and the amount of B that is unconverted at the same time is

$$B_0 - \frac{b}{a + b} x. \tag{2.4.2}$$

Thus, according to the law of mass action, the chemical reaction is governed by the differential equation

$$\frac{dx}{dt} = K\left(A_0 - \frac{a}{a + b} x\right)\left(B_0 - \frac{b}{a + b} x\right), \tag{2.4.3}$$

where K is a constant.   In order to facilitate the remaining calculations, we factor out the term $a/(a+b)$ from the first parentheses and the term $b/(a+b)$ from the second parentheses.   Then (2.4.3) can be written as

$$\frac{dx}{dt} = k(\alpha - x)(\beta - x), \tag{2.4.4}$$

where k is a constant and

$$\alpha = \frac{a+b}{a}\,A_0, \quad \beta = \frac{a+b}{b}\,B_0. \tag{2.4.5}$$

The differential equation (2.4.4) is easily solved using separation of variables.   In the case $\alpha \neq \beta$, we find (see Exercise 2.4.1)

$$x(t) = \frac{\alpha\beta[1 - e^{k(\alpha-\beta)t}]}{\beta - \alpha e^{k(\alpha-\beta)t}}, \tag{2.4.6}$$

where we have imposed the initial condition $x(0) = 0$.   If we assume that the chemicals have been chosen so that $\beta > \alpha$, then taking the limit as $t \to \infty$, we see that

$$\lim_{t \to \infty} x(t) = \alpha.$$

This is a reasonable result since the reaction will cease when we run out of either A or B; from (2.4.1), (2.4.2), and (2.4.5), this will occur when

$$x = \alpha \quad \text{or} \quad x = \beta,$$

respectively, depending on which is smaller.   Thus, since we have assumed that $\beta > \alpha$, the appropriate value is $x = \alpha$.

Example 2.4.1   In a certain chemical reaction 5 g of C are formed when 2 g of A react with 3 g of B.   Initially there are 10 g of A and 24 g of B present, and after 5 min 10 g of C has been produced.   Determine the amount of C that is produced in 15 min.

Solution   The chemicals A and B combine in the ratio $2:3$, so that the amounts of A and B required to produce x grams of C are $\frac{2}{5}x$ and $\frac{3}{5}x$, respectively.   Thus, according to the law of mass action, the differential equation governing the behavior of $x(t)$ is

$$\frac{dx}{dt} = K\left(10 - \frac{2}{5}x\right)\left(24 - \frac{3}{5}x\right),$$

or, equivalently,

$$\frac{dx}{dt} = k(25 - x)(40 - x).$$

Separating the variables yields

$$\frac{1}{(25 - x)(40 - x)}\,dx = k\,dt,$$

which can be written as

$$\frac{1}{15}\left(\frac{1}{25-x} - \frac{1}{40-x}\right) dx = k\, dt.$$

Integrating and simplifying, we obtain

$$\ln\left(\frac{40-x}{25-x}\right) = 15kt + \ln c.$$

Exponentiating both sides yields

$$\frac{40-x}{25-x} = ce^{15kt}.$$

The initial condition $x(0) = 0$ implies that $c = \frac{8}{5}$, so that

$$\frac{40-x}{25-x} = \frac{8}{5}e^{15kt}. \tag{2.4.7}$$

Further, $x(5) = 10$ requires that

$$e^{75k} = \frac{5}{4},$$

so that

$$k = \frac{1}{75}\ln\left(\frac{5}{4}\right).$$

Substituting into (2.4.7) yields

$$\frac{40-x}{25-x} = \frac{8}{5}e^{(t/5)\ln(5/4)}. \tag{2.4.8}$$

Solving for x we obtain

$$x = 200\,\frac{e^{(t/5)\ln(5/4)} - 1}{8e^{(t/5)\ln(5/4)} - 5}.$$

Thus,

$$x(15) = 200\,\frac{e^{3\ln(5/4)} - 1}{8e^{3\ln(5/4)} - 5} \approx 17.94 \text{ g.}$$

EXERCISES 2.4

1. Solve the differential equation (2.4.4) with the initial condition $x(0) = 0$, when $\alpha \neq \beta$. If $\alpha > \beta$, determine $\lim_{t\to\infty} x(t)$.

2. Solve the differential equation (2.4.4) when $\alpha = \beta$ and $x(0) = 0$. Determine $\lim_{t\to\infty} x(t)$.

3. In a certain chemical reaction 9 g of C are formed when 6 g of A combine with 3 g of B. Initially there are 20 g of both A and B, and after 10 min 15 g of C has been produced. Determine the amount of C that is produced in 20 min.

4. Chemicals A and B combine in the ratio 2 : 3. Initially there are 10 g of A and 15 g of B present, and after 5 min 10 g of C has been produced. Determine the amount of C that has been produced in 30 min. How long will it take for the reaction to be 50% complete?

5. The chemicals A and B combine in the ratio 3 : 5 in producing the chemical C. If we have 15 g of A, use the law of mass action to determine the minimum amount of B required to produce 30 g of C.

6. In a certain chemical reaction one molecule of A combines with one molecule of B to form one molecule of C. Let $x(t)$ be the number of molecules of C present at time t. If there are initially A_0 molecules of A and B_0 molecules of B present and the rate of change of x is proportional to the product of the number of molecules of A and B remaining at time t, show that the differential equation governing the reaction is

$$\frac{dx}{dt} = k(A_0 - x)(B_0 - x).$$

7. The differential equation governing a trimolecular chemical reaction is

$$\frac{dx}{dt} = k(\alpha - x)(\beta - x)(\gamma - x),$$

where k, α, β, γ are constants. Solve this differential if α, β, γ are all distinct and $x(0) = 0$.

2.5   MIXING PROBLEMS

The application of differential equations that we consider in this section gives an example of how mathematical models of physical problems are often constructed.

Statement of the problem: Consider the situation depicted in Figure 2.5.1. A tank initially contains V_0 liters of a solution in which is dissolved A_0 grams of a certain chemical. Solution containing c_1 grams/liter of the same chemical flows into the tank at a constant rate of r_1 liters/minute, and the mixture flows out at a constant rate of r_2 liters/minute. We assume that the mixture is kept uniform by stirring. Then at any time t, the concentration of chemical in the tank, $c_2(t)$, is the same throughout the tank and is given by

$$c_2 = \frac{A(t)}{V(t)}, \qquad (2.5.1)$$

where $V(t)$ denotes the volume of solution in the tank at time t. We wish to determine the amount of chemical in the tank, $A(t)$, at time t.

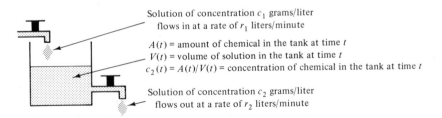

Solution of concentration c_1 grams/liter
flows in at a rate of r_1 liters/minute

$A(t)$ = amount of chemical in the tank at time t
$V(t)$ = volume of solution in the tank at time t
$c_2(t) = A(t)/V(t)$ = concentration of chemical in the tank at time t

Solution of concentration c_2 grams/liter
flows out at a rate of r_2 liters/minute

Figure 2.5.1   A mixing problem.

Mathematical formulation: The two variables in the problem are $V(t)$ and $A(t)$. To determine how they change with time we first consider their change during a short time interval, Δt minutes.

In time Δt, $r_1 \Delta t$ liters of solution flow into the tank, whereas $r_2 \Delta t$ liters flow out. Thus during the time interval Δt, the change in the volume of solution in the tank is

$$\Delta V = r_1 \Delta t - r_2 \Delta t = (r_1 - r_2)\, \Delta t. \qquad (2.5.2)$$

Since the concentration of chemical in the inflow is c_1 grams/liter (assumed constant), it follows that in the time interval Δt the amount of chemical that flows into the tank is $c_1 r_1 \Delta t$. Similarly, the amount of chemical that flows out in this same time interval is approximately $c_2 r_2 \Delta t$.[1] Thus, the total change in the amount of chemical in the tank during the time interval Δt, denoted by ΔA, is approximately

$$\Delta A \approx c_1 r_1 \Delta t - c_2 r_2 \Delta t = (c_1 r_1 - c_2 r_2)\, \Delta t. \qquad (2.5.3)$$

Dividing (2.5.2) and (2.5.3) by Δt yields

$$\frac{\Delta V}{\Delta t} = (r_1 - r_2), \qquad (2.5.4)$$

$$\frac{\Delta A}{\Delta t} \approx (c_1 r_1 - c_2 r_2), \qquad (2.5.5)$$

respectively. These equations describe the rate of change of V and A over the short, but finite, time interval Δt. To determine the instantaneous rates of change of V and A, we take the limit as $\Delta t \to 0$ to obtain

$$\frac{dV}{dt} = r_1 - r_2, \qquad (2.5.6)$$

$$\frac{dA}{dt} = c_1 r_1 - \frac{A}{V} r_2, \qquad (2.5.7)$$

where we have substituted for c_2 from (2.5.1). We can integrate (2.5.6) directly, since r_1 and r_2 are constants, to obtain

$$V(t) = (r_1 - r_2)t + V_0, \qquad (2.5.8)$$

where V_0 is an integration constant. Substituting for V into (2.5.7) and rearranging terms yields the following *linear* equation for $A(t)$:

$$\frac{dA}{dt} + \frac{r_2}{(r_1 - r_2)t + V_0} A = c_1 r_1. \qquad (2.5.9)$$

This differential equation can be solved, subject to the initial condition $A(0) = A_0$, to determine the behavior of $A(t)$.

REMARK   Do *not* memorize (2.5.9). You will be expected to derive it in a specific problem.

---

[1] This is only an approximation, since c_2 is *not* constant over the time interval Δt. The approximation will become more accurate as $\Delta t \to 0$.

Example 2.5.1   A tank contains 8 L of water in which is dissolved 32 g of chemical. A solution containing 2 g/L of the chemical flows into the tank at a rate of 4 L/min, and the well-stirred mixture flows out at a rate of 2 L/min. Determine the amount of chemical in the tank after 20 min. What is the concentration of chemical in the tank at that time?

Solution   We are given $r_1 = 4$ L/min; $r_2 = 2$ L/min; $c_1 = 2$ g/L; $V(0) = 8$; $A(0) = 32$ g. We must find $A(20)$ and $A(20)/V(20)$.
    Now,

$$\Delta V = r_1\, \Delta t - r_2\, \Delta t$$

implies that

$$\frac{dV}{dt} = 2.$$

Integrating this equation and imposing the initial condition that $V(0) = 8$ yields
$$V = 2(t + 4). \tag{2.5.10}$$

Further,

$$\Delta A \approx c_1 r_1\, \Delta t - c_2 r_2\, \Delta t$$

implies that

$$\frac{dA}{dt} = (8 - 2c_2);$$

that is, since $c_2 = A/V$,

$$\frac{dA}{dt} = 8 - 2\frac{A}{V}.$$

Thus, substituting for V from (2.5.10), we must solve
$$\frac{dA}{dt} + \frac{1}{t + 4}A = 8. \tag{2.5.11}$$

This first-order linear equation has integrating factor

$$I = e^{\int 1/(t+4)\, dt} = t + 4.$$

Consequently (2.5.11) can be written in the integrable form

$$\frac{d}{dt}[(t + 4)A] = 8(t + 4),$$

which can be integrated directly to yield
$$(t + 4)A = 4(t + 4)^2 + c,$$

so that,

$$A = \frac{1}{(t + 4)}[4(t + 4)^2 + c].$$

Imposing the given initial condition $A(0) = 32$ g implies that $c = 64$, so that

$$A = \frac{4}{(t + 4)}[(t + 4)^2 + 16].$$

Thus,

$$A(20) = \frac{1}{6}[(24)^2 + 16]$$

$$= \frac{296}{3} \, g,$$

and, from (2.5.10),

$$\frac{A(20)}{V(20)} = \frac{1}{48} \cdot \frac{296}{3} = \frac{37}{18} \, g/L.$$

EXERCISES 2.5

1. A container initially contains 10 L of water, in which is dissolved 20 g of salt. A solution containing 4 g/L of salt flows into the container at a rate of 2 L/min, and the well-stirred mixture flows out at a rate of 1 L/min. How much salt is in the tank after 40 min?

2. A tank initially contains 600 L of solution in which is dissolved 1500 g of chemical. A solution containing 5 g/L of the chemical flows into the tank at a rate of 6 L/min, and the well-stirred mixture flows out at a rate of 3 L/min. Determine the concentration of chemical in the tank after 1 h.

3. A tank whose volume is 40 L initially contains 20 L of water. A solution containing 10 g/L of salt flows into the tank at a rate of 4 L/min, and the well-stirred mixture flows out at a rate of 2 L/min. How much salt is in the tank just before the solution overflows.

4. A tank whose volume is 200 L is initially half full of a solution that contains 100 g of chemical. A solution containing 0.5 g/L of the same chemical flows into the tank at a rate of 6 L/min, and the well-stirred mixture flows out at a rate of 4 L/min. Determine the concentration of chemical in the tank just before the solution overflows.

5. A container initially contains 10 L of a salt solution. Water flows into the container at a rate of 3 L/min, and the well-stirred mixture flows out at a rate of 2 L/min. After 5 min the concentration of salt in the container is 0.2 g/L. Find:
(a) The amount of salt in the container initially.

(b) The volume of solution in the container when the concentration of salt is 0.1 g/L.

6. A tank initially contains 20 L of water. A solution containing 1 g/L of chemical flows into the tank at a rate of 3 L/min, and the mixture flows out at a rate of 2 L/min.
(a) Set up and solve the initial value problem for $A(t)$, the amount of chemical in the tank at time t.
(b) Determine the time when the concentration of chemical in the tank reaches 0.5 g/L.

7. A tank initially contains w liters of a solution, in which is dissolved A_0 grams of chemical. A solution containing k grams/liter of the same chemical flows into the tank at a rate of r liters/minute, and the mixture flows out at the same rate.
(a) Show that the amount of chemical, $A(t)$, in the tank at time t is

$$A(t) = e^{-rt/w} \, [kw(e^{rt/w} - 1) + A_0].$$

(b) Show that the concentration of chemical in the tank eventually approaches k grams/liter. Is this result reasonable?

8. Consider the double mixing problem depicted in Figure 2.5.2 on the following page.
(a) Show that the differential equations for determining $A_1(t)$ and $A_2(t)$ are

$$\frac{dA_1}{dt} + \frac{r_2}{(r_1 - r_2)t + V_1} A_1 = c_1 r_1,$$

$$\frac{dA_2}{dt} + \frac{r_3}{(r_2 - r_3)t + V_2} A_2 = \frac{r_2 A_1}{(r_1 - r_2)t + V_1},$$

where V_1 and V_2 are constants.

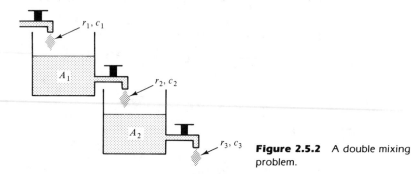

Figure 2.5.2   A double mixing problem.

(b) Consider the case in which $r_1 = 6$ L/min, $r_2 = 4$ L/min, $r_3 = 3$ L/min, and $c_1 = 0.5$ g/L. If the first tank initially holds 40 L of water in which 4 g of chemical is dissolved, whereas the second tank initially contains 20 g of chemical dissolved in 20 L of water, determine the amount of chemical in the second tank after 10 min.

2.6   SIMPLE ELECTRIC CIRCUITS

An important application of first- (and second-) order linear differential equations arises from the analysis of simple electric circuits. The most basic electric circuit is obtained by connecting the ends of a wire to the terminals of a battery or generator. This causes a flow of charge $q(t)$ through the wire, thereby producing a current $i(t)$, defined to be the rate of change of charge.[1]   Thus

$$i(t) = \frac{dq}{dt}. \tag{2.6.1}$$

In practice a circuit will contain several components that oppose the flow of charge. As current passes through these components, work has to be done; hence there is a loss of energy, which is described by the resulting voltage drop across each component. For the circuits that we will consider, the behavior of the current in the circuit is governed by Kirchoff's second law, which can be stated as follows:

> *Kirchoff's Second Law*
>
> The sum of the voltage drops around a closed circuit is zero.

To apply this law we need to know the relationship between the current passing through each component in the circuit and the resulting voltage drop. The components of most interest to us are resistors, capacitors, and inductors. We briefly describe each of these next.

1. *Resistors:* As its name suggests, a resistor is a component that, due to its composition, directly resists the flow of charge through it. According to Ohm's law, the voltage drop, ΔV_R, between the ends of a resistor is directly proportional to the current that is passing through it, that is,

$$\Delta V_R = iR, \tag{2.6.2}$$

---

[1] In SI units, charge is measured in coulombs (C) and current is measured in amperes (A).

where the constant of proportionality, R, is called the **resistance** of the resistor. The units of resistance are ohms (Ω).

2. *Capacitors:* A capacitor can be thought of as a component that stores charge and thereby opposes the passage of current. If $q(t)$ denotes the charge on the capacitor at time t, then the drop in voltage, ΔV_c, as current passes through it is directly proportional to $q(t)$. It is usual to express this law in the form

$$\Delta V_c = \frac{1}{C} q, \tag{2.6.3}$$

where the constant C is called the **capacitance** of the capacitor. The units of capacitance are farads (F).

3. *Inductors:* The third component of interest to us is an inductor. This can be considered as a component that opposes any *change* in the current flowing through it. The drop in voltage as current passes through an inductor is directly proportional to the rate at which the current is changing. We write this as

$$\Delta V_L = L \frac{di}{dt}, \tag{2.6.4}$$

where the constant L is called the **inductance** of the inductor measured in units of henrys (H).

4. *EMF:* The final component in our circuits is a source of voltage that produces an electromotive force (EMF). We can think of this as providing the force that drives the charge through the circuit. As current passes through the battery there is a voltage gain, which we denote by $E(t)$ volts (V) [that is, a voltage drop of $-E(t)$ volts].

A circuit containing all these components is shown in Figure 2.6.1.

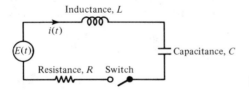

Figure 2.6.1 A simple *RLC* circuit.

According to Kirchoff's second law, the sum of the voltage drops at any instant must be zero. Applying this to the preceding *RLC* circuit, we obtain

$$\Delta V_R + \Delta V_C + \Delta V_L - E(t) = 0.$$

Substituting into this equation from (2.6.2)–(2.6.4) and rearranging yields the basic differential equation for an *RLC* circuit:

$$L \frac{di}{dt} + iR + \frac{1}{C} q = E. \tag{2.6.5}$$

We consider three cases separately.

CASE 1: AN *RL* CIRCUIT   In the case when there is no capacitor present, we have what is referred to as an *RL* circuit.   The differential equation (2.6.5) therefore reduces to

$$\frac{di}{dt} + \frac{R}{L} i = \frac{1}{L} E(t).$$

(2.6.6)

This is a first-order linear differential equation, which determines the current in the circuit at any time t.

CASE 2: AN *RC* CIRCUIT   Now consider the case when there is no inductor present in the circuit.   Setting $L = 0$ in (2.6.5) yields

$$i + \frac{1}{RC} q = \frac{E}{R}.$$

(2.6.7)

In this equation we have two unknowns, $q(t)$ and $i(t)$.   Substituting from (2.6.1) for $i(t) = dq/dt$, we obtain the following differential equation for $q(t)$:

$$\frac{dq}{dt} + \frac{1}{RC} q = \frac{E}{R}.$$

(2.6.8)

In this case we solve the linear differential equation (2.6.8) to obtain the charge $q(t)$ on the plates of the capacitor.   The current in the circuit can then be obtained from

$$i = \frac{dq}{dt}$$

by differentiation.

CASE 3: AN *RLC* CIRCUIT   In the general case we must consider all three components to be present in the circuit.   Substituting from (2.6.1) into (2.6.5) yields the following differential equation for determining the charge on the capacitor:

$$\frac{d^2q}{dt^2} + \frac{R}{L} \frac{dq}{dt} + \frac{1}{LC} q = \frac{1}{L} E(t).$$

(2.6.9)

This is a second-order linear differential equation with constant coefficients.   The techniques of Chapter 1 cannot, in general, be used to solve such a differential equation.   We will return to solve this type of differential equation in Chapter 9.

The differential equations (2.6.6) and (2.6.8) are first-order linear.   Thus, once the applied EMF, $E(t)$, has been specified, these equations can be solved using the technique of Section 1.5.   The two most important forms for $E(t)$ are

$$E(t) = E_0 \quad \text{and} \quad E(t) = E_0 \sin \omega t,$$

where E_0 and ω are constants.   The first of these corresponds to a source of EMF such as a battery.   The resulting current is called a **direct current** (DC).   The second form of EMF oscillates between $\pm E_0$.   The resulting current in the circuit is called an **alternating current** (AC).

Example 2.6.1 Determine the current in an RL circuit if the applied EMF is constant and the initial current is zero.

Solution If we let E_0 denote the constant value of the EMF, then we must solve the initial value problem

$$\frac{di}{dt} + \frac{R}{L}i = \frac{E_0}{L}, \quad i(0) = 0. \tag{2.6.10}$$

An integrating factor for (2.6.10) is $I = e^{(R/L)t}$, so that (2.6.10) can be written in the form

$$\frac{d}{dt}[e^{(R/L)t}i] = \frac{E_0}{L}e^{(R/L)t}.$$

Integrating both sides and simplifying the result yields

$$i(t) = \frac{E_0}{R} + c_1 e^{-(R/L)t}. \tag{2.6.11}$$

The initial condition $i(0) = 0$ is satisfied if and only if $c_1 = -E_0/R$. Consequently, the solution of the initial value problem is

$$i(t) = \frac{E_0}{R}[1 - e^{-(R/L)t}].$$

We see that the exponential contribution dies out rapidly with time and that the circuit soon settles down to the steady state

$$i(t) \approx \frac{E_0}{R}.$$

The behavior of $i(t)$ is shown in Figure 2.6.2.

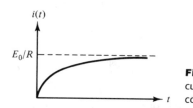

Figure 2.6.2 The behavior of the current in a simple RL circuit with constant EMF and zero initial current.

Example 2.6.2 Determine the current in an RL circuit if the applied EMF is $E(t) = E_0 \cos \omega t$, where E_0 and ω are constants.

Solution Substituting into (2.6.6) for $E(t)$ yields the differential equation

$$\frac{di}{dt} + \frac{R}{L}i = \frac{E_0}{L}\cos \omega t,$$

which we write as

$$\frac{di}{dt} + ai = \frac{E_0}{L}\cos \omega t, \tag{2.6.12}$$

where $a = R/L$. An integrating factor for (2.6.12) is $I = e^{at}$, so that we must solve

$$\frac{d}{dt}(e^{at}i) = \frac{E_0}{L}e^{at}\cos \omega t. \qquad (2.6.13)$$

Recalling the standard integral

$$\int e^{at}\cos \omega t \, dt = \frac{1}{a^2 + \omega^2}e^{at}(a \cos \omega t + \omega \sin \omega t),$$

it follows that (2.6.13) can be integrated directly to yield

$$e^{at}i = \frac{E_0}{L(a^2 + \omega^2)}e^{at}(a \cos \omega t + \omega \sin \omega t) + c,$$

where c is an integration constant. Thus

$$i = \frac{E_0}{L(a^2 + \omega^2)}(a \cos \omega t + \omega \sin \omega t) + ce^{-at}. \qquad (2.6.14)$$

Imposing the initial condition $i(0) = 0$, we find

$$c = -\frac{E_0 a}{L(a^2 + \omega^2)},$$

so that

$$i = \frac{E_0}{L(a^2 + \omega^2)}(a \cos \omega t + \omega \sin \omega t - ae^{-at}). \qquad (2.6.15)$$

This solution can be written in the form

$$i = i_S + i_T,$$

where

$$i_S = \frac{E_0}{R^2 + L^2\omega^2}(R \cos \omega t + L\omega \sin \omega t), \qquad i_T = -\frac{RE_0}{R^2 + L^2\omega^2}e^{-(R/L)t},$$

and we have substituted for $a = R/L$. The term i_T decays exponentially with time and is referred to as the **transient part** of the solution. As $t \to \infty$, the solution (2.6.15) approaches the **steady-state solution** i_S.

Finally, we consider an example that illustrates the procedure for solving the differential equation governing the behavior of an RC circuit.

Example 2.6.3 Consider the RC circuit in which $R = 0.5 \ \Omega$, $C = 0.1$ F, and $E_0 = 20$ V. Given that the capacitor has zero initial charge, determine the current in the circuit after 0.25 s.

Solution In this case we first solve (2.6.8) for $q(t)$ and then determine the current in the circuit by differentiating the result. Substituting for R, C, and E into (2.6.8) yields

$$\frac{dq}{dt} + 20q = 40,$$

which has general solution

$$q = 2 + ce^{-20t},$$

where c is an integration constant. Imposing the initial condition $q(0) = 0$ yields $c = -2$, so that

$$q = 2(1 - e^{-20t}).$$

Differentiating this expression for q yields the current in the circuit:

$$i = \frac{dq}{dt} = 40e^{-20t}.$$

Consequently,

$$i(0.25) = 40e^{-5} \approx 0.27 \text{ A}.$$

EXERCISES 2.6

1. Consider the RL circuit in which $R = 4\ \Omega$, $L = 0.1$ H, and $E(t) = 20$ V. If there is no current flowing initially, determine the current in the circuit for $t \geq 0$.

2. Consider the RC circuit in which $R = 5\ \Omega$, $C = \frac{1}{50}$ F, and $E(t) = 100$ V. If the capacitor is uncharged initially, determine the current in the circuit for $t \geq 0$.

3. An RL circuit has $E(t) = 10 \sin 4t$ volts. If $R = 2\ \Omega$, $L = \frac{2}{3}$ H, and there is no current flowing initially, determine the current for $t \geq 0$.

4. Consider the RC circuit with $R = 2\ \Omega$, $C = \frac{1}{8}$ F, and $E(t) = 10 \cos 3t$ volts. If $q(0) = 1$ C, determine the current in the circuit for $t \geq 0$.

5. Consider the general RC circuit with $E(t) = 0$. Suppose that $q(0) = 5$ C. Determine the charge

on the capacitor for $t > 0$. What happens as $t \to \infty$? Is this reasonable?

6. Determine the current in an RC circuit if the capacitor has zero charge initially and the driving EMF is $E = E_0$, where E_0 is a constant. Make a sketch depicting the change in the charge $q(t)$ on the capacitor with time and show that $q(t)$ approaches a constant value as t increases. What happens to the current in the circuit as $t \to \infty$?

7. Determine the current flowing in an RL circuit if the applied EMF is $E(t) = E_0 \sin \omega t$, where E_0 and ω are constants. Identify the transient part of the solution and the steady-state solution.

8. Determine the current flowing in an RC circuit if the capacitor is initially uncharged and the driving EMF is given by $E(t) = E_0 e^{-at}$, where E_0 and a are constants.

9. Consider the special case of the RLC circuit in which the resistance is negligible and the driving EMF is zero. In this case the differential equation governing the charge on the capacitor is

$$\frac{d^2q}{dt^2} + \frac{1}{LC}q = 0. \tag{i}$$

If the capacitor has an initial charge of q_0 coulombs and there is no current flowing initially, determine the charge on the capacitor for $t > 0$ and the corresponding current in the circuit. (*Hint:* See Section 1.10.)

10. Repeat problem 9 in the case when the driving EMF is $E(t) = E_0$, a constant.

3

Matrices

A recurring problem throughout the remainder of the text will be to find the solution properties of a system of linear algebraic equations such as

$$x_1 + x_2 - x_3 = 1,$$
$$3x_1 + 2x_2 - 4x_3 = 2.$$

Developing efficient procedures for determining the solution of such a system will require a way to represent the system. In this chapter we introduce the appropriate mathematical objects for doing this, namely, matrices. It should be noted, however, that matrices have a much broader usage than just solving systems of linear equations; for example, we will see in Chapter 11 how they give a natural framework for formulating and solving systems of linear differential equations.

3.1 DEFINITIONS AND NOTATION

We are probably all familiar with the idea of a matrix for storing information. For example, a road map that relates distances between cities by listing the appropriate mileages in rows and columns is an example of a matrix (see Figure 3.1.1).

	Los Angeles	San Diego	San Francisco	Santa Barbara
Los Angeles	0	115	390	95
San Diego	115	0	505	210
San Francisco	390	505	0	335
Santa Barbara	95	210	335	0

Figure 3.1.1 A mileage chart is an everyday example of a matrix.

The following definition gives a precise mathematical description of a matrix and also introduces the notation that we will be using.

Definition 3.1.1:   An $m \times n$ **matrix** is a rectangular array of numbers of the form

$$\overset{\longleftarrow \text{ } n \text{ columns } \longrightarrow}{A = \begin{bmatrix} a_{11} & a_{12} & a_{13} & \cdots & a_{1n} \\ a_{21} & a_{22} & a_{23} & \cdots & a_{2n} \\ a_{31} & a_{32} & a_{33} & \cdots & a_{3n} \\ \vdots & \vdots & \vdots & & \vdots \\ a_{m1} & a_{m2} & a_{m3} & \cdots & a_{mn} \end{bmatrix}} \updownarrow m \text{ rows}$$

A is said to have dimensions $m \times n$.   A typical element of A is a_{ij}, the element in the ith row and the jth column.

As in the preceding definition, we usually denote a matrix by an uppercase letter such as A or B.   However, if we wish to emphasize the elements of the matrix, then we will write $A = [a_{ij}]$, or just $[a_{ij}]$.

We see directly that the two matrices

$$\begin{bmatrix} 1 & 1 & -1 \\ 3 & 2 & -4 \end{bmatrix} \text{ and } \begin{bmatrix} 1 \\ 2 \end{bmatrix}$$

completely determine the system of linear equations given at the beginning of the chapter, since the first matrix defines the coefficients of the unknowns in the system, whereas the second matrix contains the numbers appearing on the right-hand side of the system.   Similarly, the matrix

$$\begin{bmatrix} 0 & 115 & 390 & 95 \\ 115 & 0 & 505 & 210 \\ 390 & 505 & 0 & 335 \\ 95 & 210 & 335 & 0 \end{bmatrix}$$

represents the mileage chart displayed in Figure 3.1.1.

Example 3.1.1 The following are examples of a 2×3 and a 3×3 matrix, respectively:

$$A = \begin{bmatrix} 1 & 2 & 4 \\ 1 & -3 & 5 \end{bmatrix}, \qquad B = \begin{bmatrix} 2 & -1 & 3 \\ 1 & 1 & -1 \\ 0 & 0 & 1 \end{bmatrix}.$$

An $n \times n$ matrix is called a **square matrix**, since it has the same number of rows as columns. If A is a square matrix, then the elements a_{ii}, $1 \leq i \leq n$, make up the **main diagonal**, or **leading diagonal**, of the matrix (see Figure 3.1.2 for the 3×3 case).

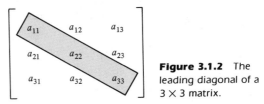

Figure 3.1.2 The leading diagonal of a 3×3 matrix.

The sum of the main diagonal elements of an $n \times n$ matrix A is called the **trace** of A and is denoted by $\mathrm{tr}(A)$. Thus,

$$\mathrm{tr}(A) = a_{11} + a_{22} + \cdots + a_{nn}.$$

Of special interest to us in the future will be $1 \times n$ and $n \times 1$ matrices. For this reason we give them special names.

Definition 3.1.2: A $1 \times n$ matrix is called a **row vector**. An $n \times 1$ matrix is called a **column vector**.

REMARK   If we wish to emphasize the number of elements in a row vector or a column vector, we will refer to such vectors as row n-vectors and column n-vectors respectively.

Example 3.1.2 $\mathbf{a} = [1 \quad -1 \quad 3 \quad 4 \quad 0]$ is a row 5-vector, whereas $\mathbf{b} = \begin{bmatrix} 2 \\ 0 \\ 1 \end{bmatrix}$ is a column 3-vector.

NOTE   As in the preceding example, we usually denote a column or row vector by a lowercase letter in **bold** print.

Associated with an $m \times n$ matrix are m row n-vectors and n column m-vectors, which are referred to as the *row vectors* of the matrix and the *column vectors* of the matrix, respectively.

Example 3.1.3   Associated with the matrix $A = \begin{bmatrix} -2 & 1 & 3 & 4 \\ 1 & 2 & 1 & 1 \\ 3 & -1 & 2 & 5 \end{bmatrix}$ are the row

vectors $[-2 \quad 1 \quad 3 \quad 4]$, $[1 \quad 2 \quad 1 \quad 1]$, $[3 \quad -1 \quad 2 \quad 5]$ and the column vectors

$$\begin{bmatrix} -2 \\ 1 \\ 3 \end{bmatrix}, \quad \begin{bmatrix} 1 \\ 2 \\ -1 \end{bmatrix}, \quad \begin{bmatrix} 3 \\ 1 \\ 2 \end{bmatrix}, \quad \begin{bmatrix} 4 \\ 1 \\ 5 \end{bmatrix}.$$

Conversely, if $\mathbf{a}_1, \mathbf{a}_2, \ldots, \mathbf{a}_n$ are each column m-vectors, then we let $[\mathbf{a}_1, \mathbf{a}_2, \ldots, \mathbf{a}_n]$ denote the $m \times n$ matrix whose column vectors are $\mathbf{a}_1, \mathbf{a}_2, \ldots, \mathbf{a}_n$. Similarly, if $\mathbf{b}_1, \mathbf{b}_2, \ldots, \mathbf{b}_m$ are each row n-vectors, then we write

$$\begin{bmatrix} \mathbf{b}_1 \\ \mathbf{b}_2 \\ \vdots \\ \mathbf{b}_m \end{bmatrix}$$

for the $m \times n$ matrix with row vectors $\mathbf{b}_1, \mathbf{b}_2, \ldots, \mathbf{b}_m$.

Example 3.1.4   If $\mathbf{a}_1 = \begin{bmatrix} 1 \\ 3 \end{bmatrix}$, $\mathbf{a}_2 = \begin{bmatrix} 4 \\ 5 \end{bmatrix}$, and $\mathbf{a}_3 = \begin{bmatrix} -1 \\ 7 \end{bmatrix}$, then

$$[\mathbf{a}_1, \mathbf{a}_2, \mathbf{a}_3] = \begin{bmatrix} 1 & 4 & -1 \\ 3 & 5 & 7 \end{bmatrix}.$$

EXERCISES 3.1

1. If $A = \begin{bmatrix} 1 & -2 & 3 & 2 \\ 7 & -6 & 5 & -1 \\ 0 & 2 & -3 & -4 \end{bmatrix}$, determine a_{31}, a_{24}, a_{14}, a_{32}, a_{21}, and a_{34}.

In problems 2–6 write out the matrix with the given elements. In each case specify the dimension of the matrix.

2. $a_{11} = 1$, $a_{21} = -1$, $a_{12} = 5$, $a_{22} = 3$.

3. $a_{11} = 2$, $a_{12} = 1$, $a_{13} = -1$, $a_{21} = 0$, $a_{22} = 4$, $a_{23} = -2$.

4. $a_{11} = -1$, $a_{41} = -5$, $a_{31} = 1$, $a_{21} = 1$.

5. $a_{11} = 1$, $a_{31} = 2$, $a_{42} = -1$, $a_{32} = 7$, $a_{13} = -2$, $a_{23} = 0$, $a_{33} = 4$, $a_{21} = 3$, $a_{41} = -4$, $a_{12} = -3$, $a_{22} = 6$, $a_{43} = 5$.

6. $a_{12} = -1$, $a_{13} = 2$, $a_{23} = 3$, $a_{ji} = -a_{ij}$, $1 \le i \le 3$, $1 \le j \le 3$.

In problems 7–9 determine tr(A) for the given matrix.

7. $A = \begin{bmatrix} 1 & 0 \\ 2 & 3 \end{bmatrix}$.

8. $A = \begin{bmatrix} 1 & 2 & -1 \\ 3 & 2 & -2 \\ 7 & 5 & -3 \end{bmatrix}$.

9. $A = \begin{bmatrix} 2 & 0 & 1 \\ 3 & 2 & 5 \\ 0 & 1 & -5 \end{bmatrix}$.

In problems 10–12 write out the column vectors and row vectors of the given matrix.

10. $A = \begin{bmatrix} 1 & -1 \\ 3 & 5 \end{bmatrix}$.

11. $A = \begin{bmatrix} 1 & 3 & -4 \\ -1 & -2 & 5 \\ 2 & 6 & 7 \end{bmatrix}$.

12. $A = \begin{bmatrix} 2 & 10 & 6 \\ 5 & -1 & 3 \end{bmatrix}$.

13. If $\mathbf{a}_1 = [1 \quad 2]$, $\mathbf{a}_2 = [3 \quad 4]$, and $\mathbf{a}_3 = [5 \quad 1]$, write the matrix $A = \begin{bmatrix} \mathbf{a}_1 \\ \mathbf{a}_2 \\ \mathbf{a}_3 \end{bmatrix}$ and determine the column vectors of A.

14. If

$$\mathbf{b}_1 = \begin{bmatrix} 2 \\ -1 \\ 4 \end{bmatrix}, \mathbf{b}_2 = \begin{bmatrix} 5 \\ 7 \\ -6 \end{bmatrix}, \mathbf{b}_3 = \begin{bmatrix} 0 \\ 0 \\ 0 \end{bmatrix}, \text{ and } \mathbf{b}_4 = \begin{bmatrix} 1 \\ 2 \\ 3 \end{bmatrix},$$

write the matrix $B = [\mathbf{b}_1, \quad \mathbf{b}_2, \quad \mathbf{b}_3, \quad \mathbf{b}_4]$ and determine the row vectors of B.

15. If $\mathbf{a}_1, \mathbf{a}_2, \ldots, \mathbf{a}_p$ are each column q-vectors, what are the dimensions of the matrix that has $\mathbf{a}_1, \mathbf{a}_2, \ldots, \mathbf{a}_p$ as its column vectors?

3.2   MATRIX ALGEBRA

In the previous section we gave a mathematical definition of a matrix. The real power of these objects can be realized only after we have defined the basic algebraic operations of addition, subtraction, and multiplication on them. We first need to define what is meant by equality of matrices.

> *Definition 3.2.1:* Two matrices A and B are **equal,** written $A = B$, if and only if
>
> **(a)** They both have the same dimensions.
> **(b)** All corresponding elements in the matrices are equal.

ADDITION AND SUBTRACTION OF MATRICES AND MULTIPLICATION OF A MATRIX BY A SCALAR

Addition (and subtraction) of matrices is defined only for matrices of the same dimension.

> *Definition 3.2.2:* If A and B are both $m \times n$ matrices, then we define the **sum** of A and B, denoted by $A + B$, to be the $m \times n$ matrix whose elements are obtained by adding *corresponding* elements of A and B. Thus, if $A = [a_{ij}]$ and $B = [b_{ij}]$ and $C = A + B$, then C has elements $c_{ij} = a_{ij} + b_{ij}$.

Example 3.2.1

$$\begin{bmatrix} 2 & -1 & 3 \\ 4 & -5 & 0 \end{bmatrix} + \begin{bmatrix} -1 & 0 & 5 \\ -5 & 2 & 7 \end{bmatrix} = \begin{bmatrix} 1 & -1 & 8 \\ -1 & -3 & 7 \end{bmatrix}.$$

Clearly, if A and B are both $m \times n$ matrices, then

$$A + B = B + A \qquad \text{(Matrix addition is commutative)},$$
$$A + (B + C) = (A + B) + C \qquad \text{(Matrix addition is associative)}.$$

In much of our work we will need to use complex as well as real numbers.[1]   In order to save some writing we introduce the term **scalar** to mean a real or complex number.

Definition 3.2.3:   If A is an $m \times n$ matrix and s is a scalar, then we let sA denote the matrix obtained by multiplying every element of A by s. This procedure is called **scalar multiplication**. In index notation, if $A = [a_{ij}]$, then $sA = [sa_{ij}]$.

Example 3.2.2   If $A = \begin{bmatrix} 2 & -1 \\ 4 & 6 \end{bmatrix}$, then $5A = \begin{bmatrix} 10 & -5 \\ 20 & 30 \end{bmatrix}$.

Example 3.2.3   If $A = \begin{bmatrix} 1+i & i \\ 2+3i & 4 \end{bmatrix}$ and $s = 1 - 2i$, where $i = \sqrt{-1}$, find sA.

Solution

$$sA = \begin{bmatrix} (1-2i)(1+i) & (1-2i)i \\ (1-2i)(2+3i) & (1-2i)4 \end{bmatrix} = \begin{bmatrix} 3-i & 2+i \\ 8-i & 4-8i \end{bmatrix}.$$

Definition 3.2.4:   We define **subtraction** of two matrices of the *same dimensions* by

$$A - B = A + (-1)B;$$

that is, we subtract corresponding elements.

Further properties satisfied by our operations of matrix addition and multiplication by a scalar are

$$1 \cdot A = A,$$
$$s(A + B) = sA + sB,$$
$$(s + t)A = sA + tA,$$
$$s(tA) = (st)A = (ts)A = t(sA),$$

for any scalars s, t.

MULTIPLICATION OF MATRICES

Our definition of scalar multiplication was essentially the only possibility if, in the case when s is a positive integer, we want sA to be the same matrix as we obtain when A is added to itself s times. We now define how to multiply two matrices together. In this case the multiplication operation is by no means obvious and so we will build up to the general case by considering two special cases.

---

[1] A review of complex numbers is given in Appendix 1.

CASE 1 If **a** is a row n-vector and **x** is a column n-vector, then their product **ax** is the 1×1 matrix defined by

$$\mathbf{ax} = \overrightarrow{[a_1 \quad a_2 \quad \cdots \quad a_n]} \begin{bmatrix} x_1 \\ x_2 \\ \vdots \\ x_n \end{bmatrix} = [a_1 x_1 + a_2 x_2 + \cdots + a_n x_n].$$

Example 3.2.4 If $\mathbf{a} = [2 \quad -1 \quad 3 \quad 5]$ and $\mathbf{x} = \begin{bmatrix} 3 \\ 2 \\ -3 \\ 4 \end{bmatrix}$, then

$$\mathbf{ax} = \overrightarrow{[2 \quad -1 \quad 3 \quad 5]} \begin{bmatrix} 3 \\ 2 \\ -3 \\ 4 \end{bmatrix} = [(2)(3) + (-1)(2) + (3)(-3) + (5)(4)] = [15].$$

CASE 2 Now suppose that A is an $m \times n$ matrix and that **x** is a column n-vector. The product $A\mathbf{x}$ is defined to be the $m \times 1$ matrix whose ith element is obtained by multiplying the ith row vector of A by **x** (see Figure 3.2.1). The ith row vector of A, $\mathbf{a}_i$, is given by

$$\mathbf{a}_i = [a_{i1} \quad a_{i2} \quad \cdots \quad a_{in}],$$

so that the ith element of $A\mathbf{x}$ is

$$(A\mathbf{x})_i = a_{i1} x_1 + a_{i2} x_2 + \cdots + a_{in} x_n,$$

which can be written as

$$(A\mathbf{x})_i = \sum_{k=1}^{n} a_{ik} x_k, \quad 1 \le i \le m. \tag{3.2.1}$$

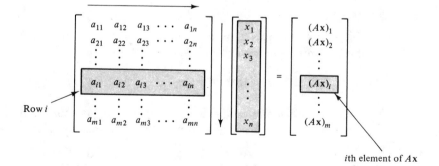

Figure 3.2.1 Multiplication of an $m \times n$ matrix by a column n-vector.

Example 3.2.5 Find $A\mathbf{x}$ if $A = \begin{bmatrix} 2 & 3 & -1 \\ 1 & 4 & -6 \\ 5 & -2 & 0 \end{bmatrix}$ and $\mathbf{x} = \begin{bmatrix} 7 \\ -3 \\ 1 \end{bmatrix}$.

Solution

$$Ax = \begin{bmatrix} 2 & 3 & -1 \\ 1 & 4 & -6 \\ 5 & -2 & 0 \end{bmatrix} \begin{bmatrix} 7 \\ -3 \\ 1 \end{bmatrix} = \begin{bmatrix} 4 \\ -11 \\ 41 \end{bmatrix}.$$

Finally we consider the general case.   If the second matrix in the product has more than one column vector, we simply multiply each of its column vectors in turn by the first matrix using the preceding rule.

Example 3.2.6   If $A = \begin{bmatrix} 1 & -1 & 2 \\ 1 & 1 & -1 \end{bmatrix}$ and $B = \begin{bmatrix} 2 & 4 \\ 1 & 3 \\ 5 & 2 \end{bmatrix}$, then

$$AB = \begin{bmatrix} 1 & -1 & 2 \\ 1 & 1 & -1 \end{bmatrix} \begin{bmatrix} 2 & 4 \\ 1 & 3 \\ 5 & 2 \end{bmatrix}$$

$$= \begin{bmatrix} \{(1)(2) + (-1)(1) + (2)(5)\} & \{(1)(4) + (-1)(3) + (2)(2)\} \\ \{(1)(2) + (1)(1) + (-1)(5)\} & \{(1)(4) + (1)(3) + (-1)(2)\} \end{bmatrix}$$

$$= \begin{bmatrix} 11 & 5 \\ -2 & 5 \end{bmatrix}.$$

The preceding discussion is summarized in the following definition.

Definition 3.2.5:

1. If $A = [a_{ij}]$ is an $m \times n$ matrix and $\mathbf{b} = [b_i]$ is a column n-vector, then $A\mathbf{b}$ is the column m-vector with elements

$$(A\mathbf{b})_i = \sum_{k=1}^{n} a_{ik}b_k, \qquad 1 \le i \le m.$$

2. If A is an $m \times n$ matrix and $B = [\mathbf{b}_1, \mathbf{b}_2, \ldots, \mathbf{b}_p]$ is an $n \times p$ matrix, then the product of A and B is the $m \times p$ matrix, denoted by AB, defined by

$$AB = [A\mathbf{b}_1, A\mathbf{b}_2, A\mathbf{b}_3, \ldots, A\mathbf{b}_p]. \qquad (3.2.2)$$

IMPORTANT REMARK   In order for the product AB to be defined, we see from the preceding definition that A and B must satisfy

number of columns of A = number of rows of B.

In such a case if C represents the product matrix AB, then the relationship between the dimensions of the matrices is

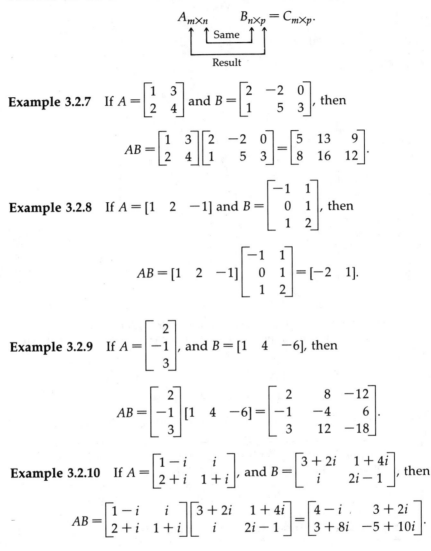

Example 3.2.7 If $A = \begin{bmatrix} 1 & 3 \\ 2 & 4 \end{bmatrix}$ and $B = \begin{bmatrix} 2 & -2 & 0 \\ 1 & 5 & 3 \end{bmatrix}$, then

$$AB = \begin{bmatrix} 1 & 3 \\ 2 & 4 \end{bmatrix}\begin{bmatrix} 2 & -2 & 0 \\ 1 & 5 & 3 \end{bmatrix} = \begin{bmatrix} 5 & 13 & 9 \\ 8 & 16 & 12 \end{bmatrix}.$$

Example 3.2.8 If $A = [1 \quad 2 \quad -1]$ and $B = \begin{bmatrix} -1 & 1 \\ 0 & 1 \\ 1 & 2 \end{bmatrix}$, then

$$AB = [1 \quad 2 \quad -1]\begin{bmatrix} -1 & 1 \\ 0 & 1 \\ 1 & 2 \end{bmatrix} = [-2 \quad 1].$$

Example 3.2.9 If $A = \begin{bmatrix} 2 \\ -1 \\ 3 \end{bmatrix}$, and $B = [1 \quad 4 \quad -6]$, then

$$AB = \begin{bmatrix} 2 \\ -1 \\ 3 \end{bmatrix}[1 \quad 4 \quad -6] = \begin{bmatrix} 2 & 8 & -12 \\ -1 & -4 & 6 \\ 3 & 12 & -18 \end{bmatrix}.$$

Example 3.2.10 If $A = \begin{bmatrix} 1-i & i \\ 2+i & 1+i \end{bmatrix}$, and $B = \begin{bmatrix} 3+2i & 1+4i \\ i & 2i-1 \end{bmatrix}$, then

$$AB = \begin{bmatrix} 1-i & i \\ 2+i & 1+i \end{bmatrix}\begin{bmatrix} 3+2i & 1+4i \\ i & 2i-1 \end{bmatrix} = \begin{bmatrix} 4-i & 3+2i \\ 3+8i & -5+10i \end{bmatrix}.$$

SOME PROPERTIES OF MATRIX MULTIPLICATION

The preceding rule for determining the product of two matrices tells us how to evaluate the product in practice. However, to prove results related to the matrix product, we require an expression for the ijth element of AB in terms of the elements of the matrices A and B. We now proceed to derive such an expression. According to our rule for multiplying two matrices, the ijth element of AB is obtained by multiplying the ith row vector of A by the jth column vector of B (see Figure 3.2.2). But, the ith row vector of A is

$$\mathbf{a}_i = [a_{i1} \quad a_{i2} \quad \cdots \quad a_{in}],$$

and the jth column vector of B is

$$\mathbf{b}_j = \begin{bmatrix} b_{1j} \\ b_{2j} \\ \vdots \\ b_{nj} \end{bmatrix}.$$

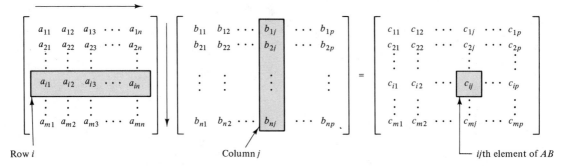

Figure 3.2.2   Multiplication of two matrices. The ijth element of the product matrix is obtained by multiplying the ith row vector of the first matrix by the jth column vector of the second matrix.

Thus, multiplying these two vectors together, we obtain

$$(AB)_{ij} = a_{i1}b_{1j} + a_{i2}b_{2j} + \cdots + a_{in}b_{nj}.$$

Expressing this using the summation notation yields the following important result.

Definition 3.2.6:   If $A = [a_{ij}]$ is an $m \times n$ matrix, $B = [b_{ij}]$ is an $n \times p$ matrix, and $C = AB$, then the elements of C are

$$c_{ij} = \sum_{k=1}^{n} a_{ik}b_{kj}, \qquad 1 \le i \le m, \quad 1 \le j \le p. \qquad (3.2.3)$$

This is called the **index form** of the matrix product.

The formula (3.2.3) for the ijth element of AB is very important and will often be required in the future. You should remember it.

We can now state and prove the basic properties of matrix multiplication.

Theorem 3.2.1:   If A, B, and C have appropriate dimensions for the operations to be performed, then

$$A(BC) = (AB)C, \qquad (3.2.4)$$
$$A(B + C) = AB + AC, \qquad (3.2.5)$$
$$(A + B)C = AC + BC. \qquad (3.2.6)$$

PROOF   The idea behind the proof of each of these results is to use the definition of matrix multiplication to show that the ijth element of the matrix on the left-hand side of each equation is equal to the ijth element of the matrix on the right-hand side. We illustrate by proving (3.2.6) and leave the proofs of the remaining properties as exercises. Suppose that A and B are $m \times n$ matrices and that C is an $n \times p$ matrix. Then, from (3.2.3),

$$[(A + B)C]_{ij} = \sum_{k=1}^{n} (a_{ik} + b_{ik})c_{kj} = \sum_{k=1}^{n} a_{ik}c_{kj} + \sum_{k=1}^{n} b_{ik}c_{kj}$$

$$= (AC)_{ij} + (BC)_{ij}$$

$$= [AC + BC]_{ij}, \qquad 1 \le i \le m, \quad 1 \le j \le p. \qquad \blacksquare$$

The preceding theorem states that matrix multiplication is associative and distributive (over addition). We now consider the question of commutativity of matrix multiplication. If A is an $m \times n$ matrix and B is an $n \times m$ matrix, we can form both of the products AB and BA. If $m \ne n$, then the matrices AB and BA will have different dimensions and so cannot be equal. It is important to realize, however, that even if $m = n$, in general (that is, except for special cases)

$$AB \ne BA.$$

This is the statement that matrix multiplication is *not* commutative. With a little bit of thought this should not be too surprising in view of the fact that the ijth element of AB is obtained by multiplying the ith row vector of $\underline{A}$ by the jth column vector of $\underline{B}$, whereas the ijth element of BA is obtained by multiplying the ith row vector of $\underline{\underline{B}}$ by the jth column vector of $\underline{\underline{A}}$. We illustrate with an example.

Example 3.2.11   If $A = \begin{bmatrix} 1 & 2 \\ -1 & 3 \end{bmatrix}$ and $B = \begin{bmatrix} 3 & 1 \\ 2 & -1 \end{bmatrix}$, find AB and BA.

Solution

$$AB = \begin{bmatrix} 1 & 2 \\ -1 & 3 \end{bmatrix}\begin{bmatrix} 3 & 1 \\ 2 & -1 \end{bmatrix} = \begin{bmatrix} 7 & -1 \\ 3 & -4 \end{bmatrix}.$$

$$BA = \begin{bmatrix} 3 & 1 \\ 2 & -1 \end{bmatrix}\begin{bmatrix} 1 & 2 \\ -1 & 3 \end{bmatrix} = \begin{bmatrix} 2 & 9 \\ 3 & 1 \end{bmatrix}.$$

Thus we see that in this example, $AB \ne BA$.

Finally, the following result will be required later in the text.

Theorem 3.2.2:   If $A = [\mathbf{a}_1, \mathbf{a}_2, \ldots, \mathbf{a}_n]$ is an $m \times n$ matrix and $\mathbf{c} = \begin{bmatrix} c_1 \\ c_2 \\ \vdots \\ c_n \end{bmatrix}$ is a column n-vector, then

$$Ac = c_1\mathbf{a}_1 + c_2\mathbf{a}_2 + \cdots + c_n\mathbf{a}_n.$$

PROOF   We must show that

$$(Ac)_i = \left(\sum_{k=1}^{n} c_k\mathbf{a}_k\right)_i, \qquad 1 \le i \le m.$$

In order to do so, we note that the ikth element of A is $a_{ik} = (\mathbf{a}_k)_i$ (the ith element of the kth column vector of A). Thus, from (3.2.1), we have

$$(Ac)_i = \sum_{k=1}^{n} a_{ik}c_k = \sum_{k=1}^{n} (\mathbf{a}_k)_i c_k = \sum_{k=1}^{n} (c_k\mathbf{a}_k)_i = \left(\sum_{k=1}^{n} c_k\mathbf{a}_k\right)_i, \qquad 1 \le i \le m,$$

as required.   ∎

EXERCISES 3.2

1. If $A = \begin{bmatrix} 1 & 2 & -1 \\ 3 & 5 & 2 \end{bmatrix}$ and $B = \begin{bmatrix} 2 & -1 & 3 \\ 1 & 4 & 5 \end{bmatrix}$, find $2A$, $-3B$, $A - 2B$, $3A + 4B$.

2. If
$$A = \begin{bmatrix} 2 & -1 & 0 \\ 3 & 1 & 2 \\ -1 & 1 & 1 \end{bmatrix}, \quad B = \begin{bmatrix} 1 & -1 & 2 \\ 3 & 0 & 1 \\ -1 & 1 & 0 \end{bmatrix},$$
$$C = \begin{bmatrix} -1 & -1 & 1 \\ 1 & 2 & 3 \\ -1 & 1 & 0 \end{bmatrix}.$$
find the matrix D such that $2A + B - 3C + 2D = A + 4C$.

3. Let $A = \begin{bmatrix} 1 & -1 & 2 \\ 3 & 1 & 4 \end{bmatrix}$, $B = \begin{bmatrix} 2 & -1 & 3 \\ 5 & 1 & 2 \\ 4 & 6 & -2 \end{bmatrix}$,
$C = \begin{bmatrix} 1 \\ -1 \\ 2 \end{bmatrix}$, $D = [2 \ -2 \ 3]$. Find, if possible, AB, BC, CA, DC, DB, AD, and CD.

In problems 4–6 determine AB for the given matrices. (In these problems i denotes $\sqrt{-1}$.)

4. $A = \begin{bmatrix} 2-i & 1+i \\ -i & 2+4i \end{bmatrix}$, $B = \begin{bmatrix} i & 1-3i \\ 0 & 4+i \end{bmatrix}$.

5. $A = \begin{bmatrix} 3+2i & 2-4i \\ 5+i & 3i-1 \end{bmatrix}$, $B = \begin{bmatrix} i-1 & 2i+3 \\ 4-3i & 1+i \end{bmatrix}$.

6. $A = \begin{bmatrix} 3-2i & i \\ -i & 1 \end{bmatrix}$,
$B = \begin{bmatrix} -1+i & 2-i & 0 \\ 1+5i & 0 & 3-2i \end{bmatrix}$.

7. Let $A = \begin{bmatrix} 1 & -1 & 2 & 3 \\ -2 & 3 & 4 & 6 \end{bmatrix}$, $B = \begin{bmatrix} 3 & 2 \\ 1 & 5 \\ 4 & -3 \\ -1 & 6 \end{bmatrix}$,
$C = \begin{bmatrix} -3 & 2 \\ 1 & -4 \end{bmatrix}$. Find ABC and CAB.

8. Let $A = \begin{bmatrix} 1 & -2 \\ 3 & 1 \end{bmatrix}$, $B = \begin{bmatrix} -1 & 2 \\ 5 & 3 \end{bmatrix}$, $C = \begin{bmatrix} 3 \\ -1 \end{bmatrix}$. Find $(2A - 3B)C$.

9. If A is an $m \times n$ matrix and C is an $r \times s$ matrix, what must the dimensions of B be in order that the product ABC is defined? Write an expression for the ijth element of ABC in terms of the elements of A, B, and C.

10. If A is an $n \times n$ matrix and k is a positive integer, we define A^k to be the matrix obtained when A is multiplied by itself k times. Find A^2, A^3, and A^4 for each A.

(a) $A = \begin{bmatrix} 1 & -1 \\ 2 & 3 \end{bmatrix}$.

(b) $A = \begin{bmatrix} 0 & 1 & 0 \\ -2 & 0 & 1 \\ 4 & -1 & 0 \end{bmatrix}$.

11. If A and B are $n \times n$ matrices, prove that
(a) $(A + B)^2 = A^2 + AB + BA + B^2$.
(b) $(A - B)^2 = A^2 - AB - BA + B^2$.

12. Find a matrix $A = \begin{bmatrix} 1 & x & z \\ 0 & 1 & y \\ 0 & 0 & 1 \end{bmatrix}$ such that

$$A^2 + \begin{bmatrix} 0 & -1 & 0 \\ 0 & 0 & -1 \\ 0 & 0 & 0 \end{bmatrix} = \begin{bmatrix} 1 & 0 & 0 \\ 0 & 1 & 0 \\ 0 & 0 & 1 \end{bmatrix}.$$

13. If $A = \begin{bmatrix} x & 1 \\ -2 & y \end{bmatrix}$, determine all values of x and y for which $A^2 = A$.

14. The Pauli spin matrices $\sigma_1, \sigma_2, \sigma_3$ are defined by

$$\sigma_1 = \begin{bmatrix} 0 & 1 \\ 1 & 0 \end{bmatrix}, \quad \sigma_2 = \begin{bmatrix} 0 & -i \\ i & 0 \end{bmatrix}, \quad \sigma_3 = \begin{bmatrix} 1 & 0 \\ 0 & -1 \end{bmatrix}.$$

Verify that they satisfy

$$\sigma_1\sigma_2 = i\sigma_3, \quad \sigma_2\sigma_3 = i\sigma_1, \quad \sigma_3\sigma_1 = i\sigma_2.$$

If A and B are $n \times n$ matrices, we define their **commutator**, denoted by $[A, B]$, by

$$[A, B] = AB - BA.$$

Thus $[A, B] = 0$ if and only if A and B commute (that is, $AB = BA$). Problems 15–18 require the commutator.

15. If $A = \begin{bmatrix} 1 & -1 \\ 2 & 1 \end{bmatrix}$ and $B = \begin{bmatrix} 3 & 1 \\ 4 & 2 \end{bmatrix}$, find $[A, B]$.

16. If $A_1 = \begin{bmatrix} 1 & 0 \\ 0 & 1 \end{bmatrix}$, $A_2 = \begin{bmatrix} 0 & 1 \\ 0 & 0 \end{bmatrix}$, and $A_3 = \begin{bmatrix} 0 & 0 \\ 1 & 0 \end{bmatrix}$, compute all of the commutators $[A_i, A_j]$, and determine which of the matrices commute.

17. If $A_1 = \frac{1}{2}\begin{bmatrix} 0 & i \\ i & 0 \end{bmatrix}$, $A_2 = \frac{1}{2}\begin{bmatrix} 0 & -1 \\ 1 & 0 \end{bmatrix}$, and $A_3 = \frac{1}{2}\begin{bmatrix} i & 0 \\ 0 & -i \end{bmatrix}$, show that

$$[A_1, A_2] = A_3, \quad [A_2, A_3] = A_1, \quad [A_3, A_1] = A_2.$$

18. If A, B, C are any $n \times n$ matrices, find an expression for $[A, [B, C]]$ in terms of products of A, B, and C, and prove the Jacobi identity

$$[A, [B, C]] + [B, [C, A]] + [C, [A, B]] = 0.$$

19. Use the index form of the matrix product to prove properties (3.2.4) and (3.2.5).

3.3 SOME IMPORTANT MATRICES

We now introduce some matrices that have special importance in matrix theory and its applications due either to their properties or particular structure.

THE ZERO MATRIX

Definition 3.3.1: The $m \times n$ **zero matrix**, denoted by 0, is the $m \times n$ matrix whose elements are all zeros.

If we wish to emphasize the dimensions of the zero matrix, then we will use the notation $0_{m \times n}$ or, in the case of the $n \times n$ zero matrix, 0_n. The following properties of the zero matrix should be clear.

PROPERTIES OF THE ZERO MATRIX   For all matrices A,

(a) $A + 0 = A$,
(b) $A - A = 0$,
(c) $A0 = 0$,
(d) $0A = 0$.

It follows from these properties that the zero matrix can be considered as taking the place of the number zero in matrix addition and matrix multiplication. One difference between matrix multiplication and the familiar real-number multiplication, however, is that the product of two nonzero matrices *can* equal zero. Thus if we have a matrix equation of the form $AB = 0$, it does not follow that either $A = 0$ or $B = 0$. We illustrate this with an example.

Example 3.3.1   If $A = \begin{bmatrix} 1 & 2 \\ 2 & 4 \end{bmatrix}$ and $B = \begin{bmatrix} -2 & 6 \\ 1 & -3 \end{bmatrix}$, then $A \neq 0$ and $B \neq 0$, but

$$AB = \begin{bmatrix} 0 & 0 \\ 0 & 0 \end{bmatrix} = 0.$$

THE IDENTITY MATRIX

One of the most important $n \times n$ matrices is the identity matrix, defined as follows.

> **Definition 3.3.2:**   The $n \times n$ **identity matrix**, I_n (or just I if the dimensions are obvious), is the $n \times n$ matrix with ones on the main diagonal and zeros elsewhere.

For example,

$$I_2 = \begin{bmatrix} 1 & 0 \\ 0 & 1 \end{bmatrix}, \qquad I_4 = \begin{bmatrix} 1 & 0 & 0 & 0 \\ 0 & 1 & 0 & 0 \\ 0 & 0 & 1 & 0 \\ 0 & 0 & 0 & 1 \end{bmatrix}.$$

The elements of I_n can be represented by the **Kronecker delta symbol**, δ_{ij}, defined by

$$\delta_{ij} = \begin{cases} 1, & \text{if } i = j, \\ 0, & \text{if } i \neq j. \end{cases}$$

We can therefore write

$$I_n = [\delta_{ij}].$$

PROPERTIES OF THE IDENTITY MATRIX

(a) $A_{m \times n} I_n = A_{m \times n}$.
(b) $I_m A_{m \times p} = A_{m \times p}$.

PROOF OF (a)

$$(AI)_{ij} = \sum_{k=1}^{n} a_{ik}\delta_{kj} = a_{ij}\delta_{jj} = a_{ij}, \qquad 1 \le i \le m, \quad 1 \le j \le n. \qquad \blacksquare$$

We leave the proof of (b) as an exercise and instead illustrate it with an example.

Example 3.3.2   If $A = \begin{bmatrix} 2 & -1 \\ 3 & 5 \\ 0 & -2 \end{bmatrix}$, then

$$I_3 A = \begin{bmatrix} 1 & 0 & 0 \\ 0 & 1 & 0 \\ 0 & 0 & 1 \end{bmatrix} \begin{bmatrix} 2 & -1 \\ 3 & 5 \\ 0 & -2 \end{bmatrix} = \begin{bmatrix} 2 & -1 \\ 3 & 5 \\ 0 & -2 \end{bmatrix} = A.$$

REMARK   It follows from properties (a) and (b) that the identity matrix can be considered as taking the place of the number 1 in matrix multiplication.

UPPER TRIANGULAR MATRICES, LOWER TRIANGULAR MATRICES, AND DIAGONAL MATRICES

Definition 3.3.3:   A square matrix A is said to be **upper triangular** if $a_{ij} = 0$ whenever $i > j$ (zeros everywhere below the main diagonal).   A square matrix A is said to be **lower triangular** if $a_{ij} = 0$ whenever $i < j$ (zeros everywhere above the main diagonal).

A schematic representation of upper and lower triangular matrices is given in Figure 3.3.1.

Upper triangular: zeros *below* the main diagonal

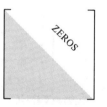

Lower triangular: zeros *above* the main diagonal

Figure 3.3.1   Schematic representation of upper and lower triangular matrices.

Example 3.3.3   The following are examples of an upper triangular and a lower triangular matrix, respectively:

$$\begin{bmatrix} 2 & 8 & 5 \\ 0 & 3 & 9 \\ 0 & 0 & -1 \end{bmatrix}, \quad \begin{bmatrix} 2 & 0 & 0 \\ 0 & 1 & 0 \\ 6 & 7 & 3 \end{bmatrix}.$$

A result that is of importance in the analysis of linear algebraic systems of equations is the following.

Theorem 3.3.1:   The product of two upper triangular matrices is an upper triangular matrix.

PROOF   Suppose that A and B are $n \times n$ upper triangular matrices.   Then, $a_{ik} = 0$ if $k < i$, and $b_{kj} = 0$ whenever $k > j$.   We wish to prove that $(AB)_{ij} = 0$ whenever $i > j$.   By definition of matrix multiplication we have

$$(AB)_{ij} = \sum_{k=1}^{n} a_{ik}b_{kj} = \sum_{k=i}^{n} a_{ik}b_{kj} \qquad \text{(since } a_{ik} = 0 \text{ if } k < i). \qquad (3.3.1)$$

But, if $i > j$, then, since $k \geq i$ in (3.3.1), it follows that $k > j$.   However this implies that $b_{kj} = 0$ (since B is upper triangular) and hence, from (3.3.1), that

$$(AB)_{ij} = 0 \text{ whenever } i > j,$$

as required.   ∎

REMARK   A similar result holds for lower triangular matrices (see problem 19).

An $n \times n$ matrix $D = [d_{ij}]$ that has all off-diagonal elements equal to zero is called a **diagonal** matrix.   Such a matrix is completely determined by giving its main diagonal elements, and this enables us to specify a diagonal matrix in the compact form

$$D = \text{diag}(d_1, d_2, \ldots, d_n),$$

where the d_i denote the diagonal elements.

Example 3.3.4 $D = \text{diag}(1, 2, 0, -3)$ is the 4×4 matrix

$$D = \begin{bmatrix} 1 & 0 & 0 & 0 \\ 0 & 2 & 0 & 0 \\ 0 & 0 & 0 & 0 \\ 0 & 0 & 0 & -3 \end{bmatrix}.$$

THE TRANSPOSE OF A MATRIX

Definition 3.3.4:   The **transpose** of an $m \times n$ matrix A is the $n \times m$ matrix obtained by interchanging row vectors and column vectors in A.   We denote the transpose of A by A^T.   In index notation, the ijth element of A^T, denoted by a_{ij}^T, is

$$a_{ij}^T = a_{ji}.$$

Example 3.3.5   If $A = \begin{bmatrix} 1 & 2 & 6 & 2 \\ 0 & 3 & 4 & 7 \end{bmatrix}$, then $A^T = \begin{bmatrix} 1 & 0 \\ 2 & 3 \\ 6 & 4 \\ 2 & 7 \end{bmatrix}$. If $A = \begin{bmatrix} 1 & -3 & 5 \\ 2 & 0 & 7 \\ 3 & 4 & 9 \end{bmatrix}$,

then $A^T = \begin{bmatrix} 1 & 2 & 3 \\ -3 & 0 & 4 \\ 5 & 7 & 9 \end{bmatrix}$.

PROPERTIES OF THE TRANSPOSE   The operation of taking the transpose of a matrix satisfies several properties, which are listed in the following theorem.

Theorem 3.3.2:   Let A and C be $m \times n$ matrices, and let B be an $n \times p$ matrix. Then

(a) $(A^T)^T = A$.

(b) $(A + C)^T = A^T + C^T$.

(c) $(AB)^T = B^T A^T$.

PROOF   The proofs of (a) and (b) are almost immediate and hence are left as exercises.   We prove (c) by showing that $[(AB)^T]_{ij} = [B^T A^T]_{ij}$.   From the definition of the transpose and the index form of the matrix product we have

$$[(AB)^T]_{ij} = (AB)_{ji} = \sum_{k=1}^{n} a_{jk} b_{ki} = \sum_{k=1}^{n} b_{ki} a_{jk} = \sum_{k=1}^{n} b_{ik}^T a_{kj}^T = [B^T A^T]_{ij},$$

$\uparrow$ $\uparrow$

Def. of   Def. of
transpose   matrix mult.

as required.

The transpose naturally picks out two important types of matrices as follows.

Definition 3.3.5:

1. A square matrix A satisfying $A^T = A$ is said to be **symmetric**.
2. A square matrix A satisfying $A^T = -A$ is said to be **skew symmetric** (or **antisymmetric**).

Example 3.3.6 $A = \begin{bmatrix} 1 & -1 & 1 & 5 \\ -1 & 2 & 2 & 6 \\ 1 & 2 & 3 & 4 \\ 5 & 6 & 4 & 9 \end{bmatrix}$ is a symmetric matrix, whereas

$B = \begin{bmatrix} 0 & -1 & -5 & 3 \\ 1 & 0 & 1 & -2 \\ 5 & -1 & 0 & 7 \\ -3 & 2 & -7 & 0 \end{bmatrix}$ is a skew-symmetric matrix.

REMARK   Notice that the main diagonal elements of the skew-symmetric matrix in the preceding example are all zeros. This is true in general since, if A is skew-symmetric, then $a_{ij} = -a_{ji}$ which implies, when $i = j$, that $a_{ii} = -a_{ii}$ or equivalently $a_{ii} = 0$.

EXERCISES 3.3

In problems 1–3 give an example of a matrix of the specified form.

1. 3×3 diagonal matrix.

2. 4×4 upper triangular matrix.

3. 4×4 skew-symmetric.

4. Prove that a symmetric upper triangular matrix is diagonal.

5. Prove property (b) of the identity matrix.

6. Determine all elements of the 3×3 skew-symmetric matrix A with $a_{21} = 1$, $a_{31} = 3$, $a_{23} = -1$.

7. A matrix that is a multiple of I_n is called an $n \times n$ **scalar** matrix. Determine the 4×4 scalar matrix whose trace is 8.

8. If A and B are $n \times n$ matrices, prove that $\text{tr}(AB) = \text{tr}(BA)$.

9. Let $A = \begin{bmatrix} 1 & -1 & 1 & 4 \\ 2 & 0 & 2 & -3 \\ 3 & 4 & -1 & 0 \end{bmatrix}$ and

$B = \begin{bmatrix} 0 & 1 \\ -1 & 2 \\ 1 & 1 \\ 2 & 1 \end{bmatrix}$, find A^T, B^T, AA^T, AB, and $B^T A^T$.

10. Let $A = \begin{bmatrix} 2 & 2 & 1 \\ 2 & 5 & 2 \\ 1 & 2 & 2 \end{bmatrix}$ and let S be the matrix

with column vectors $\mathbf{s}_1 = \begin{bmatrix} -x \\ 0 \\ x \end{bmatrix}$, $\mathbf{s}_2 = \begin{bmatrix} -y \\ y \\ -y \end{bmatrix}$, and

$\mathbf{s}_3 = \begin{bmatrix} z \\ 2z \\ z \end{bmatrix}$, where x, y, z are constants.

(a) Show that $AS = [\mathbf{s}_1, \mathbf{s}_2, 7\mathbf{s}_3]$.
(b) Find all values of x, y, z such that $S^T AS = \text{diag}(1, 1, 7)$.

11. If $A = \begin{bmatrix} 3 & -1 \\ -5 & -1 \end{bmatrix}$, calculate A^2 and verify

that A satisfies $A^2 - 2A - 8I_2 = 0_2$.

If A is an $n \times n$ matrix, then the matrices S and T defined by

$$S = \tfrac{1}{2}(A + A^T), \qquad T = \tfrac{1}{2}(A - A^T)$$

are called the symmetric and skew-symmetric parts of A, respectively. Problems 12–15 investigate the properties of S and T.

12. Use the properties of the transpose to show that S and T are symmetric and skew-symmetric, respectively.

13. Find S and T for the matrix $A = \begin{bmatrix} 1 & -5 & 3 \\ 3 & 2 & 4 \\ 7 & -2 & 6 \end{bmatrix}$.

14. If A is an $n \times n$ symmetric matrix show that $T = 0$. What is the corresponding result for skew-symmetric matrices?

15. Show that any $n \times n$ matrix can be written as the sum of a symmetric and a skew-symmetric matrix.

16. Prove that if A is an $n \times p$ matrix and $D = \text{diag}(d_1, d_2, \ldots, d_n)$, then DA is the matrix obtained by multiplying the ith row vector of A by d_i $(1 \leq i \leq n)$.

17. If $A = \begin{bmatrix} 2 & -1 & 1 \\ 0 & 4 & 3 \\ 0 & 0 & 1 \end{bmatrix}$ and $B = \begin{bmatrix} 2 & 1 & -3 \\ 0 & 1 & 5 \\ 0 & 0 & -2 \end{bmatrix}$,

find AB and thereby verify Theorem 3.3.1.

18. Use properties of the transpose to prove that
(a) (AA^T) is a symmetric matrix.
(b) $(ABC)^T = C^T B^T A^T$.

19. Prove that the product of two lower triangular matrices, A and B, is a lower triangular matrix, in two different ways:

(a) Directly in terms of the elements of A and B (as in the proof of Theorem 3.3.1).
(b) By using the result of Theorem 3.3.1 and property (c) of the transpose.

4

Solutions of Systems of Linear Equations

At some time or another we have all been faced with the problem of solving a system of linear algebraic equations. An example of such a system is

$$x_1 - x_2 = 2,$$
$$x_1 + 5x_2 = 14.$$

One approach to solving this system is to subtract the first equation from the second to eliminate x_1. The result is

$$6x_2 = 12,$$

so that

$$x_2 = 2.$$

Substituting this value of x_2 back into either of the original equations and solving for x_1 yields

$$x_1 = 4.$$

We have therefore determined the solution of the system to be $x_1 = 4$, $x_2 = 2$. In this chapter we develop systematic methods that generalize this technique for solving any linear system. Before introducing the theory we first consider an example that illustrates where linear algebraic systems may arise in the solution of differential equations.

Example Determine a particular solution of the form $y_p = A_0 + A_1 x + A_2 x^2$, where A_0, A_1, and A_2 are constants, to the differential equation

$$y'' + y' + 2y = 9 + 4x + 6x^2.$$

Solution Differentiating the given function with respect to x yields:

$$y_p' = A_1 + 2A_2 x, \qquad y_p'' = 2A_2.$$

Substituting $y = y_p$ into the left-hand side of the differential equation we obtain

$$2A_2 + (A_1 + 2A_2 x) + 2(A_0 + A_1 x + A_2 x^2) = 9 + 4x + 6x^2,$$

or, equivalently,

$$(2A_0 + A_1 + 2A_2) + (2A_1 + 2A_2)x + 2A_2 x^2 = 9 + 4x + 6x^2.$$

This equation can hold for all x if and only if A_0, A_1, A_2, satisfy the linear system

$$2A_0 + A_1 + 2A_2 = 9,$$
$$2A_1 + 2A_2 = 4,$$
$$2A_2 = 6.$$

From the third equation we have $A_2 = 3$. The second equation then implies that $A_1 = -1$, and substitution into the first equation yields $A_0 = 2$. Thus a particular solution to the given differential equation is

$$y_p = 2 - x + 3x^2.$$

This technique for determining a particular solution to a linear differential equation will be developed in detail in Chapter 9.

4.1 TERMINOLOGY AND NOTATION

The general $m \times n$ system of linear equations consists of m equations in the n unknowns $x_1, x_2, \ldots, x_n$. Such a system can be written in the form

$$
\begin{aligned}
a_{11}x_1 + a_{12}x_2 + a_{13}x_3 + \cdots + a_{1n}x_n &= b_1, \\
a_{21}x_1 + a_{22}x_2 + a_{23}x_3 + \cdots + a_{2n}x_n &= b_2, \\
&\vdots \\
a_{m1}x_1 + a_{m2}x_2 + a_{m3}x_3 + \cdots + a_{mn}x_n &= b_m,
\end{aligned}
\tag{4.1.1}
$$

where the **system coefficients** a_{ij} and the **system constants** b_j are given scalars. If

$b_j = 0$ for all j, then the system is called **homogeneous**; otherwise it is called **nonhomogeneous**.

> **Definition 4.1.1:**   By a **solution** of the system (4.1.1) we mean an ordered n-tuple of scalars, say $(c_1, c_2, \ldots, c_n)$, which, when substituted for $x_1, x_2, \ldots, x_n$ into the left-hand side of (4.1.1), yield the right-hand side.

REMARKS

1. Usually the a_{ij} and b_j will be real numbers and we will then be interested in determining only the real solutions of (4.1.1).   However, some of the problems that arise in the later chapters will require the solution of systems with complex coefficients, in which case the corresponding solutions will also be complex.

2. If $(c_1, c_2, \ldots, c_n)$ is a solution of the system (4.1.1), we will sometimes specify this solution by writing $x_1 = c_1$, $x_2 = c_2$, $\ldots$, $x_n = c_n$. For example, the ordered pair of numbers (1, 2) is a solution of the system

$$x_1 + x_2 = 3,$$
$$3x_1 - 2x_2 = -1,$$

and we could express this solution in the equivalent form $x_1 = 1$, $x_2 = 2$.

The following questions need addressing:

1. Does the system (4.1.1) have a solution?
2. If the answer to (1) is yes, then how many solutions are there?
3. How do we determine all the solutions?

To obtain an idea of the answer to questions (1) and (2), consider the special case of a system of three equations in three unknowns.   The linear system (4.1.1) then reduces to

$$a_{11}x_1 + a_{12}x_2 + a_{13}x_3 = b_1,$$
$$a_{21}x_1 + a_{22}x_2 + a_{23}x_3 = b_2,$$
$$a_{31}x_1 + a_{32}x_2 + a_{33}x_3 = b_3,$$

which can be interpreted as defining three planes in space.   An ordered triple (c_1, c_2, c_3) is a solution of this system if and only if it corresponds to the coordinates of a point of intersection of the three planes.   There are precisely four possibilities:

1. The planes have no intersection point.
2. The planes intersect in just one point.
3. The planes intersect in a line.
4. The planes are coincident.

In (1) the corresponding system has no solution, whereas in (2) the system has just one solution.   Finally, in (3) and (4) every point on the line or plane (respectively) is a solution of the linear system and hence the system has an infinite number of solutions. Cases (1) to (3) are illustrated in Figure 4.1.1. We have thus proved, geometrically, that there are only three possibilities for the solutions of a system of three equations in three unknowns. The system has no solution, it has just one solution, or it has an infinite number of solutions. The main theoretical result of this chapter is a proof that these are the only possibilities for the general $m \times n$ system (4.1.1).

Theorem 4.1.1: The system of linear equations (4.1.1) has either

(a) *no solution,* or
(b) *one solution* (a *unique solution*), or
(c) an *infinite number of solutions.*

PROOF See Section 4.3. ∎

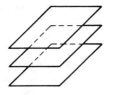

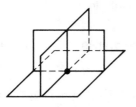

Three parallel planes (no intersection): no solution

No common intersection: no solution

Planes intersect at a point: a unique solution

Planes intersect in a line: an infinite number of solutions

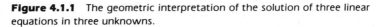

Figure 4.1.1 The geometric interpretation of the solution of three linear equations in three unknowns.

Example 4.1.1 Show that for all values of t, $(2 - 3t, 1 + t, -1 + 2t)$ is a solution of the system

$$x_1 + x_2 + x_3 = 2,$$
$$2x_1 + 8x_2 - x_3 = 13.$$

Solution Substituting for $x_1 = 2 - 3t$, $x_2 = 1 + t$, $x_3 = -1 + 2t$, we have

$$x_1 + x_2 + x_3 = (2 - 3t) + (1 + t) + (-1 + 2t) = 2,$$

so that the first equation is satisfied. Similarly,

$$2x_1 + 8x_2 - x_3 = 2(2 - 3t) + 8(1 + t) - (-1 + 2t) = 13,$$

so that the second equation is also satisfied. It follows from Definition 4.1.1 that

$$(2 - 3t, 1 + t, -1 + 2t)$$

is a solution of the given system for all values of t. In this case the system has an infinite number of solutions. Geometrically, we can interpret the two given equations as defining two planes in space. We have shown that the planes intersect in the line with parametric equations

$$x_1 = 2 - 3t, \qquad x_2 = 1 + t, \qquad x_3 = -1 + 2t.$$

A system of equations that possesses a solution is said to be **consistent**, whereas a system that has no solution is called **inconsistent**. Our problem will be first to determine whether a given system is consistent and then, in the case when we do have a consistent system, to find all its solutions.

> **Definition 4.1.2:** The set of all solutions to a system of linear equations is called the **solution set** of the system.

If we let S denote the solution set of a linear system, then Theorem 4.1.1 implies that S falls into one of the following three categories:[1]

1. S is the empty set, $\emptyset$ (if and only if the system is inconsistent).
2. S contains just one element (if and only if the system has a unique solution).
3. S contains an infinite number of elements (if and only if the system has an infinite number of solutions).

For example, the solution set of the system in Example 4.1.1 can be expressed as

$$S = \{(2 - 3t, 1 + t, -1 + 2t) : t \in \mathbf{R}\}.$$

Naturally associated with the system (4.1.1) are the following two matrices:

1. The **matrix of coefficients** $A = \begin{bmatrix} a_{11} & a_{12} & a_{13} & \cdots & a_{1n} \\ a_{21} & a_{22} & a_{23} & \cdots & a_{2n} \\ \vdots & \vdots & \vdots & & \vdots \\ a_{m1} & a_{m2} & a_{m3} & \cdots & a_{mn} \end{bmatrix}.$

2. The **augmented matrix** $A^{\#} = \begin{bmatrix} a_{11} & a_{12} & a_{13} & \cdots & a_{1n} & b_1 \\ a_{21} & a_{22} & a_{23} & \cdots & a_{2n} & b_2 \\ \vdots & \vdots & \vdots & & \vdots & \vdots \\ a_{m1} & a_{m2} & a_{m3} & \cdots & a_{mn} & b_m \end{bmatrix}.$

The augmented matrix completely characterizes a system of equations, since it contains all of the system coefficients and system constants. We will see in the fol-

---

[1] The surprising result is that if S is a nonempty finite set, then it contains only *one* element.

lowing sections that it is the relationship between A and $A^\#$ that determines the solution properties of a linear system.

Example 4.1.2 Write the system of equations with the following augmented matrix:

$$\begin{bmatrix} 1 & 2 & 9 & -1 & | & 1 \\ 2 & -3 & 7 & 4 & | & 2 \\ 1 & 3 & 5 & 0 & | & -1 \end{bmatrix}.$$

Solution

$$\begin{aligned} x_1 + 2x_2 + 9x_3 - x_4 &= 1, \\ 2x_1 - 3x_2 + 7x_3 + 4x_4 &= 2, \\ x_1 + 3x_2 + 5x_3 &= -1. \end{aligned}$$

Finally, in this section we show that the matrix of coefficients can be used to write a linear system in a very compact form as a single matrix equation. For example, the system

$$\begin{aligned} x_1 + 3x_2 - 4x_3 - x_4 &= 1, \\ 2x_1 + 5x_2 - x_3 + x_4 &= 5, \\ x_1 + 6x_3 - 7x_4 &= 3, \end{aligned}$$

can be written as the matrix equation

$$\begin{bmatrix} 1 & 3 & -4 & -1 \\ 2 & 5 & -1 & 1 \\ 1 & 0 & 6 & -7 \end{bmatrix} \begin{bmatrix} x_1 \\ x_2 \\ x_3 \\ x_4 \end{bmatrix} = \begin{bmatrix} 1 \\ 5 \\ 3 \end{bmatrix},$$

since this matrix equation is satisfied if and only if

$$\begin{bmatrix} x_1 + 3x_2 - 4x_3 - x_4 \\ 2x_1 + 5x_2 - x_3 + x_4 \\ x_1 + 6x_3 - 7x_4 \end{bmatrix} = \begin{bmatrix} 1 \\ 5 \\ 3 \end{bmatrix};$$

that is, if and only if each equation of the given system is satisfied. Similarly the general $m \times n$ system of linear equations

$$\begin{aligned} a_{11}x_1 + a_{12}x_2 + a_{13}x_3 + \cdots + a_{1n}x_n &= b_1, \\ a_{21}x_1 + a_{22}x_2 + a_{23}x_3 + \cdots + a_{2n}x_n &= b_2, \\ &\vdots \\ a_{m1}x_1 + a_{m2}x_2 + a_{m3}x_3 + \cdots + a_{mn}x_n &= b_m, \end{aligned}$$

can be written as the matrix equation

$$A\mathbf{x} = \mathbf{b}, \tag{4.1.2}$$

where A is the matrix of coefficients and

$$\mathbf{x} = \begin{bmatrix} x_1 \\ x_2 \\ \vdots \\ x_n \end{bmatrix}, \qquad \mathbf{b} = \begin{bmatrix} b_1 \\ b_2 \\ \vdots \\ b_m \end{bmatrix}.$$

We will refer to the n-vector $\mathbf{x}$ as the **vector of unknowns**, and the m-vector $\mathbf{b}$ will be referred to as the **right-hand-side vector**.

EXERCISES 4.1

In problems 1 and 2 show that the given triple of real numbers is a solution of the given system.

1. $(1, -1, 2);\ 2x_1 - 3x_2 + 4x_3 = 13,$
$\qquad\qquad\ \ x_1 + x_2 - x_3 = -2,$
$\qquad\qquad\ \ 5x_1 + 4x_2 + x_3 = 3.$

2. $(2, -3, 1);\ x_1 + x_2 - 2x_3 = -3,$
$\qquad\qquad\ 3x_1 - x_2 - 7x_3 = 2,$
$\qquad\qquad\ \ x_1 + x_2 + x_3 = 0,$
$\qquad\qquad\ 2x_1 + 2x_2 - 4x_3 = -6.$

3. Show that for all values of t, $(1 - t, 2 + 3t, 3 - 2t)$ is a solution of the linear system

$$\begin{aligned} x_1 + x_2 + x_3 &= 6, \\ x_1 - x_2 - 2x_3 &= -7, \\ 5x_1 + x_2 - x_3 &= 4. \end{aligned}$$

4. Show that for all values of s and t, $(s, s - 2t, 2s + 3t, t)$ is a solution of the linear system

$$\begin{aligned} x_1 + x_2 - x_3 + 5x_4 &= 0, \\ 2x_2 - x_3 + 7x_4 &= 0, \\ 4x_1 + 2x_2 - 3x_3 + 13x_4 &= 0. \end{aligned}$$

5. By making a sketch in the xy-plane, prove that the following linear system has no solutions.

$$\begin{aligned} 2x + 3y &= 1, \\ 2x + 3y &= 2. \end{aligned}$$

In problems 6–8, determine the coefficient matrix A, the right-hand-side vector $\mathbf{b}$, and the augmented matrix $A^\#$ of the given system.

6. $x_1 + 2x_2 - 3x_3 = 1,$
$\ \ 2x_1 + 4x_2 - 5x_3 = 2,$
$\ \ 7x_1 + 2x_2 - x_3 = 3.$

7. $x_1 + x_2 + x_3 - x_4 = 3,$
$\ \ 2x_1 + 4x_2 - 3x_3 + 7x_4 = 2.$

8. $x_1 + 2x_2 - x_3 = 0,$
$\ \ 2x_1 + 3x_2 - 2x_3 = 0,$
$\ \ 5x_1 + 6x_2 - 5x_3 = 0.$

In problems 9 and 10 write out the system of equations with the given coefficient matrix and right-hand-side vector.

9. $A = \begin{bmatrix} 1 & -1 & 2 & 3 \\ 1 & 1 & -2 & 6 \\ 3 & 1 & 4 & 2 \end{bmatrix},\ \mathbf{b} = \begin{bmatrix} 1 \\ -1 \\ 2 \end{bmatrix}.$

10. $A = \begin{bmatrix} 2 & 1 & 3 \\ 4 & -1 & 2 \\ 7 & 6 & 3 \end{bmatrix},\ \mathbf{b} = \begin{bmatrix} 1 \\ -1 \\ 2 \end{bmatrix}.$

11. Consider the $m \times n$ homogeneous system of linear equations

$$A\mathbf{x} = \mathbf{0}, \qquad\qquad (i)$$

where $\mathbf{0}$ denotes the m-vector of zeros.

(a) If $\mathbf{x} = [x_1\ x_2\ \cdots\ x_n]^T$ and $\mathbf{y} = [y_1\ y_2\ \cdots\ y_n]^T$ are solutions of (i), show that $\mathbf{z} = \mathbf{x} + \mathbf{y}$ and $\mathbf{w} = c\mathbf{x}$ are also solutions, where c is an arbitrary scalar.

(b) Is the result in (a) true when $\mathbf{x}$ and $\mathbf{y}$ are solutions of the nonhomogeneous system

$$A\mathbf{x} = \mathbf{b}?$$

4.2   ELEMENTARY ROW OPERATIONS AND ROW ECHELON MATRICES

In Section 4.3 we will develop two systematic procedures for solving a system of linear equations. Our methods will consist of reducing a given system of equations to a new system that has the same solution set as the given one but that is easier to solve. In this section we introduce the mathematical results that are necessary for developing these procedures.

ELEMENTARY ROW OPERATIONS

The first step in deriving systematic procedures for solving a linear system is to determine what operations can be performed on such a system *without altering its solution set*.

Example 4.2.1   Consider the system of equations

$$x_1 + 2x_2 + 4x_3 = 2, \tag{4.2.1}$$

$$2x_1 - 5x_2 + 3x_3 = 6, \tag{4.2.2}$$

$$4x_1 + 6x_2 - 7x_3 = 8. \tag{4.2.3}$$

If we interchange (4.2.1) and (4.2.2), for example, then the resulting system is

$$2x_1 - 5x_2 + 3x_3 = 6,$$

$$x_1 + 2x_2 + 4x_3 = 2,$$

$$4x_1 + 6x_2 - 7x_3 = 8,$$

which clearly has the same solution set as the original system. Further if we multiply (4.2.2) by 5, for example, we obtain the system

$$x_1 + 2x_2 + 4x_3 = 2,$$

$$10x_1 - 25x_2 + 15x_3 = 30,$$

$$4x_1 + 6x_2 - 7x_3 = 8,$$

which again has the same solution set as the original system. Finally, if we add twice (4.2.1) to (4.2.3), for example, we obtain the system

$$x_1 + 2x_2 + 4x_3 = 2, \tag{4.2.4}$$

$$2x_1 - 5x_2 + 3x_3 = 6, \tag{4.2.5}$$

$$(4x_1 + 6x_2 - 7x_3) + 2(x_1 + 2x_2 + 4x_3) = 8 + 2(2). \tag{4.2.6}$$

We see that if (4.2.4)–(4.2.6) are satisfied, then so are (4.2.1)–(4.2.3), and conversely. It follows that the system (4.2.4)–(4.2.6) has the same solution set as the original system (4.2.1)–(4.2.3).

More generally, a similar reasoning can be used to show that the following three types of operations can be performed on any $m \times n$ system of linear equations *without altering the solution set:*

1. Interchange equations.
2. Multiply an equation by a nonzero constant.
3. Add a multiple of one equation to another.

Since the operations (1)–(3) involve changes only in the system coefficients and constants (and not changes in the variables), they can be represented by the following operations on the augmented matrix of the system:

1. Interchange rows.
2. Multiply a row by a nonzero constant.
3. Add a multiple of one row to another row.

These three operations are called **elementary row operations**.

We will use the following notation to describe elementary row operations:

1. $R_i \leftrightarrow R_j$, meaning: interchange row i and row j.
2. $R_i \to kR_i$, meaning: multiply row i by the scalar k.
3. $R_i \to R_i + kR_j$, meaning: add k times row j to row i.

Example 4.2.2 The operations performed on the system in Example 4.2.1 can be described as follows using elementary row operations on the augmented matrix of the system:

$$\left[\begin{array}{ccc|c} 1 & 2 & 4 & 2 \\ 2 & -5 & 3 & 6 \\ 4 & 6 & -7 & 8 \end{array}\right] \xrightarrow{R_1 \leftrightarrow R_2} \left[\begin{array}{ccc|c} 2 & -5 & 3 & 6 \\ 1 & 2 & 4 & 2 \\ 4 & 6 & -7 & 8 \end{array}\right] \text{Interchange (4.2.1) and (4.2.2).}$$

$$\left[\begin{array}{ccc|c} 1 & 2 & 4 & 2 \\ 2 & -5 & 3 & 6 \\ 4 & 6 & -7 & 8 \end{array}\right] \xrightarrow{R_2 \to 5R_2} \left[\begin{array}{ccc|c} 1 & 2 & 4 & 2 \\ 10 & -25 & 15 & 30 \\ 4 & 6 & -7 & 8 \end{array}\right] \text{Multiply (4.2.2) by 5.}$$

$$\left[\begin{array}{ccc|c} 1 & 2 & 4 & 2 \\ 2 & -5 & 3 & 6 \\ 4 & 6 & -7 & 8 \end{array}\right] \xrightarrow{R_3 \to R_3 + 2R_1} \left[\begin{array}{ccc|c} 1 & 2 & 4 & 2 \\ 2 & -5 & 3 & 6 \\ 6 & 10 & 1 & 12 \end{array}\right] \text{Add 2 times (4.2.1) to (4.2.3).}$$

It is important to note that the matrices in Example 4.2.2 are *not* equal. However, they do represent linear systems that have the same solution sets. We introduce a special term for matrices that are related via elementary row operations:[1]

---

[1] It should be noted that although our discussion is motivated by the study of systems of linear equations, any results pertaining to the augmented matrix of a system will also apply to an arbitrary matrix, since such a matrix can always be considered as the augmented matrix of an appropriate linear system.

Definition 4.2.1: Let A be an $m \times n$ matrix. Any matrix obtained from A by a finite sequence of elementary row operations is said to be **row equivalent** to A.

Thus, all the matrices in the previous example are row equivalent to the augmented matrix of the system and also to each other. Since elementary row operations do not alter the solution set of a linear system, we have the following.

Theorem 4.2.1: Systems of linear equations with row-equivalent augmented matrices have the same solution sets.

Our methods for solving a system of linear equations will consist of using elementary row operations to reduce the augmented matrix of the given system to a simple form. The question that needs answering is the following: How simple a form should we aim for? In order to answer this question, consider the following system:

$$x_1 + x_2 - x_3 = 4, \qquad (4.2.7)$$

$$x_2 - 3x_3 = 5, \qquad (4.2.8)$$

$$x_3 = 2. \qquad (4.2.9)$$

A systematic way to proceed in solving this system is as follows. From (4.2.9), $x_3 = 2$. Substituting this value into (4.2.8) and solving for x_2 yields $x_2 = 5 + 6 = 11$. Finally, substituting for x_3 and x_2 into (4.2.7) and solving for x_1 we obtain $x_1 = -5$. Thus the solution to our system of equations is $(-5, 11, 2)$. This technique is called **back substitution** and could be used only because our system was in a simple form. The augmented matrix of the system is

$$\left[\begin{array}{ccc|c} 1 & 1 & -1 & 4 \\ 0 & 1 & -3 & 5 \\ 0 & 0 & 1 & 2 \end{array} \right].$$

This matrix has a simple structure, namely, 1s down the main diagonal (that is, $a_{11} = a_{22} = a_{33} = 1$) and 0s beneath it. Such a matrix is called a **unit upper triangular matrix** (the term unit coming from the fact that the main diagonal elements are all 1s). The back substitution method will work on any consistent system of linear equations with an augmented matrix of this form. Thus, if we could use elementary row operations to reduce the augmented matrix of a given system of equations to a unit upper triangular matrix, we could certainly solve the resulting system easily and hence obtain the solution of the original system. Unfortunately not all systems of equations have augmented matrices that can be reduced to a *unit* upper triangular matrix by elementary row operations. However there is a simple type of matrix, closely related to a unit upper triangular matrix, to which any matrix can be reduced by elementary row operations and which also represents a system of equations that can be solved (if it has a solution) by back substitution. This type of matrix is called a row echelon matrix and is defined as follows:

Definition 4.2.2:   An $m \times n$ matrix is called a **row echelon matrix** if it satisfies the following three conditions:

1. If there are any rows consisting entirely of zeros, they occur at the bottom of the matrix.
2. The first nonzero element in any nonzero row[1] is a one.   (This is called a **leading 1.**)
3. The leading 1 of any row (except the first) is further to the right than the leading 1 of the row above it.

Example 4.2.3   The following are examples of row echelon matrices:

$$\begin{bmatrix} 1 & -2 & 3 & 7 \\ 0 & 1 & 5 & 0 \\ 0 & 0 & 0 & 1 \end{bmatrix}, \quad \begin{bmatrix} 0 & 0 & 1 \\ 0 & 0 & 0 \\ 0 & 0 & 0 \end{bmatrix}, \quad \begin{bmatrix} 1 & -1 & 6 & 5 & 9 \\ 0 & 0 & 1 & 2 & 5 \\ 0 & 0 & 0 & 1 & 0 \\ 0 & 0 & 0 & 0 & 0 \end{bmatrix},$$

whereas

$$\begin{bmatrix} 1 & 0 & -1 \\ 0 & 1 & 2 \\ 0 & 1 & -1 \end{bmatrix} \text{ and } \begin{bmatrix} 1 & 0 & 0 \\ 0 & 0 & 0 \\ 0 & 1 & -1 \\ 0 & 0 & 1 \end{bmatrix}$$

are not row echelon matrices.

As we will illustrate in the next section, the back substitution method simplifies if the augmented matrix of a linear system is a reduced row echelon matrix, defined as follows:

Definition 4.2.3:   An $m \times n$ matrix is called a **reduced row echelon matrix** if it satisfies the following conditions:

1. It is a row echelon matrix.
2. Any *column* that contains a leading 1 has zeros everywhere else.

Example 4.2.4:   The matrices

$$\begin{bmatrix} 1 & 3 & 0 & 0 \\ 0 & 0 & 1 & 0 \\ 0 & 0 & 0 & 1 \end{bmatrix}, \quad \begin{bmatrix} 1 & -1 & 7 & 0 \\ 0 & 0 & 0 & 1 \end{bmatrix}, \quad \begin{bmatrix} 1 & 0 & 5 & 3 \\ 0 & 1 & 2 & 1 \\ 0 & 0 & 0 & 0 \end{bmatrix}, \quad \begin{bmatrix} 1 & 0 & 0 \\ 0 & 1 & 0 \\ 0 & 0 & 1 \end{bmatrix},$$

are all reduced row echelon matrices.

---

[1] A nonzero row is any row that does not consist entirely of zeros.

The basic result that will allow us to determine the solution set of any system of linear equations is the following:

Theorem 4.2.2:   Any matrix is row equivalent to a row echelon matrix.

According to this theorem, by applying an appropriate sequence of elementary row operations to any $m \times n$ matrix we can always reduce it to a row echelon matrix.   When a matrix, A, has been reduced to a row echelon matrix in this way, we say that it has been reduced to **row echelon form** and refer to the resulting matrix as a row echelon form of A.   In the case when the resulting matrix is a reduced row echelon matrix, we say that A has been reduced to **reduced row echelon form**.   The proof of Theorem 4.2.2 consists of giving an algorithm that will reduce an arbitrary matrix to a row echelon matrix after a finite sequence of elementary row operations.   Before doing so, we first illustrate the result with some examples.

Example 4.2.5   Use elementary row operations to reduce $\begin{bmatrix} 2 & 1 & -1 & 3 \\ 1 & -1 & 2 & 1 \\ -4 & 6 & -7 & 1 \\ 2 & 0 & 1 & 3 \end{bmatrix}$ to

row echelon form.

Solution   We show each step in detail.

> *Step 1:*   Put a leading 1 in the (11) position.
> This can be most easily accomplished by interchanging row 1 and row 2.

$$\begin{bmatrix} 2 & 1 & -1 & 3 \\ 1 & -1 & 2 & 1 \\ -4 & 6 & -7 & 1 \\ 2 & 0 & 1 & 3 \end{bmatrix} \xrightarrow{R_1 \leftrightarrow R_2} \begin{bmatrix} 1 & -1 & 2 & 1 \\ 2 & 1 & -1 & 3 \\ -4 & 6 & -7 & 1 \\ 2 & 0 & 1 & 3 \end{bmatrix}.$$

Leading 1

> *Step 2:*   Use the leading 1 to put 0s beneath it in column 1.
> We now add appropriate multiples of row 1 to the remaining rows in order to

put 0s beneath the leading 1.

$$\begin{matrix} R_2 \rightarrow R_2 - 2R_1 \\ R_3 \rightarrow R_3 + 4R_1 \\ \xrightarrow{\hspace{2cm}} \\ R_4 \rightarrow R_4 - 2R_1 \end{matrix} \begin{bmatrix} 1 & -1 & 2 & 1 \\ 0 & 3 & -5 & 1 \\ 0 & 2 & 1 & 5 \\ 0 & 2 & -3 & 1 \end{bmatrix}.$$

> *Step 3:*   Put a leading 1 in the (22) position.
> We could accomplish this by dividing row 2 by 3.   However, this would

introduce fractions into the matrix and thereby complicate the remaining computations.   If at all possible we should avoid the use of fractions.   In this case we can obtain our leading 1 without the use of fractions by subtracting row 3 from row 2.

$$\xrightarrow[R_2 \rightarrow R_2 - R_3]{} \begin{bmatrix} 1 & -1 & 2 & 1 \\ 0 & 1 & -6 & -4 \\ 0 & 2 & 1 & 5 \\ 0 & 2 & -3 & 1 \end{bmatrix}.$$

Leading 1

Step 4:   Use the leading 1 in the (22) position to put 0s beneath it in column 2. We now subtract appropriate multiples of row 2 from the rows beneath it.

$$\begin{matrix} R_3 \rightarrow R_3 - 2R_2 \\ \xrightarrow{\hspace{2cm}} \\ R_4 \rightarrow R_4 - 2R_2 \end{matrix} \begin{bmatrix} 1 & -1 & 2 & 1 \\ 0 & 1 & -6 & -4 \\ 0 & 0 & 13 & 13 \\ 0 & 0 & 9 & 9 \end{bmatrix}.$$

Step 5:   Put a leading 1 in the (33) position. This can be accomplished by dividing row 3 by 13.

$$\xrightarrow[R_3 \rightarrow R_3/13]{} \begin{bmatrix} 1 & -1 & 2 & 1 \\ 0 & 1 & -6 & -4 \\ 0 & 0 & 1 & 1 \\ 0 & 0 & 9 & 9 \end{bmatrix}.$$

Leading 1

Step 6:   Use the leading 1 in the (33) position to put 0s beneath it in column 3. The appropriate row operation is to subtract nine times row 3 from row 4.

$$\xrightarrow[R_4 \rightarrow R_4 - 9R_3]{} \begin{bmatrix} 1 & -1 & 2 & 1 \\ 0 & 1 & -6 & -4 \\ 0 & 0 & 1 & 1 \\ 0 & 0 & 0 & 0 \end{bmatrix}.$$

This is a row echelon matrix, and hence the given matrix has been reduced to row echelon form.

You should study the preceding example very carefully, since it illustrates the general procedure for reducing an $m \times n$ matrix to row echelon form using elementary row operations. This is a procedure that we will use repeatedly throughout the text. The idea behind reduction to row echelon form is to start at the upper left-hand corner of the matrix and proceed diagonally down the matrix. Thus at the first stage we put (if possible) a leading 1 in the (11) position and then use elementary row operations to put 0s below this leading 1 in column 1. Having accomplished this we move on to the (22) position. Using rows 2 to m we put (if possible) a leading 1 in the (22) position and use elementary row operations to put 0s below this leading 1 in column 2. Having done this we move on to the (33) position, and so on. Notice that if we want to reduce a matrix to reduced row echelon form, then we put 0s above as well as below the leading 1s at each stage.

It often happens that at the kth stage in the row-reduction procedure all of the elements $a_{ik} = 0$, $i = k, k+1, \ldots, n$. In such a situation we cannot put a leading 1 in the (kk) position. Two possibilities arise. Either the matrix is a row echelon matrix and we are finished (this happened in the preceding example—we could not put a leading 1 in the (44) position, but it did not matter since the matrix was a row echelon matrix), or we must move to the $a_{k\,k+1}$ position and proceed as before. This is illustrated in the next example.

Example 4.2.6 Reduce $\begin{bmatrix} 3 & 2 & -5 & 2 & 4 \\ 1 & 1 & -1 & 1 & 1 \\ 1 & 0 & -3 & 4 & 6 \\ 2 & 3 & 0 & 3 & 1 \end{bmatrix}$ to reduced row echelon form.

Solution We proceed as in the previous example, except we now put 0s above and below the leading 1 at each stage.

$$\begin{bmatrix} 3 & 2 & -5 & 2 & 4 \\ 1 & 1 & -1 & 1 & 1 \\ 1 & 0 & -3 & 4 & 6 \\ 2 & 3 & 0 & 3 & 1 \end{bmatrix} \xrightarrow{R_1 \leftrightarrow R_2} \begin{bmatrix} 1 & 1 & -1 & 1 & 1 \\ 3 & 2 & -5 & 2 & 4 \\ 1 & 0 & -3 & 4 & 6 \\ 2 & 3 & 0 & 3 & 1 \end{bmatrix}$$

$$\xrightarrow[\substack{R_4 \to R_4 - 2R_1}]{\substack{R_2 \to R_2 - 3R_1 \\ R_3 \to R_3 - R_1}} \begin{bmatrix} 1 & 1 & -1 & 1 & 1 \\ 0 & -1 & -2 & -1 & 1 \\ 0 & -1 & -2 & 3 & 5 \\ 0 & 1 & 2 & 1 & -1 \end{bmatrix} \xrightarrow{R_2 \to -R_2} \begin{bmatrix} 1 & 1 & -1 & 1 & 1 \\ 0 & 1 & 2 & 1 & -1 \\ 0 & -1 & -2 & 3 & 5 \\ 0 & -1 & -2 & -1 & 1 \end{bmatrix}$$

$$\xrightarrow[\substack{R_4 \to R_4 + R_2}]{\substack{R_1 \to R_1 - R_2 \\ R_3 \to R_3 + R_2}} \begin{bmatrix} 1 & 0 & -3 & 0 & 2 \\ 0 & 1 & 2 & 1 & -1 \\ 0 & 0 & 0 & 4 & 4 \\ 0 & 0 & 0 & 0 & 0 \end{bmatrix}.$$

We see that we cannot put a leading 1 in the (33) position; since the matrix is not yet in echelon form, we move to the (34) position and put a leading 1 there:

$$\xrightarrow{R_3 \to R_3/4} \begin{bmatrix} 1 & 0 & -3 & 0 & 2 \\ 0 & 1 & 2 & 1 & -1 \\ 0 & 0 & 0 & 1 & 1 \\ 0 & 0 & 0 & 0 & 0 \end{bmatrix} \xrightarrow{R_2 \to R_2 - R_3} \begin{bmatrix} 1 & 0 & -3 & 0 & 2 \\ 0 & 1 & 2 & 0 & -2 \\ 0 & 0 & 0 & 1 & 1 \\ 0 & 0 & 0 & 0 & 0 \end{bmatrix}.$$

This is a reduced row echelon matrix, and hence we are done.

We now prove Theorem 4.2.2 by presenting an algorithm that takes an arbitrary $m \times n$ matrix A and reduces it to a row echelon matrix by the application of elementary row operations. The algorithm consists of a sequence of at most m steps. At the end of each step we have a matrix that is row equivalent to A. At the beginning of step k we denote the elements of this matrix by $a_{ij}^{(k)}$. The general step is illustrated in Figure 4.2.1.

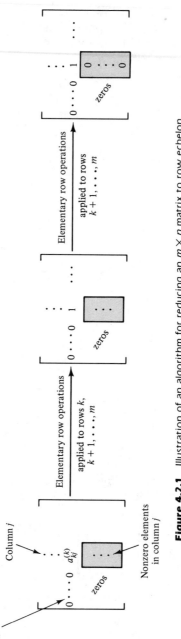

Figure 4.2.1 Illustration of an algorithm for reducing an $m \times n$ matrix to row echelon form.

---

*Algorithm for Reducing an $m \times n$ Matrix to Row
Echelon Form*

1. If $A = 0$, go to 9.
2. Set $k = 1$.
3. Let $j \geq k$ be the first column that has at least one nonzero element, $a_{ij}^{(k)}$, with $i \geq j$.
4. Use elementary row operations on rows $k, k + 1, \ldots, m$ to set $a_{kj}^{(k+1)} = 1$.
5. If $k = m$, go to 9.
6. If $k \neq m$, use elementary row operations to put 0s beneath $a_{kj}^{(k+1)}$.
$$R_i \to R_i - a_{ij}^{(k)} R_k, \qquad i = k + 1, \ldots, m.$$
7. If rows $k + 1, \ldots, m$ consist entirely of 0s, go to 9.
8. Replace k by $k + 1$ and go to 3.
9. The matrix is a row echelon matrix and we are done.

REMARK   The preceding algorithm will reduce an arbitrary $m \times n$ matrix to row echelon form. It is straightforward to extend this algorithm to reduce a matrix to reduced row echelon form. We just put 0s above and beneath the leading 1 at each step.

The techniques for solving a system of linear equations that will be developed in the next section consist of first using elementary row operations to reduce the augmented matrix of the system to row echelon form (or reduced row echelon form). Since only elementary row operations are used in reducing the augmented matrix, the resulting matrix will represent an equivalent system to the given one (that is, a system that has the same solution set as the given one). The key point behind the technique is that the equivalent system can be solved very easily (if it has a solution) using back substitution, and hence we will obtain the solution of the original system. In order to develop the theory for these techniques, we require some further results on row echelon matrices, which we now derive.

If we reduce a matrix to row echelon form, then the resulting matrix is *not* unique, since we can always add multiples of a row to any row above it while maintaining row echelon form.

Example 4.2.7 The following two matrices are row-equivalent row echelon matrices:

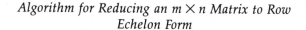

the second one being obtained from the first by performing the indicated row operations.

However, when we reduce a matrix to reduced row echelon form, we are making a particular choice of row echelon matrix, since all elements above the leading 1s are 0s. In view of this the following theorem should not be too surprising.

Theorem 4.2.3:   An $m \times n$ matrix A is row equivalent to a *unique* reduced row echelon matrix.

PROOF   A proof of this theorem can be found, for example, in G. E. Shilov, *Linear Algebra*, Dover, 1977.   ■

REMARK   We refer to this unique row-equivalent reduced row echelon matrix as the **reduced row echelon form** of A.

In reducing a row echelon matrix to its reduced row echelon form, we do not alter the number of leading 1s (we merely put 0s above each leading 1), and hence any row echelon matrix has the same number of nonzero rows as its row equivalent reduced row echelon matrix. We therefore have the following theorem.

Theorem 4.2.4:   Let A be an $m \times n$ matrix. All row echelon matrices that are row equivalent to A have the same number of nonzero rows.

PROOF   This follows directly from our preceding discussion, since all row echelon matrices that are row equivalent to A have the same number of nonzero rows as the reduced row echelon form of A.   ■

Theorem 4.2.4 associates a number with any $m \times n$ matrix A—namely, the number of nonzero rows in any row echelon form of A. As we will see in the next section, this number is fundamental in determining the solution properties of linear systems. For this reason we give it a special name:

> **Definition 4.2.4:**   The number of nonzero rows in any row echelon form of a matrix A is called the **rank** of A, and is denoted rank(A).

Example 4.2.8   Determine the rank of the matrix in the Example 4.2.6.

Solution   The row-equivalent row echelon matrix obtained in Example 4.2.6 has three nonzero rows, and hence the rank of the given matrix is 3.

EXERCISES 4.2

In problems 1–8 determine which of the given matrices are row echelon or reduced row echelon matrices.

1. $\begin{bmatrix} 1 & 0 & -1 & 0 \\ 0 & 0 & 1 & 2 \\ 0 & 0 & 0 & 0 \end{bmatrix}$.

2. $\begin{bmatrix} 1 & 0 & 2 & 5 \\ 1 & 0 & 0 & 2 \\ 0 & 1 & 1 & 0 \end{bmatrix}$.

3. $\begin{bmatrix} 1 & 0 & 0 & 0 \\ 0 & 0 & 0 & 1 \end{bmatrix}$.

4. $\begin{bmatrix} 0 & 1 \\ 1 & 0 \end{bmatrix}$.

5. $\begin{bmatrix} 1 & 1 \\ 0 & 0 \end{bmatrix}$.

6. $\begin{bmatrix} 1 & 0 & 1 & 2 \\ 0 & 0 & 1 & 1 \\ 0 & 0 & 0 & 1 \\ 0 & 0 & 0 & 0 \end{bmatrix}$.

7. $\begin{bmatrix} 0 & 0 & 0 & 0 \\ 0 & 0 & 0 & 0 \\ 0 & 0 & 0 & 0 \end{bmatrix}$.

8. $\begin{bmatrix} 0 & 1 & 0 & 0 \\ 0 & 0 & 1 & 0 \\ 0 & 0 & 0 & 1 \\ 0 & 0 & 0 & 0 \end{bmatrix}$.

In problems 9–18 use elementary row operations to reduce the given matrix to row echelon form, and hence determine the rank of each matrix.

9. $\begin{bmatrix} 2 & 1 \\ 1 & -3 \end{bmatrix}$.

10. $\begin{bmatrix} 2 & -4 \\ -4 & 8 \end{bmatrix}$.

11. $\begin{bmatrix} 2 & 1 & 4 \\ 2 & -3 & 4 \\ 3 & -2 & 6 \end{bmatrix}$.

12. $\begin{bmatrix} 0 & 1 & 3 \\ 0 & 1 & 4 \\ 0 & 3 & 5 \end{bmatrix}$.

13. $\begin{bmatrix} 2 & -1 \\ 3 & 2 \\ 2 & 5 \end{bmatrix}$.

14. $\begin{bmatrix} 2 & -1 & 3 \\ 3 & 1 & -2 \\ 2 & -2 & 1 \end{bmatrix}$.

15. $\begin{bmatrix} 2 & -1 & 3 & 4 \\ 1 & -2 & 1 & 3 \\ 1 & -5 & 0 & 5 \end{bmatrix}$.

16. $\begin{bmatrix} 2 & -2 & -1 & 3 \\ 3 & -2 & 3 & 1 \\ 1 & -1 & 1 & 0 \\ 2 & -1 & 2 & 2 \end{bmatrix}$.

17. $\begin{bmatrix} 4 & 7 & 4 & 7 \\ 3 & 5 & 3 & 5 \\ 2 & -2 & 2 & -2 \\ 5 & -2 & 5 & -2 \end{bmatrix}$.

18. $\begin{bmatrix} 2 & 1 & 3 & 4 & 2 \\ 1 & 0 & 2 & 1 & 3 \\ 2 & 3 & 1 & 5 & 7 \end{bmatrix}$.

In problems 19–25 determine the *reduced* row echelon matrix that is row equivalent to the given matrix, and hence determine the rank of the given matrix.

19. $\begin{bmatrix} 3 & 2 \\ 1 & -1 \end{bmatrix}$.

20. $\begin{bmatrix} 3 & 7 & 10 \\ 2 & 3 & -1 \\ 1 & 2 & 1 \end{bmatrix}$.

21. $\begin{bmatrix} 3 & -3 & 6 \\ 2 & -2 & 4 \\ 6 & -6 & 12 \end{bmatrix}$.

22. $\begin{bmatrix} 3 & 5 & -12 \\ 2 & 3 & -7 \\ -2 & -1 & 1 \end{bmatrix}$.

23. $\begin{bmatrix} 1 & -1 & -1 & 2 \\ 3 & -2 & 0 & 7 \\ 2 & -1 & 2 & 4 \\ 4 & -2 & 3 & 8 \end{bmatrix}$.

24. $\begin{bmatrix} 1 & -2 & 1 & 3 \\ 3 & -6 & 2 & 7 \\ 4 & -8 & 3 & 10 \end{bmatrix}$.

25. $\begin{bmatrix} 0 & 1 & 2 & 1 \\ 0 & 3 & 1 & 2 \\ 0 & 2 & 0 & 1 \end{bmatrix}$.

26. (Elementary Matrices) Consider the general

2×2 matrix $A = \begin{bmatrix} a_{11} & a_{12} \\ a_{21} & a_{22} \end{bmatrix}$.

(a) Show that multiplying A on the left by the

elementary matrix $P_{12} = \begin{bmatrix} 0 & 1 \\ 1 & 0 \end{bmatrix}$ has the effect of

interchanging row 1 with row 2.

(b) Define the elementary matrices M_1 and M_2 by

$$M_1(k) = \begin{bmatrix} k & 0 \\ 0 & 1 \end{bmatrix}, \qquad M_2(k) = \begin{bmatrix} 1 & 0 \\ 0 & k \end{bmatrix}.$$

Show that multiplying A on the left by $M_i(k)$ has the effect of multiplying the ith row of A by k.

(c) Define the elementary matrices $M_{12}(k)$ and $M_{21}(k)$ by

$$M_{12}(k) = \begin{bmatrix} 1 & k \\ 0 & 1 \end{bmatrix}, \qquad M_{21}(k) = \begin{bmatrix} 1 & 0 \\ k & 1 \end{bmatrix}.$$

Show that multiplying A on the left by $M_{ij}(k)$ has the effect of adding k times row j to row i.

[*Remark:* Notice that the elementary matrices defined in (a)–(c) can be obtained by performing the corresponding row operation on the identity matrix I_2.]

It follows from the previous question that any elementary row operation performed on a 2×2 matrix A is equivalent to multiplying A on the left

by one of the elementary matrices P_{12}, $M_i(k)$, or $M_{ij}(k)$. Further, A can be reduced to row echelon form by multiplying by an appropriate sequence of elementary matrices. In problems 27–30 determine a sequence of elementary matrices that reduces the given matrix to row echelon form.

27. $\begin{bmatrix} 1 & 0 \\ 0 & 5 \end{bmatrix}$.

28. $\begin{bmatrix} 0 & 3 \\ 1 & 2 \end{bmatrix}$.

29. $\begin{bmatrix} 2 & 3 \\ 4 & -2 \end{bmatrix}$.

30. $\begin{bmatrix} -1 & 3 \\ 2 & 7 \end{bmatrix}$.

31. Determine appropriate elementary matrices for performing elementary row operations on a 3×3 matrix.

In problems 32–34 use the elementary matrices defined in the previous problem to reduce the given matrix to row echelon form.

32. $\begin{bmatrix} 3 & -2 & 5 \\ 1 & -1 & 3 \\ 2 & -1 & 9 \end{bmatrix}$.

33. $A = \begin{bmatrix} 5 & -10 & 5 \\ 3 & -6 & 2 \\ 2 & -4 & 3 \end{bmatrix}$.

34. $A = \begin{bmatrix} 2 & 1 & 3 \\ 3 & 1 & 2 \\ 4 & 3 & 1 \end{bmatrix}$.

4.3   GAUSSIAN ELIMINATION
AND GAUSS–JORDAN ELIMINATION

We now illustrate how elementary row operations applied to the augmented matrix of a system of linear equations can be used first to determine whether a system is consistent and second, if the system is consistent, to find all of its solutions.

Example 4.3.1   Determine the solution set of

$$\begin{aligned} 3x_1 - 2x_2 + 2x_3 &= 9, \\ x_1 - 2x_2 + x_3 &= 5, \\ 2x_1 - x_2 - 2x_3 &= -1. \end{aligned} \qquad (4.3.1)$$

Solution   To illustrate the method we will work both with the augmented matrix and the equivalent system of equations. In practice you should use only the augmented matrix. We first reduce the augmented matrix to row echelon form.

$$
\begin{array}{cc}
\textit{Augmented Matrix} & \textit{Equivalent System}
\end{array}
$$

$$
\left[\begin{array}{ccc|c}
3 & -2 & 2 & 9 \\
1 & -2 & 1 & 5 \\
2 & -1 & -2 & -1
\end{array}\right]
\qquad
\begin{array}{rcr}
3x_1 - 2x_2 + 2x_3 &=& 9, \\
x_1 - 2x_2 + x_3 &=& 5, \\
2x_1 - x_2 - 2x_3 &=& -1.
\end{array}
$$

Step 1:   Put a leading 1 in the (11) position.

$$
R_1 \leftrightarrow R_2 \longrightarrow
\left[\begin{array}{ccc|c}
1 & -2 & 1 & 5 \\
3 & -2 & 2 & 9 \\
2 & -1 & -2 & -1
\end{array}\right]
\qquad
\begin{array}{rcr}
x_1 - 2x_2 + x_3 &=& 5, \\
3x_1 - 2x_2 + 2x_3 &=& 9, \\
2x_1 - x_2 - 2x_3 &=& -1.
\end{array}
$$

Step 2:   Use the leading 1 to put 0s beneath it in column 1.

$$
\begin{array}{c}
R_2 \to R_2 - 3R_1 \\
R_3 \to R_3 - 2R_1
\end{array}
\longrightarrow
\left[\begin{array}{ccc|c}
1 & -2 & 1 & 5 \\
0 & 4 & -1 & -6 \\
0 & 3 & -4 & -11
\end{array}\right]
\qquad
\begin{array}{rcr}
x_1 - 2x_2 + x_3 &=& 5, \\
4x_2 - x_3 &=& -6, \\
3x_2 - 4x_3 &=& -11.
\end{array}
$$

Step 3:   Use rows 2 and 3 to put a leading 1 in the (22) position.

$$
R_2 \to R_2 - R_3 \longrightarrow
\left[\begin{array}{ccc|c}
1 & -2 & 1 & 5 \\
0 & 1 & 3 & 5 \\
0 & 3 & -4 & -11
\end{array}\right]
\qquad
\begin{array}{rcr}
x_1 - 2x_2 + x_3 &=& 5, \\
x_2 + 3x_3 &=& 5, \\
3x_2 - 4x_3 &=& -11.
\end{array}
$$

Step 4:   Use the leading 1 in the (22) position to put 0s beneath it in column 2.

$$
R_3 \to R_3 - 3R_2 \longrightarrow
\left[\begin{array}{ccc|c}
1 & -2 & 1 & 5 \\
0 & 1 & 3 & 5 \\
0 & 0 & -13 & -26
\end{array}\right]
\qquad
\begin{array}{rcr}
x_1 - 2x_2 + x_3 &=& 5, \\
x_2 + 3x_3 &=& 5, \\
- 13x_3 &=& -26.
\end{array}
$$

Step 5:   Put a leading 1 in the (33) position.

$$
R_3 \to -R_3/13 \longrightarrow
\left[\begin{array}{ccc|c}
1 & -2 & 1 & 5 \\
0 & 1 & 3 & 5 \\
0 & 0 & 1 & 2
\end{array}\right]
\qquad
\begin{array}{rll}
x_1 - 2x_2 + x_3 = 5, & (4.3.2) \\
x_2 + 3x_3 = 5, & (4.3.3) \\
x_3 = 2. & (4.3.4)
\end{array}
$$

The augmented matrix is now in row echelon form, and the system can be solved by *back substitution* as follows. From (4.3.4), $x_3 = 2$. Substituting into (4.3.3) and solving for x_2, we find that $x_2 = -1$. Finally, substituting into (4.3.2) for x_3 and x_2 and solving for x_1 yields $x_1 = 1$. Thus our original system of equations has the unique solution, $(1, -1, 2)$, and the solution set of the system is

$$
S = \{(1, -1, 2)\}.
$$

The process of reducing the augmented matrix to row echelon form and then using back substitution to solve the equivalent system is called **Gaussian elimination**. The particular case of Gaussian elimination that arises when the augmented matrix is reduced to reduced row echelon form is called **Gauss–Jordan elimination**.

Example 4.3.2   Use Gauss–Jordan elimination to determine the solution set of

$$x_1 + 2x_2 - x_3 = 1,$$
$$2x_1 + 5x_2 - x_3 = 3,$$
$$x_1 + 3x_2 + 2x_3 = 6.$$

Solution   In this case we first reduce the augmented matrix of the system to reduced row echelon form.

$$\begin{bmatrix} 1 & 2 & -1 & | & 1 \\ 2 & 5 & -1 & | & 3 \\ 1 & 3 & 2 & | & 6 \end{bmatrix} \begin{array}{l} R_2 \to R_2 - 2R_1 \\ \xrightarrow{\hspace{1cm}} \\ R_3 \to R_3 - R_1 \end{array} \begin{bmatrix} 1 & 2 & -1 & | & 1 \\ 0 & 1 & 1 & | & 1 \\ 0 & 1 & 3 & | & 5 \end{bmatrix}$$

$$\begin{array}{l} R_1 \to R_1 - 2R_2 \\ \xrightarrow{\hspace{1cm}} \\ R_3 \to R_3 - R_2 \end{array} \begin{bmatrix} 1 & 0 & -3 & | & -1 \\ 0 & 1 & 1 & | & 1 \\ 0 & 0 & 2 & | & 4 \end{bmatrix} \begin{array}{l} R_3 \to \frac{1}{2}R_3 \\ \xrightarrow{\hspace{1cm}} \end{array} \begin{bmatrix} 1 & 0 & -3 & | & -1 \\ 0 & 1 & 1 & | & 1 \\ 0 & 0 & 1 & | & 2 \end{bmatrix}$$

$$\begin{array}{l} R_1 \to R_1 + 3R_3 \\ \xrightarrow{\hspace{1cm}} \\ R_2 \to R_2 - R_3 \end{array} \begin{bmatrix} 1 & 0 & 0 & | & 5 \\ 0 & 1 & 0 & | & -1 \\ 0 & 0 & 1 & | & 2 \end{bmatrix}.$$

The augmented matrix is now in reduced row echelon form. The equivalent system is

$$x_1 = 5,$$
$$x_2 = -1,$$
$$x_3 = 2,$$

and the solution can be read directly as $(5, -1, 2)$. The solution set is thus $\{(5, -1, 2)\}$.

We see from the previous two examples that the advantage of Gauss–Jordan elimination as compared to Gaussian elimination is that it does not require back substitution. However, the disadvantage is that reducing the augmented matrix to reduced row echelon form requires more elementary row operations than reduction to row echelon form. It can be shown, in fact, that in general Gaussian elimination is the more computationally efficient technique. The main reason for introducing the Gauss–Jordan method is its application to the computation of the inverse of an $n \times n$ matrix (see Section 4.5).

We stress the difference between Gaussian elimination and Gauss–Jordan elimination:

> *Gaussian elimination:* Use elementary row operations to reduce the augmented matrix to row echelon form and then use back substitution to solve the equivalent system.
>
> *Gauss–Jordan elimination:* Use elementary row operations to reduce the augmented matrix to *reduced row echelon form* and thereby solve the equivalent system.

REMARK   The methods are so systematic that they can be programmed easily on a computer.   Indeed many large-scale programs for solving $n \times n$ linear systems are based on the row-reduction method.

In both of the previous examples

$$\text{rank}(A) = \text{rank}(A^\#) = \text{number of unknowns in the system,}$$

and the system had a unique solution.   More generally we have the following lemma.

Lemma 4.3.1:   Consider the $m \times n$ linear system $A\mathbf{x} = \mathbf{b}$.   Let $A^\#$ denote the augmented matrix of the system.   If $\text{rank}(A) = \text{rank}(A^\#) = n$, then the system has a unique solution.

PROOF   If $\text{rank}(A) = \text{rank}(A^\#) = n$, then there are n leading 1s in any row echelon form of A and hence back substitution gives a unique solution.   ∎

There are only two other possibilities for the relationship between $\text{rank}(A)$ and $\text{rank}(A^\#)$, namely $\text{rank}(A) < \text{rank}(A^\#)$ or $\text{rank}(A) = \text{rank}(A^\#) < n$.   We must now consider what happens in these cases.

Example 4.3.3   Determine the solution set of

$$x_1 + x_2 - x_3 + x_4 = 1,$$
$$2x_1 + 3x_2 + x_3. \qquad = 4,$$
$$3x_1 + 5x_2 + 3x_3 - x_4 = 5.$$

Solution   We will reduce the augmented matrix:

$$
\begin{bmatrix}
1 & 1 & -1 & 1 & | & 1 \\
2 & 3 & 1 & 0 & | & 4 \\
3 & 5 & 3 & -1 & | & 5
\end{bmatrix}
\begin{matrix} R_2 \to R_2 - 2R_1 \\ \longrightarrow \\ R_3 \to R_3 - 3R_1 \end{matrix}
\begin{bmatrix}
1 & 1 & -1 & 1 & | & 1 \\
0 & 1 & 3 & -2 & | & 2 \\
0 & 2 & 6 & -4 & | & 2
\end{bmatrix}
$$

$$
\begin{matrix} R_3 \to R_3 - 2R_2 \\ \longrightarrow \end{matrix}
\begin{bmatrix}
1 & 1 & -1 & 1 & | & 1 \\
0 & 1 & 3 & -2 & | & 2 \\
0 & 0 & 0 & 0 & | & -2
\end{bmatrix}.
$$

The last row tells us that the system of equations has no solution (that is, it is inconsistent), since it requires

$$0x_1 + 0x_2 + 0x_3 + 0x_4 = -2,$$

which is clearly impossible.   The solution set of the system is thus the empty set $\emptyset$.

In the previous example rank$(A) = 2$, whereas rank$(A^\#) = 3$.   Thus rank$(A) <$ rank$(A^\#)$, and the corresponding system has no solution.   This result is true in general.

Lemma 4.3.2:   Consider the $m \times n$ linear system $A\mathbf{x} = \mathbf{b}$.   Let $A^\#$ denote the augmented matrix of the system.   If rank$(A) <$ rank$(A^\#)$, then the system is inconsistent.

PROOF   If rank$(A) <$ rank$(A^\#)$, then there will be one row in the reduced row echelon form of the augmented matrix whose first nonzero element arises in the last column.   Such a row corresponds to an equation of the form:

$$0x_1 + 0x_2 + \cdots + 0x_n = 1.$$

Clearly such an equation has no solution, and hence the system is inconsistent.   ∎

Now for the final possibility.

Example 4.3.4   Use Gauss–Jordan elimination to determine the solution set of

$$2x_1 - x_2 - 5x_3 = 1,$$
$$x_1 - 4x_2 - 6x_3 = -3,$$
$$3x_1 + 2x_2 - 4x_3 = 5.$$

Solution   We must reduce the augmented matrix to reduced row echelon form. This can be accomplished as follows:

$$\begin{bmatrix} 2 & -1 & -5 & | & 1 \\ 1 & -4 & -6 & | & -3 \\ 3 & 2 & -4 & | & 5 \end{bmatrix} \xrightarrow{R_2 \leftrightarrow R_1} \begin{bmatrix} 1 & -4 & -6 & | & -3 \\ 2 & -1 & -5 & | & 1 \\ 3 & 2 & -4 & | & 5 \end{bmatrix}$$

$$\xrightarrow[\substack{R_2 \to R_2 - 2R_1 \\ R_3 \to R_3 - 3R_1}]{} \begin{bmatrix} 1 & -4 & -6 & | & -3 \\ 0 & 7 & 7 & | & 7 \\ 0 & 14 & 14 & | & 14 \end{bmatrix} \xrightarrow{R_2 \to R_2/7} \begin{bmatrix} 1 & -4 & -6 & | & -3 \\ 0 & 1 & 1 & | & 1 \\ 0 & 14 & 14 & | & 14 \end{bmatrix}$$

$$\xrightarrow[\substack{R_1 \to R_1 + 4R_2 \\ R_3 \to R_3 - 14R_2}]{} \begin{bmatrix} 1 & 0 & -2 & | & 1 \\ 0 & 1 & 1 & | & 1 \\ 0 & 0 & 0 & | & 0 \end{bmatrix}.$$

This is now in reduced row echelon form, and the equivalent system is

$$x_1 \qquad - 2x_3 = 1, \tag{4.3.5}$$
$$x_2 + x_3 = 1. \tag{4.3.6}$$

The system has been reduced to its simplest form. Since we have three variables but only two equations relating them, we are free to specify one of the variables arbitrarily. The variable that we choose to specify is called a **free variable**. The remaining variables are then determined by the system of equations and are called **bound variables**. In the preceding system, we take x_3 as our free variable and set

$$x_3 = t,$$

where t can assume any real value. It follows from (4.3.6) that

$$x_2 = 1 - t.$$

Further, from (4.3.5),

$$x_1 = 1 + 2t.$$

Thus the solution set of our system of equations is

$$S = \{(1 + 2t, 1 - t, t) : t \in \mathbf{R}\}.$$

The system has an infinite number of solutions obtained by allowing the parameter t to assume all real values. For example, two particular solutions of the system are

$$(1, 1, 0) \qquad \text{(corresponding to } t = 0\text{)},$$

and

$$(-3, 3, -2) \qquad \text{(corresponding to } t = -2\text{)}.$$

REMARK The geometry of the preceding solution is as follows. The given equations can be interpreted as defining three planes in 3-space. The solution of the system then represents the coordinates of the points of intersection of the three planes. In the preceding example the planes intersect in a straight line whose parametric equations are

$$x_1 = 1 + 2t, \qquad x_2 = 1 - t, \qquad x_3 = t.$$

(See Figure 4.1.1 on page 109.)

In general the solution of a consistent $m \times n$ system of linear equations may involve more than just one free variable. Indeed, the number of free variables will depend on how many nonzero rows arise in the row echelon form of the augmented matrix, $A^{\#}$, of the system, that is on the rank of $A^{\#}$. More precisely, if $\mathrm{rank}(A^{\#}) = r^{\#}$, then the equivalent system will have only $r^{\#}$ relationships between the n variables; in general, provided the system is consistent:

$$\text{Number of free variables} = n - r^{\#}.$$

Indeed, we have the following.

Lemma 4.3.3: Consider the $m \times n$ linear system $A\mathbf{x} = \mathbf{b}$. Let $A^{\#}$ denote the augmented matrix of the system. If $\mathrm{rank}(A) = \mathrm{rank}(A^{\#}) < n$, then the system has an infinite number of solutions.

PROOF   Suppose rank$(A^\#) = $ rank$(A) = r^\#$. If $r^\# < n$, then, as discussed before, the reduced row echelon equivalent system will have only $r^\#$ equations involving the n variables, and so there will be $n - r^\#$ free variables. If we assign arbitrary values to these free variables, then the remaining $r^\#$ variables will be uniquely determined, by back substitution, from the system. Since we have one solution for each value of the free variables, in this case there are an infinite number of solutions.   ■

Example 4.3.5   Use Gaussian elimination to solve

$$x_1 - 2x_2 + 2x_3 - x_4 = 3,$$
$$3x_1 + x_2 + 6x_3 + 11x_4 = 16,$$
$$2x_1 - x_2 + 4x_3 + 4x_4 = 9.$$

Solution   A row echelon form of the augmented matrix of the system is (exercise):

$$\begin{bmatrix} 1 & -2 & 2 & -1 & 3 \\ 0 & 1 & 0 & 2 & 1 \\ 0 & 0 & 0 & 0 & 0 \end{bmatrix},$$

so that we have two free variables. The equivalent system is

$$x_1 - 2x_2 + 2x_3 - x_4 = 3, \tag{4.3.7}$$
$$x_2 + 2x_4 = 1. \tag{4.3.8}$$

Notice that we cannot choose any two variables freely. For example, from (4.3.8) we cannot specify both x_2 and x_4 independently. The bound variables should be taken as those that correspond to leading 1s in the row echelon form of $A^\#$, since these are the variables that can always be determined by back-substitution. Thus:

> Choose as free variables those variables that do not correspond to a leading 1 in the row echelon (or reduced row echelon) form of $A^\#$.

According to this rule we should choose x_3 and x_4 as free variables and so set

$$x_3 = s, \qquad x_4 = t.$$

It then follows from (4.3.8) that

$$x_2 = 1 - 2t.$$

Substitution into (4.3.7) yields

$$x_1 = 5 - 2s - 3t,$$

so that the solution set of our system is

$$S = \{(5 - 2s - 3t, 1 - 2t, s, t) : s, t \in \mathbf{R}\}.$$

Lemmas 4.3.1–4.3.3 completely characterize the solution properties of an $m \times n$ linear system. We summarize the results in a theorem.

Theorem 4.3.1: Consider the $m \times n$ linear system $A\mathbf{x} = \mathbf{b}$. Let r denote the rank of A, and let $r^{\#}$ denote the rank of the augmented matrix of the system. Then:

1. If $r < r^{\#}$ the system is inconsistent.
2. If $r = r^{\#}$ the system is consistent and:
   (a) There exists a unique solution if and only if $r^{\#} = n$.
   (b) There exist an infinite number of solutions if and only if $r^{\#} < n$.

REMARK   Notice that Theorem 4.3.1 contains the results stated in Theorem 4.1.1.

EXERCISES 4.3

In problems 1–9 use *Gaussian elimination* to determine the solution set of the given system.

1. $x_1 + 2x_2 + x_3 = 1,$
$3x_1 + 5x_2 + x_3 = 3,$
$2x_1 + 6x_2 + 7x_3 = 1.$

2. $3x_1 - x_2 = 1,$
$2x_1 + x_2 + 5x_3 = 4,$
$7x_1 - 5x_2 - 8x_3 = -3.$

3. $3x_1 + 5x_2 - x_3 = 14,$
$x_1 + 2x_2 + x_3 = 3,$
$2x_1 + 5x_2 + 6x_3 = 2.$

4. $6x_1 - 3x_2 + 3x_3 = 12,$
$2x_1 - x_2 + x_3 = 4,$
$-4x_1 + 2x_2 - 2x_3 = -8.$

5. $2x_1 - x_2 + 3x_3 = 14,$
$3x_1 + x_2 - 2x_3 = -1,$
$7x_1 + 2x_2 - 3x_3 = 3,$
$5x_1 - x_2 - 2x_3 = 5.$

6. $2x_1 - x_2 - 4x_3 = 5,$
$3x_1 + 2x_2 - 5x_3 = 8,$
$5x_1 + 6x_2 - 6x_3 = 20,$
$x_1 + x_2 - 3x_3 = -3.$

7. $x_1 + 2x_2 - x_3 + x_4 = 1,$
$2x_1 + 4x_2 - 2x_3 + 2x_4 = 2,$
$5x_1 + 10x_2 - 5x_3 + 5x_4 = 5.$

8. $x_1 + 2x_2 - x_3 + x_4 = 1,$
$2x_1 - 3x_2 + x_3 - x_4 = 2,$
$x_1 - 5x_2 + 2x_3 - 2x_4 = 1,$
$4x_1 + x_2 - x_3 + x_4 = 3.$

9. $x_1 + 2x_2 + x_3 + x_4 - 2x_5 = 3,$
$x_3 + 4x_4 - 3x_5 = 2,$
$2x_1 + 4x_2 - x_3 - 10x_4 + 5x_5 = 0.$

In problems 10–15 use *Gauss–Jordan elimination* to determine the solution set of the given system.

10. $2x_1 - x_2 - x_3 = 2,$
$4x_1 + 3x_2 - 2x_3 = -1,$
$x_1 + 4x_2 + x_3 = 4.$

11. $3x_1 + x_2 + 5x_3 = 2,$
$x_1 + x_2 - x_3 = 1,$
$2x_1 + x_2 + 2x_3 = 3.$

12. $x_1 - 2x_3 = -3,$
$3x_1 - 2x_2 - 4x_3 = -9,$
$x_1 - 4x_2 + 2x_3 = -3.$

13. $2x_1 - x_2 + 3x_3 - x_4 = 3,$
$3x_1 + 2x_2 + x_3 - 5x_4 = -6,$
$x_1 - 2x_2 + 3x_3 + x_4 = 6.$

14. $x_1 + 2x_2 - x_3 + x_4 = 1,$
$2x_1 - 3x_2 + x_3 - x_4 = 2,$
$x_1 - 5x_2 + 2x_3 - 2x_4 = 1,$
$4x_1 + x_2 - x_3 + x_4 = 3.$

15. $2x_1 - x_2 + 3x_3 + x_4 - x_5 = 11,$
$x_1 - 3x_2 - 2x_3 - x_4 - 2x_5 = 2,$
$3x_1 + x_2 - 2x_3 - x_4 + x_5 = -2,$
$x_1 + 2x_2 + x_3 + 2x_4 + 3x_5 = -3,$
$5x_1 - 3x_2 - 3x_3 + x_4 + 2x_5 = 2.$

In problems 16–20 determine the solution set of the system $A\mathbf{x} = \mathbf{b}$ for the given coefficient matrix A and right-hand-side vector $\mathbf{b}$.

16. $A = \begin{bmatrix} 1 & -3 & 1 \\ 5 & -4 & 1 \\ 2 & 4 & -3 \end{bmatrix}$, $\mathbf{b} = \begin{bmatrix} 8 \\ 15 \\ -4 \end{bmatrix}$.

17. $A = \begin{bmatrix} 1 & 0 & 5 \\ 3 & -2 & 11 \\ 2 & -2 & 6 \end{bmatrix}$, $\mathbf{b} = \begin{bmatrix} 0 \\ 2 \\ 2 \end{bmatrix}$.

18. $A = \begin{bmatrix} 0 & 1 & -1 \\ 0 & 5 & 1 \\ 0 & 2 & 1 \end{bmatrix}$, $\mathbf{b} = \begin{bmatrix} -2 \\ 8 \\ 5 \end{bmatrix}$.

19. $A = \begin{bmatrix} 1 & -1 & 0 & -1 \\ 2 & 1 & 3 & 7 \\ 3 & -2 & 1 & 0 \end{bmatrix}$, $\mathbf{b} = \begin{bmatrix} 2 \\ 2 \\ 4 \end{bmatrix}$.

20. $A = \begin{bmatrix} 1 & 1 & 0 & 1 \\ 3 & 1 & -2 & 3 \\ 2 & 3 & 1 & 2 \\ -2 & 3 & 5 & -2 \end{bmatrix}$, $\mathbf{b} = \begin{bmatrix} 2 \\ 8 \\ 3 \\ -9 \end{bmatrix}$.

21. Determine all values of the constant k for which the following system has (a) no solution, (b) an infinite number of solutions, and (c) a unique solution.

$$\begin{aligned} x_1 + 2x_2 - x_3 &= 3, \\ 2x_1 + 5x_2 + x_3 &= 7, \\ x_1 + x_2 - k^2 x_3 &= -k. \end{aligned}$$

22. Determine all values of the constants a, b for which the following system has (a) no solution, (b) a unique solution, and (c) an infinite number of solutions.

$$\begin{aligned} x_1 + x_2 - 2x_3 &= 4, \\ 3x_1 + 5x_2 - 4x_3 &= 16, \\ 2x_1 + 3x_2 - ax_3 &= b. \end{aligned}$$

23. Consider the system of linear equations

$$\begin{aligned} a_{11}x_1 + a_{12}x_2 &= b_1, \\ a_{21}x_1 + a_{22}x_2 &= b_2. \end{aligned}$$

Define Δ, Δ_1, and Δ_2 by

$$\begin{aligned} \Delta &= a_{11}a_{22} - a_{12}a_{21}, \\ \Delta_1 &= a_{22}b_1 - a_{12}b_2, \\ \Delta_2 &= a_{11}b_2 - a_{21}b_1. \end{aligned}$$

(a) Show that the given system has a unique solution if and only if $\Delta \neq 0$ and that the unique solution in this case is

$$x_1 = \frac{\Delta_1}{\Delta}, \qquad x_2 = \frac{\Delta_2}{\Delta}.$$

(b) If $\Delta = 0$ and $a_{11} \neq 0$, determine the conditions on Δ_2 that would guarantee that the system has (i) no solution, (ii) an infinite number of solutions.

(c) Interpret your results in terms of intersections of straight lines.

24. (a) An $n \times n$ system of linear equations whose matrix of coefficients is of the form

$$\begin{bmatrix} a_{11} & 0 & 0 & 0 & \cdots & 0 \\ a_{21} & a_{22} & 0 & 0 & \cdots & 0 \\ a_{31} & a_{32} & a_{33} & 0 & \cdots & 0 \\ \vdots & \vdots & \vdots & \vdots & & \vdots \\ a_{n1} & a_{n2} & a_{n3} & a_{n4} & \cdots & a_{nn} \end{bmatrix}.$$

is called a **lower triangular** system. Assuming that $a_{ii} \neq 0$ for each i, devise a method for solving such a system that is analogous to the back substitution method.

(b) Use your method from (a) to solve

$$\begin{aligned} x_1 &= 2, \\ 2x_1 - 3x_2 &= 1, \\ 3x_1 + x_2 - x_3 &= 8. \end{aligned}$$

25. Find all solutions of the following nonlinear system of equations.

$$\begin{aligned} 4x_1{}^3 + 2x_2{}^2 + 3x_3 &= 12, \\ x_1{}^3 - x_2{}^2 + x_3 &= 2, \\ 3x_1{}^3 + x_2{}^2 - x_3 &= 2. \end{aligned}$$

Does your answer contradict Theorem 4.1.1? Explain.

4.4   HOMOGENEOUS SYSTEMS OF LINEAR EQUATIONS

Many of the problems that we will meet in the future will require the solution of a homogeneous system of linear equations. The general form for such a system is

$$a_{11}x_1 + a_{12}x_2 + a_{13}x_3 + \cdots + a_{1n}x_n = 0,$$
$$a_{21}x_1 + a_{22}x_2 + a_{23}x_3 + \cdots + a_{2n}x_n = 0,$$
$$\vdots \qquad (4.4.1)$$
$$a_{m1}x_1 + a_{m2}x_2 + a_{m3}x_3 + \cdots + a_{mn}x_n = 0,$$

or, in matrix form, $A\mathbf{x} = \mathbf{0}$, where A is the coefficient matrix of the system and $\mathbf{0}$ denotes the m-vector whose elements are all zeros.

All our previous results hold for homogeneous systems, but, as we shall see, we can make some stronger statements in this case. Our main result is the following:

Theorem 4.4.1: The homogeneous linear system $A\mathbf{x} = \mathbf{0}$ is always consistent.

PROOF   The augmented matrix, $A^\#$, of a homogeneous linear system differs from that of the coefficient matrix, A, only by the addition of a column of zeros. It follows that for a homogeneous system, $\text{rank}(A^\#) = \text{rank}(A)$, and hence, from Theorem 4.3.1, such a system is necessarily consistent.   ∎

REMARKS

1. We can see directly from (4.4.1) that a homogeneous system always has the solution

$$(0, 0, \ldots , 0).$$

   This solution is referred to as the **trivial solution**. It follows that a homogeneous system either has *only* the trivial solution or it has an infinite number of solutions (one of which must be the trivial solution).

2. Once more it is worth mentioning the geometric interpretation of Theorem 4.4.1 in the case of a homogeneous system with three unknowns. We can regard each equation of such a system as defining a plane. Due to homogeneity, each plane passes through the origin, and hence the planes intersect at least at the origin.

Example 4.4.1   Determine the solution set of

$$x_1 + x_2 - x_3 = 0,$$
$$2x_1 + 3x_2 - x_3 = 0,$$
$$x_1 + 3x_2 + 4x_3 = 0.$$

Solution   We reduce the augmented matrix:

$$\begin{bmatrix} 1 & 1 & -1 & | & 0 \\ 2 & 3 & -1 & | & 0 \\ 1 & 3 & 4 & | & 0 \end{bmatrix} \xrightarrow[\substack{R_3 \to R_3 - R_1}]{\substack{R_2 \to R_2 - 2R_1}} \begin{bmatrix} 1 & 1 & -1 & | & 0 \\ 0 & 1 & 1 & | & 0 \\ 0 & 2 & 5 & | & 0 \end{bmatrix}$$

$$R_3 \rightarrow R_3 - 2R_2 \begin{bmatrix} 1 & 1 & -1 & | & 0 \\ 0 & 1 & 1 & | & 0 \\ 0 & 0 & 3 & | & 0 \end{bmatrix} \xrightarrow{R_3 \rightarrow R_3/3} \begin{bmatrix} 1 & 1 & -1 & | & 0 \\ 0 & 1 & 1 & | & 0 \\ 0 & 0 & 1 & | & 0 \end{bmatrix},$$

which implies that $x_3 = 0$, $x_2 = 0$, $x_1 = 0$ so that our given system has *only* the trivial solution $(0, 0, 0)$. The solution set is $\{(0, 0, 0)\}$.

Often we will be interested in determining whether or not a given homogeneous system has an infinite number of solutions and not in actually obtaining the solutions. The following theorem can sometimes be *helpful* in determining by inspection whether a given homogeneous system has nontrivial solutions.

Theorem 4.4.2: A homogeneous system of m linear equations in n unknowns, *with $m < n$*, has an infinite number of solutions.

PROOF   Let $r^\#$ denote the rank of the augmented matrix of the system. Then, $r^\# \leq m < n$, and so by Theorem 4.3.1 the system has an infinite number of solutions [note that we have again used rank(A) = rank(A^*) for a homogeneous system]. ∎

REMARK   If $m \geq n$, then we may or may not have nontrivial solutions, depending on whether the rank, $r^\#$, of the augmented matrix of the system satisfies $r^\# < n$ or $r^\# = n$, respectively.

Example 4.4.2   Determine the solution set of

$$x_1 - x_2 + 3x_3 + x_4 = 0,$$
$$2x_1 - x_2 + x_3 + 2x_4 = 0,$$
$$3x_1 - x_2 - x_3 + 3x_4 = 0.$$

Solution   Since we have more unknowns than equations, the preceding theorem tells us that the system has nontrivial solutions. To find the solution we must reduce the augmented matrix of the system:

$$\begin{bmatrix} 1 & -1 & 3 & 1 & | & 0 \\ 2 & -1 & 1 & 2 & | & 0 \\ 3 & -1 & -1 & 3 & | & 0 \end{bmatrix} \begin{matrix} R_2 \rightarrow R_2 - 2R_1 \\ \xrightarrow{\hspace{2cm}} \\ R_3 \rightarrow R_3 - 3R_1 \end{matrix} \begin{bmatrix} 1 & -1 & 3 & 1 & | & 0 \\ 0 & 1 & -5 & 0 & | & 0 \\ 0 & 2 & -10 & 0 & | & 0 \end{bmatrix}$$

$$\begin{matrix} R_1 \rightarrow R_1 + R_2 \\ \xrightarrow{\hspace{2cm}} \\ R_3 \rightarrow R_3 - 2R_2 \end{matrix} \begin{bmatrix} 1 & 0 & -2 & 1 & | & 0 \\ 0 & 1 & -5 & 0 & | & 0 \\ 0 & 0 & 0 & 0 & | & 0 \end{bmatrix}.$$

The matrix is now in reduced row echelon form. The equivalent system is

$$x_1 \quad - 2x_3 + x_4 = 0,$$
$$x_2 - 5x_3 \quad = 0,$$

so that there are two free variables. We set

$$x_3 = r, \qquad x_4 = s,$$

in which case $x_1 = 2r - s$, $x_2 = 5r$. It follows that the solution set of the system is $\{(2r - s, 5r, r, s) : r, s \in \mathbf{R}\}$.

We end this section with two types of systems that will be of particular interest in Chapters 8 and 11.

Example 4.4.3 Determine the solution set of $Ax = 0$ if $A = \begin{bmatrix} 0 & 2 & 3 \\ 0 & 1 & -1 \\ 0 & 3 & 7 \end{bmatrix}$.

Solution The augmented matrix of the system is

$$\begin{bmatrix} 0 & 2 & 3 & | & 0 \\ 0 & 1 & -1 & | & 0 \\ 0 & 3 & 7 & | & 0 \end{bmatrix},$$

which has reduced row echelon form (exercise)

$$\begin{bmatrix} 0 & 1 & 0 & | & 0 \\ 0 & 0 & 1 & | & 0 \\ 0 & 0 & 0 & | & 0 \end{bmatrix}.$$

The equivalent system is

$$x_2 = 0,$$
$$x_3 = 0.$$

It is tempting to conclude that the solution of the system is $x_1 = x_2 = x_3 = 0$. However, this is incorrect, since x_1 does not occur in the above system and hence it is a free variable and *not* necessarily zero. Thus the correct solution to the above system is $(r, 0, 0)$, where r is a free variable, and the solution set is $\{(r, 0, 0) : r \in \mathbf{R}\}$.

The linear systems that we have so far encountered have all had real coefficients and we have considered corresponding real solutions. The techniques that we have developed for solving linear systems are also applicable to the case when our system has complex coefficients. The corresponding solutions will also be complex. We illustrate with an example.

Example 4.4.4 Determine the solution set of

$$x_1 - ix_2 + x_3 = 0,$$
$$2ix_1 + x_2 + (1 + i)x_3 = 0,$$
$$5x_1 - 4ix_2 + (4 - i)x_3 = 0.$$

Solution We reduce the augmented matrix of the system:

$$\begin{bmatrix} 1 & -i & 1 & | & 0 \\ 2i & 1 & 1+i & | & 0 \\ 5 & -4i & 4-i & | & 0 \end{bmatrix} \xrightarrow[R_3 \to R_3 - 5R_1]{R_2 \to R_2 - 2iR_1} \begin{bmatrix} 1 & -i & 1 & | & 0 \\ 0 & -1 & 1-i & | & 0 \\ 0 & i & -(1+i) & | & 0 \end{bmatrix}$$

$$\xrightarrow{R_2 \to -R_2} \begin{bmatrix} 1 & -i & 1 & | & 0 \\ 0 & 1 & -1+i & | & 0 \\ 0 & i & -(1+i) & | & 0 \end{bmatrix} \xrightarrow{R_3 \to R_3 - iR_2} \begin{bmatrix} 1 & -i & 1 & | & 0 \\ 0 & 1 & -1+i & | & 0 \\ 0 & 0 & 0 & | & 0 \end{bmatrix}.$$

There is one free variable, which we take to be $x_3 = r$. We then have

$$x_2 = r(1-i), \qquad x_1 = -r + ir(1-i) = ir,$$

so that the solution set of the system is $\{(ir, r(1-i), r) : r \in \mathbf{C}\}$.

EXERCISES 4.4

In problems 1–11 determine the solution set of the given system.

1. $3x_1 + 2x_2 - x_3 = 0,$
   $2x_1 + x_2 + x_3 = 0,$
   $5x_1 - 4x_2 + x_3 = 0.$

2. $2x_1 + x_2 - x_3 = 0,$
   $3x_1 - x_2 + 2x_3 = 0,$
   $x_1 - x_2 - x_3 = 0,$
   $5x_1 + 2x_2 - 2x_3 = 0.$

3. $2x_1 - x_2 - x_3 = 0,$
   $5x_1 - x_2 + 2x_3 = 0,$
   $x_1 + x_2 + 4x_3 = 0.$

4. $(1+2i)x_1 + (1-i)x_2 + x_3 = 0,$
   $ix_1 + (1+i)x_2 + ix_3 = 0,$
   $2ix_1 + ix_2 + (1+3i)x_3 = 0.$

5. $3x_1 + 2x_2 + x_3 = 0,$
   $6x_1 - x_2 + 2x_3 = 0,$
   $12x_1 + 6x_2 + 4x_3 = 0.$

6. $2x_1 + x_2 - 8x_3 = 0,$
   $3x_1 - 2x_2 - 5x_3 = 0,$
   $5x_1 - 6x_2 - 3x_3 = 0,$
   $3x_1 - 5x_2 + x_3 = 0.$

7. $x_1 + (1+i)x_2 + (1-i)x_3 = 0,$
   $ix_1 + x_2 + ix_3 = 0,$
   $(1-2i)x_1 + (i-1)x_2 + (1-3i)x_3 = 0.$

8. $x_1 - x_2 + x_3 = 0,$
   $3x_2 + 2x_3 = 0,$
   $3x_1 - x_3 = 0,$
   $5x_1 + x_2 - x_3 = 0.$

9. $2x_1 - 4x_2 + 6x_3 = 0,$
   $3x_1 - 6x_2 + 9x_3 = 0,$
   $x_1 - 2x_2 + 3x_3 = 0,$
   $5x_1 - 10x_2 + 15x_3 = 0.$

10. $4x_1 - 2x_2 - x_3 - x_4 = 0,$
   $3x_1 + x_2 - 2x_3 + 3x_4 = 0,$
   $5x_1 - x_2 - 2x_3 + x_4 = 0.$

11. $2x_1 + x_2 - x_3 + x_4 = 0,$
   $x_1 + x_2 + x_3 - x_4 = 0,$
   $3x_1 - x_2 + x_3 - 2x_4 = 0,$
   $4x_1 + 2x_2 - x_3 + x_4 = 0.$

In problems 12–22 determine the solution set of the system $A\mathbf{x} = \mathbf{0}$ for the given matrix A.

12. $A = \begin{bmatrix} 2 & -1 \\ 3 & 4 \end{bmatrix}.$

13. $A = \begin{bmatrix} 1-i & 2i \\ 1+i & -2 \end{bmatrix}.$

14. $A = \begin{bmatrix} 1+i & 1-2i \\ -1+i & 2+i \end{bmatrix}.$

15. $A = \begin{bmatrix} 1 & 2 & 3 \\ 2 & -1 & 0 \\ 1 & 1 & 1 \end{bmatrix}$.

20. $A = \begin{bmatrix} 1 & -1 & 0 & 1 \\ 3 & -2 & 0 & 5 \\ -1 & 2 & 0 & 1 \end{bmatrix}$.

16. $A = \begin{bmatrix} 1 & 1 & 1 & -1 \\ -1 & 0 & -1 & 2 \\ 1 & 3 & 2 & 2 \end{bmatrix}$.

21. $A = \begin{bmatrix} 1 & 0 & -3 & 0 \\ 3 & 0 & -9 & 0 \\ -2 & 0 & 6 & 0 \end{bmatrix}$.

17. $A = \begin{bmatrix} 2-3i & 1+i & i-1 \\ 3+2i & -1+i & -1-i \\ 5-i & 2i & -2 \end{bmatrix}$.

22. $A = \begin{bmatrix} 2+i & i & 3-2i \\ i & 1-i & 4+3i \\ 3-i & 1+i & 1+5i \end{bmatrix}$.

18. $A = \begin{bmatrix} 1 & 3 & 0 \\ -2 & -3 & 0 \\ 1 & 4 & 0 \end{bmatrix}$.

23. Determine all values of the constant k for which the following system has **(a)** a unique solution and **(b)** an infinite number of solutions.

19. $A = \begin{bmatrix} 1 & 0 & 3 \\ 3 & -1 & 7 \\ 2 & 1 & 8 \\ 1 & 1 & 5 \\ -1 & 1 & -1 \end{bmatrix}$.

$$\begin{aligned} 2x_1 + x_2 - x_3 + x_4 &= 0, \\ x_1 + x_2 + x_3 - x_4 &= 0, \\ 4x_1 + 2x_2 - x_3 + x_4 &= 0, \\ 3x_1 - x_2 + x_3 + kx_4 &= 0. \end{aligned}$$

4.5   THE INVERSE OF A SQUARE MATRIX

Consider the $n \times n$ linear system

$$A\mathbf{x} = \mathbf{b}, \tag{4.5.1}$$

and suppose that there exists a matrix B such that

$$BA = I_n.$$

Then, multiplying both sides of (4.5.1) on the left by B yields

$$(BA)\mathbf{x} = B\mathbf{b},$$

that is, since $BA = I_n$,

$$\mathbf{x} = B\mathbf{b}.$$

Thus we have determined a solution of the system (4.5.1) by a matrix multiplication. In this section we investigate the question as to when, for a given $n \times n$ matrix A, there exists a matrix B satisfying[1]

$$AB = I_n \quad \text{and} \quad BA = I_n, \tag{4.5.2}$$

and derive an efficient method for determining B when it does exist. We begin this investigation by proving that there can be at most one matrix, B, satisfying (4.5.2).

---

[1] It can be shown that if A and B are $n \times n$ and $AB = I_n$, then necessarily $BA = I_n$ (see Exercise 32).

Theorem 4.5.1:   Let A be an $n \times n$ matrix. Suppose B and C are both $n \times n$ matrices satisfying

$$AB = BA = I_n, \tag{4.5.3}$$

$$AC = CA = I_n, \tag{4.5.4}$$

respectively. Then $B = C$.

PROOF   From (4.5.3) it follows that

$$C = CI_n = C(AB),$$

that is,

$$C = (CA)B = I_n B = B,$$

where we have used (4.5.4) to replace CA by I_n in the second step.   ∎

Since the identity matrix I_n plays the role of the number 1 in the multiplication of matrices, the properties given in (4.5.2) are the analogues for matrices of the properties

$$xx^{-1} = 1, \qquad x^{-1}x = 1,$$

which hold for all (nonzero) numbers x.   It is thus natural to denote the matrix B in (4.5.2) by A^{-1} and to call it the inverse of A.   The following definition introduces the appropriate terminology.

Definition 4.5.1:   Let A be an $n \times n$ matrix. If there exists a matrix A^{-1} satisfying

$$AA^{-1} = A^{-1}A = I_n,$$

then we call A^{-1} the matrix **inverse** to A, or just the inverse of A. We say that A is **nonsingular** if A^{-1} exists. If A^{-1} does not exist, then we say that A is **singular**.

REMARK   It is important to realize that A^{-1} denotes the matrix that satisfies $AA^{-1} = A^{-1}A = I_n$. It does *not* mean $1/A$, which has no meaning whatsoever.

Example 4.5.1   If $A = \begin{bmatrix} 1 & -1 & 2 \\ 2 & -3 & 3 \\ 1 & -1 & 1 \end{bmatrix}$, show that $B = \begin{bmatrix} 0 & -1 & 3 \\ 1 & -1 & 1 \\ 1 & 0 & -1 \end{bmatrix}$ is the inverse of A.

Solution   By direct multiplication we find that

$$AB = \begin{bmatrix} 1 & -1 & 2 \\ 2 & -3 & 3 \\ 1 & -1 & 1 \end{bmatrix} \begin{bmatrix} 0 & -1 & 3 \\ 1 & -1 & 1 \\ 1 & 0 & -1 \end{bmatrix} = I_3,$$

and

$$BA = \begin{bmatrix} 0 & -1 & 3 \\ 1 & -1 & 1 \\ 1 & 0 & -1 \end{bmatrix} \begin{bmatrix} 1 & -1 & 2 \\ 2 & -3 & 3 \\ 1 & -1 & 1 \end{bmatrix} = I_3.$$

Thus (4.5.2) is satisfied and hence B is indeed the inverse of A.   We therefore write

$$A^{-1} = \begin{bmatrix} 0 & -1 & 3 \\ 1 & -1 & 1 \\ 1 & 0 & -1 \end{bmatrix}.$$

The inverse of an $n \times n$ matrix satisfies the following three properties.

Theorem 4.5.2:   Let A and B be nonsingular $n \times n$ matrices.   Then

1. A^{-1} is nonsingular and $(A^{-1})^{-1} = A$.

2. AB is nonsingular and $(AB)^{-1} = B^{-1}A^{-1}$.

3. A^T is nonsingular and $(A^T)^{-1} = (A^{-1})^T$.

PROOF   The proof of each result consists of checking that the matrix on the right-hand side of the equality is indeed the required inverse.   We prove (3) and leave (1) and (2) as exercises.   Since A is nonsingular, $(A^{-1})^T$ is a well-defined matrix.   We wish to show that $(A^{-1})^T$ is the inverse of A^T, that is,

$$A^T(A^{-1})^T = I_n \quad \text{and} \quad (A^{-1})^T A^T = I_n.$$

First recall the general property of the transpose that $A^T B^T = (BA)^T$.   Using this property with $B = A^{-1}$ yields

$$A^T(A^{-1})^T = (A^{-1}A)^T = (I_n)^T = I_n.$$

Similarly,

$$(A^{-1})^T A^T = (AA^{-1})^T = (I_n)^T = I_n.$$

Consequently A^T is nonsingular and has inverse $(A^{-1})^T$.   ∎

We now return to the $n \times n$ system (4.5.1).   Our next theorem is a direct consequence of the uniqueness of A^{-1}.

Theorem 4.5.3:   If A^{-1} exists, then the $n \times n$ system of linear equations

$$Ax = b$$

has the *unique* solution

$$x = A^{-1}b.$$

Example 4.5.2   Solve the system of equations $A\mathbf{x} = \mathbf{b}$ if

$$A = \begin{bmatrix} 1 & -1 & 2 \\ 2 & -3 & 3 \\ 1 & -1 & 1 \end{bmatrix} \quad \text{and} \quad \mathbf{b} = \begin{bmatrix} 3 \\ 1 \\ 1 \end{bmatrix}.$$

Solution   We showed in the previous example that

$$A^{-1} = \begin{bmatrix} 0 & -1 & 3 \\ 1 & -1 & 1 \\ 1 & 0 & -1 \end{bmatrix}.$$

Multiplying both sides of $A\mathbf{x} = \mathbf{b}$ on the left by A^{-1} yields

$$\mathbf{x} = A^{-1}\mathbf{b} = \begin{bmatrix} 0 & -1 & 3 \\ 1 & -1 & 1 \\ 1 & 0 & -1 \end{bmatrix}\begin{bmatrix} 3 \\ 1 \\ 1 \end{bmatrix} = \begin{bmatrix} 2 \\ 3 \\ 2 \end{bmatrix},$$

so that the unique solution to the system is (2, 3, 2).

The next step in our investigation of the inverse of an $n \times n$ matrix A is to determine when A^{-1} exists. This is accomplished in the following theorem.

Theorem 4.5.4:   An $n \times n$ matrix A is nonsingular if and only if rank$(A) = n$.

PROOF   If A^{-1} exists, then $A\mathbf{x} = \mathbf{b}$ has a unique solution and so, from Theorem 4.3.1, rank$(A) = n$.

Conversely, suppose rank$(A) = n$. We must show that there exists a matrix X satisfying

$$AX = I_n = XA.$$

Let $\mathbf{e}_1, \mathbf{e}_2, \ldots, \mathbf{e}_n$ denote the column vectors of the identity matrix I_n. Then, since rank$(A) = n$, Theorem 4.3.1 implies that each of the linear systems

$$A\mathbf{x}_i = \mathbf{e}_i, \qquad i = 1, 2, \ldots, n \tag{4.5.5}$$

has a unique solution.[1]   If we let $X = [\mathbf{x}_1, \mathbf{x}_2, \ldots, \mathbf{x}_n]$, where $\mathbf{x}_1, \mathbf{x}_2, \ldots, \mathbf{x}_n$ are the unique solutions to the systems in (4.5.5), it follows that

$$A[\mathbf{x}_1, \mathbf{x}_2, \ldots, \mathbf{x}_n] = [A\mathbf{x}_1, A\mathbf{x}_2, \ldots, A\mathbf{x}_n] = [\mathbf{e}_1, \mathbf{e}_2, \ldots, \mathbf{e}_n],$$

that is,

$$AX = I_n. \tag{4.5.6}$$

---

[1] Notice that for an $n \times n$ system $A\mathbf{x} = \mathbf{b}$, if rank$(A) = n$, then so must rank$(A^\#)$.

We must also show that $XA = I_n$. Multiplying both sides of (4.5.6) on the right by A yields

$$(AX)A = A,$$

that is,

$$A(XA - I_n) = 0_n. \qquad (4.5.7)$$

Now let $\mathbf{y}_1, \mathbf{y}_2, \ldots, \mathbf{y}_n$ denote the column vectors of the $n \times n$ matrix $(XA - I_n)$. Equating corresponding column vectors on either side of (4.5.7) yields

$$A\mathbf{y}_i = \mathbf{0}, \qquad i = 1, 2, \ldots, n. \qquad (4.5.8)$$

But, by assumption, rank$(A) = n$, and so each of the systems in (4.5.8) has a unique solution, which, since the systems are homogeneous, must be the trivial solution. Thus

$$XA - I_n = 0_n,$$

and hence

$$XA = I_n. \qquad (4.5.9)$$

Equations (4.5.6) and (4.5.9) imply that $X = A^{-1}$. ∎

We now have a converse to Theorem 4.5.3.

Corollary 4.5.1: Let A be an $n \times n$ matrix. If $A\mathbf{x} = \mathbf{b}$ has a unique solution, then A^{-1} exists.

PROOF If $A\mathbf{x} = \mathbf{b}$ has a unique solution, then, from Theorem 4.3.1, rank$(A) = n$; so from the previous theorem, A^{-1} exists. ∎

Having established when A^{-1} exists, it remains only to develop a method for finding it. Assuming that rank$(A) = n$, let $\mathbf{x}_1, \mathbf{x}_2, \ldots, \mathbf{x}_n$ denote the column vectors of A^{-1}. Then, as in the proof of the previous theorem, since $AA^{-1} = I_n$ these column vectors can be determined by solving the n linear systems

$$A\mathbf{x}_i = \mathbf{e}_i, \qquad i = 1, 2, \ldots, n. \qquad (4.5.10)$$

We illustrate the method in the 3×3 case. From (4.5.10), the column vectors of A^{-1} are determined by solving the three linear systems

$$A\mathbf{x}_1 = \mathbf{e}_1, \qquad A\mathbf{x}_2 = \mathbf{e}_2, \qquad A\mathbf{x}_3 = \mathbf{e}_3.$$

The augmented matrices of these systems can be written as

$$\left[\begin{array}{c|c} A & \begin{matrix} 1 \\ 0 \\ 0 \end{matrix} \end{array}\right], \quad \left[\begin{array}{c|c} A & \begin{matrix} 0 \\ 1 \\ 0 \end{matrix} \end{array}\right], \quad \left[\begin{array}{c|c} A & \begin{matrix} 0 \\ 0 \\ 1 \end{matrix} \end{array}\right],$$

respectively. We now use elementary row operations to reduce these augmented matrices to reduced row echelon form. Since rank$(A) = 3$ by assumption, it follows that the reduced row echelon form of A is just I_3, and so we have, schematically,

$$\begin{bmatrix} A & \begin{matrix} 1 \\ 0 \\ 0 \end{matrix} \end{bmatrix} \xrightarrow{\text{Elementary row operations}} \begin{bmatrix} 1 & 0 & 0 & a_1 \\ 0 & 1 & 0 & a_2 \\ 0 & 0 & 1 & a_3 \end{bmatrix},$$

which implies that the first column vector of A^{-1} is

$$\mathbf{x}_1 = \begin{bmatrix} a_1 \\ a_2 \\ a_3 \end{bmatrix}.$$

Similarly,

$$\begin{bmatrix} A & \begin{matrix} 0 \\ 1 \\ 0 \end{matrix} \end{bmatrix} \xrightarrow{\text{Elementary row operations}} \begin{bmatrix} 1 & 0 & 0 & b_1 \\ 0 & 1 & 0 & b_2 \\ 0 & 0 & 1 & b_3 \end{bmatrix}$$

implies that the second column vector of A^{-1} is

$$\mathbf{x}_2 = \begin{bmatrix} b_1 \\ b_2 \\ b_3 \end{bmatrix}.$$

Finally,

$$\begin{bmatrix} A & \begin{matrix} 0 \\ 0 \\ 1 \end{matrix} \end{bmatrix} \xrightarrow{\text{Elementary row operations}} \begin{bmatrix} 1 & 0 & 0 & c_1 \\ 0 & 1 & 0 & c_2 \\ 0 & 0 & 1 & c_3 \end{bmatrix}$$

implies that the third column vector of A^{-1} is

$$\mathbf{x}_3 = \begin{bmatrix} c_1 \\ c_2 \\ c_3 \end{bmatrix}.$$

It follows that

$$A^{-1} = [\mathbf{x}_1, \mathbf{x}_2, \mathbf{x}_3] = \begin{bmatrix} a_1 & b_1 & c_1 \\ a_2 & b_2 & c_2 \\ a_3 & b_3 & c_3 \end{bmatrix}.$$

The key point to notice is that in solving for $\mathbf{x}_1$, $\mathbf{x}_2$, $\mathbf{x}_3$, we use the *same* elementary row operations to reduce A to I_3. We can therefore save a significant amount of work by combining the above operations as follows:

$$\begin{bmatrix} A & \begin{matrix} 1 & 0 & 0 \\ 0 & 1 & 0 \\ 0 & 0 & 1 \end{matrix} \end{bmatrix} \xrightarrow{\text{Elementary row operations}} \begin{bmatrix} 1 & 0 & 0 & a_1 & b_1 & c_1 \\ 0 & 1 & 0 & a_2 & b_2 & c_2 \\ 0 & 0 & 1 & a_3 & b_3 & c_3 \end{bmatrix}.$$

The generalization to the $n \times n$ case is immediate. We form the $n \times 2n$ matrix $[A \mid I_n]$ and reduce A to I_n using elementary row operations. Schematically,

$$\text{Elementary row operations}$$
$$[A \mid I_n] \xrightarrow{\hspace{4cm}} [I_n \mid A^{-1}].$$

This method of finding A^{-1} is called the **Gauss–Jordan technique.**

REMARK   Notice that if we are given an $n \times n$ matrix, A, we will not know rank(A), and hence we will not know whether A^{-1} exists. However, if at any stage in the row reduction of $[A \mid I_n]$ we find that rank(A) $< n$, then it will follow from Theorem 4.5.4 that A is singular and hence that A^{-1} does not exist.

Example 4.5.3   Find A^{-1} if $A = \begin{bmatrix} 1 & 1 & 3 \\ 0 & 1 & 2 \\ 3 & 5 & -1 \end{bmatrix}$.

Solution   Using the Gauss–Jordan technique, we proceed as follows:

$$\left[\begin{array}{ccc|ccc} 1 & 1 & 3 & 1 & 0 & 0 \\ 0 & 1 & 2 & 0 & 1 & 0 \\ 3 & 5 & -1 & 0 & 0 & 1 \end{array}\right] \xrightarrow{R_3 \to R_3 - 3R_1} \left[\begin{array}{ccc|ccc} 1 & 1 & 3 & 1 & 0 & 0 \\ 0 & 1 & 2 & 0 & 1 & 0 \\ 0 & 2 & -10 & -3 & 0 & 1 \end{array}\right]$$

$$\xrightarrow[R_3 \to R_3 - 2R_2]{R_1 \to R_1 - R_2} \left[\begin{array}{ccc|ccc} 1 & 0 & 1 & 1 & -1 & 0 \\ 0 & 1 & 2 & 0 & 1 & 0 \\ 0 & 0 & -14 & -3 & -2 & 1 \end{array}\right]$$

$$\xrightarrow{R_3 \to -R_3/14} \left[\begin{array}{ccc|ccc} 1 & 0 & 1 & 1 & -1 & 0 \\ 0 & 1 & 2 & 0 & 1 & 0 \\ 0 & 0 & 1 & \frac{3}{14} & \frac{1}{7} & -\frac{1}{14} \end{array}\right]$$

$$\xrightarrow[R_2 \to R_2 - 2R_3]{R_1 \to R_1 - R_3} \left[\begin{array}{ccc|ccc} 1 & 0 & 0 & \frac{11}{14} & -\frac{8}{7} & \frac{1}{14} \\ 0 & 1 & 0 & -\frac{3}{7} & \frac{5}{7} & \frac{1}{7} \\ 0 & 0 & 1 & \frac{3}{14} & \frac{1}{7} & -\frac{1}{14} \end{array}\right].$$

Thus,

$$A^{-1} = \begin{bmatrix} \frac{11}{14} & -\frac{8}{7} & \frac{1}{14} \\ -\frac{3}{7} & \frac{5}{7} & \frac{1}{7} \\ \frac{3}{14} & \frac{1}{7} & -\frac{1}{14} \end{bmatrix}.$$

We leave it as an exercise to check that $AA^{-1} = A^{-1}A = I_n$.

Example 4.5.4   Use A^{-1} to solve the system

$$x_1 + x_2 + 3x_3 = 2,$$
$$x_2 + 2x_3 = 1,$$
$$3x_1 + 5x_2 - x_3 = 4.$$

Solution The system can be written as

$$Ax = b,$$

where A is the matrix in the previous example and $b = [2 \ 1 \ 4]^T$. Since A is non-singular, the solution of the system can be written as $x = A^{-1}b$, that is, from the preceding example,

$$x = \begin{bmatrix} \frac{11}{14} & -\frac{8}{7} & \frac{1}{14} \\ -\frac{3}{7} & \frac{5}{7} & \frac{1}{7} \\ \frac{3}{14} & \frac{1}{7} & -\frac{1}{14} \end{bmatrix} \begin{bmatrix} 2 \\ 1 \\ 4 \end{bmatrix} = \begin{bmatrix} \frac{5}{7} \\ \frac{3}{7} \\ \frac{2}{7} \end{bmatrix}.$$

Thus $x_1 = \frac{5}{7}$, $x_2 = \frac{3}{7}$, $x_3 = \frac{2}{7}$, so that the solution of the system is $(\frac{5}{7}, \frac{3}{7}, \frac{2}{7})$.

Finally, we note that if we are presented with an $n \times n$ linear system $Ax = b$ with a nonsingular coefficient matrix, we now have three different techniques that could be applied to determine its solution:

1. Solve by Gaussian elimination.
2. Solve by Gauss–Jordan elimination.
3. Find A^{-1} and then obtain the solution directly as $x = A^{-1}b$.

Since the first two of these require the row reduction of an $n \times (n + 1)$ augmented matrix, whereas the computation of A^{-1} in (3) requires the row reduction of an $n \times 2n$ matrix, it should be clear that the first two techniques are superior from a computational viewpoint. The significance of (3) is more as a theoretical tool.

EXERCISES 4.5

In problems 1–3 verify, by direct multiplication, that the given matrices are inverses.

1. $A = \begin{bmatrix} 2 & -1 \\ 3 & -1 \end{bmatrix}$, $A^{-1} = \begin{bmatrix} -1 & 1 \\ -3 & 2 \end{bmatrix}$.

2. $A = \begin{bmatrix} 4 & 9 \\ 3 & 7 \end{bmatrix}$, $A^{-1} = \begin{bmatrix} 7 & -9 \\ -3 & 4 \end{bmatrix}$.

3. $A = \begin{bmatrix} 3 & 5 & 1 \\ 1 & 2 & 1 \\ 2 & 6 & 7 \end{bmatrix}$, $A^{-1} = \begin{bmatrix} 8 & -29 & 3 \\ -5 & 19 & -2 \\ 2 & -8 & 1 \end{bmatrix}$.

In problems 4–16 determine A^{-1}, if possible, using the Gauss–Jordan method. If A^{-1} exists, check your answer by verifying that $AA^{-1} = I_n$ and $A^{-1}A = I_n$.

4. $A = \begin{bmatrix} 1 & 2 \\ 1 & 3 \end{bmatrix}$.

5. $A = \begin{bmatrix} 1 & 1+i \\ 1-i & 1 \end{bmatrix}$.

6. $A = \begin{bmatrix} 1 & -i \\ i-1 & 2 \end{bmatrix}$.

7. $A = \begin{bmatrix} 0 & 0 \\ 0 & 0 \end{bmatrix}$.

8. $A = \begin{bmatrix} 1 & -1 & 2 \\ 2 & 1 & 11 \\ 4 & -3 & 10 \end{bmatrix}$.

9. $A = \begin{bmatrix} 3 & 5 & 1 \\ 1 & 2 & 1 \\ 2 & 6 & 7 \end{bmatrix}$.

10. $A = \begin{bmatrix} 0 & 1 & 0 \\ 0 & 0 & 1 \\ 0 & 1 & 2 \end{bmatrix}$.

11. $A = \begin{bmatrix} 4 & 2 & -13 \\ 2 & 1 & -7 \\ 3 & 2 & 4 \end{bmatrix}$.

12. $A = \begin{bmatrix} 1 & 2 & -3 \\ 2 & 6 & -2 \\ -1 & 1 & 4 \end{bmatrix}$.

13. $A = \begin{bmatrix} 1 & i & 2 \\ 1+i & -1 & 2i \\ 2 & 2i & 5 \end{bmatrix}$.

14. $A = \begin{bmatrix} 2 & 1 & 3 \\ 1 & -1 & 2 \\ 3 & 3 & 4 \end{bmatrix}$.

15. $A = \begin{bmatrix} 1 & -1 & 2 & 3 \\ 2 & 0 & 3 & -4 \\ 3 & -1 & 7 & 8 \\ 1 & 0 & 3 & 5 \end{bmatrix}$.

16. $A = \begin{bmatrix} 0 & -2 & -1 & -3 \\ 2 & 0 & 2 & 1 \\ 1 & -2 & 0 & 2 \\ 3 & -1 & -2 & 0 \end{bmatrix}$.

In problems 17–21 use A^{-1} to find the solution of the given system.

17. $x_1 + 3x_2 = 1,$
$2x_1 + 5x_2 = 3.$

18. $x_1 - 2ix_2 = 2,$
$(2 - i)x_1 + 4ix_2 = -i.$

19. $x_1 + x_2 - 2x_3 = -2,$
$x_2 + x_3 = 3,$
$2x_1 + 4x_2 - 3x_3 = 1.$

20. $x_1 + x_2 + 2x_3 = 12,$
$x_1 + 2x_2 - x_3 = 24,$
$2x_1 - x_2 + x_3 = -36.$

21. $3x_1 + 4x_2 + 5x_3 = 1,$
$2x_1 + 10x_2 + x_3 = 1,$
$4x_1 + x_2 + 8x_3 = 1.$

An $n \times n$ matrix is called **orthogonal** if $A^T = A^{-1}$. In problems 22–25 show that the given matrices are orthogonal.

22. $A = \begin{bmatrix} 0 & 1 \\ -1 & 0 \end{bmatrix}$.

23. $\begin{bmatrix} \frac{\sqrt{3}}{2} & \frac{1}{2} \\ -\frac{1}{2} & \frac{\sqrt{3}}{2} \end{bmatrix}$.

24. $\begin{bmatrix} \cos\alpha & \sin\alpha \\ -\sin\alpha & \cos\alpha \end{bmatrix}$.

25. $\begin{bmatrix} \frac{1}{2} & -\frac{1}{2} & \frac{1}{2} & \frac{1}{2} \\ \frac{1}{2} & \frac{1}{2} & \frac{1}{2} & -\frac{1}{2} \\ \frac{1}{2} & \frac{1}{2} & -\frac{1}{2} & \frac{1}{2} \\ \frac{1}{2} & -\frac{1}{2} & -\frac{1}{2} & -\frac{1}{2} \end{bmatrix}$.

26. Prove (1) and (2) of Theorem 4.5.2.

27. Use properties of the inverse to prove the following:
(a) If A is an $n \times n$ nonsingular *symmetric* matrix, then A^{-1} is symmetric.
(b) If A is an $n \times n$ nonsingular *skew-symmetric* matrix, then A^{-1} is skew-symmetric.

28. Prove that if A, B, C are $n \times n$ matrices satisfying $BA = I_n$ and $AC = I_n$, then $B = C$.

29. Consider the 2×2 matrix

$$A = \begin{bmatrix} a_{11} & a_{12} \\ a_{21} & a_{22} \end{bmatrix}.$$

Let $\Delta = a_{11}a_{22} - a_{12}a_{21}$ with $a_{11} \neq 0$. Show that, provided $\Delta \neq 0$,

$$A^{-1} = \frac{1}{\Delta}\begin{bmatrix} a_{22} & -a_{12} \\ -a_{21} & a_{11} \end{bmatrix}.$$

The quantity Δ is referred to as the determinant of A. We will investigate determinants in more detail in the next chapter.

30. Let A be an $n \times n$ matrix and suppose that we have to solve the p linear systems

$$A\mathbf{x}_i = \mathbf{b}_i, \qquad i = 1, 2, \ldots, p,$$

where the $\mathbf{b}_i$ are given. Devise an efficient method for solving these systems.

31. Use your method from the previous problem to solve the three linear systems

$$Ax_i = b_i, \qquad i = 1, 2, 3,$$

if

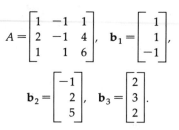

$$A = \begin{bmatrix} 1 & -1 & 1 \\ 2 & -1 & 4 \\ 1 & 1 & 6 \end{bmatrix}, \quad b_1 = \begin{bmatrix} 1 \\ 1 \\ -1 \end{bmatrix},$$

$$b_2 = \begin{bmatrix} -1 \\ 2 \\ 5 \end{bmatrix}, \quad b_3 = \begin{bmatrix} 2 \\ 3 \\ 2 \end{bmatrix}.$$

32. Let A and B be $n \times n$ matrices, and suppose that $AB = I_n$.

(a) Show that for all y, the system $Ax = y$ has solution $x = By$.

(b) Let A^* denote the reduced row echelon form of A. Use the fact that elementary row operations are reversible to show that the system $A^*x = b$ has a solution for all b.

(c) If rank$(A) < n$, prove that the linear system $A^*x = e_n$ has no solution, where e_n denotes the nth column vector of I_n. Use the result from (b) to conclude that rank$(A) = n$.

(d) Use your result from (c) and the results of this section to prove that $BA = I_n$.

33. Let A be an $m \times n$ matrix with $m \le n$.

(a) If rank$(A) = m$, prove that there exists a matrix B satisfying $AB = I_m$. Such a matrix is called a **right inverse** of A.

(b) If $A = \begin{bmatrix} 1 & 3 & 1 \\ 2 & 7 & 4 \end{bmatrix}$, determine all right inverses of A.

5

Determinants

In this chapter we introduce a basic tool in applied mathematics, namely, the determinant of an $n \times n$ matrix. The determinant can be considered as a number associated with an $n \times n$ matrix A, whose nonvanishing characterizes when the linear system $A\mathbf{x} = \mathbf{b}$ has a unique solution (or, equivalently, when A^{-1} exists). As we shall see in later chapters, the determinant also plays a fundamental role in the theory of linear differential equations and linear systems of differential equations. Other areas of applications of the determinant include coordinate geometry and function theory.

There are several ways to define a determinant. One of these, with which some of you will already be familiar, is the inductive definition. In this approach the determinant of a 1×1 matrix is given and then the determinant of an $n \times n$ matrix is defined in terms of a sum of determinants of appropriate $(n - 1) \times (n - 1)$ matrices. Although this is certainly a very simple way of introducing the determinant, we feel that it lacks motivation; it also has the drawback that the proofs of many results about determinants are rather tedious when using this definition. We prefer to use the classical definition in which a determinant is defined as an appropriate sum of products of the elements from the matrix A. The disadvantage to this approach is that it requires the introduction of some results about permutations.

However, this is more than compensated for by the gain in understanding of the determinant, and it also enables us to prove all the results that we require about determinants in a fairly straightforward manner.

5.1   PERMUTATIONS

Consider the first n positive integers $1, 2, \ldots, n$. Any arrangement of these integers in a specific order, say $(p_1, p_2, \ldots, p_n)$, is called a **permutation**.

Example 5.1.1   There are precisely six distinct permutations of the integers 1, 2, 3:

$$(1, 2, 3), (1, 3, 2), (2, 3, 1), (2, 1, 3), (3, 1, 2), (3, 2, 1).$$

More generally we have the following result:

Theorem 5.1.1:   There are precisely $n!$ distinct permutations of the integers 1, 2, \ldots, n.

PROOF   The proof is left as an exercise.   ■

The elements in the permutation $(1, 2, 3, \ldots, n)$ are said to be in their natural increasing order. We now introduce a number that describes how far from natural order a given permutation is. The elements p_i and p_j in the permutation $(p_1, p_2, \ldots, p_n)$ are said to be *inverted* if they are out of their natural order—that is, if $p_i > p_j$ with $i < j$. For example, in the permutation $(4, 2, 3, 1)$ the pairs $(4, 2), (4, 3)$, $(4, 1), (2, 1), (3, 1)$ are all out of their natural order, and so there are a total of five inversions in this permutation.

In general we let $N(p_1, p_2, \ldots, p_n)$ denote the total number of inversions in the permutation $(p_1, p_2, \ldots, p_n)$.

Example 5.1.2   Find the number of inversions in the permutations $(1, 3, 2, 4, 5)$, $(2, 4, 5, 3, 1)$.

Solution   The only pair of elements in the permutation $(1, 3, 2, 4, 5)$ that is out of natural order is $(3, 2)$, so that $N(1, 3, 2, 4, 5) = 1$.

The permutation $(2, 4, 5, 3, 1)$ has the following pairs of elements out of natural order: $(2, 1), (4, 3), (4, 1), (5, 3), (5, 1), (3, 1)$. Thus $N(2, 4, 5, 3, 1) = 6$.

REMARK   It can be shown that the number of inversions gives the minimum number of adjacent interchanges of elements in the permutation that are required to restore the permutation to its natural increasing order. This justifies the claim that the number of inversions describes how far from natural order a given permutation is. For example $N(3, 2, 1) = 3$, and the permutation $(3, 2, 1)$ can be restored to its natural order by the following sequence of adjacent interchanges:

$$(3, 2, 1) \rightarrow (3, 1, 2) \rightarrow (1, 3, 2) \rightarrow (1, 2, 3).$$

The number of inversions enables us to distinguish two different types of permutations as follows.

Even permutations: If $N(p_1, p_2, \ldots, p_n)$ is an even integer (or zero), we say that $(p_1, p_2, \ldots, p_n)$ is an even permutation. We also say that $(p_1, p_2, \ldots, p_n)$ has **even parity**.

Odd permutations: If $N(p_1, p_2, \ldots, p_n)$ is an odd integer, we say that $(p_1, p_2, \ldots, p_n)$ is an odd permutation. We also say that $(p_1, p_2, \ldots, p_n)$ has **odd parity**.

Example 5.1.3 The permutation $(4, 1, 3, 2)$ has even parity, since $N(4, 1, 3, 2) = 4$, whereas $(3, 2, 1, 4)$ is an odd permutation, since $N(3, 2, 1, 4) = 3$.

We associate a plus or a minus sign with a permutation, depending on whether it has even or odd parity, respectively. Thus the sign associated with the permutation $(p_1, p_2, \ldots, p_n)$ can be specified by the indicator $\sigma(p_1, p_2, \ldots, p_n)$ defined by:

$$\sigma(p_1, p_2, \ldots, p_n) = \begin{cases} +1 & \text{if } (p_1, p_2, \ldots, p_n) \text{ has even parity,} \\ -1 & \text{if } (p_1, p_2, \ldots, p_n) \text{ has odd parity.} \end{cases}$$

This can be expressed directly in terms of the number of inversions as

$$\sigma(p_1, p_2, \ldots, p_n) = (-1)^{N(p_1, p_2, \ldots, p_n)}.$$

Example 5.1.4 It follows from Example 5.1.2 that

$$\sigma(1, 3, 2, 4, 5) = (-1)^1 = -1,$$
$$\sigma(2, 4, 5, 3, 1) = (-1)^6 = +1.$$

Finally, the proof of some of our later results will depend upon the following lemma.

Lemma 5.1.1: If any two elements in a permutation are interchanged, then the parity of the resulting permutation is opposite to that of the original permutation.

PROOF We first show that the interchange of two adjacent terms in a permutation changes the parity. Consider an arbitrary permutation $(p_1, p_2, \ldots, p_k, p_{k+1}, \ldots, p_n)$ and suppose we interchange the adjacent elements p_k and p_{k+1}. Then

1. If $p_k > p_{k+1}$, $N(p_1, p_2, \ldots, p_{k+1}, p_k, \ldots, p_n) = N(p_1, p_2, \ldots, p_k, p_{k+1}, \ldots, p_n) - 1$,

2. If $p_k < p_{k+1}$, $N(p_1, p_2, \ldots, p_{k+1}, p_k, \ldots, p_n) = N(p_1, p_2, \ldots, p_k, p_{k+1}, \ldots, p_n) + 1$.

Thus the parity is changed in both cases. Now suppose we interchange the elements p_i and p_k in the permutation $(p_1, p_2, \ldots, p_i, \ldots, p_k, \ldots, p_n)$. We can accomplish this by successively interchanging adjacent elements. In moving p_k to the ith position, we perform $(k - i)$ interchanges, and the resulting permutation is

$$(p_1, p_2, \ldots, p_k, p_i, \ldots, p_{k-1}, p_{k+1}, \ldots, p_n).$$

We now move p_i to the kth position. This requires $[(k - i) - 1]$ interchanges. Thus, the total number of interchanges is $2(k - i) - 1$, which is always an odd integer. Since each adjacent interchange changes the parity, the permutation resulting from an odd number of adjacent interchanges has opposite parity to the original permutation. ∎

EXERCISES 5.1

1. Write out all distinct permutations of the integers 1, 2, 3, and find the number of inversions in each permutation.

2. Determine the parity of the given permutation:

(a) (2, 1, 3, 4).
(b) (1, 3, 2, 4).
(c) (1, 4, 3, 5, 2).
(d) (5, 4, 3, 2, 1).

5.2   THE DEFINITION OF A DETERMINANT

We motivate the definition of a determinant by studying the structure of the solution of an $n \times n$ system of linear equations in some special cases.

CASE 1: $n = 1$ The system

$$a_{11}x_1 = b_1, \quad \text{with coefficient matrix } A = [a_{11}]$$

has solution

$$x_1 = \frac{b_1}{a_{11}},$$

provided the 1×1 determinant, $\det(A)$, defined by

$$\det(A) = a_{11}$$

is nonzero.

CASE 2: $n = 2$ The system

$$a_{11}x_1 + a_{12}x_2 = b_1,$$
$$a_{21}x_1 + a_{22}x_2 = b_2, \quad \text{with coefficient matrix } A = \begin{bmatrix} a_{11} & a_{12} \\ a_{21} & a_{22} \end{bmatrix}$$

has solution

$$x_1 = -\frac{a_{12}b_2 - a_{22}b_1}{a_{11}a_{22} - a_{12}a_{21}}, \qquad x_2 = \frac{a_{11}b_2 - a_{21}b_1}{a_{11}a_{22} - a_{12}a_{21}},$$

provided the 2×2 determinant, $\det(A)$, defined by

$$\det(A) = a_{11}a_{22} - a_{12}a_{21} \tag{5.2.1}$$

is nonzero.

CASE 3: $n = 3$ The system

$$\begin{aligned} a_{11}x_1 + a_{12}x_2 + a_{13}x_3 &= b_1, \\ a_{21}x_1 + a_{22}x_2 + a_{23}x_3 &= b_2, \\ a_{31}x_1 + a_{32}x_2 + a_{33}x_3 &= b_3, \end{aligned} \qquad \text{with coefficient matrix } A = \begin{bmatrix} a_{11} & a_{12} & a_{13} \\ a_{21} & a_{22} & a_{23} \\ a_{31} & a_{32} & a_{33} \end{bmatrix}$$

has solution:[1]

$$x_1 = \frac{E_1}{\det(A)}, \qquad x_2 = \frac{E_2}{\det(A)}, \qquad x_3 = \frac{E_3}{\det(A)},$$

provided the 3×3 determinant, $\det(A)$, is nonzero, where

$$\det(A) = a_{11}a_{22}a_{33} + a_{12}a_{23}a_{31} + a_{13}a_{21}a_{32}$$

$$- a_{11}a_{23}a_{32} - a_{12}a_{21}a_{33} - a_{13}a_{22}a_{31}.$$

Structure: Each determinant consists of the sum of $n!$ products.   Each product term consists of one element from each row and one element from each column in the matrix of coefficients of the system.   Also, each term is assigned a plus or a minus sign.   If we look more closely at the expression for, say, the 3×3 determinant we see that the row indices of each term have been arranged in their natural order and that the column indices are each a permutation, (p_1, p_2, p_3), of 1, 2, 3. Further, the sign attached to each term coincides with the sign of the permutation of the corresponding column indices, that is, $\sigma(p_1, p_2, p_3)$.   These observations motivate the following general definition of the determinant of an $n \times n$ matrix:

Definition 5.2.1:   Let $A = [a_{ij}]$ be an $n \times n$ matrix.   We define the **determinant of A**, denoted by $\det(A)$, as follows:

$$\det(A) = \sum \sigma(p_1, p_2, \ldots, p_n) a_{1p_1} a_{2p_2} a_{3p_3} \cdots a_{np_n}, \tag{5.2.2}$$

where the summation is over the $n!$ distinct permutations $(p_1, p_2, \ldots, p_n)$ of the integers $1, 2, 3, \ldots, n$.   The determinant of an $n \times n$ matrix is said to have **order n**.

---

[1] Here E_1, E_2, and E_3, are algebraic expressions involving the a_{ij} and b_i whose exact form need not concern us at the moment.

We will also denote the determinant of

$$A = \begin{bmatrix} a_{11} & a_{12} & \cdots & a_{1n} \\ a_{21} & a_{22} & \cdots & a_{2n} \\ \vdots & \vdots & & \vdots \\ a_{n1} & a_{n2} & \cdots & a_{nn} \end{bmatrix}$$

by

$$\begin{vmatrix} a_{11} & a_{12} & \cdots & a_{1n} \\ a_{21} & a_{22} & \cdots & a_{2n} \\ \vdots & \vdots & & \vdots \\ a_{n1} & a_{n2} & \cdots & a_{nn} \end{vmatrix}.$$

Thus, for example, from (5.2.1),

$$\begin{vmatrix} a_{11} & a_{12} \\ a_{21} & a_{22} \end{vmatrix} = a_{11}a_{22} - a_{12}a_{21}.$$

Example 5.2.1   Use Definition 5.2.1 to derive the expression for the determinant of order 3.

Solution   In the 3×3 case (5.2.2) reduces to

$$\det(A) = \sum \sigma(p_1, p_2, p_3)a_{1p_1}a_{2p_2}a_{3p_3},$$

where the summation is over the 3! permutations of 1, 2, 3.   It follows that the 3! terms in this summation are

$$a_{11}a_{22}a_{33}, \quad a_{11}a_{23}a_{32}, \quad a_{12}a_{21}a_{33}, \quad a_{12}a_{23}a_{31}, \quad a_{13}a_{21}a_{32}, \quad a_{13}a_{22}a_{31},$$

so that

$$\det(A) = \sigma(1, 2, 3)a_{11}a_{22}a_{33} + \sigma(1, 3, 2)a_{11}a_{23}a_{32} + \sigma(2, 1, 3)a_{12}a_{21}a_{33}$$
$$+ \sigma(2, 3, 1)a_{12}a_{23}a_{31} + \sigma(3, 1, 2)a_{13}a_{21}a_{32} + \sigma(3, 2, 1)a_{13}a_{22}a_{31}.$$

Using our rules for determining $\sigma(p_1, p_2, p_3)$, we find

$$\sigma(1, 2, 3) = +1, \quad \sigma(1, 3, 2) = -1, \quad \sigma(2, 1, 3) = -1,$$
$$\sigma(2, 3, 1) = +1, \quad \sigma(3, 1, 2) = +1, \quad \sigma(3, 2, 1) = -1,$$

so that

$$\det(A) = \begin{vmatrix} a_{11} & a_{12} & a_{13} \\ a_{21} & a_{22} & a_{23} \\ a_{31} & a_{32} & a_{33} \end{vmatrix}$$

$$= a_{11}a_{22}a_{33} + a_{12}a_{23}a_{31} + a_{13}a_{21}a_{32} - a_{11}a_{23}a_{32} - a_{12}a_{21}a_{33} - a_{13}a_{22}a_{31}.$$

Example 5.2.2   Evaluate:

(a) $|-3|$.

(b) $\begin{vmatrix} 3 & -2 \\ 1 & 4 \end{vmatrix}$.

(c) $\begin{vmatrix} 1 & 2 & -3 \\ 4 & -1 & 2 \\ 0 & 3 & 1 \end{vmatrix}$.

Solution

(a) $|-3| = -3$.

(b) $\begin{vmatrix} 3 & -2 \\ 1 & 4 \end{vmatrix} = (3)(4) - (1)(-2) = 14$.

(c) $\begin{vmatrix} 1 & 2 & -3 \\ 4 & -1 & 2 \\ 0 & 3 & 1 \end{vmatrix} = (1)(-1)(1) + (2)(2)(0) + (-3)(4)(3) - (0)(-1)(-3)$
$$- (3)(2)(1) - (1)(4)(2)$$
$$= -51.$$

In the definition of determinant we have chosen to arrange the row indices in their natural order and sum over the distinct permutations on the column indices. This is purely convention, in the sense that we could equally well arrange the column indices in their natural order and sum over all distinct permutations of the row indices.   As we will now show, the resulting number,

$$\sum \sigma(p_1, p_2, \ldots, p_n) a_{p_11} a_{p_22} a_{p_33} \cdots a_{p_nn},$$

is the same as that given by (5.2.2). Since interchanging row and column indices is equivalent to taking the transpose of A, our claim is that $\det(A^T) = \det(A)$.   This is the content of the following theorem.

Theorem 5.2.1:   Let A be an $n \times n$ matrix.   Then $\det(A^T) = \det(A)$.

PROOF   Using Definition 5.2.1 we have

$$\det(A^T) = \sum \sigma(p_1, p_2, \ldots, p_n) a_{p_11} a_{p_22} a_{p_33} \cdots a_{p_nn}. \qquad (5.2.3)$$

Since $p_1, p_2, \ldots, p_n$ is a permutation of $1, 2, \ldots, n$, it follows that, by rearranging terms,

$$a_{p_11} a_{p_22} a_{p_33} \cdots a_{p_nn} = a_{1q_1} a_{2q_2} a_{3q_3} \cdots a_{nq_n}, \qquad (5.2.4)$$

for an appropriate permutation $(q_1, q_2, \ldots, q_n)$.   Further, since $\sigma(1, 2, \ldots, n) = 1$, Lemma 5.1.1 implies that

CHAP. 5: Determinants

$$N(p_1, p_2, \ldots, p_n) = \text{number of interchanges in changing } (1, 2, \ldots, n) \text{ to}$$
$$(p_1, p_2, \ldots, p_n)$$
$$= \text{number of interchanges in changing } (p_1, p_2, \ldots, p_n) \text{ to}$$
$$(1, 2, \ldots, n)$$
$$= \text{number of interchanges in changing } (1, 2, \ldots, n) \text{ to}$$
$$(q_1, q_2, \ldots, q_n)$$
$$= N(q_1, q_2, \ldots, q_n).$$

Thus,

$$\sigma(p_1, p_2, \ldots, p_n) = \sigma(q_1, q_2, \ldots, q_n). \tag{5.2.5}$$

Substituting from (5.2.4) and (5.2.5) into (5.2.3), we have

$$\det(A^T) = \sum \sigma(q_1, q_2, \ldots, q_n) a_{1q_1} a_{2q_2} a_{3q_3} \cdots a_{nq_n}$$
$$= \det(A). \qquad \blacksquare$$

EXERCISES 5.2

1. Use the definition of determinant to derive the general expression for the determinant of A if

$$A = \begin{bmatrix} a_{11} & a_{12} \\ a_{21} & a_{22} \end{bmatrix}.$$

In problems 2–5 evaluate the determinant of the given matrix.

2. $A = \begin{bmatrix} 1 & -1 \\ 2 & 3 \end{bmatrix}.$

3. $A = \begin{bmatrix} 2 & -1 \\ 6 & -3 \end{bmatrix}.$

4. $A = \begin{bmatrix} 1 & -1 & 0 \\ 2 & 3 & 6 \\ 0 & 2 & -1 \end{bmatrix}.$

5. $A = \begin{bmatrix} 2 & 1 & 5 \\ 4 & 2 & 3 \\ 9 & 5 & 1 \end{bmatrix}.$

6. (a) Write out all the 24 distinct permutations of the integers 1, 2, 3, 4, and determine the parity of each permutation. Hence derive the expres-

sion for the determinant of order 4.
(b) Use the result of (a) to evaluate

$$\begin{vmatrix} 1 & -1 & 0 & 1 \\ 3 & 0 & 2 & 5 \\ 2 & 1 & 0 & 3 \\ 9 & -1 & 2 & 1 \end{vmatrix}.$$

7. (a) If $A = \begin{bmatrix} a_{11} & a_{12} \\ a_{21} & a_{22} \end{bmatrix}$ and c is a constant, show that $\det(cA) = c^2 \det(A)$.
(b) Use the definition of determinant to prove that if A is an $n \times n$ matrix and c is a constant, then $\det(cA) = c^n \det(A)$.

8. Determine whether the following are terms in the determinant of order 5. If they are, determine whether the permutation of the column indices has even or odd parity and hence find whether the terms have a plus or a minus sign attached to them:
(a) $a_{11} a_{25} a_{33} a_{42} a_{54}.$
(b) $a_{11} a_{23} a_{34} a_{43} a_{52}.$
(c) $a_{11} a_{32} a_{24} a_{43} a_{55}.$
9. Determine the values of the indices p and q such that the following are terms in the determinant of order 4. In each case determine the number of inversions in the permutation of the

column indices and hence find the appropriate sign that should be attached to each term.

(a) $a_{13}a_{p4}a_{32}a_{2q}$.

(b) $a_{21}a_{3q}a_{p2}a_{43}$.

(c) $a_{3q}a_{p4}a_{13}a_{42}$.

10. If A is the general $n \times n$ matrix, determine the sign attached to the term

$$a_{1n}a_{2n-1}a_{3n-2} \cdots a_{n1}$$

that arises in $\det(A)$.

11. Consider the schematic given in Figure 5.2.1. Show that taking the product of the ele-

ments joined by each arrow and attaching the indicated sign to the result yields the six terms in the determinant of the 3×3 matrix $A = [a_{ij}]$.

In problems 12–14, evaluate the given determinant using the schematic given in Figure 5.2.1.

12. $\begin{vmatrix} 2 & 3 & -1 \\ 1 & 4 & 1 \\ 3 & 1 & 6 \end{vmatrix}$.

13. $\begin{vmatrix} 3 & 2 & 6 \\ 2 & 1 & -1 \\ -1 & 1 & 4 \end{vmatrix}$.

14. $\begin{vmatrix} 2 & 3 & 6 \\ 0 & 1 & 2 \\ 1 & 5 & 0 \end{vmatrix}$.

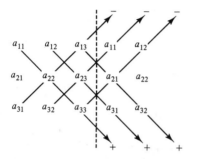

Figure 5.2.1 A schematic for obtaining the determinant of a 3×3 matrix $A = [a_{ij}]$.

5.3   PROPERTIES OF DETERMINANTS

Evaluating a determinant of order n using the definition given in the previous section is not very practical since the number of terms in the determinant grows as $n!$ (for example, a determinant of order 10 contains 3,628,800 terms). In the next two sections we develop better techniques for evaluating determinants. The following theorem suggests one way to proceed.

Theorem 5.3.1: If A is an $n \times n$ upper or lower triangular matrix, then

$$\det(A) = a_{11}a_{22}a_{33} \cdots a_{nn} = \prod_{i=1}^{n} a_{ii}. \tag{5.3.1}$$

PROOF   We use the definition of determinant to prove the result in the upper triangular case. From (5.2.2),

$$\det(A) = \sum \sigma(p_1, p_2, \ldots, p_n)a_{1p_1}a_{2p_2}a_{3p_3} \cdots a_{np_n}. \tag{5.3.2}$$

If A is upper triangular, then $a_{ij} = 0$ whenever $i > j$, and thus the only nonzero terms in (5.3.2) are those with $p_i \geq i$ for all i. Since all the p_i must be distinct, the only possibility is (start at the right-hand end and work toward the left)

$$p_i = i, \qquad 1 \leq i \leq n,$$

and so (5.3.2) reduces to the single term

$$\det(A) = \sigma(1, 2, \ldots, n) a_{11} a_{22} \cdots a_{nn}.$$

Since $\sigma(1, 2, \ldots, n) = 1$, it follows that

$$\det(A) = a_{11} a_{22} \cdots a_{nn}.$$

The result for lower triangular matrices now follows directly since $\det(A) = \det(A^T)$ and A^T is upper triangular if A is lower triangular. ∎

Example 5.3.1   According to the previous theorem,

$$\begin{vmatrix} 2 & 5 & -1 & 3 \\ 0 & -1 & 0 & 4 \\ 0 & 0 & 7 & 8 \\ 0 & 0 & 0 & 5 \end{vmatrix} = (2)(-1)(7)(5) = -70.$$

We now recall from Chapter 4 that any matrix can be reduced to row echelon form by a sequence of elementary row operations. In the particular case of an $n \times n$ matrix, any row echelon form will be upper triangular. The preceding theorem suggests that we should consider how elementary row operations performed on a matrix, A, alter the value of $\det(A)$.

Elementary Row Operations and Determinants

Let A be an $n \times n$ matrix.

P1. If B is a matrix obtained by interchanging two distinct rows of A, then

$$\det(A) = -\det(B).$$

P2. If B is the matrix obtained by dividing any row of A by a nonzero scalar k, then

$$\det(A) = k \det(B).$$

P3. If B is the matrix obtained by adding a multiple of any row of A to a different row of A, then

$$\det(A) = \det(B).$$

REMARK   The main use of P2 is that it enables us to factor a common multiple of any row out of the determinant. For example:

$$\begin{vmatrix} -5 & 20 \\ 3 & -2 \end{vmatrix} \overset{R_1 \to \frac{1}{5} R_1}{=} 5 \begin{vmatrix} -1 & 4 \\ 3 & -2 \end{vmatrix} = 5(2 - 12) = -50.$$

$$\underset{\det(A)}{\uparrow} \qquad \underset{k}{\uparrow} \quad \underset{\det(B)}{\uparrow}$$

The proofs of P1 and P3 are given at the end of the section, whereas the proof of P2 is left as an exercise.

We now illustrate how the preceding properties, together with Theorem 5.3.1, can be used to evaluate a determinant. The basic idea is the same as that for Gaussian elimination. We use elementary row operations to reduce the determinant to upper triangular form and then use Theorem 5.3.1 to evaluate the resulting determinant.

WARNING   When using the preceding properties to simplify a determinant, you must remember to take account of any change in the value of the determinant that arises from the operations that you have performed on it.

Example 5.3.2   Evaluate $\begin{vmatrix} 2 & -1 & 3 & 7 \\ 1 & -2 & 4 & 3 \\ 3 & 4 & 2 & -1 \\ 2 & -2 & 8 & -4 \end{vmatrix}$.

Solution

$$\begin{vmatrix} 2 & -1 & 3 & 7 \\ 1 & -2 & 4 & 3 \\ 3 & 4 & 2 & -1 \\ 2 & -2 & 8 & -4 \end{vmatrix} \begin{array}{c} R_1 \leftrightarrow R_2 \\ = \\ R_4 \to \frac{1}{2}R_4 \end{array} -2 \begin{vmatrix} 1 & -2 & 4 & 3 \\ 2 & -1 & 3 & 7 \\ 3 & 4 & 2 & -1 \\ 1 & -1 & 4 & -2 \end{vmatrix}$$

$$\begin{array}{c} R_2 \to R_2 - 2R_1 \\ R_3 \to R_3 - 3R_1 \\ = \\ R_4 \to R_4 - R_1 \end{array} -2 \begin{vmatrix} 1 & -2 & 4 & 3 \\ 0 & 3 & -5 & 1 \\ 0 & 10 & -10 & -10 \\ 0 & 1 & 0 & -5 \end{vmatrix} \begin{array}{c} R_2 \leftrightarrow R_4 \\ = \\ R_3 \to \frac{1}{10}R_3 \end{array} 20 \begin{vmatrix} 1 & -2 & 4 & 3 \\ 0 & 1 & 0 & -5 \\ 0 & 1 & -1 & -1 \\ 0 & 3 & -5 & 1 \end{vmatrix}$$

$$\begin{array}{c} R_3 \to R_3 - R_2 \\ = \\ R_4 \to R_4 - 3R_2 \end{array} 20 \begin{vmatrix} 1 & -2 & 4 & 3 \\ 0 & 1 & 0 & -5 \\ 0 & 0 & -1 & 4 \\ 0 & 0 & -5 & 16 \end{vmatrix} \begin{array}{c} R_4 \to R_4 - 5R_3 \\ = \end{array} 20 \begin{vmatrix} 1 & -2 & 4 & 3 \\ 0 & 1 & 0 & -5 \\ 0 & 0 & -1 & 4 \\ 0 & 0 & 0 & -4 \end{vmatrix}$$

$$= 80.$$

In the previous section we proved the following general property of determinants:

P4. If A is an $n \times n$ matrix, then $\det(A^T) = \det(A)$.

The main importance of this property is the implication that any results that hold for the rows of a determinant also hold for columns. In particular, elementary column operations can also be used to reduce a determinant to upper triangular form. Once more we must be careful to take account of any changes in the value of the determinant that result from the applied operations. We illustrate with an example.

Example 5.3.3   Evaluate $\begin{vmatrix} 6 & 10 & 3 & 4 \\ 3 & 6 & -1 & 2 \\ 9 & 20 & 5 & 4 \\ 15 & 34 & 3 & 8 \end{vmatrix}$.

Solution

$$\begin{vmatrix} 6 & 10 & 3 & 4 \\ 3 & 6 & -1 & 2 \\ 9 & 20 & 5 & 4 \\ 15 & 34 & 3 & 8 \end{vmatrix} \begin{matrix} C_1 \to \frac{1}{3}C_1 \\ C_2 \to \frac{1}{2}C_2 \\ = 3 \cdot 2 \cdot 2 \\ C_4 \to \frac{1}{2}C_4 \end{matrix} \begin{vmatrix} 2 & 5 & 3 & 2 \\ 1 & 3 & -1 & 1 \\ 3 & 10 & 5 & 2 \\ 5 & 17 & 3 & 4 \end{vmatrix}$$

$$\begin{matrix} R_1 \leftrightarrow R_2 \\ = \end{matrix} -12 \begin{vmatrix} 1 & 3 & -1 & 1 \\ 2 & 5 & 3 & 2 \\ 3 & 10 & 5 & 2 \\ 5 & 17 & 3 & 4 \end{vmatrix} \begin{matrix} R_2 \to R_2 - 2R_1 \\ R_3 \to R_3 - 3R_1 \\ = \\ R_4 \to R_4 - 5R_1 \end{matrix} -12 \begin{vmatrix} 1 & 3 & -1 & 1 \\ 0 & -1 & 5 & 0 \\ 0 & 1 & 8 & -1 \\ 0 & 2 & 8 & -1 \end{vmatrix}$$

$$\begin{matrix} R_3 \to R_3 + R_2 \\ = \\ R_4 \to R_4 + 2R_2 \end{matrix} -12 \begin{vmatrix} 1 & 3 & -1 & 1 \\ 0 & -1 & 5 & 0 \\ 0 & 0 & 13 & -1 \\ 0 & 0 & 18 & -1 \end{vmatrix}$$

$$\begin{matrix} C_3 \to C_3 + 18C_4 \\ = \end{matrix} -12 \begin{vmatrix} 1 & 3 & 17 & 1 \\ 0 & -1 & 5 & 0 \\ 0 & 0 & -5 & -1 \\ 0 & 0 & 0 & -1 \end{vmatrix} = -12(1)(-1)(-5)(-1) = 60.$$

FURTHER PROPERTIES OF DETERMINANTS

In addition to elementary row and column operations, the following properties can be useful in evaluating a determinant.

Let A and B be $n \times n$ matrices.

P5. Let $\mathbf{a}_1, \mathbf{a}_2, \ldots, \mathbf{a}_n$ denote the row vectors of A. If the ith row vector of A is the sum of two row vectors—for example, $\mathbf{a}_i = \mathbf{b}_i + \mathbf{c}_i$—then

$$\det(A) = \det(B) + \det(C),$$

where

$$B = \begin{bmatrix} \mathbf{a}_1 \\ \vdots \\ \mathbf{a}_{i-1} \\ \mathbf{b}_i \\ \mathbf{a}_{i+1} \\ \vdots \\ \mathbf{a}_n \end{bmatrix} \quad \text{and} \quad C = \begin{bmatrix} \mathbf{a}_1 \\ \vdots \\ \mathbf{a}_{i-1} \\ \mathbf{c}_i \\ \mathbf{a}_{i+1} \\ \vdots \\ \mathbf{a}_n \end{bmatrix}.$$

The corresponding property is also true for columns.

P6. If A has a row (or column) of zeros, then $\det(A) = 0$.

P7. If two rows (or columns) of A are the same, then $\det(A) = 0$.

P8. $\det(AB) = \det(A)\det(B)$.

The proofs of P5 and P7 are given at the end of the section. Property P6 follows directly from the definition of determinant. The proof of P8 is more difficult and, except for some special cases treated in the exercises, is omitted (see, for example, I. N. Herstein, *Topics in Algebra*, John Wiley, 1975).

The property that usually gives the most difficulty is P5. We explicitly indicate its use in the following example.

Example 5.3.4 Use property P5 to express the following determinant as a sum of four determinants:

$$\begin{vmatrix} a_1 + b_1 & c_1 + d_1 \\ a_2 + b_2 & c_2 + d_2 \end{vmatrix}.$$

Solution Applying P5 to row 1 yields

$$\begin{vmatrix} a_1 + b_1 & c_1 + d_1 \\ a_2 + b_2 & c_2 + d_2 \end{vmatrix} = \begin{vmatrix} a_1 & c_1 \\ a_2 + b_2 & c_2 + d_2 \end{vmatrix} + \begin{vmatrix} b_1 & d_1 \\ a_2 + b_2 & c_2 + d_2 \end{vmatrix}.$$

We now apply P5 to row 2 in both of the determinants on the right-hand side to obtain

$$\begin{vmatrix} a_1 + b_1 & c_1 + d_1 \\ a_2 + b_2 & c_2 + d_2 \end{vmatrix} = \begin{vmatrix} a_1 & c_1 \\ a_2 & c_2 \end{vmatrix} + \begin{vmatrix} a_1 & c_1 \\ b_2 & d_2 \end{vmatrix} + \begin{vmatrix} b_1 & d_1 \\ a_2 & c_2 \end{vmatrix} + \begin{vmatrix} b_1 & d_1 \\ b_2 & d_2 \end{vmatrix}.$$

Notice that we could also have applied P5 to the columns of the given determinant.

Example 5.3.5   Evaluate:

$$
\text{(a)} \quad \begin{vmatrix} 1 & 2 & -3 & 1 \\ -2 & 4 & 6 & 2 \\ -3 & -6 & 9 & 3 \\ 2 & 11 & -6 & 4 \end{vmatrix}. \qquad \text{(b)} \quad \begin{vmatrix} 2-4x & -4 & 2 \\ 5+3x & 3 & -3 \\ 1-2x & -2 & 1 \end{vmatrix}.
$$

Solution

$$
\text{(a)} \quad \begin{vmatrix} 1 & 2 & -3 & 1 \\ -2 & 4 & 6 & 2 \\ -3 & -6 & 9 & 3 \\ 2 & 11 & -6 & 4 \end{vmatrix} = -3 \begin{vmatrix} 1 & 2 & 1 & 1 \\ -2 & 4 & -2 & 2 \\ -3 & -6 & -3 & 3 \\ 2 & 11 & 2 & 4 \end{vmatrix} = 0,
$$

since column 1 = column 3.

$$
\text{(b)} \quad \begin{vmatrix} 2-4x & -4 & 2 \\ 5+3x & 3 & -3 \\ 1-2x & -2 & 1 \end{vmatrix} \quad \underset{\substack{\uparrow \\ \text{P5 applied to column 1}}}{=} \quad \begin{vmatrix} 2 & -4 & 2 \\ 5 & 3 & -3 \\ 1 & -2 & 1 \end{vmatrix} + \begin{vmatrix} -4x & -4 & 2 \\ 3x & 3 & -3 \\ -2x & -2 & 1 \end{vmatrix}
$$

$$
= 2 \begin{vmatrix} 1 & -2 & 1 \\ 5 & 3 & -3 \\ 1 & -2 & 1 \end{vmatrix} + x \begin{vmatrix} -4 & -4 & 2 \\ 3 & 3 & -3 \\ -2 & -2 & 1 \end{vmatrix}
$$

$$
= 0,
$$

since $R_1 = R_3$ in the first determinant and $C_1 = C_2$ in the second determinant.

Example 5.3.6

If $A = \begin{bmatrix} \sin\varphi & \cos\varphi \\ -\cos\varphi & \sin\varphi \end{bmatrix}$ and $B = \begin{bmatrix} \cos\theta & -\sin\theta \\ \sin\theta & \cos\theta \end{bmatrix}$, show that $\det(AB) = 1$.

Solution   Using P8, we have

$$
\det(AB) = \det(A)\det(B) = (\sin^2\varphi + \cos^2\varphi)(\cos^2\theta + \sin^2\theta) = 1.
$$

Example 5.3.7   Find all x satisfying $\begin{vmatrix} x^2 & x & 1 \\ 1 & 1 & 1 \\ 4 & 2 & 1 \end{vmatrix} = 0.$

Solution   If we expanded the determinant, then we would have a quadratic equation in x.   Thus there are at most two distinct values of x that satisfy the equation. By inspection the determinant vanishes when $x = 1$ (since then $R_1 = R_2$) and when $x = 2$ (since then $R_1 = R_3$).   Consequently the two values of x satisfying the given equation are $x = 1, 2$.

PROOFS OF THE PROPERTIES OF DETERMINANTS

We now prove several of the properties of determinants.

PROOF OF P1   Let B be the matrix obtained by interchanging row r with row s in A. Then the elements of B are related to those of A as follows:

$$b_{ij} = \begin{cases} a_{ij} & \text{if } i \neq r, s, \\ a_{sj} & \text{if } i = r, \\ a_{rj} & \text{if } i = s. \end{cases}$$

Thus, from Definition 5.2.1,

$$\det(B) = \sum \sigma(p_1, p_2, \ldots, p_n) b_{1p_1} b_{2p_2} b_{3p_3} \cdots b_{np_n}$$

$$= \sum \sigma(p_1, p_2, \ldots, p_r, \ldots, p_s, \ldots, p_n) a_{1p_1} a_{2p_2} \cdots a_{sp_r} \cdots a_{rp_s} \cdots a_{np_n}.$$

Interchanging p_r and p_s in $\sigma(p_1, p_2, \ldots, p_r, \ldots, p_s, \ldots, p_n)$ and recalling from Lemma 5.1.1 that such an interchange has the effect of changing the parity, we obtain

$$\det(B) = -\sum \sigma(p_1, p_2, \ldots, p_s, \ldots, p_r, \ldots, p_n) a_{1p_1} a_{2p_2} \cdots a_{rp_s} \cdots a_{sp_r} \cdots a_{np_n},$$

where we have also rearranged the terms so that the row suffixes are in their natural increasing order. The sum on the right-hand side of this equation is just $\det(A)$, so that

$$\det(B) = -\det(A). \qquad \blacksquare$$

We prove properties P5 and P7 next, since they simplify the proof of P3.

PROOF OF P5   The elements of A are

$$a_{kj} = \begin{cases} a_{kj}, & \text{if } k \neq i, \\ b_{ij} + c_{ij}, & \text{if } k = i. \end{cases}$$

Thus, from Definition 5.2.1,

$$\det(A) = \sum \sigma(p_1, p_2, \ldots, p_n) a_{1p_1} a_{2p_2} a_{3p_3} \cdots a_{np_n}$$

$$= \sum \sigma(p_1, p_2, \ldots, p_n) a_{1p_1} a_{2p_2} \cdots a_{i-1p_{i-1}} (b_{ip_i} + c_{ip_i}) a_{i+1p_{i+1}} \cdots a_{np_n}$$

$$= \sum \sigma(p_1, p_2, \ldots, p_n) a_{1p_1} a_{2p_2} \cdots a_{i-1p_{i-1}} b_{ip_i} a_{i+1p_{i+1}} \cdots a_{np_n}$$

$$+ \sum \sigma(p_1, p_2, \ldots, p_n) a_{1p_1} a_{2p_2} \cdots a_{i-1p_{i-1}} c_{ip_i} a_{i+1p_{i+1}} \cdots a_{np_n}$$

$$= \det(B) + \det(C). \qquad \blacksquare$$

PROOF OF P7   Suppose rows i and j in A are the same.   Then if we interchange these rows, the matrix—and hence its determinant—is unaltered.   However, according to P1, the determinant of the resulting matrix is $-\det(A)$.   Thus we have

$$\det(A) = -\det(A),$$

which implies that

$$\det(A) = 0.$$ ∎

PROOF OF P3   Let $A = [\mathbf{a}_1, \mathbf{a}_2, \ldots, \mathbf{a}_n]^T$, and let B be the matrix obtained from A when k times row j of A is added to row i of A.   Then $B = [\mathbf{a}_1, \mathbf{a}_2, \ldots, \mathbf{a}_i + k\mathbf{a}_j, \ldots, \mathbf{a}_n]^T$, so that, using P5,

$$\det(B) = \det[\mathbf{a}_1, \mathbf{a}_2, \ldots, \underset{\underset{\text{ith row vector of }B}{\uparrow}}{\mathbf{a}_i + k\mathbf{a}_j}, \ldots, \mathbf{a}_n]^T$$

$$= \det[\mathbf{a}_1, \mathbf{a}_2, \ldots, \mathbf{a}_n]^T + \det[\mathbf{a}_1, \mathbf{a}_2, \ldots, \underset{\underset{\text{ith row vector}}{\uparrow}}{k\mathbf{a}_j}, \ldots, \mathbf{a}_n]^T.$$

By P2 we can factor out k from row i of the second determinant on the right-hand side.   If we do this it follows that row i and row j of the resulting determinant are the same and so, from P7, the value of the second determinant is zero.   Thus,

$$\det(B) = \det[\mathbf{a}_1, \mathbf{a}_2, \ldots, \mathbf{a}_n]^T = \det(A),$$

as required. ∎

EXERCISES 5.3

In problems 1–12 reduce the given determinant to upper triangular form and then evaluate.

1. $\begin{vmatrix} 1 & 2 & 3 \\ 2 & 6 & 4 \\ 3 & -5 & 2 \end{vmatrix}.$

2. $\begin{vmatrix} 2 & -1 & 4 \\ 3 & 2 & 1 \\ -2 & 1 & 4 \end{vmatrix}.$

3. $\begin{vmatrix} 2 & 1 & 3 \\ -1 & 2 & 6 \\ 4 & 1 & 12 \end{vmatrix}.$

4. $\begin{vmatrix} 0 & 1 & -2 \\ -1 & 0 & 3 \\ 2 & -3 & 0 \end{vmatrix}.$

5. $\begin{vmatrix} 3 & 7 & 1 \\ 5 & 9 & -6 \\ 2 & 1 & 3 \end{vmatrix}.$

6. $\begin{vmatrix} 1 & -1 & 2 & 4 \\ 3 & 1 & 2 & 4 \\ -1 & 1 & 3 & 2 \\ 2 & 1 & 4 & 2 \end{vmatrix}.$

7. $\begin{vmatrix} 2 & 32 & 1 & 4 \\ 26 & 104 & 26 & -13 \\ 2 & 56 & 2 & 7 \\ 1 & 40 & 1 & 5 \end{vmatrix}.$

8. $\begin{vmatrix} 0 & 1 & -1 & 1 \\ -1 & 0 & 1 & 1 \\ 1 & -1 & 0 & 1 \\ -1 & -1 & -1 & 0 \end{vmatrix}.$

9. $\begin{vmatrix} 2 & 1 & 3 & 5 \\ 3 & 0 & 1 & 2 \\ 4 & 1 & 4 & 3 \\ 5 & 2 & 5 & 3 \end{vmatrix}.$

10. $\begin{vmatrix} 2 & -1 & 3 & 4 \\ 7 & 1 & 2 & 3 \\ -2 & 4 & 8 & 6 \\ 6 & -6 & 18 & -24 \end{vmatrix}.$

11. $\begin{vmatrix} 7 & -1 & 3 & 4 \\ 14 & 2 & 4 & 6 \\ 21 & 1 & 3 & 4 \\ -7 & 4 & 5 & 8 \end{vmatrix}.$

12. $\begin{vmatrix} 3 & 7 & 1 & 2 & 3 \\ 1 & 1 & -1 & 0 & 1 \\ 4 & 8 & -1 & 6 & 6 \\ 3 & 7 & 0 & 9 & 4 \\ 8 & 16 & -1 & 8 & 12 \end{vmatrix}.$

13. If $A = \begin{bmatrix} 1 & -1 & 2 \\ 3 & 1 & 4 \\ 0 & 1 & 3 \end{bmatrix}$, find $\det(A)$, and use properties of determinants to find $\det(A^T)$ and $\det(-2A)$.

14. If $A = \begin{bmatrix} 1 & -1 \\ 2 & 3 \end{bmatrix}$ and $B = \begin{bmatrix} 1 & 2 \\ -2 & 4 \end{bmatrix}$, evaluate $\det(AB)$ and verify P8.

15. If $A = \begin{bmatrix} \cosh x & \sinh x \\ \sinh x & \cosh x \end{bmatrix}$ and

$B = \begin{bmatrix} \cosh y & \sinh y \\ \sinh y & \cosh y \end{bmatrix}$, evaluate $\det(AB)$.

16. If $A = \begin{bmatrix} a_{11} & a_{12} \\ a_{21} & a_{22} \end{bmatrix}$ and $B = \begin{bmatrix} b_{11} & b_{12} \\ b_{21} & b_{22} \end{bmatrix}$, prove that $\det(AB) = \det(A)\det(B)$.

In problems 17–19 use properties of determinants to show that $\det(A) = 0$ for the given matrix A.

17. $A = \begin{bmatrix} 3 & 2 & 1 \\ 6 & 4 & -1 \\ 9 & 6 & 2 \end{bmatrix}.$

18. $A = \begin{bmatrix} 1 & -3 & 1 \\ 2 & -1 & 7 \\ 3 & 1 & 13 \end{bmatrix}.$

19. $A = \begin{bmatrix} 1+3a & 1 & 3 \\ 1+2a & 1 & 2 \\ 2 & 2 & 0 \end{bmatrix}.$

20. Without expanding the determinant, determine all values of x for which $\det(A) = 0$ if

$$A = \begin{bmatrix} 1 & -1 & x \\ 2 & 1 & x^2 \\ 4 & -1 & x^3 \end{bmatrix}.$$

21. Use *only* properties P5, P1, and P2 to show that

$$\begin{vmatrix} \alpha x - \beta y & \beta x - \alpha y \\ \beta x + \alpha y & \alpha x + \beta y \end{vmatrix} = (x^2 + y^2) \begin{vmatrix} \alpha & \beta \\ \beta & \alpha \end{vmatrix}.$$

22. Use properties P5, P1, and P2 to find the value of $\alpha\beta\gamma$ such that

$$\begin{vmatrix} a_1 + \beta b_1 & b_1 + \gamma c_1 & c_1 + \alpha a_1 \\ a_2 + \beta b_2 & b_2 + \gamma c_2 & c_2 + \alpha a_2 \\ a_3 + \beta b_3 & b_3 + \gamma c_3 & c_3 + \alpha a_3 \end{vmatrix} = 0,$$

for all values of a_i, b_i, c_i.

23. If A is a nonsingular $n \times n$ matrix, prove that $\det(A) \neq 0$ and

$$\det(A^{-1}) = \frac{1}{\det(A)}.$$

24. An $n \times n$ matrix A that satisfies $A^T = A^{-1}$ is called an **orthogonal matrix**. Prove that if A is orthogonal, then $\det(A) = \pm 1$.

25. **(a)** Use the definition of determinant to prove that if A is an $n \times n$ lower triangular matrix, then

$$\det(A) = a_{11} a_{22} a_{33} \cdots a_{nn} = \prod_{i=1}^{n} a_{ii}.$$

(b) Evaluate the following determinant by first

reducing it to lower triangular form and then using the result from (a):

$$\begin{vmatrix} 2 & -1 & 3 & 5 \\ 1 & 2 & 2 & 1 \\ 3 & 0 & 1 & 4 \\ 1 & 2 & 0 & 1 \end{vmatrix}.$$

26. Prove property P2 of determinants.

27. Prove property P8 in the particular case when B is an $n \times n$ *diagonal* matrix.

28. Let A and B be $n \times n$ upper triangular matrices. From Theorem 3.3.1 the product AB is also an upper triangular matrix.
(a) Prove that the diagonal elements of AB are just the products of the corresponding diagonal elements of A and B; that is, $(AB)_{ii} = a_{ii}b_{ii}$, $1 \le i \le n$.
(b) Use your result from (a) to prove property P8 for *upper* triangular matrices A and B.

29. Show that $\begin{vmatrix} x & y & 1 \\ x_1 & y_1 & 1 \\ x_2 & y_2 & 1 \end{vmatrix} = 0$ represents the equation of the straight line through the distinct points (x_1, y_1) and (x_2, y_2).

30. Without expanding the determinant, show that

$$\begin{vmatrix} 1 & x & x^2 \\ 1 & y & y^2 \\ 1 & z & z^2 \end{vmatrix} = (y - z)(z - x)(x - y).$$

31. If A is an $n \times n$ *skew-symmetric* matrix with n odd, prove that $\det(A) = 0$.

32. Let $A = [\mathbf{a}_1, \mathbf{a}_2, \dots, \mathbf{a}_n]$ be an $n \times n$ matrix, and let $\mathbf{b} = c_1\mathbf{a}_1 + c_2\mathbf{a}_2 + \dots + c_n\mathbf{a}_n$, where $c_1, c_2, \dots, c_n$ are constants. If B_k denotes the matrix obtained from A by replacing its kth column vector by $\mathbf{b}$, prove that

$$\det(B_k) = c_k\det(A), \qquad k = 1, 2, \dots, n.$$

5.4   MINORS AND COFACTORS

We now obtain an alternative method for evaluating a determinant. The basic idea is that we can reduce a determinant of order n to a sum of determinants of order $(n - 1)$. Continuing in this manner it is possible to express any given determinant as the sum of $(n!)/2$ determinants of order 2. This method is the one most frequently used to evaluate a given determinant by hand, although the triangulation procedure of the previous section involves less work in general. We first require two preliminary definitions.

> **Definition 5.4.1:**   Let A be an $n \times n$ matrix. The **minor**, M_{ij}, of the element a_{ij} is the determinant of the matrix obtained by deleting the ith row vector and jth column vector of A.

REMARK   Notice that M_{ij} is a determinant of order $(n - 1)$ if A is an $n \times n$ matrix.

Example 5.4.1   If $A = \begin{bmatrix} a_{11} & a_{12} & a_{13} \\ a_{21} & a_{22} & a_{23} \\ a_{31} & a_{32} & a_{33} \end{bmatrix}$, then, for example, $M_{23} = \begin{vmatrix} a_{11} & a_{12} \\ a_{31} & a_{32} \end{vmatrix}$.

Example 5.4.2   Determine all minors of $A = \begin{bmatrix} 2 & 1 & 3 \\ -1 & 4 & -2 \\ 3 & 1 & 5 \end{bmatrix}$.

Solution Using Definition 5.4.1, we have

$$M_{11} = \begin{vmatrix} 4 & -2 \\ 1 & 5 \end{vmatrix} = 22, \qquad M_{12} = \begin{vmatrix} -1 & -2 \\ 3 & 5 \end{vmatrix} = 1, \qquad M_{13} = \begin{vmatrix} -1 & 4 \\ 3 & 1 \end{vmatrix} = -13,$$

$$M_{21} = \begin{vmatrix} 1 & 3 \\ 1 & 5 \end{vmatrix} = 2, \qquad M_{22} = \begin{vmatrix} 2 & 3 \\ 3 & 5 \end{vmatrix} = 1, \qquad M_{23} = \begin{vmatrix} 2 & 1 \\ 3 & 1 \end{vmatrix} = -1,$$

$$M_{31} = \begin{vmatrix} 1 & 3 \\ 4 & -2 \end{vmatrix} = -14, \qquad M_{32} = \begin{vmatrix} 2 & 3 \\ -1 & -2 \end{vmatrix} = -1, \qquad M_{33} = \begin{vmatrix} 2 & 1 \\ -1 & 4 \end{vmatrix} = 9.$$

Definition 5.4.2: Let A be an $n \times n$ matrix. The **cofactor**, A_{ij}, of the element a_{ij} is defined by

$$A_{ij} = (-1)^{i+j} M_{ij},$$

where M_{ij} is the minor of a_{ij}.

The appropriate sign in the cofactor A_{ij} is easy to remember, since it alternates in the following manner:

$$\begin{vmatrix} + & - & + & - & + & \cdots \\ - & + & - & + & - & \cdots \\ + & - & + & - & + & \cdots \\ \vdots & \vdots & \vdots & \vdots & \vdots & \end{vmatrix}.$$

Example 5.4.3 Determine the cofactors of the matrix $A = \begin{bmatrix} 2 & 1 & 3 \\ -1 & 4 & -2 \\ 3 & 1 & 5 \end{bmatrix}$.

Solution We have already determined the minors of A in the previous example. It follows that

$$
\begin{aligned}
A_{11} &= +M_{11} = 22, & A_{12} &= -M_{12} = -1, & A_{13} &= +M_{13} = -13, \\
A_{21} &= -M_{21} = -2, & A_{22} &= +M_{22} = 1, & A_{23} &= -M_{23} = 1, \\
A_{31} &= +M_{31} = -14, & A_{32} &= -M_{32} = 1, & A_{33} &= +M_{33} = 9.
\end{aligned}
$$

Example 5.4.4 If $A = \begin{bmatrix} a_{11} & a_{12} \\ a_{21} & a_{22} \end{bmatrix}$, show that $\det(A) = a_{11}A_{11} + a_{12}A_{12}$.

Solution In this case,

$$A_{11} = +|a_{22}| = a_{22}, \qquad A_{12} = -|a_{21}| = -a_{21},$$

and hence

$$a_{11}A_{11} + a_{12}A_{12} = a_{11}a_{22} - a_{12}a_{21} = \det(A).$$

The preceding example is a special case of the following important theorem.

Theorem 5.4.1 (Cofactor Expansion Theorem):   Let A be an $n \times n$ matrix. If we multiply the elements in any row (or column) of A by their cofactors, then the sum of the resulting products is det(A). Thus

1. If we expand along row i,

$$\det(A) = a_{i1}A_{i1} + a_{i2}A_{i2} + \cdots + a_{in}A_{in} = \sum_{j=1}^{n} a_{ij}A_{ij}.$$

2. If we expand along column j,

$$\det(A) = a_{1j}A_{1j} + a_{2j}A_{2j} + \cdots + a_{nj}A_{nj} = \sum_{i=1}^{n} a_{ij}A_{ij}.$$

PROOF   It follows from the definition of determinant that det(A) can be written in the form

$$\det(A) = a_{i1}\hat{A}_{i1} + a_{i2}\hat{A}_{i2} + \cdots + a_{in}\hat{A}_{in}, \tag{5.4.1}$$

where the coefficients $\hat{A}_{ij}$ contain no elements from row i or column j. We must show that

$$\hat{A}_{ij} = A_{ij},$$

the cofactor of a_{ij}.

Consider first a_{11}. From Definition 5.2.1 (page 151) the terms in det(A) that contain a_{11} are given by

$$a_{11} \sum \sigma(1, p_2, p_3, \ldots, p_n) a_{2p_2} a_{3p_3} \cdots a_{np_n},$$

where the summation is over the $(n-1)!$ distinct permutations of the integers 2, 3, $\ldots$, n. Thus

$$\hat{A}_{11} = \sum \sigma(1, p_2, p_3, \ldots, p_n) a_{2p_2} a_{3p_3} \cdots a_{np_n}.$$

However, this summation is just the minor M_{11} and, since $A_{11} = M_{11}$, we have shown that the coefficient of a_{11} in the expansion of det(A) is indeed the cofactor A_{11}. Now consider the element a_{ij}. By successively interchanging adjacent rows and columns in A, we can move a_{ij} into the (11) position *without altering the relative positions of the other rows and columns of A.* We let A' denote the resulting matrix. Obtaining A' from A requires $(i-1)$ row interchanges and $(j-1)$ column interchanges. Thus the total number of interchanges required to obtain A' from A is $(i + j - 2)$, and hence

$$\det(A) = (-1)^{i+j-2}\det(A') = (-1)^{i+j}\det(A').$$

Now for the key point: The coefficient of a_{ij} in det(A) must be $(-1)^{i+j}$ times the coefficient of a_{ij} in det(A'). But a_{ij} occurs in the (11) position of A', and hence—as we showed previously—its coefficient in $\det(A')$ is M'_{11}. Since the relative positions of the remaining rows in A have not been altered, it follows that $M'_{11} = M_{ij}$, and hence the coefficient of a_{ij} in $\det(A')$ is M_{ij}. Consequently, the coefficient of a_{ij} in $\det(A)$ is $(-1)^{i+j}M_{ij} = A_{ij}$. Applying this result to the elements $a_{i1}, a_{i2}, \ldots, a_{in}$ and comparing with (5.4.1) yields

$$\hat{A}_{ij} = A_{ij}, \qquad j = 1, 2, \ldots, n,$$

and hence the theorem is established for expansion along a row. The result for expansion along a column follows directly since $\det(A^T) = \det(A)$. ∎

Example 5.4.5 Evaluate $\begin{vmatrix} 2 & 3 & 4 \\ 1 & -1 & 1 \\ 6 & 3 & 0 \end{vmatrix}$ along (a) row 1, (b) column 3.

Solution

(a) $\begin{vmatrix} 2 & 3 & 4 \\ 1 & -1 & 1 \\ 6 & 3 & 0 \end{vmatrix} = 2\begin{vmatrix} -1 & 1 \\ 3 & 0 \end{vmatrix} - 3\begin{vmatrix} 1 & 1 \\ 6 & 0 \end{vmatrix} + 4\begin{vmatrix} 1 & -1 \\ 6 & 3 \end{vmatrix} = -6 + 18 + 36 = 48.$

(b) $\begin{vmatrix} 2 & 3 & 4 \\ 1 & -1 & 1 \\ 6 & 3 & 0 \end{vmatrix} = 4\begin{vmatrix} 1 & -1 \\ 6 & 3 \end{vmatrix} - 1\begin{vmatrix} 2 & 3 \\ 6 & 3 \end{vmatrix} + 0 = 36 + 12 + 0 = 48.$

Notice that (b) was easier than (a) because of the zero in column 3. Whenever you use the cofactor expansion method to evaluate a determinant, always choose the row or column carefully in order to minimize the amount of work.

Example 5.4.6 Evaluate $\begin{vmatrix} 3 & 0 & -1 & 0 \\ 0 & 5 & 8 & 2 \\ 2 & 7 & 5 & 4 \\ 1 & 6 & 7 & 0 \end{vmatrix}$.

Solution In this case it is easiest to use either row 1 or column 4. We choose row 1

$\begin{vmatrix} 3 & 0 & -1 & 0 \\ 0 & 5 & 8 & 2 \\ 2 & 7 & 5 & 4 \\ 1 & 6 & 7 & 0 \end{vmatrix} = 3\begin{vmatrix} 5 & 8 & 2 \\ 7 & 5 & 4 \\ 6 & 7 & 0 \end{vmatrix} + (-1)\begin{vmatrix} 0 & 5 & 2 \\ 2 & 7 & 4 \\ 1 & 6 & 0 \end{vmatrix}$

$= 3[2(49 - 30) - 4(35 - 48) + 0] - [0 - 2(0 - 12) + (20 - 14)]$

$= 240.$

We now have two computational methods for evaluating determinants, namely, the triangulation technique of the previous section and the cofactor expansion theorem. In evaluating a given determinant it is usually most efficient (and least error prone) to use a combination of the two techniques. More specifically, we use the properties of determinants to set all except one element in a row or column equal to zero and then use the cofactor expansion theorem on the corresponding row or column. We illustrate with an example.

Example 5.4.7   Evaluate $\begin{vmatrix} 1 & 3 & -1 & 2 \\ 1 & 4 & 1 & 3 \\ -1 & 2 & 1 & 4 \\ 2 & 1 & 8 & 6 \end{vmatrix}$.

Solution

$$\begin{vmatrix} 1 & 3 & -1 & 2 \\ 1 & 4 & 1 & 3 \\ -1 & 2 & 1 & 4 \\ 2 & 1 & 8 & 6 \end{vmatrix} \begin{matrix} R_2 \to R_2 - R_1 \\ R_3 \to R_3 + R_1 \\ = \\ R_4 \to R_4 - 2R_1 \end{matrix} \begin{vmatrix} 1 & 3 & -1 & 2 \\ 0 & 1 & 2 & 1 \\ 0 & 5 & 0 & 6 \\ 0 & -5 & 10 & 2 \end{vmatrix}$$

$$= \begin{vmatrix} 1 & 2 & 1 \\ 5 & 0 & 6 \\ -5 & 10 & 2 \end{vmatrix} \begin{matrix} R_3 \to R_3 - 5R_1 \\ = \end{matrix} \begin{vmatrix} 1 & 2 & 1 \\ 5 & 0 & 6 \\ -10 & 0 & -3 \end{vmatrix} = -90.$$

Cofactor expansion along column 1 | Cofactor expansion along column 2

Finally we prove a result that will be required in the next section. In order to motivate this result we consider a particular example.

Example 5.4.8   If $A = \begin{bmatrix} a_{11} & a_{12} \\ a_{21} & a_{22} \end{bmatrix}$, show that $a_{11}A_{21} + a_{12}A_{22} = 0$.

Solution   We have $A_{21} = -a_{12}$, $A_{22} = a_{11}$, so that

$$a_{11}A_{21} + a_{12}A_{22} = a_{11}(-a_{12}) + a_{12}a_{11} = 0,$$

as required.

The previous example shows that, for a 2×2 matrix, if we multiply the elements of row 1 by the cofactors of row 2 and sum the results, then the value of the sum is zero. The following theorem generalizes this result to an arbitrary $n \times n$ matrix.

Theorem 5.4.2:   If the elements in the ith row (or column) of an $n \times n$ matrix A are multiplied by the cofactors of a different row (or column), then the sum of the resulting products is zero. That is,

1. If we use the elements of row i and the cofactors of row j then

$$\sum_{k=1}^{n} a_{ik}A_{jk} = 0, \qquad i \neq j. \tag{5.4.2}$$

2. If we use the elements of column i and the cofactors of column j then

$$\sum_{k=1}^{n} a_{ki}A_{kj} = 0, \qquad i \neq j. \tag{5.4.3}$$

PROOF   We prove (5.4.2). Adding row i to row $j(i \neq j)$ and expanding the resulting determinant along row j yields

$$\det(A) = \sum_{k=1}^{n} (a_{jk} + a_{ik})A_{jk} = \sum_{k=1}^{n} a_{jk}A_{jk} + \sum_{k=1}^{n} a_{ik}A_{jk},$$

that is,

$$\det(A) = \det(A) + \sum_{k=1}^{n} a_{ik}A_{jk},$$

which implies that

$$\sum_{k=1}^{n} a_{ik}A_{jk} = 0, \qquad i \neq j.$$

The result (5.4.3) can be proved similarly.   ■

REMARK   We can combine the previous result with that of Theorem 5.4.1 to obtain

$$\sum_{k=1}^{n} a_{ik}A_{jk} = \delta_{ij}\det(A), \qquad \sum_{k=1}^{n} a_{ki}A_{kj} = \delta_{ij}\det(A) \tag{5.4.4}$$

where δ_{ij} is the Kronecker delta.

EXERCISES 5.4

In problems 1–3 determine all minors and cofactors of the given matrix.

1. $A = \begin{bmatrix} 1 & -3 \\ 2 & 4 \end{bmatrix}$.

2. $A = \begin{bmatrix} 1 & -1 & 2 \\ 3 & -1 & 4 \\ 2 & 1 & 5 \end{bmatrix}$.

3. $A = \begin{bmatrix} 2 & 10 & 3 \\ 0 & -1 & 0 \\ 4 & 1 & 5 \end{bmatrix}$.

4. If $A = \begin{bmatrix} 1 & 3 & -1 & 2 \\ 3 & 4 & 1 & 2 \\ 7 & 1 & 4 & 6 \\ 5 & 0 & 1 & 2 \end{bmatrix}$, determine the minors

M_{12}, M_{31}, M_{23}, and M_{42} and the corresponding cofactors.

In problems 5–10 use the cofactor expansion theorem to evaluate the given determinant along the specified row or column.

5. $\begin{vmatrix} 1 & -2 \\ 1 & 3 \end{vmatrix}$, row 1.

6. $\begin{vmatrix} -1 & 2 & 3 \\ 1 & 4 & -2 \\ 3 & 1 & 4 \end{vmatrix}$, column 3.

7. $\begin{vmatrix} 2 & 1 & -4 \\ 7 & 1 & 3 \\ 1 & 5 & -2 \end{vmatrix}$, row 2.

8. $\begin{vmatrix} 3 & 1 & 4 \\ 7 & 1 & 2 \\ 2 & 3 & -5 \end{vmatrix}$, column 1.

9. $\begin{vmatrix} 0 & 2 & -3 \\ -2 & 0 & 5 \\ 3 & -5 & 0 \end{vmatrix}$, row 3.

10. $\begin{vmatrix} 1 & -2 & 3 & 0 \\ 4 & 0 & 7 & -2 \\ 0 & 1 & 3 & 4 \\ 1 & 5 & -2 & 0 \end{vmatrix}$, column 4.

In problems 11–20 evaluate the given determinant using the techniques of this section.

11. $\begin{vmatrix} 1 & 0 & -2 \\ 3 & 1 & -1 \\ 7 & 2 & 5 \end{vmatrix}$.

12. $\begin{vmatrix} -1 & 2 & 3 \\ 0 & 1 & 4 \\ 2 & -1 & 3 \end{vmatrix}$.

13. $\begin{vmatrix} 2 & -1 & 3 \\ 5 & 2 & 1 \\ 3 & -3 & 7 \end{vmatrix}$.

14. $\begin{vmatrix} 0 & -2 & 1 \\ 2 & 0 & -3 \\ -1 & 3 & 0 \end{vmatrix}$.

15. $\begin{vmatrix} 1 & 0 & -1 & 0 \\ 0 & 1 & 0 & -1 \\ -1 & 0 & -1 & 0 \\ 0 & 1 & 0 & 1 \end{vmatrix}$.

16. $\begin{vmatrix} 2 & -1 & 3 & 1 \\ 1 & 4 & -2 & 3 \\ 0 & 2 & -1 & 0 \\ 1 & 3 & -2 & 4 \end{vmatrix}$.

17. $\begin{vmatrix} 3 & 5 & 2 & 6 \\ 2 & 3 & 5 & -5 \\ 7 & 5 & -3 & -16 \\ 9 & -6 & 27 & -12 \end{vmatrix}$.

18. $\begin{vmatrix} 2 & -7 & 4 & 3 \\ 5 & 5 & -3 & 7 \\ 6 & 2 & 6 & 3 \\ 4 & 2 & -4 & 5 \end{vmatrix}$.

19. $\begin{vmatrix} 2 & 0 & -1 & 3 & 0 \\ 0 & 3 & 0 & 1 & 2 \\ 0 & 1 & 3 & 0 & 4 \\ 1 & 0 & 1 & -1 & 0 \\ 3 & 0 & 2 & 0 & 5 \end{vmatrix}$.

20. $\begin{vmatrix} 3 & -1 & 4 & 2 & 1 \\ -1 & 2 & 3 & 5 & 2 \\ 1 & -1 & 0 & 3 & 4 \\ -2 & 1 & 4 & 4 & 0 \\ 5 & -8 & 3 & 1 & -1 \end{vmatrix}$.

21. If $A = \begin{bmatrix} 0 & x & y & z \\ -x & 0 & 1 & -1 \\ -y & -1 & 0 & 1 \\ -z & 1 & -1 & 0 \end{bmatrix}$, show that

$\det(A) = (x + y + z)^2$.

22. (a) Consider the 3×3 *Vandermonde* determinant $V(r_1, r_2, r_3)$ defined by

$$V(r_1, r_2, r_3) = \begin{vmatrix} 1 & 1 & 1 \\ r_1 & r_2 & r_3 \\ r_1^2 & r_2^2 & r_3^2 \end{vmatrix}.$$

Show that

$$V(r_1, r_2, r_3) = (r_2 - r_1)(r_3 - r_1)(r_3 - r_2).$$

(b) More generally, show that the $n \times n$ Vandermonde determinant $V(r_1, r_2, \ldots, r_n)$

$$= \begin{vmatrix} 1 & 1 & 1 & \cdots & 1 \\ r_1 & r_2 & r_3 & \cdots & r_n \\ r_1^2 & r_2^2 & r_3^2 & \cdots & r_n^2 \\ \vdots & \vdots & \vdots & \cdots & \vdots \\ r_1^{n-1} & r_2^{n-1} & r_3^{n-1} & \cdots & r_n^{n-1} \end{vmatrix},$$

has value

$$V(r_1, r_2, \ldots, r_n) = \prod_{1 \le i < m \le n} (r_m - r_i).$$

5.5   DETERMINANTS AND A^{-1}

So far we have introduced the determinant of an $n \times n$ matrix but have not discussed its use.   The two main applications of determinants are in characterizing when a matrix is singular or nonsingular and also in the analysis of $n \times n$ linear systems.   In this section we consider the first of these applications and return to study $n \times n$ linear systems in Section 5.6.   Our results will show the relationship between rank(A), det(A), the existence of A^{-1}, and the solution of $n \times n$ linear systems.

The main result of this section is the following.

Theorem 5.5.1:   An $n \times n$ matrix is nonsingular if and only if det(A) $\neq 0$.

PROOF   It follows from Theorem 4.5.4 (page 140) that A is nonsingular if and only if rank(A) $= n$.   Thus, if we let A^* denote the reduced row echelon form of A, then A is nonsingular if and only if det(A^*) $\neq 0$.   But, A^* is obtained from A by performing a sequence of elementary row operations, and so, from properties P1–P3 of determinants, det(A) is just a nonzero multiple of det(A^*)—that is,

$$\det(A) = k \det(A^*),$$

where k is a *nonzero* constant.   Thus A is nonsingular if and only if det(A) $\neq 0$.   ∎

Example 5.5.1   Determine whether the following matrices are nonsingular.

(a) $A = \begin{bmatrix} 1 & -1 & 3 \\ 2 & 4 & -2 \\ 3 & 5 & 7 \end{bmatrix}$.

(b) $B = \begin{bmatrix} 2 & -3 & 1 \\ 1 & 5 & -2 \\ 4 & 7 & -3 \end{bmatrix}$.

Solution

(a) det(A) $= 52$ implies that A is nonsingular.
(b) det(B) $= 0$ implies that B is singular.

If det(A) $\neq 0$ then, as we have just shown, A^{-1} exists, and we could find it using the Gauss–Jordan method.   We now derive an alternative method for finding A^{-1} that is based upon determinants.   In order to do so we need the following definition.

Definition 5.5.1: If we replace every element in an $n \times n$ matrix by its cofactor, the resulting matrix is called the **matrix of cofactors**. The transpose of the matrix of cofactors is called the (classical[1]) **adjoint** of A and is denoted by adj(A). Thus

$$[\text{adj}(A)]_{ij} = A_{ji}.$$

Example 5.5.2 Determine adj(A) if $A = \begin{bmatrix} 2 & 0 & -3 \\ -1 & 5 & 4 \\ 3 & -2 & 0 \end{bmatrix}$.

Solution   We first determine the cofactors of A.

$$A_{11} = \begin{vmatrix} 5 & 4 \\ -2 & 0 \end{vmatrix} = 8, \quad A_{12} = -\begin{vmatrix} -1 & 4 \\ 3 & 0 \end{vmatrix} = 12, \quad A_{13} = \begin{vmatrix} -1 & 5 \\ 3 & -2 \end{vmatrix} = -13,$$

$$A_{21} = -\begin{vmatrix} 0 & -3 \\ -2 & 0 \end{vmatrix} = 6, \quad A_{22} = \begin{vmatrix} 2 & -3 \\ 3 & 0 \end{vmatrix} = 9, \quad A_{23} = -\begin{vmatrix} 2 & 0 \\ 3 & -2 \end{vmatrix} = 4,$$

$$A_{31} = \begin{vmatrix} 0 & -3 \\ 5 & 4 \end{vmatrix} = 15, \quad A_{32} = -\begin{vmatrix} 2 & -3 \\ -1 & 4 \end{vmatrix} = -5, \quad A_{33} = \begin{vmatrix} 2 & 0 \\ -1 & 5 \end{vmatrix} = 10.$$

Thus,

$$\text{matrix of cofactors} = \begin{bmatrix} 8 & 12 & -13 \\ 6 & 9 & 4 \\ 15 & -5 & 10 \end{bmatrix},$$

and hence

$$\text{adj}(A) = \begin{bmatrix} 8 & 6 & 15 \\ 12 & 9 & -5 \\ -13 & 4 & 10 \end{bmatrix}.$$

We can now prove the following:

Theorem 5.5.2 (The Adjoint Method for Computing A^{-1}): If det(A) $\neq 0$, then

$$A^{-1} = \frac{1}{\det(A)} \text{adj}(A). \tag{5.5.1}$$

---

[1] The term *classical* is used to distinguish adj(A) from the **Hermitian adjoint**, A^H, which is the transpose of the matrix obtained when every element in A is replaced by its complex conjugate: $A^H = \overline{A}^T$.

PROOF   From (5.4.4) we have

$$\sum_{k=1}^{n} a_{ik} A_{jk} = \delta_{ij} \det(A),$$

which can be written as

$$\frac{1}{\det(A)} \sum_{k=1}^{n} a_{ik} [\text{adj}(A)]_{kj} = \delta_{ij}, \tag{5.5.2}$$

provided $\det(A) \neq 0$. But, using the index form of the matrix product (page 96) and recalling that $I_n = [\delta_{ij}]$, (5.5.2) is simply the index form of:

$$\frac{1}{\det(A)} A \, \text{adj}(A) = I_n,$$

which implies, due to the uniqueness of A^{-1}, that

$$A^{-1} = \frac{1}{\det(A)} \text{adj}(A). \qquad \blacksquare$$

REMARK   Notice that (5.5.1) gives a formula for the elements of A^{-1} in terms of the elements of adj(A).

Example 5.5.3   For the matrix in the previous example,

$$\det(A) = 55,$$

so that

$$A^{-1} = \frac{1}{55} \begin{bmatrix} 8 & 6 & 15 \\ 12 & 9 & -5 \\ -13 & 4 & 10 \end{bmatrix}.$$

In Chapter 11, we will find that the solution of a system of *differential* equations can be expressed naturally in terms of matrix functions—that is, matrices whose elements are functions defined on some interval. Certain problems will require us to find the inverse of such matrix functions and, for small systems, the adjoint method is very quick.

Example 5.5.4   Find A^{-1} if $A = \begin{bmatrix} e^{2t} & e^{-t} \\ 3e^{2t} & 6e^{-t} \end{bmatrix}$.

Solution   In this case, $\det(A) = 3e^t$, and

$$\text{adj}(A) = \begin{bmatrix} 6e^{-t} & -e^{-t} \\ -3e^{2t} & e^{2t} \end{bmatrix},$$

so that

$$A^{-1} = \begin{bmatrix} 2e^{-2t} & -\frac{1}{3}e^{-2t} \\ \\ -e^t & \frac{1}{3}e^t \end{bmatrix}.$$

EXERCISES 5.5

In problems 1–7 determine whether the given
matrix is singular or nonsingular.

1. $\begin{bmatrix} 2 & 1 \\ 3 & 2 \end{bmatrix}.$

2. $\begin{bmatrix} -1 & 1 \\ 1 & -1 \end{bmatrix}.$

3. $\begin{bmatrix} 2 & 6 & -1 \\ 3 & 5 & 1 \\ 2 & 0 & 1 \end{bmatrix}.$

4. $\begin{bmatrix} -1 & 2 & 3 \\ 5 & -2 & 1 \\ 8 & -2 & 5 \end{bmatrix}.$

5. $\begin{bmatrix} 1 & 0 & 2 & -1 \\ 3 & -2 & 1 & 4 \\ 2 & 1 & 6 & 2 \\ 1 & -3 & 4 & 0 \end{bmatrix}.$

6. $\begin{bmatrix} 1 & 1 & 1 & 1 \\ -1 & 1 & -1 & 1 \\ 1 & 1 & -1 & -1 \\ -1 & 1 & 1 & -1 \end{bmatrix}.$

7. $\begin{bmatrix} 1 & 2 & -3 & 5 \\ -1 & 2 & -3 & 6 \\ 2 & 3 & -1 & 4 \\ 1 & -2 & 3 & -6 \end{bmatrix}.$

In problems 8–17 find, if possible, **(a)** the matrix
of cofactors, **(b)** adj(A), and **(c)** A^{-1}.

8. $A = \begin{bmatrix} 3 & 1 \\ 4 & 5 \end{bmatrix}.$

9. $A = \begin{bmatrix} -1 & -2 \\ 4 & 1 \end{bmatrix}.$

10. $A = \begin{bmatrix} 5 & 2 \\ -15 & -6 \end{bmatrix}.$

11. $A = \begin{bmatrix} 2 & -3 & 0 \\ 2 & 1 & 5 \\ 0 & -1 & 2 \end{bmatrix}.$

12. $A = \begin{bmatrix} -2 & 3 & -1 \\ 2 & 1 & 5 \\ 0 & 2 & 3 \end{bmatrix}.$

13. $A = \begin{bmatrix} 1 & -1 & 2 \\ 3 & -1 & 4 \\ 5 & 1 & 7 \end{bmatrix}.$

14. $A = \begin{bmatrix} 0 & 1 & 2 \\ -1 & -1 & 3 \\ 1 & -2 & 1 \end{bmatrix}.$

15. $A = \begin{bmatrix} 2 & -3 & 5 \\ 1 & 2 & 1 \\ 0 & 7 & -1 \end{bmatrix}.$

16. $A = \begin{bmatrix} 1 & 1 & 1 & 1 \\ -1 & 1 & -1 & 1 \\ 1 & 1 & -1 & -1 \\ -1 & 1 & 1 & -1 \end{bmatrix}.$

17. $A = \begin{bmatrix} 1 & 0 & 3 & 5 \\ -2 & 1 & 1 & 3 \\ 3 & 9 & 0 & 2 \\ 2 & 0 & 3 & -1 \end{bmatrix}.$

18. Determine all values of x, y, z for which

$$A = \begin{bmatrix} 1 & x & y \\ -x & 0 & z \\ -y & -z & 0 \end{bmatrix}$$ is nonsingular. Use the ad-

joint method to find A^{-1}.

19. Find the element in the second row and third column of A^{-1} if

$$A = \begin{bmatrix} 1 & 0 & 1 & 0 \\ 2 & -1 & 1 & 3 \\ 0 & 1 & -1 & 2 \\ -1 & 1 & 2 & 0 \end{bmatrix}.$$

In problems 20–22 find A^{-1}.

20. $A = \begin{bmatrix} 3e^t & e^{2t} \\ 2e^t & 2e^{2t} \end{bmatrix}.$

21. $A = \begin{bmatrix} e^t\sin 2t & -e^{-t}\cos 2t \\ e^t\cos 2t & e^{-t}\sin 2t \end{bmatrix}.$

22. $A = \begin{bmatrix} e^t & te^t & e^{-2t} \\ e^t & 2te^t & e^{-2t} \\ e^t & te^t & 2e^{-2t} \end{bmatrix}.$

5.6 DETERMINANTS AND n × n LINEAR SYSTEMS

We now illustrate the significant role that determinants play in the analysis of $n \times n$ linear systems. Consider the general $n \times n$ system of linear equations

$$a_{11}x_1 + a_{12}x_2 + \cdots + a_{1n}x_n = b_1,$$
$$a_{21}x_1 + a_{22}x_2 + \cdots + a_{2n}x_n = b_2, \qquad (5.6.1)$$
$$\vdots$$
$$a_{n1}x_1 + a_{n2}x_2 + \cdots + a_{nn}x_n = b_n,$$

which we can write as $A\mathbf{x} = \mathbf{b}$, where A denotes the matrix of coefficients of the system and $\mathbf{b} = [b_1, b_2, \ldots, b_n]^T$.

We first prove an important general result.

Theorem 5.6.1: The $n \times n$ system of equations $A\mathbf{x} = \mathbf{b}$ has a unique solution if and only if

$$\det(A) \neq 0.$$

PROOF From Theorem 4.5.3 and Corollary 4.5.1, an $n \times n$ system $A\mathbf{x} = \mathbf{b}$ has a unique solution if and only if A^{-1} exists—that is, from Theorem 5.5.1, if and only if $\det(A) \neq 0$. ■

If an $n \times n$ system of equations is *nonhomogeneous, and* $\det(A) = 0$, then it follows from the results of Chapter 4 and the previous theorem that the system either has an infinite number of solutions or else it is inconsistent, although we could not tell which without reducing the augmented matrix. However, if the system is *homogeneous* then it is necessarily consistent, and hence $\det(A)$ completely characterizes the solution properties of the system.

Corollary 5.6.1: An $n \times n$ *homogeneous* system of linear equations, with coefficient matrix A, has an infinite number of solutions if and only if

$$\det(A) = 0,$$

and hence has only the trivial solution if and only if

$$\det(A) \neq 0.$$

PROOF   From Theorem 5.6.1, the system has a unique solution if and only if $\det(A) \neq 0$, and hence, since a homogeneous system is always consistent, there are an infinite number of solutions if and only if $\det(A) = 0$.   ■

REMARK   The preceding corollary is *very* important, since often we are interested only in determining the solution properties of a homogeneous linear system and not actually in finding the solutions themselves. We will refer back to this corollary on several occasions during the remainder of the text.

Example 5.6.1   Determine all k for which the system

$$10x_1 + kx_2 - x_3 = 0,$$
$$kx_1 + x_2 - x_3 = 0,$$
$$2x_1 + x_2 - 3x_3 = 0,$$

has nontrivial solutions.

Solution   The determinant of the matrix of coefficients of the system is

$$\det(A) = \begin{vmatrix} 10 & k & -1 \\ k & 1 & -1 \\ 2 & 1 & -3 \end{vmatrix} \begin{array}{c} C_1 \to C_1 - 2C_2 \\ = \\ C_3 \to C_3 + 3C_2 \end{array} \begin{vmatrix} 10 - 2k & k & -1 + 3k \\ k - 2 & 1 & 2 \\ 0 & 1 & 0 \end{vmatrix}$$

$$= -[2(10 - 2k) - (k - 2)(3k - 1)]$$
$$= 3(k^2 - k - 6)$$
$$= 3(k - 3)(k + 2).$$

From the previous theorem the system has nontrivial solutions if and only if $\det(A) = 0$, that is, if and only if $k = 3$ or $k = -2$.

Now let us return to system (5.6.1). According to Theorem 5.6.1, assuming that $\det(A) \neq 0$, the system has a unique solution, which we could find using Gaussian elimination. We now derive an alternative method, known as *Cramer's rule*, which is also useful, at least theoretically.

Consider the $n \times n$ linear system $A\mathbf{x} = \mathbf{b}$ and let B_k denote the matrix obtained by replacing the kth column vector of A with $\mathbf{b}$. Thus:

$$B_k = \begin{bmatrix} a_{11} & a_{12} & \cdots & b_1 & \cdots & a_{1n} \\ a_{21} & a_{22} & \cdots & b_2 & \cdots & a_{2n} \\ \vdots & \vdots & & \vdots & & \vdots \\ a_{n1} & a_{n2} & \cdots & b_n & \cdots & a_{nn} \end{bmatrix}, \qquad k = 1, 2, \ldots, n,$$

$$\uparrow$$
$$\text{column } k$$

or, in column vector notation,

$$B_k = [\mathbf{a}_1, \mathbf{a}_2, \ldots, \mathbf{a}_{k-1}, \mathbf{b}, \mathbf{a}_{k+1}, \ldots, \mathbf{a}_n].$$

Example 5.6.2 If $A = \begin{bmatrix} 2 & -1 & 3 \\ 5 & 1 & -2 \\ 4 & -3 & 6 \end{bmatrix}$ and $\mathbf{b} = \begin{bmatrix} 9 \\ 8 \\ 7 \end{bmatrix}$, then

$$B_1 = \begin{bmatrix} 9 & -1 & 3 \\ 8 & 1 & -2 \\ 7 & -3 & 6 \end{bmatrix}, \qquad B_2 = \begin{bmatrix} 2 & 9 & 3 \\ 5 & 8 & -2 \\ 4 & 7 & 6 \end{bmatrix}, \qquad B_3 = \begin{bmatrix} 2 & -1 & 9 \\ 5 & 1 & 8 \\ 4 & -3 & 7 \end{bmatrix}.$$

The key point to notice about B_k is that the cofactors of the elements in the kth column coincide with the corresponding cofactors of A. Thus expanding $\det(B_k)$ along the kth column using the cofactor expansion theorem yields

$$\det(B_k) = A_{1k}b_1 + A_{2k}b_2 + \cdots + A_{nk}b_n = \sum_{i=1}^{n} A_{ik}b_i. \qquad (5.6.2)$$

We can now prove Cramer's rule.

Theorem 5.6.2 (Cramer's Rule): If $\det(A) \neq 0$, the unique solution to the $n \times n$ system $A\mathbf{x} = \mathbf{b}$ is $(x_1, x_2, \ldots, x_n)$, where

$$x_k = \frac{\det(B_k)}{\det(A)}, \qquad k = 1, 2, \ldots, n. \qquad (5.6.3)$$

PROOF If $\det(A) \neq 0$, then the system $A\mathbf{x} = \mathbf{b}$ has the unique solution

$$\mathbf{x} = A^{-1}\mathbf{b}, \qquad (5.6.4)$$

where, from Theorem 5.5.2, we can write

$$A^{-1} = \frac{1}{\det(A)} \text{adj}(A). \qquad (5.6.5)$$

If we let $\mathbf{x} = [x_1\ x_2 \cdots x_n]^T$ and $\mathbf{b} = [b_1\ b_2 \cdots b_n]^T$ and recall that $[\text{adj}(A)]_{ij} = A_{ji}$, then substitution from (5.6.5) into (5.6.4) and use of the index form of the matrix product yields:

$$x_k = \sum_{i=1}^{n} (A^{-1})_{ki} b_i = \sum_{i=1}^{n} \frac{1}{\det(A)} [\text{adj}(A)]_{ki} b_i$$

$$= \frac{1}{\det(A)} \sum_{i=1}^{n} A_{ik} b_i, \qquad k = 1, 2, \ldots, n.$$

Thus, substituting from (5.6.2),

$$x_k = \frac{\det(B_k)}{\det(A)}, \qquad k = 1, 2, \ldots, n$$

as required. ∎

REMARK   Cramer's rule in general requires more work than the Gaussian elimination method and is also restricted to $n \times n$ systems with nonsingular coefficient matrix. However it is a powerful theoretical tool, since it gives us a formula for the solution of an $n \times n$ system provided $\det(A) \neq 0$.

Example 5.6.3 Solve

$$3x_1 + 2x_2 - x_3 = 4,$$
$$x_1 + x_2 - 5x_3 = -3,$$
$$-2x_1 - x_2 + 4x_3 = 0.$$

Solution

$$\det(A) = \begin{vmatrix} 3 & 2 & -1 \\ 1 & 1 & -5 \\ -2 & -1 & 4 \end{vmatrix} \begin{matrix} R_1 \rightarrow R_1 - 3R_2 \\ = \\ R_3 \rightarrow R_3 + 2R_2 \end{matrix} \begin{vmatrix} 0 & -1 & 14 \\ 1 & 1 & -5 \\ 0 & 1 & -6 \end{vmatrix} = 8.$$

$$\det(B_1) = \begin{vmatrix} 4 & 2 & -1 \\ -3 & 1 & -5 \\ 0 & -1 & 4 \end{vmatrix} \begin{matrix} C_3 \rightarrow C_3 + 4C_2 \\ = \end{matrix} \begin{vmatrix} 4 & 2 & 7 \\ -3 & 1 & -1 \\ 0 & -1 & 0 \end{vmatrix} = 17.$$

$$\det(B_2) = \begin{vmatrix} 3 & 4 & -1 \\ 1 & -3 & -5 \\ -2 & 0 & 4 \end{vmatrix} \begin{matrix} C_3 \rightarrow C_3 + 2C_1 \\ = \end{matrix} \begin{vmatrix} 3 & 4 & 5 \\ 1 & -3 & -3 \\ -2 & 0 & 0 \end{vmatrix} = -6.$$

$$\det(B_3) = \begin{vmatrix} 3 & 2 & 4 \\ 1 & 1 & -3 \\ -2 & -1 & 0 \end{vmatrix} \begin{matrix} C_1 \rightarrow C_1 - 2C_2 \\ = \end{matrix} \begin{vmatrix} -1 & 2 & 4 \\ -1 & 1 & -3 \\ 0 & -1 & 0 \end{vmatrix} = 7.$$

Thus, from (5.6.3) we have $x_1 = \frac{17}{8}$, $x_2 = -\frac{6}{8} = -\frac{3}{4}$, and $x_3 = \frac{7}{8}$, so that the solution to the system is $(\frac{17}{8}, -\frac{3}{4}, \frac{7}{8})$.

EXERCISES 5.6

In problems 1–4 use Cramer's rule to solve the given linear system.

1. $2x_1 - 3x_2 = 2,$
    $x_1 + 2x_2 = 4.$

2. $3x_1 - 2x_2 + x_3 = 4,$
    $x_1 + x_2 - x_3 = 2,$
    $2x_1 \qquad + x_3 = 1.$

3. $x_1 - 3x_2 + x_3 = 0,$
    $x_1 + 4x_2 - x_3 = 0,$
    $2x_1 + x_2 - 3x_3 = 0.$

4. $x_1 - 2x_2 + 3x_3 - x_4 = 1,$
    $2x_1 \qquad + x_3 \qquad = 2,$
    $x_1 + x_2 \qquad - x_4 = 0,$
    $x_2 - 2x_3 + x_4 = 3.$

5. Use Cramer's rule to determine x_1 and x_2 if

$$e^t x_1 + e^{-2t} x_2 = 3 \sin t,$$
$$e^t x_1 - 2e^{-2t} x_2 = 4 \cos t.$$

6. Determine the value of x_2 such that

$$\begin{aligned} x_1 + 4x_2 - 2x_3 + x_4 &= 2, \\ 2x_1 + 9x_2 - 3x_3 - 2x_4 &= 5, \\ x_1 + 5x_2 \qquad\quad - x_4 &= 3, \\ 3x_1 + 14x_2 + 7x_3 - 2x_4 &= 6. \end{aligned}$$

7. Determine all values of λ for which the system

$$\begin{aligned} x_1 + 2x_2 + x_3 &= \lambda x_1, \\ 2x_1 + x_2 + x_3 &= \lambda x_2, \\ x_1 + x_2 + 2x_3 &= \lambda x_3, \end{aligned}$$

has *nontrivial* solutions. Obtain the solution set when $\lambda = 1$.

8. Find all solutions of the system

$$\begin{aligned} (b+c)x_1 + a(x_2 + x_3) &= a, \\ (c+a)x_2 + b(x_3 + x_1) &= b, \\ (a+b)x_3 + c(x_1 + x_2) &= c, \end{aligned}$$

where a, b, and c are constants. Make sure you consider all cases (that is, those when there is a unique solution, an infinite number of solutions, and no solution).

5.7   SUMMARY OF RESULTS

In this section we summarize the results derived in this chapter.

DEFINITION OF A DETERMINANT

If A is an $n \times n$ matrix, then the determinant of A is defined by

$$\det(A) = \sum \sigma(p_1, p_2, \ldots, p_n) a_{1p_1} a_{2p_2} a_{3p_3} \cdots a_{np_n},$$

where the summation is over the $n!$ distinct permutations $(p_1, p_2 \ldots, p_n)$ of the integers $1, 2, \ldots, n$.

PROPERTIES OF DETERMINANTS

Let A and B be $n \times n$ matrices.

P1. If B is a matrix obtained by interchanging two distinct rows (or columns) of A, then $\det(A) = -\det(B)$.

P2. If B is the matrix obtained by dividing any row (or column) of A by a nonzero scalar k, then $\det(A) = k \det(B)$.

P3. If B is the matrix obtained by adding a multiple of any row (or column) of A to another row (or column) of A, then $\det(A) = \det(B)$.

P4. $\det(A^T) = \det(A)$.

P5. Let $\mathbf{a}_1, \mathbf{a}_2, \ldots, \mathbf{a}_n$ denote the row vectors of A. If the ith row vector of A is the sum of two row vectors—for example, $\mathbf{a}_i = \mathbf{b}_i + \mathbf{c}_i$—then $\det(A) = \det(B) + \det(C)$, where $B = [\mathbf{a}_1, \mathbf{a}_2, \ldots, \mathbf{a}_{i-1}, \mathbf{b}_i, \mathbf{a}_{i+1}, \ldots, \mathbf{a}_n]^T$ and $C = [\mathbf{a}_1, \mathbf{a}_2, \ldots, \mathbf{a}_{i-1}, \mathbf{c}_i, \mathbf{a}_{i+1}, \ldots, \mathbf{a}_n]^T$. The corresponding property is also true for columns.

P6. If A has a row (or column) of zeros, then $\det(A) = 0$.

P7. If two rows (or columns) of A are the same, then $\det(A) = 0$.

P8. $\det(AB) = \det(A)\det(B)$.

EVALUATION OF DETERMINANTS

We have introduced four different ways to evaluate a determinant.

1. Use the definition.
2. Use the preceding properties to reduce a determinant to upper triangular form and then use Theorem 5.3.1 to evaluate the resulting determinant.
3. Use the cofactor expansion theorem.
4. Use properties of determinants to set all of the elements except one in a row or column to zero and then use the cofactor expansion theorem.

APPLICATIONS OF DETERMINANTS

1. An $n \times n$ matrix is nonsingular if and only if $\det(A) \neq 0$.
2. If A is nonsingular, then A^{-1} can be expressed directly in terms of determinants as

$$A^{-1} = \frac{1}{\det(A)}\,\text{adj}(A),$$

where $\text{adj}(A)$ denotes the adjoint of A.

3. An $n \times n$ linear system $A\mathbf{x} = \mathbf{b}$ has a unique solution if and only if $\det(A) \neq 0$.
4. An $n \times n$ *homogeneous* linear system $A\mathbf{x} = \mathbf{0}$ has an infinite number of solutions if and only if $\det(A) = 0$.
5. Cramer's rule: If $\det(A) \neq 0$, then the unique solution to $A\mathbf{x} = \mathbf{b}$ is $(x_1, x_2, \ldots, x_n)$, where $x_k = \dfrac{\det(B_k)}{\det(A)}$, $k = 1, 2, \ldots, n$.

6

Vector Spaces

INTRODUCTION AND MOTIVATION

The main aim of this text is to study linear mathematics. So far we have considered two different types of linear problems. In Chapter 1 we defined what is meant by a linear differential equation, whereas in Chapter 4 we studied systems of linear algebraic equations. The theory underlying the solution of both of these problems can be considered as a special case of a general framework for linear problems. In order to illustrate this framework we consider two examples.

It is straightforward to show that the linear algebraic system $A\mathbf{x} = \mathbf{0}$, where

$$A = \begin{bmatrix} 1 & -1 & 2 \\ 2 & -2 & 4 \\ 3 & -3 & 6 \end{bmatrix}$$

has solution set

$$S = \{(r - 2s, r, s) : r, s \in \mathbf{R}\}.$$

Geometrically we can interpret each solution as defining the coordinates of a point in space or, equivalently, as the geometric vector with components

$$\mathbf{x} = (r - 2s, \, r, \, s).$$

Using the standard operations of vector addition and multiplication of a vector by a real number, it follows that $\mathbf{x}$ can be written in the form

$$\mathbf{x} = r(1, \, 1, \, 0) + s(-2, \, 0, \, 1).$$

The key point to notice for the present discussion is that we have expressed every solution of the given linear system as a *linear combination* of the two basic solutions (see Figure 6.1.1)

$$\mathbf{v}_1 = (1, \, 1, \, 0), \qquad \mathbf{v}_2 = (-2, \, 0, \, 1).$$

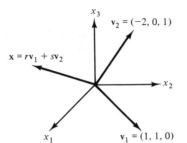

Figure 6.1.1 Two basic solutions to $A\mathbf{x} = \mathbf{0}$, and an example of an arbitrary solution to the system.

Now consider the second-order *linear* differential equation

$$y'' + y = 0.$$

It is easily verified (by direct substitution) that

$$y_1 = \cos x, \qquad y_2 = \sin x,$$

are solutions of the differential equation on any interval I. Further, it is shown in the exercise at the end of this section that *every* solution of the differential equation is of the form

$$y = c_1 \cos x + c_2 \sin x,$$

where c_1 and c_2 are arbitrary real numbers. Thus once more the solution of a linear problem has been expressed as a linear combination of two basic solutions.

The similarities between the above two problems are quite striking. In both cases we have a set of "vectors," V (in the first problem the vectors are ordered triples of numbers, whereas in the second problem they are functions that are at least twice differentiable on an interval I), and a *linear* vector equation defined in V. Further, in both cases, all solutions to the given equation could be expressed as a linear combination of two particular solutions.

In the next two chapters we develop this way of formulating linear problems in terms of an abstract set of vectors, V, and a linear equation defined on V. We will find that many problems fit into this framework and that the solutions to these problems can be expressed as linear combinations of a certain number (not necessarily two) of basic solutions.   In particular, we will be able to derive the theory underlying linear differential equations and linear systems of differential equations as special cases of the general framework.

Before proceeding any further we give a word of encouragement to the more application-oriented reader.   It will probably seem at times that the ideas we are introducing are rather esoteric and that the formalism is pure mathematical abstraction.   However, in addition to the inherent mathematical beauty of the formalism, the ideas that it incorporates pervade many areas of applied mathematics, particularly engineering mathematics and mathematical physics, where the problems under investigation are often linear in nature.   Indeed the linear algebra introduced in the next few chapters should be considered an extremely important addition to your mathematical repertoire, perhaps on a par with the more concrete ideas of elementary calculus.

EXERCISES 6.1

1. Consider the differential equation $y'' + y = 0$.
(a) Verify that $y = c_1 \cos x + c_2 \sin x$ is a solution of the differential equation, on any interval, for all values of the constants c_1, c_2.
(b) Show that the initial value problem

$$y'' + y = 0, \qquad y(x_0) = y_0, \qquad y'(x_0) = y_1$$

has a unique solution for all values of the constants x_0, y_0, y_1.

This justifies the claim made in the text that all solutions of the differential equation $y'' + y = 0$ are of the form $y = c_1 \cos x + c_2 \sin x$.

6.2   DEFINITION OF A VECTOR SPACE

In order to motivate the definition for which we are aiming, we begin by considering the set of all geometric vectors (directed line segments), which we denote by V. In previous courses you will have defined how to add two vectors and how to multiply a vector by a real number (see Figure 6.2.1).

These operations satisfy many properties, the most important of which are the following:

1. If **u** and **v** are elements of V, then $\mathbf{u} + \mathbf{v}$ and $r\mathbf{u}$ are elements of V for all real numbers r.   We say that V is *closed* under addition and *closed* under multiplication by a real number.

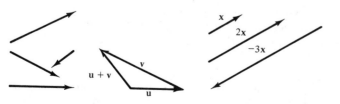

Elements of V Addition in V Scalar multiplication in V **Figure 6.2.1**   Geometric vectors.

2. The operation of vector addition is commutative and associative, that is, for all **u**, **v**, **w** in V,

$$\mathbf{u} + \mathbf{v} = \mathbf{v} + \mathbf{u}$$

and

$$\mathbf{u} + (\mathbf{v} + \mathbf{w}) = (\mathbf{u} + \mathbf{v}) + \mathbf{w}.$$

3. There exists a vector called the zero vector, denoted by **0**, with the property

$$\mathbf{u} + \mathbf{0} = \mathbf{u}$$

for all **u** in V.

4. For any **u** in V, there is a vector $-\mathbf{u}$ in V with the property that

$$\mathbf{u} + (-\mathbf{u}) = \mathbf{0}.$$

We call $-\mathbf{u}$ the additive inverse of **u**.

5. If **u** and **v** are arbitrary vectors in V, and r and s are real numbers, then

$$r(\mathbf{u} + \mathbf{v}) = r\mathbf{u} + r\mathbf{v}, \qquad (r + s)\mathbf{u} = r\mathbf{u} + s\mathbf{u},$$

$$(rs)\mathbf{u} = r(s\mathbf{u}), \qquad\qquad 1\mathbf{u} = \mathbf{u}.$$

If you analyze other sets of objects for which operations of addition and multiplication by a real number are defined (for example, the set of all real numbers, the set of all $n \times n$ matrices, the set of all real-valued functions defined on an interval), you will find that many of them also satisfy the preceding properties. We are thus led to focus our attention on the operations of addition and scalar multiplication and their associated properties rather than on the particular set of objects under consideration. This is where the idea of a vector space arises. Informally a vector space consists of a set of objects, V, whose elements are called vectors, together with a rule for adding vectors in V and a rule for multiplying vectors in V by a real number. Further, these addition and multiplication operations must satisfy the same properties as the addition and multiplication operations for geometric vectors. Actually what we have just described is more correctly called a *real* vector space, since we have assumed that the multiplication operation in V is defined only for real numbers. The formulation of some of the problems in Chapters 8 and 11 will require the slightly more general idea of a complex vector space, which arises when the multiplication operation in V is defined for complex numbers. In order to give a unified definition of both types of vector space, we let F denote the set of scalars for which the multiplication operation in V is defined. Thus, for us, F will either be the set of all real numbers, **R**, or the set of all complex numbers, **C**.[1] With these comments in mind we are now ready to give the formal definition of a vector space.

---

[1] In more advanced treatments of vector space theory, the scalars are elements of a mathematical structure called a *field* (see, for example, I. N. Herstein, *Topics in Algebra*, John Wiley, 1975). The set of all real numbers and the set of all complex numbers are particular examples of fields.

Definition 6.2.1: A nonempty set V, whose elements are called vectors, is said to form a **vector space** (over F) provided the following conditions are satisfied:

(a) *Addition*: There is a rule that associates with each pair of vectors $\mathbf{u}$, $\mathbf{v}$ in V another vector $\mathbf{u} + \mathbf{v}$ in V, called the *sum* of $\mathbf{u}$ and $\mathbf{v}$. The vector addition operation must satisfy the following axioms:

Addition Axioms

A1. For all $\mathbf{u}$, $\mathbf{v}$, and $\mathbf{w}$ in V:

$$\mathbf{u} + \mathbf{v} = \mathbf{v} + \mathbf{u}, \qquad \text{(Addition is commutative.)}$$
$$(\mathbf{u} + \mathbf{v}) + \mathbf{w} = \mathbf{u} + (\mathbf{v} + \mathbf{w}). \qquad \text{(Addition is associative.)}$$

A2. There is a vector, denoted by $\mathbf{0}$, in V such that

$$\mathbf{u} + \mathbf{0} = \mathbf{u}, \qquad \text{for every } \mathbf{u} \text{ in } V.$$

We refer to this vector as the **zero vector** in V.

A3. For each vector $\mathbf{u}$ in V, there is a vector, denoted by $-\mathbf{u}$, in V such that

$$\mathbf{u} + (-\mathbf{u}) = \mathbf{0}.$$

$-\mathbf{u}$ is called the **additive inverse** of $\mathbf{u}$.

(b) *Scalar multiplication*: There is a rule that associates with each vector $\mathbf{v}$ in V and each scalar s in F a vector $s\mathbf{v}$ in V. The scalar multiplication operation must satisfy the following axiom:

Scalar Multiplication Axiom

A4. For all vectors $\mathbf{u}$, $\mathbf{v}$ in V and all scalars r, s, t in F,

$$1\mathbf{u} = \mathbf{u}, \qquad\qquad (st)\mathbf{u} = s(t\mathbf{u}),$$
$$r(\mathbf{u} + \mathbf{v}) = r\mathbf{u} + r\mathbf{v}, \qquad (s + t)\mathbf{u} = s\mathbf{u} + t\mathbf{u}.$$

IMPORTANT REMARKS:

1. The key point to note from the preceding definition is that a vector space has four components:
   (a) A nonempty set of vectors, V.
   (b) A set of scalars F (either **R** or **C**).
   (c) An addition operation, $+$, defined on V.
   (d) A scalar multiplication operation, $\cdot$, defined on V.

   If we wish to emphasize this point, we will use the notation $(V, F, +, \cdot)$ to denote these four components. Thus $(V, F, +, \cdot)$ forms a vector space if and only if all parts of Definition 6.2.1 are satisfied.

Terminology

If the set of scalars F is the set of all real numbers, we will often refer to $(V, \mathbf{R}, +, \cdot)$ as a **real vector space**, whereas if the set of scalars F is the set of all complex numbers, then we will refer to $(V, \mathbf{C}, +, \cdot)$ as a **complex vector space**.

---

2. As we have indicated in the preceding definition, we will use bold print to denote vectors in a given vector space. In handwriting it is strongly advised that you denote vectors either as $\vec{v}$ or as $\underset{\sim}{\mathbf{v}}$. This will avoid any confusion between vectors in V and scalars in F.

3. When we deal with a familiar vector space, we will use the usual notation for vectors in the space. For example, the set of all real-valued functions defined on an interval forms a vector space (see Example 6.2.2) and we will denote the vectors in this vector space by $f, g, \ldots$. Similarly, when we are dealing with the vector space of all $m \times n$ matrices, we will denote the vectors in the vector space by A, B, and so forth.

EXAMPLES OF VECTOR SPACES

1. The set of all real numbers together with the usual operations of addition and multiplication is a real vector space.

2. The set of all complex numbers is a complex vector space when we use the usual operations of addition and multiplication by a complex number.

3. The set of all $m \times n$ matrices with real entries is a real vector space when we use the usual operations of addition of matrices and multiplication of matrices by a real number (here the vectors are $m \times n$ matrices, A, B, etc.).

Example 6.2.1 Let V be the set of all 2×2 matrices with real elements. Show that V, together with the usual operations of addition and multiplication by a real number, forms a real vector space.

Solution If A and B are in V (that is, are 2×2 matrices with real entries), then $A + B$ and cA are in V for all real numbers c. We now check the axioms.

A1. This axiom is satisfied, since, as we have shown in Chapter 3 (page 91), matrix addition is associative and commutative.

A2. If A is any vector in V and $0_2 = \begin{bmatrix} 0 & 0 \\ 0 & 0 \end{bmatrix}$, then

$$A + 0_2 = A.$$

Thus 0_2 is the zero vector in V.

A3. If $A = \begin{bmatrix} a & b \\ c & d \end{bmatrix}$, then $-A = \begin{bmatrix} -a & -b \\ -c & -d \end{bmatrix}$, since $A + (-A) = \begin{bmatrix} 0 & 0 \\ 0 & 0 \end{bmatrix} = 0_2$.

A4. These are basic properties of matrix algebra (see page 92).

Thus V, together with the given operations, forms a real vector space.

REMARK We will denote the set of all $n \times n$ matrices with real elements by $M_n(\mathbf{R})$. The preceding example shows that $M_2(\mathbf{R})$ is a vector space. It is not too difficult to show that, for any positive integer n, $M_n(\mathbf{R})$ is one also.

Example 6.2.2 Let V be the set of all real-valued functions defined on an interval I. Define addition and scalar multiplication in V in the usual manner; that is, if f and g are in V and r is any *real* number, then $f + g$ and rf are defined by

$$(f + g)(x) = f(x) + g(x), \qquad (rf)(x) = rf(x) \qquad \text{for all } x \text{ in I.}$$

Show that $(V, \mathbf{R}, +, \cdot)$ forms a real vector space.

Solution If f and g are in V and c is any real number, then $f + g$ and cf are also in V. We now check the axioms.

A1. Properties of addition of functions.

A2. The zero vector in V is the zero function, o, defined by $o(x) = 0$ for all x in I, since

$$(f + o)(x) = f(x) + o(x) = f(x) + 0 = f(x), \qquad \text{for all } f \text{ in } V \text{ and all } x \text{ in I.}$$

A3. If f is any function in V, then $-f$ is defined by $(-f)(x) = -[f(x)]$ for all x in I, since

$$[f + (-f)](x) = f(x) + (-f)(x) = f(x) - f(x) = 0 \qquad \text{for all } x \text{ in I,}$$

which implies that $f + (-f) = o$, the zero function.

A4. Each of these properties is easily proved. As an example we show that $(s + t)f = sf + tf$ for all real numbers s and t. Let f be in V and let s and t be arbitrary real numbers. Then, for all x in I,

$$[(s + t)f](x) = (s + t)f(x) = sf(x) + tf(x) = (sf)(x) + (tf)(x) = [(sf) + (tf)](x),$$

and so $(s + t)f = sf + tf$. The other properties are proved similarly.

It follows that V is a real vector space.

THE VECTOR SPACE $\mathbf{R}^n$

We now introduce one of the most important real vector spaces. In order to define this vector space we must give a set of objects, together with operations of addition and scalar multiplication defined on the set, which satisfy Definition 6.2.1. We first define the set of objects.

> *Definition 6.2.2:*   Let $\mathbf{R}^n$ denote the set of all ordered n-tuples of real numbers. That is,
>
> $$\mathbf{R}^n = \{(x_1, x_2, \dots, x_n) : x_i \in \mathbf{R}\}.$$
>
> We refer to the elements in $\mathbf{R}^n$ as **vectors in $\mathbf{R}^n$**.

We are used to dealing with $\mathbf{R}^2$ and $\mathbf{R}^3$ from coordinate geometry, where we interpret the ordered pairs and triples of numbers as the coordinates of points in the plane and in 3-space respectively, or as the components of geometric vectors. Also, it follows from Definition 4.1.1 (page 108) that any real solution of an $m \times n$ linear system is a vector in $\mathbf{R}^n$.

Example 6.2.3   Give examples of vectors in $\mathbf{R}^3$, $\mathbf{R}^5$, and $\mathbf{R}^8$.

Solution

$(1, -\frac{1}{2}, 2), (3, \pi, -10), (31, 23, -177)$ are vectors in $\mathbf{R}^3$.

$(1, 2, 3, 4, 5), (0, 0, 0, 0, 0), (-23, 13, 17, -5, 1)$ are vectors in $\mathbf{R}^5$.

$(14, 13, -10, 0, 25, 1, 22, -1), (1, 0, 1, 0, 0, -1, 29, -3)$ are vectors in $\mathbf{R}^8$.

We now define operations of addition and scalar multiplication for vectors in $\mathbf{R}^n$. It is important to note that when dealing with $\mathbf{R}^n$, our scalars are always *real numbers.*

> *Definition 6.2.3:*   Let $\mathbf{x} = (x_1, x_2, \dots, x_n)$ and $\mathbf{y} = (y_1, y_2, \dots, y_n)$ be vectors in $\mathbf{R}^n$ and let k be an arbitrary real number. We define **addition** and **scalar multiplication** in $\mathbf{R}^n$ by
>
> $$\mathbf{x} + \mathbf{y} = (x_1, x_2, \dots, x_n) + (y_1, y_2, \dots, y_n) = (x_1 + y_1, x_2 + y_2, \dots, x_n + y_n),$$
> $$k\mathbf{x} = k(x_1, x_2, \dots, x_n) = (kx_1, kx_2, \dots, kx_n),$$
>
> respectively.

Example 6.2.4   If $\mathbf{x} = (1, -1, 0, 2)$ and $\mathbf{y} = (0, -1, 4, 5)$, find $\mathbf{x} + \mathbf{y}$ and $2\mathbf{x} + 3\mathbf{y}$.

Solution

$\mathbf{x} + \mathbf{y} = (1, -2, 4, 7);\quad 2\mathbf{x} + 3\mathbf{y} = (2, -2, 0, 4) + (0, -3, 12, 15) = (2, -5, 12, 19).$

Notice that we have a natural one-to-one correspondence between vectors in $\mathbf{R}^n$, row n-vectors, and column n-vectors, defined by:

$$(x_1, x_2, \ldots, x_n) \leftrightarrow [x_1 \ x_2 \cdots x_n] \leftrightarrow \begin{bmatrix} x_1 \\ x_2 \\ \vdots \\ x_n \end{bmatrix}.$$

Further, this correspondence is preserved under the operations of addition and scalar multiplication defined for vectors in $\mathbf{R}^n$ and for row and column vectors. For this reason we will often treat vectors in $\mathbf{R}^n$, row n-vectors, and column n-vectors as if they are just different representations of the same basic object. For example, a solution of an $m \times n$ linear system $A\mathbf{x} = \mathbf{b}$ can and will be written in either of the equivalent forms

$$\mathbf{x} = (x_1, x_2, \ldots, x_n) \quad \text{or} \quad \mathbf{x} = \begin{bmatrix} x_1 \\ x_2 \\ \vdots \\ x_n \end{bmatrix}.$$

It is now a straightforward procedure to check that $\mathbf{R}^n$ together with the preceding defined operations of addition and scalar multiplication form a (real) vector space. We leave the full verification as an exercise and simply write the zero vector and additive inverse in the vector space:

Zero vector: $\mathbf{0} = (0, 0, \ldots, 0)$.

Additive inverse: If $\mathbf{x} = (x_1, x_2, \ldots, x_n)$, then $-\mathbf{x} = (-x_1, -x_2, \ldots, -x_n)$.

The next example illustrates the fact that a vector space consists of more than just a set of vectors—the operations defined on those vectors are an intrinsic part of the definition. If we change either of these operations while maintaining the same set of vectors, the result will, in general, not give a vector space.

Example 6.2.5   Consider $\mathbf{R}^2$, the set of all ordered pairs of real numbers. Define addition and multiplication by a real number in $\mathbf{R}^2$ (denoted by $\oplus$ and $\cdot$, respectively), as follows:

$$(x_1, x_2) \oplus (y_1, y_2) = (x_1 + y_1, x_2 + y_2),$$

$$c \cdot (x_1, x_2) = (cx_1, x_2).$$

Show that $(\mathbf{R}^2, \mathbf{R}, \oplus, \cdot)$ does not form a vector space.

Solution   We have to show only that one of the properties in Definition 6.2.1 is not satisfied. Since we have altered only the scalar multiplication operation and not the addition operation in $\mathbf{R}^2$, the only axiom that can fail is A4. If axiom A4 were satisfied, then it would follow that

$$(a + b) \cdot (x_1, x_2) = [a \cdot (x_1, x_2)] \oplus [b \cdot (x_1, x_2)]$$

for all (x_1, x_2) in $\mathbf{R}^2$. However, according to the scalar multiplication operation given here,

$$(a + b) \cdot (x_1, x_2) = ((a + b)x_1, x_2),$$

whereas

$$[a \cdot (x_1, x_2)] \oplus [b \cdot (x_1, x_2)] = (ax_1, x_2) \oplus (bx_1, x_2) = ((a+b)x_1, 2x_2).$$

Thus,

$$(a+b) \cdot (x_1, x_2) \neq [a \cdot (x_1, x_2)] \oplus [b \cdot (x_1, x_2)],$$

and so $(\mathbf{R}^2, \mathbf{R}, \oplus, \cdot)$ does not form a vector space, since axiom A4 is not satisfied.

THE VECTOR SPACE $\mathbf{C}^n$

The vector spaces that we have considered so far have been real vector spaces.   We now introduce our most important complex vector space.

Definition 6.2.4:   Let $\mathbf{C}^n$ denote the set of all ordered n-tuples of complex numbers.   That is,

$$\mathbf{C}^n = \{(z_1, z_2, \ldots, z_n) : z_i \in \mathbf{C}\}.$$

We refer to the elements in $\mathbf{C}^n$ as **vectors in $\mathbf{C}^n$**.

A typical vector in $\mathbf{C}^n$ is $(z_1, z_2, \ldots, z_n)$, where each z_i is a complex number.

Example 6.2.6   The following are examples of vectors in $\mathbf{C}^2$, and $\mathbf{C}^4$, respectively:

$$\mathbf{u} = (\tfrac{2}{3} - \tfrac{3}{7}i, -\tfrac{1}{2} + i), \qquad \mathbf{v} = (5 + 7i, 2 - i, 3 + 4i, -9 - 17i).$$

In order to obtain a vector space, we must define appropriate operations of addition and multiplication by a scalar.   Based on our operations in $\mathbf{R}^n$, the following definition should not be too surprising:

Definition 6.2.5:   Let $\mathbf{u} = (u_1, u_2, \ldots, u_n)$ and $\mathbf{v} = (v_1, v_2, \ldots, v_n)$ be vectors in $\mathbf{C}^n$ and let k be an arbitrary complex number.   We define **addition** and **scalar multiplication** in $\mathbf{C}^n$ by

$$\mathbf{u} + \mathbf{v} = (u_1, u_2, \ldots, u_n) + (v_1, v_2, \ldots, v_n)$$
$$= (u_1 + v_1, u_2 + v_2, \ldots, u_n + v_n),$$
$$k\mathbf{u} = k(u_1, u_2, \ldots, u_n) = (ku_1, ku_2, \ldots, ku_n),$$

respectively.

Example 6.2.7   If $\mathbf{u} = (1 - 3i, 2 + 4i)$, $\mathbf{v} = (-2 + 4i, 5 - 6i)$, and $k = 2 + i$, find $\mathbf{u} + k\mathbf{v}$.

Solution

$$\begin{aligned}
\mathbf{u} + k\mathbf{v} &= (1 - 3i, 2 + 4i) + (2 + i)(-2 + 4i, 5 - 6i) \\
&= (1 - 3i, 2 + 4i) + (-8 + 6i, 16 - 7i) \\
&= (-7 + 3i, 18 - 3i).
\end{aligned}$$

It is straightforward to show that $\mathbf{C}^n$ together with the preceding operations of addition and scalar multiplication form a (complex) vector space.

FURTHER PROPERTIES OF VECTOR SPACES

The main reason for formalizing the definition of a vector space is that any results that we can prove based solely on the definition will then apply to all vector spaces (for example, we do not have to prove separate results for geometric vectors, $n \times n$ matrices, vectors in $\mathbf{R}^n$ or $\mathbf{C}^n$, or real-valued functions). The next theorem lists three results that can be proved using the vector space axioms. We will require the first two of these in the proof of Theorem 6.3.1, whereas the third result will be used in Section 7.2. If these results had not been consequences of the definition of a vector space, then they would have been put in as additional axioms.

Theorem 6.2.1:   Let V be a vector space.

1. $0\mathbf{u} = \mathbf{0}$ for all $\mathbf{u} \in V$.
2. $-\mathbf{u} = (-1)\mathbf{u}$ for all $\mathbf{u} \in V$.
3. $c\mathbf{0} = \mathbf{0}$ for all scalars c.

PROOF   We prove only property 1. The proofs of (2) and (3) are left as exercises. Let $\mathbf{u}$ be an arbitrary element in a vector space V. Then

$$0 = 0 + 0$$

implies that

$$0\mathbf{u} = (0 + 0)\mathbf{u}.$$

Using axiom A4 this can be written as

$$0\mathbf{u} = 0\mathbf{u} + 0\mathbf{u}.$$

Adding $-0\mathbf{u}$ to both sides of this equation yields

$$(0\mathbf{u}) + (-0\mathbf{u}) = [0\mathbf{u} + 0\mathbf{u}] + (-0\mathbf{u}),$$

that is, since addition in a vector space is associative (axiom A1),

$$(0\mathbf{u}) + (-0\mathbf{u}) = 0\mathbf{u} + [0\mathbf{u} + (-0\mathbf{u})].$$

Using axiom A3 this can be written as

$$\mathbf{0} = 0\mathbf{u} + \mathbf{0},$$

that is, using axiom A2,

$$\mathbf{0} = 0\mathbf{u}. \qquad \blacksquare$$

REMARK   Notice that the preceding proof required the use of each of the vector space axioms.   This indicates that there is no excess baggage in the definition.

We end this section with a list of the most important vector spaces that will be required throughout the remainder of the text.   In each case the addition and scalar multiplication operations are the usual ones associated with the set of vectors.

1. $\mathbf{R}^n$, the (real) vector space of all ordered n-tuples of real numbers.
2. $\mathbf{C}^n$, the (complex) vector space of all ordered n-tuples of complex numbers.
3. $M_n(\mathbf{R})$, the (real) vector space of all $n \times n$ matrices with real elements.
4. $C^k(\mathrm{I})$, the vector space of all real-valued functions that are continuous and have (at least) k continuous derivatives on I.   We will show that this set of vectors is a (real) vector space in the next section.
5. P_n, the vector space of all polynomials of degree less than n with real coefficients, that is,

$$P_n = \{p(x) = a_0 + a_1 x + a_2 x^2 + \cdots + a_{n-1} x^{n-1} : a_0, a_1, \ldots, a_{n-1} \in \mathbf{R}\}.$$

We leave the verification that P_n forms a (real) vector space as an exercise.

EXERCISES 6.2

In problems 1–21 determine whether the given set of vectors forms a *real* vector space.   In each case the operations $+$ and $\cdot$ are the usual ones associated with the set of objects.   For each vector space you must explicitly exhibit the zero vector and the additive inverse of a general vector. If the set and operations do not form a vector space, you must explain why not.

1. $M_3(\mathbf{R})$.

2. $\mathbf{R}^2$.

3. The set of all vectors in $\mathbf{R}^2$ of the form $(x, -x)$.

4. The set of all real numbers of the form $a + b\sqrt{5}$, where a and b are real numbers.

5. The set consisting only of the real number zero.

6. The set of all functions defined on the interval $(-\infty, \infty)$ that are solutions to the linear differential equation $y'' + y = 0$.

7. The set of all functions defined on the interval $(-\infty, \infty)$ that are solutions to the differential equation $y'' + y = x$.

8. $M_n(\mathbf{R})$.

9. The set of all column 3-vectors with real elements.

10. The set of all row 2-vectors with real elements.

11. The set of all solutions to the homogeneous $m \times n$ linear system $A\mathbf{x} = \mathbf{0}$, where A is a fixed $m \times n$ matrix.

12. The set of all solutions to the nonhomogeneous linear system $A\mathbf{x} = \mathbf{b}$, where A is a fixed matrix and $\mathbf{b}$ ($\neq \mathbf{0}$) is a fixed vector.

13. The set of all *nonsingular* 2×2 matrices.

14. The set of all functions defined on the interval $[a, b]$ and satisfying $f(a) = 0$.

15. The set of all functions defined on the interval $[a, b]$ and satisfying $f(a) = 1$.

16. The set of all functions of the form $f(x) = ax$, where a is a constant.

17. The set of all functions of the form $f(x) = ax + b$, where a and b are constants, $b \neq 0$.

18. P_3, the set of all polynomials of degree less than 3 with real coefficients.

19. The set of all polynomials of degree exactly 2—that is, polynomials of the form $p(x) = a_0 + a_1 x + a_2 x^2$, $a_2 \neq 0$.

20. The set of all functions of the form $f(x) = a/x$ on the interval $(0, \infty)$, where a is a constant.

21. The set of all convergent infinite series.

In problems 22 and 23 determine whether the given set of objects forms a *complex* vector space. In each case the operations of + and · are the usual ones associated with the set of objects.

22. $\mathbf{C}^2$.

23. $M_2(\mathbf{C})$, the set of all 2×2 matrices with complex elements.

24. Is $\mathbf{C}^3$ a *real* vector space?

25. On $\mathbf{R}^2$ define the operations of addition and multiplication by a real number ($\oplus$ and · respectively) as follows:

$$\mathbf{u} \oplus \mathbf{v} = \mathbf{u} - \mathbf{v},$$

$$k \cdot \mathbf{u} = -k\mathbf{u},$$

where the operations on the right-hand side are the usual ones. Which of the axioms for a vector space are satisfied by $\mathbf{R}^2$ with the operations $\oplus$ and ·?

26. On $\mathbf{R}^2$ define the operation of addition by

$$(x_1, x_2) \oplus (y_1, y_2) = (x_1 y_1, x_2 y_2).$$

Do axioms A2 and A3 in the definition of a vector space hold?

27. On $M_2(\mathbf{R})$ define the operation of addition by

$$A \oplus B = AB.$$

Determine which axioms for a vector space are satisfied by $M_2(\mathbf{R})$ with this addition operation and the usual scalar multiplication operation.

28. In $M_2(\mathbf{R})$ define the operations of addition and multiplication by a real number ($\oplus$ and ·, respectively) as follows:

$$A \oplus B = -(A + B),$$

$$k \cdot A = -kA,$$

where the operations on the right-hand side are the usual ones associated with $M_2(\mathbf{R})$. Determine which of the axioms for a vector space are satisfied by $M_2(\mathbf{R})$ with the operations $\oplus$ and ·.

29. Consider the set of all 2×2 matrices with real entries. Define operations of addition and multiplication by a real number on this set that do not yield a vector space.

30. Prove that if $\mathbf{v}_1, \mathbf{v}_2, \ldots, \mathbf{v}_k$ are vectors in a vector space V and $c_1, c_2, \ldots, c_k$ are scalars, then

$$c_1 \mathbf{v}_1 + c_2 \mathbf{v}_2 + \cdots + c_k \mathbf{v}_k$$

is a vector in V.

31. Prove property 2 of Theorem 6.2.1.

32. Prove property 3 of Theorem 6.2.1.

6.3   SUBSPACES

In many of the problems that we encounter in the future we will be working with a subset of vectors, S, from a vector space V and will need to determine whether that subset is itself a vector space (under the same operation of addition and scalar multiplication as in V). For example, we have shown that the set of all functions defined on an interval forms a vector space, V. When we solve a differential equation on an interval I, the set of all its solutions is a subset of V. The theory behind solving certain linear differential equations will follow directly once we have shown that the subset of all solutions is itself a vector space. We will have to develop some more vector space theory before we can fill in the details, however; Example 6.3.7 treats a special case.

The importance of this type of problem leads to the following.

Definition 6.3.1: Let S be a nonempty subset of a vector space V. If S is itself a vector space under the same operations of addition and scalar multiplication as in V, then we say that S is a **subspace** of V.

Example 6.3.1   Let V be the vector space $\mathbf{R}^3$ and let S be the set of vectors of the form $(x, y, 0)$. Then S is a subspace of V, since Definition 6.2.1 is satisfied in S when we use the same operations of addition and multiplication by a real number in S as in $\mathbf{R}^3$. Geometrically we can interpret $\mathbf{R}^3$ as the whole of 3-space, in which case the subspace of vectors of the form $(x, y, 0)$ corresponds to the set of points in the xy-plane. See Figure 6.3.1.

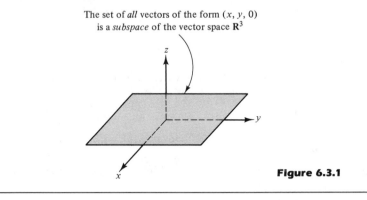

The set of *all* vectors of the form $(x, y, 0)$
is a *subspace* of the vector space $\mathbf{R}^3$

Figure 6.3.1

For a subset, S, of a vector space V to be a subspace of V, it must satisfy Definition 6.2.1 and hence certainly must be closed under addition and scalar multiplication. That is, whenever we add two vectors in S, the resulting vector must also be in S, and whenever we multiply a vector in S by a scalar in F, the result must be a vector in S. The following theorem establishes that provided these closure properties are satisfied, then axioms A1–A4 necessarily hold in S, and hence S is a subspace of V. This result will be used on several occasions during the remainder of the text (see Figure 6.3.2 for an illustration).

Theorem 6.3.1:   Let S be a nonempty subset of a vector space V. Then S is a subspace of V if and only if S is closed under the operations of addition and scalar multiplication in V—that is, if and only if

1. $\mathbf{u} + \mathbf{v}$ is in S whenever $\mathbf{u}$ and $\mathbf{v}$ are in S
and
2. $c\mathbf{u}$ is in S for all vectors $\mathbf{u}$ in S and all scalars c.

PROOF   If S is a subspace of V, then it is a vector space and hence it is certainly closed under addition and scalar multiplication. Conversely, assume (1) and (2) are satisfied. Then S is closed under addition and scalar multiplication. We must prove that axioms A1–A4 hold when we restrict to vectors in S. Consider first A1 and A4. These are properties of the addition and scalar multiplication operations and, hence, since we use the same operations in S as in V, axioms A1 and A4 necessarily hold when we restrict to vectors in S. That A2 and A3 also hold for vectors in S can be proved as follows. If $\mathbf{u}$ is in S, then, since S is closed under scalar multiplication, both $(-1)\mathbf{u}$ and $0\mathbf{u}$ are in S. Thus, from Theorem 6.2.1, $-\mathbf{u}$ and $\mathbf{0}$ are in S, and hence A2 and A3 are satisfied.   ■

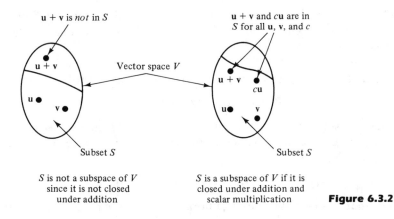

u + v is *not* in S

u + v and *c***u** are in
S for all **u**, **v**, and *c*

Vector space *V*

u + v

*c***u**

u

v

Subset S

Subset S

S is not a subspace of *V*
since it is not closed
under addition

S is a subspace of *V* if it is
closed under addition and
scalar multiplication

Figure 6.3.2

In determining whether a subset S of a vector space V is a subspace of V, we must keep clear in our minds what the given vector space is and what conditions on the vectors in V restrict them to lie in the subset S. This is most easily done by expressing S in set notation as follows:

$$S = \{\mathbf{v} \in V : \text{conditions on } \mathbf{v}\}.$$

↗ ↖

Vector space   Conditions restricting **v** to S

Example 6.3.2 If $V = M_2(\mathbf{R})$—the vector space of all 2×2 matrices with real elements—and S is the subset of V consisting of those 2×2 matrices with $a_{11} = a_{22} = 0$, show that S is a subspace of V.

Solution   We first express the given subset in set notation.

$$S = \{A \in M_2(\mathbf{R}) : A = \begin{bmatrix} 0 & a \\ b & 0 \end{bmatrix}, \text{ where } a \text{ and } b \text{ are real numbers}\}.$$

If A and B are two vectors in S, then

$$A = \begin{bmatrix} 0 & a \\ b & 0 \end{bmatrix} \quad \text{and} \quad B = \begin{bmatrix} 0 & c \\ d & 0 \end{bmatrix}$$

for some real numbers a, b, c, d. Thus

$$A + B = \begin{bmatrix} 0 & a \\ b & 0 \end{bmatrix} + \begin{bmatrix} 0 & c \\ d & 0 \end{bmatrix} = \begin{bmatrix} 0 & a+c \\ b+d & 0 \end{bmatrix},$$

so that $A + B$ is in S and hence S is closed under addition. Further, if k is an arbitrary real number,

$$kA = k\begin{bmatrix} 0 & a \\ b & 0 \end{bmatrix} = \begin{bmatrix} 0 & ka \\ kb & 0 \end{bmatrix},$$

which is once more a vector in S. It follows that S is also closed under scalar multiplication; hence, from Theorem 6.3.1, since S is clearly nonempty, S is a subspace of V.

Example 6.3.3   Let V be the vector space of all real-valued functions defined on an interval $[a, b]$, and let S denote the set of all functions in V that satisfy $f(a) = 0$. Show that S is a subspace of V.

Solution   We first express S in set notation.   For the present problem we have

$$S = \{f \in V : f(a) = 0\},$$

which is nonempty since it contains, for example, the zero function. We now check for closure under addition and scalar multiplication.   If f and g are in S, then $f(a) = 0$, and $g(a) = 0$.   Thus

$$(f + g)(a) = f(a) + g(a) = 0 + 0 = 0,$$

which implies that S is closed under addition.   Further, if c is any real number,

$$(cf)(a) = cf(a) = c0 = 0,$$

so that S is also closed under scalar multiplication.   Theorem 6.3.1 therefore implies that S *is* a subspace of V.

Example 6.3.4   Show that the set of all solutions to the $m \times n$ homogeneous linear system $A\mathbf{x} = \mathbf{0}$ is a subspace of $\mathbf{R}^n$.

Solution   The subset of $\mathbf{R}^n$ of interest is

$$S = \{\mathbf{x} \in \mathbf{R}^n : A\mathbf{x} = \mathbf{0}\},$$

that is, of all the vectors in $\mathbf{R}^n$, we consider only those that are solutions of the homogeneous linear system $A\mathbf{x} = \mathbf{0}$.   This subset is necessarily nonempty, since $A\mathbf{x} = \mathbf{0}$ always admits the trivial solution $\mathbf{x} = \mathbf{0}$.   Let $\mathbf{x}$ and $\mathbf{y}$ be vectors in S.   This means that

$$A\mathbf{x} = \mathbf{0} \quad \text{and} \quad A\mathbf{y} = \mathbf{0}.$$

Thus

$$A(\mathbf{x} + \mathbf{y}) = A\mathbf{x} + A\mathbf{y} = \mathbf{0} + \mathbf{0} = \mathbf{0},$$

which implies that $\mathbf{x} + \mathbf{y}$ is in S and hence S is closed under addition.   Further, if c is any real number,

$$A(c\mathbf{x}) = cA\mathbf{x} = c\mathbf{0} = \mathbf{0},$$

so that S is also closed under scalar multiplication; hence S *is* a subspace of $\mathbf{R}^n$.

Example 6.3.5   Let S be the subset of $\mathbf{R}^3$ that consists of all vectors of the form $(t, 1, -t)$, where t is a real number.   Show that S is *not* a subspace of $\mathbf{R}^3$.

Solution   In this case we have

$$S = \{\mathbf{x} \in \mathbf{R}^3 : \mathbf{x} = (t, 1, -t), \text{ for some } t \in \mathbf{R}\}.$$

To show that S is not a subspace of $\mathbf{R}^3$, we show that it is not closed under addition. (Note, however, that S does not contain the zero vector and so directly cannot form a vector space.) If $\mathbf{x}$ and $\mathbf{y}$ are vectors in S, then $\mathbf{x} = (r, 1, -r)$ and $\mathbf{y} = (s, 1, -s)$, for some real numbers r and s. Thus

$$\mathbf{x} + \mathbf{y} = (r, 1, -r) + (s, 1, -s) = (r+s, 2, -r-s).$$

That is,

$$\mathbf{x} + \mathbf{y} = (t, 2, -t),$$

where $t = r + s$. Hence $\mathbf{x} + \mathbf{y}$ is not in S since it is not of the form $(t, 1, -t)$. It follows from Theorem 6.3.1 that S is *not* a subspace of $\mathbf{R}^3$.

We now introduce the vector space of primary importance in the study of linear ordinary differential equations. This vector space arises as a subspace of the vector space of all functions that are defined on an interval I.

Example 6.3.6 Let V denote the vector space of all functions that are defined on an interval I, and let $C^k(\text{I})$ denote the set of all functions that are continuous and have (at least) k continuous derivatives on the interval I. Show that $C^k(\text{I})$ is a subspace of V.

Solution In this case

$$C^k(\text{I}) = \{f \in V : f, f', f'', \ldots, f^{(k)} \text{ exist and are continuous on I}\}.$$

It follows from the properties of derivatives that if we add two functions in $C^k(\text{I})$, the result is a function in $C^k(\text{I})$, and similarly if we multiply a function in $C^k(\text{I})$ by a scalar, then the result is a function in $C^k(\text{I})$. Thus Theorem 6.2.1 implies that $C^k(\text{I})$ is a subspace of V.

REMARK In general k will be a positive integer. We use the notation $C^\infty(\text{I})$ to denote the vector space of all functions that have continuous derivatives of all orders on I.

Our final example indicates how vector spaces naturally arise in differential equation theory.

Example 6.3.7 Let S be the set of all solutions to the linear differential equation

$$y'' + ay' + by = 0,$$

that are defined on an interval I (here we will assume that a and b are real constants). Show that S is a subspace of $C^2(\text{I})$.

Solution Any solution to the given differential equation is certainly in $C^2(\text{I})$, and hence the set of all solutions is a subset of $C^2(\text{I})$. Indeed, we can write

$$S = \{y \in C^2(\text{I}) : y'' + ay' + by = 0\}.$$

We now show that S is, in fact, a subspace of $C^2(I)$. S is nonempty since, for example, the zero function, $y(x) = 0$ for all x in I, is a solution of $y'' + ay' + by = 0$ and therefore is in S.   Now let y_1 and y_2 be in S.   Thus

$$y_1'' + ay_1' + by_1 = 0 \quad \text{and} \quad y_2'' + ay_2' + by_2 = 0.$$

We must show that $y_1 + y_2$ and cy_1 are also in S.   We proceed as follows:

$$(y_1 + y_2)'' + a(y_1 + y_2)' + b(y_1 + y_2) = (y_1'' + y_2'') + (ay_1' + ay_2') + b(y_1 + y_2)$$
$$= (y_1'' + ay_1' + by_1) + (y_2'' + ay_2' + by_2)$$
$$= 0.$$

Thus S is closed under addition.   Further, for all real numbers c,

$$(cy_1)'' + a(cy_1)' + b(cy_1) = cy_1'' + acy_1' + bcy_1$$
$$= c(y_1'' + ay_1' + by_1) = c(0) = 0,$$

so that S is also closed under scalar multiplication; hence S is a subspace of $C^2(I)$.

Note that in the preceding example we did not have to *find* the solutions of the differential equation in order to show that the set of all solutions forms a subspace of $C^2(I)$.

EXERCISES 6.3

In problems 1–17 express S in set notation and then determine whether it is a subspace of the given vector space V.

1. $V = \mathbf{R}^3$ and S is the set of all vectors in V of the form $(0, x_2, x_3)$.

2. $V = \mathbf{R}^2$ and S is the set of all vectors (x_1, x_2) in V satisfying $3x_1 + 2x_2 = 0$.

3. $V = \mathbf{R}^3$ and S is the set of all vectors (x_1, x_2, x_3) in V satisfying $x_3 = 2x_1 - x_2$.

4. $V = \mathbf{R}^4$ and S is the set of vectors of the form $(x_1, 0, x_3, 2)$.

5. $V = \mathbf{R}^3$ and S is the set of all vectors (x_1, x_2, x_3) in V satisfying $x_1 + x_2 + x_3 = 1$.

6. $V = \mathbf{R}^n$ and S is the set of all solutions to the nonhomogeneous linear system $A\mathbf{x} = \mathbf{b}$, where A is a fixed $m \times n$ matrix and $\mathbf{b}$ $(\neq \mathbf{0})$ is a fixed vector.

7. $V = M_2(\mathbf{R})$ and S is the subset of all 2×2 matrices with $\det(A) = 1$.

8. V is the vector space of all real-valued functions defined on the interval $[a, b]$ and S is the subset of V consisting of those functions satisfying $f(a) = f(b)$.

9. V is the vector space of all real-valued functions defined on the interval $[a, b]$ and S is the subset of V consisting of those functions satisfying $f(a) = 1$.

10. V is the vector space of all real-valued functions defined on the interval $(-\infty, \infty)$ and S is the subset of V consisting of those functions satisfying $f(-x) = -f(x)$, for all x in $(-\infty, \infty)$.

11. $V = \mathbf{R}^3$ and S consists of all vectors in $\mathbf{R}^3$ of the form $c\mathbf{x}$, where $\mathbf{x}$ is a fixed nonzero vector and c is an arbitrary real number.   Describe S geometrically.

12. $V = C^2(I)$ and S is the subset of V consisting of those functions satisfying the differential equation $f'' + 2f' - f = 0$ on I.

13. $V = C^2(I)$ and S is the subset of V consisting of those functions satisfying the differential equation $f'' + 2f' - f = 1$ on I.

14. $V = M_n(\mathbf{R})$ and S is the subset of all *nonsingular* matrices.

15. $V = M_2(\mathbf{R})$ and S is the subset of all 2×2 *symmetric* matrices.

16. $V = M_2(\mathbf{R})$ and S is the subset of all 2×2 *skew-symmetric* matrices.

17. $V = \mathbf{R}^2$ and S consists of all vectors $\mathbf{x} = (x_1, x_2)$ satisfying $x_1^2 - x_2^2 = 0$.

18. Describe geometrically all possible subspaces of $\mathbf{R}^3$.

19. Let S denote the set of all vectors lying in the plane $3x + 2y - z = 4$. Is S a subspace of $\mathbf{R}^3$?

20. Let $\mathbf{u}$ be a fixed geometric vector, and let S denote the set of all vectors that are perpendicular to $\mathbf{u}$. Show that S is a subspace of the vector space of all geometric vectors and give a geometric interpretation of S.

6.4 LINEAR COMBINATIONS AND SPANNING SETS

Suppose we are given a set of vectors $\{\mathbf{v}_1, \mathbf{v}_2, \ldots, \mathbf{v}_k\}$ in a vector space V. The only way in which we can obtain new vectors from this set is by applying the addition and scalar multiplication operations in V. Specifically, if $c_1, c_2, \ldots, c_k$ are scalars, then the only way we can combine the vectors $\mathbf{v}_1, \mathbf{v}_2, \ldots, \mathbf{v}_k$ is in the following *linear* manner:

$$c_1 \mathbf{v}_1 + c_2 \mathbf{v}_2 + \cdots + c_k \mathbf{v}_k.$$

Since V is closed under the operations of addition and scalar multiplication it follows that the preceding linear combination of $\mathbf{v}_1, \mathbf{v}_2, \ldots, \mathbf{v}_k$ is also a vector in V. We formalize this idea in a definition.

Definition 6.4.1: A vector $\mathbf{v}$ in a vector space V is said to be a **linear combination** of the vectors $\mathbf{v}_1, \mathbf{v}_2, \ldots, \mathbf{v}_k$ in V if there exist scalars $c_1, c_2, \ldots, c_k$ such that

$$\mathbf{v} = c_1 \mathbf{v}_1 + c_2 \mathbf{v}_2 + \cdots + c_k \mathbf{v}_k = \sum_{i=1}^{k} c_i \mathbf{v}_i.$$

Example 6.4.1 The vector $\mathbf{v} = (2, -1)$ in $\mathbf{R}^2$ can be written as a linear combination of the vectors $\mathbf{v}_1 = (1, 0)$, $\mathbf{v}_2 = (0, 1)$, since

$$\mathbf{v} = 2\mathbf{v}_1 - \mathbf{v}_2.$$

In fact any vector, $\mathbf{x} = (x_1, x_2)$, in $\mathbf{R}^2$ can be written as a linear combination of $\mathbf{v}_1 = (1, 0)$ and $\mathbf{v}_2 = (0, 1)$ as follows:

$$\mathbf{x} = (x_1, x_2) = x_1(1, 0) + x_2(0, 1) = x_1 \mathbf{v}_1 + x_2 \mathbf{v}_2.$$

In this case we say that $\mathbf{v}_1$ and $\mathbf{v}_2$ *span* $\mathbf{R}^2$.

Generalizing the previous example to an arbitrary vector space leads to the following definition.

> **Definition 6.4.2:** Let $v_1, v_2, \ldots, v_k$ be vectors in a vector space V. If *every* vector in V can be written as a linear combination of $v_1, v_2, \ldots, v_k$ we say that V is **spanned** by $v_1, v_2, \ldots, v_k$ and call the set of vectors $\{v_1, v_2, \ldots, v_k\}$ a **spanning set** for V.

We illustrate this important idea with several examples.

Example 6.4.2   If $v_1 = (1, 0, 0)$, $v_2 = (0, 1, 0)$, and $v_3 = (0, 0, 1)$, then $\mathbf{R}^3$ is spanned by v_1, v_2, v_3, since any vector (a, b, c) in $\mathbf{R}^3$ can be written as

$$(a, b, c) = a(1, 0, 0) + b(0, 1, 0) + c(0, 0, 1).$$

Example 6.4.3   Show that $\mathbf{R}^2$ is spanned by the vectors $v_1 = (1, 1)$ and $v_2 = (1, -1)$.

Solution   Let $x = (x_1, x_2)$ be an arbitrary vector in $\mathbf{R}^2$. We must show that there exist constants c_1, c_2 such that

$$c_1 v_1 + c_2 v_2 = (x_1, x_2). \tag{6.4.1}$$

Substituting for v_1 and v_2, we require c_1 and c_2 to satisfy

$$c_1(1, 1) + c_2(1, -1) = (x_1, x_2),$$

that is,

$$c_1 + c_2 = x_1,$$
$$c_1 - c_2 = x_2.$$

The determinant of the matrix of coefficients of this system is

$$\begin{vmatrix} 1 & 1 \\ 1 & -1 \end{vmatrix} = -2 \neq 0,$$

and hence the system has a unique solution for any x_1, x_2. Thus (6.4.1) can be satisfied for any vector $x = (x_1, x_2)$, and so the given vectors span $\mathbf{R}^2$. Indeed, solving the preceding system for c_1 and c_2 yields

$$c_1 = \tfrac{1}{2}(x_1 + x_2), \qquad c_2 = \tfrac{1}{2}(x_1 - x_2),$$

so that

$$x = (x_1, x_2) = \tfrac{1}{2}(x_1 + x_2)v_1 + \tfrac{1}{2}(x_1 - x_2)v_2.$$

Example 6.4.4   Let S be the subspace of $\mathbf{R}^3$ consisting of all vectors of the form $(c_1, 2c_2 - 5c_1, 3c_2)$, where c_1 and c_2 are real numbers. Determine a set of vectors that spans S.

Solution In this case we have

$$S = \{\mathbf{x} \in \mathbf{R}^3 : \mathbf{x} = (c_1, 2c_2 - 5c_1, 3c_2),\ c_1, c_2 \in \mathbf{R}\}$$
$$= \{\mathbf{x} \in \mathbf{R}^3 : \mathbf{x} = (c_1, -5c_1, 0) + (0, 2c_2, 3c_2),\ c_1, c_2 \in \mathbf{R}\}$$
$$= \{\mathbf{x} \in \mathbf{R}^3 : \mathbf{x} = c_1(1, -5, 0) + c_2(0, 2, 3),\ c_1, c_2 \in \mathbf{R}\}.$$

Thus every vector in S can be written as a linear combination of the vectors

$$\mathbf{v}_1 = (1, -5, 0), \qquad \mathbf{v}_2 = (0, 2, 3),$$

and hence a spanning set for S is $\{\mathbf{v}_1, \mathbf{v}_2\}$.

Now let $\mathbf{v}_1, \mathbf{v}_2, \ldots, \mathbf{v}_k$ be vectors in a vector space V, and let S denote the subset of V consisting of *all* linear combinations of these vectors, that is,

$$S = \{\mathbf{v} \in V : \mathbf{v} = c_1\mathbf{v}_1 + c_2\mathbf{v}_2 + \cdots + c_k\mathbf{v}_k,\ \text{for some scalars } c_1, c_2, \ldots, c_k\}$$
$$= \{\mathbf{v} \in V : \mathbf{v} = \sum_{i=1}^{k} c_i\mathbf{v}_i,\ \text{for some scalars } c_1, c_2, \ldots, c_k\}. \tag{6.4.2}$$

The next theorem establishes that S is a *subspace* of V.

Theorem 6.4.1: Let $\mathbf{v}_1, \mathbf{v}_2, \ldots, \mathbf{v}_k$ be vectors in a vector space V and let S be the set of all linear combinations of these vectors. Then S forms a subspace of V.

PROOF We must show that S is closed under addition and closed under scalar multiplication. If $\mathbf{u}$ and $\mathbf{v}$ are in S, then, from (6.4.2),

$$\mathbf{u} = \sum_{i=1}^{k} a_i\mathbf{v}_i \quad \text{and} \quad \mathbf{v} = \sum_{i=1}^{k} b_i\mathbf{v}_i,$$

for some scalars a_i, b_i. Thus

$$\mathbf{u} + \mathbf{v} = \sum_{i=1}^{k} a_i\mathbf{v}_i + \sum_{i=1}^{k} b_i\mathbf{v}_i = \sum_{i=1}^{k} (a_i + b_i)\mathbf{v}_i, \qquad c\mathbf{u} = c\sum_{i=1}^{k} a_i\mathbf{v}_i = \sum_{i=1}^{k} (ca_i)\mathbf{v}_i,$$

so that $\mathbf{u} + \mathbf{v}$ and $c\mathbf{u}$ are in S; hence, from Theorem 6.3.1, S is a subspace of V. ∎

The previous theorem showed that S, the set of all linear combinations of the vectors $\mathbf{v}_1, \mathbf{v}_2, \ldots, \mathbf{v}_k$, is a subspace of V. Since, by definition, each vector in S can be written as a linear combination of $\mathbf{v}_1, \mathbf{v}_2, \ldots, \mathbf{v}_k$, it follows that S is spanned by $\mathbf{v}_1, \mathbf{v}_2, \ldots, \mathbf{v}_k$. Due to the importance of this subspace, we introduce the following terminology and notation:

Definition 6.4.3: Let $\mathbf{v}_1, \mathbf{v}_2, \ldots, \mathbf{v}_k$ be vectors in a vector space V. The set of all linear combinations of these vectors is called the **subspace spanned by** $\mathbf{v}_1, \mathbf{v}_2, \ldots, \mathbf{v}_k$ and is denoted by $\text{span}\{\mathbf{v}_1, \mathbf{v}_2, \ldots, \mathbf{v}_k\}$.

Example 6.4.5   If $V = \mathbf{R}^2$ and $\mathbf{v}_1 = (-1, 1)$, determine span$\{\mathbf{v}_1\}$.

Solution   By Definition 6.4.3,

$$\text{span}\{\mathbf{v}_1\} = \{\mathbf{x} \in \mathbf{R}^2 : \mathbf{x} = c_1\mathbf{v}_1, c_1 \in \mathbf{R}\}$$
$$= \{\mathbf{x} \in \mathbf{R}^2 : \mathbf{x} = c_1(-1, 1), c_1 \in \mathbf{R}\}$$
$$= \{\mathbf{x} \in \mathbf{R}^2 : \mathbf{x} = (-c_1, c_1), c_1 \in \mathbf{R}\}.$$

Geometrically this is the line through the origin with parametric equations $x = -c_1$, $y = c_1$, so that the Cartesian equation of the line is $y = -x$ (see Figure 6.4.1).

Example 6.4.6   If $V = \mathbf{R}^3$ and $\mathbf{v}_1 = (1, 0, 1)$, $\mathbf{v}_2 = (0, 1, 1)$, determine the subspace of $\mathbf{R}^3$ spanned by $\mathbf{v}_1$ and $\mathbf{v}_2$.   Does $\mathbf{v} = (1, 1, -1)$ lie in this subspace?

Solution

$$\text{span}\{\mathbf{v}_1, \mathbf{v}_2\} = \{\mathbf{x} \in \mathbf{R}^3 : \mathbf{x} = c_1\mathbf{v}_1 + c_2\mathbf{v}_2, c_1, c_2 \in \mathbf{R}\}$$
$$= \{\mathbf{x} \in \mathbf{R}^3 : \mathbf{x} = c_1(1, 0, 1) + c_2(0, 1, 1), c_1, c_2 \in \mathbf{R}\}$$
$$= \{\mathbf{x} \in \mathbf{R}^3 : \mathbf{x} = (c_1, c_2, c_1 + c_2), c_1, c_2 \in \mathbf{R}\}$$

If $\mathbf{v} = (1, 1, -1)$, then we see that this is not of the form $(c_1, c_2, c_1 + c_2)$, and hence $\mathbf{v}$ does not lie in span$\{\mathbf{v}_1, \mathbf{v}_2\}$.   Geometrically, span$\{\mathbf{v}_1, \mathbf{v}_2\}$ is the plane through the origin determined by the two given vectors $\mathbf{v}_1$ and $\mathbf{v}_2$.   It has parametric equations $x = c_1$, $y = c_2$, $z = c_1 + c_2$, which implies that its Cartesian equation is $z = x + y$.   Thus the fact that $\mathbf{v}$ is not in span$\{\mathbf{v}_1, \mathbf{v}_2\}$ means that it does not lie in this plane.   The subspace is depicted in Figure 6.4.2.

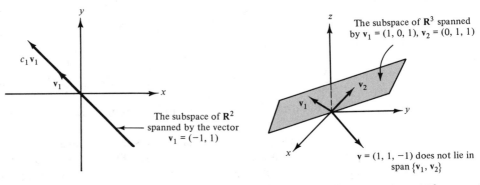

Figure 6.4.1

The subspace of $\mathbf{R}^2$ spanned by the vector $\mathbf{v}_1 = (-1, 1)$

The subspace of $\mathbf{R}^3$ spanned by $\mathbf{v}_1 = (1, 0, 1)$, $\mathbf{v}_2 = (0, 1, 1)$

$\mathbf{v} = (1, 1, -1)$ does not lie in span $\{\mathbf{v}_1, \mathbf{v}_2\}$

Figure 6.4.2   The subspace of $\mathbf{R}^3$ spanned by $\mathbf{v}_1 = (1, 0, 1)$ and $\mathbf{v}_2 = (0, 1, 1)$ is the plane $z = x + y$.

Example 6.4.7   If $V = M_2(\mathbf{R})$ and $A_1 = \begin{bmatrix} 1 & 0 \\ 0 & 0 \end{bmatrix}$, $A_2 = \begin{bmatrix} 0 & 1 \\ 1 & 0 \end{bmatrix}$, and

$A_3 = \begin{bmatrix} 0 & 0 \\ 0 & 1 \end{bmatrix}$, determine span$\{A_1, A_2, A_3\}$.

Solution By definition we have:

span$\{A_1, A_2, A_3\}$

$$= \{A \in M_2(\mathbf{R}) : A = c_1 A_1 + c_2 A_2 + c_3 A_3, \, c_1, c_2, c_3 \in \mathbf{R}\}$$

$$= \left\{ A \in M_2(\mathbf{R}) : A = c_1 \begin{bmatrix} 1 & 0 \\ 0 & 0 \end{bmatrix} + c_2 \begin{bmatrix} 0 & 1 \\ 1 & 0 \end{bmatrix} + c_3 \begin{bmatrix} 0 & 0 \\ 0 & 1 \end{bmatrix}, \, c_1, c_2, c_3 \in \mathbf{R} \right\}$$

$$= \left\{ A \in M_2(\mathbf{R}) : A = \begin{bmatrix} c_1 & c_2 \\ c_2 & c_3 \end{bmatrix} \right\}.$$

This is the set of all real *symmetric* 2×2 matrices.

Example 6.4.8 We have shown in Section 6.1 that the set of all solutions to the linear differential equation $y'' + y = 0$ is

$$S = \{y \in C^2(\mathrm{I}) : y = c_1 \cos x + c_2 \sin x\}.$$

In terms of the ideas introduced in this section, we can therefore write

$$S = \mathrm{span}\{\cos x, \sin x\};$$

that is, the solution set is the *subspace* of $C^2(\mathrm{I})$ spanned by $y_1 = \cos x$, $y_2 = \sin x$. We will develop this way of looking at the solution of linear differential equations further in the later chapters.

EXERCISES 6.4

In problems 1 and 2 determine whether the given vectors span $\mathbf{R}^2$.

1. $v_1 = (1, 2)$, $v_2 = (0, -1)$.

2. $v_1 = (2, 1)$, $v_2 = (-4, -2)$.

3. Show that $v_1 = (2, -1)$, $v_2 = (3, 2)$ span $\mathbf{R}^2$ and express the vector $v = (5, -7)$ as a linear combination of v_1, v_2.

4. Show that $v_1 = (-1, 3, 2)$, $v_2 = (1, -2, 1)$, $v_3 = (2, 1, 1)$ span $\mathbf{R}^3$ and express $x = (x_1, x_2, x_3)$ as a linear combination of v_1, v_2, v_3.

5. Show that $v_1 = (1, 1)$, $v_2 = (-1, 2)$, $v_3 = (1, 4)$ span $\mathbf{R}^2$. Do v_1 and v_2 span $\mathbf{R}^2$ also?

6. Let S be the subspace of $\mathbf{R}^3$ consisting of all vectors of the form

$$x = (c_1, c_2, c_2 - 2c_1).$$

Show that S is spanned by $v_1 = (1, 0, -2)$, $v_2 = (0, 1, 1)$.

7. Let S be the subspace of $\mathbf{R}^4$ consisting of all vectors of the form

$$v = (x_1, x_2, x_2 - x_1, x_1 - 2x_2).$$

Determine a set of vectors that spans S.

8. Let S be the subspace of $\mathbf{R}^3$ consisting of all solutions to the linear equation

$$x_1 - 2x_2 - x_3 = 0.$$

Determine a set of vectors that spans S.

9. Let S be the set of all solutions to the linear system $Ax = 0$, where $A = \begin{bmatrix} 1 & 2 & 3 \\ 1 & 3 & 4 \\ 2 & 4 & 6 \end{bmatrix}$. Determine a vector that spans S.

10. Let S be the subspace of $M_2(\mathbf{R})$ consisting of all *symmetric* 2×2 matrices with real elements.

Show that S is spanned by the matrices

$$A_1 = \begin{bmatrix} 1 & 0 \\ 0 & 0 \end{bmatrix}, \quad A_2 = \begin{bmatrix} 0 & 0 \\ 0 & 1 \end{bmatrix},$$

$$A_3 = \begin{bmatrix} 0 & 1 \\ 1 & 0 \end{bmatrix}.$$

11. Let S be the subspace of $M_2(\mathbf{R})$ consisting of all *skew-symmetric* 2×2 matrices with real elements. Determine a matrix that spans S.

12. Let S be the subset of $M_2(\mathbf{R})$ consisting of all matrices of the form $\begin{bmatrix} a_{11} & a_{12} \\ 0 & a_{22} \end{bmatrix}$.

(a) Show that S is a subspace of $M_2(\mathbf{R})$.
(b) Determine a set of 2×2 matrices that spans S.

13. Let S be the subspace of $\mathbf{R}^3$ spanned by the vector $\mathbf{x} = (1, -1, 2)$. Show that the geometric interpretation of S is a line through the origin, and find its equation.

In problems 14 and 15 determine span$\{\mathbf{v}_1, \mathbf{v}_2\}$ for the given vectors in $\mathbf{R}^3$ and interpret it geometrically.

14. $\mathbf{v}_1 = (1, -1, 2)$, $\mathbf{v}_2 = (2, -1, 3)$.
15. $\mathbf{v}_1 = (1, 2, -1)$, $\mathbf{v}_2 = (-2, -4, 2)$.
16. Let S be the subspace of $\mathbf{R}^3$ spanned by the vectors $\mathbf{v}_1 = (1, 1, -1)$, $\mathbf{v}_2 = (2, 1, 3)$, $\mathbf{v}_3 = (-2, -2, 2)$. Show that S is also spanned by $\mathbf{v}_1$ and $\mathbf{v}_2$ only.

In problems 17–19 determine whether the given vector $\mathbf{v}$ lies in span$\{\mathbf{v}_1, \mathbf{v}_2\}$.

17. $\mathbf{v}_1 = (1, -1, 2)$, $\mathbf{v}_2 = (2, 1, 3)$, $\mathbf{v} = (3, 3, 4)$ in $\mathbf{R}^3$.

18. $\mathbf{v}_1 = (-1, 1, 2)$, $\mathbf{v}_2 = (3, 1, -4)$, $\mathbf{v} = (5, 3, -6)$ in $\mathbf{R}^3$.

19. $\mathbf{v}_1 = (3, 1, 2)$, $\mathbf{v}_2 = (-2, -1, 1)$, $\mathbf{v} = (1, 1, -2)$ in $\mathbf{R}^3$.

20. Let $p_1(x) = x - 4$, and $p_2(x) = x^2 - x + 3$ be vectors in P_3. Determine whether

$$p(x) = 2x^2 - x + 2$$

is in span$\{p_1, p_2\}$.

21. Consider   the   vectors $A_1 = \begin{bmatrix} 1 & -1 \\ 2 & 0 \end{bmatrix}$,

$A_2 = \begin{bmatrix} 0 & 1 \\ -2 & 1 \end{bmatrix}$,   and $A_3 = \begin{bmatrix} 3 & 0 \\ 1 & 2 \end{bmatrix}$ in $M_2(\mathbf{R})$.

Determine span$\{A_1, A_2, A_3\}$.

22. Consider   the   vectors $A_1 = \begin{bmatrix} 1 & 2 \\ -1 & 3 \end{bmatrix}$ and

$A_2 = \begin{bmatrix} -2 & 1 \\ 1 & -1 \end{bmatrix}$ in $M_2(\mathbf{R})$.   Find span$\{A_1, A_2\}$

and determine whether $B = \begin{bmatrix} 3 & 1 \\ -2 & 4 \end{bmatrix}$ lies in this subspace.

23. Let $V = C^\infty(I)$ and let S be the subspace of V spanned by the functions

$$f(x) = \cosh x, \ g(x) = \sinh x.$$

(a) Give an expression for a general vector in S.
(b) Show that S is also spanned by the functions $h(x) = e^x$, $j(x) = e^{-x}$.

6.5   LINEAR DEPENDENCE AND INDEPENDENCE

This is one of the most important sections in the chapter, since the ideas and terminology introduced here will be used throughout the remainder of the text.
     One of the basic problems in linear algebra is to determine whether a given set of vectors $\{\mathbf{v}_1, \mathbf{v}_2, \ldots, \mathbf{v}_k\}$ is dependent in the sense that at least one of the vectors can be written as a linear combination of the remaining vectors in the set. This idea is formulated mathematically in the following definition.

Definition 6.5.1:   A set of vectors $\{v_1, v_2, \ldots, v_k\}$ in a vector space V is said to be **linearly dependent** if there exist scalars $c_1, c_2, \ldots, c_k$, *not all zero*, such that

$$c_1 v_1 + c_2 v_2 + \cdots + c_k v_k = 0.$$

The next theorem shows that Definition 6.5.1 has indeed captured the idea of dependence.

Theorem 6.5.1:   The set of vectors $\{v_1, v_2, \ldots, v_k\}$ in a vector space V is linearly dependent if and only if at least one vector in the set is a linear combination of the others.

PROOF   If $\{v_1, v_2, \ldots, v_k\}$ is a linearly dependent set of vectors, then there exist scalars $c_1, c_2, \ldots, c_k$, not all zero, such that

$$c_1 v_1 + c_2 v_2 + \cdots + c_k v_k = 0.$$

Suppose that $c_i \neq 0$. Then we can express v_i as a linear combination of the other vectors as follows:

$$v_i = -\frac{1}{c_i}[c_1 v_1 + c_2 v_2 + \cdots + c_{i-1} v_{i-1} + c_{i+1} v_{i+1} + \cdots + c_k v_k].$$

Conversely, suppose that one of the vectors, say v_j, can be expressed as a linear combination of the remaining vectors, that is,

$$v_j = c_1 v_1 + c_2 v_2 + \cdots + c_{j-1} v_{j-1} + c_{j+1} v_{j+1} + \cdots + c_k v_k.$$

Then,

$$c_1 v_1 + c_2 v_2 + \cdots + c_{j-1} v_{j-1} - v_j + c_{j+1} v_{j+1} + \cdots + c_k v_k = 0.$$

Since the coefficient of $v_j = -1 \neq 0$, the set of vectors $\{v_1, v_2, \ldots, v_k\}$ is linearly dependent. ∎

Although Definition 6.5.1 applies to vectors in an arbitrary vector space, as usual we can obtain a geometric interpretation by considering vectors in R^3. In particular, two vectors x_1 and x_2 in R^3 are linearly dependent if and only if one is a scalar multiple of the other. Geometrically the vectors must be colinear.

If the set of vectors $\{v_1, v_2, \ldots, v_k\}$ is not linearly dependent, then it is said to be *linearly independent*.

Definition 6.5.2:   The set of vectors $\{v_1, v_2, \ldots, v_k\}$ in a vector space V is **linearly independent** if the *only* values of the scalars $c_1, c_2, \ldots, c_k$ for which

$$c_1 v_1 + c_2 v_2 + \cdots + c_k v_k = 0$$

are $c_1 = c_2 = \cdots = c_k = 0.$

REMARKS

1. We can use the result of Theorem 3.2.2 (page 97) to write the homogeneous linear system $A\mathbf{c} = \mathbf{0}$ as

$$c_1\mathbf{a}_1 + c_2\mathbf{a}_2 + \cdots + c_n\mathbf{a}_n = \mathbf{0},$$

where $\mathbf{a}_1, \mathbf{a}_2, \ldots, \mathbf{a}_n$ denote the column vectors of A. It follows that the problem of determining whether a set of vectors in an arbitrary vector space is linearly independent or linearly dependent can be considered as a generalization of the problem of determining whether $A\mathbf{c} = \mathbf{0}$ has just the trivial solution or an infinite number of solutions respectively.

2. Even though linear dependence and linear independence is a property of the set of vectors $\{\mathbf{v}_1, \mathbf{v}_2, \ldots, \mathbf{v}_k\}$, we will often talk about the vectors $\mathbf{v}_1, \mathbf{v}_2, \ldots, \mathbf{v}_k$ being linearly dependent or independent.

Example 6.5.1

(a) In $\mathbf{R}^3$ the vectors $\mathbf{v}_1 = (1, 3, 1)$, $\mathbf{v}_2 = (1, -1, 3)$, and $\mathbf{v}_3 = (1, 7, -1)$ are linearly dependent, since $2\mathbf{v}_1 - \mathbf{v}_2 - \mathbf{v}_3 = \mathbf{0}$.

(b) The vectors $(1, 0, 0)$, $(0, 1, 0)$, $(0, 0, 1)$ in $\mathbf{R}^3$ are linearly independent, since

$$c_1(1, 0, 0) + c_2(0, 1, 0) + c_3(0, 0, 1) = (0, 0, 0)$$

is satisfied if and only if $c_1 = 0$, $c_2 = 0$, $c_3 = 0$.

(c) In $C^\infty(I)$, $f_1(x) = \sin^2 x$, $f_2(x) = \cos^2 x$, $f_3(x) = 1$ are linearly dependent, since

$$f_1 + f_2 - f_3 = 0.$$

Example 6.5.2   Determine whether the following vectors are linearly dependent or linearly independent in $M_2(\mathbf{R})$:

$$A_1 = \begin{bmatrix} 1 & -1 \\ 2 & 0 \end{bmatrix}, \quad A_2 = \begin{bmatrix} 2 & 1 \\ 0 & 3 \end{bmatrix}, \quad A_3 = \begin{bmatrix} 1 & -1 \\ 2 & 1 \end{bmatrix}.$$

Solution   The condition for determining linear dependence or independence, namely,

$$c_1 A_1 + c_2 A_2 + c_3 A_3 = 0,$$

is equivalent, in this case, to

$$c_1\begin{bmatrix} 1 & -1 \\ 2 & 0 \end{bmatrix} + c_2\begin{bmatrix} 2 & 1 \\ 0 & 3 \end{bmatrix} + c_3\begin{bmatrix} 1 & -1 \\ 2 & 1 \end{bmatrix} = \begin{bmatrix} 0 & 0 \\ 0 & 0 \end{bmatrix},$$

which is satisfied if and only if

$$
\begin{aligned}
c_1 + 2c_2 + c_3 &= 0, \\
-c_1 + c_2 - c_3 &= 0, \\
2c_1 \qquad\quad + 2c_3 &= 0, \\
3c_2 + c_3 &= 0.
\end{aligned}
$$

The reduced row echelon form of the augmented matrix is (exercise)

$$\left[\begin{array}{ccc|c} 1 & 0 & 0 & 0 \\ 0 & 1 & 0 & 0 \\ 0 & 0 & 1 & 0 \\ 0 & 0 & 0 & 0 \end{array}\right],$$

which implies that the system only has the trivial solution $c_1 = c_2 = c_3 = 0$. It follows from Definition 6.5.2 that the given vectors are therefore linearly independent.

Example 6.5.3 Recall that P_n denotes the vector space of all polynomials of degree less than n. Determine whether the vectors $p_1(x) = 1 + x$, $p_2(x) = 3 - 2x$ are linearly dependent or linearly independent in P_2.

Solution In this case the condition for determining linear dependence or independence is

$$c_1 p_1(x) + c_2 p_2(x) = 0.$$

This is satisfied if and only if

$$(c_1 + 3c_2) + (c_1 - 2c_2)x = 0$$

for all x—that is, if and only if

$$c_1 + 3c_2 = 0, \qquad c_1 - 2c_2 = 0.$$

This homogeneous system of linear equations has only the trivial solution $c_1 = c_2 = 0$, and hence the given vectors are linearly independent.

As the preceding examples suggest, in a general vector space in order to tell whether a given set of vectors $\{v_1, v_2, \dots, v_k\}$ is linearly independent or linearly dependent, we must set up the equations

$$c_1 v_1 + c_2 v_2 + \cdots + c_k v_k = 0 \tag{6.5.1}$$

and see whether there are nontrivial solutions (this always reduces to solving a homogeneous linear system). However, if we are given n vectors in $\mathbf{R}^m$, then the following result tells us how to determine directly whether the vectors are linearly dependent or independent.

Theorem 6.5.2: Let $v_1, v_2, \dots, v_n$ be vectors in $\mathbf{R}^m$, and let $D = [v_1, v_2, \dots, v_n]$.[1]

1. If $n > m$, the vectors are linearly dependent.
2. If $n = m$, the vectors are linearly dependent if and only if $\det(D) = 0$.
3. If $n < m$, the vectors are linearly dependent if and only if $\mathrm{rank}(D) < n$.

---

[1] Note that we are switching between vectors in $\mathbf{R}^m$ and column m-vectors. See Section 6.2 for justification.

PROOF   The condition for determining linear dependence or independence

$$c_1 v_1 + c_2 v_2 + \cdots + c_n v_n = 0$$

can be written in matrix form as

$$[v_1, v_2, \ldots, v_n]c = 0, \qquad (6.5.2)$$

where $c = [c_1, c_2, \ldots, c_n]^T$. This is an $m \times n$ homogeneous system of linear equations with coefficient matrix

$$D = [v_1, v_2, \ldots, v_n].$$

From our previous results for such systems, we can conclude that

1. If $n > m$, the system has an infinite number of solutions (Theorem 4.4.2, page 134) and hence the vectors are linearly dependent.
2. If $n = m$, the system (6.5.2) is $n \times n$ and, hence, from Corollary 5.6.1 (page 175) has an infinite number of solutions if and only if $\det(D) = 0$.
3. If $n < m$, then the system has an infinite number of solutions if and only if (Theorem 4.3.1, page 131) rank$(D) < n$ (recall that for a homogeneous system the rank of the coefficient matrix and the rank of the augmented matrix coincide).   ∎

Example 6.5.4   Determine whether the following vectors are linearly dependent or linearly independent:
   (a) $v_1 = (1, 0, 0)$, $v_2 = (0, 2, 3)$, $v_3 = (4, 5, 6)$, $v_4 = (1, -1, 2)$.
   (b) $v_1 = (1, -1, 2, 3)$, $v_2 = (1, 1, 0, 1)$, $v_3 = (1, 5, -4, -3)$.
   (c) $v_1 = (1, 0, 1, 0)$, $v_2 = (1, 1, 0, 0)$, $v_3 = (1, 0, 0, 1)$, $v_4 = (1, -1, 2, 1)$.

Solution
   (a) The vectors v_1, v_2, v_3, v_4 are necessarily linearly dependent, since we have four vectors in $\mathbf{R}^3$.
   (b) In this case we must determine the rank of the matrix

$$D = [v_1, v_2, v_3] = \begin{bmatrix} 1 & 1 & 1 \\ -1 & 1 & 5 \\ 2 & 0 & -4 \\ 3 & 1 & -3 \end{bmatrix}.$$

The reduced row echelon form of this matrix is (exercise):

$$\begin{bmatrix} 1 & 0 & -2 \\ 0 & 1 & 3 \\ 0 & 0 & 0 \\ 0 & 0 & 0 \end{bmatrix},$$

so that rank$(D) = 2$. Since our vectors are from $\mathbf{R}^4$, it follows from Theorem 6.5.2 that they are linearly dependent.

(c) In this case we have four vectors in $\mathbf{R}^4$, and hence we can use the determinant. It is straightforward to show that

$$\det(D) = \det[\mathbf{v}_1, \mathbf{v}_2, \mathbf{v}_3, \mathbf{v}_4] = \begin{vmatrix} 1 & 1 & 1 & 1 \\ 0 & 1 & 0 & -1 \\ 1 & 0 & 0 & 2 \\ 0 & 0 & 1 & 1 \end{vmatrix} = -1.$$

Since this determinant is nonzero, it follows from Theorem 6.5.2 that the given vectors are linearly independent.

EXERCISES 6.5

In problems 1–9 determine whether the given vectors are linearly independent or linearly dependent in $\mathbf{R}^n$.

1. $(1, -1), (1, 1)$.
2. $(2, -1), (3, 2), (0, 1)$.
3. $(1, -1, 0), (0, 1, -1), (1, 1, 1)$.
4. $(1, 2, 3), (1, -1, 2), (1, -4, 1)$.
5. $(-1, 1, 2), (0, 2, -1), (3, 1, 2), (-1, -1, 1)$.
6. $(1, -1, 2, 3), (2, -1, 1, -1), (-1, 1, 1, 1)$.
7. $(-2, 4, -6), (3, -6, 9)$.
8. $(1, -1, 2), (2, 1, 0)$.
9. $(2, -1, 0, 1), (1, 0, -1, 2), (0, 3, 1, 2), (-1, 1, 2, 1)$.

10. Determine all values of the constant k for which the vectors

$$(1, 1, k), (0, 2, k), (1, k, 6)$$

are linearly dependent in $\mathbf{R}^3$.

In problems 11 and 12, determine all values of the constant k for which the given vectors are linearly independent in $\mathbf{R}^4$.

11. $(1, 0, 1, k), (-1, 0, k, 1), (2, 0, 1, 3)$.
12. $(1, 1, 0, -1), (1, k, 1, 1), (2, 1, k, 1), (-1, 1, 1, k)$.

In problems 13–15 determine whether the given vectors are linearly independent in $M_2(\mathbf{R})$.

13. $A_1 = \begin{bmatrix} 1 & 1 \\ 0 & 1 \end{bmatrix}, A_2 = \begin{bmatrix} 2 & -1 \\ 0 & 1 \end{bmatrix}, A_3 = \begin{bmatrix} 3 & 6 \\ 0 & 4 \end{bmatrix}$.

14. $A_1 = \begin{bmatrix} 2 & -1 \\ 3 & 4 \end{bmatrix}, A_2 = \begin{bmatrix} -1 & 2 \\ 1 & 3 \end{bmatrix}$.

15. $A_1 = \begin{bmatrix} 1 & 0 \\ 1 & 2 \end{bmatrix}, A_2 = \begin{bmatrix} -1 & 1 \\ 2 & 1 \end{bmatrix}, A_3 = \begin{bmatrix} 2 & 1 \\ 5 & 7 \end{bmatrix}$.

16. Determine whether the following vectors are linearly independent in P_2:
(a) $p_1(x) = 1 - x, p_2(x) = 1 + x$.
(b) $p_1(x) = 2 + 3x, p_2(x) = 4 + 6x$.

17. Show that the vectors $p_1(x) = a + bx$, $p_2(x) = c + dx$ are linearly independent in P_2 if and only if the constants a, b, c, d satisfy $ad - bc \neq 0$.

18. If $f_1(x) = \cos 2x, f_2(x) = \sin^2 x, f_3(x) = \cos^2 x$, determine whether f_1, f_2, f_3 are linearly dependent or linearly independent in $C^\infty(-\infty, \infty)$.

19. Let $\mathbf{v}_1$ and $\mathbf{v}_2$ be linearly independent vectors in a vector space V, and let $\mathbf{x} = \alpha \mathbf{v}_1 + \mathbf{v}_2$, $\mathbf{y} = \mathbf{v}_1 + \alpha \mathbf{v}_2$, where α is a constant. Determine all values of α for which $\mathbf{x}$ and $\mathbf{y}$ are linearly independent.

20. If $\mathbf{v}_1$ and $\mathbf{v}_2$ are linearly independent vectors in a vector space V and $\mathbf{u}_1, \mathbf{u}_2, \mathbf{u}_3$ are each linear combinations of them, prove that $\mathbf{u}_1, \mathbf{u}_2, \mathbf{u}_3$ are linearly dependent.

21. Let $\mathbf{v}_1, \mathbf{v}_2, \ldots, \mathbf{v}_m$ be linearly independent vectors in a vector space V, and suppose that the vectors $\mathbf{u}_1, \mathbf{u}_2, \ldots, \mathbf{u}_n$ are each linear combinations of them. It follows that we can write

$$\mathbf{u}_k = \sum_{i=1}^m a_{ik} \mathbf{v}_i, \quad k = 1, 2, \ldots, n,$$

for appropriate constants a_{ik}.

(a) If $n > m$, prove that the vectors $\mathbf{u}_1, \mathbf{u}_2, \ldots,$ $\mathbf{u}_n$ are necessarily linearly dependent in V.
(b) If $m = n$, prove that the vectors $\mathbf{u}_1, \mathbf{u}_2, \ldots,$ $\mathbf{u}_n$ are linearly independent in V if and only if $\det[a_{ij}] \neq 0$.
(c) If $n < m$, prove that the vectors $\mathbf{u}_1, \mathbf{u}_2, \ldots,$ $\mathbf{u}_n$ are linearly independent in V if and only if rank$(A) = n$, where $A = [a_{ij}]$.
(d) Which theorem from this section do these results generalize?

22. Prove that any set of vectors that includes the zero vector is linearly dependent.

6.6   LINEAR INDEPENDENCE OF FUNCTIONS: THE WRONSKIAN

When we resume the study of differential equations in Chapter 9 we will be interested in obtaining solutions that are linearly independent on some interval. In this section we derive a criterion for determining whether a given set of functions is linearly dependent or linearly independent on an interval I. Before doing so, we first specialize the general definition of linear independence given in the previous section to the case of a set of functions defined on an interval I.

> *Definition 6.6.1:*   The functions $f_1, f_2, \ldots, f_n$ are **linearly independent on an interval** I if and only if the only values of the scalars $c_1, c_2, \ldots, c_n$ such that
>
> $$c_1 f_1(x) + c_2 f_2(x) + \cdots + c_n f_n(x) = 0 \qquad (6.6.1)$$
>
> *for all x in* I, *are* $c_1 = c_2 = \cdots = c_n = 0$.

The main point to notice is that condition (6.6.1) must hold for all x in I.

Example 6.6.1   Determine whether f and g defined by $f(x) = \sin x$, $g(x) = \cos x$ are linearly independent on $(-\infty, \infty)$.

Solution   If

$$c_1 \sin x + c_2 \cos x = 0, \qquad (6.6.2)$$

for all x in $(-\infty, \infty)$ then, differentiating, we must also have

$$c_1 \cos x - c_2 \sin x = 0, \qquad (6.6.3)$$

for all x in $(-\infty, \infty)$. However (6.6.2), (6.6.3) is a homogeneous linear system for c_1, c_2. The determinant of the matrix of coefficients is

$$\begin{vmatrix} \sin x & \cos x \\ \cos x & -\sin x \end{vmatrix} = -1.$$

It follows that the only solution to (6.6.2), (6.6.3) is the trivial solution $c_1 = c_2 = 0$, and hence the given functions are linearly independent on $(-\infty, \infty)$.

The method used in the preceding example to prove linear independence can be generalized to any (finite) set of sufficiently smooth functions. The first step in this generalization is to make the following definition.

Definition 6.6.2: Let $f_1, f_2, \ldots, f_n$ be functions in $C^{n-1}(I)$. The **Wronskian** of these functions is the $n \times n$ determinant defined by

$$W[f_1, f_2, \ldots, f_n](x) = \begin{vmatrix} f_1(x) & f_2(x) & \cdots & f_n(x) \\ f_1'(x) & f_2'(x) & \cdots & f_n'(x) \\ \vdots & \vdots & & \vdots \\ f_1^{(n-1)}(x) & f_2^{(n-1)}(x) & \cdots & f_n^{(n-1)}(x) \end{vmatrix}.$$

REMARK   Notice that the Wronskian is a function defined on I. Also note that the value of the Wronskian depends on the order of the functions. For example,

$$W[f_1, f_2, \ldots, f_n](x) = -W[f_2, f_1, \ldots, f_n](x).$$

Example 6.6.2   If $f_1(x) = x$, $f_2(x) = x^2$, $f_3(x) = x^3$ on $(-\infty, \infty)$, then

$$W[f_1, f_2, f_3](x) = \begin{vmatrix} x & x^2 & x^3 \\ 1 & 2x & 3x^2 \\ 0 & 2 & 6x \end{vmatrix} = x(12x^2 - 6x^2) - (6x^3 - 2x^3) = 2x^3.$$

We can now state and prove the main result of this section.

Theorem 6.6.1:   Let $f_1, f_2, \ldots, f_n$ be functions in $C^{n-1}(I)$. If $W[f_1, f_2, \ldots, f_n](x)$ is nonzero at any point in I, then $f_1, f_2, \ldots, f_n$ are linearly independent on I.

PROOF   The proof is a generalization of the technique used in Example 6.6.1. If

$$c_1 f_1(x) + c_2 f_2(x) + \cdots + c_n f_n(x) = 0$$

for all x in I, then, differentiating $(n-1)$ times yields the linear system

$$\begin{aligned} c_1 f_1(x) &+ c_2 f_2(x) &+ \cdots + & c_n f_n(x) &= 0, \\ c_1 f_1'(x) &+ c_2 f_2'(x) &+ \cdots + & c_n f_n'(x) &= 0, \\ & & \vdots & & \\ c_1 f_1^{(n-1)}(x) &+ c_2 f_2^{(n-1)}(x) &+ \cdots + & c_n f_n^{(n-1)}(x) &= 0, \end{aligned}$$

where the unknowns in the system are $c_1, c_2, \ldots, c_n$. The determinant of the matrix of coefficients of this system is just $W[f_1, f_2, \ldots, f_n](x)$. Thus if $W[f_1, f_2, \ldots, f_n](x_0) \neq 0$ for some x_0 in I, then the determinant of the matrix of coefficients of the system is nonzero at that point, and hence the only solution to the system is the trivial solution $c_1 = c_2 = \cdots = c_n = 0$; that is, the given functions are linearly independent on I. ∎

REMARKS

1. Notice that we require $W[f_1, f_2, \ldots, f_n](x)$ to be nonzero at only one point in I for the functions to be linearly independent on I.
2. The logical equivalent of the preceding theorem is

   If $f_1, f_2, \ldots, f_n$ are linearly dependent on I then $W[f_1, f_2, \ldots, f_n](x) = 0$ at every point of I.

   This does *not* say that if $W[f_1, f_2, \ldots, f_n](x) = 0$ at every point of I, then $f_1, f_2, \ldots, f_n$ are linearly dependent on I.

   When $W[f_1, f_2, \ldots, f_n](x) = 0$ at every point of I, we cannot conclude linear dependence or linear independence from Theorem 6.6.1.

Example 6.6.3   Determine whether the following functions are linearly dependent or linearly independent on $I = (-\infty, \infty)$:

(a) $f_1(x) = e^x$, $f_2(x) = x^2 e^x$.

(b) $f_1(x) = x$, $f_2(x) = x + x^2$, $f_3(x) = 2x - x^2$.

(c) $f_1(x) = x^2$, $-\infty < x < \infty$, $f_2(x) = \begin{cases} 2x^2, & \text{if } x \geq 0 \\ -x^2, & \text{if } x < 0 \end{cases}$

Solution

(a) $W[f_1, f_2](x) = \begin{vmatrix} e^x & x^2 e^x \\ e^x & e^x(x^2 + 2x) \end{vmatrix} = e^{2x} \begin{vmatrix} 1 & x^2 \\ 1 & x^2 + 2x \end{vmatrix} = 2xe^{2x}.$

Since $W[f_1, f_2](x) \neq 0$ on $(-\infty, \infty)$ (except at $x = 0$), the functions are linearly independent on $(-\infty, \infty)$.

(b) $W[f_1, f_2, f_3](x) = \begin{vmatrix} x & x + x^2 & 2x - x^2 \\ 1 & 1 + 2x & 2 - 2x \\ 0 & 2 & -2 \end{vmatrix}$

$= x[(-2)(1 + 2x) - 2(2 - 2x)] - [(-2)(x + x^2) - 2(2x - x^2)]$

$= 0.$

Thus, no conclusion can be drawn from Theorem 6.6.1, and we must return to the definition of linear independence.   For the given functions we have

$$c_1 f_1 + c_2 f_2 + c_3 f_3 = 0$$

for all x if and only if

$$c_1 x + c_2(x + x^2) + c_3(2x - x^2) = 0$$

for all x, that is, if and only if

$$(c_1 + c_2 + 2c_3)x + (c_2 - c_3)x^2 = 0,$$

for all x. Since the functions x and x^2 are linearly independent on $(-\infty, \infty)$ [their Wronskian is $W[x, x^2](x) = x^2$], it follows that the constants c_1, c_2, c_3 must satisfy

$$c_1 + c_2 + 2c_3 = 0,$$

$$c_2 - c_3 = 0.$$

This is a homogeneous system of two equations in three unknowns, and hence it has an infinite number of solutions. It follows that the given functions are linearly dependent on $(-\infty, \infty)$. Indeed, by inspection, we see that

$$3f_1 - f_2 - f_3 = 0.$$

(c) If $x \geq 0$, then

$$W[f_1, f_2](x) = \begin{vmatrix} x^2 & 2x^2 \\ 2x & 4x \end{vmatrix} = 0,$$

whereas if $x < 0$, then

$$W[f_1, f_2](x) = \begin{vmatrix} x^2 & -x^2 \\ 2x & -2x \end{vmatrix} = 0.$$

Thus $W[f_1, f_2](x) = 0$ for all x in $(-\infty, \infty)$, and so no conclusion can be drawn from Theorem 6.6.1. In this case, however, the condition

$$c_1 f_1(x) + c_2 f_2(x) = 0$$

is satisfied on $(-\infty, \infty)$ if and only if

$$c_1 x^2 + 2c_2 x^2 = 0, \qquad \text{for } x \geq 0,$$

$$c_1 x^2 - c_2 x^2 = 0, \qquad \text{for } x < 0.$$

Thus on $(-\infty, \infty)$ c_1 and c_2 must satisfy

$$c_1 + 2c_2 = 0,$$

$$c_1 - c_2 = 0,$$

which implies that $c_1 = c_2 = 0$; hence the given functions are linearly independent on $(-\infty, \infty)$ [see Figure 6.6.1 for a sketch of $y = f_1(x)$ and $y = f_2(x)$].

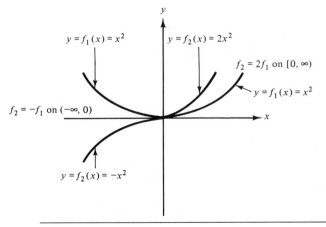

Figure 6.6.1 Two functions that are linearly *independent* on $(-\infty, \infty)$ but whose Wronskian is identically zero on that interval.

REMARK   Notice that the functions in (c) of the previous example are linearly *dependent* on the interval $(-\infty, 0)$, since on this interval we have

$$f_1 + f_2 = 0.$$

Also they are linearly *dependent* on the interval $[0, \infty)$, since on this interval we have

$$2f_1 - f_2 = 0.$$

The key point to realize is that there is no set of *nonzero* constants c_1, c_2 for which

$$c_1 f_1 + c_2 f_2 = 0,$$

for all x in $(-\infty, \infty)$.   Once more this emphasizes the importance of the role played by the interval when discussing linear dependence and linear independence of functions.

It might appear at this stage that the usefulness of the Wronskian is questionable, since if $W[f_1, f_2, \ldots, f_n]$ vanishes on an interval I, then no conclusion can be drawn as to the linear dependence or linear independence of the functions $f_1, f_2, \ldots, f_n$ on I.   However, the real power of the Wronskian is in its application to solutions of linear differential equations of the form

$$a_0(x)y^{(n)} + a_1(x)y^{(n-1)} + \cdots + a_{n-1}(x)y' + a_n(x)y = 0. \qquad (6.6.4)$$

In Chapter 9 we will show that if we have n functions that are *solutions of an equation of the form* (6.6.4) on an interval I, then if the Wronskian of these functions is identically zero on I, the functions are linearly dependent on I.   Thus the Wronskian does indeed completely characterize the linear dependence or linear independence of solutions of such equations.   This is a fundamental result in the theory of linear differential equations.

Example 6.6.4   Show that $y_1 = e^{2x}$, $y_2 = e^{-x}$ are linearly independent solutions of

$$y'' - y' - 2y = 0 \qquad (6.6.5)$$

on any interval.

Solution   We first check that y_1 and y_2 are solutions.   We have

$$y_1'' - y_1' - 2y_1 = e^{2x}(4 - 2 - 2) = 0,$$

so that y_1 is a solution of (6.6.5) on $(-\infty, \infty)$.   Similarly,

$$y_2'' - y_2' - 2y_2 = e^{-x}(1 + 1 - 2) = 0,$$

so that y_2 is also a solution of (6.6.5) on $(-\infty, \infty)$.   Further, the Wronskian of y_1 and y_2 is

$$W[y_1, y_2](x) = \begin{vmatrix} e^{2x} & e^{-x} \\ 2e^{2x} & -e^{-x} \end{vmatrix} = -3e^x.$$

Since the Wronskian is nonzero for all x, Theorem 6.6.1 implies that the functions are linearly independent on any interval.

EXERCISES 6.6

In problems 1–13 determine whether the given functions are linearly dependent or linearly independent on the given interval I.

1. $f_1(x) = 1, f_2(x) = x, f_3(x) = x^2, I = (-\infty, \infty)$.

2. $f_1(x) = 1, f_2(x) = x, f_3(x) = 2x - 1, I = (-\infty, \infty)$.

3. $f_1(x) = e^x, f_2(x) = e^{-x}, f_3(x) = \cosh x, I = (-\infty, \infty)$.

4. $f_1(x) = \sin x, f_2(x) = \cos x, f_3(x) = \tan x, I = \left(-\frac{\pi}{2}, \frac{\pi}{2}\right)$.

5. $f_1(x) = 1/x, f_2(x) = x, I = (0, \infty)$.

6. $f_1(x) = 1, f_2(x) = 3x, f_3(x) = x^2 - 1, I = (-\infty, \infty)$.

7. $f_1(x) = \sin 2x, f_2(x) = \sin x, I = (-\infty, \infty)$.

8. $f_1(x) = 3x, f_2(x) = x(2x + 1), f_3(x) = x(x - 1), I = (-\infty, \infty)$.

9. $f_1(x) = \ln x, f_2 = x \ln x, I = (0, \infty)$.

10. $f_1(x) = e^{2x}, f_2(x) = e^{3x}, f_3(x) = e^{-x}, I = (-\infty, \infty)$.

11. $f_1(x) = \begin{cases} 5x^3, & \text{if } x \ge 0, \\ -3x^3, & \text{if } x < 0, \end{cases}$

$f_2(x) = 2x^3, I = (-\infty, \infty)$.

12. $f_1(x) = \begin{cases} x^2, & \text{if } x \ge 0, \\ 3x^3, & \text{if } x < 0, \end{cases}$

$f_2(x) = 7x^2, I = (-\infty, \infty)$.

13. $f_1(x) = \cos 2x, f_2(x) = \sin^2 x, f_3(x) = 1, I = (-\infty, \infty)$.

14. Show that the functions $f_1(x) = 1, f_2(x) = x, f_3(x) = x^3$ are linearly independent on any interval and make a sketch of the Wronskian $W[f_1, f_2, f_3](x)$. Also sketch the Wronskians $W[f_2, f_1, f_3](x)$ and $W[f_2, f_3, f_1](x)$. How are these Wronskians related?

15. Consider the functions

$$f_1(x) = x, f_2(x) = \begin{cases} x, & \text{if } x \ge 0, \\ -x, & \text{if } x < 0. \end{cases}$$

(a) Show that f_2 is not in $C^1(-\infty, \infty)$.
(b) Show that f_1 and f_2 are linearly dependent on $(-\infty, 0)$ and $[0, \infty)$ and that they are linearly independent on $(-\infty, \infty)$. Justify your results by making a sketch showing both functions.

16. Show that the functions

$$f(x) = \begin{cases} (x - 1), & \text{if } x \ge 1, \\ 2(x - 1), & \text{if } x < 1, \end{cases} \quad g(x) = 2x, \quad h(x) = 3$$

are linearly independent on $(-\infty, \infty)$. Determine all intervals on which they are linearly dependent.

17. Determine whether the functions

$$f_1(x) = x, \qquad f_2(x) = \begin{cases} x, & \text{if } x \ne 0, \\ 1, & \text{if } x = 0, \end{cases}$$

are linearly dependent or linearly independent on $I = (-\infty, \infty)$.

18. (a) Show that the functions $f_0(x) = 1, f_1(x) = x, f_2(x) = x^2, f_3 = x^3$, are linearly independent on any interval.
(b) If $f_k(x) = x^k, k = 0, 1, 2, \ldots, n$, show that the functions $f_0, f_1, \ldots, f_n$ are linearly independent on any interval for all fixed n.

19. Consider the set of vectors $\{p_1(x), p_2(x), \ldots, p_n(x)\}$ in P_n defined by

$$p_1(x) = a_{11} + a_{21}x + a_{31}x^2 + \cdots + a_{n1}x^{n-1},$$
$$p_2(x) = a_{12} + a_{22}x + a_{32}x^2 + \cdots + a_{n2}x^{n-1},$$
$$\vdots$$
$$p_n(x) = a_{1n} + a_{2n}x + a_{3n}x^2 + \cdots + a_{nn}x^{n-1}.$$

Show that this set of vectors is linearly independent on any interval if and only if $\det(A) \ne 0$, where

$$A = \begin{bmatrix} a_{11} & a_{12} & \cdots & a_{1n} \\ a_{21} & a_{22} & \cdots & a_{2n} \\ \vdots & \vdots & & \vdots \\ a_{n1} & a_{n2} & \cdots & a_{nn} \end{bmatrix}.$$

In problems 20–23 show that the given functions are *linearly independent* solutions of the given differential equation on the interval I.

20. $y_1 = e^x, y_2 = e^{-3x}, y'' + 2y' - 3y = 0, I = (-\infty, \infty)$.

21. $y_1 = \cos 2x, y_2 = \sin 2x, y'' + 4y = 0, I = (-\infty, \infty)$.

22. $y_1 = e^{ax}, y_2 = e^{bx}, y'' - (a + b)y' + aby = 0, a \ne b$ constants, $I = (-\infty, \infty)$.

23. $y_1 = x^{-3}, y_2 = x^2, x^2y'' + 2xy' - 6y = 0, I = (0, \infty)$.

24. (a) Show that the functions

$$f_1(x) = e^{r_1 x}, \qquad f_2(x) = e^{r_2 x}, \qquad f_3(x) = e^{r_3 x},$$

have Wronskian

$$W[f_1, f_2, f_3](x)$$

$$= e^{(r_1 + r_2 + r_3)x} \begin{vmatrix} 1 & 1 & 1 \\ r_1 & r_2 & r_3 \\ r_1^2 & r_2^2 & r_3^2 \end{vmatrix}$$

$$= e^{(r_1 + r_2 + r_3)x}(r_3 - r_1)(r_3 - r_2)(r_2 - r_1),$$

and hence determine the conditions on r_1, r_2, r_3, such that f_1, f_2, f_3 are linearly independent on any interval.

(b) More generally show that the functions

$$f_1(x) = e^{r_1 x}, f_2(x) = e^{r_2 x}, \ldots, f_n(x) = e^{r_n x},$$

are linearly independent on any interval if and only if all the r_i are distinct. [*Hint:* Show that the Wronskian of the given functions is a multiple of the $n \times n$ Vandermonde determinant (see Exercise 22 in Section 5.4.)]

6.7   BASES AND DIMENSION

In working with geometric vectors in three space, it is usual to represent vectors as a linear combination of the basic unit vectors **i**, **j**, **k**. That is, we write

$$\mathbf{v} = v_1 \mathbf{i} + v_2 \mathbf{j} + v_3 \mathbf{k},$$

where (v_1, v_2, v_3) are the components of **v**. More generally, if $\mathbf{v}_1, \mathbf{v}_2, \ldots, \mathbf{v}_k$ span a vector space V, then any vector in V can be expressed as a linear combination of them. In this section we are interested in determining the *minimum* number of vectors that span V. If $\{\mathbf{v}_1, \mathbf{v}_2, \ldots, \mathbf{v}_k\}$ is a linearly dependent spanning set, then (at least) one of the vectors in the set can be expressed in terms of the others, and hence there exists a smaller set of vectors that span V. We might therefore suspect that in order for $\{\mathbf{v}_1, \mathbf{v}_2, \ldots, \mathbf{v}_k\}$ to be a minimal spanning set, the key property that we require is that, in addition to spanning V, $\{\mathbf{v}_1, \mathbf{v}_2, \ldots, \mathbf{v}_k\}$ is also linearly independent. This is indeed correct, although its proof will require some preliminary results. We begin with a fundamental definition in vector space theory.

> **Definition 6.7.1:**   A set of vectors $\{\mathbf{v}_1, \mathbf{v}_2, \ldots, \mathbf{v}_k\}$ in a vector space V is called a **basis** for V if
>
> 1. The vectors are linearly independent.
> 2. The vectors span V.
>
> A vector space V is called **finite-dimensional** if it has a basis consisting of a finite number of vectors.

The vectors $(1, 0, 0), (0, 1, 0), (0, 0, 1)$ are linearly independent (see Example 6.5.1) and span $\mathbf{R}^3$ (see Example 6.4.2). Thus $\{(1, 0, 0), (0, 1, 0), (0, 0, 1)\}$ is a basis for $\mathbf{R}^3$. We call this basis the **standard**, or **natural**, basis for $\mathbf{R}^3$. The corresponding geometric vectors are the familiar **i**, **j**, **k** mentioned previously. See Figure 6.7.1.

Example 6.7.1 Show that the vectors $\mathbf{v}_1 = (1, -1, 1)$, $\mathbf{v}_2 = (2, 1, 3)$, and $\mathbf{v}_3 = (3, 1, -2)$ form a basis for $\mathbf{R}^3$.

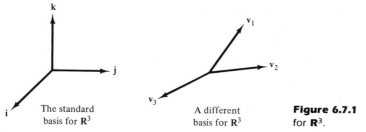

The standard basis for $\mathbf{R}^3$ A different basis for $\mathbf{R}^3$ **Figure 6.7.1** *Two different bases for $\mathbf{R}^3$.*

Solution We must show that the given vectors are linearly independent and that they span $\mathbf{R}^3$. We have:

$$\det[\mathbf{v}_1, \mathbf{v}_2, \mathbf{v}_3] = \begin{vmatrix} 1 & 2 & 3 \\ -1 & 1 & 1 \\ 1 & 3 & -2 \end{vmatrix} = -19,$$

so that, from Theorem 6.5.2, the vectors are certainly linearly independent. Further, if $\mathbf{x}$ is an arbitrary vector in $\mathbf{R}^3$, the linear system

$$c_1\mathbf{v}_1 + c_2\mathbf{v}_2 + c_3\mathbf{v}_3 = \mathbf{x}$$

can be written as the matrix equation

$$[\mathbf{v}_1, \mathbf{v}_2, \mathbf{v}_3]\mathbf{c} = \mathbf{x}.$$

Since $\det[\mathbf{v}_1, \mathbf{v}_2, \mathbf{v}_3] \neq 0$, it follows from Theorem 5.6.1 (page 175) that this system has a unique solution for any and all vectors $\mathbf{x}$. Thus any vector in $\mathbf{R}^3$ can be written as a linear combination of $\mathbf{v}_1, \mathbf{v}_2, \mathbf{v}_3$, and hence these vectors span $\mathbf{R}^3$. Since $\mathbf{v}_1, \mathbf{v}_2, \mathbf{v}_3$ are linearly independent and span $\mathbf{R}^3$, they form a basis for $\mathbf{R}^3$.

Example 6.7.2 The following vectors in $M_2(\mathbf{R})$,

$$A_1 = \begin{bmatrix} 1 & 0 \\ 0 & 0 \end{bmatrix}, \quad A_2 = \begin{bmatrix} 0 & 1 \\ 0 & 0 \end{bmatrix}, \quad A_3 = \begin{bmatrix} 0 & 0 \\ 1 & 0 \end{bmatrix}, \quad A_4 = \begin{bmatrix} 0 & 0 \\ 0 & 1 \end{bmatrix},$$

are linearly independent. Further, an arbitrary vector,

$$A = \begin{bmatrix} a & b \\ c & d \end{bmatrix},$$

in $M_2(\mathbf{R})$ can be written as

$$\begin{bmatrix} a & b \\ c & d \end{bmatrix} = aA_1 + bA_2 + cA_3 + dA_4.$$

It follows that A_1, A_2, A_3, A_4 also span $M_2(\mathbf{R})$, and hence $\{A_1, A_2, A_3, A_4\}$ is a basis for $M_2(\mathbf{R})$.

Example 6.7.3 Determine a basis for P_3.

Solution In set notation we have:

$$P_3 = \{p(x) = a_0 + a_1x + a_2x^2 : a_0, a_1, a_2 \in \mathbf{R}\},$$

so that every vector in P_3 can be written as

$$p(x) = a_0 p_0(x) + a_1 p_1(x) + a_2 p_2(x),$$

where

$$p_0(x) = 1, \qquad p_1(x) = x, \qquad p_2(x) = x^2.$$

Thus p_0, p_1, p_2 span P_3. Further, these vectors have Wronskian

$$W[p_0, p_1, p_2](x) = \begin{vmatrix} 1 & x & x^2 \\ 0 & 1 & 2x \\ 0 & 0 & 2 \end{vmatrix} = 2 \neq 0,$$

which implies that p_0, p_1, p_2 are linearly independent on any interval. It follows from Definition 6.7.1 that $\{p_0, p_1, p_2\}$ is a basis for P_3.

We know from Definition 6.7.1 that if we have a basis for a vector space V, then, since the vectors in the basis span V, any vector in V can be expressed as a linear combination of the basis vectors. The following theorem establishes that there is only one way in which we can do this.

Theorem 6.7.1:   If V is a finite-dimensional vector space with basis $\{v_1, v_2, \ldots, v_n\}$, then any vector v in V can be written *uniquely* as a linear combination of $v_1, v_2, \ldots, v_n$.

PROOF   Since $v_1, v_2, \ldots, v_n$ span V, it follows that any vector v in V can be expressed as a linear combination of $v_1, v_2, \ldots, v_n$. That is,

$$v = \sum_{i=1}^n a_i v_i, \qquad \text{for some scalars } a_1, a_2, \ldots, a_n. \tag{6.7.1}$$

Suppose also that

$$v = \sum_{i=1}^n b_i v_i, \qquad \text{for some scalars } b_1, b_2, \ldots, b_n. \tag{6.7.2}$$

We wish to show that $a_i = b_i$ for each i. Subtracting (6.7.2) from (6.7.1) yields

$$\sum_{i=1}^n (a_i - b_i) v_i = 0. \tag{6.7.3}$$

But $v_1, v_2, \ldots, v_n$ are linearly independent, and so (6.7.3) implies that we must have

$$a_i - b_i = 0, \qquad i = 1, 2, \ldots, n,$$

that is,

$$a_i = b_i, \qquad i = 1, 2, \ldots, n. \qquad \blacksquare$$

Example 6.7.4   The vector $x = (x_1, x_2, x_3)$ in $\mathbf{R}^3$ can be written in terms of the standard basis for $\mathbf{R}^3$ as

$$x = (x_1, x_2, x_3) = x_1(1, 0, 0) + x_2(0, 1, 0) + x_3(0, 0, 1).$$

If $\{\mathbf{v}_1, \mathbf{v}_2, \ldots, \mathbf{v}_n\}$ is a basis for V, and $\mathbf{v}$ is any vector in V, then the unique constants $c_1, c_2, \ldots, c_k$ such that

$$\mathbf{v} = c_1\mathbf{v}_1 + c_2\mathbf{v}_2 + \cdots + c_k\mathbf{v}_k$$

are called the **components of v relative to the basis** $\{\mathbf{v}_1, \mathbf{v}_2, \ldots, \mathbf{v}_k\}$.

Example 6.7.5 Determine the components of the vector $\mathbf{v} = (1, 7)$ relative to the basis $\mathbf{v}_1 = (1, 2)$, $\mathbf{v}_2 = (3, 1)$ in $\mathbf{R}^2$.

Solution We must determine constants c_1, c_2 such that

$$c_1\mathbf{v}_1 + c_2\mathbf{v}_2 = \mathbf{v},$$

that is, such that

$$c_1(1, 2) + c_2(3, 1) = (1, 7).$$

This requires

$$c_1 + 3c_2 = 1,$$
$$2c_1 + c_2 = 7.$$

The solution of this system is $(4, -1)$, so that the components of $\mathbf{v}$ relative to the basis $\{\mathbf{v}_1, \mathbf{v}_2\}$ are $c_1 = 4$, $c_2 = -1$ (see Figure 6.7.2). Thus

$$\mathbf{v} = 4\mathbf{v}_1 - \mathbf{v}_2.$$

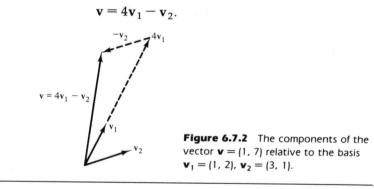

Figure 6.7.2 The components of the vector $\mathbf{v} = (1, 7)$ relative to the basis $\mathbf{v}_1 = (1, 2)$, $\mathbf{v}_2 = (3, 1)$.

The following theorem and its corollaries are fundamental results in vector space theory.

Theorem 6.7.2 If a finite-dimensional vector space has a basis consisting of m vectors, then any set of more than m vectors is linearly dependent.

PROOF Let $\{\mathbf{v}_1, \mathbf{v}_2, \ldots, \mathbf{v}_m\}$ be a basis for V, and consider an arbitrary set of vectors in V, say $\{\mathbf{u}_1, \mathbf{u}_2, \ldots, \mathbf{u}_n\}$, with $n > m$. We wish to prove that $\{\mathbf{u}_1, \mathbf{u}_2, \ldots, \mathbf{u}_n\}$ is necessarily linearly dependent.

Since $\{\mathbf{v}_1, \mathbf{v}_2, \ldots, \mathbf{v}_m\}$ is a basis for V, it follows that each $\mathbf{u}_j$ can be written as a linear combination of $\mathbf{v}_1, \mathbf{v}_2, \ldots, \mathbf{v}_m$. Thus there exist constants a_{ij} such that

$$\mathbf{u}_j = \sum_{i=1}^{m} a_{ij}\mathbf{v}_i, \qquad j = 1, 2, \ldots, n.$$

To prove that $\{\mathbf{u}_1, \mathbf{u}_2, \ldots, \mathbf{u}_n\}$ is linearly dependent, we must show that there exist nonzero constants $c_1, c_2, \ldots, c_n$ such that

$$\sum_{j=1}^{n} c_j\mathbf{u}_j = \mathbf{0}. \tag{6.7.4}$$

But, (6.7.4) holds if and only if

$$\sum_{j=1}^{n} c_j \left\{ \sum_{i=1}^{m} a_{ij}\mathbf{v}_i \right\} = \mathbf{0},$$

that is, if and only if

$$\sum_{i=1}^{m} \left\{ \sum_{j=1}^{n} a_{ij}c_j \right\} \mathbf{v}_i = \mathbf{0}. \tag{6.7.5}$$

Since $\{\mathbf{v}_1, \mathbf{v}_2, \ldots, \mathbf{v}_m\}$ is a linearly independent set of vectors, (6.7.5) is satisfied if and only if

$$\sum_{j=1}^{n} a_{ij}c_j = 0, \qquad i = 1, 2, \ldots, m. \tag{6.7.6}$$

This is an $m \times n$ homogeneous system of linear equations with $m < n$ and hence, from Theorem 4.4.2 (page 134), has nontrivial solutions for $c_1, c_2, \ldots, c_n$. It therefore follows from (6.7.4) that the set of vectors $\{\mathbf{u}_1, \mathbf{u}_2, \ldots, \mathbf{u}_n\}$ is linearly dependent. ∎

Corollary 6.7.1:   All bases in a finite-dimensional vector space V contain the same number of vectors.

PROOF   Suppose $\{\mathbf{v}_1, \mathbf{v}_2, \ldots, \mathbf{v}_n\}$ and $\{\mathbf{u}_1, \mathbf{u}_2, \ldots, \mathbf{u}_m\}$ are two bases for V. From Theorem 6.7.2 we cannot have $m > n$ (otherwise $\{\mathbf{u}_1, \mathbf{u}_2, \ldots, \mathbf{u}_m\}$ would be a linearly dependent set and hence could not form a basis) or $n > m$ (otherwise $\{\mathbf{v}_1, \mathbf{v}_2, \ldots, \mathbf{v}_n\}$ would be a linearly dependent set and hence could not form a basis). It follows that we must have $m = n$. ∎

We can now prove that any basis provides a minimal spanning set for V.

Corollary 6.7.2:   If a finite-dimensional vector space has a basis consisting of n vectors, then any spanning set must contain at least n vectors.

PROOF   Let $\{\mathbf{v}_1, \mathbf{v}_2 \ldots, \mathbf{v}_k\}$ be a spanning set for V, and assume that $k < n$. Then $\{\mathbf{v}_1, \mathbf{v}_2, \ldots, \mathbf{v}_k\}$ must be linearly dependent otherwise it would be a basis consisting of less than n vectors, which would contradict the previous corollary. Since $\{\mathbf{v}_1, \mathbf{v}_2, \ldots, \mathbf{v}_k\}$ is linearly dependent, at least one of the vectors in the set, say $\mathbf{v}_j$, can be written as a linear combination of the other vectors in the set. Hence $\mathbf{v}_j$ can be removed from the set and the resulting subset is also a spanning set for V. This process can be repeated (if necessary) until we obtain a linearly independent spanning set—a basis—containing less than n vectors. However, the existence of such a basis contradicts Corollary 6.7.1 and, consequently, we must have $k \geq n$.   ∎

Once more we emphasize the reason for introducing the idea of a basis. If $\{\mathbf{v}_1, \mathbf{v}_2, \ldots, \mathbf{v}_k\}$ is a basis for V, then any vector $\mathbf{v}$ in V can be written uniquely as

$$\mathbf{v} = c_1 \mathbf{v}_1 + c_2 \mathbf{v}_2 + \cdots + c_k \mathbf{v}_k,$$

for appropriate scalars $c_1, c_2, \ldots, c_k$. Thus, knowledge of a basis is equivalent to knowing all the vectors in a vector space. Since a basis is a minimal spanning set, it gives the minimum number of vectors that are required to define all vectors in V.

The number of vectors in a basis for a vector space is clearly a fundamental property of the vector space. For this reason it is given a special name.

Definition 6.7.2:   The **dimension** of a finite-dimensional vector space, written $\dim[V]$, is the number of vectors in a basis for V. If V consists only of the zero vector, then we define its dimension to be zero.

REMARK   We say that the dimension of the world that we live in is three for the very reason that the maximum number of independent directions that we can perceive is three. If a vector space has a basis containing n vectors, then, from Theorem 6.7.2, the maximum number of vectors in any linearly independent set is n. Thus we see that the terminology *dimension* used in an arbitrary vector space is once more a natural generalization of a familiar idea.

Example 6.7.6   It follows from Examples 6.7.1, 6.7.2, and 6.7.3 that $\dim[\mathbf{R}^3] = 3$, $\dim[M_2(\mathbf{R})] = 4$, and $\dim[P_3] = 3$, respectively. More generally, it is easily shown that

$$\dim[\mathbf{R}^n] = n, \qquad \dim[M_n(\mathbf{R})] = n^2, \quad \text{and} \quad \dim[P_n] = n.$$

For example, a basis for $M_n(\mathbf{R})$ is given by the n^2 linearly independent matrices $\{I_{ij}\}$, where I_{ij} denotes the $n \times n$ matrix which has a 1 in the ijth position and 0s elsewhere [see Example 6.7.2 for the case of $M_2(\mathbf{R})$]. Similarly, a basis for P_n is given by the n linearly independent polynomials $\{1, x, x^2, \ldots, x^{n-1}\}$. We leave it as an exercise to determine a basis for $\mathbf{R}^n$.

Example 6.7.7   Determine a basis for the subspace of $\mathbf{R}^3$ that consists of all solutions of the equation $x_1 + 2x_2 - x_3 = 0$.

Solution   The subspace of interest is

$$S = \{(x_1, x_2, x_3) \in \mathbf{R}^3 : x_1 + 2x_2 - x_3 = 0\}.$$

To obtain the solution of the equation $x_1 + 2x_2 - x_3 = 0$, we set $x_2 = r$, $x_3 = s$, in which case $x_1 = -2r + s$.   It follows that

$$S = \{x \in \mathbf{R}^3 : x = (-2r + s, r, s), r, s \in \mathbf{R}\}.$$

We can write this in the equivalent form

$$S = \{x \in \mathbf{R}^3 : x = r(-2, 1, 0) + s(1, 0, 1), r, s \in \mathbf{R}\}.$$

If we let $v_1 = (-2, 1, 0)$ and $v_2 = (1, 0, 1)$, then we see that

$$S = \text{span}\{v_1, v_2\}.$$

Further, v_1 and v_2 are linearly independent, and hence they form a basis for S. Since this basis consists of two vectors, it follows that $\dim[S] = 2$.

Example 6.7.8   Determine a basis for the set of all solutions, S, to the differential equation

$$y'' + y = 0,$$

and hence find $\dim[S]$.

Solution   We have already shown in Example 6.4.8 that the solution set of the given differential equation on any interval I is

$$S = \{y \in C^2(I) : y = c_1 \cos x + c_2 \sin x\} = \text{span}\{y_1, y_2\}$$

where

$$y_1 = \cos x, \qquad y_2 = \sin x.$$

Further, the Wronskian of these solutions is

$$W[y_1, y_2](x) = \begin{vmatrix} \cos x & \sin x \\ -\sin x & \cos x \end{vmatrix} = 1 \neq 0,$$

so that y_1 and y_2 are linearly independent on I.   Thus $\{\cos x, \sin x\}$ is a basis for S, and hence $\dim[S] = 2$.

If a vector space has dimension n, then the maximum number of vectors in any linearly independent set is n.   We might therefore suspect that *any* set of n linearly independent vectors is a basis for V.   This is indeed true and is proved in the following theorem.

Theorem 6.7.3:   If $\dim[V] = n$, then *any* set of n linearly independent vectors in V is a basis for V.

PROOF   Let $v_1, v_2, \ldots, v_n$ be n linearly independent vectors in V.   We must show that they span V.   If v is an arbitrary vector in V, then, from Theorem 6.7.2, the set of vectors $\{v, v_1, v_2, \ldots, v_n\}$ is linearly dependent, and so there exist scalars $c_0, c_1, c_2, \ldots, c_n$, not all zero, such that

$$c_0 v + c_1 v_1 + \cdots + c_n v_n = 0. \tag{6.7.7}$$

Since $\mathbf{v}_1, \mathbf{v}_2, \ldots, \mathbf{v}_n$ are linearly independent, it follows that necessarily $c_0 \neq 0$, and so, from (6.7.7),

$$\mathbf{v} = -\frac{1}{c_0}(c_1\mathbf{v}_1 + c_2\mathbf{v}_2 + \cdots + c_n\mathbf{v}_n),$$

that is, any vector can be written as a linear combination of $\mathbf{v}_1, \mathbf{v}_2, \ldots, \mathbf{v}_n$, and hence $\{\mathbf{v}_1, \mathbf{v}_2, \ldots, \mathbf{v}_n\}$ is a basis for V. ■

REMARK   Theorem 6.7.3 is perhaps the most important result of the section. We will use it repeatedly in later chapters.

Example 6.7.9   Show that the vectors $\mathbf{v}_1 = (1, -1, 2)$, $\mathbf{v}_2 = (1, 0, -1)$, $\mathbf{v}_3 = (2, 1, 1)$ form a basis for $\mathbf{R}^3$.

Solution $\det[\mathbf{v}_1, \mathbf{v}_2, \mathbf{v}_3] = 6 \neq 0$, and so Theorem 6.5.2 implies that $\mathbf{v}_1, \mathbf{v}_2, \mathbf{v}_3$ are three linearly independent vectors in $\mathbf{R}^3$. Since $\dim[\mathbf{R}^3] = 3$, it follows from the previous theorem that $\mathbf{v}_1, \mathbf{v}_2, \mathbf{v}_3$ form a basis for $\mathbf{R}^3$.

Example 6.7.10   Show that $p_1(x) = 1 + x$, $p_2(x) = 2 - 2x + x^2$, $p_3(x) = 1 + x^2$ form a basis for P_3.

Solution   Since $\dim[P_3] = 3$, all that is required is to show that p_1, p_2, p_3 are linearly independent. The Wronskian of the given polynomials is

$$W[p_1, p_2, p_3](x) = \begin{vmatrix} 1 + x & 2 - 2x + x^2 & 1 + x^2 \\ 1 & -2 + 2x & 2x \\ 0 & 2 & 2 \end{vmatrix} = -6 \neq 0.$$

Since the Wronskian is nonzero, the vectors are indeed linearly independent and hence they do form a basis for P_3.

In our final theorem of this section we establish that a basis for a subspace of a vector space V can be extended to a basis for V. This will be required in Chapter 7.

Theorem 6.7.4:   Let S ($\neq \{0\}$) be a subspace of a finite-dimensional vector space V. Any basis for S is part of a basis for V.

PROOF   Let $\{\mathbf{v}_1, \mathbf{v}_2, \ldots, \mathbf{v}_k\}$ be a basis for S. If $V = \mathrm{span}\{\mathbf{v}_1, \mathbf{v}_2, \ldots, \mathbf{v}_k\}$, then $\{\mathbf{v}_1, \mathbf{v}_2, \ldots, \mathbf{v}_k\}$ is a basis for V and the theorem is proved. If $V \neq \mathrm{span}\{\mathbf{v}_1, \mathbf{v}_2, \ldots, \mathbf{v}_k\}$, then there exists at least one vector, say $\mathbf{v}_{k+1}$, in V that is not in $\mathrm{span}\{\mathbf{v}_1, \mathbf{v}_2, \ldots, \mathbf{v}_k\}$. Thus the set $\{\mathbf{v}_1, \mathbf{v}_2, \ldots, \mathbf{v}_k, \mathbf{v}_{k+1}\}$ is linearly independent. If $V = \mathrm{span}\{\mathbf{v}_1, \mathbf{v}_2, \ldots, \mathbf{v}_k, \mathbf{v}_{k+1}\}$, then $\{\mathbf{v}_1, \mathbf{v}_2, \ldots, \mathbf{v}_k, \mathbf{v}_{k+1}\}$ is a basis for V and we are done. If $V \neq \mathrm{span}\{\mathbf{v}_1, \mathbf{v}_2, \ldots, \mathbf{v}_k, \mathbf{v}_{k+1}\}$, then we can repeat the procedure to obtain the linearly independent set $\{\mathbf{v}_1, \mathbf{v}_2, \ldots, \mathbf{v}_k, \mathbf{v}_{k+1}, \mathbf{v}_{k+2}\}$. This process will terminate when we have a linearly independent set consisting of n vectors, where n is the dimension of V, and the resulting set of vectors will be a basis for V. (Why?) ■

REMARK The process used in proving the previous theorem is referred to as **extending a basis.**

It is important to realize that not all vector spaces are finite-dimensional. A vector space in which we can find an arbitrarily large number of linearly independent vectors is said to be **infinite-dimensional**. Indeed, the vector space of primary importance in differential equation theory, $C^n(I)$, is an infinite-dimensional vector space, as we now show.

Example 6.7.11 Show that the vector space $C^n(I)$ is an infinite-dimensional vector space.

Solution Consider the functions $1, x, x^2, \cdots, x^k$. For each fixed k the Wronskian of these functions is nonzero, and hence the functions are linearly independent on I. Since we can choose k arbitrarily, it follows that there is an arbitrarily large number of linearly independent vectors in $C^n(I)$, and hence $C^n(I)$ is infinite-dimensional.

It follows from the previous example that, in particular, the vector space $C^2(I)$ is an infinite-dimensional vector space. We will show in Chapter 9, however, that the set of all solutions to any linear differential equation of the form

$$y'' + a(x)y' + b(x)y = 0, \tag{6.7.8}$$

where a and b are continuous functions on an interval I, forms a *two*-dimensional subspace of $C^2(I)$ (see Example 6.7.8 for a special case). It will then follow immediately from Theorem 6.7.3 that any pair of solutions to (6.7.8), say y_1, y_2, that are linearly independent on I will form a basis for the set of all solutions, and hence every solution of the differential equation can be written as

$$y = c_1y_1 + c_2y_2,$$

for appropriate values of the constants c_1, c_2. Thus finding all solutions of the differential equation will be reduced to finding just two linearly independent solutions on I.

EXERCISES 6.7

1. Let S be the subspace of $\mathbf{R}^3$ that consists of all solutions of the equation $x_1 - 3x_2 + x_3 = 0$. Determine a basis for S and hence find dim$[S]$.

2. $A = \begin{bmatrix} 1 & 3 \\ -2 & -6 \end{bmatrix}$.

In problems 2–4 let S be the subspace of $\mathbf{R}^n$ consisting of all solutions to the system $A\mathbf{x} = \mathbf{0}$ for the given matrix A. Determine a basis for S and hence find dim$[S]$.

3. $A = \begin{bmatrix} 1 & -1 & 4 \\ 2 & 3 & -2 \\ 1 & 2 & -2 \end{bmatrix}$.

4. $A = \begin{bmatrix} 1 & -1 & 2 & 3 \\ 2 & -1 & 3 & 4 \\ 1 & 0 & 1 & 1 \\ 3 & -1 & 4 & 5 \end{bmatrix}$.

5. Let S be the subspace of $\mathbf{R}^3$ spanned by the vectors

$\mathbf{v}_1 = (1, 0, 1)$, $\mathbf{v}_2 = (0, 1, 1)$, $\mathbf{v}_3 = (2, 0, 2)$.

Determine a basis for S, and hence find $\dim[S]$.

6. Let S be the subspace of $\mathbf{R}^3$ consisting of all vectors of the form $(r, r - 2s, 3s - 5r)$, where r and s are real numbers. Determine a basis for S, and hence find $\dim[S]$.

7. Let S be the vector space consisting of the set of all linear combinations of the functions $f_1(x) = e^x$, $f_2(x) = e^{-x}$, $f_3(x) = \cosh(x)$. Determine a basis for S, and hence find $\dim[S]$.

In problems 8–12 determine whether the given vectors form a basis for $\mathbf{R}^n$.

8. $(1, 1)$, $(-1, 1)$.

9. $(1, 2, 1)$, $(3, -1, 2)$, $(1, 1, -1)$.

10. $(1, -1, 1)$, $(2, 5, -2)$, $(3, 11, -5)$.

11. $(1, 1, -1, 2)$, $(1, 0, 1, -1)$, $(2, -1, 1, -1)$.

12. $(1, 1, 0, 2)$, $(2, 1, 3, -1)$, $(-1, 1, 1, -2)$, $(2, -1, 1, 2)$.

13. Determine all values of the constant k for which the vectors

$(0, -1, 0, k)$, $(1, 0, 1, 0)$, $(0, 1, 1, 0)$, $(k, 0, 2, 1)$

form a basis for $\mathbf{R}^4$.

14. If $\mathbf{v}_1 = (1, 1)$, $\mathbf{v}_2 = (-1, 1)$, show that $\{\mathbf{v}_1, \mathbf{v}_2\}$ is a basis for $\mathbf{R}^2$. Determine the components of $\mathbf{v} = (5, -1)$ relative to the basis $\{\mathbf{v}_1, \mathbf{v}_2\}$.

15. Show that $\mathbf{v}_1 = (2, 1)$, $\mathbf{v}_2 = (3, 1)$ form a basis for $\mathbf{R}^2$ and determine the components of each of the standard basis vectors $\mathbf{e}_1 = (1, 0)$, $\mathbf{e}_2 = (0, 1)$ relative to the basis $\{\mathbf{v}_1, \mathbf{v}_2\}$.

16. If $\mathbf{v}_1 = (0, 6, 3)$, $\mathbf{v}_2 = (3, 0, 3)$, $\mathbf{v}_3 = (6, -3, 0)$ show that $\{\mathbf{v}_1, \mathbf{v}_2, \mathbf{v}_3\}$ is a basis for $\mathbf{R}^3$ and determine the components of an arbitrary vector $\mathbf{x} = (x_1, x_2, x_3)$ relative to the basis $\{\mathbf{v}_1, \mathbf{v}_2, \mathbf{v}_3\}$.

17. Let

$p_1(x) = 1 + x$, $p_2(x) = x(x - 1)$,

$p_3(x) = 1 + 2x^2$.

Show that $\{p_1, p_2, p_3\}$ is a basis for P_3, and determine the components of

$$p(x) = a_0 + a_1 x + a_2 x^2$$

relative to this basis.

18. Determine all values of the constant α for which

$p_1(x) = 1 + \alpha x^2$, $\quad p_2(x) = 1 + x + x^2$,
$$p_3(x) = 2 + x$$

form a basis for P_3.

19. Determine a basis for P_4 and hence prove that $\dim[P_4] = 4$.

20. The Legendre polynomial of degree n, $p_n(x)$, is defined to be the polynomial solution of the differential equation

$$(1 - x^2)y'' - 2xy' + n(n + 1)y = 0,$$

which has been normalized so that $p_n(1) = 1$. The first three Legendre polynomials are $p_0(x) = 1$, $p_1(x) = x$, $p_2(x) = \frac{1}{2}(3x^2 - 1)$. Show that $\{p_0, p_1, p_2\}$ is a basis for P_3, and determine the components of $p(x) = x^2 - x + 2$ relative to this basis.

21. Show that the vectors

$A_1 = \begin{bmatrix} -1 & 1 \\ 0 & 1 \end{bmatrix}$, $\quad A_2 = \begin{bmatrix} 1 & 3 \\ -1 & 0 \end{bmatrix}$,

$A_3 = \begin{bmatrix} 1 & 0 \\ 1 & 2 \end{bmatrix}$, $\quad A_4 = \begin{bmatrix} 0 & 1 \\ 2 & 3 \end{bmatrix}$,

form a basis for $M_2(\mathbf{R})$.

In problems 22 and 23, $\mathrm{Sym}_n(\mathbf{R})$ and $\mathrm{Skew}_n(\mathbf{R})$ denote the vector spaces consisting of all real $n \times n$ matrices which are symmetric and skew-symmetric, respectively.

22. Find a basis for $\mathrm{Sym}_2(\mathbf{R})$ and $\mathrm{Skew}_2(\mathbf{R})$ and show that

$\dim[\mathrm{Sym}_2(\mathbf{R})] + \dim[\mathrm{Skew}_2(\mathbf{R})] = \dim[M_2(\mathbf{R})]$.

23. Determine the dimensions of $\mathrm{Sym}_n(\mathbf{R})$ and $\mathrm{Skew}_n(\mathbf{R})$, and show that

$\dim[\mathrm{Sym}_n(\mathbf{R})] + \dim[\mathrm{Skew}_n(\mathbf{R})] = \dim[M_n(\mathbf{R})]$.

24. Let V be a finite-dimensional vector space of dimension n. If S is a subspace of V with $\dim[S] = n$, prove that $S = V$.

6.8   INNER PRODUCT SPACES

In this section we extend the familiar idea of the dot product for geometric vectors to a general vector space, V, and show how this leads to the definition of the concepts of length and orthogonality for vectors in V.   In order to motivate our definitions we begin with a brief review of the dot product.

Consider two geometric vectors $\mathbf{x} = (x_1, x_2, x_3)$ and $\mathbf{y} = (y_1, y_2, y_3)$ in 3-space.   The dot product of these vectors is defined geometrically by

$$\mathbf{x} \cdot \mathbf{y} = \| \mathbf{x} \| \, \| \mathbf{y} \| \cos \theta, \tag{6.8.1}$$

where $\| \mathbf{x} \|$ and $\| \mathbf{y} \|$ denote the lengths of $\mathbf{x}$ and $\mathbf{y}$, respectively, and $0 \le \theta \le \pi$ is the angle between them (see Figure 6.8.1).

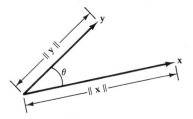

Figure 6.8.1   Two geometric vectors.

The dot product can also be defined algebraically in terms of the components of $\mathbf{x}$ and $\mathbf{y}$ (relative to the standard basis in $\mathbf{R}^3$) as

$$\mathbf{x} \cdot \mathbf{y} = x_1 y_1 + x_2 y_2 + x_3 y_3. \tag{6.8.2}$$

Taking $\mathbf{y} = \mathbf{x}$ in (6.8.1) and (6.8.2) yields

$$\| \mathbf{x} \|^2 = \mathbf{x} \cdot \mathbf{x} = x_1^2 + x_2^2 + x_3^2,$$

so that the length of a geometric vector is given in terms of the dot product by

$$\| \mathbf{x} \| = \sqrt{\mathbf{x} \cdot \mathbf{x}} = \sqrt{x_1^2 + x_2^2 + x_3^2}.$$

Further, from (6.8.1), the angle between two nonzero vectors $\mathbf{x}$ and $\mathbf{y}$ is

$$\cos \theta = \frac{\mathbf{x} \cdot \mathbf{y}}{\| \mathbf{x} \| \, \| \mathbf{y} \|},$$

which implies that $\mathbf{x}$ and $\mathbf{y}$ are orthogonal (perpendicular) if and only if

$$\mathbf{x} \cdot \mathbf{y} = 0.$$

In a general vector space we do not have a geometrical picture to guide us in defining the dot product, and hence our definitions must be purely algebraic.   We begin by considering the vector space $\mathbf{R}^n$, since there is a natural way to extend (6.8.2) in this case.   Before proceeding we note that from now on we will use the standard terms *inner product* and *norm* in place of *dot product* and *length*, respectively.

> **Definition 6.8.1:** *Let* $\mathbf{x} = (x_1, x_2, \ldots, x_n)$ *and* $\mathbf{y} = (y_1, y_2, \ldots, y_n)$ *be vectors in* $\mathbf{R}^n$. We define the standard **inner product**, $<\mathbf{x}, \mathbf{y}>$, in $\mathbf{R}^n$ by
>
> $$<\mathbf{x}, \mathbf{y}> = x_1 y_1 + x_2 y_2 + x_3 y_3 + \cdots + x_n y_n.$$
>
> The **norm** of $\mathbf{x}$ is
>
> $$\|\mathbf{x}\| = \sqrt{<\mathbf{x}, \mathbf{x}>} = \sqrt{x_1^2 + x_2^2 + \cdots + x_n^2}.$$

For $n > 3$ we cannot picture these very easily, and we certainly could not measure the norm of a vector in, say, $\mathbf{R}^4$ using a ruler. We have used a concrete geometric idea to motivate an abstract algebraic definition.

Example 6.8.1 If $\mathbf{x} = (1, -1, 0, 2, 4)$ and $\mathbf{y} = (2, 1, 1, 3, 0)$, then

$$<\mathbf{x}, \mathbf{y}> = 2 - 1 + 0 + 6 + 0 = 7,$$
$$\|\mathbf{x}\| = \sqrt{1 + 1 + 0 + 4 + 16} = \sqrt{22},$$
$$\|\mathbf{y}\| = \sqrt{4 + 1 + 1 + 9 + 0} = \sqrt{15}.$$

BASIC PROPERTIES OF THE STANDARD INNER PRODUCT IN $\mathbf{R}^n$

In the case of $\mathbf{R}^n$, the definition of an inner product was a natural extension of the familiar dot product in $\mathbf{R}^3$. To generalize this definition to an arbitrary vector space, we first isolate the most important properties of the standard inner product in $\mathbf{R}^n$. We can consider the standard inner product in $\mathbf{R}^n$ as being a mapping that associates with any two vectors $\mathbf{x} = (x_1, x_2, \ldots, x_n)$ and $\mathbf{y} = (y_1, y_2, \ldots, y_n)$ in $\mathbf{R}^n$ the real number

$$<\mathbf{x}, \mathbf{y}> = x_1 y_1 + x_2 y_2 + x_3 y_3 + \cdots + x_n y_n.$$

Further, this mapping satisfies the following properties.
For all $\mathbf{x}$, $\mathbf{y}$, and $\mathbf{z}$ in $\mathbf{R}^n$ and all real numbers c:

1. $<\mathbf{x}, \mathbf{x}> \geq 0$, and $<\mathbf{x}, \mathbf{x}> = 0$ if and only if $\mathbf{x} = \mathbf{0}$.
2. $<\mathbf{y}, \mathbf{x}> = <\mathbf{x}, \mathbf{y}>$.
3. $<c\mathbf{x}, \mathbf{y}> = c<\mathbf{x}, \mathbf{y}>$.
4. $<\mathbf{x} + \mathbf{y}, \mathbf{z}> = <\mathbf{x}, \mathbf{z}> + <\mathbf{y}, \mathbf{z}>$.

These properties all follow directly from the definition of the standard inner product. For example, to prove (1) we proceed as follows. From Definition 6.8.1,

$$<\mathbf{x}, \mathbf{x}> = x_1^2 + x_2^2 + \cdots + x_n^2.$$

Since this is a sum of squares of real numbers it is necessarily nonnegative. Further, $<\mathbf{x}, \mathbf{x}> = 0$ if and only if $x_1 = x_2 = \cdots = x_n = 0$—that is, if and only if $\mathbf{x} = \mathbf{0}$.

Similarly,

$$<\mathbf{y}, \mathbf{x}> = y_1x_1 + y_2x_2 + \cdots + y_nx_n = x_1y_1 + x_2y_2 + \cdots + x_ny_n = <\mathbf{x}, \mathbf{y}>,$$

so that (2) is satisifed. We leave the verification of properties 3 and 4 as exercises.

DEFINITION OF A REAL INNER PRODUCT SPACE

We now use (1)–(4) as the basic defining properties of an inner product in a real vector space.

Definition 6.8.2:   Let V be a real vector space. An **inner product** in V is a mapping that associates with each pair of vectors $\mathbf{x}$, $\mathbf{y}$ in V a real number, denoted by $<\mathbf{x}, \mathbf{y}>$, and satisfying the following properties.
    For all $\mathbf{x}$, $\mathbf{y}$, and $\mathbf{z}$ in V and all real numbers c,

1. $<\mathbf{x}, \mathbf{x}> \geq 0$, and $<\mathbf{x}, \mathbf{x}> = 0$ if and only if $\mathbf{x} = \mathbf{0}$.
2. $<\mathbf{y}, \mathbf{x}> = <\mathbf{x}, \mathbf{y}>$.
3. $<c\mathbf{x}, \mathbf{y}> = c<\mathbf{x}, \mathbf{y}>$.
4. $<\mathbf{x} + \mathbf{y}, \mathbf{z}> = <\mathbf{x}, \mathbf{z}> + <\mathbf{y}, \mathbf{z}>$.

The **norm** of $\mathbf{x}$ is defined in terms of the inner product by

$$\|\mathbf{x}\| = \sqrt{<\mathbf{x}, \mathbf{x}>}.$$

A real vector space together with an inner product defined in it is called a **real inner product space**.

Thus $\mathbf{R}^n$ together with the inner product defined in Definition 6.8.1 is an example of a real inner product space.

Recall that $C^0[a, b]$ denotes the vector space of all functions that are continuous on the interval $[a, b]$. An important inner product space arises when we define the following inner product in $C^0[a, b]$:

$$<f, g> = \int_a^b f(x)g(x)\, dx \tag{6.8.3}$$

for all f and g in $C^0[a, b]$. To show that this mapping does indeed define an inner product in $C^0[a, b]$ we must show that properties (1)–(4) of Definition 6.8.2 are satisfied. Consider property 1. If f is in $C^0[a, b]$, then

$$<f, f> = \int_a^b [f(x)]^2\, dx.$$

Since the integrand is a nonnegative function, it follows that $<f, f> \geq 0$, and, further, the integral vanishes if and only if $f(x) = 0$, for all x in $[a, b]$. Thus property 1 of Definition 6.8.2 is satisfied. The remaining properties are just standard properties of the integral, and hence we leave the verification of their validity as exercises.

Example 6.8.2 Find the inner product of the following functions in $C^0[0, 1]$:

$$f(x) = 8x \quad \text{and} \quad g(x) = x^2 - 1.$$

Also find $\|f\|$ and $\|g\|$.

Solution From (6.8.3),

$$<f, g> = \int_0^1 8x(x^2 - 1)\, dx = [2x^4 - 4x^2]_0^1 = -2.$$

$$\|f\| = \sqrt{\int_0^1 64x^2\, dx} = \frac{8}{\sqrt{3}}.$$

$$\|g\| = \sqrt{\int_0^1 (x^2 - 1)^2\, dx} = \sqrt{\int_0^1 (x^4 - 2x^2 + 1)\, dx}$$

$$= \sqrt{\frac{8}{15}}.$$

COMPLEX INNER PRODUCTS

The preceding discussion has been concerned with real vector spaces. To general-ize the definition of an inner product to a complex vector space, we first consider the case of C^n. It might be thought that an appropriate inner product for C^n could be obtained by using Definition 6.8.1. However, one of our main reasons for introducing an inner product is so that we can obtain a concept of "length" of a vector. If we tried applying the R^n inner product directly, say, to the vector $\mathbf{u} = (1 - i, 1 + i)$, we would obtain

$$\|\mathbf{u}\|^2 = (1 - i)^2 + (1 + i)^2 = 0,$$

that is, a nonzero vector having zero "length." To rectify this situation, we extend the definition of an inner product in C^n as follows.

Definition 6.8.3: If $\mathbf{u} = (u_1, u_2, \ldots, u_n)$ and $\mathbf{v} = (v_1, v_2, \ldots, v_n)$ are vec-tors in C^n, we define the standard **inner product** in C^n by[1]

$$<\mathbf{u}, \mathbf{v}> = u_1\bar{v}_1 + u_2\bar{v}_2 + \cdots + u_n\bar{v}_n.$$

The **norm** of $\mathbf{u}$ is defined to be the real number

$$\|\mathbf{u}\| = \sqrt{<\mathbf{u}, \mathbf{u}>} = \sqrt{|u_1|^2 + |u_2|^2 + \cdots + |u_n|^2}.$$

---

[1] Recall that if $u = a + ib$, then $\bar{u} = a - ib$ and $|u|^2 = u\bar{u} = (a + ib)(a - ib) = a^2 + b^2$. See Appen-dix 1.

REMARK   The preceding inner product is a mapping that associates with the two vectors $\mathbf{u} = (u_1, u_2, \ldots, u_n)$ and $\mathbf{v} = (u_1, u_2, \ldots, u_n)$ in $\mathbf{C}^n$ the *scalar*

$$\langle \mathbf{u}, \mathbf{v}\rangle = u_1\bar{v}_1 + u_2\bar{v}_2 + \cdots + u_n\bar{v}_n.$$

In general $\langle \mathbf{u}, \mathbf{v}\rangle$ is a complex number. However, the key point to notice is that the norm of $\mathbf{u}$ is always a *real* number, even though the separate components of $\mathbf{u}$ are complex numbers.

Example 6.8.3   If $\mathbf{u} = (1 + 2i, 2 - 3i)$ and $\mathbf{v} = (2 - i, 3 + 4i)$ find $\langle \mathbf{u}, \mathbf{v}\rangle$ and $\|\mathbf{u}\|$.

Solution   Using Definition 6.8.3, we have

$$\langle \mathbf{u}, \mathbf{v}\rangle = (1 + 2i)(2 + i) + (2 - 3i)(3 - 4i) = 5i - 6 - 17i = -6(1 + 2i).$$
$$\|\mathbf{u}\| = \sqrt{\langle \mathbf{u}, \mathbf{u}\rangle} = \sqrt{(1 + 2i)(1 - 2i) + (2 - 3i)(2 + 3i)} = \sqrt{18} = 3\sqrt{2}.$$

The standard inner product in $\mathbf{C}^n$ satisfies properties 1, 3, and 4 of Definition 6.8.2 but not property 2. We now derive the appropriate generalization of this property when we are in $\mathbf{C}^n$.

Let $\mathbf{u} = (u_1, u_2, \ldots, u_n)$ and $\mathbf{v} = (v_1, v_2, \ldots, v_n)$ be vectors in $\mathbf{C}^n$. Then, from Definition 6.8.3,

$$\langle \mathbf{u}, \mathbf{v}\rangle = u_1\bar{v}_1 + u_2\bar{v}_2 + \cdots + u_n\bar{v}_n.$$

Taking the conjugate of both sides of this equation yields

$$\overline{\langle \mathbf{u}, \mathbf{v}\rangle} = \overline{u_1\bar{v}_1} + \overline{u_2\bar{v}_2} + \cdots + \overline{u_n\bar{v}_n} = v_1\bar{u}_1 + v_2\bar{u}_2 + \cdots + v_n\bar{u}_n = \langle \mathbf{v}, \mathbf{u}\rangle.$$

Thus

$$\langle \mathbf{v}, \mathbf{u}\rangle = \overline{\langle \mathbf{u}, \mathbf{v}\rangle}.$$

We now use the properties satisfied by the standard inner product in $\mathbf{C}^n$ to define an inner product in an arbitrary (that is, real or complex) vector space.

Definition 6.8.4:   Let V be a vector space. An **inner product** in V is a mapping that associates with each pair of vectors $\mathbf{u}, \mathbf{v}$ in V a scalar, denoted by $\langle \mathbf{u}, \mathbf{v}\rangle$, and that satisfies the following properties.
For all $\mathbf{u}, \mathbf{v}$, and $\mathbf{w}$ in V and all scalars c,

1. $\langle \mathbf{u}, \mathbf{u}\rangle \geq 0$, and $\langle \mathbf{u}, \mathbf{u}\rangle = 0$ if and only if $\mathbf{u} = \mathbf{0}$.
2. $\langle \mathbf{v}, \mathbf{u}\rangle = \overline{\langle \mathbf{u}, \mathbf{v}\rangle}$.
3. $\langle c\mathbf{u}, \mathbf{v}\rangle = c\langle \mathbf{u}, \mathbf{v}\rangle$.
4. $\langle \mathbf{u} + \mathbf{v}, \mathbf{w}\rangle = \langle \mathbf{u}, \mathbf{w}\rangle + \langle \mathbf{v}, \mathbf{w}\rangle$.

The **norm** of **u** is defined in terms of the inner product by

$$\|\mathbf{u}\| = \sqrt{<\mathbf{u}, \mathbf{u}>}.$$

A vector space together with an inner product defined in it is called an **inner product space**.

REMARKS

1. If V is a complex vector space we will sometimes refer to V together with its associated inner product as a **complex inner product space**.

2. Notice that the properties in the above definition *do* reduce to those in Definition 6.8.2 in the case that V is a *real* vector space, since in such a case the complex conjugates are superfluous.

Example 6.8.4   Use properties 2 and 3 of Definition 6.8.4 to prove that in an inner product space $<\mathbf{u}, c\mathbf{v}> = \bar{c}<\mathbf{u}, \mathbf{v}>$ for all vectors **u**, **v**, and all scalars c.

Solution   From property 2 we have

$$<\mathbf{u}, c\mathbf{v}> = \overline{<c\mathbf{v}, \mathbf{u}>},$$

so that, using property 3,

$$<\mathbf{u}, c\mathbf{v}> = \bar{c}\,\overline{<\mathbf{v}, \mathbf{u}>} = \bar{c}<\mathbf{u}, \mathbf{v}>,$$

where we have once more used property 2 in the final step. Notice that in the particular case of a real vector space the above result reduces to

$$<\mathbf{u}, c\mathbf{v}> = c<\mathbf{u}, \mathbf{v}>,$$

since in such a case the scalars are real numbers.

The dot product is useful in $\mathbf{R}^3$ because it allows us to measure lengths of vectors. More importantly, it tells us whether two geometric vectors are orthogonal. We now abstract this idea of orthogonality to *define* the corresponding property for vectors in any inner product space.

Definition 6.8.5:   Let V be an inner product space.

1. Two vectors **x** and **y** in V are said to be **orthogonal** if and only if

$$<\mathbf{x}, \mathbf{y}> = 0.$$

2. A set of nonzero vectors $\{\mathbf{x}_1, \mathbf{x}_2, \ldots, \mathbf{x}_k\}$ in V is said to form an **orthogonal set** of vectors if

$$<\mathbf{x}_i, \mathbf{x}_j> = 0, \qquad \text{whenever } i \neq j,$$

(that is, each vector is orthogonal to every other vector in the set).

3. A vector $\mathbf{x}$ in V is called a **unit vector** if $\|\mathbf{x}\| = 1$.

4. An orthogonal set of *unit* vectors is called an **orthonormal set** of vectors. Thus, $\{\mathbf{x}_1, \mathbf{x}_2, \dots, \mathbf{x}_k\}$ in V is an orthonormal set if and only if
   (a) $<\mathbf{x}_i, \mathbf{x}_j> = 0$, whenever $i \neq j$,
   (b) $<\mathbf{x}_i, \mathbf{x}_i> = 1$, $i = 1, 2, \dots, k$.

These two conditions can be written compactly in terms of the Kronecker delta symbol as

$$<\mathbf{x}_i, \mathbf{x}_j> = \delta_{ij}, \qquad i, j = 1, 2, \dots, k.$$

REMARKS

1. Note that the inner products occurring in (1)–(4) of the preceding definition will depend upon in which inner product space we are working.

2. If $\mathbf{u}$ is any nonzero vector, then $\mathbf{u}/\|\mathbf{u}\|$ is a unit vector.

Example 6.8.5 Show that $\{(-2, 1, 3, 0), (0, -3, 1, -6), (-2, -4, 0, 2)\}$ forms an orthogonal set of vectors in $\mathbf{R}^4$ and use it to construct an orthonormal set of vectors in $\mathbf{R}^4$.

Solution   Let $\mathbf{x}_1 = (-2, 1, 3, 0)$, $\mathbf{x}_2 = (0, -3, 1, -6)$, $\mathbf{x}_3 = (-2, -4, 0, 2)$.   Then

$$<\mathbf{x}_1, \mathbf{x}_2> = 0, \qquad <\mathbf{x}_1, \mathbf{x}_3> = 0, \qquad <\mathbf{x}_2, \mathbf{x}_3> = 0,$$

so that the given vectors do form an orthogonal set.   We now divide each vector by its norm in order to obtain unit vectors.   It follows that an orthonormal set of vectors is

$$\left\{ \frac{1}{\sqrt{14}}\mathbf{x}_1, \frac{1}{\sqrt{46}}\mathbf{x}_2, \frac{1}{2\sqrt{6}}\mathbf{x}_3 \right\}.$$

Example 6.8.6   Repeat the previous example for the set of vectors

$$\{(3 - i, i), (1 - i, -2 - 4i)\}$$

in $\mathbf{C}^2$.

Solution   Let $\mathbf{x}_1 = (3 - i, i)$ and $\mathbf{x}_2 = (1 - i, -2 - 4i)$.   Then

$$<\mathbf{x}_1, \mathbf{x}_2> = [(3 - i)(1 + i) + i(-2 + 4i)] = [(4 + 2i) - (4 + 2i)] = 0,$$

so that $\mathbf{x}_1$ and $\mathbf{x}_2$ are orthogonal.   Further,

$$\|\mathbf{x}_1\| = \sqrt{(3 - i)(3 + i) + i(-i)} = \sqrt{11},$$

$$\|\mathbf{x}_2\| = \sqrt{(1 - i)(1 + i) + (-2 - 4i)(-2 + 4i)} = \sqrt{22},$$

which implies that an orthonormal set of vectors in $\mathbf{C}^2$ is

$$\left\{ \frac{1}{\sqrt{11}}\mathbf{x}_1, \frac{1}{\sqrt{22}}\mathbf{x}_2 \right\}.$$

Example 6.8.7   Show that the functions $f_1(x) = 1$, $f_2(x) = \sin x$, $f_3(x) = \cos x$ are orthogonal in $C^0[-\pi, \pi]$, and use them to construct an orthonormal set of functions in $C^0[-\pi, \pi]$.

Solution   In this case we have

$$<f_1, f_2> = \int_{-\pi}^{\pi} \sin x \, dx = [-\cos x]_{-\pi}^{\pi} = 0,$$

$$<f_1, f_3> = \int_{-\pi}^{\pi} \cos x \, dx = [\sin x]_{-\pi}^{\pi} = 0,$$

$$<f_2, f_3> = \int_{-\pi}^{\pi} \sin x \cos x \, dx = \left[\frac{1}{2} \sin^2 x\right]_{-\pi}^{\pi} = 0,$$

so that the functions are certainly orthogonal on $[-\pi, \pi]$.   Taking the norm of each function we obtain

$$\|f_1\| = \sqrt{\int_{-\pi}^{\pi} 1 \, dx} = \sqrt{2\pi},$$

$$\|f_2\| = \sqrt{\int_{-\pi}^{\pi} \sin^2 x \, dx} = \sqrt{\int_{-\pi}^{\pi} \frac{1 - \cos 2x}{2} \, dx} = \sqrt{\pi},$$

$$\|f_3\| = \sqrt{\int_{-\pi}^{\pi} \cos^2 x \, dx} = \sqrt{\int_{-\pi}^{\pi} \frac{1 + \cos 2x}{2} \, dx} = \sqrt{\pi}.$$

Thus an orthonormal set of functions on $[-\pi, \pi]$ is

$$\left\{ \frac{1}{\sqrt{2\pi}}, \frac{1}{\sqrt{\pi}} \sin x, \frac{1}{\sqrt{\pi}} \cos x \right\}.$$

EXERCISES 6.8

In problems 1–4 determine whether the given set of vectors is an orthogonal set in $\mathbf{R}^n$. For those that are, determine a corresponding orthonormal set of vectors.

1. $\{(2, -1, 1), (1, 1, -1), (0, 1, 1)\}$.

2. $\{(1, 3, -1, 1), (-1, 1, 1, -1), (1, 0, 2, 1)\}$.

3. $\{(1, 2, -1, 0), (1, 0, 1, 2), (-1, 1, 1, 0), (1, -1, -1, 0)\}$.

4. $\{(1, 2, -1, 0, 3), (1, 1, 0, 2, -1), (4, 2, -4, -5, -4)\}$.

5. Determine all nonzero vectors $\mathbf{x}_3$ such that $\{\mathbf{x}_1, \mathbf{x}_2, \mathbf{x}_3\}$ is an orthogonal set, where $\mathbf{x}_1 = (1, 2, 3)$, $\mathbf{x}_2 = (1, 1, -1)$. Hence obtain an orthonormal set of vectors in $\mathbf{R}^3$.

In problems 6 and 7 show that the given set of vectors is an orthogonal set in $\mathbf{C}^n$, and hence determine a corresponding orthonormal set of vectors in each case.

6. $\{(1 - i, 3 + 2i), (2 + 3i, 1 - i)\}$.

7. $\{(1-i, 1+i, i), (0, i, 1-i), (-3+3i, 2+2i, 2i)\}$.

8. Consider the vectors $x_1 = (1-i, \ 1+2i)$, $x_2 = (2+i, \ z)$ in C^2. Determine the complex number z such that $\{x_1, x_2\}$ is an orthogonal set of vectors and hence obtain an orthonormal set of vectors in C^2.

Consider the inner product space $C^0[a, b]$, where the inner product is defined in Equation (6.8.3). In problems 9–11 find $<f_1, f_2>$, $\|f_1\|$, and $\|f_2\|$ for the given functions. Also determine whether the functions are orthogonal.

9. $f_1(x) = 1, f_2(x) = x^2$ in $C^0[-1, 1]$.

10. $f_1(x) = x^2, f_2(x) = x^3$ in $C^0[-1, 1]$.

11. $f_1(x) = e^x, f_2(x) = e^{-x}$ in $C^0[0, 1]$.

In problems 12–13 show that the given functions in $C^0[-1, 1]$ are orthogonal and use them to construct an orthonormal set of functions in $C^0[-1, 1]$.

12. $f_1(x) = 1, f_2(x) = \sin \pi x, f_3(x) = \cos \pi x$.

13. $f_1(x) = 1, \quad f_2(x) = x, \quad f_3(x) = \frac{1}{2}(3x^2 - 1)$. These are the Legendre polynomials, which arise as solutions of the Legendre differential equation

$$(1 - x^2)y'' - 2xy' + n(n+1)y = 0,$$

when $n = 0, 1, 2$, respectively (see Chapter 13).

In problems 14–15 show that the given functions are orthonormal on $[-1, 1]$.

14. $f_1(x) = \sin \pi x, f_2(x) = \sin 2\pi x$, $f_3(x) = \sin 3\pi x$. (Hint: $\sin a \sin b = \frac{1}{2}[\cos(a+b) - \cos(a-b)]$.)

15. $f_1(x) = \cos \pi x, f_2(x) = \cos 2\pi x$, $f_3(x) = \cos 3\pi x$.

16. Let $A = \begin{bmatrix} a_{11} & a_{12} \\ a_{21} & a_{22} \end{bmatrix}$ and $B = \begin{bmatrix} b_{11} & b_{12} \\ b_{21} & b_{22} \end{bmatrix}$ be arbitrary vectors in $M_2(\mathbf{R})$. Prove that the mapping

$$<A, B> = a_{11}b_{11} + a_{12}b_{12} + a_{21}b_{21} + a_{22}b_{22} \quad \text{(i)}$$

defines an inner product in $M_2(\mathbf{R})$ (that is, verify properties 1–4 of Definition 6.8.2).

17. If $A = \begin{bmatrix} 2 & -1 \\ 3 & 5 \end{bmatrix}$ and $B = \begin{bmatrix} 3 & 1 \\ -1 & 2 \end{bmatrix}$, use the inner product (i) defined in problem 16 to determine $<A, B>$, $\|A\|$, and $\|B\|$.

18. Let $A_1 = \begin{bmatrix} 1 & 1 \\ -1 & 2 \end{bmatrix}$, $A_2 = \begin{bmatrix} -1 & 1 \\ 2 & 1 \end{bmatrix}$, and $A_3 = \begin{bmatrix} -1 & -3 \\ 0 & 2 \end{bmatrix}$. Use the inner product (i) defined in problem 16 to determine all matrices $A_4 = \begin{bmatrix} a & b \\ c & d \end{bmatrix}$ such that $\{A_1, A_2, A_3, A_4\}$ is an orthogonal set of matrices in $M_2(\mathbf{R})$.

19. Let $p_1(x) = a + bx$ and $p_2(x) = c + dx$ be arbitrary vectors in P_2. Determine a mapping $<p_1, p_2>$ that defines an inner product on P_2.

Consider the vector space $\mathbf{R}^2$. Define the mapping $<, >$ by

$$<x, y> = 2x_1y_1 + x_1y_2 + x_2y_1 + 2x_2y_2 \quad \text{(ii)}$$

for all vectors $x = (x_1, x_2)$ and $y = (y_1, y_2)$ in $\mathbf{R}^2$. This mapping is required for problems 20–23.

20. Show that (ii) defines an inner product on $\mathbf{R}^2$ (that is, verify properties 1–4 of Definition 6.8.2).

In problems 21–23 determine whether the given vectors are orthogonal when we use **(a)** the inner product (ii), and **(b)** the standard inner product in $\mathbf{R}^2$.

21. $x = (1, 0), y = (-1, 2)$.

22. $x = (2, -1), y = (3, 6)$.

23. $x = (1, -2), y = (2, 1)$.

24. Consider the vector space $\mathbf{R}^2$. Define the mapping $<, >$ by

$$<x, y> = x_1y_1 - x_2y_2, \quad \text{(iii)}$$

for all vectors $x = (x_1, x_2), y = (y_1, y_2)$. Show that all the properties in Definition 6.8.2 except 1 are satisfied by (iii).

The mapping (iii) is called a **pseudo–inner product** on $\mathbf{R}^2$ and, when generalized to $\mathbf{R}^4$, is of fundamental importance in Einstein's special relativity theory.

25. Using (iii), determine all nonzero vectors satisfying $<x, x> = 0$. Such vectors are called **null vectors**.

26. Using (iii), determine all vectors satisfying $<x, x> < 0$. Such vectors are called **timelike vectors**.

27. Using (iii), determine all vectors satisfying $<x, x> > 0$. Such vectors are called **spacelike** vectors.

28. Make a sketch of $\mathbf{R}^2$ and indicate the position of the null, timelike, and spacelike vectors.

29. Let V be a real inner product space.
(a) Prove that for all $\mathbf{x}$, $\mathbf{y}$ in V

$$\|\mathbf{x} + \mathbf{y}\|^2 = \|\mathbf{x}\|^2 + 2<\mathbf{x}, \mathbf{y}> + \|\mathbf{y}\|^2.$$

(*Hint:* $\|\mathbf{x} + \mathbf{y}\|^2 = <\mathbf{x} + \mathbf{y}, \mathbf{x} + \mathbf{y}>$.)
(b) Prove that for all $\mathbf{x}$, $\mathbf{y}$ in V
(i) $\|\mathbf{x} + \mathbf{y}\|^2 - \|\mathbf{x} - \mathbf{y}\|^2 = 4<\mathbf{x}, \mathbf{y}>$,
(ii) $\|\mathbf{x} + \mathbf{y}\|^2 + \|\mathbf{x} - \mathbf{y}\|^2 = 2(\|\mathbf{x}\|^2 + \|\mathbf{y}\|^2)$.

30. Let V be a complex inner product space. Prove that for all $\mathbf{u}$, $\mathbf{v}$ in V

$$\|\mathbf{u} + \mathbf{v}\|^2 = \|\mathbf{u}\|^2 + 2\ \mathrm{Re}\{<\mathbf{u}, \mathbf{v}>\} + \|\mathbf{v}\|^2,$$

where Re denotes the real part of a complex number.

31. Let $\{\mathbf{v}_1, \mathbf{v}_2, \ldots, \mathbf{v}_k\}$ be an *orthogonal* set of *nonzero* vectors in an inner product space. Show that this set of vectors is necessarily linearly independent. (*Hint:* Write out the general condition for determining linear dependence or independence, and take the inner product with $\mathbf{v}_j$.)

6.9   THE GRAM–SCHMIDT ORTHOGONALIZATION PROCEDURE

When using geometric vectors in practice, we often express all such vectors in terms of the basis vectors $\mathbf{i}, \mathbf{j}, \mathbf{k}$. These basis vectors have very special properties, namely, they are mutually orthogonal unit vectors. In this section we consider the possibility of choosing an analogous basis in an arbitrary (finite-dimensional) inner product space. We show that, provided we can find a basis for such a space, we can always construct a basis that consists of mutually orthogonal unit vectors. The following definition introduces the appropriate terminology.

Definition 6.9.1:   A basis $\{\mathbf{v}_1, \mathbf{v}_2, \ldots, \mathbf{v}_m\}$ for a (finite-dimensional) inner product space is called an **orthogonal** basis if

$$<\mathbf{v}_i, \mathbf{v}_j> = 0 \qquad \text{whenever } i \neq j,$$

and is called an **orthonormal** basis if

$$<\mathbf{v}_i, \mathbf{v}_j> = \delta_{ij}, \quad i, j = 1, 2, \ldots, m.$$

Once more the motivation behind the result that we wish to derive comes from studying geometric vectors in $\mathbf{R}^3$. If $\mathbf{v}_1$ and $\mathbf{v}_2$ are any two linearly independent vectors in $\mathbf{R}^3$, then the **orthogonal projection of $\mathbf{v}_2$ on $\mathbf{v}_1$** is the vector $\mathbf{P}(\mathbf{v}_2, \mathbf{v}_1)$ shown in Figure 6.9.1.

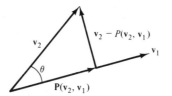

Figure 6.9.1   Obtaining an orthogonal basis for a subspace of $\mathbf{R}^3$.

We see from the figure that the vectors $\mathbf{u}_1 = \mathbf{v}_1$ and $\mathbf{u}_2 = \mathbf{v}_2 - \mathbf{P}(\mathbf{v}_2, \mathbf{v}_1)$ form an orthogonal basis for the subspace of $\mathbf{R}^3$ spanned by $\mathbf{v}_1$ and $\mathbf{v}_2$. To understand the generalization of this result to the case of an arbitrary inner product space it is useful to derive an expression for $\mathbf{P}(\mathbf{v}_2, \mathbf{v}_1)$ in terms of the inner product in $\mathbf{R}^3$. We see from Figure 6.9.1 that the norm (length) of $\mathbf{P}(\mathbf{v}_2, \mathbf{v}_1)$ is

$$\|\mathbf{P}(\mathbf{v}_2, \mathbf{v}_1)\| = \|\mathbf{v}_2\|\cos\theta,$$

so that

$$\mathbf{P}(\mathbf{v}_2, \mathbf{v}_1) = \|\mathbf{v}_2\|\cos\theta \, \frac{\mathbf{v}_1}{\|\mathbf{v}_1\|},$$

which we can write as

$$\mathbf{P}(\mathbf{v}_2, \mathbf{v}_1) = \frac{\|\mathbf{v}_2\| \, \|\mathbf{v}_1\|}{\|\mathbf{v}_1\|^2} (\cos\theta)\mathbf{v}_1. \tag{6.9.1}$$

If we recall that the dot product of the vectors $\mathbf{v}_1$, $\mathbf{v}_2$ is defined by

$$\mathbf{v}_2 \cdot \mathbf{v}_1 = \|\mathbf{v}_2\| \, \|\mathbf{v}_1\|\cos\theta,$$

it follows from (6.9.1) that

$$\mathbf{P}(\mathbf{v}_2, \mathbf{v}_1) = \frac{\mathbf{v}_2 \cdot \mathbf{v}_1}{\|\mathbf{v}_1\|^2}\mathbf{v}_1,$$

that is, using the notation introduced in the previous section,

$$\mathbf{P}(\mathbf{v}_2, \mathbf{v}_1) = \frac{\langle \mathbf{v}_2, \mathbf{v}_1 \rangle}{\|\mathbf{v}_1\|^2}\mathbf{v}_1. \tag{6.9.2}$$

More generally, let $\mathbf{v}_1$, $\mathbf{v}_2$ be linearly independent vectors in an *arbitrary* inner product space V. We now show that the preceding method for determining an orthogonal basis in $\mathbf{R}^3$ can also be applied in V to obtain an orthogonal basis $\{\mathbf{u}_1, \mathbf{u}_2\}$ for the subspace of V spanned by $\mathbf{v}_1$ and $\mathbf{v}_2$. In order for $\mathbf{u}_1$ and $\mathbf{u}_2$ to remain in the subspace spanned by $\mathbf{v}_1$ and $\mathbf{v}_2$, they must be linear combinations of $\mathbf{v}_1$, $\mathbf{v}_2$. We take $\mathbf{u}_1 = \mathbf{v}_1$ and let $\mathbf{u}_2 = \mathbf{v}_2 + \lambda\mathbf{u}_1$, where λ is a scalar. We wish to choose λ such that $\mathbf{u}_1$ and $\mathbf{u}_2$ are orthogonal, that is, such that $\langle \mathbf{u}_2, \mathbf{u}_1 \rangle = 0$. But this latter condition is satisfied if and only if

$$\langle \mathbf{v}_2 + \lambda\mathbf{u}_1, \mathbf{u}_1 \rangle = 0.$$

Using the properties of the inner product, we can write the above condition as

$$\langle \mathbf{v}_2, \mathbf{u}_1 \rangle + \lambda \langle \mathbf{u}_1, \mathbf{u}_1 \rangle = 0,$$

that is,

$$\langle \mathbf{v}_2, \mathbf{u}_1 \rangle + \lambda \|\mathbf{u}_1\|^2 = 0.$$

Solving for λ yields

$$\lambda = -\frac{\langle \mathbf{v}_2, \mathbf{u}_1 \rangle}{\|\mathbf{u}_1\|^2}.$$

Thus $\{\mathbf{u}_1, \mathbf{u}_2\}$ will form an orthogonal basis for the subspace spanned by $\mathbf{v}_1, \mathbf{v}_2$ if we choose $\mathbf{u}_1 = \mathbf{v}_1$ and

$$\mathbf{u}_2 = \mathbf{v}_2 - \frac{<\mathbf{v}_2, \mathbf{u}_1>}{\|\mathbf{u}_1\|^2} \mathbf{u}_1.$$

If we use (6.9.2) to *define* the orthogonal projection of $\mathbf{v}_2$ on $\mathbf{v}_1$ in an arbitrary inner product space, we see that $\mathbf{u}_2$ is obtained by subtracting $\mathbf{P}(\mathbf{v}_2, \mathbf{u}_1)$ from $\mathbf{v}_2$. This is exactly the same formula as that obtained above in $\mathbf{R}^3$.

Now suppose that we are given a set of linearly independent vectors $\{\mathbf{v}_1, \mathbf{v}_2, \ldots, \mathbf{v}_m\}$ in an inner product space V. In a similar manner we can construct an orthogonal basis for the subspace spanned by these vectors. We let $\mathbf{u}_1 = \mathbf{v}_1$ and then successively define the remaining orthogonal vectors $\mathbf{u}_2, \mathbf{u}_3, \ldots, \mathbf{u}_m$ by subtracting off appropriate projections on the previous vectors. The resulting procedure is called the **Gram–Schmidt orthogonalization procedure**. The formal statement of the result is as follows.

Theorem 6.9.1: Let $\{\mathbf{v}_1, \mathbf{v}_2, \ldots, \mathbf{v}_m\}$ be linearly independent vectors in an inner product space V. Then an *orthogonal basis* for the subspace of V spanned by these vectors is $\{\mathbf{u}_1, \mathbf{u}_2, \ldots, \mathbf{u}_m\}$, where

$$\mathbf{u}_1 = \mathbf{v}_1,$$

$$\mathbf{u}_2 = \mathbf{v}_2 - \frac{<\mathbf{v}_2, \mathbf{u}_1>}{\|\mathbf{u}_1\|^2} \mathbf{u}_1, \qquad \leftarrow \mathbf{v}_2 - \mathbf{P}(\mathbf{v}_2, \mathbf{u}_1)$$

$$\mathbf{u}_3 = \mathbf{v}_3 - \frac{<\mathbf{v}_3, \mathbf{u}_1>}{\|\mathbf{u}_1\|^2} \mathbf{u}_1 - \frac{<\mathbf{v}_3, \mathbf{u}_2>}{\|\mathbf{u}_2\|^2} \mathbf{u}_2, \qquad \leftarrow \mathbf{v}_3 - \mathbf{P}(\mathbf{v}_3, \mathbf{u}_1) - \mathbf{P}(\mathbf{v}_3, \mathbf{u}_2)$$

$$\vdots$$

$$\mathbf{u}_i = \mathbf{v}_i - \sum_{k=1}^{i-1} \frac{<\mathbf{v}_i, \mathbf{u}_k>}{\|\mathbf{u}_k\|^2} \mathbf{u}_k, \qquad \leftarrow \text{General term } \mathbf{v}_i - \sum_{k=1}^{i-1} \mathbf{P}(\mathbf{v}_i, \mathbf{u}_k)$$

$$\vdots$$

$$\mathbf{u}_m = \mathbf{v}_m - \frac{<\mathbf{v}_m, \mathbf{u}_1>}{\|\mathbf{u}_1\|^2} \mathbf{u}_1 - \frac{<\mathbf{v}_m, \mathbf{u}_2>}{\|\mathbf{u}_2\|^2} \mathbf{u}_2 - \cdots - \frac{<\mathbf{v}_m, \mathbf{u}_{m-1}>}{\|\mathbf{u}_{m-1}\|^2} \mathbf{u}_{m-1}.$$

PROOF (*Outline*) The proof is constructional. We begin by choosing $\mathbf{u}_1 = \mathbf{v}_1$, and

$$\mathbf{u}_2 = \mathbf{v}_2 - \frac{<\mathbf{v}_2, \mathbf{u}_1>}{\|\mathbf{u}_1\|^2} \mathbf{u}_1.$$

We have already shown above that these two vectors are orthogonal. Now use mathematical induction. Assume we have k orthogonal vectors $\mathbf{u}_1, \mathbf{u}_2, \ldots, \mathbf{u}_k$, and let

$$\mathbf{u}_{k+1} = \mathbf{v}_{k+1} + \sum_{i=1}^{k} \lambda_i \mathbf{u}_i.$$

The orthogonality conditions $\langle \mathbf{u}_{k+1}, \mathbf{u}_i \rangle = 0$ for $1 \le i \le k$ imply that

$$\lambda_i = -\frac{\langle \mathbf{v}_{k+1}, \mathbf{u}_i \rangle}{\|\mathbf{u}_i\|^2}, \qquad i = 1, 2, \ldots, k,$$

and so lead to the formula given.   ∎

REMARKS

1. If $\{\mathbf{u}_1, \mathbf{u}_2, \ldots, \mathbf{u}_m\}$ is an orthogonal basis for an inner product space V, then

$$\left\{ \frac{1}{\|\mathbf{u}_1\|} \mathbf{u}_1, \frac{1}{\|\mathbf{u}_2\|} \mathbf{u}_2, \ldots, \frac{1}{\|\mathbf{u}_m\|} \mathbf{u}_m \right\}$$

   is an orthonormal basis for V.

2. In applying the Gram–Schmidt procedure, you must remember to use the appropriate inner product in the space that you are dealing with.

Example 6.9.1   Obtain an orthogonal basis for the subspace of $\mathbf{R}^4$ spanned by $\mathbf{v}_1 = (1, 0, 1, 0)$, $\mathbf{v}_2 = (1, 1, 1, 1)$, $\mathbf{v}_3 = (-1, 2, 0, 1)$.

Solution   Using the Gram–Schmidt procedure, we choose

$$\mathbf{u}_1 = \mathbf{v}_1 = (1, 0, 1, 0),$$

Thus

$$\langle \mathbf{v}_2, \mathbf{u}_1 \rangle = 1 + 0 + 1 + 0 = 2, \qquad \|\mathbf{u}_1\|^2 = 2,$$

so that

$$\mathbf{u}_2 = \mathbf{v}_2 - \frac{\langle \mathbf{v}_2, \mathbf{u}_1 \rangle}{\|\mathbf{u}_1\|^2} \mathbf{u}_1 = (1, 1, 1, 1) - \frac{2}{2}(1, 0, 1, 0) = (0, 1, 0, 1).$$

Continuing, we have

$$\langle \mathbf{v}_3, \mathbf{u}_1 \rangle = -1 + 0 + 0 + 0 = -1,$$

$$\langle \mathbf{v}_3, \mathbf{u}_2 \rangle = 0 + 2 + 0 + 1 = 3,$$

and

$$\|\mathbf{u}_2\|^2 = \langle \mathbf{u}_2, \mathbf{u}_2 \rangle = 2,$$

so that

$$\mathbf{u}_3 = \mathbf{v}_3 - \frac{\langle \mathbf{v}_3, \mathbf{u}_1 \rangle}{\|\mathbf{u}_1\|^2} \mathbf{u}_1 - \frac{\langle \mathbf{v}_3, \mathbf{u}_2 \rangle}{\|\mathbf{u}_2\|^2} \mathbf{u}_2$$

$$= (-1, 2, 0, 1) + \frac{1}{2}(1, 0, 1, 0) - \frac{3}{2}(0, 1, 0, 1)$$

$$= \frac{1}{2}(-1, 1, 1, -1).$$

Thus an orthogonal basis for the subspace spanned by $\mathbf{v}_1$, $\mathbf{v}_2$, $\mathbf{v}_3$ is

$$\left\{ (1, 0, 1, 0),\ (0, 1, 0, 1),\ \frac{1}{2}(-1, 1, 1, -1) \right\},$$

and an orthonormal basis is

$$\left\{ \frac{\sqrt{2}}{2}(1, 0, 1, 0),\ \frac{\sqrt{2}}{2}(0, 1, 0, 1),\ \frac{1}{2}(-1, 1, 1, -1) \right\}.$$

Example 6.9.2 Determine an orthogonal basis for the subspace of $C^0[-1, 1]$ spanned by the functions $f_1(x) = x$, $f_2(x) = x^3$, $f_3(x) = x^5$.

Solution In this case we let g_1, g_2, g_3 denote the orthogonal basis vectors, and apply the Gram–Schmidt process using the appropriate inner product in $C^0[-1, 1]$. Thus $g_1(x) = x$, and

$$g_2(x) = f_2(x) - \frac{<f_2, g_1>}{||g_1||^2} g_1(x). \tag{6.9.3}$$

Applying the inner product in $C^0[-1, 1]$ we have

$$<f_2, g_1> = \int_{-1}^{1} f_2(x)\, g_1(x)\, dx = \int_{-1}^{1} x^4\, dx = \frac{2}{5},$$

$$||g_1||^2 = <g_1, g_1> = \int_{-1}^{1} x^2\, dx = \frac{2}{3}.$$

Thus, substituting into (6.9.3),

$$g_2(x) = x^3 - \frac{2}{5} \cdot \frac{3}{2} x,$$

that is,

$$g_2(x) = \frac{1}{5} x(5x^2 - 3). \tag{6.9.4}$$

We now compute $g_3(x)$. From Theorem 6.9.1 we have

$$g_3(x) = f_3(x) - \frac{<f_3, g_1>}{||g_1||^2} g_1(x) - \frac{<f_3, g_2>}{||g_2||^2} g_2(x). \tag{6.9.5}$$

We first evaluate the required inner products.

$$<f_3, g_1> = \int_{-1}^{1} f_3(x)\, g_1(x)\, dx = \int_{-1}^{1} x^6\, dx = \frac{2}{7}.$$

$$<f_3, g_2> = \int_{-1}^{1} f_3(x)\, g_2(x)\, dx = \frac{1}{5} \int_{-1}^{1} x^6(5x^2 - 3)\, dx = \frac{1}{5}\left(\frac{10}{9} - \frac{6}{7}\right) = \frac{16}{315}.$$

$$||g_2||^2 = \int_{-1}^{1} [g_2(x)]^2\, dx = \frac{1}{25} \int_{-1}^{1} x^2(5x^2 - 3)^2\, dx = \frac{1}{25} \int_{-1}^{1} (25x^6 - 30x^4 + 9x^2)\, dx$$

$$= \frac{8}{175}.$$

Substituting into (6.9.5) yields

$$g_3(x) = x^5 - \frac{3}{7}x - \frac{2}{9}x(5x^2 - 3) = \frac{1}{63}(63x^5 - 70x^3 + 15x).$$

Thus an orthogonal basis for the subspace of $C^0[-1, 1]$ spanned by f_1, f_2, f_3 is

$$\left\{ x, \frac{1}{5}x(5x^2 - 3), \frac{1}{63}x(63x^4 - 70x^2 + 15) \right\}.$$

EXERCISES 6.9

In problems 1–6 use the Gram–Schmidt process to determine an ortho*normal* basis for the subspace of $\mathbf{R}^n$ spanned by the given set of vectors.

1. $\{(1, -1, -1), (2, 1, -1)\}$.
2. $\{(2, 1, -2), (1, 3, -1)\}$.
3. $\{(-1, 1, 1, 1), (1, 2, 1, 2)\}$.
4. $\{(1, 0, -1, 0), (1, 1, -1, 0), (-1, 1, 0, 1)\}$.
5. $\{(1, 2, 0, 1), (2, 1, 1, 0), (1, 0, 2, 1)\}$.
6. $\{(1, 1, -1, 0), (-1, 0, 1, 1), (2, -1, 2, 1)\}$.

In problems 7 and 8 determine an ortho*normal* basis for the subspace of $\mathbf{C}^3$ spanned by the given set of vectors. Make sure that you use the appropriate inner product in $\mathbf{C}^3$.

7. $\{(1 - i, 0, i), (1, 1 + i, 0)\}$.
8. $\{(1 + i, i, 2 - i), (1 + 2i, 1 - i, i)\}$.

In problems 9–11 determine an orthogonal basis for the subspace of $C^0[a, b]$ spanned by the given vectors, for the given interval $[a, b]$.

9. $f_1(x) = 1, f_2(x) = x, f_3(x) = x^2, a = 0, b = 1$.
10. $f_1(x) = 1, f_2(x) = x^2, f_3(x) = x^4$, $a = -1, b = 1$.
11. $f_1(x) = 1, f_2(x) = \sin x, f_3(x) = \cos x, a = -\frac{\pi}{2}$, $b = \frac{\pi}{2}$.

On $M_2(\mathbf{R})$ define the inner product $<A, B>$ by

$$<A, B> = a_{11}b_{11} + a_{12}b_{12} + a_{21}b_{21} + a_{22}b_{22} \quad \text{(i)}$$

for all matrices

$$A = \begin{bmatrix} a_{11} & a_{12} \\ a_{21} & a_{22} \end{bmatrix} \text{ and } B = \begin{bmatrix} b_{11} & b_{12} \\ b_{21} & b_{22} \end{bmatrix}.$$

In problems 12 and 13 use the inner product (i) in the Gram–Schmidt procedure to determine an orthogonal basis for the subspace of $M_2(\mathbf{R})$ spanned by the given matrices.

12. $A_1 = \begin{bmatrix} 1 & -1 \\ 2 & 1 \end{bmatrix}, A_2 = \begin{bmatrix} 2 & -3 \\ 4 & 1 \end{bmatrix}.$

13. $A_1 = \begin{bmatrix} 0 & 1 \\ 1 & 0 \end{bmatrix}, A_2 = \begin{bmatrix} 0 & 1 \\ 1 & 1 \end{bmatrix}, A_3 = \begin{bmatrix} 1 & 1 \\ 1 & 0 \end{bmatrix}.$

Also identify the subspace of $M_2(\mathbf{R})$ spanned by A_1, A_2, A_3.

On P_n define the inner product $<p_1, p_2>$ by

$$<p_1, p_2> = a_0b_0 + a_1b_1 + \cdots + a_{n-1}b_{n-1} \text{(ii)}$$

for all polynomials $p_1(x) = a_0 + a_1x + \cdots + a_{n-1}x^{n-1}, p_2(x) = b_0 + b_1x + \cdots + b_{n-1}x^{n-1}$. In problems 14 and 15, use the inner product (ii) to determine an orthogonal basis for the subspace of P_n spanned by the given polynomials.

14. $p_1(x) = 1 - 2x + 2x^2, p_2(x) = 2 - x - x^2$.

15. $p_1(x) = 1 + x^2, p_2(x) = 2 - x + x^3, p_3(x) = 2x^2 - x$.

16. Let $\{\mathbf{u}_1, \mathbf{u}_2, \mathbf{v}\}$ be linearly independent vectors in an inner product space V, and suppose that $\mathbf{u}_1$ and $\mathbf{u}_2$ are orthogonal. Define the vector $\mathbf{u}_3$ in V by

$$\mathbf{u}_3 = \mathbf{v} + \lambda\mathbf{u}_1 + \mu\mathbf{u}_2,$$

where λ and μ are scalars. Derive the values of λ and μ such that $\{\mathbf{u}_1, \mathbf{u}_2, \mathbf{u}_3\}$ is an orthogonal basis for the subspace of V spanned by $\{\mathbf{u}_1, \mathbf{u}_2, \mathbf{v}\}$.

6.10   SUMMARY OF RESULTS

In this chapter we have derived some basic results in linear algebra regarding vector spaces. These results form the framework for much of linear mathematics. Following are listed some of the highlights of the chapter.

THE DEFINITION OF A VECTOR SPACE

A vector space consists of the following four different components:

1. A set of vectors V.
2. A set of scalars F (either the set of all real numbers $\mathbf{R}$ or the set of all complex numbers $\mathbf{C}$).
3. A rule, $+$, for adding vectors in V.
4. A rule, $\cdot$, for multiplying vectors in V by scalars in F.

Then $(V, F, +, \cdot)$ forms a vector space if and only if Definition 6.2.1 is satisfied. If F is the set of all real numbers, then $(V, \mathbf{R}, +, \cdot)$ is called a *real* vector space, whereas if F is the set of all complex numbers, then $(V, \mathbf{C}, +, \cdot)$ is called a *complex* vector space. Since it is usually quite clear what the addition and scalar multiplication operations are, we usually specify a vector space simply by giving the set of vectors V. The vector spaces that we have dealt with are the following:

$\mathbf{R}^n$, the (real) vector space of all ordered n-tuples of real numbers.

$\mathbf{C}^n$, the (complex) vector space of all ordered n-tuples of complex numbers.

$M_n(\mathbf{R})$, the (real) vector space of all $n \times n$ matrices with real elements.

$C^k(\mathrm{I})$, the vector space of all real-valued functions that are continuous and have (at least) k continuous derivatives on I.

P_n, the vector space of all polynomials of degree less than n with real coefficients.

SUBSPACES

Usually the vector space that underlies a given problem is known; for example, it may be $\mathbf{R}^n$ or $C^n(\mathrm{I})$. However the solution of a given problem in general only involves a subset of vectors from this vector space. The question that then arises is whether this subset of vectors is itself a vector space under the same operations of addition and scalar multiplication as in V. In order to answer this question we have the basic Theorem 6.3.1, which tells us that *a nonempty subset of a vector space is a subspace if and only if it is closed under addition and closed under scalar multiplication.*

SPANNING SETS

A set of vectors $\{\mathbf{v}_1, \mathbf{v}_2, \ldots, \mathbf{v}_k\}$ in a vector space V is said to *span* V if *every* vector in V can be written as a linear combination of $\mathbf{v}_1, \mathbf{v}_2, \ldots, \mathbf{v}_k$, that is, if for any $\mathbf{v} \in V$

there exist scalars $c_1, c_2, \ldots c_k$ such that

$$\mathbf{v} = c_1\mathbf{v}_1 + c_2\mathbf{v}_2 + \cdots + c_k\mathbf{v}_k.$$

Conversely, given a set of vectors $\{\mathbf{v}_1, \mathbf{v}_2, \ldots, \mathbf{v}_k\}$ in a vector space V, we can form the set of *all* vectors that can be written as a linear combination of $\mathbf{v}_1, \mathbf{v}_2, \ldots, \mathbf{v}_k$. This set of vectors is a subspace of V called the *subspace spanned by* $\{\mathbf{v}_1, \mathbf{v}_2, \ldots, \mathbf{v}_k\}$ and denoted by $\mathrm{span}\{\mathbf{v}_1, \mathbf{v}_2, \ldots, \mathbf{v}_k\}$. Thus

$$\mathrm{span}\{\mathbf{v}_1, \mathbf{v}_2, \ldots, \mathbf{v}_k\} = \{\mathbf{v} \in V : \mathbf{v} = c_1\mathbf{v}_1 + c_2\mathbf{v}_2 + \cdots + c_k\mathbf{v}_k\}.$$

LINEAR DEPENDENCE AND LINEAR INDEPENDENCE

Let $\{\mathbf{v}_1, \mathbf{v}_2, \ldots, \mathbf{v}_k\}$ be a set of vectors in a vector space V, and consider the vector equation

$$c_1\mathbf{v}_1 + c_2\mathbf{v}_2 + \cdots + c_k\mathbf{v}_k = \mathbf{0}. \qquad (6.10.1)$$

Clearly this equation will hold if $c_1 = c_2 = \cdots = c_k = 0$. The question of interest is whether there are nonzero values of the scalars $c_1, c_2, \ldots, c_k$ such that (6.10.1) holds. This leads to the following two ideas:

Linear dependence: There exist scalars $c_1, c_2, \ldots, c_k$, *not all zero,* such that (6.10.1) holds.

Linear independence: The *only* values of the scalars $c_1, c_2, \ldots, c_k$ such that (6.10.1) holds are $c_1 = c_2 = \cdots = c_k = 0$.

To determine whether a set of vectors is linearly dependent or linearly independent, we usually have to use (6.10.1). However, if the vectors are from $\mathbf{R}^n$ (or $\mathbf{C}^n$), then we can use Theorem 6.5.2, whereas for vectors in $C^{n-1}(\mathrm{I})$, the Wronskian can be useful.

BASES AND DIMENSION

A set of linearly independent vectors that spans a vector space V is called a *basis* for V. If $\{\mathbf{v}_1, \mathbf{v}_2, \ldots, \mathbf{v}_k\}$ is a basis for V, then any vector in V can be written uniquely as

$$\mathbf{v} = c_1\mathbf{v}_1 + c_2\mathbf{v}_2 + \cdots + c_k\mathbf{v}_k.$$

for appropriate values of the scalars $c_1, c_2, \ldots, c_k$.

1. All bases in a finite-dimensional vector space contain the same number of vectors, and this number is called the *dimension* of V, denoted by $\dim[V]$.

2. We can view the dimension of a finite-dimensional vector space in two different ways. First, it gives the minimum number of vectors that span V. Alternatively, we can regard $\dim[V]$ as determining the maximum number of vectors that a linearly independent set can contain.

3. If $\dim[V] = n$, then *any* set of n linearly independent vectors in V forms a basis for V.

INNER PRODUCT SPACES

An inner product is a mapping that associates with any two vectors, $\mathbf{u}$, $\mathbf{v}$, in a vector space V a scalar that we denote by $<\mathbf{u}, \mathbf{v}>$. This mapping must satisfy the properties given in Definition 6.8.4. The main reason for introducing the idea of an inner product is that it enables us to extend the familiar idea of orthogonality of vectors in $\mathbf{R}^3$ to a general vector space. Thus $\mathbf{u}$ and $\mathbf{v}$ are said to be orthogonal in an inner product space if and only if

$$<\mathbf{u}, \mathbf{v}> = 0.$$

THE GRAM–SCHMIDT ORTHOGONALIZATION PROCEDURE

The Gram–Schmidt procedure is a process that takes a set of linearly independent vectors $\{\mathbf{v}_1, \mathbf{v}_2, \ldots, \mathbf{v}_k\}$ in an inner product space V and returns an *orthogonal* basis $\{\mathbf{u}_1, \mathbf{u}_2, \ldots, \mathbf{u}_k\}$ for $\text{span}\{\mathbf{v}_1, \mathbf{v}_2, \ldots, \mathbf{v}_k\}$.

7

Linear Transformations

DEFINITION OF A LINEAR TRANSFORMATION

In the previous chapter we introduced an appropriate framework for studying *linear* mathematics, namely, the vector space. In order for us to be able to formulate applied problems within this framework, we must now study mappings (or functions) between vector spaces. This should be considered as a generalization of the familiar idea of a function f of a real variable, such as $f(x) = 2x$.

Definition 7.1.1: Let V and W be vector spaces. A **mapping**, T, from V into W is a rule that associates with each vector $\mathbf{x}$ in V exactly one vector $\mathbf{y}$ in W. We write $T(\mathbf{x}) = \mathbf{y}$ or $T\mathbf{x} = \mathbf{y}$.

REMARKS

1. The vector space V is called the **domain** of the mapping, and the set of vectors that we can map onto in W is called the **range** of the transformation (see Figure 7.7.1).

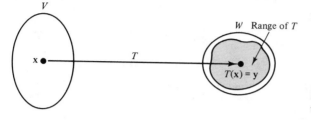

Figure 7.1.1 A mapping between two vector spaces.

2. We often write $T : V \rightarrow W$ to denote a mapping from V into W.

The basic operations of addition and scalar multiplication in a vector space enable us to combine vectors only linearly—that is, to form linear combinations of vectors. In keeping with the aim of studying linear mathematics, it is natural to restrict attention to mappings that preserve such linear combinations of vectors in the following sense:

$$T(c_1\mathbf{v}_1 + c_2\mathbf{v}_2 + \cdots + c_n\mathbf{v}_n) = c_1 T(\mathbf{v}_1) + c_2 T(\mathbf{v}_2) + \cdots + c_n T(\mathbf{v}_n) \qquad (7.1.1)$$

for all vectors $\mathbf{v}_1, \mathbf{v}_2, \cdots, \mathbf{v}_n$ in V and all scalars $c_1, c_2, \cdots, c_n$. The most general type of mapping that does this is called a linear transformation and is defined precisely as follows.

Definition 7.1.2: Let V and W be vector spaces.[1] A mapping $T : V \rightarrow W$ is called a **linear transformation** (or **linear operator**) from V into W if it has the following properties:

1. $T(\mathbf{x} + \mathbf{y}) = T(\mathbf{x}) + T(\mathbf{y})$ for *all* $\mathbf{x}, \mathbf{y}$ in V.
2. $T(c\mathbf{x}) = cT(\mathbf{x})$ for *all* $\mathbf{x}$ in V and *all* scalars c.

REMARKS

1. We leave it as an exercise to verify that a linear transformation does indeed satisfy (7.1.1) and hence preserves linear combinations of vectors.
2. A mapping that does not satisfy Definition 7.1.2 is called a **nonlinear transformation**.

Example 7.1.1   Define $T : C^1(\mathrm{I}) \rightarrow C^0(\mathrm{I})$ by

$$T(f) = f'.$$

---

[1] V and W must either be both real vector spaces or both complex vector spaces in order that we use the same scalars in both spaces.

Show that T is a linear transformation.

Solution   If f and g are in $C^1(I)$, and c is a real number, then
$$T(f+g) = (f+g)' = f' + g' = T(f) + T(g),$$
and
$$T(cf) = (cf)' = cf' = cT(f).$$
Thus T satisfies both properties of Definition 7.1.2 and hence is a linear transformation.

Example 7.1.2   Define $T:\mathbf{R}^2 \to \mathbf{R}^2$ by $T(x_1, x_2) = (x_1 - x_2, x_1 + x_2)$. Show that T is a linear transformation.

Solution   Let $\mathbf{x} = (x_1, x_2)$ and $\mathbf{y} = (y_1, y_2)$ be arbitrary vectors in $\mathbf{R}^2$. Then
$$\begin{aligned} T(\mathbf{x} + \mathbf{y}) &= T[(x_1, x_2) + (y_1, y_2)] \\ &= T[(x_1 + y_1, x_2 + y_2)] = ((x_1 + y_1) - (x_2 + y_2), (x_1 + y_1) + (x_2 + y_2)) \\ &= ((x_1 - x_2) + (y_1 - y_2), (x_1 + x_2) + (y_1 + y_2)) \\ &= (x_1 - x_2, x_1 + x_2) + (y_1 - y_2, y_1 + y_2) \\ &= T(x_1, x_2) + T(y_1, y_2) = T(\mathbf{x}) + T(\mathbf{y}). \end{aligned}$$
Similarly, if c is an arbitrary real number,
$$\begin{aligned} T(c\mathbf{x}) &= T[c(x_1, x_2)] = T(cx_1, cx_2) = (cx_1 - cx_2, cx_1 + cx_2) \\ &= c(x_1 - x_2, x_1 + x_2) = cT(x_1, x_2) = cT(\mathbf{x}). \end{aligned}$$
Thus T is a linear transformation.

Example 7.1.3   Define $S:M_n(\mathbf{R}) \to M_n(\mathbf{R})$ by $S(A) = A^T$. Show that S is a linear transformation.

Solution   Let A and B be any two matrices in $M_n(\mathbf{R})$. Then, using the properties of the transpose,
$$S(A + B) = (A + B)^T = A^T + B^T = S(A) + S(B),$$
and
$$S(cA) = (cA)^T = cA^T = cS(A),$$
for any real number c. Since both conditions of Definition 7.1.2 are satisfied, it follows that S is a linear transformation.

The following example introduces one of the most important linear transformations.

Example 7.1.4   Let A be an $m \times n$ matrix and define $T:\mathbf{R}^n \to \mathbf{R}^m$ by $T(\mathbf{x}) = A\mathbf{x}$. Show that T is a linear transformation.

Solution From the definition of T and properties of matrix multiplication it follows that

$$T(\mathbf{x} + \mathbf{y}) = A(\mathbf{x} + \mathbf{y}) = A\mathbf{x} + A\mathbf{y} = T(\mathbf{x}) + T(\mathbf{y}),$$

$$T(c\mathbf{x}) = A(c\mathbf{x}) = cA\mathbf{x} = cT(\mathbf{x}).$$

Thus T is a linear transformation.

Example 7.1.5 Explicitly write out the transformation $T:\mathbf{R}^3 \to \mathbf{R}^2$ defined by $T(\mathbf{x}) = A\mathbf{x}$, if $A = \begin{bmatrix} 1 & 3 & -1 \\ 2 & 1 & 5 \end{bmatrix}$.

Solution For any $\mathbf{x} = (x_1, x_2, x_3)$ in $\mathbf{R}^3$ we have

$$T(\mathbf{x}) = A\mathbf{x} = \begin{bmatrix} 1 & 3 & -1 \\ 2 & 1 & 5 \end{bmatrix} \begin{bmatrix} x_1 \\ x_2 \\ x_3 \end{bmatrix} = \begin{bmatrix} x_1 + 3x_2 - x_3 \\ 2x_1 + x_2 + 5x_3 \end{bmatrix},$$

so that

$$T(x_1, x_2, x_3) = (x_1 + 3x_2 - x_3, 2x_1 + x_2 + 5x_3).$$

REMARK Notice in the preceding example how we have once more switched between vectors in $\mathbf{R}^3$ and column 3-vectors.

Example 7.1.6 Define $T:\mathbf{R}^2 \to \mathbf{R}^2$ by $T(x_1, x_2) = (x_1 + 1, x_2 - 2)$. Show that T is a nonlinear transformation.

Solution We only need to show that one of the properties in Definition 7.1.2 is not satisfied. Let $\mathbf{x} = (x_1, x_2)$ and $\mathbf{y} = (y_1, y_2)$ be arbitrary vectors in $\mathbf{R}^2$. Then

$$T(\mathbf{x} + \mathbf{y}) = T(x_1 + y_1, x_2 + y_2) = (x_1 + y_1 + 1, x_2 + y_2 - 2),$$

whereas

$$T(\mathbf{x}) + T(\mathbf{y}) = (x_1 + 1, x_2 - 2) + (y_1 + 1, y_2 - 2) = (x_1 + y_1 + 2, x_2 + y_2 - 4).$$

Thus we see that $T(\mathbf{x} + \mathbf{y}) \neq T(\mathbf{x}) + T(\mathbf{y})$, and hence T is *not* a linear transformation. We leave it as an exercise to show that $T(c\mathbf{x}) \neq cT(\mathbf{x})$ in this case either.

Now suppose that $T:V \to W$ is a linear transformation, where V is a finite-dimensional vector space. Let $\{\mathbf{v}_1, \mathbf{v}_2, \ldots, \mathbf{v}_n\}$ be a basis for V. Then, for any $\mathbf{v}$ in V, there are scalars $c_1, c_2, \ldots, c_n$ such that

$$\mathbf{v} = c_1\mathbf{v}_1 + c_2\mathbf{v}_2 + \cdots + c_n\mathbf{v}_n.$$

Thus applying T to $\mathbf{v}$ and using the fact that T preserves linear combinations, we have

$$T(\mathbf{v}) = T(c_1\mathbf{v}_1 + c_2\mathbf{v}_2 + \cdots + c_n\mathbf{v}_n) = c_1T(\mathbf{v}_1) + c_2T(\mathbf{v}_2) + \cdots + c_nT(\mathbf{v}_n).$$

This shows that if we know how the vectors in a basis for V transform, then we can determine the transformation of any vector in V.   Once more this emphasizes the usefulness of the concept of a basis for a vector space.   We illustrate with an example.

Example 7.1.7   Let $T:\mathbf{R}^2 \to \mathbf{R}^2$ be a *linear* transformation.   Suppose that the basis vectors $\mathbf{v}_1 = (1, 0)$, $\mathbf{v}_2 = (0, 1)$ are transformed as follows:

$$T(\mathbf{v}_1) = (3, -2), \qquad T(\mathbf{v}_2) = (1, 5).$$

Determine $T(\mathbf{x})$ for an arbitrary vector $\mathbf{x} = (x_1, x_2)$.

Solution   We first express $\mathbf{x} = (x_1, x_2)$ in terms of the basis $\{\mathbf{v}_1, \mathbf{v}_2\}$.   In this case we have

$$\mathbf{x} = x_1(1, 0) + x_2(0, 1) = x_1\mathbf{v}_1 + x_2\mathbf{v}_2.$$

Applying T to $\mathbf{x}$ yields:

$$T(\mathbf{x}) = T(x_1\mathbf{v}_1 + x_2\mathbf{v}_2) = x_1T(\mathbf{v}_1) + x_2T(\mathbf{v}_2),$$

that is,

$$T(\mathbf{x}) = x_1(3, -2) + x_2(1, 5).$$

Thus

$$T(\mathbf{x}) = (3x_1 + x_2, -2x_1 + 5x_2).$$

For example,

$$T(2, -1) = (5, -9), \qquad T(-3, 0) = (-9, 6).$$

We now generalize to linear transformations some ideas that are familiar from functions of a single variable.   We first need the following definition.

Definition 7.1.3:   If $T_1:V \to W$ and $T_2:V \to W$ are two linear transformations, then we say that T_1 **equals** T_2 and write $T_1 = T_2$ if and only if $T_1(\mathbf{v}) = T_2(\mathbf{v})$ for all $\mathbf{v}$ in V.

We can now define how to add two linear transformations and how to multiply a linear transformation by a scalar.

Definition 7.1.4:   Let $T_1:V \to W$ and $T_2:V \to W$ be linear transformations, and let c be a scalar.   We define the **sum** $T_1 + T_2$ and the **scalar product** cT_1 by

$$(T_1 + T_2)(\mathbf{v}) = T_1(\mathbf{v}) + T_2(\mathbf{v}), \qquad \text{for all } \mathbf{v} \text{ in } V,$$
$$(cT_1)(\mathbf{v}) = cT_1(\mathbf{v}), \qquad \text{for all } \mathbf{v} \text{ in } V,$$

respectively.

Theorem 7.1.1:   If $T_1:V \rightarrow W$ and $T_2:V \rightarrow W$ are linear transformations and c is a scalar, then $T_1 + T_2$ and cT_1 are also linear transformations.

PROOF   We must show that the transformations $T_1 + T_2$ and cT_1 satisfy both properties of Definition 7.1.2.   Consider first $T_1 + T_2$.   Let $\mathbf{v}_1$ and $\mathbf{v}_2$ be arbitrary vectors in V.   Then

$$(T_1 + T_2)(\mathbf{v}_1 + \mathbf{v}_2) = T_1(\mathbf{v}_1 + \mathbf{v}_2) + T_2(\mathbf{v}_1 + \mathbf{v}_2) = T_1(\mathbf{v}_1) + T_1(\mathbf{v}_2) + T_2(\mathbf{v}_1) + T_2(\mathbf{v}_2)$$

$$= T_1(\mathbf{v}_1) + T_2(\mathbf{v}_1) + T_1(\mathbf{v}_2) + T_2(\mathbf{v}_2)$$

$$= (T_1 + T_2)(\mathbf{v}_1) + (T_1 + T_2)(\mathbf{v}_2),$$

so that property 1 of Definition 7.1.2 is satisfied.   Further, if k is any scalar, then

$$(T_1 + T_2)(k\mathbf{v}) = T_1(k\mathbf{v}) + T_2(k\mathbf{v}) = kT_1(\mathbf{v}) + kT_2(\mathbf{v})$$

$$= k[T_1(\mathbf{v}) + T_2(\mathbf{v})] = k(T_1 + T_2)(\mathbf{v}),$$

so that property 2 of Definition 7.1.2 is also satisfied.   It follows that $T_1 + T_2$ is a linear transformation.

The proof that cT_1 is a linear transformation is similar and hence we leave it as an exercise.   ∎

Example 7.1.8   Let $A = \begin{bmatrix} 2 & 1 \\ -1 & 3 \end{bmatrix}$ and $B = \begin{bmatrix} 1 & -1 \\ 2 & 1 \end{bmatrix}$.   Define $T_1:\mathbf{R}^2 \rightarrow \mathbf{R}^2$ and $T_2:\mathbf{R}^2 \rightarrow \mathbf{R}^2$ by

$$T_1(\mathbf{x}) = A\mathbf{x}, \qquad T_2(\mathbf{x}) = B\mathbf{x}.$$

Find (a) $(T_1 + T_2)$, (b) (cT_1).

Solution

(a) Let $\mathbf{x} = (x_1, x_2)$ be an arbitrary vector in $\mathbf{R}^2$.   Then

$$(T_1 + T_2)(\mathbf{x}) = T_1(\mathbf{x}) + T_2(\mathbf{x}) = A\mathbf{x} + B\mathbf{x} = (A + B)\mathbf{x}.$$

Substituting for the given matrices yields

$$(T_1 + T_2)(\mathbf{x}) = \begin{bmatrix} 3 & 0 \\ 1 & 4 \end{bmatrix}\begin{bmatrix} x_1 \\ x_2 \end{bmatrix} = \begin{bmatrix} 3x_1 \\ x_1 + 4x_2 \end{bmatrix}.$$

Thus

$$(T_1 + T_2)(x_1, x_2) = (3x_1, x_1 + 4x_2).$$

(b) Let c be an arbitrary real number.   Then

$$(cT_1)(\mathbf{x}) = cT_1(\mathbf{x}) = cA\mathbf{x}.$$

Substituting for A we obtain

$$(cT_1)(\mathbf{x}) = c\begin{bmatrix} 2 & 1 \\ -1 & 3 \end{bmatrix}\begin{bmatrix} x_1 \\ x_2 \end{bmatrix} = c\begin{bmatrix} 2x_1 + x_2 \\ -x_1 + 3x_2 \end{bmatrix} = \begin{bmatrix} 2cx_1 + cx_2 \\ -cx_1 + 3cx_2 \end{bmatrix}.$$

Thus

$$(cT_1)(x_1, x_2) = (2cx_1 + cx_2, -cx_1 + 3cx_2).$$

Finally in this section, let V and W be vector spaces and consider the set of all linear transformations from V into W. We denote this set by $L(V, W)$. Definition 7.1.4 provides us with a way of adding elements in this set and multiplying elements in the set by a scalar. Further, it follows from Theorem 7.1.1 that $L(V, W)$ is closed under both of these operations. It is natural to ask, therefore, whether $L(V, W)$ together with the previously defined operations forms a vector space. The following theorem gives an affirmative answer to this question.

Theorem 7.1.2: Let V and W be vector spaces. The set of all linear transformations from V into W, $L(V, W)$, together with the operations of addition and scalar multiplication defined in Definition 7.1.4, forms a vector space.

PROOF   We need to check that all the vector space axioms are satisfied. We leave the verification of Axioms A1 and A4 as exercises and simply give the zero vector in $L(V, W)$ and the additive inverse of a general vector.

The zero vector in $L(V, W)$ is the *zero transformation* $O:V \rightarrow W$ defined by[1]

$$O(\mathbf{v}) = \mathbf{0} \qquad \text{for all } \mathbf{v} \text{ in } V,$$

where $\mathbf{0}$ denotes the zero vector in W. To show that O is indeed the zero vector in $L(V, W)$, let T be any transformation in $L(V, W)$. Then

$$(T + O)(\mathbf{v}) = T(\mathbf{v}) + O(\mathbf{v}) = T(\mathbf{v}) + \mathbf{0} = T(\mathbf{v}) \qquad \text{for all } \mathbf{v} \in V,$$

so that $T + O = T$.

The additive inverse of the transformation $T \in L(V, W)$ is the linear transformation $-T$ defined by $-T = (-1)T$, since

$$[T + (-T)](\mathbf{v}) = T(\mathbf{v}) + (-T)(\mathbf{v}) = T(\mathbf{v}) + (-1)T(\mathbf{v}) = T(\mathbf{v}) - T(\mathbf{v}) = \mathbf{0} \text{ for all } \mathbf{v} \text{ in } V,$$

so that $T + (-T) = O$. ∎

REMARK   Notice that the "vectors" in $L(V, W)$ are themselves linear transformations.

EXERCISES 7.1

In problems 1–6 determine whether the given mapping $T:\mathbf{R}^m \rightarrow \mathbf{R}^n$ defines a linear transformation.

1. $T(x_1, x_2) = (x_1 + 2x_2, 2x_1 - x_2)$.
2. $T(x_1, x_2) = (x_1, x_2 - x_1, x_2)$.

3. $T(x_1, x_2, x_3) = (x_1 + x_2, x_3, 1)$.
4. $T(x_1, x_2) = x_1 + x_2$.
5. $T(x_1, x_2, x_3) = x_1^2$.
6. $T(x_1, x_2, x_3) = (x_1 + 3x_2 + x_3, x_1 - x_2)$.

---

[1] We leave the verification that the zero transformation is a *linear* transformation as an exercise.

In problems 7–10 determine the matrix A such that the given transformation can be written in the form $T(\mathbf{x}) = A\mathbf{x}$.

7. $T(x_1, x_2) = (3x_1 - 2x_2, x_1 + 5x_2)$.

8. $T(x_1, x_2) = (x_1 + 3x_2, 2x_1 - 7x_2, x_1)$.

9. $T(x_1, x_2, x_3) = (x_1 - x_2 + x_3, x_3 - x_1)$.

10. $T(x_1, x_2, x_3) = x_1 + 5x_2 - 3x_3$.

11. Define the mapping $T:C^0[a, b] \to \mathbf{R}$ by $T(f) = \displaystyle\int_a^b f(x)\, dx$. Show that T is a linear transformation.

12. Define $T:M_n(\mathbf{R}) \to M_n(\mathbf{R})$ by $T(A) = AB - BA$, where B is a fixed $n \times n$ matrix. Show that T is a linear transformation.

13. Let V be a real inner product space and let $\mathbf{u}$ be a fixed (nonzero) vector in V. Define $T:V \to \mathbf{R}$ by

$$T(\mathbf{v}) = <\mathbf{u},\ \mathbf{v}>$$

(that is, the inner product of $\mathbf{u}$ with $\mathbf{v}$). Show that T is a linear transformation.

14. Consider the mapping $S:M_n(\mathbf{R}) \to M_n(\mathbf{R})$ defined by $S(A) = A - A^T$. Show that S is a linear transformation.

15. Recall that the sum of the diagonal elements of an $n \times n$ matrix A is called the *trace* of A, denoted by $\mathrm{tr}(A)$. Thus

$$\mathrm{tr}(A) = \sum_{i=1}^{n} a_{ii}.$$

Define $T:M_n(\mathbf{R}) \to \mathbf{R}$ by $T(A) = \mathrm{tr}(A)$. Show that T is a linear transformation.

16. (a) Define $T:M_2(\mathbf{R}) \to \mathbf{R}$ by $T(A) = \det(A)$. Determine whether T is a linear transformation. **(b)** Define $T:M_n(\mathbf{R}) \to \mathbf{R}$ by $T(A) = \det(A)$. Determine all values of n for which T is a linear transformation.

17. Let $T:\mathbf{R}^2 \to \mathbf{R}^3$ be the *linear* transformation defined by

$$T(1, 0) = (1, -1, 2), \qquad T(0, 1) = (2, 1, -1).$$

Show that for all $(x_1, x_2) \in \mathbf{R}^2$,

$$T(x_1, x_2) = (x_1 + 2x_2, -x_1 + x_2, 2x_1 - x_2).$$

18. (a) Show that the vectors $\mathbf{v}_1 = (1, 1)$, $\mathbf{v}_2 = (1, -1)$ form a basis for $\mathbf{R}^2$.

(b) Let $T:\mathbf{R}^2 \to \mathbf{R}^2$ be the *linear* transformation satisfying

$$T(\mathbf{v}_1) = (2, 3), \qquad T(\mathbf{v}_2) = (-1, 1),$$

where $\mathbf{v}_1$ and $\mathbf{v}_2$ are the basis vectors given in (a). Find $T(x_1, x_2)$ for an arbitrary vector (x_1, x_2) in $\mathbf{R}^2$. What is $T(4, -2)$?

19. Let $T:P_3 \to P_3$ be the *linear* transformation satisfying

$$T(1) = x + 1, \qquad T(x) = x^2 - 1, \qquad T(x^2) = 3x + 2.$$

Determine $T(ax^2 + bx + c)$, where a, b, c are arbitrary real numbers.

20. Let $T:V \to V$ be a linear transformation, and suppose that

$$T(2\mathbf{v}_1 + 3\mathbf{v}_2) = \mathbf{v}_1 + \mathbf{v}_2, \quad T(\mathbf{v}_1 + \mathbf{v}_2) = 3\mathbf{v}_1 - \mathbf{v}_2.$$

Find $T(\mathbf{v}_1)$ and $T(\mathbf{v}_2)$.

21. Let $T:P_3 \to P_3$ be the *linear* transformation satisfying

$$T(x^2 - 1) = x^2 + x - 3, \qquad T(2x) = 4x,$$

$$T(3x + 2) = 2(x + 3).$$

Find $T(1)$, $T(x)$, and $T(x^2)$, and hence show that

$$T(ax^2 + bx + c) = ax^2 - (a - 2b + 2c)x + 3c,$$

where a, b, c are arbitrary real numbers.

22. Let $\{\mathbf{v}_1, \mathbf{v}_2\}$ be a basis for the vector space V. If $T:V \to V$ is the linear transformation satisfying

$$T(\mathbf{v}_1) = 3\mathbf{v}_1 - \mathbf{v}_2, \qquad T(\mathbf{v}_2) = \mathbf{v}_1 + 2\mathbf{v}_2,$$

find $T(\mathbf{v})$ for an arbitrary vector in V.

23. Consider the linear transformations $T_1:\mathbf{R}^2 \to \mathbf{R}^2$, $T_2:\mathbf{R}^2 \to \mathbf{R}^2$ defined by

$$T_1(\mathbf{x}) = A\mathbf{x}, \quad \text{where } A = \begin{bmatrix} 2 & -1 \\ 3 & 2 \end{bmatrix},$$

$$T_2(\mathbf{x}) = B\mathbf{x}, \quad \text{where } B = \begin{bmatrix} 1 & 2 \\ 2 & -1 \end{bmatrix}.$$

Find: $T_1 + T_2$, $5T_1$, $2T_1 - 3T_2$.

24. Define the linear transformations $T_1:P_2 \to P_3$, $T_2:P_2 \to P_3$ by

$$T_1(ax + b) = ax^2 + (a - b)x + b,$$

$$T_2(ax + b) = bx^2 + (b - a)x + a.$$

Determine $T_1 + T_2$.

25. Let $\{\mathbf{v}_1, \mathbf{v}_2\}$ be a basis for the vector space V. If $T_1:V \to V$ and $T_2:V \to V$ are linear transformations satisfying

$$T_1(\mathbf{v}_1) = 2\mathbf{v}_1 - \mathbf{v}_2, \qquad T_1(\mathbf{v}_2) = \mathbf{v}_1 + 5\mathbf{v}_2,$$

$$T_2(\mathbf{v}_1) = \mathbf{v}_1 + \mathbf{v}_2, \qquad T_2(\mathbf{v}_2) = \mathbf{v}_1 - 3\mathbf{v}_2,$$

find $(T_1 + T_2)(\mathbf{v})$, for an arbitrary vector $\mathbf{v}$ in V.

26. Complete the proof of Theorem 7.1.1 by showing that if T is a linear transformation then so also is cT, where c is a scalar.

27. Complete the proof of Theorem 7.1.2 by verifying that the vector space Axioms A1 and A4 are satisfied in $L(V, W)$ with the given operations of addition and scalar multiplication.

7.2   THE KERNEL AND RANGE OF A LINEAR TRANSFORMATION

In this section we show how linear transformations naturally give rise to *linear equations* defined in a vector space. We introduce this idea by reformulating two familiar problems in terms of linear transformations.

Let A be an $m \times n$ matrix. One of the problems that we have studied in detail is that of finding the solution set, S, of the homogeneous linear system $A\mathbf{x} = \mathbf{0}$. This problem can be reformulated in linear transformation terms as follows. We can use A to define a linear transformation $T:\mathbf{R}^n \to \mathbf{R}^m$ by $T(\mathbf{x}) = A\mathbf{x}$. Then, determining the solution set to $A\mathbf{x} = \mathbf{0}$ is equivalent to finding all vectors $\mathbf{x}$ in $\mathbf{R}^n$ satisfying the linear equation

$$T(\mathbf{x}) = \mathbf{0}.$$

We can therefore write

$$S = \{\mathbf{x} \in \mathbf{R}^n : T(\mathbf{x}) = \mathbf{0}\}.$$

Now consider solving the linear differential equation $y'' + y = 0$. If we define the linear transformation[1] $T:C^2(I) \to C^0(I)$ by

$$T(y) = y'' + y,$$

then we see that finding all solutions to the given differential equation is equivalent to finding all vectors y in $C^2(I)$ satisfying the linear equation

$$T(y) = 0.$$

Denoting the set of all solutions to $y'' + y = 0$ by S, it follows that

$$S = \{y \in C^2(I) : T(y) = 0\}.$$

These problems arise as special cases of the following general linear problem.

> Let $T:V \to W$ be a linear transformation. Find *all* vectors $\mathbf{x} \in V$ satisfying the linear equation $T(\mathbf{x}) = \mathbf{0}$.

This discussion suggests that the set of vectors that are mapped to the zero vector by a linear transformation plays a fundamental role in the formulation and

---

[1] We leave the verification that T is indeed a *linear* transformation as an exercise.

analysis of linear problems. The following definition introduces the standard terminology that we will use for this set of vectors.

Definition 7.2.1: Let $T:V \to W$ be a linear transformation. The set of *all* vectors $\mathbf{x} \in V$ such that $T(\mathbf{x}) = \mathbf{0}$ is called the **kernel**, or **null space**, of T and is denoted by Ker(T). Thus,

$$\mathrm{Ker}(T) = \{\mathbf{x} \in V : T(\mathbf{x}) = \mathbf{0}\}.$$

The next example establishes that Ker(T) is necessarily a *nonempty* set.

Example 7.2.1   If $T:V \to W$ is a linear transformation, prove that $\mathbf{0}$ is in Ker(T).

Solution   We must prove that $T(\mathbf{0}) = \mathbf{0}$ (note that the zero vector appearing on the left-hand side is the zero vector in V, whereas the zero vector appearing on the right-hand side is the zero vector in W). Recall from Theorem 6.2.1 (page 191) that if $\mathbf{v}$ is any vector in V, then we can write $\mathbf{0} = 0\mathbf{v}$. Thus,

$$T(\mathbf{0}) = T(0\mathbf{v}) = 0T(\mathbf{v}) = \mathbf{0}.$$

Another fundamental set of vectors associated with a linear transformation $T:V \to W$ is the range of T. In words this is the set of vectors in W that are obtained when we allow T to act on all the vectors in V. More precisely we have the following.

Definition 7.2.2: Let $T:V \to W$ be a linear transformation. The set of *all* vectors $\mathbf{y}$ in W such that $T(\mathbf{x}) = \mathbf{y}$ for at least one $\mathbf{x}$ in V is called the **range** of T and is denoted by Rng(T). Thus

$$\mathrm{Rng}(T) = \{\mathbf{y} \in W : T(\mathbf{x}) = \mathbf{y} \text{ for at least one } \mathbf{x} \text{ in } V\}.$$

See Figure 7.2.1.

If A is an $m \times n$ matrix, then the range of the linear transformation $T(\mathbf{x}) = A\mathbf{x}$ is

$$\mathrm{Rng}(T) = \{\mathbf{y} \in \mathbf{R}^m : A\mathbf{x} = \mathbf{y}, \text{ for some } \mathbf{x} \text{ in } \mathbf{R}^n\}.$$

The system $A\mathbf{x} = \mathbf{y}$ is nonhomogeneous (when $\mathbf{y} \neq \mathbf{0}$); hence, in general, there will be some vectors $\mathbf{y}$ for which this system has a solution and some for which it does not have a solution. The range of T consists of all vectors $\mathbf{y} \in \mathbf{R}^m$ for which $A\mathbf{x} = \mathbf{y}$

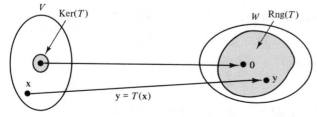

Figure 7.2.1 A pictorial representation of Ker(T) and Rng(T). Every vector in Ker(T) is mapped to the zero vector in W.

has a solution, that is (using the terminology introduced in Chapter 4), the set of all **y** for which $A\mathbf{x} = \mathbf{y}$ is consistent.

Thus, summarizing:

If $T:\mathbf{R}^n \rightarrow \mathbf{R}^m$ is defined by $T(\mathbf{x}) = A\mathbf{x}$, then

$$\text{Ker}(T) = \{\mathbf{x} \in \mathbf{R}^n : A\mathbf{x} = \mathbf{0}\},$$

which is the set of all solutions to the homogeneous linear system $A\mathbf{x} = \mathbf{0}$. Further,

$$\text{Rng}(T) = \{\mathbf{y} \in \mathbf{R}^m : A\mathbf{x} = \mathbf{y}, \text{ for some } \mathbf{x} \text{ in } \mathbf{R}^n\},$$

which is the set of vectors in $\mathbf{R}^m$ for which the nonhomogeneous system $A\mathbf{x} = \mathbf{y}$ is consistent.

It follows from Example 6.3.4 that, for the preceding transformation, $\text{Ker}(T)$ is in fact a *subspace* of $\mathbf{R}^n$, and it is easily shown that $\text{Rng}(T)$ is a subspace of $\mathbf{R}^m$. The importance of $\text{Ker}(T)$ and $\text{Rng}(T)$ stems from the fact that they are subspaces in general. Indeed we have the following basic theorem.

Theorem 7.2.1:   If $T:V \rightarrow W$ is a linear transformation, then

1. $\text{Ker}(T)$ is a subspace of V.
2. $\text{Rng}(T)$ is a subspace of W.

PROOF   It follows from Example 7.2.1 that $\text{Ker}(T)$ and $\text{Rng}(T)$ are necessarily nonempty subsets of V and W, respectively. We must show that both these subsets are closed under addition and closed under scalar multiplication in the appropriate vector space.

1. If **u** and **v** are in $\text{Ker}(T)$ and c is any scalar, then

$$T(\mathbf{u} + \mathbf{v}) = T(\mathbf{u}) + T(\mathbf{v}) = \mathbf{0} + \mathbf{0} = \mathbf{0},$$

so that $\text{Ker}(T)$ is closed under addition. Further,

$$T(c\mathbf{u}) = cT(\mathbf{u}) = c\mathbf{0} = \mathbf{0},$$

so that $\text{Ker}(T)$ is also closed under scalar multiplication. Thus $\text{Ker}(T)$ is a subspace of V.

2. If **v** and **w** are in $\text{Rng}(T)$, then $\mathbf{v} = T(\mathbf{x})$, $\mathbf{w} = T(\mathbf{y})$ for some **x**, **y** in V. Thus

$$\mathbf{v} + \mathbf{w} = T(\mathbf{x}) + T(\mathbf{y}) = T(\mathbf{x} + \mathbf{y}),$$

that is, $\mathbf{v} + \mathbf{w}$ is in $\text{Rng}(T)$. Further, if c is any scalar, then

$$c\mathbf{v} = cT(\mathbf{x}) = T(c\mathbf{x}),$$

so that $c\mathbf{v}$ is in $\text{Rng}(T)$. Hence $\text{Rng}(T)$ is a subspace of W. ∎

Example 7.2.2 Consider the linear transformation $T: \mathbf{R}^2 \to \mathbf{R}^2$ defined by $T(\mathbf{x}) = A\mathbf{x}$, where $A = \begin{bmatrix} 1 & 2 \\ -1 & -2 \end{bmatrix}$. Find $\text{Ker}(T)$, $\text{Rng}(T)$, and their dimensions.

Solution

$$T(\mathbf{x}) = A\mathbf{x} = \begin{bmatrix} x_1 + 2x_2 \\ -x_1 - 2x_2 \end{bmatrix},$$

so that

$$\text{Ker}(T) = \{(x_1, x_2) \in \mathbf{R}^2 : x_1 + 2x_2 = 0\}.$$

If we set $x_2 = r$, then we can write the solution of $x_1 + 2x_2 = 0$ in the form $(-2r, r)$, so that

$$\text{Ker}(T) = \{\mathbf{x} \in \mathbf{R}^2 : \mathbf{x} = r(-2, 1), r \in \mathbf{R}\}.$$

A basis for $\text{Ker}(T)$ is $\{(-2, 1)\}$, and hence $\dim[\text{Ker}(T)] = 1$. Geometrically, $\text{Ker}(T)$ is the line through the origin determined by the vector $(-2, 1)$.

We now find $\text{Rng}(T)$. For the given transformation, we have

$$\text{Rng}(T) = \{(y_1, y_2) \in \mathbf{R}^2 : x_1 + 2x_2 = y_1, -x_1 - 2x_2 = y_2\}.$$

We must therefore determine the values of y_1, y_2 for which the system

$$x_1 + 2x_2 = y_1$$
$$-x_1 - 2x_2 = y_2$$

is consistent.   Adding these two equations yields the consistency requirement that

$$y_1 + y_2 = 0.$$

Thus

$$\text{Rng}(T) = \{(y_1, y_2) \in \mathbf{R}^2 : y_1 + y_2 = 0\}.$$

Introducing the free variable $y_2 = s$, we can express $\text{Rng}(T)$ in the form

$$\text{Rng}(T) = \{\mathbf{y} \in \mathbf{R}^2 : \mathbf{y} = s(-1, 1), s \in \mathbf{R}\}.$$

Consequently a basis for $\text{Rng}(T)$ is $\{(-1, 1)\}$, and hence $\dim[\text{Rng}(T)] = 1$. Geometrically $\text{Rng}(T)$ is also a line in $\mathbf{R}^2$.   (See Figure 7.2.2.)

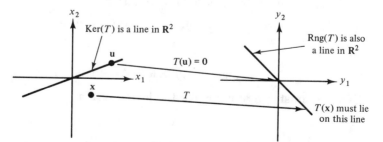

Figure 7.2.2   The kernel and range of the linear transformation in Example 7.2.2.

If the matrix is more complicated than in the previous example, then we have to reduce the augmented matrix of the system $A\mathbf{x} = \mathbf{y}$ to determine $\mathrm{Rng}(T)$. This is illustrated in the following example.

Example 7.2.3 Consider the linear transformation $T:\mathbf{R}^3 \to \mathbf{R}^3$ defined by

$T(\mathbf{x}) = A\mathbf{x}$, where $A = \begin{bmatrix} 1 & 1 & 1 \\ 2 & 3 & 1 \\ 3 & 5 & 1 \end{bmatrix}$. Find $\mathrm{Ker}(T)$ and $\mathrm{Rng}(T)$, and give a geometric

interpretation of each of them.

Solution   In this case

$$\mathrm{Ker}(T) = \{\mathbf{x} \in \mathbf{R}^3 : A\mathbf{x} = \mathbf{0}\},$$

that is, the set of all solutions to the homogeneous system

$$x_1 + x_2 + x_3 = 0,$$
$$2x_1 + 3x_2 + x_3 = 0,$$
$$3x_1 + 5x_2 + x_3 = 0.$$

The reduced row echelon form of the augmented matrix of this system is

$$\begin{bmatrix} 1 & 0 & 2 & | & 0 \\ 0 & 1 & -1 & | & 0 \\ 0 & 0 & 0 & | & 0 \end{bmatrix},$$

so that the solution of the system can be written as

$$(-2r, r, r),$$

where r is a free variable.   It follows that

$$\mathrm{Ker}(T) = \{\mathbf{x} \in \mathbf{R}^3 : \mathbf{x} = r(-2, 1, 1), r \in \mathbf{R}\}.$$

Geometrically this is a line in $\mathbf{R}^3$.   [It is the subspace of $\mathbf{R}^3$ spanned by the vector $(-2, 1, 1)$.]

For the given transformation we have

$$\mathrm{Rng}(T) = \{\mathbf{y} \in \mathbf{R}^3 : A\mathbf{x} = \mathbf{y} \text{ for at least one } \mathbf{x} \in \mathbf{R}^3\},$$

that is,

$$\mathrm{Rng}(T) = \{\mathbf{y} \in \mathbf{R}^3 : A\mathbf{x} = \mathbf{y} \text{ is consistent}\}.$$

Thus we must determine when the linear system

$$x_1 + x_2 + x_3 = y_1,$$
$$2x_1 + 3x_2 + x_3 = y_2,$$
$$3x_1 + 5x_2 + x_3 = y_3,$$

is consistent.   Reducing the augmented matrix of the system yields

$$\begin{bmatrix} 1 & 1 & 1 \\ 2 & 3 & 1 \\ 3 & 5 & 1 \end{bmatrix} \begin{matrix} y_1 \\ y_2 \\ y_3 \end{matrix} \begin{matrix} R_2 \to R_2 - 2R_1 \\ \xrightarrow{} \\ R_3 \to R_3 - 3R_1 \end{matrix} \begin{bmatrix} 1 & 1 & 1 \\ 0 & 1 & -1 \\ 0 & 2 & -2 \end{bmatrix} \begin{matrix} y_1 \\ y_2 - 2y_1 \\ y_3 - 3y_1 \end{matrix}$$

$$\xrightarrow{R_3 \to R_3 - 2R_2} \begin{bmatrix} 1 & 1 & 1 \\ 0 & 1 & -1 \\ 0 & 0 & 0 \end{bmatrix} \begin{matrix} y_1 \\ y_2 - 2y_1 \\ y_1 - 2y_2 + y_3 \end{matrix}.$$

The augmented matrix is now in row echelon form, and the last row tells us directly that the system is consistent if and only if

$$y_1 - 2y_2 + y_3 = 0. \tag{7.2.1}$$

Thus,

$$\text{Rng}(T) = \{(y_1, y_2, y_3) \in \mathbf{R}^3 : y_1 - 2y_2 + y_3 = 0\}.$$

Equation (7.2.1) involves three unknowns, and hence there are two free variables. Setting

$$y_2 = s, \qquad y_3 = t,$$

it follows that

$$y_1 = 2s - t.$$

We can therefore express the range of T in the form

$$\text{Rng}(T) = \{\mathbf{y} \in \mathbf{R}^3 : \mathbf{y} = (2s - t, s, t), \ s, \ t \in \mathbf{R}\},$$

or, equivalently,

$$\text{Rng}(T) = \{\mathbf{y} \in \mathbf{R}^3 : \mathbf{y} = s(2, 1, 0) + t(-1, 0, 1), \ s, \ t \in \mathbf{R}\}.$$

We see that the range of T is the subspace of $\mathbf{R}^3$ spanned by the linearly independent vectors $(2, 1, 0)$, $(-1, 0, 1)$, and hence $\dim[\text{Rng}(T)] = 2$.   From (7.2.1) the geometrical interpretation of $\text{Rng}(T)$ is that it represents a plane in $\mathbf{R}^3$.   See Figure 7.2.3.

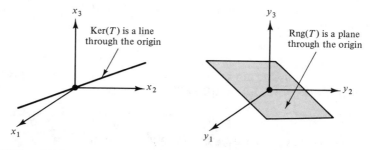

Figure 7.2.3   The kernel and range of the transformation in Example 7.2.3.

Example 7.2.4 Consider the linear transformation $T:P_3 \rightarrow P_2$ defined by

$$T(ax^2 + bx + c) = (4a - c)x + (2b - c).$$

Determine $\text{Ker}(T)$, $\text{Rng}(T)$, and their dimensions.

Solution From Definition 7.2.1,

$$\text{Ker}(T) = \{p \in P_3 : T(p) = 0\}$$
$$= \{ax^2 + bx + c : (4a - c)x + (2b - c) = 0 \text{ for all } x\}.$$

But,

$$(4a - c)x + (2b - c) = 0$$

for all x if and only if a, b, and c satisfy

$$4a \quad - c = 0,$$
$$2b - c = 0.$$

Solving this system yields

$$a = r, \qquad b = 2r, \qquad c = 4r,$$

where r is a free variable, so that

$$\text{Ker}(T) = \{r(x^2 + 2x + 4) : r \in \mathbf{R}\}.$$

Thus $\text{Ker}(T)$ is spanned by the nonzero vector $p_1(x) = x^2 + 2x + 4$, and hence $\dim[\text{Ker}(T)] = 1$. The range of T is

$$\text{Rng}(T) = \{\alpha x + \beta \in P_2 : T(p) = \alpha x + \beta, \text{ for some } p \in P_3\}$$
$$= \{\alpha x + \beta \in P_2 : T(ax^2 + bx + c) = \alpha x + \beta, \text{ for some } a, b, c \in \mathbf{R}\}$$
$$= \{\alpha x + \beta : (4a - c)x + (2b - c) = \alpha x + \beta \text{ for some } a, b, c \in \mathbf{R}\}$$
$$= \{\alpha x + \beta : 4a - c = \alpha, 2b - c = \beta \text{ is consistent}\}.$$

The system

$$4a \quad - c = \alpha,$$
$$2b - c = \beta$$

is consistent for all values of α, β [$\text{rank}(A^\#) = \text{rank}(A)$] and hence every vector in P_2 is the image of some vector in P_3. It follows that $\text{Rng}(T) = P_2$, and hence $\dim[\text{Rng}(T)] = 2$.

The linear transformation $T:P_3 \rightarrow P_2$ in the previous example has $\dim[\text{Ker}(T)] = 1$ and $\dim[\text{Rng}(T)] = 2$, so that

$$\dim[\text{Ker}(T)] + \dim[\text{Rng}(T)] = \dim[P_3].$$

This is a special case of the following general result, which is one of the more important theorems in linear algebra.

Theorem 7.2.2: If $T: V \rightarrow W$ is a linear transformation and V is finite-dimensional, then

$$\dim[\text{Ker}(T)] + \dim[\text{Rng}(T)] = \dim[V].$$

PROOF   Suppose that $\dim[V] = n$. There are three different cases to consider. If $\text{Ker}(T) = V$, then $\text{Rng}(T) = \{0\}$, and so

$$\dim[\text{Ker}(T)] + \dim[\text{Rng}(T)] = n + 0 = n,$$

as required.

Now consider the case when $\dim[\text{Ker}(T)] = k$, where $0 < k < n$. Let $\{\mathbf{v}_1, \mathbf{v}_2, \ldots, \mathbf{v}_k\}$ be a basis for $\text{Ker}(T)$. Then, using Theorem 6.7.4 (page 223), we can extend this basis to a basis for V, which we denote by $\{\mathbf{v}_1, \mathbf{v}_2, \ldots, \mathbf{v}_k, \mathbf{v}_{k+1}, \ldots, \mathbf{v}_n\}$. We prove that $\{T(\mathbf{v}_{k+1}), T(\mathbf{v}_{k+2}), \ldots, T(\mathbf{v}_n)\}$ is a basis for $\text{Rng}(T)$. Let $\mathbf{w}$ be any vector in $\text{Rng}(T)$. Then $\mathbf{w} = T(\mathbf{v})$ for some $\mathbf{v} \in V$. Consequently, there exist scalars $c_1, c_2, \ldots, c_n$ such that

$$\mathbf{w} = T(c_1\mathbf{v}_1 + c_2\mathbf{v}_2 + \cdots + c_n\mathbf{v}_n)$$
$$= c_1 T(\mathbf{v}_1) + c_2 T(\mathbf{v}_2) + \cdots + c_n T(\mathbf{v}_n).$$

Since $\mathbf{v}_1, \mathbf{v}_2, \ldots, \mathbf{v}_k$ are in $\text{Ker}(T)$, this reduces to

$$\mathbf{w} = c_{k+1} T(\mathbf{v}_{k+1}) + c_{k+2} T(\mathbf{v}_{k+2}) + \cdots + c_n T(\mathbf{v}_n).$$

Thus

$$\text{Rng}(T) = \text{span}\{T(\mathbf{v}_{k+1}), T(\mathbf{v}_{k+2}), \ldots, T(\mathbf{v}_n)\}.$$

We must also show that $\{T(\mathbf{v}_{k+1}), T(\mathbf{v}_{k+2}), \ldots, T(\mathbf{v}_n)\}$ is linearly independent. Suppose that

$$d_{k+1} T(\mathbf{v}_{k+1}) + d_{k+2} T(\mathbf{v}_{k+2}) + \cdots + d_n T(\mathbf{v}_n) = \mathbf{0}, \qquad (7.2.2)$$

where $d_{k+1}, d_{k+2}, \ldots, d_n$ are scalars. Then, using the linearity of T,

$$T(d_{k+1}\mathbf{v}_{k+1} + d_{k+2}\mathbf{v}_{k+2} + \cdots + d_n\mathbf{v}_n) = \mathbf{0},$$

which implies that the vector $d_{k+1}\mathbf{v}_{k+1} + d_{k+2}\mathbf{v}_{k+2} + \cdots + d_n\mathbf{v}_n$ is in $\text{Ker}(T)$. Consequently there exist scalars $d_1, d_2, \ldots, d_k$ such that

$$d_{k+1}\mathbf{v}_{k+1} + d_{k+2}\mathbf{v}_{k+2} + \cdots + d_n\mathbf{v}_n = d_1\mathbf{v}_1 + d_2\mathbf{v}_2 + \cdots + d_k\mathbf{v}_k,$$

that is, such that

$$d_1\mathbf{v}_1 + d_2\mathbf{v}_2 + \cdots + d_k\mathbf{v}_k - (d_{k+1}\mathbf{v}_{k+1} + d_{k+2}\mathbf{v}_{k+2} + \cdots + d_n\mathbf{v}_n) = \mathbf{0}.$$

But the vectors $\mathbf{v}_1, \mathbf{v}_2, \ldots, \mathbf{v}_k, \mathbf{v}_{k+1}, \ldots, \mathbf{v}_n$ are linearly independent, and so we must have

$$d_1 = d_2 = \cdots = d_k = d_{k+1} = \cdots = d_n = 0.$$

Thus, from (7.2.2), $T(\mathbf{v}_{k+1}), T(\mathbf{v}_{k+2}), \ldots, T(\mathbf{v}_n)$ are linearly independent, and therefore $\{T(\mathbf{v}_{k+1}), T(\mathbf{v}_{k+2}), \ldots, T(\mathbf{v}_n)\}$ is a basis for $\text{Rng}(T)$. Since there are $n - k$ vectors in this basis, it follows that $\dim[\text{Rng}(T)] = n - k$. Consequently,

$$\dim[\text{Ker}(T)] + \dim[\text{Rng}(T)] = k + (n - k) = n,$$

as required.

Finally, if $\text{Ker}(T) = \{0\}$, let $\{v_1, v_2, \ldots, v_n\}$ be any basis for V. By a similar argument to that used above (see problem 15), it follows that $\{T(v_1), T(v_2), \ldots, T(v_n)\}$ is a basis for $\text{Rng}(T)$, and so once more we have

$$\dim[\text{Ker}(T)] + \dim[\text{Rng}(T)] = n. \qquad \blacksquare$$

The final example in this section gives a further indication of the role vector spaces play in the theory of linear differential equations. This will be developed fully in Chapter 9.

Example 7.2.5 Define the *linear* transformation $L : C^2(I) \to C^0(I)$ by $Ly = y'' + ay' + by$, where a prime denotes differentiation with respect to x and a, b are (real) constants. Describe $\text{Ker}(L)$.

Solution   In this case,

$$\text{Ker}(L) = \{y \in C^2(I) : Ly = 0\} = \{y \in C^2(I) : y'' + ay' + by = 0\}.$$

Thus, $\text{Ker}(L)$ consists of the set of all solutions of the linear differential equation

$$y'' + ay' + by = 0. \qquad (7.2.3)$$

From Theorem 7.2.1, this set of solutions is a subspace of $C^2(I)$ and hence is itself a vector space (we showed this by brute force in Section 6.3). What do you think the dimension of this vector space is? We will prove in Chapter 9 that the dimension is, in fact, 2. Thus, accepting this fact, it follows that any solution of the differential equation (7.2.3) is of the form

$$y = c_1 y_1 + c_2 y_2,$$

where y_1, y_2 are any two *linearly independent* solutions of (7.2.3). Thus our vector space and linear transformation results have reduced the problem of finding all solutions of (7.2.3) to that of finding just two linearly independent solutions.

EXERCISES 7.2

1. Consider $T : \mathbf{R}^3 \to \mathbf{R}^2$ defined by $T(\mathbf{x}) = A\mathbf{x}$, where $A = \begin{bmatrix} 1 & -1 & 2 \\ 1 & -2 & -3 \end{bmatrix}$. Determine whether the following vectors are in $\text{Ker}(T)$.

(a) $(7, 5, -1)$.
(b) $(-21, -15, 2)$.
(c) $(35, 25, -5)$.

In problems 2–6 find $\text{Ker}(T)$ and $\text{Rng}(T)$ and give a geometrical description of each. Also

find $\dim[\text{Ker}(T)]$ and $\dim[\text{Rng}(T)]$ and verify Theorem 7.2.2.

2. $T : \mathbf{R}^2 \to \mathbf{R}^2$ defined by $T(\mathbf{x}) = A\mathbf{x}$, where $A = \begin{bmatrix} 3 & 6 \\ 1 & 2 \end{bmatrix}$.

3. $T : \mathbf{R}^3 \to \mathbf{R}^3$ defined by $T(\mathbf{x}) = A\mathbf{x}$, where $A = \begin{bmatrix} 1 & -1 & 0 \\ 0 & 1 & 2 \\ 2 & -1 & 1 \end{bmatrix}$.

4. $T:\mathbf{R}^3 \to \mathbf{R}^3$ defined by $T(\mathbf{x}) = A\mathbf{x}$, where

$$A = \begin{bmatrix} 1 & -2 & 1 \\ 2 & -3 & -1 \\ 5 & -8 & -1 \end{bmatrix}.$$

5. $T:\mathbf{R}^3 \to \mathbf{R}^2$ defined by $T(\mathbf{x}) = A\mathbf{x}$, where

$$A = \begin{bmatrix} 1 & -1 & 2 \\ -3 & 3 & -6 \end{bmatrix}.$$

6. $T:\mathbf{R}^3 \to \mathbf{R}^2$ defined by $T(\mathbf{x}) = A\mathbf{x}$, where

$$A = \begin{bmatrix} 1 & 3 & 2 \\ 2 & 6 & 5 \end{bmatrix}.$$

7. Define the linear transformation $L:C^2(I) \to C^0(I)$ by $L(f) = f''$. Determine a basis for Ker(L) and hence find dim[Ker(L)].

8. Consider the linear transformation $T:\mathbf{R}^3 \to \mathbf{R}$ defined by

$$T(\mathbf{v}) = <\mathbf{u}, \mathbf{v}>$$

(that is, the inner product of $\mathbf{u}$ with $\mathbf{v}$), where $\mathbf{u}$ is a fixed vector in $\mathbf{R}^3$.
(a) Find Ker(T) and dim[Ker(T)], and interpret geometrically.
(b) Find Rng(T) and dim[Rng(T)].

9. Consider the linear transformation $S:M_n(\mathbf{R}) \to M_n(\mathbf{R})$ defined by $S(A) = A - A^T$, where A is an $n \times n$ matrix.
(a) Find Ker(S) and describe it.
(b) In the particular case when A is a 2×2 matrix, determine a basis for Ker(S) and hence find its dimension.

10. Consider the linear transformation $T:M_n(\mathbf{R}) \to M_n(\mathbf{R})$ defined by

$$T(A) = AB - BA,$$

where B is a fixed $n \times n$ matrix. Find Ker(T) and describe it.

11. Consider the *linear* transformation $T:P_3 \to P_3$ defined by

$$T(ax^2 + bx + c) = ax^2 + (a + 2b + c)x + (3a - 2b - c),$$

where a, b, c are arbitrary constants.
(a) Show that Ker(T) consists of all polynomials of the form $b(x - 2)$ and hence find its dimension.
(b) Find Rng(T) and its dimension.

12. Consider the *linear* transformation $T:P_3 \to P_2$ defined by

$$T(ax^2 + bx + c) = (a + b) + (b - c)x,$$

where a, b, c are arbitrary real numbers. Determine Ker(T), Rng(T), and their dimensions.

13. Consider the *linear* transformation $T:P_2 \to P_3$ defined by

$$T(ax + b) = (b - a) + (2b - 3a)x + bx^2.$$

Determine Ker(T), Rng(T), and their dimensions.

14. Let $\{\mathbf{v}_1, \mathbf{v}_2, \mathbf{v}_3\}$ and $\{\mathbf{w}_1, \mathbf{w}_2\}$ be bases for the (real) vector spaces V and W respectively, and let $T:V \to W$ be the linear transformation satisfying

$$T(\mathbf{v}_1) = 2\mathbf{w}_1 - \mathbf{w}_2, \qquad T(\mathbf{v}_2) = \mathbf{w}_1 - \mathbf{w}_2,$$
$$T(\mathbf{v}_3) = \mathbf{w}_1 + 2\mathbf{w}_2.$$

Find Ker(T), Rng(T), and their dimensions.

15. Let $T:V \to W$ be a linear transformation, and suppose that $\dim[V] = n$. If Ker(T) = {0} and $\{\mathbf{v}_1, \mathbf{v}_2, \dots, \mathbf{v}_n\}$ is any basis for V, prove that $\{T(\mathbf{v}_1), T(\mathbf{v}_2), \dots, T(\mathbf{v}_n)\}$ is a basis for Rng(T). (This fills in the missing details in the proof of Theorem 7.2.2.)

7.3   INVERSE TRANSFORMATIONS

Recall that if f and g are two real-valued functions, then the composition of f with g, denoted by $f \circ g$, is the function defined by $(f \circ g)(x) = f(g(x))$, whenever $g(x)$ is in the domain of f. Further, if there exists a function f^{-1} with the property that $(f \circ f^{-1})(x) = (f^{-1} \circ f)(x) = x$ for all x in some interval, then we say that f^{-1} is the inverse of f. We now introduce the corresponding ideas for linear transformations between two vector spaces. We must first define how to compose two linear transformations.

> **Definition 7.3.1:** Let $T_1:U \to V$ and $T_2:V \to W$ be two linear transformations.[1] We define the **composition** (or **product**) $T_2T_1:U \to W$ by
>
> $$(T_2T_1)(\mathbf{u}) = T_2(T_1(\mathbf{u})) \qquad \text{for all } \mathbf{u} \text{ in } U.$$
>
> See Figure 7.3.1.

REMARK   Note that for linear transformations we use the product notation T_2T_1 rather than the composition notation $T_2 \circ T_1$.

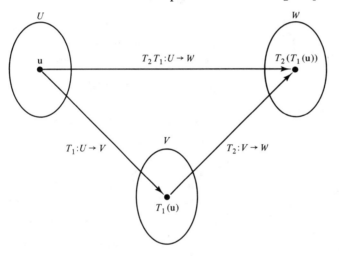

Figure 7.3.1 The composition of two linear transformations.

Our first result establishes that T_2T_1 is a *linear* transformation.

> **Theorem 7.3.1:** Let $T_1:U \to V$ and $T_2:V \to W$ be linear transformations. Then $T_2T_1:U \to W$ is a linear transformation.

PROOF   Let $\mathbf{v}_1, \mathbf{v}_2$ be arbitrary vectors in U, and let c be a scalar. We must prove that

$$(T_2T_1)(\mathbf{v}_1 + \mathbf{v}_2) = (T_2T_1)(\mathbf{v}_1) + (T_2T_1)(\mathbf{v}_2), \qquad (7.3.1)$$

and that

$$(T_2T_1)(c\mathbf{v}_1) = c(T_2T_1)(\mathbf{v}_1). \qquad (7.3.2)$$

Consider first (7.3.1). We have

---

[1] U, V, and W must be either all real vector spaces or all complex vector spaces.

$$(T_2T_1)(\mathbf{v}_1 + \mathbf{v}_2) = T_2(T_1(\mathbf{v}_1 + \mathbf{v}_2)) = T_2(T_1(\mathbf{v}_1) + T_1(\mathbf{v}_2))$$

↑   ↑

Definition of T_2T_1 Using the linearity of T_1

$$= T_2(T_1(\mathbf{v}_1)) + T_2(T_1(\mathbf{v}_2)) = (T_2T_1)(\mathbf{v}_1) + (T_2T_1)(\mathbf{v}_2),$$

↑   ↑

Using the linearity of T_2 Definition of T_2T_1

so that (7.3.1) is satisfied. We leave the proof of (7.3.2) as an exercise. ∎

Example 7.3.1 Let $T_1:\mathbf{R}^3 \to \mathbf{R}^2$ and $T_2:\mathbf{R}^2 \to \mathbf{R}^2$ be defined by $T_1(\mathbf{x}) = A\mathbf{x}$, $T_2(\mathbf{x}) = B\mathbf{x}$ respectively, where $A = \begin{bmatrix} 1 & -1 & 3 \\ 2 & 3 & -1 \end{bmatrix}$ and $B = \begin{bmatrix} 1 & 5 \\ -1 & 2 \end{bmatrix}$. Find T_2T_1.

Solution Let $\mathbf{x}$ be an arbitrary vector in $\mathbf{R}^3$. Then,

$$(T_2T_1)(\mathbf{x}) = T_2(T_1(\mathbf{x})) = T_2(A\mathbf{x}) = B(A\mathbf{x}).$$

Thus, if $\mathbf{x} = (x_1, x_2, x_3)$, then

$$(T_2T_1)(\mathbf{x}) = \begin{bmatrix} 1 & 5 \\ -1 & 2 \end{bmatrix}\begin{bmatrix} 1 & -1 & 3 \\ 2 & 3 & -1 \end{bmatrix}\begin{bmatrix} x_1 \\ x_2 \\ x_3 \end{bmatrix} = \begin{bmatrix} 1 & 5 \\ -1 & 2 \end{bmatrix}\begin{bmatrix} x_1 - x_2 + 3x_3 \\ 2x_1 + 3x_2 - x_3 \end{bmatrix}$$

$$= \begin{bmatrix} 11x_1 + 14x_2 - 2x_3 \\ 3x_1 + 7x_2 - 5x_3 \end{bmatrix}.$$

Hence,

$$(T_2T_1)(x_1, x_2, x_3) = (11x_1 + 14x_2 - 2x_3, 3x_1 + 7x_2 - 5x_3).$$

Notice that T_2T_1 is indeed a linear transformation from $\mathbf{R}^3$ into $\mathbf{R}^2$.

PROPERTIES OF THE COMPOSITION

The operation of composition satisfies many familiar properties, the most important of which are the following:

Theorem 7.3.2: Let T_1, T_2, T_3 be linear transformations. Then, provided the transformations are defined, we have

1. $T_1(T_2T_3) = (T_1T_2)T_3$.
2. $T_1(T_2 + T_3) = T_1T_2 + T_1T_3$.
3. $(T_1 + T_2)T_3 = T_1T_3 + T_2T_3$.

PROOF Consider (1). By Definition 7.3.1 we have:

$$[T_1(T_2T_3)](\mathbf{v}) = T_1[(T_2T_3)(\mathbf{v})] = T_1[T_2(T_3(\mathbf{v}))], \qquad (7.3.3)$$

whereas

$$[(T_1T_2)T_3](\mathbf{v}) = (T_1T_2)(T_3(\mathbf{v})) = T_1[T_2(T_3(\mathbf{v}))], \qquad (7.3.4)$$

and the result follows directly from (7.3.3) and (7.3.4). The proofs of (2) and (3) are left as exercises. ∎

If $T_1:V \to W$ and $T_2:W \to V$ are linear transformations, then both of the product transformations $T_1 T_2$ and $T_2 T_1$ are defined, and it is natural to ask whether they are equal. The answer is that, in general, $T_1 T_2 \neq T_2 T_1$. When V and W are different vector spaces this is obvious, since $T_2 T_1:V \to V$, whereas $T_1 T_2:W \to W$. However, even if V and W are the same vector space, $T_1 T_2 \neq T_2 T_1$ in general (that is except for special cases). You should draw a picture similar to Figure 7.3.1 to convince yourselves of this.

Example 7.3.2 Let $T_1:\mathbf{R}^n \to \mathbf{R}^n$ and $T_2:\mathbf{R}^n \to \mathbf{R}^n$ be defined by $T_1(\mathbf{x}) = A\mathbf{x}$, $T_2(\mathbf{x}) = B\mathbf{x}$, where A and B are $n \times n$ matrices. Then

$$(T_1 T_2)(\mathbf{x}) = T_1(T_2(\mathbf{x})) = T_1(B\mathbf{x}) = A(B\mathbf{x}) = (AB)\mathbf{x},$$

whereas

$$(T_2 T_1)(\mathbf{x}) = T_2(T_1(\mathbf{x})) = T_2(A\mathbf{x}) = B(A\mathbf{x}) = (BA)\mathbf{x}.$$

We see that in general $T_1 T_2 \neq T_2 T_1$, since matrix multiplication is not, in general, commutative.

We now turn our attention to inverse transformations, and ask the following question.

For a given linear transformation $T:V \to W$, does there exist a linear transformation $T^{-1}:W \to V$ with the properties

$$(T^{-1}T)(\mathbf{v}) = \mathbf{v}, \qquad \text{for all } \mathbf{v} \in V,$$
$$(TT^{-1})(\mathbf{w}) = \mathbf{w}, \qquad \text{for all } \mathbf{w} \in W?$$

Before we can answer this question, we require two important definitions for linear transformations:

Definition 7.3.2: A linear transformation $T:V \to W$ is said to be

1. **One-to-one** if distinct elements in V are mapped to distinct elements in W —that is,

    If $\mathbf{x} \neq \mathbf{y}$, then $T(\mathbf{x}) \neq T(\mathbf{y})$ for all $\mathbf{x}$ and $\mathbf{y}$ in V.

    See Figure 7.3.2.
2. **Onto** if the range of T is the whole of W—that is, if *every* $\mathbf{w}$ in W is the image of some $\mathbf{v}$ in V. See Figure 7.3.3.

Example 7.3.3 Define $T:\mathbf{R}^3 \to \mathbf{R}^2$ by $T(\mathbf{x}) = A\mathbf{x}$, where $A = \begin{bmatrix} 1 & 1 & -2 \\ 3 & 4 & 1 \end{bmatrix}$. Determine whether T is one-to-one or onto.

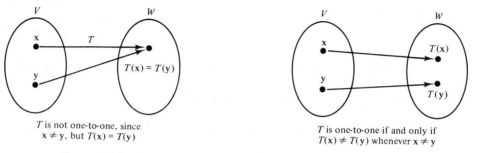

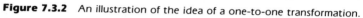

Figure 7.3.2 An illustration of the idea of a one-to-one transformation.

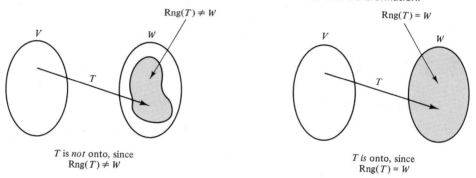

Figure 7.3.3 Illustration of the idea of an onto transformation.

Solution We compute Rng(T). Recall from the previous section that this is the set of vectors in $\mathbf{R}^2$ for which the linear system $A\mathbf{x} = \mathbf{y}$ is consistent. This system is

$$x_1 + x_2 - 2x_3 = y_1,$$
$$3x_1 + 4x_2 + x_3 = y_2.$$

The reduced row echelon form of the augmented matrix of this system is

$$\begin{bmatrix} 1 & 0 & -9 & | & 4y_1 - y_2 \\ 0 & 1 & 7 & | & y_2 - 3y_1 \end{bmatrix},$$

which implies that the system is consistent for all $\mathbf{y} = (y_1, y_2)$ in $\mathbf{R}^2$. Thus Rng(T) $= \mathbf{R}^2$; hence, from Definition 7.3.2, T is onto. Further, in this case the system $A\mathbf{x} = \mathbf{y}$ has an infinite number of solutions for each $\mathbf{y}$; that is, each $\mathbf{y}$ in $\mathbf{R}^2$ is the image of (infinitely) many $\mathbf{x}$ in $\mathbf{R}^3$. This implies that T is *not* one-to-one.

The following theorem can be helpful in determining whether a given linear transformation is one-to-one.

Theorem 7.3.3: Let $T:V \rightarrow W$ be a linear transformation. Then T is one-to-one if and only if Ker(T) consists only of the zero vector.

PROOF   Since T is a linear transformation, we have $T(\mathbf{0}) = \mathbf{0}$ (see Example 7.2.3). Thus if T is one-to-one, there can be no other $\mathbf{x}$ satisfying $T(\mathbf{x}) = \mathbf{0}$, and so $\text{Ker}(T) = \{\mathbf{0}\}$.

Conversely, suppose $\text{Ker}(T) = \{\mathbf{0}\}$.   If $\mathbf{x} \neq \mathbf{y}$ then $\mathbf{x} - \mathbf{y} \neq \mathbf{0}$ and, since $\text{Ker}(T) = \{\mathbf{0}\}$, $T(\mathbf{x} - \mathbf{y}) \neq \mathbf{0}$.   That is, $T(\mathbf{x}) - T(\mathbf{y}) \neq \mathbf{0}$ or, equivalently, $T(\mathbf{x}) \neq T(\mathbf{y})$. Thus if $\text{Ker}(T) = \{\mathbf{0}\}$, then T is one-to-one.   ∎

REMARK   We now have a complete characterization of one-to-one and onto in terms of the kernel and range of T.

Consider the linear transformation $T:V \to W$.

T is one-to-one if and only if $\text{Ker}(T) = \{\mathbf{0}\}$.

T is onto if and only if $\text{Rng}(T) = W$.

Example 7.3.4   Consider the linear transformation $T:P_3 \to P_3$ defined by

$$T(ax^2 + bx + c) = ax^2 + (b - 2c)x + (a - b + 2c).$$

Determine whether T is one-to-one or onto.

Solution   To determine whether T is one-to-one, we find $\text{Ker}(T)$.   For the given transformation we have

$$\text{Ker}(T) = \{p \in P_3 : T(p) = 0\}$$

$$= \{ax^2 + bx + c : T(ax^2 + bx + c) = 0, \text{ for all } x\}$$

$$= \{ax^2 + bx + c : ax^2 + (b - 2c)x + (a - b + 2c) = 0, \text{ for all } x\}.$$

But,

$$ax^2 + (b - 2c)x + (a - b + 2c) = 0$$

for all real x if and only if

$$a = 0, \qquad b - 2c = 0, \qquad a - b + 2c = 0.$$

These equations are satisfied if and only if

$$a = 0, \qquad b = 2c,$$

so that

$$\text{Ker}(T) = \{ax^2 + bx + c \in P_3 : a = 0, b = 2c\}$$

$$= \{p \in P_3 : p(x) = c(2x + 1), c \in \mathbf{R}\}.$$

It follows that the kernel of T does not consist of just the zero vector, and hence Theorem 7.3.3 implies that T is *not* one-to-one.   Rather than computing $\text{Rng}(T)$ to determine whether T is onto, we use Theorem 7.2.2.   The vectors in $\text{Ker}(T)$ consist of all scalar multiples of the vector $p_1(x) = 2x + 1$.   Thus p_1 forms a basis for

Ker(T), and hence dim[Ker(T)] = 1. Further, we have seen in Chapter 6 that dim[P_3] = 3. Thus applying the result of Theorem 7.2.2 we have

$$1 + \dim[\text{Rng}(T)] = 3,$$

which implies that

$$\dim[\text{Rng}(T)] = 2.$$

Since the dimension of P_3 is 3, it follows that Rng(T) $\neq P_3$, and hence T is *not* onto.

We can now answer our question as to the existence of T^{-1}. If $T:V \to W$ is both one-to-one and onto, then for each $\mathbf{w}$ in W, there is a *unique* $\mathbf{v}$ in V such that $T(\mathbf{v}) = \mathbf{w}$. Thus we can define a transformation $T^{-1}:W \to V$ by (see Figure 7.3.4)

$$T^{-1}(\mathbf{w}) = \mathbf{v} \quad \text{if and only if} \quad \mathbf{w} = T(\mathbf{v}).$$

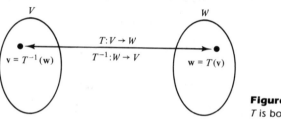

Figure 7.3.4 T^{-1} exists if and only if T is both one-to-one and onto.

This transformation satisfies the basic properties of an inverse, namely,

$$T^{-1}(T(\mathbf{v})) = \mathbf{v} \qquad \text{for all } \mathbf{v} \text{ in } V,$$
$$T(T^{-1}(\mathbf{w})) = \mathbf{w} \qquad \text{for all } \mathbf{w} \text{ in } W,$$

and so we call T^{-1} the *inverse* transformation. Again we stress that T^{-1} exists if and only if T is both one-to-one *and* onto.

REMARK   We leave it as an exercise to show that T^{-1} *is a linear* transformation.

Definition 7.3.3:   Let $T:V \to W$ be a linear transformation. If T is both one-to-one and onto, then the linear transformation $T^{-1}:W \to V$ defined by

$$T^{-1}(\mathbf{w}) = \mathbf{v} \quad \text{if and only if} \quad \mathbf{w} = T(\mathbf{v})$$

is called the **inverse** transformation to T.

Since neither of the transformations in the previous two examples is one-to-one, it follows that T^{-1} does not exist in either case.

For linear transformations $T:\mathbf{R}^n \to \mathbf{R}^m$ defined by $T(\mathbf{x}) = A\mathbf{x}$, we can characterize when T^{-1} exists and, when T^{-1} does exist, we can find it.

Theorem 7.3.4: Consider the linear transformation $T:\mathbf{R}^n \rightarrow \mathbf{R}^m$ defined by $T(\mathbf{x}) = A\mathbf{x}$. T^{-1} exists if and only if A is a nonsingular $n \times n$ matrix. In such a case T^{-1} is defined by

$$T^{-1}(\mathbf{y}) = A^{-1}\mathbf{y}.$$

PROOF   We first show that if $m \neq n$, then T^{-1} cannot exist. Suppose that $m < n$. Then Theorem 4.4.2 (page 134) implies that the $m \times n$ system of linear equations $A\mathbf{x} = \mathbf{0}$ has an infinite number of solutions, and hence $\text{Ker}(T) \neq \{\mathbf{0}\}$. Thus T is not one-to-one in this case, and so T^{-1} does not exist. Now suppose that $m > n$. Then, from Theorem 7.2.2, we have

$$\dim[\text{Ker}(T)] + \dim[\text{Rng}(T)] = n,$$

which implies that $\dim[\text{Rng}(T)] \leq n < m$. It follows that $\text{Rng}(T) \neq \mathbf{R}^m$, and hence T is not onto. Thus T^{-1} cannot exist in this case either. Finally consider the case in which $m = n$. Then T is one-to-one if and only if $\text{Ker}(T) = \{\mathbf{0}\}$—that is, if and only if $A\mathbf{x} = \mathbf{0}$ has only the trivial solution. But, from Corollary 5.6.1 (page 175), this is true if and only if $\det(A) \neq 0$. Further, if $\det(A) \neq 0$ then the system

$$A\mathbf{x} = \mathbf{y}$$

has the unique solution $\mathbf{x} = A^{-1}\mathbf{y}$ for any $\mathbf{y} \in \mathbf{R}^n$. It follows that T is also onto, and hence T^{-1} exists. It is defined by

$$T^{-1}(\mathbf{y}) = A^{-1}\mathbf{y} \qquad \text{for all } \mathbf{y} \text{ in } \mathbf{R}^n,$$

since then $(TT^{-1})(\mathbf{x}) = \mathbf{x} = (T^{-1}T)(\mathbf{x})$ for all $\mathbf{x}$ in $\mathbf{R}^n$. ∎

Example 7.3.5   Define $T:\mathbf{R}^2 \rightarrow \mathbf{R}^2$ by $T(\mathbf{x}) = A\mathbf{x}$, where $A = \begin{bmatrix} 2 & -1 \\ 3 & 1 \end{bmatrix}$. Show that T^{-1} exists and find it.

Solution   In this case $\det(A) = 5$, so that T^{-1} exists and is defined by $T^{-1}(\mathbf{y}) = A^{-1}\mathbf{y}$. It is easily shown that

$$A^{-1} = \begin{bmatrix} \frac{1}{5} & \frac{1}{5} \\ -\frac{3}{5} & \frac{2}{5} \end{bmatrix}.$$

Example 7.3.6   Define the linear transformation $T:P_2 \rightarrow P_2$ by

$$T(ax + b) = (a + b)x + (b - a).$$

Show that T^{-1} exists and find it.

Solution   We first show that T is both one-to-one and onto. In this case,

$$\text{Ker}(T) = \{ax + b : (a + b)x + (b - a) = 0, \text{ for all } x\}.$$

The only values of a and b for which

$$(a + b)x + (b - a) = 0$$

for all real x are

$$a = b = 0,$$

so that Ker(T) consists only of the zero polynomial $p(x) = 0$ for all x. It follows from Theorem 7.3.3 that T is one-to-one. Using Theorem 7.2.2 we have

$$\dim[\text{Rng}(T)] = \dim[P_2],$$

and since Rng(T) is a subspace of P_2, it follows that we must have

$$\text{Rng}(T) = P_2.$$

Thus, T is also onto. Since T is both one-to-one and onto, T^{-1} exists. From the definition of T it follows that

$$T^{-1}((a+b)x + b - a) = ax + b. \tag{7.3.5}$$

This does not determine T^{-1}, since we must specify $T^{-1}(Ax + B)$ for an arbitrary polynomial $Ax + B$. In order to do so we let

$$a + b = A,$$
$$-a + b = B,$$

in (7.3.5) and solve these equations to express a and b in terms of A and B. The result is

$$a = \tfrac{1}{2}(A - B), \quad b = \tfrac{1}{2}(A + B).$$

so that, substituting into (7.3.5),

$$T^{-1}(Ax + B) = \tfrac{1}{2}(A - B)x + \tfrac{1}{2}(A + B).$$

EXERCISES 7.3

1. Let $T_1:\mathbf{R}^2 \to \mathbf{R}^2$ and $T_2:\mathbf{R}^2 \to \mathbf{R}^2$ be linear transformations defined by $T_1(\mathbf{x}) = A\mathbf{x}$, $T_2(\mathbf{x}) = B\mathbf{x}$, where $A = \begin{bmatrix} -1 & 2 \\ 3 & 1 \end{bmatrix}$, $B = \begin{bmatrix} 1 & 5 \\ -2 & 0 \end{bmatrix}$. Find T_1T_2 and T_2T_1. Does $T_1T_2 = T_2T_1$?

2. Let $T_1:\mathbf{R}^2 \to \mathbf{R}^2$ be defined by $T_1(\mathbf{x}) = A\mathbf{x}$, where $A = \begin{bmatrix} 1 & -1 \\ 3 & 2 \end{bmatrix}$, and let $T_2:\mathbf{R}^2 \to \mathbf{R}$ be defined by $T_2(\mathbf{x}) = B\mathbf{x}$, where $B = [-1 \quad 1]$. Find T_2T_1. Does T_1T_2 exist? Explain.

3. Consider the linear transformations $T_1:\mathbf{R}^2 \to \mathbf{R}^2$, $T_2:\mathbf{R}^2 \to \mathbf{R}^2$ defined by

$$T_1(\mathbf{x}) = A\mathbf{x}, \quad \text{where } A = \begin{bmatrix} 1 & -1 \\ 2 & -2 \end{bmatrix},$$

$$T_2(\mathbf{x}) = B\mathbf{x}, \quad \text{where } B = \begin{bmatrix} 2 & 1 \\ 3 & -1 \end{bmatrix}.$$

Find Ker(T_1), Ker(T_2), Ker(T_1T_2), and Ker(T_2T_1).

4. Let $T_1:M_n(\mathbf{R}) \to M_n(\mathbf{R})$ and $T_2:M_n(\mathbf{R}) \to M_n(\mathbf{R})$ be linear transformations defined by $T_1(A) = (A - A^T)$, $T_2(A) = (A + A^T)$. Show that T_2T_1 is the zero transformation.

5. Define $T_1:C^1[a,b] \to C^0[a,b]$, $T_2:C^0[a,b] \to C^1[a,b]$ by

$$T_1(f) = f', \quad [T_2(f)](x) = \int_a^x f(t)\, dt, \quad a \le x \le b.$$

(a) If $f(x) = \sin(x - a)$, find $[T_1(f)](x)$, $[T_2(f)](x)$, and show that, for the given function, $[T_1T_2](f) = [T_2T_1](f) = f$.

(b) Show that for general functions f and g,

$$[T_1 T_2](f) = f, \qquad \{[T_2 T_1](g)\}(x) = g(x) - g(a).$$

6. Let $\{v_1, v_2\}$ be a basis for the vector space V, and suppose that $T_1 : V \to V$ and $T_2 : V \to V$ are the linear transformations satisfying

$$T_1(v_1) = v_1 - v_2, \qquad T_1(v_2) = 2v_1 + v_2,$$
$$T_2(v_1) = v_1 + 2v_2, \qquad T_2(v_2) = 3v_1 - v_2.$$

Determine $(T_2 T_1)(v)$ for an arbitrary vector in V.

7. Let $T_1 : U \to V$ and $T_2 : V \to W$ both be linear transformations. Complete the proof of Theorem 7.3.1 by showing that $(T_2 T_1)(cv_1) = c(T_2 T_1)(v_1)$.

8. Prove properties 2 and 3 of Theorem 7.3.2.

In problems 9–11 find Ker(T) and Rng(T), and hence determine whether the given transformation is one-to-one or onto. If T^{-1} exists, find it.

9. $T(x) = Ax$, where $A = \begin{bmatrix} 4 & 2 \\ 1 & 3 \end{bmatrix}$.

10. $T(x) = Ax$, where $A = \begin{bmatrix} 1 & 2 \\ -2 & -4 \end{bmatrix}$.

11. $T(x) = Ax$, where $A = \begin{bmatrix} 1 & 2 & -1 \\ 2 & 5 & 1 \end{bmatrix}$.

12. Let V be a vector space and define $T : V \to V$ by $T(x) = \lambda x$, where λ is a (nonzero) scalar. Show that T is a linear transformation that is one-to-one and onto, and find T^{-1}.

13. Define $T : P_2 \to P_2$ by

$$T(ax + b) = (2b - a)x + (b + a).$$

Show that T is both one-to-one and onto, and find T^{-1}.

14. Define $T : P_3 \to P_2$ by:

$$T(ax^2 + bx + c) = (a - b)x + c.$$

Determine whether T is one-to-one or onto. Does T^{-1} exist?

15. Let $\{v_1, v_2\}$ be a basis for the vector space V, and suppose that $T : V \to V$ is a linear transformation. If $T(v_1) = v_1 + 2v_2$ and $T(v_2) = 2v_1 - 3v_2$, show that T is one-to-one and onto and find T^{-1}.

16. Let v_1, v_2 be a basis for the vector space V, and suppose that $T_1 : V \to V$ and $T_2 : V \to V$ are the linear transformations satisfying

$$T_1(v_1) = v_1 + v_2, \qquad T_1(v_2) = v_1 - v_2,$$
$$T_2(v_1) = \tfrac{1}{2}(v_1 + v_2), \qquad T_2(v_2) = \tfrac{1}{2}(v_1 - v_2).$$

Find $(T_1 T_2)(v)$ and $(T_2 T_1)(v)$ for an arbitrary vector in V, and show that $T_2 = T_1^{-1}$.

17. If $T : V \to W$ is an invertible linear transformation (that is, T^{-1} exists) show that $T^{-1} : W \to V$ is also a linear transformation.

18. Let V be a finite-dimensional vector space, and let $T : V \to W$ be a linear transformation that is both one-to-one and onto. Prove that $\dim[V] = \dim[W]$. (*Hint:* Use Theorems 7.2.2 and 7.3.3.)

19. Let $T : V \to W$ be a linear transformation. Prove that if $\dim[W] < \dim[V]$, then T cannot be one-to-one.

20. Let $T : V \to W$ be a linear transformation. Prove that if $\dim[W] > \dim[V]$, then T cannot be onto.

21. Prove that if $T : V \to V$ is a one-to-one linear transformation, then T^{-1} exists.

22. Let $T : V \to W$ be a linear transformation and suppose that $\dim[W] = n$. Prove that if $\dim[\text{Rng}(T)] = n$, then T is onto.

23. Let $T_1 : V \to V$, and $T_2 : V \to V$, be linear transformations and suppose that T_2 is one-to-one. If

$$(T_1 T_2)(v) = v, \qquad \text{for all } v \text{ in } V,$$

prove that

$$(T_2 T_1)(v) = v, \qquad \text{for all } v \text{ in } V.$$

7.4   SUMMARY OF RESULTS

In this chapter we have considered mappings $T : V \to W$ between vector spaces V and W that satisfy the basic linearity properties

$$T(\mathbf{x} + \mathbf{y}) = T(\mathbf{x}) + T(\mathbf{y}), \qquad \text{for all } \mathbf{x} \text{ and } \mathbf{y} \text{ in } V,$$
$$T(c\mathbf{x}) = cT(\mathbf{x}), \qquad \text{for all } \mathbf{x} \text{ in } V \text{ and all scalars } c.$$

We now list some of the key definitions and theorems for linear transformations.
    We have identified the following two important subsets of vectors associated with a linear transformation:

1. The *kernel* of T, denoted by Ker(T). This is the set of all vectors in V that are mapped to the zero vector in W.
2. The *range* of T, denoted by Rng(T). This is the set of vectors in W that we obtain when we allow T to act on every vector in V.

The key results about Ker(T) and Rng(T) are as follows.
    Let $T:V \to W$ be a linear transformation. Then

1. Ker(T) is a subspace of V.
2. Rng(T) is a subspace of W.
3. If V is finite-dimensional, $\dim[\text{Ker}(T)] + \dim[\text{Rng}(T)] = \dim[V]$.
4. T is one-to-one if and only if Ker(T) = $\{\mathbf{0}\}$.
5. T is onto if and only if $[\text{Rng}(T)] = W$.

Finally, if $T:V \to W$ is a linear transformation, then the inverse transformation $T^{-1}:W \to V$ exists if and only if T is both one-to-one and onto.

8

Eigenvalues
and Eigenvectors

8.1   MOTIVATION

In order to motivate the problem to be studied in this chapter, we recall from Chapter 1 that the differential equation

$$\frac{dx}{dt} = ax,$$

(8.1.1)

where a is a constant, has general solution

$$x = ce^{at}.$$

(8.1.2)

Now consider the pair of linear differential equations

$$\frac{dx_1}{dt} = a_{11}x_1 + a_{12}x_2,$$

(8.1.3)

$$\frac{dx_2}{dt} = a_{21}x_1 + a_{22}x_2,$$

(8.1.4)

where x_1 and x_2 are functions of the independent variable t, and the a_{ij} are constants. By a solution to this system we mean a pair of functions

$$x_1 = f_1(t), \qquad x_2 = f_2(t)$$

that satisfy (8.1.3) and (8.1.4) simultaneously. Based on the solution (8.1.2) to the differential equation (8.1.1), we might suspect that (8.1.3) and (8.1.4) have a solution of the form

$$x_1 = e^{\lambda t} v_1, \qquad x_2 = e^{\lambda t} v_2,$$

where λ, v_1, v_2 are constants. Substituting these expressions for x_1 and x_2 into (8.1.3) and (8.1.4) yields the following conditions that λ, v_1, and v_2 must satisfy in order that we have a solution:

$$\lambda e^{\lambda t} v_1 = e^{\lambda t}(a_{11} v_1 + a_{12} v_2),$$

$$\lambda e^{\lambda t} v_2 = e^{\lambda t}(a_{21} v_1 + a_{22} v_2).$$

Dividing both equations by $e^{\lambda t}$ and rearranging terms yields

$$a_{11} v_1 + a_{12} v_2 = \lambda v_1,$$

$$a_{21} v_1 + a_{22} v_2 = \lambda v_2,$$

which can be written as the single matrix equation

$$\boxed{A\mathbf{v} = \lambda \mathbf{v},}$$

(8.1.5)

where

$$A = \begin{bmatrix} a_{11} & a_{12} \\ a_{21} & a_{22} \end{bmatrix} \quad \text{and} \quad \mathbf{v} = \begin{bmatrix} v_1 \\ v_2 \end{bmatrix}.$$

Thus we have shown that

$$\mathbf{x} = e^{\lambda t} \mathbf{v}$$

is a solution to the system of differential equations (8.1.3) and (8.1.4), provided λ and $\mathbf{v}$ satisfy (8.1.5). In Chapter 11 we will pursue this technique for determining solutions to general linear systems of differential equations. In the present chapter, however, we focus our attention on the mathematical problem of finding all scalars λ and all nonzero vectors $\mathbf{v}$ satisfying (8.1.5) for a given $n \times n$ matrix A. This is a fundamental problem that arises in many areas of pure mathematics such as geometry, algebra, and differential equations as well as most areas of applied mathematics.

8.2 THE EIGENVALUE-EIGENVECTOR PROBLEM

The discussion in the previous section leads us to analyze the following mathematical problem:

For a given $n \times n$ matrix A, find all scalars λ and all *nonzero* vectors $\mathbf{v}$ such that

$$A\mathbf{v} = \lambda\mathbf{v}. \tag{8.2.1}$$

We begin by introducing the appropriate terminology.

> **Definition 8.2.1:** Let A be an $n \times n$ matrix. Any values of λ for which $A\mathbf{v} = \lambda\mathbf{v}$ has nontrivial solutions are called **eigenvalues** of A. The corresponding *nonzero* vectors $\mathbf{v}$ are called **eigenvectors** of A.

REMARK   Eigenvalues and eigenvectors are also often referred to as characteristic values and characteristic vectors, respectively.

To formulate the eigenvalue-eigenvector problem within the vector space framework, we will interpret the matrix A as defining a linear transformation $T:\mathbf{C}^n \rightarrow \mathbf{C}^n$ in the usual manner—that is, $T(\mathbf{v}) = A\mathbf{v}$. In most of our problems A and λ will both be real, which will enable us to restrict attention to $\mathbf{R}^n$, although we will sometimes require the more general vector space $\mathbf{C}^n$.

Example 8.2.1   Show that $\mathbf{v} = (1, 3)$ is an eigenvector of the matrix $A = \begin{bmatrix} 1 & 1 \\ -3 & 5 \end{bmatrix}$

corresponding to the eigenvalue $\lambda = 4$.

Solution   For the given vector we have[1]

$$A\mathbf{v} = \begin{bmatrix} 1 & 1 \\ -3 & 5 \end{bmatrix}\begin{bmatrix} 1 \\ 3 \end{bmatrix} = \begin{bmatrix} 4 \\ 12 \end{bmatrix} = 4\begin{bmatrix} 1 \\ 3 \end{bmatrix},$$

that is, $A\mathbf{v} = 4\mathbf{v}$. It follows that $\mathbf{v}$ is an eigenvector of A corresponding to the eigenvalue $\lambda = 4$.

It is always helpful to have a geometric interpretation of the problem under consideration. The eigenvectors of an $n \times n$ matrix A are those nonzero vectors that are mapped into themselves, up to a constant multiple, by the linear transformation $T(\mathbf{v}) = A\mathbf{v}$. This is illustrated for the case of $\mathbf{R}^2$ in Figure 8.2.1.

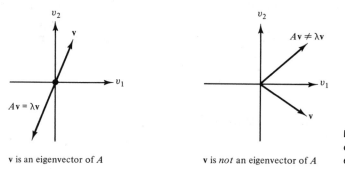

Figure 8.2.1   A geometrical description of the eigenvalue-eigenvector problem.

v is an eigenvector of A v is *not* an eigenvector of A

---

[1] Notice that once more we will switch between vectors in $\mathbf{R}^n$ and column n-vectors.

REMARK   If $A\mathbf{v} = \lambda\mathbf{v}$ and c is an arbitrary scalar, then

$$A(c\mathbf{v}) = cA\mathbf{v} = c(\lambda\mathbf{v}) = \lambda(c\mathbf{v}).$$

Thus, if $\mathbf{v}$ is an eigenvector of A, then so is $\mathbf{w} = c\mathbf{v}$ for any nonzero scalar c. It follows that eigenvectors corresponding to a given eigenvalue are defined only up to scalar multiples (this should be clear, geometrically, from Figure 8.2.1).

SOLUTION OF THE PROBLEM

The solution of the eigenvalue-eigenvector problem hinges on the observation that (8.2.1) can be written in the equivalent form

$$(A - \lambda I)\mathbf{v} = \mathbf{0}, \tag{8.2.2}$$

where I denotes the identity matrix. Thus the eigenvalues of A are those values of λ for which the $n \times n$ linear system (8.2.2) has nontrivial solutions and the eigenvectors are the corresponding solutions. But, according to Corollary 5.6.1 (page 175), the system (8.2.2) has nontrivial solutions if and only if

$$\det(A - \lambda I) = 0.$$

Thus to solve the eigenvalue-eigenvector problem we proceed as follows:

---

1. Find all scalars λ such that $\det(A - \lambda I) = 0$. These are the eigenvalues of A.

2. If $\lambda_1, \lambda_2, \ldots, \lambda_k$ are the *distinct* eigenvalues obtained in (1), then solve the k systems of linear equations

$$(A - \lambda_i I)\mathbf{v}_i = 0, \qquad i = 1, 2, 3, \ldots, k,$$

to find all eigenvectors $\mathbf{v}_i$ corresponding to each eigenvalue.

For a given $n \times n$ matrix A, the degree n polynomial $p(\lambda)$ defined by

$$p(\lambda) = \det(A - \lambda I)$$

is called the **characteristic polynomial of** A, and the equation

$$p(\lambda) = 0$$

is called the **characteristic equation of** A. It follows from (1) that the eigenvalues of A are the roots of the characteristic equation.

Example 8.2.2   Determine the characteristic polynomial and characteristic equation of $A = \begin{bmatrix} a_{11} & a_{12} \\ a_{21} & a_{22} \end{bmatrix}$.

Solution   By definition the characteristic polynomial of A is

$$p(\lambda) = \det(A - \lambda I) = \begin{vmatrix} a_{11} - \lambda & a_{12} \\ a_{21} & a_{22} - \lambda \end{vmatrix}$$

$$= \lambda^2 - (a_{11} + a_{22})\lambda + (a_{11}a_{22} - a_{12}a_{21}).$$

Thus the characteristic equation of A is

$$\lambda^2 - (a_{11} + a_{22})\lambda + (a_{11}a_{22} - a_{12}a_{21}) = 0.$$

Before considering some examples we recall that a quadratic equation

$$a\lambda^2 + b\lambda + c = 0,$$

always has two roots (not necessarily distinct) given by the **quadratic formula**

$$\lambda = \frac{-b \pm \sqrt{b^2 - 4ac}}{2a}.$$

This is sometimes required in determining the eigenvalues of a matrix.

SOME BASIC EXAMPLES

REAL EIGENVALUES AND EIGENVECTORS   We now consider several examples in order to illustrate some of the possibilities that arise in the eigenvalue-eigenvector problem.   To begin with, we will consider the simplest case of a real matrix with real eigenvalues.   In this case the eigenvectors can always be chosen to have real components, and hence they are vectors in $\mathbf{R}^n$.

Example 8.2.3   Find all eigenvalues and eigenvectors of $A = \begin{bmatrix} 1 & 2 \\ 4 & 3 \end{bmatrix}$.

Solution   The linear system for determining the eigenvalues and eigenvectors is $(A - \lambda I)\mathbf{v} = \mathbf{0}$, that is,

$$\begin{bmatrix} 1 - \lambda & 2 \\ 4 & 3 - \lambda \end{bmatrix} \begin{bmatrix} v_1 \\ v_2 \end{bmatrix} = \begin{bmatrix} 0 \\ 0 \end{bmatrix}. \tag{8.2.3}$$

This system has nontrivial solutions if and only if $\det(A - \lambda I) = 0$, so that λ must satisfy

$$\begin{vmatrix} 1 - \lambda & 2 \\ 4 & 3 - \lambda \end{vmatrix} = 0.$$

Expanding the determinant we obtain the characteristic equation

$$\lambda^2 - 4\lambda - 5 = 0,$$

which can be factored to yield

$$(\lambda + 1)(\lambda - 5) = 0.$$

Consequently the eigenvalues of A are $\lambda_1 = -1$, $\lambda_2 = 5$.

Eigenvectors   The corresponding eigenvectors are obtained by successively substituting the preceding eigenvalues into (8.2.3) and solving the resulting system.
$\lambda = -1$: Substituting $\lambda = -1$ into (8.2.3) yields the system

$$2v_1 + 2v_2 = 0,$$

$$4v_1 + 4v_2 = 0.$$

There is only one independent equation, $v_1 + v_2 = 0$, so that the system has an infinite number of solutions given by $(-r, r)$, where r is a free variable. It follows that the eigenvectors corresponding to $\lambda = -1$ are those vectors in $\mathbf{R}^2$ of the form

$$\mathbf{v} = r(-1, 1),$$

where r is any *nonzero* real number. Notice that there is only one *linearly independent* eigenvector corresponding to the eigenvalue $\lambda = -1$, which we may take as $\mathbf{v}_1 = (-1, 1)$. All the other eigenvectors (corresponding to $\lambda = -1$) are scalar multiples of $\mathbf{v}_1$.
$\lambda = 5$: Substituting $\lambda = 5$ into (8.2.3) yields the system

$$-4v_1 + 2v_2 = 0,$$

$$4v_1 - 2v_2 = 0,$$

which has solution $(s, 2s)$, where s is a free variable. It follows that the eigenvectors corresponding to $\lambda = 5$ are those vectors in $\mathbf{R}^2$ of the form

$$\mathbf{v} = s(1, 2),$$

where s is any (nonzero) real number. Once more we note that there is only one *linearly independent* eigenvector corresponding to the eigenvalue $\lambda = 5$, which we may take as $\mathbf{v}_2 = (1, 2)$. All the other eigenvectors (corresponding to $\lambda = 5$) are scalar multiples of $\mathbf{v}_2$.
Notice that the eigenvectors $\mathbf{v}_1 = (-1, 1)$ and $\mathbf{v}_2 = (1, 2)$ (which correspond to different eigenvalues) are linearly independent in $\mathbf{R}^2$. This is shown in Figure 8.2.2.

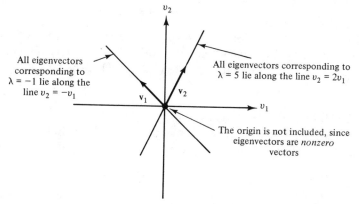

Figure 8.2.2 Eigenvectors corresponding to different eigenvalues are linearly independent.

Example 8.2.4   Find all eigenvalues and eigenvectors of $A = \begin{bmatrix} 1 & -2 & 0 \\ 0 & -1 & 0 \\ 4 & -4 & -1 \end{bmatrix}$.

Solution   In this case the system $(A - \lambda I)\mathbf{v} = \mathbf{0}$ assumes the form

$$\begin{bmatrix} 1-\lambda & -2 & 0 \\ 0 & -1-\lambda & 0 \\ 4 & -4 & -1-\lambda \end{bmatrix} \begin{bmatrix} v_1 \\ v_2 \\ v_3 \end{bmatrix} = \begin{bmatrix} 0 \\ 0 \\ 0 \end{bmatrix}. \tag{8.2.4}$$

This system has nontrivial solutions if and only if $\det(A - \lambda I) = 0$—that is, using the cofactor expansion theorem along column 3, if and only if

$$(\lambda + 1)(\lambda - 1)(\lambda + 1) = 0.$$

This can be written as

$$(\lambda - 1)(\lambda + 1)^2 = 0,$$

so that the eigenvalues are

$$\lambda = 1, -1, -1.$$

There are just two *distinct* eigenvalues ($\lambda = -1$ is a *repeated* eigenvalue with *multiplicity* 2),

$$\lambda_1 = 1, \qquad \lambda_2 = -1.$$

Eigenvectors
$\lambda = 1$:   Substitution into (8.2.4) leads to the system

$$- 2v_2 \qquad\quad = 0,$$
$$- 2v_2 \qquad\quad = 0,$$
$$4v_1 - 4v_2 - 2v_3 = 0.$$

If we set $v_3 = 2r$, then the solution of this system is $(r, 0, 2r)$, so that the eigenvectors corresponding to $\lambda = 1$ are those vectors in $\mathbf{R}^3$ of the form

$$\mathbf{v} = r(1, 0, 2),$$

where r is an arbitrary (nonzero) real number.   Again we notice that there is only one *linearly independent* eigenvector, which we may take as $\mathbf{v}_1 = (1, 0, 2)$.
$\lambda = -1$:   Substitution into (8.2.4) yields the single equation

$$v_1 - v_2 = 0,$$

for the *three* unknowns v_1, v_2, v_3.   Thus there are two free variables in this case. Setting $v_2 = s$, $v_3 = t$, it follows that $v_1 = s$, so that the eigenvectors of A corresponding to the eigenvalue $\lambda = -1$ are those nonzero vectors in $\mathbf{R}^3$ of the form

$$\mathbf{v} = (s, s, t),$$

that is,

$$\mathbf{v} = s(1,\ 1,\ 0) + t(0,\ 0,\ 1).$$

We see that there are two linearly independent eigenvectors corresponding to $\lambda = -1$, which we may take as $\mathbf{v}_2 = (1,\ 1,\ 0)$ (obtained by setting $s = 1$ and $t = 0$) and $\mathbf{v}_3 = (0,\ 0,\ 1)$ (obtained by setting $s = 0$, and $t = 1$). All other eigenvectors corresponding to $\lambda = -1$ are obtained by taking nontrivial linear combinations of $\mathbf{v}_2$ and $\mathbf{v}_3$. This is illustrated in Figure 8.2.3.

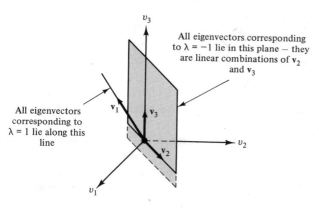

Figure 8.2.3 *Geometrical description of the eigenvectors in Example 8.2.4.*

Example 8.2.5 Find all eigenvalues and eigenvectors of $A = \begin{bmatrix} 2 & -1 & 0 \\ 1 & 0 & 0 \\ 0 & 0 & 3 \end{bmatrix}$.

Solution In this case the system $(A - \lambda I)\mathbf{v} = \mathbf{0}$ is

$$\begin{bmatrix} 2 - \lambda & -1 & 0 \\ 1 & -\lambda & 0 \\ 0 & 0 & 3 - \lambda \end{bmatrix} \begin{bmatrix} v_1 \\ v_2 \\ v_3 \end{bmatrix} = \begin{bmatrix} 0 \\ 0 \\ 0 \end{bmatrix}, \tag{8.2.5}$$

which has nontrivial solutions if and only if $\det(A - \lambda I) = 0$—that is, if and only if

$$(\lambda - 3)(\lambda - 1)^2 = 0.$$

Thus the eigenvalues are $\lambda = 3,\ 1,\ 1$, and again we have a repeated eigenvalue. The distinct eigenvalues are

$$\lambda_1 = 3, \qquad \lambda_2 = 1.$$

Eigenvectors
$\lambda = 3$: Substitution into (8.2.5) yields the system

$$-v_1 - v_2 = 0,$$
$$v_1 - 3v_2 = 0,$$

which has solution $(0,\ 0,\ r)$, where r is a free variable. Thus the eigenvectors corresponding to $\lambda = 3$ are those vectors in $\mathbf{R}^3$ of the form

$$\mathbf{v} = r(0,\ 0,\ 1), \qquad r \neq 0.$$

There is only one *linearly independent* eigenvector, which we can take as $\mathbf{v}_1 = (0, 0, 1)$.

$\lambda = 1$: Substitution into (8.2.5) yields

$$v_1 - v_2 \quad\quad = 0,$$
$$2v_3 = 0,$$

which has solution $(s, s, 0)$, where s is a free variable. Thus the eigenvectors corresponding to $\lambda = 1$ are of the form $\mathbf{v} = s(1, 1, 0)$, and once more there is only one linearly independent eigenvector, for example, $\mathbf{v}_2 = (1, 1, 0)$. In this case we say that A is *defective*, since it has an eigenvalue of multiplicity 2 but only *one* corresponding *linearly independent* eigenvector (compare with Example 8.2.4).

COMPLEX EIGENVALUES AND EIGENVECTORS   The matrices and eigenvalues in the previous examples have all been *real*, and this enabled us to regard the eigenvectors as being vectors in $\mathbf{R}^n$. We now consider the case when some or all the eigenvalues and eigenvectors are complex. The steps in determining the eigenvalues and eigenvectors do not change, although they are a little more complicated algebraically, since the equations determining the eigenvectors will have complex coefficients. If the matrix A has only *real* elements, then the following theorem can save some work in determining any complex eigenvectors.

Theorem 8.2.1   Let A be an $n \times n$ matrix with *real* elements. If λ is a complex eigenvalue of A with corresponding eigenvector $\mathbf{v}$, then $\bar{\lambda}$ is an eigenvalue of A with corresponding eigenvector $\bar{\mathbf{v}}$.

PROOF   If $A\mathbf{v} = \lambda\mathbf{v}$ then $\overline{A\mathbf{v}} = \overline{\lambda\mathbf{v}}$, which implies that[1] $A\bar{\mathbf{v}} = \bar{\lambda}\bar{\mathbf{v}}$ if A has real entries. ∎

REMARK   According to the previous theorem, if we find the eigenvectors of a real matrix A corresponding to a complex eigenvalue λ, then we can obtain the eigenvectors corresponding to the eigenvalue $\bar{\lambda}$ *without having to solve a linear system.*

Example 8.2.6   Find all eigenvalues and eigenvectors of $A = \begin{bmatrix} -2 & -6 \\ 3 & 4 \end{bmatrix}$.

Solution   In this case the system $(A - \lambda I)\mathbf{v} = \mathbf{0}$ is

$$\begin{bmatrix} -2 - \lambda & -6 \\ 3 & 4 - \lambda \end{bmatrix}\begin{bmatrix} v_1 \\ v_2 \end{bmatrix} = \begin{bmatrix} 0 \\ 0 \end{bmatrix}, \tag{8.2.6}$$

which has nontrivial solutions if and only if $\det(A - \lambda I) = 0$—that is, if and only if

$$\lambda^2 - 2\lambda + 10 = 0.$$

Thus the eigenvalues of A are

$$\lambda = 1 + 3i, \quad\quad 1 - 3i.$$

---

[1] If $A = [a_{ij}]$, then, by definition, $\bar{A} = [\bar{a}_{ij}]$.

Since these are complex eigenvalues, we take the underlying vector space as being $\mathbf{C}^2$, and hence any scalars that arise in the solution of the problem will be complex.

Eigenvectors

$\lambda = 1 + 3i$: Substitution into (8.2.6) yields the system

$$(-3 - 3i)v_1 - \qquad 6v_2 = 0, \qquad\qquad (8.2.7)$$

$$3v_1 + (3 - 3i)v_2 = 0. \qquad\qquad (8.2.8)$$

We know that this system has an infinite number of solutions, and hence there is only one independent equation, which we may take as (8.2.8).[1] If we set $v_2 = r$, then $v_1 = (-(1 - i)r, r)$, so that the eigenvectors of A corresponding to $\lambda = 1 + 3i$ are those vectors in $\mathbf{C}^2$ of the form

$$\mathbf{v} = r(-(1 - i), 1),$$

where r is an arbitrary (nonzero) *complex* number.

$\lambda = 1 - 3i$: From Theorem 8.2.1, the eigenvectors in this case are those vectors in $\mathbf{C}^2$ of the form $\mathbf{v} = s(-(1 + i), 1)$, where s is an arbitrary (nonzero) complex number.

Notice that the eigenvectors corresponding to different eigenvalues are linearly independent vectors in $\mathbf{C}^2$. For example, a set of linearly independent eigenvectors is $\{(-(1 - i), 1), (-(1 + i), 1)\}$.

Finally, we remark that a common mistake in solving for the eigenvalues of a matrix A is first to reduce A to row echelon form and then find the eigenvalues of the resulting matrix. It is important to realize that this is not permissible, since the eigenvalues of the resulting matrix will *not* coincide with the eigenvalues of A. (The eigenvalues of a row echelon matrix can only be either 0 or 1.)

EXERCISES 8.2

In problems 1–3 use (8.2.1) to show that λ and $\mathbf{v}$ are an eigenvalue-eigenvector pair for the given matrix A.

1. $\lambda = 4$, $\mathbf{v} = (1, 1)$, $A = \begin{bmatrix} 1 & 3 \\ 2 & 2 \end{bmatrix}$.

2. $\lambda = 3$, $\mathbf{v} = (2, 1, -1)$, $A = \begin{bmatrix} 1 & -2 & -6 \\ -2 & 2 & -5 \\ 2 & 1 & 8 \end{bmatrix}$.

3. $\lambda = -2$, $\mathbf{v} = \alpha(1, 0, -3) + \beta(4, -3, 0)$,

$A = \begin{bmatrix} 1 & 4 & 1 \\ 3 & 2 & 1 \\ 3 & 4 & -1 \end{bmatrix}$, where α and β are constants.

4. Given that $\mathbf{v}_1 = (1, -2)$ and $\mathbf{v}_2 = (1, 1)$ are eigenvectors of $A = \begin{bmatrix} 4 & 1 \\ 2 & 3 \end{bmatrix}$, determine the eigenvalues of A.

In problems 5–23 determine all eigenvalues and corresponding eigenvectors of the given matrix.

---

[1] To see that (8.2.7) and (8.2.8) are equivalent, multiply (8.2.8) by $-(1 + i)$.

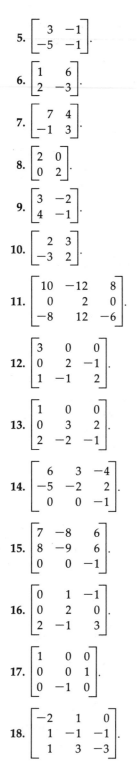

5. $\begin{bmatrix} 3 & -1 \\ -5 & -1 \end{bmatrix}$.

6. $\begin{bmatrix} 1 & 6 \\ 2 & -3 \end{bmatrix}$.

7. $\begin{bmatrix} 7 & 4 \\ -1 & 3 \end{bmatrix}$.

8. $\begin{bmatrix} 2 & 0 \\ 0 & 2 \end{bmatrix}$.

9. $\begin{bmatrix} 3 & -2 \\ 4 & -1 \end{bmatrix}$.

10. $\begin{bmatrix} 2 & 3 \\ -3 & 2 \end{bmatrix}$.

11. $\begin{bmatrix} 10 & -12 & 8 \\ 0 & 2 & 0 \\ -8 & 12 & -6 \end{bmatrix}$.

12. $\begin{bmatrix} 3 & 0 & 0 \\ 0 & 2 & -1 \\ 1 & -1 & 2 \end{bmatrix}$.

13. $\begin{bmatrix} 1 & 0 & 0 \\ 0 & 3 & 2 \\ 2 & -2 & -1 \end{bmatrix}$.

14. $\begin{bmatrix} 6 & 3 & -4 \\ -5 & -2 & 2 \\ 0 & 0 & -1 \end{bmatrix}$.

15. $\begin{bmatrix} 7 & -8 & 6 \\ 8 & -9 & 6 \\ 0 & 0 & -1 \end{bmatrix}$.

16. $\begin{bmatrix} 0 & 1 & -1 \\ 0 & 2 & 0 \\ 2 & -1 & 3 \end{bmatrix}$.

17. $\begin{bmatrix} 1 & 0 & 0 \\ 0 & 0 & 1 \\ 0 & -1 & 0 \end{bmatrix}$.

18. $\begin{bmatrix} -2 & 1 & 0 \\ 1 & -1 & -1 \\ 1 & 3 & -3 \end{bmatrix}$.

19. $\begin{bmatrix} 2 & -1 & 3 \\ 3 & 1 & 0 \\ 2 & -1 & 3 \end{bmatrix}$.

20. $\begin{bmatrix} 5 & 0 & 0 \\ 0 & 5 & 0 \\ 0 & 0 & 5 \end{bmatrix}$.

21. $\begin{bmatrix} 0 & 2 & 2 \\ 2 & 0 & 2 \\ 2 & 2 & 0 \end{bmatrix}$.

22. $\begin{bmatrix} 1 & 2 & 3 & 4 \\ 4 & 3 & 2 & 1 \\ 4 & 5 & 6 & 7 \\ 7 & 6 & 5 & 4 \end{bmatrix}$.

23. $\begin{bmatrix} 0 & 1 & 0 & 0 \\ -1 & 0 & 0 & 0 \\ 0 & 0 & 0 & -1 \\ 0 & 0 & 1 & 0 \end{bmatrix}$.

24. Find all eigenvalues and corresponding eigenvectors of

$$A = \begin{bmatrix} 1+i & 0 & 0 \\ 2-2i & 1-i & 0 \\ 2i & 0 & 1 \end{bmatrix}.$$

Note that the eigenvectors do not occur in complex conjugate pairs. Does this contradict Theorem 8.2.1? Explain.

25. Consider the matrix $A = \begin{bmatrix} 1 & -1 \\ 2 & 4 \end{bmatrix}$.

(a) Show that the characteristic polynomial of A is $p(\lambda) = \lambda^2 - 5\lambda + 6$.

(b) Show that A satisfies its characteristic equation. That is, $A^2 - 5A + 6I_2 = 0_2$. (This result is known as the Cayley–Hamilton theorem and is true for a general $n \times n$ matrix.)

(c) Use the result from (b) to find A^{-1}. [Hint: Multiply the equation in (b) by A^{-1}.]

26. Let $A = \begin{bmatrix} 1 & 2 \\ 2 & -2 \end{bmatrix}$.

(a) Determine all eigenvalues of A.

(b) Reduce A to row echelon form and determine the eigenvalues of the resulting matrix. Are these the same as the eigenvalues of A?

27. If $v_1 = (1, -1)$ and $v_2 = (2, 1)$ are eigenvectors of the matrix A corresponding to the eigenvalues $\lambda_1 = 2$, $\lambda_2 = -3$, respectively, find $A(3v_1 - v_2)$.

28. Let $v_1 = (1, -1, 1)$, $v_2 = (2, 1, 3)$, $v_3 = (-1, -1, 2)$ be eigenvectors of the matrix A that correspond to the eigenvalues $\lambda_1 = 2$, $\lambda_2 = -2$, $\lambda_3 = 3$, respectively, and let $v = (5, 0, 3)$.
(a) Express v as a linear combination of v_1, v_2, v_3.
(b) Find Av.

29. If v_1, v_2, v_3 are eigenvectors of A corresponding to the eigenvalue λ, and c_1, c_2, c_3 are scalars (not all zero), show that $c_1v_1 + c_2v_2 + c_3v_3$ is also an eigenvector of A corresponding to the eigenvalue λ.

30. Prove that the eigenvalues of an upper (or lower) triangular matrix are just the diagonal elements of the matrix.

31. Let A be an $n \times n$ nonsingular matrix. Show that if λ is an eigenvalue of A, then $1/\lambda$ is an eigenvalue of A^{-1}.

32. Let A be an $n \times n$ matrix. Prove that $\lambda = 0$ is an eigenvalue of A if and only if $\det(A) = 0$.

33. Let A be an $n \times n$ matrix. Prove that A and A^T have the same eigenvalues.
[*Hint:* Prove that $\det(A^T - \lambda I) = \det(A - \lambda I)$.]

34. Let A be an $n \times n$ real matrix with complex eigenvalue $\lambda = a + ib$, $b \neq 0$ and let $v = r + is$ be a corresponding eigenvector of A.
(a) Prove that r and s are *nonzero* vectors in $\mathbf{R}^n$.
(b) Prove that r and s are *linearly independent* vectors in $\mathbf{R}^n$.

8.3 GENERAL RESULTS FOR EIGENVALUES AND EIGENVECTORS

In this section we look more closely at the relationship between the eigenvalues and eigenvectors of an $n \times n$ matrix. Our aim is to formalize many of the ideas introduced via the examples of the previous section.

For a given $n \times n$ matrix $A = [a_{ij}]$, the characteristic polynomial $p(\lambda)$ assumes the form

$$p(\lambda) = \det(A - \lambda I) = \begin{vmatrix} a_{11} - \lambda & a_{12} & a_{13} & \cdots & a_{1n} \\ a_{21} & a_{22} - \lambda & a_{23} & \cdots & a_{2n} \\ \vdots & \vdots & \vdots & & \vdots \\ a_{n1} & a_{n2} & a_{n3} & \cdots & a_{nn} - \lambda \end{vmatrix}.$$

The expansion of this determinant would yield a polynomial of degree n in λ with leading coefficient $(-1)^n$. It follows that $p(\lambda)$ can be written in the form

$$p(\lambda) = (-1)^n\lambda^n + b_1\lambda^{n-1} + b_2\lambda^{n-2} + \cdots + b_n,$$

where $b_1, b_2, \ldots, b_n$ are scalars. Since we consider the underlying vector space to be $\mathbf{C}^n$, the fundamental theorem of algebra guarantees that $p(\lambda)$ will have precisely n zeros (not necessarily distinct), and hence A will have n eigenvalues. If we let λ_1, $\lambda_2, \ldots, \lambda_k$ denote the *distinct* eigenvalues of A, then $p(\lambda)$ can be factored as

$$p(\lambda) = (-1)^n(\lambda - \lambda_1)^{m_1}(\lambda - \lambda_2)^{m_2}(\lambda - \lambda_3)^{m_3} \cdots (\lambda - \lambda_k)^{m_k},$$

where, since $p(\lambda)$ has degree n, $m_1 + m_2 + m_3 + \cdots + m_k = n$. Thus, associated with each eigenvalue λ_i is a number m_i, called the **multiplicity** of λ_i.

We now focus our attention on the eigenvectors of A.

> **Definition 8.3.1:**   Let A be an $n \times n$ matrix.   For a given eigenvalue λ_i let E_i denote the set of all eigenvectors corresponding to λ_i together with the zero vector.   Then E_i is called the **eigenspace** of A corresponding to the eigenvalue λ_i.

REMARKS

1. Two alternative, but equivalent, ways to define the eigenspace E_i are as follows:
   (a) E_i is the solution set of the homogeneous linear system $(A - \lambda_i I)\mathbf{v} = \mathbf{0}$.
   (b) E_i is the kernel of the linear transformation $T_i : \mathbf{C}^n \to \mathbf{C}^n$ defined by $T_i(\mathbf{v}) = (A - \lambda_i I)\mathbf{v}$.
2. It is important to notice that there is one eigenspace associated with each eigenvalue of A.

Example 8.3.1   Determine all eigenspaces for the matrix $A = \begin{bmatrix} 1 & 2 \\ 4 & 3 \end{bmatrix}$.

Solution   We have already computed the eigenvalues and eigenvectors of A in Example 8.2.3.   The eigenvalues of A are $\lambda_1 = -1$ and $\lambda_2 = 5$.   The eigenvectors corresponding to $\lambda_1 = -1$ are of the form $\mathbf{v} = r(-1, 1)$.   Thus the eigenspace corresponding to $\lambda_1 = -1$ is

$$E_1 = \{\mathbf{v} \in \mathbf{R}^2 : \mathbf{v} = r(-1, 1), \, r \in \mathbf{R}\}.$$

The eigenvectors corresponding to the eigenvalue $\lambda_2 = 5$ are of the form $\mathbf{v} = s(1, 2)$, so that the eigenspace corresponding to $\lambda_2 = 5$ is

$$E_2 = \{\mathbf{v} \in \mathbf{R}^2 : \mathbf{v} = s(1, 2), \, s \in \mathbf{R}\}.$$

We have one main result for eigenspaces.

Theorem 8.3.1:   Let λ_i be an eigenvalue of A of multiplicity m_i and let E_i denote the corresponding eigenspace.   Then

1. For each i, E_i is a vector space.
2. If n_i denotes the dimension of E_i then $1 \leq n_i \leq m_i$ for each i.   In words, the dimension of the eigenspace corresponding to λ_i is less than or equal to the multiplicity of λ_i.

PROOF   As remarked before, E_i is the kernel of the linear transformation $T_i : \mathbf{C}^n \to \mathbf{C}^n$ defined by $T(\mathbf{v}) = (A - \lambda_i I)\mathbf{v}$ and hence is a vector space.   The proof of (2) requires more advanced ideas about linear transformations and is therefore omitted (see, for example, G. E. Shilov, *Linear Algebra*, Dover Publications, 1977).   ∎

REMARK   For a given eigenvalue λ_i, the terminology that is often used is to refer to m_i as the **algebraic** multiplicity of λ_i, and n_i as the **geometric** multiplicity of λ_i.

Example 8.3.2 Determine all eigenspaces and their dimensions for the matrix

$$A = \begin{bmatrix} 3 & -1 & 0 \\ 0 & 2 & 0 \\ -1 & 1 & 2 \end{bmatrix}.$$

Solution A straightforward calculation yields the characteristic polynomial $p(\lambda) = -(\lambda - 2)^2(\lambda - 3)$, so that the eigenvalues of A are $\lambda_1 = 2$ (with multiplicity 2) and $\lambda_2 = 3$. The eigenvectors corresponding to $\lambda_1 = 2$ are determined by solving

$$v_1 - v_2 = 0$$

and hence are of the form

$$\mathbf{v} = (r, r, s) = r(1, 1, 0) + s(0, 0, 1),$$

where r and s cannot be zero simultaneously (since eigenvectors are nonzero vectors). Thus the eigenspace corresponding to $\lambda_1 = 2$ is

$$E_1 = \{\mathbf{v} \in \mathbf{R}^3 : \mathbf{v} = r(1, 1, 0) + s(0, 0, 1), r, s \in \mathbf{R}\}.$$

We see that the linearly independent vectors $\mathbf{v}_1 = (1, 1, 0)$, $\mathbf{v}_2 = (0, 0, 1)$ form a basis for the eigenspace E_1 and hence the dimension of this eigenspace is 2 (that is, $n_1 = 2$).

It is easily shown that the eigenvectors corresponding to the eigenvalue $\lambda_2 = 3$ are of the form

$$\mathbf{v} = t(1, 0, -1),$$

where t is a nonzero real number. Thus the eigenspace corresponding to $\lambda_2 = 3$ is

$$E_2 = \{\mathbf{v} \in \mathbf{R}^3 : \mathbf{v} = t(1, 0, -1), t \in \mathbf{R}\}.$$

It follows that $\mathbf{v}_3 = (1, 0, -1)$ forms a basis for this eigenspace, and hence the dimension of the eigenspace is 1 (that is, $n_2 = 1$).

The eigenspaces E_1 and E_2 are sketched in Figure 8.3.1.

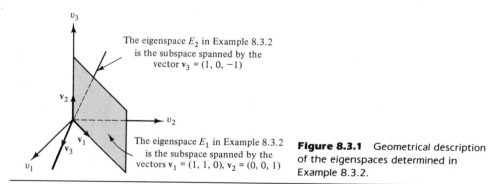

The eigenspace E_2 in Example 8.3.2 is the subspace spanned by the vector $v_3 = (1, 0, -1)$

The eigenspace E_1 in Example 8.3.2 is the subspace spanned by the vectors $v_1 = (1, 1, 0)$, $v_2 = (0, 0, 1)$

Figure 8.3.1 Geometrical description of the eigenspaces determined in Example 8.3.2.

We now consider the relationship between eigenvectors corresponding to *distinct* eigenvalues. There is one key theorem which has already been illustrated in the previous section.

Theorem 8.3.2:   Eigenvectors corresponding to *distinct* eigenvalues are linearly independent.

PROOF   We use induction to prove the result.   Let $\lambda_1, \lambda_2, \ldots, \lambda_m$ be distinct eigenvalues of A with corresponding eigenvectors $\mathbf{v}_1, \mathbf{v}_2, \ldots, \mathbf{v}_m$.   It is certainly true that $\{\mathbf{v}_1\}$ is linearly independent.   Now suppose that $\{\mathbf{v}_1, \mathbf{v}_2 \ldots, \mathbf{v}_k\}$ is linearly independent for some $k < m$, and consider the set $\{\mathbf{v}_1, \mathbf{v}_2 \ldots, \mathbf{v}_k, \mathbf{v}_{k+1}\}$.   We wish to show that this set of vectors is linearly independent.   Consider

$$c_1 \mathbf{v}_1 + c_2 \mathbf{v}_2 + \cdots + c_k \mathbf{v}_k + c_{k+1} \mathbf{v}_{k+1} = \mathbf{0}. \qquad (8.3.1)$$

Multiplying both sides on the left by A yields, using $A\mathbf{v}_i = \lambda_i \mathbf{v}_i$,

$$c_1 \lambda_1 \mathbf{v}_1 + c_2 \lambda_2 \mathbf{v}_2 + \cdots + c_k \lambda_k \mathbf{v}_k + c_{k+1} \lambda_{k+1} \mathbf{v}_{k+1} = \mathbf{0}. \qquad (8.3.2)$$

But, from (8.3.1),

$$c_{k+1} \mathbf{v}_{k+1} = -(c_1 \mathbf{v}_1 + c_2 \mathbf{v}_2 + \cdots + c_k \mathbf{v}_k),$$

so that (8.3.2) can be written as

$$c_1 \lambda_1 \mathbf{v}_1 + c_2 \lambda_2 \mathbf{v}_2 + \cdots + c_k \lambda_k \mathbf{v}_k - \lambda_{k+1}(c_1 \mathbf{v}_1 + c_2 \mathbf{v}_2 + \cdots + c_k \mathbf{v}_k) = \mathbf{0}.$$

that is,

$$c_1(\lambda_1 - \lambda_{k+1})\mathbf{v}_1 + c_2(\lambda_2 - \lambda_{k+1})\mathbf{v}_2 + \cdots + c_k(\lambda_k - \lambda_{k+1})\mathbf{v}_k = \mathbf{0}.$$

Since $\mathbf{v}_1, \mathbf{v}_2, \ldots, \mathbf{v}_k$ are linearly independent, this implies that

$$c_1(\lambda_1 - \lambda_{k+1}) = 0, \qquad c_2(\lambda_2 - \lambda_{k+1}) = 0, \ldots, c_k(\lambda_k - \lambda_{k+1}) = 0;$$

since the λ_i are distinct,

$$c_1 = c_2 = \cdots = c_k = 0.$$

But now, since $\mathbf{v}_{k+1} \neq \mathbf{0}$, it follows from (8.3.1) that $c_{k+1} = 0$ also, and so $\mathbf{v}_1, \mathbf{v}_2, \ldots, \mathbf{v}_k, \mathbf{v}_{k+1}$ are linearly independent.

We have shown that the desired result is true for $\{\mathbf{v}_1, \mathbf{v}_2, \ldots, \mathbf{v}_k, \mathbf{v}_{k+1}\}$ whenever it is true for $\{\mathbf{v}_1, \mathbf{v}_2, \ldots, \mathbf{v}_k\}$, and, since the result is true for a single eigenvector, it is true for $\{\mathbf{v}_1, \mathbf{v}_2, \ldots, \mathbf{v}_k\}$, $1 \leq k \leq m$.   ∎

Since the dimension of $\mathbf{R}^n$ (or $\mathbf{C}^n$) is n, the maximum number of linearly independent eigenvectors that A can have is n.   In such a case we say that A possesses a *complete* set of eigenvectors.   The following definition introduces the appropriate terminology.

Definition 8.3.2:   An $n \times n$ matrix that has n linearly independent eigenvectors is said to have a **complete set of eigenvectors** and we say that A is *non-defective*.   If A does *not* have a complete set of eigenvectors, then we say that it is *defective*.

If A is nondefective, then any set of n linearly independent eigenvectors of A is a basis for $\mathbf{R}^n$ (or $\mathbf{C}^n$).   Such a basis is sometimes referred to as an *eigenbasis* of A.

Example 8.3.3   For the matrix in the previous example a complete set of eigenvectors is $\{(1, 0, 0), (0, 0, 1), (1, 0, -1)\}$. Thus the matrix is nondefective.

Example 8.3.4   Determine whether $A = \begin{bmatrix} 4 & -1 \\ 1 & 2 \end{bmatrix}$ is defective or nondefective.

Solution   The characteristic polynomial of A is

$$p(\lambda) = (4 - \lambda)(2 - \lambda) + 1 = \lambda^2 - 6\lambda + 9 = (\lambda - 3)^2.$$

Thus $\lambda_1 = 3$ is an eigenvalue of multiplicity 2. The eigenvectors of A are determined by solving

$$v_1 - v_2 = 0,$$

so that the eigenspace corresponding to $\lambda_1 = 3$ is

$$E_1 = \{\mathbf{v} \in \mathbf{R}^2 : \mathbf{v} = r(1, 1),\ r \in \mathbf{R}\}.$$

Thus, $\dim[E_1] = 1$, which implies that A is defective.

The next result is a direct consequence of the previous theorem.

Theorem 8.3.3:   If an $n \times n$ matrix A has n *distinct* eigenvalues, then it possesses a complete set of eigenvectors.

PROOF   If A has n distinct eigenvalues, then from our previous theorem it has n linearly independent eigenvectors.   ∎

Note that if A does *not* have n distinct eigenvalues, it *may* still have a complete set of eigenvectors (in Example 8.3.2 there are only two distinct eigenvalues, but there are three linearly independent eigenvectors and hence a complete set). The general result is as follows.

Theorem 8.3.4:   An $n \times n$ matrix A possesses a complete set of eigenvectors if and only if the dimension of each eigenspace is the same as the multiplicity of the corresponding eigenvalue—that is, if and only if $n_i = m_i$ for each i.

PROOF   We leave the proof as an exercise. (*Hint:* Remember that the m_i must satisfy $m_1 + m_2 + \cdots + m_k = n$.)   ∎

EXERCISES 8.3

In problems 1–15 determine the multiplicity of each eigenvalue and a basis for each eigenspace of the given matrix. Hence determine the dimension of each eigenspace and state whether the matrix is defective or nondefective.

1. $\begin{bmatrix} 1 & 4 \\ 2 & 3 \end{bmatrix}$.

2. $\begin{bmatrix} 3 & 0 \\ 0 & 3 \end{bmatrix}$.

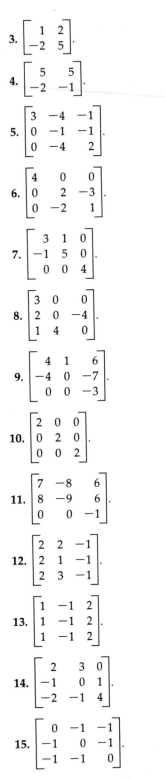

3. $\begin{bmatrix} 1 & 2 \\ -2 & 5 \end{bmatrix}$.

4. $\begin{bmatrix} 5 & 5 \\ -2 & -1 \end{bmatrix}$.

5. $\begin{bmatrix} 3 & -4 & -1 \\ 0 & -1 & -1 \\ 0 & -4 & 2 \end{bmatrix}$.

6. $\begin{bmatrix} 4 & 0 & 0 \\ 0 & 2 & -3 \\ 0 & -2 & 1 \end{bmatrix}$.

7. $\begin{bmatrix} 3 & 1 & 0 \\ -1 & 5 & 0 \\ 0 & 0 & 4 \end{bmatrix}$.

8. $\begin{bmatrix} 3 & 0 & 0 \\ 2 & 0 & -4 \\ 1 & 4 & 0 \end{bmatrix}$.

9. $\begin{bmatrix} 4 & 1 & 6 \\ -4 & 0 & -7 \\ 0 & 0 & -3 \end{bmatrix}$.

10. $\begin{bmatrix} 2 & 0 & 0 \\ 0 & 2 & 0 \\ 0 & 0 & 2 \end{bmatrix}$.

11. $\begin{bmatrix} 7 & -8 & 6 \\ 8 & -9 & 6 \\ 0 & 0 & -1 \end{bmatrix}$.

12. $\begin{bmatrix} 2 & 2 & -1 \\ 2 & 1 & -1 \\ 2 & 3 & -1 \end{bmatrix}$.

13. $\begin{bmatrix} 1 & -1 & 2 \\ 1 & -1 & 2 \\ 1 & -1 & 2 \end{bmatrix}$.

14. $\begin{bmatrix} 2 & 3 & 0 \\ -1 & 0 & 1 \\ -2 & -1 & 4 \end{bmatrix}$.

15. $\begin{bmatrix} 0 & -1 & -1 \\ -1 & 0 & -1 \\ -1 & -1 & 0 \end{bmatrix}$.

In problems 16–20 determine whether the given matrix is defective or nondefective.

16. $A = \begin{bmatrix} 2 & 3 \\ 2 & 1 \end{bmatrix}$, characteristic polynomial $p(\lambda) = (\lambda + 1)(\lambda - 4)$.

17. $A = \begin{bmatrix} 6 & 5 \\ -5 & -4 \end{bmatrix}$, characteristic polynomial $p(\lambda) = (\lambda - 1)^2$.

18. $A = \begin{bmatrix} 1 & -2 \\ 5 & 3 \end{bmatrix}$, characteristic polynomial $p(\lambda) = \lambda^2 - 4\lambda + 13$.

19. $A = \begin{bmatrix} 1 & -3 & 1 \\ -1 & -1 & 1 \\ -1 & -3 & 3 \end{bmatrix}$, characteristic polynomial $p(\lambda) = -(\lambda - 2)^2(\lambda + 1)$.

20. $A = \begin{bmatrix} -1 & 2 & 2 \\ -4 & 5 & 2 \\ -4 & 2 & 5 \end{bmatrix}$, characteristic polynomial $p(\lambda) = (3 - \lambda)^3$.

In problems 21–25 determine a basis for each eigenspace of A and sketch the eigenspaces.

21. $A = \begin{bmatrix} 2 & 1 \\ 3 & 4 \end{bmatrix}$.

22. $A = \begin{bmatrix} 2 & 3 \\ 0 & 2 \end{bmatrix}$.

23. $A = \begin{bmatrix} 5 & 0 \\ 0 & 5 \end{bmatrix}$.

24. $A = \begin{bmatrix} 3 & 1 & -1 \\ 1 & 3 & -1 \\ -1 & -1 & 3 \end{bmatrix}$. [You may assume that A has characteristic polynomial $p(\lambda) = (5 - \lambda)(\lambda - 2)^2$.]

25. $\begin{bmatrix} -3 & 1 & 0 \\ -1 & -1 & 2 \\ 0 & 0 & -2 \end{bmatrix}$.

26. The matrix $A = \begin{bmatrix} 2 & -2 & 3 \\ 1 & -1 & 3 \\ 1 & -2 & 4 \end{bmatrix}$ has eigenvalues $\lambda_1 = 1$, $\lambda_2 = 3$.

(a) Determine a basis for the eigenspace E_1 and then use the Gram–Schmidt process to obtain an orthogonal basis for E_1.

(b) Are the vectors in E_1 orthogonal to those in E_2?

27. Repeat the previous question for

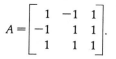

$$A = \begin{bmatrix} 1 & -1 & 1 \\ -1 & 1 & 1 \\ 1 & 1 & 1 \end{bmatrix}.$$

(A has eigenvalues $\lambda_1 = 2$, $\lambda_2 = -1$.) What is special about the matrix in this problem? (See Section 8.5 for a generalization.)

28. The matrix $A = \begin{bmatrix} a & b & c \\ a & b & c \\ a & b & c \end{bmatrix}$ has eigenvalues 0,

0, $(a + b + c)$. Determine all values of the constants a, b, c for which A is nondefective.

29. Consider the characteristic polynomial of an $n \times n$ matrix A, namely,

$$p(\lambda) = \det(A - \lambda I) =$$

$$\begin{vmatrix} a_{11} - \lambda & a_{12} & a_{13} & \cdots & a_{1n} \\ a_{21} & a_{22} - \lambda & a_{23} & \cdots & a_{2n} \\ \vdots & \vdots & \vdots & & \vdots \\ a_{n1} & a_{n2} & a_{n3} & \cdots & a_{nn} - \lambda \end{vmatrix}, \quad \text{(i)}$$

which can be written in either of the forms:

$$p(\lambda) = (-1)^n \lambda^n + b_1 \lambda^{n-1} + b_2 \lambda^{n-2} + \cdots + b_n, \quad \text{(ii)}$$

$$p(\lambda) = (\lambda_1 - \lambda)(\lambda_2 - \lambda)(\lambda_3 - \lambda) \cdots (\lambda_n - \lambda), \quad \text{(iii)}$$

where $\lambda_1, \lambda_2, \ldots, \lambda_n$ are the eigenvalues of A.

(a) Use (i) and (ii) to show that

$$b_1 = (-1)^{n-1}(a_{11} + a_{22} + \cdots + a_{nn}), \quad b_n = \det(A).$$

Recall that the quantity $a_{11} + a_{22} + \cdots + a_{nn}$ is called the *trace* of the matrix A, denoted by $\text{tr}(A)$.

(b) Use (ii) and (iii) to show that

$$b_1 = (-1)^{n-1}(\lambda_1 + \lambda_2 + \cdots + \lambda_n),$$

$$b_n = \lambda_1 \lambda_2 \cdots \lambda_n.$$

(c) Use your results from (a) and (b) to show that:

$$\det(A) = \text{product of the eigenvalues of } A.$$

$$\text{tr}(A) = \text{sum of the eigenvalues of } A.$$

(d) Find the sum and the product of the eigenvalues of

$$A = \begin{bmatrix} 12 & 11 & 9 & -7 \\ 2 & 3 & -5 & 6 \\ 10 & 8 & 5 & 4 \\ 1 & 0 & 3 & 4 \end{bmatrix}.$$

30. Let E_i denote the eigenspace of A corresponding to the eigenvalue λ_i. Use Theorem 6.3.1 to prove that E_i is a subspace of $\mathbf{C}^n$.

31. Let $\mathbf{v}_1$ and $\mathbf{v}_2$ be eigenvectors of A corresponding to the distinct eigenvalues λ_1, λ_2, respectively. Prove that $\mathbf{v}_1$ and $\mathbf{v}_2$ are linearly independent. (Model your proof on the general case considered in Theorem 8.3.3.)

32. (a) Let E_i denote the eigenspace of A corresponding to the eigenvalue λ_i. If $\{\mathbf{v}_1\}$ is a basis for E_1 and $\{\mathbf{v}_2, \mathbf{v}_3\}$ is a basis for E_2, prove that $\{\mathbf{v}_1, \mathbf{v}_2, \mathbf{v}_3\}$ is linearly independent.

(b) Extend the result of (a) to the case when $\{\mathbf{v}_{i1}, \mathbf{v}_{i2}, \ldots, \mathbf{v}_{im_i}\}$ is a basis for E_i, $i = 1, 2, \ldots k$.

33. Prove Theorem 8.3.4.

8.4   DIAGONALIZATION

We begin by presenting some motivation behind the problem that we will be studying in this and the next section.

Let $T: V \rightarrow V$ be a linear transformation and suppose that V has finite dimension n. Let $\{\mathbf{e}_1, \mathbf{e}_2, \ldots, \mathbf{e}_n\}$ be a basis for V. If we apply T to the basis vector $\mathbf{e}_1$, then the resulting vector $T(\mathbf{e}_1)$ will also be in V, and hence there exist scalars a_{11}, $a_{21}, \ldots, a_{n1}$ such that

$$T(\mathbf{e}_1) = a_{11}\mathbf{e}_1 + a_{21}\mathbf{e}_2 + \cdots + a_{n1}\mathbf{e}_n = \sum_{i=1}^{n} a_{i1}\mathbf{e}_i.$$

Similarly, if we successively apply T to the remaining vectors in the basis, the result can be written as

$$T(\mathbf{e}_k) = \sum_{i=1}^{n} a_{ik}\mathbf{e}_i, \qquad k = 1, 2, \ldots, n, \qquad (8.4.1)$$

for appropriate scalars a_{ik}. This defines an $n \times n$ matrix $A = [a_{ik}]$. Further, for any $\mathbf{v}$ in V, we have

$$\mathbf{v} = \sum_{k=1}^{n} v_k\mathbf{e}_k,$$

and so, since T is a linear transformation,

$$T(\mathbf{v}) = T\left(\sum_{k=1}^{n} v_k\mathbf{e}_k\right) = \sum_{k=1}^{n} T(v_k\mathbf{e}_k) = \sum_{k=1}^{n} v_k T(\mathbf{e}_k).$$

Substituting for $T(\mathbf{e}_k)$ from (8.4.1) yields

$$T(\mathbf{v}) = \sum_{k=1}^{n}\left(v_k\sum_{i=1}^{n} a_{ik}\mathbf{e}_i\right) = \sum_{i=1}^{n}\left(\sum_{k=1}^{n} a_{ik}v_k\right)\mathbf{e}_i = \sum_{i=1}^{n}(A\mathbf{v})_i\mathbf{e}_i.$$

This shows that the linear transformation T is completely determined by the $n \times n$ matrix A. This matrix is called the matrix of T *relative to the basis* $\{\mathbf{e}_1, \mathbf{e}_2, \ldots, \mathbf{e}_n\}$. If we use a different basis in V, say $\{\mathbf{f}_1, \mathbf{f}_2, \ldots, \mathbf{f}_n\}$, then

$$\mathbf{v} = \sum_{k=1}^{n} v_k^*\mathbf{f}_k,$$

where $v_1^*, v_2^*, \ldots, v_n^*$ denote the components of $\mathbf{v}$ relative to the basis $\{\mathbf{f}_1, \mathbf{f}_2, \ldots, \mathbf{f}_n\}$. Thus

$$T(\mathbf{v}) = \sum_{k=1}^{n} v_k^* T(\mathbf{f}_k)$$
$$= \sum_{i=1}^{n}(B\mathbf{v}^*)_i\mathbf{f}_i,$$

where the $n \times n$ matrix $B = [b_{ik}]$ is defined by

$$T(\mathbf{f}_k) = \sum_{i=1}^{n} b_{ik}\mathbf{f}_i, \qquad k = 1, 2, \ldots, n.$$

The key point to note is that the matrix representing T is dependent on the basis that we are using in V. It can be shown, however, that matrices representing the same linear transformation relative to different bases are related in the following way:[1]

$$B = S^{-1}AS,$$

where S is a nonsingular matrix that gives the relationship between the two bases. This motivates the following definition.

---

[1] A proof of this result is outlined in problem 22 at the end of the section.

Definition 8.4.1:   Let A and B be $n \times n$ matrices. A is said to be *similar* to B if there exists a nonsingular matrix S such that

$$B = S^{-1}AS.$$

Example 8.4.1   If $A = \begin{bmatrix} -4 & -10 \\ 3 & 7 \end{bmatrix}$ and $B = \begin{bmatrix} 1 & -1 \\ 0 & 2 \end{bmatrix}$, show that $B = S^{-1}AS$, where $S = \begin{bmatrix} 2 & -3 \\ -1 & 2 \end{bmatrix}$.

Solution   It is easily shown that $S^{-1} = \begin{bmatrix} 2 & 3 \\ 1 & 2 \end{bmatrix}$, so that

$$S^{-1}AS = \begin{bmatrix} 2 & 3 \\ 1 & 2 \end{bmatrix} \begin{bmatrix} -4 & -10 \\ 3 & 7 \end{bmatrix} \begin{bmatrix} 2 & -3 \\ -1 & 2 \end{bmatrix}$$

$$= \begin{bmatrix} 2 & 3 \\ 1 & 2 \end{bmatrix} \begin{bmatrix} 2 & -8 \\ -1 & 5 \end{bmatrix} = \begin{bmatrix} 1 & -1 \\ 0 & 2 \end{bmatrix},$$

that is,

$$S^{-1}AS = B.$$

The preceding analysis showed that a linear transformation $T:V \rightarrow V$ can be represented by a matrix and that the matrix is dependent on the basis that is being used in V. The question that naturally arises is the following:

Can we choose a basis so that the matrix representing T has a simple form?

Before answering this question we must first determine how simple a form we should aim for. The following theorem helps us decide.

Theorem 8.4.1:   Similar matrices have the same eigenvalues.

PROOF   If A is similar to B, then $B = S^{-1}AS$ for some nonsingular matrix S. Thus,

$$\det(B - \lambda I) = \det(S^{-1}AS - \lambda I) = \det(S^{-1}AS - \lambda S^{-1}S)$$

$$= \det[S^{-1}(A - \lambda I)S] = \det(S^{-1})\det(S)\det(A - \lambda I) = \det(A - \lambda I).$$

Consequently A and B have the same characteristic polynomial and hence the same eigenvalues.   ∎

Returning now to the question just posed, suppose that A and B represent the same linear transformation, T, relative to different bases. Then A and B are similar and hence have the same eigenvalues, say $\lambda_1, \lambda_2, \ldots, \lambda_n$. The simplest matrix that has these eigenvalues is the diagonal matrix $D = \text{diag}(\lambda_1, \lambda_2, \ldots, \lambda_n)$. Thus the simplest possible representation of a linear transformation is by a diagonal matrix, and hence we now ask the following question:

For a given square matrix A, does there exist a nonsingular matrix S such that

$$S^{-1}AS = \text{diag}(\lambda_1, \lambda_2, \ldots, \lambda_n)?$$

Equivalently,

When is A similar to a diagonal matrix?

The next theorem answers this question and also gives a matrix S that accomplishes the diagonalization.

Theorem 8.4.2:   An $n \times n$ matrix A is similar to a diagonal matrix if and only if A has n linearly independent eigenvectors $\mathbf{v}_1, \mathbf{v}_2, \ldots, \mathbf{v}_n$.   In such a case if we let

$$S = [\mathbf{v}_1, \mathbf{v}_2, \ldots, \mathbf{v}_n]$$

then

$$S^{-1}AS = \text{diag}(\lambda_1, \lambda_2, \ldots, \lambda_n),$$

where $\lambda_1, \lambda_2, \ldots, \lambda_n$ are the eigenvalues of A (not necessarily distinct) corresponding to the eigenvectors $\mathbf{v}_1, \mathbf{v}_2, \ldots, \mathbf{v}_n$.

PROOF   Let $D = \text{diag}(\lambda_1, \lambda_2, \ldots, \lambda_n)$, where the λ_i are scalars.   Then A is similar to D if and only if there exists a *nonsingular* matrix $S = [\mathbf{v}_1, \mathbf{v}_2, \ldots, \mathbf{v}_n]$ such that

$$S^{-1}AS = D,$$

that is, if and only if

$$AS = SD,$$

or, equivalently,

$$[A\mathbf{v}_1, A\mathbf{v}_2, \ldots, A\mathbf{v}_n] = [\lambda_1\mathbf{v}_1, \lambda_2\mathbf{v}_2, \ldots, \lambda_n\mathbf{v}_n].$$

Equating corresponding column vectors, we must have

$$A\mathbf{v}_1 = \lambda_1\mathbf{v}_1, A\mathbf{v}_2 = \lambda_2\mathbf{v}_2, \ldots, A\mathbf{v}_n = \lambda_n\mathbf{v}_n.$$

Thus $\mathbf{v}_1, \mathbf{v}_2, \ldots, \mathbf{v}_n$ must be eigenvectors of A corresponding to the eigenvalues $\lambda_1, \lambda_2, \ldots, \lambda_n$.   Further, since $\det(S) \neq 0$, it follows that the eigenvectors are linearly independent.   ∎

Definition 8.4.2:   An $n \times n$ matrix that is similar to a diagonal matrix is said to be **diagonalizable**.

The preceding theorem thus states that an $n \times n$ matrix is diagonalizable if and only if it is nondefective.

REMARK   It follows that the matrices in Examples 8.2.3, 8.2.4, 8.2.6, and 8.3.2 *are* diagonalizable, whereas the matrices in Example 8.2.5 and 8.3.4 *are not* diagonalizable.   As an exercise you should write an appropriate matrix S in each of Examples 8.2.3, 8.2.4, 8.2.6, 8.3.2, together with diagonal matrices to which the given matrices are similar.

As we have seen, not all matrices are diagonalizable. However, combining the previous theorem with Theorem 8.3.3 yields the following.

Theorem 8.4.3: If an $n \times n$ matrix has n distinct eigenvalues, then it is diagonalizable.

Example 8.4.2 Show that $A = \begin{bmatrix} 1 & -2 & -1 \\ 0 & 2 & -1 \\ 0 & -2 & 1 \end{bmatrix}$ is diagonalizable, and find a matrix S such that $S^{-1}AS = \text{diag}(\lambda_1, \lambda_2, \lambda_3)$.

Solution The characteristic polynomial in this case is $p(\lambda) = -\lambda(\lambda - 1)(\lambda - 3)$. Thus the eigenvalues are $\lambda = 0, 1, 3$. Since these eigenvalues are distinct, the previous theorem guarantees that A is diagonalizable. To find S we must determine the eigenvectors of A. A straightforward calculation yields the following linearly independent eigenvectors:

$$\lambda_1 = 0 \Rightarrow v_1 = (4, 1, 2); \quad \lambda_2 = 1 \Rightarrow v_2 = (1, 0, 0); \quad \lambda_3 = 3 \Rightarrow v_3 = (1, -2, 2),$$

where we have only chosen one eigenvector corresponding to each eigenvalue. It follows from Theorem 8.4.2 that the matrix

$$S = [v_1, v_2, v_3] = \begin{bmatrix} 4 & 1 & 1 \\ 1 & 0 & -2 \\ 2 & 0 & 2 \end{bmatrix}$$

satisfies $S^{-1}AS = \text{diag}(0, 1, 3)$. (Note that the ordering of the eigenvalues along the diagonal corresponds to the ordering of the eigenvectors in S.)

Example 8.4.3 Repeat the previous example for the matrix $A = \begin{bmatrix} 1 & 0 & -1 \\ -1 & 2 & -1 \\ 0 & 0 & 2 \end{bmatrix}$.

Solution In this case the characteristic polynomial is $p(\lambda) = -(\lambda - 2)^2(\lambda - 1)$, so that the eigenvalues of A are $\lambda = 2, 2, 1$. We must compute the eigenvectors in this case before we can determine whether or not A is diagonalizable. It is easy to show that the following are linearly independent eigenvectors:

$$\lambda = 2 \Rightarrow v_1 = (0, 1, 0), \quad v_2 = (-1, 0, 1); \quad \lambda = 1 \Rightarrow v_3 = (1, 1, 0).$$

It follows that A has a complete set of eigenvectors and hence is diagonalizable. If we set

$$S = \begin{bmatrix} 0 & -1 & 1 \\ 1 & 0 & 1 \\ 0 & 1 & 0 \end{bmatrix},$$

then

$$S^{-1}AS = \text{diag}(2, 2, 1).$$

Finally in this section we note that if a matrix is defective then, from Theorem 8.4.2, there does *not* exist a matrix S such that $S^{-1}AS = \text{diag}(\lambda_1, \lambda_2, \ldots, \lambda_n)$. Equivalently, there is no basis relative to which the corresponding linear transformation is represented by a diagonal matrix. In more advanced treatments of the eigenvalue-eigenvector problem, the possible "simple" forms of a defective matrix are discussed. Some special cases are considered in problems 19–21.

EXERCISES 8.4

In problems 1–12 determine whether the given matrix is diagonalizable. Where possible, find a matrix S such that $S^{-1}AS = \text{diag}(\lambda_1, \lambda_2, \ldots, \lambda_n)$.

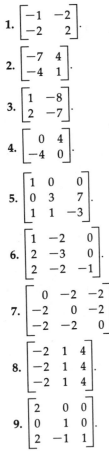

1. $\begin{bmatrix} -1 & -2 \\ -2 & 2 \end{bmatrix}$.

2. $\begin{bmatrix} -7 & 4 \\ -4 & 1 \end{bmatrix}$.

3. $\begin{bmatrix} 1 & -8 \\ 2 & -7 \end{bmatrix}$.

4. $\begin{bmatrix} 0 & 4 \\ -4 & 0 \end{bmatrix}$.

5. $\begin{bmatrix} 1 & 0 & 0 \\ 0 & 3 & 7 \\ 1 & 1 & -3 \end{bmatrix}$.

6. $\begin{bmatrix} 1 & -2 & 0 \\ 2 & -3 & 0 \\ 2 & -2 & -1 \end{bmatrix}$.

7. $\begin{bmatrix} 0 & -2 & -2 \\ -2 & 0 & -2 \\ -2 & -2 & 0 \end{bmatrix}$.

8. $\begin{bmatrix} -2 & 1 & 4 \\ -2 & 1 & 4 \\ -2 & 1 & 4 \end{bmatrix}$.

9. $\begin{bmatrix} 2 & 0 & 0 \\ 0 & 1 & 0 \\ 2 & -1 & 1 \end{bmatrix}$.

10. $\begin{bmatrix} 4 & 0 & 0 \\ 3 & -1 & -1 \\ 0 & 2 & 1 \end{bmatrix}$.

11. $\begin{bmatrix} 0 & 2 & -1 \\ -2 & 0 & -2 \\ 1 & 2 & 0 \end{bmatrix}$.

12. $\begin{bmatrix} 1 & -2 & 0 \\ -2 & 1 & 0 \\ 0 & 0 & 3 \end{bmatrix}$.

13. Prove the following properties for similar matrices:
(a) If A is similar to B, then B is similar to A.
(b) A is similar to itself.
(c) If A is similar to B and B is similar to C, then A is similar to C.

14. If A is similar to B, prove that A^T is similar to B^T.

15. In Theorem 8.4.1 we proved that similar matrices have the same eigenvalues. This problem investigates the relationship between their eigenvectors. Let **v** be an eigenvector of A corresponding to the eigenvalue λ. Prove that if $B = S^{-1}AS$, then $S^{-1}\mathbf{v}$ is an eigenvector of B corresponding to the eigenvalue λ.

16. Let A be nondefective and let S be such that $S^{-1}AS = \text{diag}(\lambda_1, \lambda_2, \ldots, \lambda_n)$, where all λ_i are nonzero.
(a) Prove that A^{-1} exists.
(b) Prove that $S^{-1}A^{-1}S = \text{diag}\left(\dfrac{1}{\lambda_1}, \dfrac{1}{\lambda_2}, \ldots, \dfrac{1}{\lambda_n}\right)$.

17. Let A be nondefective and let S be such that $S^{-1}AS = \text{diag}(\lambda_1, \lambda_2, \ldots, \lambda_n)$.
(a) Prove that $Q^{-1}A^TQ = \text{diag}(\lambda_1, \lambda_2, \ldots, \lambda_n)$, where $Q = (S^T)^{-1}$. (This establishes that A^T is also nondefective.)
(b) If C denotes the matrix of cofactors of S, prove that the column vectors of C are linearly independent eigenvectors of A^T. (*Hint:* Use the adjoint method to determine S^{-1}.)

18. If $A = \begin{bmatrix} -2 & 4 \\ 1 & 1 \end{bmatrix}$, determine S such that $S^{-1}AS = \text{diag}(-3, 2)$, and use the result from previous problem to determine all eigenvectors of A^T.

Problems 19–21 deal with the generalization of the diagonalization problem to defective matrices.

19. Let A be a 2×2 *defective* matrix. It follows from Theorem 8.4.2 that A is not diagonalizable. However, it can be shown that A is similar to the (elementary) Jordan matrix $J_1(\lambda) = \begin{bmatrix} \lambda & 1 \\ 0 & \lambda \end{bmatrix}$.

Thus there exists a matrix $S = [\mathbf{v}_1, \mathbf{v}_2]$, such that
$$S^{-1}AS = J_1(\lambda).$$
Prove that $\mathbf{v}_1$ and $\mathbf{v}_2$ must satisfy
$$(A - \lambda I)\mathbf{v}_1 = \mathbf{0}, \qquad \text{(i)}$$
$$(A - \lambda I)\mathbf{v}_2 = \mathbf{v}_1. \qquad \text{(ii)}$$
Equation (i) is the statement that $\mathbf{v}_1$ must be an eigenvector of A corresponding to the eigenvalue λ. Any vectors that satisfy (ii) are called **generalized eigenvectors** of A.

20. Show that $A = \begin{bmatrix} 2 & 1 \\ -1 & 4 \end{bmatrix}$ is defective and use the previous problem to determine a matrix S such that
$$S^{-1}AS = \begin{bmatrix} 3 & 1 \\ 0 & 3 \end{bmatrix}.$$

21. Let λ be an eigenvalue of the 3×3 matrix A of multiplicity 3, and suppose the corresponding eigenspace has dimension 1. It can be shown that in this case there exists a matrix $S = [\mathbf{v}_1, \mathbf{v}_2, \mathbf{v}_3]$ such that
$$S^{-1}AS = \begin{bmatrix} \lambda & 1 & 0 \\ 0 & \lambda & 1 \\ 0 & 0 & \lambda \end{bmatrix}.$$

Prove that $\mathbf{v}$, $\mathbf{v}_2$, $\mathbf{v}_3$ must satisfy
$$(A - \lambda I)\mathbf{v}_1 = \mathbf{0},$$
$$(A - \lambda I)\mathbf{v}_2 = \mathbf{v}_1,$$
$$(A - \lambda I)\mathbf{v}_3 = \mathbf{v}_2.$$

22. Let $\{\mathbf{e}_1, \mathbf{e}_2, \ldots, \mathbf{e}_n\}$ and $\{\mathbf{f}_1, \mathbf{f}_2, \ldots, \mathbf{f}_n\}$ be bases for a vector space V and let $T:V \to V$ be a linear transformation. Define the $n \times n$ matrices $A = [a_{ik}]$, $B = [b_{ik}]$ by
$$T(\mathbf{e}_k) = \sum_{i=1}^{n} a_{ik}\mathbf{e}_i, \qquad k = 1, 2, \ldots, n, \quad \text{(iii)}$$
$$T(\mathbf{f}_k) = \sum_{i=1}^{n} b_{ik}\mathbf{f}_i, \qquad k = 1, 2, \ldots, n. \quad \text{(iv)}$$
Expressing each of the basis vectors $\mathbf{f}_1, \mathbf{f}_2, \ldots, \mathbf{f}_n$ in terms of the basis $\{\mathbf{e}_1, \mathbf{e}_2, \ldots, \mathbf{e}_n\}$ yields
$$\mathbf{f}_i = \sum_{j=1}^{n} s_{ji}\mathbf{e}_j, \qquad i = 1, 2, \ldots, n. \quad \text{(v)}$$
for appropriate scalars s_{ji}. Thus the matrix $S = [s_{ji}]$ describes the relationship between the two bases.
(a) Prove that S is nonsingular. (*Hint:* Use the fact that $\mathbf{f}_1, \mathbf{f}_2, \ldots, \mathbf{f}_n$ are linearly independent.)
(b) Use (iv) and (v) to show that
$$T(\mathbf{f}_k) = \sum_{j=1}^{n}\left(\sum_{i=1}^{n} s_{ji}b_{ik}\right)\mathbf{e}_j, \quad k = 1, 2, \ldots, n,$$
or, equivalently,
$$T(\mathbf{f}_k) = \sum_{i=1}^{n}\left(\sum_{j=1}^{n} s_{ij}b_{jk}\right)\mathbf{e}_i, \quad k = 1, 2, \ldots, n. \quad \text{(vi)}$$
(c) Use (v) and (iii) to show that
$$T(\mathbf{f}_k) = \sum_{i=1}^{n}\left(\sum_{j=1}^{n} a_{ij}s_{jk}\right)\mathbf{e}_i, \quad k = 1, 2, \ldots, n. \quad \text{(vii)}$$
(d) Use (vi) and (vii) together with the linear independence of $\mathbf{e}_1, \mathbf{e}_2, \ldots, \mathbf{e}_n$ to show that
$$\sum_{j=1}^{n} s_{ij}b_{jk} = \sum_{j=1}^{n} a_{ij}s_{jk}, \quad 1 \le i \le n, \ 1 \le k \le n,$$
and hence that
$$SB = AS.$$
Finally conclude that A and B are related by
$$B = S^{-1}AS.$$

8.5   EIGENVALUES AND EIGENVECTORS OF REAL SYMMETRIC MATRICES

Recall that a square matrix A is called *symmetric* if $A^T = A$. We now investigate the eigenvalue-eigenvector problem for real symmetric matrices (that is, symmetric matrices with *real* elements), since such matrices often arise in applications. The following lemma will be useful in proving our results.

> *Lemma 8.5.1:*   Let $\mathbf{v}_1$ and $\mathbf{v}_2$ be vectors in $\mathbf{R}^n$ (or $\mathbf{C}^n$) and let A be an $n \times n$ real symmetric matrix. Then
> $$\langle A\mathbf{v}_1, \mathbf{v}_2\rangle = \langle \mathbf{v}_1, A\mathbf{v}_2\rangle$$

PROOF   The key to the proof is to note that the inner product of two vectors in $\mathbf{R}^n$ or $\mathbf{C}^n$ can be written in matrix form as

$$[\langle \mathbf{x}, \mathbf{y}\rangle] = \mathbf{x}^T\overline{\mathbf{y}}.$$

Applying this to the vectors $A\mathbf{v}_1$, $\mathbf{v}_2$ yields

$$[\langle A\mathbf{v}_1, \mathbf{v}_2\rangle] = (A\mathbf{v}_1)^T\overline{\mathbf{v}}_2$$
$$= \mathbf{v}_1^T A^T \overline{\mathbf{v}}_2$$
$$= \mathbf{v}_1^T \overline{(A\mathbf{v}_2)}$$
$$= [\langle \mathbf{v}_1, A\mathbf{v}_2\rangle],$$

from which the result follows directly.   ■

BASIC RESULTS FOR REAL SYMMETRIC MATRICES

We summarize the special properties satisfied by the eigenvalues and eigenvectors of a real symmetric matrix in a theorem.

> *Theorem 8.5.1:*   Let A be a real symmetric matrix. Then
> 1. All eigenvalues of A are real.
> 2. Real eigenvectors of A that correspond to *distinct* eigenvalues are *orthogonal*.
> 3. A is nondefective.
> 4. A possesses a complete set of *orthonormal* eigenvectors.

PROOF

1. Suppose that $(\lambda_1, \mathbf{v}_1)$ are an eigenvalue-eigenvector pair, that is,

$$A\mathbf{v}_1 = \lambda_1\mathbf{v}_1. \qquad (8.5.1)$$

We must prove that

$$\lambda_1 = \overline{\lambda}_1. \qquad (8.5.2)$$

Taking the inner product of both sides of (8.5.1) with the vector $\mathbf{v}_1$ yields

$$\langle A\mathbf{v}_1, \mathbf{v}_1 \rangle = \langle \lambda_1 \mathbf{v}_1, \mathbf{v}_1 \rangle;$$

that is, using the properties of the inner product and recalling that by definition $\langle \mathbf{v}_1, \mathbf{v}_1 \rangle = \| \mathbf{v}_1 \|^2$,

$$\langle A\mathbf{v}_1, \mathbf{v}_1 \rangle = \lambda_1 \| \mathbf{v}_1 \|^2. \tag{8.5.3}$$

Taking the complex conjugate of (8.5.3) yields (remember that $\| \mathbf{v}_1 \|$ is a *real* number),

$$\overline{\langle A\mathbf{v}_1, \mathbf{v}_1 \rangle} = \bar{\lambda}_1 \| \mathbf{v}_1 \|^2;$$

that is, using the fact that $\overline{\langle \mathbf{u}, \mathbf{v} \rangle} = \langle \mathbf{v}, \mathbf{u} \rangle$,

$$\langle \mathbf{v}_1, A\mathbf{v}_1 \rangle = \bar{\lambda}_1 \| \mathbf{v}_1 \|^2. \tag{8.5.4}$$

Subtracting (8.5.4) from (8.5.3) and using Lemma 8.5.1 with $\mathbf{v}_2 = \mathbf{v}_1$ yields

$$(\lambda_1 - \bar{\lambda}_1) \| \mathbf{v}_1 \|^2 = 0. \tag{8.5.5}$$

However, since $\mathbf{v}_1$ is an eigenvector of A it follows that it is necessarily a *non-zero* vector. It therefore follows from (8.5.5) that we must have

$$\lambda_1 = \bar{\lambda}_1.$$

2. It follows from (1) that all eigenvalues of A are necessarily real. We can therefore take the underlying vector space as being $\mathbf{R}^n$ so that the corresponding eigenvectors of A will also be real. Let $(\lambda_1, \mathbf{v}_1)$, $(\lambda_2, \mathbf{v}_2)$ be two such real eigenvalue-eigenvector pairs with $\lambda_1 \neq \lambda_2$. Then, in addition to (8.5.1) we also have:

$$A\mathbf{v}_2 = \lambda_2 \mathbf{v}_2. \tag{8.5.6}$$

We must prove that

$$\langle \mathbf{v}_1, \mathbf{v}_2 \rangle = 0.$$

Taking the inner product of (8.5.1) with $\mathbf{v}_2$ and the inner product of (8.5.6) with $\mathbf{v}_1$ and using $\langle \mathbf{v}_1, \lambda_2 \mathbf{v}_2 \rangle = \lambda_2 \langle \mathbf{v}_1, \mathbf{v}_2 \rangle$ yields, respectively,

$$\langle A\mathbf{v}_1, \mathbf{v}_2 \rangle = \lambda_1 \langle \mathbf{v}_1, \mathbf{v}_2 \rangle \qquad \langle \mathbf{v}_1, A\mathbf{v}_2 \rangle = \lambda_2 \langle \mathbf{v}_1, \mathbf{v}_2 \rangle.$$

Subtracting the second equation from the first and using Lemma 8.5.1, we obtain

$$(\lambda_1 - \lambda_2)\langle \mathbf{v}_1, \mathbf{v}_2 \rangle = 0.$$

Since $\lambda_1 - \lambda_2 \neq 0$ by assumption, it follows that $\langle \mathbf{v}_1, \mathbf{v}_2 \rangle = 0$.

3. The proof of this property requires some deeper results from linear algebra than we have encountered and hence is omitted. See, for example, G. E. Shilov, *Linear Algebra*, Dover Publications, 1977.

4. Eigenvectors corresponding to distinct eigenvalues are orthogonal from (2). Now suppose that the eigenvalue λ_i has multiplicity m_i. Then, since A possesses a complete set of eigenvectors [from (3)], it follows from Theorem 8.3.4

that we can find m_i linearly independent eigenvectors corresponding to λ_i. These vectors span the eigenspace corresponding to λ_i and hence we can use the Gram–Schmidt process to find m_i *orthogonal* eigenvectors (corresponding to λ_i) that span this eigenspace (see Figure 8.5.1 for an illustration of the 3×3 case). Proceeding in this manner for each eigenvalue, we obtain a complete set of orthogonal eigenvectors, and hence by normalization we can find an orthonormal set.   ∎

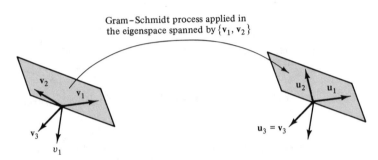

Gram–Schmidt process applied in
the eigenspace spanned by $\{\mathbf{v}_1, \mathbf{v}_2\}$

Figure 8.5.1   Construction of a complete set of *orthogonal* eigenvectors $\{\mathbf{u}_1, \mathbf{u}_2,$ $\mathbf{u}_3\}$ for a real symmetric 3×3 matrix that has an eigenvalue of multiplicity 2.

Example 8.5.1   Find a complete orthonormal set of eigenvectors for

$$A = \begin{bmatrix} 2 & 2 & 1 \\ 2 & 5 & 2 \\ 1 & 2 & 2 \end{bmatrix}.$$

Solution   Since A is real and symmetric, a complete orthonormal set of eigenvectors exists. We first find the eigenvalues of A. The characteristic polynomial of A is

$$
\begin{aligned}
\det(A - \lambda I) &= \begin{vmatrix} 2 - \lambda & 2 & 1 \\ 2 & 5 - \lambda & 2 \\ 1 & 2 & 2 - \lambda \end{vmatrix} \\
&= (2 - \lambda)[(5 - \lambda)(2 - \lambda) - 4] - 2[4 - 2\lambda - 2] + [4 + \lambda - 5] \\
&= (2 - \lambda)[\lambda^2 - 7\lambda + 6] + 5(\lambda - 1) \\
&= (2 - \lambda)(\lambda - 6)(\lambda - 1) + 5(\lambda - 1) \\
&= -(\lambda - 1)(\lambda^2 - 8\lambda + 7) \\
&= -(\lambda - 1)(\lambda - 1)(\lambda - 7) \\
&= -(\lambda - 1)^2(\lambda - 7).
\end{aligned}
$$

Thus A has eigenvalues

$$\lambda = 1, 1, 7.$$

Eigenvectors

$\lambda = 1$:   In this case the system $(A - \lambda I)\mathbf{v} = 0$ reduces to the single equation

$$v_1 + 2v_2 + v_3 = 0,$$

which has solution $(-2r - s, r, s)$, so that the corresponding eigenvectors are

$$\mathbf{v} = (-2r - s, r, s) = r(-2, 1, 0) + s(-1, 0, 1). \tag{8.5.7}$$

Two *linearly independent* eigenvectors corresponding to the eigenvalue $\lambda = 1$ are

$$\mathbf{v}_1 = (-1, 0, 1), \qquad \mathbf{v}_2 = (-2, 1, 0). \tag{8.5.8}$$

The eigenvectors $\mathbf{v}_1$ and $\mathbf{v}_2$ are *not* orthogonal. However we can use the Gram–Schmidt process to obtain a pair of orthogonal eigenvectors corresponding to $\lambda = 1$. We take $\mathbf{u}_1 = \mathbf{v}_1$—that is, $\mathbf{u}_1 = (-1, 0, 1)$. Then $\mathbf{u}_2$ is given by

$$\mathbf{u}_2 = \mathbf{v}_2 - \frac{\langle \mathbf{v}_2, \mathbf{u}_1 \rangle}{\| \mathbf{u}_1 \|^2} \mathbf{u}_1 = (-2, 1, 0) - \frac{2}{2}(-1, 0, 1),$$

that is,

$$\mathbf{u}_2 = (-1, 1, -1).$$

Thus the vectors $\mathbf{u}_1$, $\mathbf{u}_2$ are orthogonal eigenvectors corresponding to $\lambda = 1$ (they form an orthogonal basis for the eigenspace corresponding to $\lambda = 1$). Two orthonormal eigenvectors corresponding to $\lambda = 1$ are thus

$$\mathbf{w}_1 = \frac{\mathbf{u}_1}{\| \mathbf{u}_1 \|} = \left(-\frac{1}{\sqrt{2}}, 0, \frac{1}{\sqrt{2}} \right), \tag{8.5.9}$$

$$\mathbf{w}_2 = \frac{\mathbf{u}_2}{\| \mathbf{u}_2 \|} = \left(-\frac{1}{\sqrt{3}}, \frac{1}{\sqrt{3}}, -\frac{1}{\sqrt{3}} \right). \tag{8.5.10}$$

$\lambda = 7$: In this case the system $(A - \lambda I)\mathbf{v} = 0$ is

$$-5v_1 + 2v_2 + \ v_3 = 0,$$
$$2v_1 - 2v_2 + 2v_3 = 0,$$
$$v_1 + 2v_2 - 5v_3 = 0.$$

Solving using Gaussian elimination yields $(t, 2t, t)$, so that the eigenvectors corresponding to $\lambda = 7$ are

$$\mathbf{v} = t(1, 2, 1), \tag{8.5.11}$$

and a *unit* eigenvector corresponding to $\lambda = 7$ is

$$\mathbf{w}_3 = \frac{\mathbf{v}}{\| \mathbf{v} \|} = \left(\frac{1}{\sqrt{6}}, \frac{2}{\sqrt{6}}, \frac{1}{\sqrt{6}} \right). \tag{8.5.12}$$

Notice that $\mathbf{w}_3$ is orthogonal to both $\mathbf{w}_1$ and $\mathbf{w}_2$, as guaranteed by Theorem 8.5.1.

It follows from (8.5.9), (8.5.10), and (8.5.12) that a complete set of orthonormal eigenvectors for the given matrix A is $\{\mathbf{w}_1, \mathbf{w}_2, \mathbf{w}_3\}$, that is,

$$\left\{\left(-\frac{1}{\sqrt{2}}, 0, \frac{1}{\sqrt{2}}\right), \left(-\frac{1}{\sqrt{3}}, \frac{1}{\sqrt{3}}, -\frac{1}{\sqrt{3}}\right), \left(\frac{1}{\sqrt{6}}, \frac{2}{\sqrt{6}}, \frac{1}{\sqrt{6}}\right)\right\}.$$

Since a real symmetric matrix possesses a complete set of eigenvectors, it follows that such a matrix is necessarily diagonalizable, that is,

$$S^{-1}AS = \text{diag}(\lambda_1, \lambda_2, \ldots, \lambda_n),$$

where S is a matrix whose column vectors are any complete set of eigenvectors for A. We now show that there is a preferred choice for S when A is real and symmetric. This requires a definition.

> **Definition 8.5.1:** A real nonsingular matrix A is called **orthogonal** if
> $$A^{-1} = A^T.$$

Example 8.5.2 Show that the following matrix is an orthogonal matrix:

$$A = \begin{bmatrix} \frac{1}{2} & -\frac{1}{2} & \frac{1}{2} & \frac{1}{2} \\ \frac{1}{2} & \frac{1}{2} & \frac{1}{2} & -\frac{1}{2} \\ \frac{1}{2} & \frac{1}{2} & -\frac{1}{2} & \frac{1}{2} \\ \frac{1}{2} & -\frac{1}{2} & -\frac{1}{2} & -\frac{1}{2} \end{bmatrix}.$$

Solution For the given matrix we have

$$AA^T = \begin{bmatrix} \frac{1}{2} & -\frac{1}{2} & \frac{1}{2} & \frac{1}{2} \\ \frac{1}{2} & \frac{1}{2} & \frac{1}{2} & -\frac{1}{2} \\ \frac{1}{2} & \frac{1}{2} & -\frac{1}{2} & \frac{1}{2} \\ \frac{1}{2} & -\frac{1}{2} & -\frac{1}{2} & -\frac{1}{2} \end{bmatrix} \begin{bmatrix} \frac{1}{2} & \frac{1}{2} & \frac{1}{2} & \frac{1}{2} \\ -\frac{1}{2} & \frac{1}{2} & \frac{1}{2} & -\frac{1}{2} \\ \frac{1}{2} & \frac{1}{2} & -\frac{1}{2} & -\frac{1}{2} \\ \frac{1}{2} & -\frac{1}{2} & \frac{1}{2} & -\frac{1}{2} \end{bmatrix}$$

$$= I_4.$$

Similarly $A^TA = I_4$, so that $A^T = A^{-1}$. Consequently A is an orthogonal matrix.

If we look more closely at the above example, we see that the column vectors of A and the row vectors of A are *orthonormal* vectors. The next theorem establishes that this is a basic characterizing property of all orthogonal matrices.

> ***Theorem 8.5.2:*** A real $n \times n$ matrix A is an orthogonal matrix if and only if the row (or column) vectors are orthogonal unit vectors.

PROOF The proof is left as an exercise.   ∎

We now return to the diagonalization problem for a real symmetric matrix. From (4) of Theorem 8.5.1, such a matrix possesses a complete set of *orthonormal* eigenvectors. It follows that if we use *this* complete set for the column vectors of S, then these column vectors are orthonormal; hence, from Theorem 8.5.2, the matrix S is an orthogonal matrix (that is, $S^{-1} = S^T$). We have therefore proved the following result.

Theorem 8.5.3: If A is a real symmetric matrix and $S = [s_1, s_2, \ldots, s_n]$, where $\{s_1, s_2, \ldots, s_n\}$ is a complete set of *orthonormal* eigenvectors for A, then

$$S^TAS = \operatorname{diag}(\lambda_1, \lambda_2, \ldots, \lambda_n).$$

Example 8.5.3 We showed in Example 8.5.1 that a complete set of *orthonormal*

eigenvectors for the real symmetric matrix $A = \begin{bmatrix} 2 & 2 & 1 \\ 2 & 5 & 2 \\ 1 & 2 & 2 \end{bmatrix}$ is

$$\left\{ \left(-\frac{1}{\sqrt{2}}, 0, \frac{1}{\sqrt{2}} \right), \left(-\frac{1}{\sqrt{3}}, \frac{1}{\sqrt{3}}, -\frac{1}{\sqrt{3}} \right), \left(\frac{1}{\sqrt{6}}, \frac{2}{\sqrt{6}}, \frac{1}{\sqrt{6}} \right) \right\},$$

corresponding to the eigenvalues $\lambda = 1, 1, 7$, respectively. It follows from the previous theorem that if we set

$$S = \begin{bmatrix} -\dfrac{1}{\sqrt{2}} & -\dfrac{1}{\sqrt{3}} & \dfrac{1}{\sqrt{6}} \\ 0 & \dfrac{1}{\sqrt{3}} & \dfrac{2}{\sqrt{6}} \\ \dfrac{1}{\sqrt{2}} & -\dfrac{1}{\sqrt{3}} & \dfrac{1}{\sqrt{6}} \end{bmatrix},$$

then

$$S^TAS = \operatorname{diag}(1, 1, 7).$$

REMARK   Notice that Theorem 8.5.3 requires the use of a complete *orthonormal* set of eigenvectors. Often the mistake is made of forgetting to normalize the orthogonal vectors.

EXERCISES 8.5

In problems 1–12, determine an orthogonal matrix S such that $S^TAS = \operatorname{diag}(\lambda_1, \lambda_2, \ldots, \lambda_n)$, where A denotes the given matrix.

1. $\begin{bmatrix} 2 & 2 \\ 2 & -1 \end{bmatrix}$.

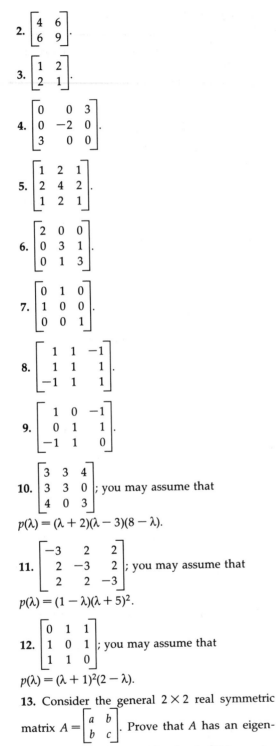

2. $\begin{bmatrix} 4 & 6 \\ 6 & 9 \end{bmatrix}$.

3. $\begin{bmatrix} 1 & 2 \\ 2 & 1 \end{bmatrix}$.

4. $\begin{bmatrix} 0 & 0 & 3 \\ 0 & -2 & 0 \\ 3 & 0 & 0 \end{bmatrix}$.

5. $\begin{bmatrix} 1 & 2 & 1 \\ 2 & 4 & 2 \\ 1 & 2 & 1 \end{bmatrix}$.

6. $\begin{bmatrix} 2 & 0 & 0 \\ 0 & 3 & 1 \\ 0 & 1 & 3 \end{bmatrix}$.

7. $\begin{bmatrix} 0 & 1 & 0 \\ 1 & 0 & 0 \\ 0 & 0 & 1 \end{bmatrix}$.

8. $\begin{bmatrix} 1 & 1 & -1 \\ 1 & 1 & 1 \\ -1 & 1 & 1 \end{bmatrix}$.

9. $\begin{bmatrix} 1 & 0 & -1 \\ 0 & 1 & 1 \\ -1 & 1 & 0 \end{bmatrix}$.

10. $\begin{bmatrix} 3 & 3 & 4 \\ 3 & 3 & 0 \\ 4 & 0 & 3 \end{bmatrix}$; you may assume that
$p(\lambda) = (\lambda + 2)(\lambda - 3)(8 - \lambda)$.

11. $\begin{bmatrix} -3 & 2 & 2 \\ 2 & -3 & 2 \\ 2 & 2 & -3 \end{bmatrix}$; you may assume that
$p(\lambda) = (1 - \lambda)(\lambda + 5)^2$.

12. $\begin{bmatrix} 0 & 1 & 1 \\ 1 & 0 & 1 \\ 1 & 1 & 0 \end{bmatrix}$; you may assume that
$p(\lambda) = (\lambda + 1)^2(2 - \lambda)$.

13. Consider the general 2×2 real symmetric matrix $A = \begin{bmatrix} a & b \\ b & c \end{bmatrix}$. Prove that A has an eigenvalue of multiplicity two if and only if it is a scalar matrix (that is a matrix of the form dI_2, where d is a constant).

14. **(a)** Let A be an $n \times n$ real symmetric matrix. Prove that if λ is an eigenvalue of A of multiplicity n, then A is a scalar matrix. (*Hint:* Prove that there exists an orthogonal matrix S such that $S^T A S = \lambda I_n$, and then solve for A.)
(b) State and prove the corresponding result for general $n \times n$ matrices.

15. A 2×2 real symmetric matrix A has two distinct eigenvalues, λ_1 and λ_2. If $v_1 = (1, 2)$ is an eigenvector of A corresponding to the eigenvalue λ_1, determine an eigenvector corresponding to λ_2.

16. A 2×2 real symmetric matrix A has two distinct eigenvalues, λ_1 and λ_2.
(a) If $v_1 = (a, b)$ is an eigenvector of A corresponding to the eigenvalue λ_1, determine an eigenvector corresponding to λ_2, and hence find an orthogonal matrix S such that $S^T A S = \text{diag}(\lambda_1, \lambda_2)$.
(b) Use your result from (a) to find A. (Your answer will involve λ_1, λ_2, a, and b.)

17. A 3×3 real symmetric matrix A has eigenvalues λ_1 and λ_2 (multiplicity 2).
(a) If $v_1 = (1, -1, 1)$ spans the eigenspace E_1, determine a basis for E_2 and hence find an orthogonal matrix S such that $S^T A S = \text{diag}(\lambda_1, \lambda_2, \lambda_2)$.
(b) Use your result from (a) to find A.

18. Prove that a real $n \times n$ matrix A is orthogonal if and only if its row (or column) vectors are orthogonal unit vectors.

19. Prove that if A and B are $n \times n$ orthogonal matrices, then AB is also an orthogonal matrix.

Problems 20–22 deal with the eigenvalue-eigenvector problem for $n \times n$ real *skew-symmetric* matrices.

20. Let A be an $n \times n$ real skew-symmetric matrix.
(a) Prove that for all v_1 and v_2 in $\mathbb{C}^n$,

$$\langle Av_1, v_2 \rangle = -\langle v_1, Av_2 \rangle,$$

where $\langle, \rangle$ denotes the standard inner product in $\mathbb{C}^n$. (*Hint:* See Lemma 8.5.1.)
(b) Prove that all *nonzero* eigenvalues of A are pure imaginary $(\lambda = -\bar{\lambda})$. [*Hint:* Model your proof after that of (1) in Theorem 8.5.1.]

21. It follows from the previous problem that the only real eigenvalue that a real skew-symmetric matrix can possess is $\lambda = 0$. Use this to prove that if A is an $n \times n$ real skew-symmetric matrix, *with n odd,* then A necessarily has zero as one of its eigenvalues.

22. Determine all eigenvalues and corresponding eigenvectors of the matrix

$$A = \begin{bmatrix} 0 & 4 & -4 \\ -4 & 0 & -2 \\ 4 & 2 & 0 \end{bmatrix}.$$

8.6 SUMMARY OF RESULTS

In this chapter we have introduced one of the fundamental problems in mathematics, the eigenvalue-eigenvector problem. In this final section we list the important results of the chapter.

1. For a given $n \times n$ matrix A, the eigenvalue-eigenvector problem consists of determining all scalars λ and all *nonzero* vectors $\mathbf{v}$ such that

$$A\mathbf{v} = \lambda \mathbf{v}.$$

2. The eigenvalues are the roots of the characteristic polynomial

$$p(\lambda) = \det(A - \lambda I) = 0 \qquad (8.6.1)$$

and the eigenvectors are obtained by solving the linear systems

$$(A - \lambda I)\mathbf{v} = \mathbf{0}, \qquad (8.6.2)$$

when λ assumes the values obtained in (8.6.1).

3. If A is a matrix with real elements, then complex eigenvalues and eigenvectors occur in conjugate pairs.

4. Associated with each eigenvalue λ is a vector space called the eigenspace of λ. This is the set of all eigenvectors corresponding to λ together with the zero vector. Equivalently, it can be considered as the set of *all* solutions to the linear system (8.6.2).

5. If m denotes the multiplicity of the eigenvalue λ and n denotes the dimension of the corresponding eigenspace, then

$$1 \le n \le m.$$

6. Eigenvectors corresponding to distinct eigenvalues are linearly independent.

7. An $n \times n$ matrix that has n linearly independent eigenvectors is said to have a *complete set of eigenvectors,* and we call such a matrix *nondefective.*

8. Two $n \times n$ matrices A and B are said to be *similar* if there exists a matrix S such that

$$B = S^{-1}AS.$$

We can consider such matrices as representing the same linear transformation relative to different bases.

9. A matrix that is similar to a diagonal matrix is said to be diagonalizable. We have shown that A is diagonalizable if and only if it is nondefective.

10. If A is nondefective and $S = [\mathbf{v}_1, \mathbf{v}_2, \ldots, \mathbf{v}_n]$, where $\mathbf{v}_1, \mathbf{v}_2, \ldots, \mathbf{v}_n$ are linearly independent eigenvectors of A, then

$$S^{-1}AS = \text{diag}(\lambda_1, \lambda_2, \ldots, \lambda_n),$$

where $\lambda_1, \lambda_2, \ldots, \lambda_n$ are the eigenvalues of A corresponding to the eigenvectors $\mathbf{v}_1, \mathbf{v}_2, \ldots, \mathbf{v}_n$.

11. If A is a *real symmetric* matrix, then
    (a) All eigenvalues of A are real.
    (b) Eigenvectors corresponding to different eigenvalues are *orthogonal*.
    (c) A is nondefective.
    (d) A has a complete set of orthonormal eigenvectors, say $\mathbf{s}_1, \mathbf{s}_2, \ldots, \mathbf{s}_n$.
    (e) If we let $S = [\mathbf{s}_1, \mathbf{s}_2, \ldots, \mathbf{s}_n]$, where $\mathbf{s}_1, \mathbf{s}_2, \ldots, \mathbf{s}_n$ are the orthonormal eigenvectors determined in (d), then S is an orthogonal matrix ($S^{-1} = S^T$), and hence, from (10), for this matrix,

$$S^T A S = \text{diag}(\lambda_1, \lambda_2, \ldots, \lambda_n).$$

Linear Ordinary Differential Equations

9.1   INTRODUCTION

In Chapter 1 we studied first-order differential equations of the form

$$\frac{dy}{dx} = f(x, y)$$

and were able to develop special techniques for rewriting certain equations as

$$\frac{d}{dx} [g(x, y)] = F(x).$$

The solution, $y = y(x)$, of the differential equation could then be obtained by a direct integration. These techniques, however, cannot be extended to differential equations of higher order, and so more sophisticated methods are required. We now return to develop such methods for *linear* differential equations of arbitrary order.   The general form for a linear differential equation of order n is[1]

$$a_0(x)y^{(n)} + a_1(x)y^{(n-1)} + \cdots + a_{n-1}(x)y' + a_n(x)y = F(x). \qquad (9.1.1)$$

We have postponed a discussion of these differential equations until now so that the underlying theory could be developed as an application of the vector space results obtained in Chapters 6 and 7.   In order to apply these results, we will first reformulate the problem of determining all solutions to (9.1.1) in terms of a linear operator problem.   Indeed we will show that (9.1.1) can be written as

$$Ly = F,$$

for an appropriate linear transformation L.   The key theoretical result, proved in section 9.3, is that the kernel of L has dimension n if the differential equation has order n.   It will then follow directly from the results of Chapter 6 that any solution to

$$Ly = 0 \qquad (9.1.2)$$

can be expressed as

$$y = c_1 y_1 + c_2 y_2 + \cdots + c_n y_n,$$

where $\{y_1, y_2, \ldots, y_n\}$ is any set of linearly independent solutions to $Ly = 0$.   We will also show that all solutions to (9.1.1) can be written as

$$y = c_1 y_1 + c_2 y_2 + \cdots + c_n y_n + y_p,$$

where $y = y_p$ is any particular solution to (9.1.1).   Consequently determining all solutions to (9.1.1) will be reduced to determining n linearly independent solutions of (9.1.2) together with one particular solution of (9.1.1).   The fact that we will be able to derive these results in a few pages is an illustration of the power of the vector space techniques.

Having completely set up the underlying theory for linear differential equations, we will then turn our attention to obtaining the solutions whose existence is guaranteed from the theory.   We will find, perhaps somewhat surprisingly, that the determination of these solutions can usually be reduced to an algebraic problem, as opposed to a calculus problem.

This chapter is the focal point of the text.   It is here, particularly in Section 9.3, that you should finally appreciate the power and use of the abstract mathematical framework that has been developed in Chapters 6 and 7.   Furthermore, the solution techniques that we derive in this chapter are fundamental in the theory and applications of differential equations.

---

[1] You are encouraged to reread Section 1.1 in order to familiarize yourself once more with the general terminology used in differential equation theory.

9.2   LINEAR DIFFERENTIAL OPERATORS

In order to apply our vector space results in deriving the underlying theory for linear differential equations, we need to formulate the differential equations as operator equations.   In this section we define the appropriate linear operators.

We first recall that $C^n(I)$ denotes the vector space of all functions that have (at least) n continuous derivatives on the interval I.   [$C^0(I)$ denotes the set of all functions that are continuous on I.]

We now define the *linear* transformation $D:C^1(I) \to C^0(I)$ by

$$Df = f' \qquad \text{for all } f \text{ in } C^1(I),$$

that is,

$$Df(x) = f'(x) \qquad \text{for all } f \text{ in } C^1(I) \text{ and all } x \text{ in I.}$$

D is called the **derivative** operator.

We have already seen (Example 7.1.1, page 245) that D *is a linear* transformation.   Higher-order derivative operators are defined by composition, as follows:

$$D^2:C^2(I) \to C^0(I), D^2f = D(Df), \qquad \text{for all } f \text{ in } C^2(I),$$

$$D^3:C^3(I) \to C^0(I), D^3f = D(D^2f), \qquad \text{for all } f \text{ in } C^3(I),$$

$$\vdots \qquad\qquad \vdots \qquad\qquad\qquad \vdots$$

$$D^n:C^n(I) \to C^0(I), D^nf = D(D^{n-1}f), \qquad \text{for all } f \text{ in } C^n(I).$$

Thus,

$$\boxed{D^nf = f^{(n)}} \quad ,$$

the nth derivative of f.

Example 9.2.1   If $f(x) = x^2 + \sin x$, then

$$(Df)(x) = 2x + \cos x,$$

$$(D^2f)(x) = D(Df)(x) = 2 - \sin x,$$

$$(D^3f)(x) = D(D^2f)(x) = -\cos x,$$

and so on.

We will be interested in operators that are obtained by taking linear combinations of the derivative operators.   For example, if a_0, a_1, a_2 are functions defined on an interval I, then we define the operator

$$L = a_0D^2 + a_1D + a_2$$

by

$$Lf = a_0D^2f + a_1Df + a_2f = a_0f'' + a_1f' + a_2f$$

for all functions f in $C^2(I)$. It is easily shown that

$$L(f+g) = Lf + Lg, \qquad L(cf) = cLf,$$

for all f in $C^2(I)$ and all real numbers c, so that L is indeed a linear transformation between the vector spaces $C^2(I)$ and $C^0(I)$.

More generally we have the following definition.

Definition 9.2.1:   Let $a_0, a_1, \ldots, a_n$ be given functions that are continuous on an interval I, with a_0 not identically zero on I. By a **linear differential operator of order** n, $L:C^n(I) \to C^0(I)$, we mean the operator

$$L = a_0 D^n + a_1 D^{n-1} + \cdots + a_{n-1}D + a_n,$$

defined by

$$Lf = a_0 f^{(n)} + a_1 f^{(n-1)} + \cdots + a_{n-1}f' + a_n f \qquad \text{for all } f \text{ in } C^n(I).$$

We leave the verification that L is a linear transformation as an exercise.

Example 9.2.2   Let $L_1 = D + x^2$, $L_2 = x^2 D - x$. Find $L_1 f$ and $L_2 f$ if $f(x) = 3x$.

Solution   For any differentiable function f we have

$$L_1 f = (D + x^2)f = f' + x^2 f,$$
$$L_2 f = (x^2 D - x)f = x^2 f' - xf.$$

Thus

$$L_1(3x) = 3 + 3x^3.$$

Similarly,

$$L_2(3x) = 3x^2 - 3x^2 = 0,$$

which implies that $f(x) = 3x$ is in $\text{Ker}(L_2)$.

If we have two linear differential operators, one $L_1:C^n(I) \to C^0(I)$ of order n and the other $L_2:C^m(I) \to C^0(I)$ of order m, then we can compose the two operators to obtain two new linear differential operators of order $m+n$, namely, $L_1 L_2:C^{m+n}(I) \to C^0(I)$ and $L_2 L_1:C^{m+n}(I) \to C^0(I)$.[1]   Since composition of linear operators is not, in general, commutative, it follows that $L_1 L_2 \neq L_2 L_1$ in general. We illustrate this important point with an example.

Example 9.2.3   If L_1 and L_2 are the same as in the previous example, find $L_2 L_1$ and $L_1 L_2$ and show that $L_2 L_1 \neq L_1 L_2$.

---

[1] We assume that the functions arising in the definitions of L_1 and L_2 are sufficiently smooth for the operations to be performed.

Solution

$$(L_2L_1)f = L_2(L_1f) = L_2(f' + x^2f) = x^2(f' + x^2f)' - x(f' + x^2f)$$
$$= x^2f'' + x^4f' + 2x^3f - xf' - x^3f$$
$$= x^2f'' + (x^4 - x)f' + x^3f = [x^2D^2 + (x^4 - x)D + x^3]f,$$

so that

$$L_2L_1 = x^2D^2 + (x^4 - x)D + x^3. \tag{9.2.1}$$

Similarly,

$$(L_1L_2)f = L_1(L_2f) = L_1(x^2f' - xf) = (x^2f' - xf)' + x^2(x^2f' - xf)$$
$$= x^2f'' + 2xf' - xf' - f + x^4f' - x^3f = x^2f'' + (x^4 + x)f' - (x^3 + 1)f,$$

so that

$$L_1L_2 = x^2D^2 + (x^4 + x)D - (x^3 + 1). \tag{9.2.2}$$

We see from (9.2.1) and (9.2.2) that $L_2L_1 \neq L_1L_2$.

A linear differential operator of order n in which $a_0, a_1, \ldots, a_n$ are real *constants* is called a **polynomial differential operator of order** n and is denoted by $P(D)$. Thus, the general form for a polynomial differential operator of order n is

$$P(D) = a_0D^n + a_1D^{n-1} + \cdots + a_{n-1}D + a_n,$$

where $a_0, a_1, \ldots, a_n$ are real *constants* and $a_0 \neq 0$. If $P_1(D)$ and $P_2(D)$ are two polynomial differential operators, then we can always form the operators $P_1(D)P_2(D)$ and $P_2(D)P_1(D)$. Further, since the coefficients in each operator are constants, it follows that

$$P_1(D)P_2(D) = P_2(D)P_1(D).$$

Thus *polynomial differential operators commute.*

Example 9.2.4 If $P_1(D) = D - 2$, and $P_2(D) = D - 1$, verify that

$$P_1(D)P_2(D) = P_2(D)P_1(D).$$

Solution For any f in $C^2(I)$

$$[P_1(D)P_2(D)]f = (D - 2)(f' - f) = f'' - 3f' + 2f = (D^2 - 3D + 2)f,$$

so that

$$P_1(D)P_2(D) = D^2 - 3D + 2.$$

Similarly,

$$[P_2(D)P_1(D)]f = (D - 1)(f' - 2f) = f'' - 3f' + 2f = (D^2 - 3D + 2)f.$$

Thus,

$$P_2(D)P_1(D) = D^2 - 3D + 2.$$

Consequently,

$$P_1(D)P_2(D) = P_2(D)P_1(D).$$

Finally, in this section we indicate the role that differential operators are going to play in the development of the theory underlying the solution of linear differential equations.

Example 9.2.5 If $L:C^1(0, \infty) \rightarrow C^0(0, \infty)$ is defined by $L = xD + 1$, find Ker(L).

Solution

$$\text{Ker}(L) = \{f \in C^1(0, \infty) : Lf = 0\}.$$

But, $Lf = 0$ if and only if $xDf + f = 0$—that is, if and only if $xf' + f = 0$. This first-order linear differential equation has general solution $f(x) = cx^{-1}$. Since all solutions of the differential equation are of this form, we have

$$\text{Ker}(L) = \{f(x) = cx^{-1} : c \in \mathbf{R}, x \in (0, \infty)\}.$$

In the previous example Ker(L) coincided with the set of all solutions to the differential equation $Ly = 0$, where $L = xD + 1$. More generally, if L is any linear differential operator of order n, then a function y is a solution of the differential equation $Ly = 0$ if and only if y belongs to Ker(L). It is this result that will enable us in the next section to use the vector space ideas from Chapters 6 and 7 to derive the theory underlying the solution of linear differential equations.

EXERCISES 9.2

In problems 1–3 find Lf for the given differential operator if $f(x) = 2x - 3e^{2x}$.

1. $L = D - x$.

2. $L = D^2 - x^2D + x$.

3. $L = D^3 - 2xD^2$.

In problems 4 and 5 verify that the given function is in the kernel of L.

4. $f(x) = x^{-2}$, $L = x^2D^2 + 2xD - 2$.

5. $f(x) = \sin x^2$, $L = D^2 - x^{-1}D + 4x^2$.

6. Find Ker(L) if $L = D - 2x$.

In problems 7–9 find L_1L_2 and L_2L_1 for the given differential operators, and determine whether $L_1L_2 = L_2L_1$.

7. $L_1 = xD + 1$, $L_2 = D - 2x^2$.

8. $L_1 = D + x$, $L_2 = 2D + (2x - 1)$.

9. $L_1 = xD^2 - x^2$, $L_2 = xD$.

10. If $L_1 = a_0(x)D + a_1(x)$, determine all differential operators of the form $L_2 = b_0(x)D + b_1(x)$ such that $L_1L_2 = L_2L_1$ (you may assume that a_0 and b_0 are nonzero on the interval of interest).

11. If $P_1(D) = aD + b$ and $P_2(D) = cD + d$, where a, b, c, d are arbitrary constants, verify that $P_1(D)P_2(D) = P_2(D)P_1(D)$.

In problems 12 and 13 verify that the given function is in Ker($P(D)$) for all values of the constants c_1, c_2.

12. $f(x) = c_1e^{2x} + c_2e^{-3x}$, $P(D) = D^2 + D - 6$.

13. $f(x) = e^{-x}(c_1 + c_2x)$, $P(D) = D^2 + 2D + 1$.

14. Find Ker($P(D)$) if $P(D) = D - r$, where r is a constant.

9.3   GENERAL THEORY OF LINEAR DIFFERENTIAL EQUATIONS

In this section we apply the vector space results from Chapters 6 and 7 to develop the theory underlying the solution of linear differential equations. The ease with which these results are obtained is an indication of the power of the vector space framework that we have constructed.

We recall from Chapter 1 that a differential equation of the form

$$a_0(x)y^{(n)} + a_1(x)y^{(n-1)} + \cdots + a_{n-1}(x)y' + a_n(x)y = F(x) \qquad (9.3.1)$$

where $a_0, a_1, \ldots, a_n$, and F are functions specified on an interval I is called a linear differential equation of order n. If F is identically zero on I, then the differential equation is called **homogeneous**; otherwise it is called **nonhomogeneous**. We will assume that $a_0(x)$ is nonzero in I, in which case we can divide (9.3.1) by a_0 and redefine the remaining functions to obtain the **standard form**

$$y^{(n)} + a_1(x)y^{(n-1)} + \cdots + a_{n-1}(x)y' + a_n(x)y = F(x). \qquad (9.3.2)$$

In order to develop the theory for linear differential equations we need to consider the corresponding initial value problem (see Definition 1.1.6, page 8), namely,

$$y^{(n)} + a_1(x)y^{(n-1)} + \cdots + a_{n-1}(x)y' + a_n(x)y = F(x),$$

$$y(x_0) = \alpha_1, \, y'(x_0) = \alpha_2, \, \ldots, \, y^{(n-1)}(x_0) = \alpha_n,$$

where $\alpha_1, \alpha_2, \ldots, \alpha_n$ are arbitrary constants and x_0 is any point in I. The key result that we require is the following existence-uniqueness theorem.

Theorem 9.3.1:   Let $a_1, \ldots, a_n$, F be functions that are continuous on an interval I. Then, for any x_0 in I, the initial value problem

$$y^{(n)} + a_1(x)y^{(n-1)} + \cdots + a_{n-1}(x)y' + a_n(x)y = F(x),$$

$$y(x_0) = \alpha_1, \, y'(x_0) = \alpha_2, \, \ldots, \, y^{(n-1)}(x_0) = \alpha_n,$$

has a unique solution on I.

PROOF   The proof of this theorem requires concepts from advanced calculus and is best left for a second course in differential equations. See, for example, E. A. Coddington and N. Levinson, *Theory of Differential Equations*, McGraw-Hill, 1955. ∎

The differential equation (9.3.2) is said to be **regular** on I if the functions $a_1, \ldots, a_n$, F are continuous on I. In developing the theory for linear differential equations we will always assume that our differential equations are regular on the interval of interest so that the existence-uniqueness theorem can be applied on that interval.

HOMOGENEOUS LINEAR DIFFERENTIAL EQUATIONS

We first consider the nth-order linear homogeneous differential equation

$$y^{(n)} + a_1(x)y^{(n-1)} + \cdots + a_{n-1}(x)y' + a_n(x)y = 0 \qquad (9.3.3)$$

on an interval I. This differential equation can be written as the operator equation

$$Ly = 0,$$

where $L:C^n(I) \to C^0(I)$ is the nth-order linear differential operator

$$L = D^n + a_1 D^{n-1} + \cdots + a_{n-1}D + a_n.$$

If we let S denote the set of all solutions to the differential equation (9.3.3), then we can write

$$S = \{y \in C^n(I) : Ly = 0\},$$

that is,

$$S = \mathrm{Ker}(L).$$

In Chapter 7 we proved that the kernel of any linear operator $T:V \to W$ is a subspace of V. It follows directly from this result that S, the set of all solutions to (9.3.3), is a subspace of $C^n(I)$. We will refer to this subspace as the **solution space** of the differential equation. If we can determine the dimension of S, then we will know how many linearly independent solutions are required to span the solution space. This is dealt with in the following theorem.

Theorem 9.3.2:   The set of all solutions to the regular nth-order homogeneous linear differential equation

$$y^{(n)} + a_1(x)y^{(n-1)} + \cdots + a_{n-1}(x)y' + a_n(x)y = 0$$

on an interval I forms a vector space of dimension n.

PROOF   The given differential equation can be written in operator form as

$$Ly = 0.$$

We have already shown above that the set of all solutions to this differential equation is a vector space. To prove that the dimension of the solution space is n, we must determine a basis consisting of n solutions. For simplicity we provide the details only for the case $n = 2$.

Let y_1, y_2 be the unique solutions of the initial value problems

$$Ly_1 = 0, \qquad y_1(x_0) = 1, \qquad y_1'(x_0) = 0,$$
$$Ly_2 = 0, \qquad y_2(x_0) = 0, \qquad y_2'(x_0) = 1,$$

respectively, where

$$L = D^2 + a_1(x)D + a_0(x).$$

The Wronskian of these solutions at x_0 is $W[y_1, y_2](x_0) = \det(I_2) = 1 \neq 0$, so that the solutions are linearly independent on I.   In order to form a basis for the solution space, y_1 and y_2 must also span the solution space.   Let $y = u(x)$ be any solution of the differential equation $Ly = 0$ on I, and suppose that

$$u(x_0) = c_1, \ u'(x_0) = c_2,$$

where c_1, c_2 are constants.   Then $y = u(x)$ is the unique solution to the initial value problem:

$$\begin{cases} Ly = 0, \\ y(x_0) = c_1, \ y'(x_0) = c_2. \end{cases} \tag{9.3.4}$$

However, if we define

$$w(x) = c_1 y_1(x) + c_2 y_2(x),$$

then $w(x)$ also satisfies the initial value problem (9.3.4).   Thus, by uniqueness, we must have

$$u(x) = w(x),$$

that is,

$$u(x) = c_1 y_1(x) + c_2 y_2(x).$$

Therefore, we have shown that any solution to $Ly = 0$ can be written as a linear combination of the linearly independent solutions y_1, y_2, and hence these solutions do span the solution space.   It follows that $\{y_1, y_2\}$ is a basis for the solution space and, since the basis consists of 2 vectors, the dimension of this solution space is 2. The extension of the above proof to arbitrary n is left as an exercise.   ∎

It follows from the previous theorem and Theorem 6.7.3 (page 222), that *any* set of n *linearly independent* solutions, say $\{y_1, y_2, \ldots, y_n\}$, to

$$y^{(n)} + a_1(x)y^{(n-1)} + \cdots + a_{n-1}(x)y' + a_n(x)y = 0 \tag{9.3.5}$$

forms a basis for the solution space of this differential equation.   Thus any solution of the differential equation can be written as

$$y = c_1 y_1 + c_2 y_2 + \cdots + c_n y_n, \tag{9.3.6}$$

for appropriate constants $c_1, c_2, \ldots, c_n$.   In keeping with our terminology from Chapter 1, we refer to (9.3.6) as the **general solution** of (9.3.5).

Example 9.3.1   Show that $y_1 = e^{3x}$, $y_2 = e^{4x}$ are solutions of

$$y'' - 7y' + 12y = 0 \tag{9.3.7}$$

on $(-\infty, \infty)$ and hence find the general solution.   Also find the solution satisfying the initial conditions $y(0) = 0$, $y'(0) = -1$.

Solution   We first verify that y_1 and y_2 are solutions of the given differential equation on $(-\infty, \infty)$.   Direct substitution yields

$$y_1'' - 7y_1' + 12y_1 = (9 - 21 + 12)e^{3x} = 0,$$

so that $y_1 = e^{3x}$ is a solution of (9.3.7). Similarly,

$$y_2'' - 7y_2' + 12y_2 = (16 - 28 + 12)e^{4x} = 0,$$

which implies that $y_2 = e^{4x}$ is also a solution of (9.3.7). Further, the Wronskian of y_1 and y_2 is $W[y_1, y_2](x) = e^{7x}$, so that y_1 and y_2 are linearly independent on $(-\infty, \infty)$. It follows from Theorem 9.3.2 that a basis for the set of all solutions to the differential equation is $\{e^{3x}, e^{4x}\}$. Consequently the general solution of the differential equation is

$$y = c_1 e^{3x} + c_2 e^{4x}.$$

Imposing the given initial conditions yields the following equations for c_1 and c_2:

$$c_1 + c_2 = 0,$$
$$3c_1 + 4c_2 = -1.$$

Solving for c_1 and c_2 we obtain $c_1 = 1, c_2 = -1$, so that the required unique solution is

$$y(x) = e^{3x} - e^{4x}.$$

In the previous example we were able to determine that the solutions y_1, y_2 were linearly independent on $(-\infty, \infty)$, since their Wronskian was nonzero on $(-\infty, \infty)$. What would have happened if their Wronskian had been identically zero on $(-\infty, \infty)$? Based on Theorem 6.6.1 (page 211), we would not have been able to draw any conclusion as to the linear dependence or linear independence of the solutions. We now show, however, that *when dealing with solutions of an nth-order homogeneous linear differential equation*, if the Wronskian of the solutions is zero at any point in I, then the solutions are linearly dependent on I.

Theorem 9.3.3:   Let $y_1, y_2, \ldots, y_n$, be solutions of the regular nth-order differential equation $Ly = 0$ on an interval I, and let $W[y_1, y_2, \ldots, y_n](x)$ denote their Wronskian. If $W[y_1, y_2, \ldots, y_n](x_0) = 0$ at any point in I, then $y_1, y_2, \ldots, y_n$ are linearly dependent on I.

PROOF   We provide details for the case $n = 2$, and leave the extension to arbitrary n as an exercise. Once more the proof depends on the existence-uniqueness theorem. Let x_0 be a point in I at which $W[y_1, y_2](x_0) = 0$, and consider the linear system

$$c_1 y_1(x_0) + c_2 y_2(x_0) = 0,$$
$$c_1 y_1'(x_0) + c_2 y_2'(x_0) = 0,$$

where the unknowns are c_1, c_2. The determinant of the matrix of coefficients of this system is $W[y_1, y_2](x_0) = 0$, so that the system has nontrivial solutions. Let (α_1, α_2) be one such *nontrivial* solution, and define the function $u(x)$ by

$$u(x) = \alpha_1 y_1(x) + \alpha_2 y_2(x).$$

Then $y = u(x)$ satisfies the initial value problem

$$Ly = 0, \qquad y(x_0) = 0, \qquad y'(x_0) = 0.$$

However, $y(x) = 0$ also satisfies the preceding initial value problem; hence, by uniqueness, we must have $u(x) = 0$—that is,

$$\alpha_1 y_1(x) + \alpha_2 y_2(x) = 0,$$

where at least one of α_1, α_2 is nonzero. Consequently, y_1, y_2 are linearly dependent on I. ∎

Combining this result with Theorem 6.6.1 we have the following corollary.

Corollary 9.3.1: If y_1, y_2, ..., y_n are solutions of $Ly = 0$ on I, then their Wronskian is either identically zero on I (if and only if the solutions are linearly dependent on I) or their Wronskian is *never* zero on I (if and only if the solutions are linearly independent on I).

Thus, the vanishing or nonvanishing of the Wronskian on an interval I completely characterizes the linear dependence or independence on I of *solutions* to $Ly = 0$.

Example 9.3.2 Show that $y_1 = \cos 2x$, $y_2 = 3(1 - 2\sin^2 x)$ are solutions of the differential equation $y'' + 4y = 0$ on $(-\infty, \infty)$. Determine whether they are linearly independent on $(-\infty, \infty)$.

Solution It is easily shown by direct substitution that

$$y_1'' + 4y_1 = 0 \quad \text{and} \quad y_2'' + 4y_2 = 0,$$

so that y_1 and y_2 are solutions of the given differential equation on $(-\infty, \infty)$. To determine whether they are linearly independent on $(-\infty, \infty)$, we compute their Wronskian.

$$W[y_1, y_2](x) = \begin{vmatrix} \cos 2x & 3(1 - 2\sin^2 x) \\ -2\sin 2x & -12\sin x \cos x \end{vmatrix}$$

$$= -6\sin 2x \cos 2x + 6\sin 2x \cos 2x$$

$$= 0,$$

so that, from Theorem 9.3.3, the functions are linearly dependent on $(-\infty, \infty)$. Indeed, since $\cos 2x = 1 - 2\sin^2 x$, we have $y_2 = 3y_1$—that is, $3y_1 - y_2 = 0$. We leave it as an exercise to verify that a second linearly independent solution of the given differential equation is $y_3 = \sin 2x$, so that the general solution of $y'' + 4y = 0$ is

$$y = c_1 \cos 2x + c_2 \sin 2x.$$

NONHOMOGENEOUS LINEAR DIFFERENTIAL EQUATIONS

We now consider the nonhomogeneous linear differential equation

$$y^{(n)} + a_1(x)y^{(n-1)} + \cdots + a_{n-1}(x)y' + a_n(x)y = F(x), \qquad (9.3.8)$$

where $F(x)$ is not identically zero on the interval of interest. If we set $F(x) = 0$ in (9.3.8), we obtain the **associated homogeneous equation**

$$y^{(n)} + a_1(x)y^{(n-1)} + \cdots + a_{n-1}(x)y' + a_n(x)y = 0. \qquad (9.3.9)$$

Equations (9.3.8) and (9.3.9) can be written in operator form as

$$Ly = F \quad \text{and} \quad Ly = 0,$$

respectively, where

$$L = D^n + a_1(x)D^{n-1} + \cdots + a_{n-1}(x)D + a_n(x).$$

The main theoretical result for nonhomogeneous linear differential equations is given in the following theorem.

Theorem 9.3.4:   Let $y_1, y_2, \ldots, y_n$ be linearly independent solutions to (9.3.9) on an interval I, and let $y = y_p$ be any *particular* solution to (9.3.8) on I. Then every solution to (9.3.8) on I is of the form

$$y = c_1 y_1 + c_2 y_2 + \cdots + c_n y_n + y_p$$

for appropriate constants $c_1, c_2, \ldots, c_n$.

PROOF   Since $y = y_p$ satisfies (9.3.8) we have

$$Ly_p = F. \qquad (9.3.10)$$

Let $y = u$ be any solution of (9.3.8).   Then

$$Lu = F. \qquad (9.3.11)$$

Subtracting (9.3.10) from (9.3.11) and using the fact that L is a linear operator yields

$$L(u - y_p) = 0.$$

Thus $y = u - y_p$ is a solution of the associated homogeneous equation $Ly = 0$ and therefore can be written as

$$u - y_p = c_1 y_1 + c_2 y_2 + \cdots + c_n y_n$$

for appropriately chosen $c_1, c_2, \ldots, c_n$.   Consequently,

$$u = c_1 y_1 + c_2 y_2 + \cdots + c_n y_n + y_p. \qquad \blacksquare$$

According to the previous theorem, the **general solution** of the nonhomogeneous differential equation $Ly = F$ is of the form

$$y = y_c + y_p,$$

where

$$y_c = c_1 y_1 + c_2 y_2 + \cdots + c_n y_n$$

is the general solution of the associated homogeneous equation $Ly = 0$ and y_p is a particular solution of $Ly = F$. We refer to y_c as the **complementary function** for $Ly = F$.

Example 9.3.3 Find the general solution of

$$y'' - 4y = 32. \tag{9.3.12}$$

Solution It is easily verified that $y_1 = e^{2x}$, $y_2 = e^{-2x}$ are linearly independent solutions to the associated homogeneous equation $y'' - 4y = 0$, so that the complementary function is

$$y_c = c_1 e^{2x} + c_2 e^{-2x}.$$

Further, we see by inspection that $y_p = -8$ is a particular solution to (9.3.12). Consequently the general solution to (9.3.12) is

$$y = y_c + y_p = c_1 e^{2x} + c_2 e^{-2x} - 8.$$

The following example gives an illustration of a solution technique that we will develop in the next several sections for solving certain types of linear differential equations.

Example 9.3.4 Consider the differential equation

$$y'' - 2y' - 3y = 15e^{4x}. \tag{9.3.13}$$

(a) Determine all values of the constant r such that $y = e^{rx}$ is a solution of the associated homogeneous equation. Hence find the complementary function for (9.3.13).

(b) Determine the value of the constant A_0 such that $y_p = A_0 e^{4x}$ is a particular solution of (9.3.13).

(c) Find the general solution of (9.3.13) on $(-\infty, \infty)$.

Solution

(a) Differentiating $y = e^{rx}$ twice with respect to x yields

$$y' = re^{rx}, \qquad y'' = r^2 e^{rx}.$$

Thus, $y'' - 2y' - 3y = 0$ if and only if

$$e^{rx}(r^2 - 2r - 3) = 0.$$

Consequently, r must satisfy

$$r^2 - 2r - 3 = 0,$$

that is,

$$(r - 3)(r + 1) = 0.$$

It follows that $r = 3$ or -1, and so the corresponding solutions to the associated homogeneous equation are

$$y_1 = e^{3x}, \qquad y_2 = e^{-x}.$$

Further,

$$W[y_1, y_2](x) = \begin{vmatrix} e^{3x} & e^{-x} \\ 3e^{3x} & -e^{-x} \end{vmatrix} = -4e^{2x}.$$

Since this is nonzero, the solutions are linearly independent on $(-\infty, \infty)$, so that the general solution of the associated homogeneous equation is

$$y_c = c_1 e^{3x} + c_2 e^{-x}.$$

This is the complementary function for (9.3.13).
  (b) Differentiating $y_p = A_0 e^{4x}$ twice yields

$$y_p' = 4A_0 e^{4x}, \qquad y_p'' = 16A_0 e^{4x}.$$

Substituting into (9.3.13) it follows that y_p is indeed a solution, provided

$$5A_0 e^{4x} = 15e^{4x},$$

so that

$$A_0 = 3.$$

Consequently, a particular solution to (9.3.13) is

$$y_p = 3e^{4x}.$$

  (c) The general solution of (9.3.13) on $(-\infty, \infty)$ is

$$y = y_c + y_p = c_1 e^{3x} + c_2 e^{-x} + 3e^{4x}.$$

Finally, we will require the following result.

Theorem 9.3.5:   If $y = u_p$ and $y = v_p$ are particular solutions of $Ly = f(x)$, and $Ly = g(x)$, respectively, then $y = u_p + v_p$ is a solution of $Ly = f(x) + g(x)$.

PROOF

$$L(u_p + v_p) = Lu_p + Lv_p = f(x) + g(x). \qquad \blacksquare$$

We have now derived the fundamental theory for linear differential equations. In the remainder of the chapter we focus our attention on developing techniques for finding the solutions whose existence is guaranteed by our theory.

EXERCISES 9.3

In problems 1 and 2 write the given nonhomogeneous differential equation as an operator equation, and give the associated homogeneous differential equation.

1. $y''' + x^2 y'' - (\sin x)y' + e^x y = x^3$.

2. $y'' + 4xy' - 6x^2 y = x^2 \sin x$.

3. Use the existence-uniqueness theorem to prove that the only solution to the initial value problem

$$y'' + x^2 y' + e^x y = 0, \qquad y(0) = 0, \ y'(0) = 0,$$

is the trivial solution $y(x) = 0$.

4. Use the existence-uniqueness theorem to formulate and prove a general theorem regarding the solution of the initial value problem

$$Ly = 0, \ y(x_0) = 0, \ y'(x_0) = 0, \ \ldots, \ y^{(n-1)}(x_0) = 0.$$

In problems 5–7 show that the given functions are solutions of the given differential equation on $(-\infty, \infty)$ and use them to determine a basis for the solution space of the differential equation. Hence obtain the general solution of the given differential equation.

5. $y'' - y' - 6y = 0$, $y_1 = e^{3x}$, $y_2 = e^{-2x}$.

6. $y'' + 4y = 0$, $y_1 = \cos 2x$, $y_2 = \sin 2x$.

7. $y'' + 2y' + y = 0$, $y_1 = e^{-x}$, $y_2 = xe^{-x}$, $y_3 = e^{-x}(x + 2)$.

8. Determine all values of the constant r such that $y = e^{rx}$ is a solution of the differential equation

$$y'' - 4y' + 3y = 0. \qquad \text{(i)}$$

Hence find the general solution of (i) on $(-\infty, \infty)$.

9. Determine all values of the constant r such that $y = x^r$ is a solution of

$$x^2 y'' + 3xy' - 8y = 0, \qquad x > 0. \qquad \text{(ii)}$$

Hence find the general solution of (ii) on $(0, \infty)$.

10. Consider the differential equation

$$y'' + y' - 6y = 18e^{3x}. \qquad \text{(iii)}$$

(a) Determine all values of the constant r such that $y = e^{rx}$ is a solution of the associated homogeneous equation on $(-\infty, \infty)$. Hence determine the complementary function for (iii).

(b) Determine the value of the constant A_0 such that $y_p = A_0 e^{3x}$ is a particular solution of (iii).
(c) Use your results from (a) and (b) to determine the general solution to (iii).

11. Consider the differential equation

$$y'' + y' - 2y = 4x^2. \qquad \text{(iv)}$$

(a) Determine all values of the constant r such that $y = e^{rx}$ is a solution of the associated homogeneous equation on $(-\infty, \infty)$. Hence determine the complementary function for (iv).
(b) Determine the values of the constants a_0, a_1, a_2 such that

$$y_p = a_0 + a_1 x + a_2 x^2$$

is a particular solution of (iv).
(c) Use your results from (a) and (b) to determine the general solution to (iv).

12. Let y_1 and y_2 be solutions of

$$y'' + a_1(x)y' + a_2(x)y = 0,$$

on an interval I. Show that the Wronskian $W(x) = y_1 y_2' - y_2 y_1'$ satisfies the first-order linear differential equation

$$W'(x) + a_1(x)W(x) = 0,$$

and hence show that (Abel's formula)

$$W(x) = c \, \exp\left[-\int a_1(x) \, dx\right].$$

It follows that $W(x)$ is either identically zero on I ($c = 0$) or never zero on I ($c \neq 0$).

13. Extend the proof of Theorem 9.3.2 to arbitrary n.

14. Extend the proof of Theorem 9.3.3 to arbitrary n.

15. Let $T:V \to W$ be a linear transformation, and suppose that $\{v_1, v_2, \ldots, v_n\}$ is a basis for $\text{Ker}(T)$. Prove that every solution of the operator equation

$$T(v) = w \qquad \text{(v)}$$

is of the form

$$v = c_1 v_1 + c_2 v_2 + \cdots + c_n v_n + v_p,$$

where v_p is any particular solution of (v).

9.4   LINEAR DIFFERENTIAL EQUATIONS WITH CONSTANT COEFFICIENTS

In the next few sections we develop techniques for solving linear differential equations that have constant coefficients. These are differential equations that can be written in the form

$$y^{(n)} + a_1 y^{(n-1)} + \cdots + a_{n-1} y' + a_n y = F(x),$$

where $a_1, a_2, \ldots, a_n$ are real *constants*. To determine the general solution of this differential equation we will require its complementary function. We therefore begin by analyzing the associated homogeneous equation

$$y^{(n)} + a_1 y^{(n-1)} + \cdots + a_{n-1} y' + a_n y = 0,$$

which can be written as the operator equation

$$P(D)y = 0, \tag{9.4.1}$$

where $P(D)$ is the polynomial differential operator

$$P(D) = D^n + a_1 D^{n-1} + \cdots + a_{n-1} D + a_n. \tag{9.4.2}$$

To motivate a solution technique for (9.4.1), we note that the first-order differential equation

$$(D + a)y = 0$$

has general solution (separate the variables)

$$y = c_1 e^{-ax},$$

where c_1 is an arbitrary constant. This suggests that the differential equation (9.4.1) may also have such solutions. In order to investigate this possibility we consider the effect that the polynomial differential operator (9.4.2) has on a function of the form $y = e^{rx}$. Using the linearity of $P(D)$ it follows that

$$P(D)(e^{rx}) = D^n(e^{rx}) + a_1 D^{n-1}(e^{rx}) + \cdots + a_{n-1} D(e^{rx}) + a_n(e^{rx}).$$

Thus, since $D^k(e^{rx}) = r^k e^{rx}$, we can write

$$P(D)(e^{rx}) = e^{rx}(r^n + a_1 r^{n-1} + \cdots + a_{n-1} r + a_n) = e^{rx} P(r), \tag{9.4.3}$$

where

$$P(r) = r^n + a_1 r^{n-1} + \cdots + a_{n-1} r + a_n.$$

The real polynomial $P(r)$ is called the **auxiliary polynomial** associated with the differential operator $P(D)$.[1] It can be determined directly from $P(D)$ by simply replacing D by r. However, the distinction between $P(D)$ and $P(r)$ must be kept clearly in mind. The former defines a *linear differential operator*, whereas the latter defines a *polynomial function*. It follows from (9.4.3) that $y = e^{rx}$ is a solution of the

---

[1] Here *real polynomial* means a polynomial with real constant coefficients.

differential equation $P(D)y = 0$ whenever r is a zero of the auxiliary polynomial—that is, a root of the **auxiliary equation**

$$P(r) = r^n + a_1 r^{n-1} + \cdots + a_{n-1} r + a_n = 0.$$

We state this fundamental result in a theorem.

Theorem 9.4.1:   The function $y = e^{ax}$ is a solution of the differential equation $P(D)y = 0$ whenever $r = a$ is a real root of the auxiliary equation $P(r) = 0$.

Example 9.4.1   Determine the general solution of the differential equation

$$(D^2 - D - 2)y = 0.$$

Solution   The auxiliary polynomial is

$$P(r) = r^2 - r - 2 = (r - 2)(r + 1).$$

It follows that $r_1 = 2$, $r_2 = -1$ are two roots of the auxiliary equation; hence, from Theorem 9.4.1, two solutions of the differential equation are

$$y_1 = e^{2x}, \qquad y_2 = e^{-x}.$$

Further, the Wronskian of these solutions is

$$W[y_1, y_2](x) = \begin{vmatrix} e^{2x} & e^{-x} \\ 2e^{2x} & -e^{-x} \end{vmatrix} = -3e^x.$$

Since this Wronskian is never zero the solutions are linearly independent on any interval, and hence $\{y_1, y_2\}$ is a basis for the set of all solutions.   It follows that the general solution of the differential equation is

$$y = c_1 e^{2x} + c_2 e^{-x}.$$

The auxiliary polynomial corresponding to the differential operator

$$P(D) = D^n + a_1 D^{n-1} + \cdots + a_{n-1} D + a_n$$

is

$$P(r) = r^n + a_1 r^{n-1} + \cdots + a_{n-1} r + a_n.$$

This can be factored as

$$P(r) = (r - r_1)(r - r_2) \cdots (r - r_n), \tag{9.4.4}$$

where $r_1, r_2, \ldots, r_n$ are the roots of the auxiliary equation. Since $P(r)$ corresponds to a unique polynomial differential operator $P(D)$, it follows that $P(D)$ can also be factored as

$$P(D) = (D - r_1)(D - r_2) \cdots (D - r_n), \tag{9.4.5}$$

and, since polynomial differential operators commute, the order of the factors does not matter.   In general some (or all) of the r_i may be complex, in which case (9.4.4) and (9.4.5) will contain complex linear terms.   We recall, however, that complex roots of a real polynomial always occur in conjugate pairs.   We now use this fact to show that, by combining any complex conjugate linear terms, $P(D)$ can be factored into a product of real linear and irreducible quadratic terms.[1]

Theorem 9.4.2:   Any polynomial differential operator can be factored into linear and irreducible quadratic terms with *real* coefficients.

PROOF   Complex coefficients can only occur when we have a complex root in the auxiliary equation.   We show that this situation gives rise to quadratic factors with real coefficients.   Suppose α is a complex root of the auxiliary equation.   Then so also is $\overline{\alpha}$.   Consequently (9.4.5) will contain the factor $(D - \alpha)(D - \overline{\alpha})$.   But, writing $\alpha = a + ib$, we have

$$(D - \alpha)(D - \overline{\alpha}) = [D - (a + ib)][D - (a - ib)] = D^2 - 2aD + (a^2 + b^2),$$

which is an irreducible quadratic factor with real coefficients.   ∎

Example 9.4.2   Factor the polynomial differential operator

$$P(D) = D^3 - 3D^2 + 7D - 5$$

into linear and irreducible quadratic terms.

Solution   The auxiliary polynomial is $P(r) = r^3 - 3r^2 + 7r - 5$.   By inspection we see that $P(1) = 0$, so that $r = 1$ is a zero of P.   It follows that $(r - 1)$ is a factor of $P(r)$, and it is easily shown that

$$P(r) = (r - 1)(r^2 - 2r + 5).$$

The discriminant of the quadratic factor is $-16 < 0$, so that this term is irreducible. It follows that the given polynomial differential operator can be factored as

$$P(D) = (D - 1)(D^2 - 2D + 5)$$

and that it cannot be reduced to a product of *real* linear factors.

If the auxiliary equation $P(r) = 0$ has n *real* and *distinct* roots, say $r_1, r_2, \ldots,$ r_n, then Theorem 9.4.1 implies that $e^{r_1 x}, e^{r_2 x}, \ldots, e^{r_n x}$ are n solutions to $P(D)y = 0$, and it can be shown (see problem 24) that these solutions are linearly independent on any interval.   We can therefore state the following theorem.

Theorem 9.4.3:   Let $P(D)$ be a polynomial differential operator of order n.   If the auxiliary equation $P(r) = 0$ has n real and distinct roots $r_1, r_2, \ldots, r_n$, then $e^{r_1 x}$,

---

[1] Recall that the quadratic $ax^2 + bx + c$ has **discriminant** $b^2 - 4ac$ and that the roots of the quadratic are given by

$$x = \frac{-b \pm \sqrt{b^2 - 4ac}}{2a}.$$

If $b^2 - 4ac < 0$, then the quadratic has no real factors and is said to be **irreducible**.

$e^{r_2 x}, \ldots, e^{r_n x}$ are solutions to $P(D)y = 0$, which are linearly independent on any interval. Hence the general solution to $P(D)y = 0$ in this case is

$$y = c_1 e^{r_1 x} + c_2 e^{r_2 x} + \cdots + c_n e^{r_n x}.$$

Example 9.4.3 Find the general solution to

$$D(D + 5)(D - 7)y = 0. \tag{9.4.6}$$

Solution The auxiliary equation is $r(r + 5)(r - 7) = 0$, with roots $r_1 = 0, r_2 = -5$, $r_3 = 7$. It follows directly from Theorem 9.4.3 that $y_1 = 1, y_2 = e^{-5x}, y_3 = e^{7x}$ are linearly independent solutions to (9.4.6), so that the general solution is

$$y = c_1 + c_2 e^{-5x} + c_3 e^{7x}.$$

Example 9.4.4 Determine a basis for the solution space to

$$y''' - y'' - 14y' + 24y = 0, \tag{9.4.7}$$

and hence find the general solution.

Solution In this case the auxiliary polynomial is

$$P(r) = r^3 - r^2 - 14r + 24.$$

By trial and error we find that $P(2) = 0$, so that $(r - 2)$ is a factor of $P(r)$. Consequently we can write

$$P(r) = (r - 2)(r^2 + r - 12).$$

The quadratic term can also be factored to yield

$$P(r) = (r - 2)(r - 3)(r + 4).$$

The roots of the auxiliary equation are $r_1 = 2, r_2 = 3, r_3 = -4$, and hence Theorem 9.4.3 implies that a basis for the solution space of the differential equation (9.4.7) is $\{e^{2x}, e^{3x}, e^{-4x}\}$. Hence, (9.4.7) has general solution

$$y = c_1 e^{2x} + c_2 e^{-4x} + c_3 e^{3x}.$$

Notice in the preceding examples we did not have to check linear independence of the solutions, since that is guaranteed by the previous theorem.

If the auxiliary equation corresponding to a polynomial differential operator of order n, $P(D)$, has *fewer* than n distinct real roots, then Theorem 9.4.1 will not yield enough solutions to span the solution space of the differential equation $P(D)y = 0$. This will occur when a real root has multiplicity greater than one, or when $P(r)$ has complex roots. In order to deal with the complex case, we must first introduce some results about complex-valued functions. This is the subject of the next section. In Section 9.6 we will return to give a unified treatment of the complete solution of the differential equation

$$P(D)y = 0$$

in all cases.

EXERCISES 9.4

In problems 1 and 2 determine the polynomial differential operator $P(D)$ such that the given differential equation can be written as $P(D)y = 0$.

1. $y'' - 3y' + 6y = 0$.

2. $y''' - 2y'' + 4y = 0$.

In problems 3–6 find the auxiliary polynomial for the given polynomial differential operator and factor $P(D)$ into linear and irreducible quadratic terms.

3. $P(D) = D^2 + D - 12$.

4. $P(D) = D^3 + D^2 - 6D$.

5. $P(D) = D^3 - 3D^2 + 3D - 1$.

6. $P(D) = D^3 - 3D^2 + 4D - 2$.

In problems 7–9 determine the polynomial differential operator of lowest order, with real coefficients, whose auxiliary polynomial has the following numbers among its zeros.

7. $r_1 = 1$, $r_2 = -2$.

8. $r_1 = 3$, $r_2 = 1$, $r_3 = 0$.

9. $r_1 = 1$, $r_2 = 2 + i$.

In problems 10–17 use Theorem 9.4.3 to determine the general solution of the given differential equation.

10. $(D^2 - 1)y = 0$.

11. $(D^2 - D - 6)y = 0$.

12. $(D + 4)(D - 2)(D + 3)y = 0$.

13. $y'' - 5y' + 6y = 0$.

14. $y'' - 3y' - 4y = 0$.

15. $D(D - 1)y = 0$.

16. $y''' - y'' - 4y' + 4y = 0$.

17. $(D + 4)(D - 2)(D^2 - 9)y = 0$.

In problems 18–20 determine the constant coefficient linear homogeneous differential equation of lowest order whose solution space is spanned by the given functions.

18. $y_1 = e^{2x}$, $y_2 = e^{3x}$.

19. $y_1 = e^x$, $y_2 = e^{-x}$.

20. $y_1 = e^{2x}$, $y_2 = e^{-2x}$, $y_3 = e^{3x}$.

21. Determine a differential equation that has general solution $y = c_1 e^{3x} + c_2 e^{-4x}$, where c_1 and c_2 are arbitrary constants.

22. Try to find the general solution to $y'' - 2y' + y = 0$.

23. (a) Consider the differential equation

$$y'' + by' + cy = 0, \qquad (i)$$

where b and c are constants. Suppose that the auxiliary equation has the two real and distinct roots r_1 and r_2. What conditions must these roots satisfy in order that every solution of (i) satisfies

$$\lim_{x \to \infty} y(x) = 0?$$

(b) Formulate a similar result to that in (a) for the general nth-order differential equation $P(D)y = 0$.

24. Let $r_1, r_2, \ldots, r_n$ be *distinct* real numbers. Prove that the set of functions

$$\{e^{r_1 x}, e^{r_2 x}, \ldots, e^{r_n x}\}$$

is linearly independent on any interval. [*Hint:* The condition for determining linear dependence or linear independence is

$$c_1 e^{r_1 x} + c_2 e^{r_2 x} + \cdots + c_n e^{r_n x} = 0.$$

Divide by $e^{r_1 x}$ and differentiate the result to eliminate c_1. Continue in this manner to obtain $(r_n - r_1)(r_n - r_2) \cdots (r_n - r_{n-1})c_n = 0$.]

9.5   COMPLEX-VALUED FUNCTIONS

We saw in the previous section that $y = e^{rx}$ is a solution of the homogeneous differential equation $P(D)y = 0$ whenever r is a real root of the auxiliary equation $P(r) = 0$. In order to extend this result to the case when r is a complex root of $P(r) = 0$, we must first introduce some results from the theory of complex-valued functions. No attempt at rigor is made. We are essentially just listing the results that will be required in order to determine the solutions of our differential equations.

A function $w(x)$ of the form

$$w(x) = u(x) + iv(x)$$

where u and v are real-valued functions of a real variable x (and $i^2 = -1$) is called a **complex-valued function** of a real variable ($w:\mathbf{R} \to \mathbf{C}$). An example of such a function is

$$w(x) = 3 \cos 2x + 4i \sin 3x.$$

It is easily shown that the set of all complex-valued functions defined on an interval I forms a vector space (addition and scalar multiplication are defined in the usual manner).

THE COMPLEX EXPONENTIAL FUNCTION

The extension of Theorem 9.4.1 to the case of complex roots of $P(r) = 0$ will be straightforward once we have attached a meaning to a function of the form e^{rx}, where r is a complex number. We now indicate how such a function is defined.

Recall that for all real x, the function e^x has the Maclaurin expansion

$$e^x = \sum_{n=0}^{\infty} \frac{1}{n!} x^n.$$

It is also possible to discuss convergence of infinite series of complex numbers. We define e^{ib}, where b is a real number, by

$$e^{ib} = \sum_{n=0}^{\infty} \frac{1}{n!} (ib)^n = \left[1 + ib + \frac{1}{2!} (ib)^2 + \frac{1}{3!} (ib)^3 + \cdots + \frac{1}{n!} (ib)^n + \cdots \right].$$

Factoring the even and odd powers of b and using the formulas

$$i^{2k} = (-1)^k, \qquad i^{2k+1} = (-1)^k i,$$

yields

$$e^{ib} = \left[1 - \frac{1}{2!} b^2 + \frac{1}{4!} b^4 + \cdots + \frac{(-1)^k}{(2k)!} b^{2k} + \cdots \right]$$
$$+ i \left[b - \frac{1}{3!} b^3 + \frac{1}{5!} b^5 + \cdots + \frac{(-1)^k}{(2k+1)!} b^{2k+1} + \cdots \right],$$

that is,

$$e^{ib} = \underbrace{\sum_{n=0}^{\infty} (-1)^n \frac{b^{2n}}{(2n)!}}_{\cos b} + i \underbrace{\sum_{n=0}^{\infty} (-1)^n \frac{b^{2n+1}}{(2n+1)!}}_{\sin b}.$$

The two series appearing in this equation are, respectively, the Maclaurin series expansions of $\cos b$ and $\sin b$, which converge for all real b. Thus we have shown that

$$\boxed{e^{ib} = \cos b + i \sin b}\,,$$

(9.5.1)

which is called **Euler's formula**.   It is now natural to *define* e^{a+ib} by

$$e^{a+ib} = e^{a}e^{ib} = e^{a}(\cos b + i \sin b),$$

(9.5.2)

where a and b are any real numbers.

A function of the form $f(x) = e^{rx}$, where $r = a + ib$ and x is a real variable, is called a **complex exponential function**.   Replacing ib by ibx in (9.5.1) and $a + ib$ by $(a + ib)x$ in (9.5.2) yields the following important formulas:

$$e^{ibx} = \cos bx + i \sin bx, \qquad e^{(a+ib)x} = e^{ax}(\cos bx + i \sin bx).$$

Replacing i with $-i$ in the above formulas, we obtain:

$$e^{-ibx} = \cos bx - i \sin bx, \qquad e^{(a-ib)x} = e^{ax}(\cos bx - i \sin bx).$$

Example 9.5.1   Express $e^{(3-5i)x}$ in terms of trigonometric functions.

Solution

$$e^{(3-5i)x} = e^{3x}(\cos 5x - i \sin 5x).$$

The preceding definition of $e^{(a+ib)x}$ also enables us to attach a meaning to x^{a+ib}.   We recall that for nonrational r, and positive x, x^r is defined by

$$x^r = e^{r \ln x}.$$

We now extend this definition to the case when r is complex and therefore define

$$\boxed{x^{a+ib} = e^{(a+ib)\ln x}.}$$

Using Euler's formula, this can be written as

$$x^{a+ib} = x^{a}e^{ib \ln x} = x^{a}[\cos(b \ln x) + i \sin(b \ln x)].$$

For example,

$$x^{2+3i} = x^{2}[\cos(3 \ln x) + i \sin(3 \ln x)].$$

DIFFERENTIATION OF COMPLEX-VALUED FUNCTIONS

We now return to the general complex-valued function $w(x) = u(x) + iv(x)$.   If $u'(x)$ and $v'(x)$ exist, then we define the derivative of w by

$$w'(x) = u'(x) + iv'(x).$$

Higher-order derivatives are defined similarly.

In particular, we have the following important result.

$$\frac{d}{dx}(e^{rx}) = re^{rx} \text{ when } r \text{ is complex.}$$

This coincides with the usual formula for the derivative of e^{rx} when r is a real number.   To prove this formula, we proceed as follows.   If $r = a + ib$, then

$$e^{rx} = e^{(a+ib)x} = e^{ax}(\cos bx + i \sin bx).$$

Differentiating with respect to x using the product rule yields:

$$\frac{d}{dx}(e^{rx}) = ae^{ax}(\cos bx + i \sin bx) + be^{ax}(-\sin bx + i \cos bx)$$
$$= ae^{ax}(\cos bx + i \sin bx) + ibe^{ax}(\cos bx + i \sin bx)$$
$$= (a + ib)e^{ax}(\cos bx + i \sin bx) = re^{rx},$$

as required.

Similarly, it can be shown that

$$\frac{d}{dx}(x^r) = rx^{r-1} \text{ when } r \text{ is complex.}$$

We now introduce two results about complex-valued solutions to linear differential equations.   These will be required in the remaining sections.   As usual, we let

$$L = D^n + a_1(x)D^{n-1} + \cdots + a_{n-1}(x)D + a_n(x)$$

where $a_1, a_2, \ldots, a_n$ are real-valued functions (this includes the case of polynomial differential operators).

Theorem 9.5.1:   Let $y(x) = u(x) + iv(x)$ be a complex-valued solution to $Ly = f(x) + ig(x)$ on an interval I.   Then

$$Lu = f(x) \quad \text{and} \quad Lv = g(x),$$

on I.

PROOF   By assumption,

$$L(u + iv) = f(x) + ig(x),$$

that is, using the linearity of L,

$$Lu + iLv = f(x) + ig(x).$$

Equating real and imaginary parts on both sides of this equation yields

$$Lu = f(x) \qquad Lv = g(x),$$

as required.   ∎

Applying the preceding theorem in the special case when $f = g = 0$ yields the following corollary.

Corollary 9.5.1:   The real and imaginary parts of a complex-valued solution to $Ly = 0$ on an interval I are themselves real-valued solutions to this differential equation on I.

Example 9.5.2   Show that $y = \cos x + i \sin x$ is a solution of the differential equation $y'' + y = 0$, and hence determine the general solution.

Solution   By direct substitution it is easily shown that $y'' + y = 0$. Taking the real and imaginary parts of this complex-valued solution, it follows that $y_1 = \cos x$ and $y_2 = \sin x$ are two real-valued solutions. Since

$$W[y_1, y_2](x) = \begin{vmatrix} \cos x & \sin x \\ -\sin x & \cos x \end{vmatrix} = 1 \neq 0,$$

the solutions are linearly independent on any interval and hence the general solution to $y'' + y = 0$ is $y = c_1 \cos x + c_2 \sin x$.

Example 9.5.3   Determine all values of the constant r such that $y = e^{rx}$ is a solution of

$$y'' + 2y' + 10y = 0. \tag{9.5.3}$$

Hence find the general solution.

Solution   Substituting $y = e^{rx}$ into (9.5.3) yields the auxiliary equation

$$r^2 + 2r + 10 = 0.$$

This has roots

$$r = -1 \pm 3i.$$

Consider the root $r = -1 + 3i$. The corresponding complex-valued solution to (9.5.3) is

$$y = e^{(-1+3i)x} = e^{-x}(\cos 3x + i \sin 3x).$$

Taking the real and imaginary parts of this complex-valued solution yields the two real-valued solutions

$$y_1 = e^{-x}\cos 3x, \qquad y_2 = e^{-x}\sin 3x.$$

The Wronskian of these solutions is

$$W[y_1, y_2](x) = \begin{vmatrix} e^{-x}\cos 3x & e^{-x}\sin 3x \\ -e^{-x}(\cos 3x + 3\sin 3x) & e^{-x}(3\cos 3x - \sin 3x) \end{vmatrix} = 3e^{-2x} \neq 0$$

so that y_1 and y_2 are linearly independent on any interval. Consequently the general solution of (9.5.3) is

$$y = c_1 e^{-x}\cos 3x + c_2 e^{-x}\sin 3x = e^{-x}(c_1 \cos 3x + c_2 \sin 3x).$$

Finally we remark that Theorem 6.6.1 (page 211) is also valid for complex-valued functions. Thus a set of complex-valued functions $\{f_1, f_2, \ldots, f_k\}$ is linearly independent on an interval I if their Wronskian is nonzero at any point in I.

EXERCISES 9.5

In problems 1–8 express the given complex-valued function in the form $f(x) + ig(x)$ for appropriate real-valued functions f and g.

1. e^{2ix}.

2. $e^{(3+4i)x}$.

3. e^{-5ix}.

4. $e^{-(2+i)x}$.

5. x^{2-i}.

6. x^{3i}.

7. x^{-1+2i}.

8. $x^{2i}e^{(3+4i)x}$.

9. Derive the famous mathematical formula $e^{i\pi} + 1 = 0$.

10. Show that

$$\cos bx = \tfrac{1}{2}(e^{ibx} + e^{-ibx})$$

and

$$\sin bx = \tfrac{1}{2i}(e^{bx} - e^{-ibx}).$$

[A comparison of these formulas with the corresponding formulas

$$\cosh bx = \tfrac{1}{2}(e^{bx} + e^{-bx}),$$

$$\sinh bx = \tfrac{1}{2}(e^{bx} - e^{-bx}),$$

indicates why the trigonometric and hyperbolic functions satisfy similar identities.]

In problems 11–13 use the result of problem 10 to express the given functions in terms of complex exponential functions.

11. $\sin 4x$.

12. $\cos 8x$.

13. $\tan x$.

14. Use the result of problem 10 to verify the identity $\sin^2 x + \cos^2 x = 1$.

15. Show that $y = e^{(1+2i)x}$ is a complex-valued solution of

$$y'' - 2y' + 5y = 0. \qquad (i)$$

Hence obtain two linearly independent real-valued solutions of (i) and thereby find its general solution.

In problems 16–19 determine all values (real or complex) of the constant r for which $y = e^{rx}$ is a solution of the given differential equation. Hence obtain *linearly independent* real-valued solutions of the differential equation on $(-\infty, \infty)$.

16. $y'' + y = 0$.

17. $y'' - 4y' + 5y = 0$.

18. $y'' + 6y' + 13y = 0$.

19. $(D-1)(D^2+4)y = 0$.

In problems 20–23 determine all values of the constant r such that $y = x^r$ is a solution of the given differential equation. Hence find two *linearly independent* real-valued solutions on $(0, \infty)$.

20. $x^2y'' + xy' + 4y = 0$.

21. $x^2y'' + xy' + 25y = 0$.

22. $x^2y'' - xy' + 10y = 0$.

23. $x^2y'' + 5xy' + 8y = 0$.

24. Determine the value of the constant A_0 such that $y = A_0e^{ix}$ is a complex-valued solution of

$$y'' + 4y = 6e^{ix}.$$

Using the result from Theorem 9.5.1, find a real-valued solution to each of:

$$y'' + 4y = 6\cos x, \qquad y'' + 4y = 6\sin x.$$

25. Prove that the set of all complex-valued functions defined on an interval I forms a vector space.

9.6   SOLUTION OF CONSTANT COEFFICIENT LINEAR HOMOGENEOUS DIFFERENTIAL EQUATIONS

In this section we develop fully the solution technique introduced in Section 9.4 for solving a differential equation of the form

$$P(D)y = 0, \tag{9.6.1}$$

where $P(D)$ is a polynomial differential operator.   We begin by extending Theorem 9.4.1 to the case of complex roots of the auxiliary equation.

Theorem 9.6.1:   If r is a (real or complex) root of $P(r) = 0$, then a solution of the differential equation $P(D)y = 0$ is $y = e^{rx}$.

PROOF   The proof is identical to that of Theorem 9.4.1, since $D^k(e^{rx}) = r^k e^{rx}$ holds when r is complex.   ■

We first restrict our attention to second-order differential equations, since they can be analyzed very simply and are the equations of most interest in applications.   The generalization of the solution technique to arbitrary order is straightforward.
    Consider the differential equation

$$y'' + a_1 y' + a_2 y = 0, \tag{9.6.2}$$

where a_1 and a_2 are real constants.   The auxiliary equation is $r^2 + a_1 r + a_2 = 0$, with roots

$$r_1 = \frac{-a_1 + \sqrt{a_1^2 - 4a_2}}{2}, \qquad r_2 = \frac{-a_1 - \sqrt{a_1^2 - 4a_2}}{2}.$$

Thus (9.6.2) can be written in factored form as

$$(D - r_1)(D - r_2)y = 0. \tag{9.6.3}$$

Three subcases arise depending on whether r_1 and r_2 are real and distinct, real and equal, or complex conjugate.   We consider these three subcases separately below.

CASE 1: REAL AND DISTINCT ROOTS   If r_1 and r_2 are real and distinct ($r_1 \neq r_2$), it follows directly from Theorem 9.4.3 that a basis for the solution space is $\{e^{r_1 x}, e^{r_2 x}\}$, and hence the general solution to (9.6.2) is

$$\boxed{y = c_1 e^{r_1 x} + c_2 e^{r_2 x}.}$$

CASE 2: REPEATED REAL ROOTS   In this case $r_1 = r_2$, and (9.6.3) reduces to

$$(D - r_1)^2 y = 0. \tag{9.6.4}$$

It follows from Theorem 9.6.1 that one solution to (9.6.4) is

$$y_1 = e^{r_1 x}. \tag{9.6.5}$$

If y_2 is a second linearly independent solution to (9.6.4), then the ratio y_2/y_1 must be nonconstant; that is, we must have

$$y_2 = u(x)y_1,$$

for some nonconstant function $u(x)$. We now determine $u(x)$ by substituting into (9.6.4).[1] We have

$$(D - r_1)^2(uy_1) = (D - r_1)(D - r_1)(ue^{r_1 x}) = (D - r_1)(u'e^{r_1 x}) = u''e^{r_1 x}.$$

Thus, $y_2 = u(x)y_1$ is a solution of (9.6.4) provided u satisfies

$$u'' = 0,$$

which implies that

$$u = A + Bx,$$

where A and B are constants. Choosing $A = 0$, $B = 1$, we obtain the solution

$$y_2 = xe^{r_1 x}.$$

Further, the Wronskian of y_1 and y_2 is

$$W[y_1, y_2](x) = \begin{vmatrix} e^{r_1 x} & xe^{r_1 x} \\ r_1 e^{r_1 x} & e^{r_1 x}(1 + r_1 x) \end{vmatrix} = e^{2r_1 x} \neq 0,$$

so that $\{y_1, y_2\}$ is linearly independent on any interval. Consequently a basis for the solution space of the differential equation (9.6.4) is $\{e^{r_1 x}, xe^{r_1 x}\}$, and the general solution is, therefore,

$$\boxed{y = c_1 e^{r_1 x} + c_2 x e^{r_1 x}.}$$

CASE 3: COMPLEX CONJUGATE ROOTS   In this case $r_2 = \bar{r}_1$, so that (9.6.3) can be written as

$$(D - r_1)(D - \bar{r}_1)y = 0. \tag{9.6.6}$$

Writing $r_1 = a + ib$ ($b \neq 0$), it follows from Theorem 9.6.1 that a complex-valued solution to (9.6.6) corresponding to the root $r = r_1$ is

$$w = e^{r_1 x} = e^{(a+ib)x} = e^{ax}(\cos bx + i \sin bx).$$

Taking the real and imaginary parts of this complex-valued solution and using the result of Corollary 9.5.1 yields the real-valued solutions

$$y_1 = e^{ax}\cos bx, \qquad y_2 = e^{ax}\sin bx.$$

The Wronskian of these solutions is (exercise) $W[y_1, y_2](x) = be^{2ax} \neq 0$, so that the solutions are linearly independent on any interval. It follows that a basis for

---

[1] This method of determining a second linearly independent solution to a second-order linear differential equation is called *reduction of order* and will be developed fully in Section 9.11.

the solution space of the differential equation is $\{e^{ax}\cos bx, e^{ax}\sin bx\}$ and hence the general solution of (9.6.6) is

$$y = c_1 e^{ax}\cos bx + c_2 e^{ax}\sin bx.$$

We write this in the equivalent form

$$\boxed{y = e^{ax}(c_1\cos bx + c_2\sin bx).}$$

The question that naturally arises at this point is whether the complex-valued solution corresponding to $r = \bar{r}_1$ leads to any further *linearly independent* real-valued solutions. Since we know from our general theory that the dimension of the solution space of the differential equation (9.6.6) is two, it follows that there cannot be any further *linearly independent* real-valued solutions. Indeed, we leave it as an exercise to show that the real-valued solutions obtained by taking the real and imaginary parts of the conjugate solution $y = e^{(a-ib)x}$ are

$$Y_1 = e^{ax}\cos bx, \qquad Y_2 = -e^{ax}\sin bx,$$

which clearly depend on y_1 and y_2. We see that the linearly independent *real-valued* solutions derived in this case are really associated with the *pair* of complex conjugate roots. More precisely, substituting for $r_1 = a + ib$ into (9.6.6), we obtain

$$(D^2 - 2aD + a^2 + b^2)y = 0.$$

The auxiliary polynomial is a real irreducible quadratic, and it is this real quadratic that gives rise to the two *real-valued* linearly independent solutions

$$y_1 = e^{ax}\cos bx, \qquad y_2 = e^{ax}\sin bx.$$

This discussion is summarized in Table 9.6.1.

TABLE 9.6.1

Summary
Linearly independent solutions of
$$y'' + a_1 y' + a_2 y = 0$$
are determined as follows:

Roots of the Auxiliary Equation	Linearly Independent Solutions of DE
Real distinct: $r_1 \neq r_2$	$y_1 = e^{r_1 x}$, $y_2 = e^{r_2 x}$.
Real repeated: $r_1 = r_2$	$y_1 = e^{r_1 x}$, $y_2 = x e^{r_1 x}$.
Complex conjugate: $r_1 = a + ib$, $r_2 = a - ib$	$y_1 = e^{ax}\cos bx$, $y_2 = e^{ax}\sin bx$.

Example 9.6.1   Find the general solution of $y'' - 6y' + 25y = 0$.

Solution   The auxiliary equation is $r^2 - 6r + 25 = 0$, with roots $r = 3 \pm 4i$. Consequently, two linearly independent real-valued solutions of the differential equation are:

$$y_1 = e^{3x}\cos 4x, \qquad y_2 = e^{3x}\sin 4x,$$

and hence the general solution of the differential equation is

$$y = e^{3x}(c_1\cos 4x + c_2\sin 4x).$$

Example 9.6.2 Solve the initial value problem

$$y'' - 4y' + 4y = 0, \qquad y(0) = 1, \qquad y'(0) = 4.$$

Solution The auxiliary equation is $r^2 - 4r + 4 = 0$; that is, $(r - 2)^2 = 0$. Thus $r = 2$ is a repeated root, and hence two linearly independent solutions of the given differential equation are

$$y_1 = e^{2x}, \qquad y_2 = xe^{2x}.$$

Consequently the general solution is

$$y = e^{2x}(c_1 + c_2 x).$$

The initial condition $y(0) = 1$ implies that

$$c_1 = 1.$$

Thus

$$y = e^{2x}(1 + c_2 x).$$

Differentiating this expression yields

$$y' = 2e^{2x}(1 + c_2 x) + c_2 e^{2x},$$

so that the second initial condition requires $c_2 = 2$. Hence the unique solution to the initial value problem is $y = e^{2x}(1 + 2x)$.

GENERALIZATION TO HIGHER ORDER

Now consider the general nth-order equation

$$y^{(n)} + a_1 y^{(n-1)} + \cdots + a_{n-1}y' + a_n y = 0, \tag{9.6.7}$$

or, equivalently,

$$(D^n + a_1 D^{n-1} + \cdots + a_{n-1}D + a_n)y = 0.$$

Let $r_1, r_2, \ldots, r_k$ denote the *distinct* roots of the auxiliary equation, so that the auxiliary polynomial can be factored as

$$P(r) = (r - r_1)^{m_1}(r - r_2)^{m_2} \cdots (r - r_k)^{m_k},$$

where m_i denotes the multiplicity of the root r_i and $m_1 + m_2 + \cdots + m_k = n$. Then (9.6.7) can be written in the equivalent form

$$(D - r_1)^{m_1}(D - r_2)^{m_2} \cdots (D - r_k)^{m_k} \, y = 0. \tag{9.6.8}$$

The first result that we require is the following.

Theorem 9.6.2: If $P(D) = P_1(D)P_2(D) \cdots P_k(D)$, where each $P_i(D)$ is a polynomial differential operator, then any solution of $P_i(D)y = 0$ is also a solution of $P(D)y = 0$.

PROOF   Suppose $P_i(D)u = 0$. Then, since we can change the order of the factors in a polynomial differential operator (with constant coefficients), it follows that

$$P(D)u = P_1(D) \cdots P_{i-1}(D)P_{i+1}(D) \cdots P_k(D)P_i(D)u$$

$$= 0. \qquad \blacksquare$$

Applying this theorem to (9.6.8) we see that any solutions to

$$(D - r_i)^{m_i} y = 0 \qquad (9.6.9)$$

will also be solutions to the full differential equation. Thus we first focus our attention on differential equations of the form (9.6.9). A set of linearly independent solutions to this differential equation is given in the following theorem.

Theorem 9.6.3:   The differential equation $(D - r)^m y = 0$, where m is a positive integer and r is a real or complex number, has the following m solutions, which are linearly independent on any interval:

$$e^{rx}, \, xe^{rx}, \, x^2 e^{rx}, \, \ldots, \, x^{m-1}e^{rx}.$$

PROOF   Consider the effect of applying the differential operator $(D - a)^m$ to the function $e^{rx}u(x)$, where u is an arbitrary (but sufficiently smooth) function. When $m = 1$, we obtain

$$(D - r)(e^{rx}u) = e^{rx}u' + re^{rx}u - re^{rx}u.$$

Thus

$$(D - r)(e^{rx}u) = e^{rx}u'.$$

Repeating this procedure yields

$$(D - r)^2(e^{rx}u) = (D - r)(e^{rx}u') = e^{rx}u''.$$

The general result is

$$(D - r)^m(e^{rx}u) = e^{rx}D^m(u). \qquad (9.6.10)$$

Choosing $u(x) = x^k$ and using the fact that $D^m(x^k) = 0$, for $k = 0, 1, \ldots, m - 1$, it follows from (9.6.10) that

$$(D - r)^m(e^{rx}x^k) = 0, \qquad k = 0, 1, \ldots, m - 1,$$

and hence $e^{rx}, \, xe^{rx}, \, x^2 e^{rx}, \, \ldots, \, x^{m-1}e^{rx}$ are solutions of the differential equation $(D - r)^m y = 0$. We now prove that these solutions are linearly independent on any interval. We must show that

$$c_1 e^{rx} + c_2 xe^{rx} + c_3 x^2 e^{rx} + \cdots + c_m x^{m-1}e^{rx} = 0$$

for x in any interval if and only if $c_1 = c_2 = \cdots = c_m = 0$. Dividing by e^{rx} we obtain the equivalent expression

$$c_1 + c_2 x + c_3 x^2 + \cdots + c_m x^{m-1} = 0.$$

Since the set of functions $\{1, x, x^2, \ldots, x^{m-1}\}$ is linearly independent on any interval (see problem 18 in Section 6.6), it follows that we must have

$c_1 = c_2 = \cdots = c_m = 0$, and hence the given functions are also linearly independent on any interval. ∎

We now apply the results of the previous two theorems to the differential equation

$$(D - r_1)^{m_1}(D - r_2)^{m_2} \cdots (D - r_k)^{m_k} y = 0. \tag{9.6.11}$$

The solutions that are obtained due to a term of the form $(D - r)^m$ depend on whether r is a real or a complex number. We consider the two cases separately.

1. Each term of the form $(D - r)^m$, where r is *real*, contributes the linearly independent solutions

$$e^{rx}, xe^{rx}, \ldots, x^{m-1}e^{rx}.$$

2. $(D - r)^m$, $r = a + ib$, $b \neq 0$. In this case the term $(D - \bar{r})^m$ must also occur in (9.6.11). These complex conjugate terms contribute the complex-valued solutions

$$e^{(a \pm ib)x}, xe^{(a \pm ib)x}, x^2e^{(a \pm ib)x}, \ldots, x^{m-1}e^{(a \pm ib)x},$$

which we can write, using Euler's formula, as

$$e^{ax}(\cos bx \pm i \sin bx), xe^{ax}(\cos bx \pm i \sin bx), x^2e^{ax}(\cos bx \pm i \sin bx), \ldots,$$

$$x^{m-1}e^{ax}(\cos bx \pm i \sin bx).$$

Taking the real and imaginary parts of these complex-valued solutions yields the following $2m$ *real-valued* solutions to (9.6.11)

$$e^{ax}\cos bx, xe^{ax}\cos bx, x^2e^{ax}\cos bx, \ldots, x^{m-1}e^{ax}\cos bx,$$

$$e^{ax}\sin bx, xe^{ax}\sin bx, x^2e^{ax}\sin bx, \ldots, x^{m-1}e^{ax}\sin bx.$$

We leave the proof that these solutions are linearly independent on any interval as an exercise.

By considering each term in (9.6.1) successively, we can obtain n real-valued solutions to $P(D)y = 0$. The proof that these solutions are linearly independent on any interval is tedious, not particularly instructive and hence is omitted (see, for example, W. Kaplan, *Differential Equations*, Addison-Wesley, 1958).

We now summarize our results.

Theorem 9.6.4:   Consider the differential equation

$$P(D)y = 0. \tag{9.6.12}$$

Let $r_1, r_2, \ldots, r_k$ be the distinct roots of the auxiliary equation, so that

$$P(r) = (r - r_1)^{m_1}(r - r_2)^{m_2} \cdots (r - r_k)^{m_k},$$

where m_i denotes the multiplicity of the root $r = r_i$.

1. If r_i is *real*, then the functions $e^{r_i x}, xe^{r_i x}, \ldots, x^{m_i-1}e^{r_i x}$ are linearly independent solutions of (9.6.12) on any interval.

2. If r_i is *complex*, say $r_i = a + ib(a, b$ real, $b \neq 0)$ then the functions

$$e^{ax}\cos bx, \ xe^{ax}\cos bx, \ \ldots, \ x^{m_i-1}e^{ax}\cos bx,$$

$$e^{ax}\sin bx, \ xe^{ax}\sin bx, \ \ldots, \ x^{m_i-1}e^{ax}\sin bx,$$

corresponding to the conjugate roots $r = a \pm ib$, are linearly independent solutions of (9.6.12) on any interval.

3. The n real-valued solutions $y_1, y_2, \ldots, y_n$ of (9.6.12) that are obtained by considering the distinct roots $r_1, r_2, \ldots, r_k$ are linearly independent on any interval, and hence the general solution of (9.6.12) is

$$y = c_1 y_1 + c_2 y_2 + \cdots + c_n y_n.$$

Example 9.6.3   Find the general solution of $y''' + 2y'' + 3y' + 2y = 0$.

Solution   The auxiliary polynomial is $P(r) = r^3 + 2r^2 + 3r + 2$, which can be factored into $P(r) = (r + 1)(r^2 + r + 2)$.   The roots of the auxiliary equation are, therefore,

$$r = -1 \quad \text{and} \quad r = \frac{-1 \pm i\sqrt{7}}{2}.$$

It follows that three linearly independent solutions to the given differential equation are

$$y_1 = e^{-x}, \quad y_2 = e^{-x/2}\cos\left(\frac{\sqrt{7}x}{2}\right), \quad y_3 = e^{-x/2}\sin\left(\frac{\sqrt{7}x}{2}\right),$$

so that the general solution is

$$y = c_1 e^{-x} + c_2 e^{-x/2}\cos\left(\frac{\sqrt{7}x}{2}\right) + c_3 e^{-x/2}\sin\left(\frac{\sqrt{7}x}{2}\right).$$

Example 9.6.4   Find the general solution to

$$(D - 3)(D^2 + 2D + 2)^2 y = 0. \tag{9.6.13}$$

Solution   The auxiliary polynomial is

$$P(r) = (r - 3)(r^2 + 2r + 2)^2,$$

so that the roots of the auxiliary equation are $r = 3$, and $r = -1 \pm i$ (multiplicity 2). The corresponding linearly independent solutions to (9.6.13) are

$$y_1 = e^{3x}, \quad y_2 = e^{-x}\cos x, \quad y_3 = xe^{-x}\cos x, \quad y_4 = e^{-x}\sin x, \quad y_5 = xe^{-x}\sin x,$$

and hence the general solution to (9.6.13) is

$$y = c_1 e^{3x} + e^{-x}(c_2\cos x + c_3 x \cos x + c_4\sin x + c_5 x \sin x).$$

Example 9.6.5   Find the general solution of $D^3(D - 2)^2(D^2 + 1)^2 y = 0$.

Solution The auxiliary polynomial is $P(r) = r^3(r-2)^2(r^2+1)^2$, with zeros $r = 0$ (multiplicity 3), $r = 2$ (multiplicity 2), and $r = \pm i$ (multiplicity 2). We therefore obtain the following linearly independent solutions to the given differential equation:

$$y_1 = 1, \qquad y_2 = x, \qquad y_3 = x^2, \qquad y_4 = e^{2x}, \qquad y_5 = xe^{2x},$$

$$y_6 = \cos x, \qquad y_7 = x \cos x, \qquad y_8 = \sin x, \qquad y_9 = x \sin x.$$

It follows that the general solution of the differential equation is

$$y = c_1 + c_2 x + c_3 x^2 + c_4 e^{2x} + c_5 x e^{2x} + (c_6 + c_7 x)\cos x + (c_8 + c_9 x)\sin x.$$

EXERCISES 9.6

In problems 1–20 find the general solution of the given differential equation.

1. $y'' - y' - 2y = 0$.
2. $y'' - 6y' + 9y = 0$.
3. $y'' + 6y' + 25y = 0$.
4. $(D^2 - 4D - 5)y = 0$.
5. $(D+2)^2 y = 0$.
6. $(D-1)(D^2 - 2D + 1)y = 0$.
7. $(D^2 - 6D + 34)y = 0$.
8. $(D^2 - 2)y = 0$.
9. $y'' + 10y' + 25y = 0$.
10. $y'' + 2y' + 2y = 0$.
11. $(D-2)(D^2 - 16)y = 0$.
12. $(D^2 + 3)(D+1)^2 y = 0$.
13. $y'' + 8y' + 20y = 0$.
14. $y'' - 2y' - 8y = 0$.
15. $y''' - 2y'' - 4y' + 8y = 0$.
16. $(D^2 + 4)(D+1)y = 0$.
17. $D^2(D-1)y = 0$.
18. $(D-1)^3(D^2 + 9)y = 0$.
19. $(D^2 - 2D + 2)^2(D^2 - 1)y = 0$.
20. $y''' + 8y'' + 22y' + 20y = 0$.

In problems 21–23 solve the given initial value problem.

21. $y'' + y' - 6y = 0$, $y(0) = 3$, $y'(0) = 1$.
22. $y'' - 8y' + 16y = 0$, $y(0) = 2$, $y'(0) = 7$.
23. $(D^2 - 4D + 5)y = 0$, $y(0) = 3$, $y'(0) = 5$.

24. Solve the initial value problem

$$y'' - 2my' + (m^2 + k^2)y = 0, \quad y(0) = 0, \quad y'(0) = k,$$

where m and k are positive constants.

25. Find the general solution of

$$y'' - 2my' + (m^2 - k^2)y = 0$$

(m, k positive constants) and show that it can be written in the form

$$y = e^{mx}(c_1\cosh kx + c_2\sinh kx).$$

26. An object of mass m is attached to one end of a spring, and the other end of the unstretched spring is attached to a fixed wall (see Figure 9.6.1). The object is pulled to the right a distance x_0 and released from rest. Assuming that the forces resisting the motion are proportional to the velocity of the object, an application of Hooke's law and Newton's second law of motion yields the following initial value problem that governs the motion of the object:

$$\frac{d^2x}{dt^2} + 2c\frac{dx}{dt} + k^2 x = 0, \quad x(0) = x_0, \quad \frac{dx}{dt}(0) = 0,$$

(using appropriate units) where c and k are positive constants.

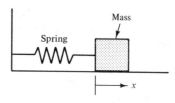

Figure 9.6.1

(a) Assuming that $c^2 < k^2$, solve the preceding initial value problem to obtain

$$x = \left(\frac{x_0}{\omega}\right) e^{-ct}(\omega \cos \omega t + c \sin \omega t),$$

where $\omega = \sqrt{k^2 - c^2}$.

(b) Show that the solution in (a) can be written in the form

$$x = \left(\frac{kx_0}{\omega}\right) e^{-ct}\sin(\omega t + \phi),$$

where $\phi = \tan^{-1}(\omega/c)$, and hence sketch the graph of x against t. Is the predicted motion reasonable?

27. Consider the *partial differential equation* (Laplace's equation)

$$\frac{\partial^2 u}{\partial x^2} + \frac{\partial^2 u}{\partial y^2} = 0. \tag{i}$$

(a) Show that the substitution $u(x, y) = e^{x/\alpha} f(\xi)$, where $\xi = \beta x - \alpha y$, (α, β positive constants) reduces (i) to the differential equation

$$\frac{d^2 f}{d\xi^2} + 2p\frac{df}{d\xi} + \frac{q}{\alpha^2}f = 0, \tag{ii}$$

where

$$p = \frac{\beta}{\alpha(\alpha^2 + \beta^2)}, \quad q = \frac{1}{(\alpha^2 + \beta^2)}. \tag{iii}$$

[*Hint:* Use the chain rule, for example,

$$\frac{\partial f}{\partial x} = \frac{df}{d\xi}\frac{\partial \xi}{\partial x},$$

etc.]

(b) Solve (ii) and hence find the corresponding solution of (i). [*Hint:* In solving (ii) you will need to use (iii) in order to obtain a "simple" form of solution.]

In problems 28–31, determine the constant coefficient homogeneous linear differential equation of lowest order whose solution space is spanned by the given set of functions.

28. $\{e^x, xe^x\}$.

29. $\{\sin 2x, \cos 2x\}$.

30. $\{e^{-x}\cos 3x, e^{-x}\sin 3x\}$.

31. $\{e^x\cos 2x, e^x\sin 2x, e^{3x}\}$.

In problems 32–35, determine the constant coefficient homogeneous linear differential equation of lowest order that has the given function as solution.

32. $y = xe^{2x}$.

33. $y = 1 + \sin 2x$.

34. $y = e^{3x}\cos 4x$.

35. $y = xe^x\cos x$.

36. Consider the differential equation

$$y'' + a_1 y' + a_2 y = 0, \tag{iv}$$

where a_1 and a_2 are real constants.
(a) If the auxiliary equation has real roots r_1, r_2, what conditions on these roots would guarantee that every solution of (iv) satisfies

$$\lim_{x \to +\infty} y(x) = 0?$$

(b) If the auxiliary equation has complex conjugate roots $r = a \pm ib$, what conditions on these roots would guarantee that every solution of (iv) satisfies

$$\lim_{x \to +\infty} y(x) = 0?$$

(c) If a_1 and a_2 are positive, prove that $\lim_{x \to +\infty} y(x) = 0$ for every solution of (iv).

(d) If $a_1 > 0$ and $a_2 = 0$, prove that all solutions to (iv) approach a constant value as $t \to +\infty$.
(e) If $a_1 = 0$, and $a_2 > 0$, prove that all solutions of (iv) remain bounded as $t \to +\infty$.

37. Consider $P(D)y = 0$. What conditions on the roots of the auxiliary equation would guarantee that every solution of the differential equation satisfies

$$\lim_{x \to +\infty} y(x) = 0?$$

38. Prove that the functions

$e^{ax}\cos bx, \ xe^{ax}\cos bx, \ x^2 e^{ax}\cos bx, \ \ldots,$

$$x^{m-1}e^{ax}\cos bx,$$

$e^{ax}\sin bx, \ xe^{ax}\sin bx, \ x^2 e^{ax}\sin bx, \ \ldots,$

$$x^{m-1}e^{ax}\sin bx,$$

are linearly independent on $(-\infty, \infty)$. [*Hint*: Show that the condition for determining linear dependence or independence can be written as

$$P(x)\cos bx + Q(x)\sin bx = 0$$

where $P(x) = c_1 + c_2x + \cdots + c_mx^{m-1}$

and $Q(x) = d_1 + d_2x + \cdots + d_mx^{m-1}$.

Then show that this implies $P(n\pi/b) = 0$, $Q((2n+1)\pi/b) = 0$ for all integers n, which means that P and Q must both be the zero polynomial.]

9.7   THE METHOD OF UNDETERMINED COEFFICIENTS: ANNIHILATORS

According to Theorem 9.3.4, the general solution to the nonhomogeneous differential equation

$$P(D)y = F(x) \tag{9.7.1}$$

is of the form

$$y = y_c + y_p,$$

where y_c is the general solution of the associated homogeneous equation and y_p is a particular solution of (9.7.1). We have seen in the previous section how y_c can be obtained. We now turn our attention to determining a particular solution y_p. In this section we develop a method that can be applied whenever $F(x)$ has certain special forms. This technique can be introduced quite simply as follows.

Consider the differential equation (9.7.1), and suppose that there is a polynomial differential operator $Q(D)$ such that

$$Q(D)F = 0.$$

Then, operating on (9.7.1) with $Q(D)$ yields the homogeneous differential equation

$$Q(D)P(D)y = 0. \tag{9.7.2}$$

The key point to realize is the following. Every solution of (9.7.1) is also a solution of (9.7.2). Thus the general solution of (9.7.2) must contain a particular solution of (9.7.1). However, (9.7.2) is just a linear homogeneous constant coefficient differential equation whose general solution can be obtained using the method of the previous section. Before developing this technique fully, we consider a simple example.

Example 9.7.1   Determine the general solution to

$$(D^2 - 1)y = 16e^{3x}. \tag{9.7.3}$$

Solution   We first obtain the complementary function. The auxiliary polynomial is

$$P(r) = r^2 - 1 = (r - 1)(r + 1),$$

so that

$$y_c = c_1e^x + c_2e^{-x}.$$

In this case $F(x) = 16e^{3x}$ and so we require a polynomial differential operator $Q(D)$ such that $Q(D)F = 0$.   Since $(D - a)(e^{ax}) = 0$, it follows that we can choose

$$Q(D) = D - 3,$$

Operating on (9.7.3) with $Q(D)$ yields the homogeneous differential equation

$$(D - 3)(D^2 - 1)y = 0,$$

which has general solution

$$y = \underbrace{c_1 e^x + c_2 e^{-x}}_{\text{Complementary function}} + \underbrace{A_0 e^{3x}}_{\text{Trial solution}}.$$

This solution must contain a particular solution to (9.7.3) for appropriate values of the constants c_1, c_2, and A_0.   However, the first two terms coincide with the complementary function and hence cannot contribute to a particular solution of (9.7.3).   It follows that (9.7.3) must have a solution of the form

$$y_p = A_0 e^{3x}. \tag{9.7.4}$$

We call y_p a **trial solution.**   It contains one "undetermined coefficient," A_0.   In order to determine the appropriate value of A_0, we substitute the trial solution into (9.7.3).   Differentiating (9.7.4) twice yields

$$y_p' = 3A_0 e^{3x}, \qquad y_p'' = 9A_0 e^{3x},$$

so that y_p is a solution of (9.7.3) if and only if A_0 satisfies

$$(9A_0 - A_0)\, e^{3x} = 16e^{3x}.$$

Thus

$$A_0 = 2,$$

so that a particular solution to (9.7.3) is

$$y_p = 2e^{3x}.$$

Consequently, the general solution of (9.7.3) is

$$y = c_1 e^x + c_2 e^{-x} + 2e^{3x}.$$

This technique for determining a particular solution is called the **method of undetermined coefficients**.   It is only applicable to differential equations that satisfy the following conditions.

---

1. The differential equation has constant coefficients and hence is of the form

$$P(D)y = F(x).$$

2. There exists a polynomial differential operator $Q(D)$ such that

$$Q(D)F = 0. \tag{9.7.5}$$

The previous example contains the essence of the technique. It remains for us to determine only what types of functions satisfy (9.7.5). We first introduce some terminology.

Definition 9.7.1: A polynomial differential operator $Q(D)$ is said to **annihilate** a function $F(x)$ if

$$Q(D)F(x) = 0.$$

REMARK   In terms of our linear transformation ideas, we can say that $Q(D)$ annihilates F if and only if F is in $\text{Ker}[Q(D)]$.

Example 9.7.2   Show that $Q(D) = D - 5$ annihilates $F(x) = 16e^{5x}$.

Solution

$$Q(D)(16e^{5x}) = (D - 5)(16e^{5x}) = 16D(e^{5x}) - 80e^{5x} = 80e^{5x} - 80e^{5x} = 0.$$

More generally, a polynomial differential operator $Q(D)$ annihilates F if and only if $y = F(x)$ is a solution of

$$Q(D)y = 0.$$

Thus, the only types of functions that can be annihilated by a polynomial differential operator are those that arise as solutions of a homogeneous constant coefficient linear differential equation. From our results of the previous section it follows that $F(x)$ must be one of the following forms:

1. $F(x) = cx^k e^{ax}$,
2. $F(x) = cx^k e^{ax}\sin bx$,
3. $F(x) = cx^k e^{ax}\cos bx$,
4. sums of (1)–(3),

where a, b, c are real numbers and k is a nonnegative integer. We now derive appropriate polynomial differential operators that annihilate each of these functions. Consider first the case when $F(x) = x^k e^{ax}$, where a is a real number. The differential equation

$$(D - a)^{k+1}y = 0,$$

where a is a real number and k is a nonnegative integer, has the real-valued solutions

$$e^{ax}, xe^{ax}, \ldots, x^k e^{ax}.$$

An alternative way to state this is as follows:

---

1. $Q(D) = (D - a)^{k+1}$ annihilates each of the functions

$$e^{ax}, \ xe^{ax}, \ \ldots, \ x^k e^{ax},$$

and hence it also annihilates

$$F(x) = (a_0 + a_1 x + \cdots + a_k x^k) e^{ax}$$

for all values of the constants $a_0, a_1, \ldots, a_k$.

REMARK   Note the special case of (1) that arises when $a = 0$, namely,

---

$Q(D) = D^{k+1}$ annihilates $F(x) = a_0 + a_1 x + \cdots + a_k x^k$.

Example 9.7.3   Determine a polynomial differential operator that annihilates the given function.
  (a) $F(x) = 2e^{3x}$.
  (b) $F(x) = e^{2x}(3 + 4x)$.
  (c) $F(x) = 3e^{-x} + x$.
  (d) $F(x) = 5e^{4x} + 6x^2 e^{-7x}$.

Solution
  (a) $Q(D) = D - 3$.
  (b) $Q(D) = (D - 2)^2$.
  (c) Since $Q_1(D) = D + 1$ annihilates $3e^{-x}$ whereas $Q_2(D) = D^2$ annihilates x, it follows that $Q(D) = D^2(D + 1)$ annihilates $F(x) = 3e^{-x} + x$.
  (d) $Q(D) = (D - 4)(D + 7)^3$.

Now consider the functions $e^{ax}\cos bx$, $e^{ax}\sin bx$, where a and b are real numbers. These functions arise as linearly independent (real-valued) solutions of the differential equation $(D - \alpha)(D - \overline{\alpha})y = 0$, where $\alpha = a + ib$. Expanding the polynomial differential operator we have

$$[D^2 - 2aD + (a^2 + b^2)]y = 0.$$

Thus

---

2. $Q(D) = D^2 - 2aD + (a^2 + b^2)$ annihilates both of the functions

$$e^{ax}\cos bx, \qquad e^{ax}\sin bx,$$

and hence it annihilates

$$F(x) = e^{ax}(a_0\cos bx + b_0\sin bx)$$

for all values of the constants a_0, b_0.
In particular,

$$Q(D) = D^2 + b^2$$

annihilates the functions $\cos bx$ and $\sin bx$.

Example 9.7.4   Determine a polynomial differential operator that annihilates the given function.

(a) $F(x) = 3e^x \cos 2x$.
(b) $F(x) = 4 \sin 3x + 5 \cos 3x$.
(c) $F(x) = 2e^{-x}\sin 4x - e^{2x}\cos x$.

Solution

(a) In this case $a = 1$ and $b = 2$, so that $Q(D) = D^2 - 2D + 5$.
(b) $Q(D) = D^2 + 9$.
(c) $Q_1(D) = D^2 + 2D + 17$ annihilates $2e^{-x}\sin 4x$, whereas $Q_2(D) = D^2 - 4D + 5$ annihilates $e^{2x}\cos x$. It follows that $Q(D) = (D^2 + 2D + 17)(D^2 - 4D + 5)$ annihilates $F(x) = 2e^{-x}\sin 4x - e^{2x}\cos x$.

Finally, the functions

$$e^{ax}\cos bx,\ xe^{ax}\cos bx,\ x^2 e^{ax}\cos bx,\ \dots,\ x^k e^{ax}\cos bx,$$
$$e^{ax}\sin bx,\ xe^{ax}\sin bx,\ x^2 e^{ax}\sin bx,\ \dots,\ x^k e^{ax}\sin bx,$$

arise as linearly independent (real-valued) solutions of the differential equation

$$[D^2 - 2aD + (a^2 + b^2)]^{k+1}\, y = 0.$$

Equivalently we can state that

---

3. $Q(D) = [D^2 - 2aD + (a^2 + b^2)]^{k+1}$ annihilates each of the functions

$$e^{ax}\cos bx,\ xe^{ax}\cos bx,\ x^2 e^{ax}\cos bx,\ \dots,\ x^k e^{ax}\cos bx,$$

$$e^{ax}\sin bx,\ xe^{ax}\sin bx,\ x^2 e^{ax}\sin bx,\ \dots,\ x^k e^{ax}\sin bx,$$

and hence, for all values of the constants $a_0, a_1, \dots, a_k, b_0, b_1, \dots, b_k$ it annihilates

$$F(x) = (a_0 + a_1 x + \cdots + a_k x^k)e^{ax}\cos bx + (b_0 + b_1 x + \cdots + b_k x^k)e^{ax}\sin bx.$$

Example 9.7.5   Determine a polynomial differential operator that annihilates the given function.

(a) $F(x) = x \cos x$.
(b) $F(x) = x^2 e^x \sin 3x$.
(c) $F(x) = 2x \sin 2x + 3e^{2x}\cos 5x$.

Solution

(a) $Q(D) = (D^2 + 1)^2$.
(b) In this case we have $a = 1$, $b = 3$, $k = 2$, so that $Q(D) = (D^2 - 2D + 10)^3$.
(c) $Q(D) = (D^2 + 4)^2(D^2 - 4D + 29)$.

Terminology

The polynomial differential operator of lowest order with real coefficients that annihilates $F(x)$ will be called the **annihilator** of $F(x)$.

Summarizing the preceding results, we have the following:

Theorem 9.7.1:   Let $a_0, a_1, \ldots, a_k, b_0, b_1, \ldots, b_k, a, b, c$ be real numbers, and let k be a nonnegative integer.

1. The annihilator of

$$F(x) = (a_0 + a_1 x + \cdots + a_k x^k) e^{ax}$$

is $Q(D) = (D - a)^{k+1}$.

2. The annihilator of

$$F(x) = (a_0 + a_1 x + \cdots + a_k x^k) e^{ax} \sin bx + (b_0 + b_1 x + \cdots + b_k x^k) e^{ax} \cos bx,$$

is $Q(D) = [D^2 - 2aD + (a^2 + b^2)]^{k+1}$.

3. If $F(x)$ is a sum of functions of the forms given in (1) and (2), then the annihilator of $F(x)$ is the corresponding *product* of the annihilators in (1) and (2).

We now consider several examples that illustrate the method of undetermined coefficients.

Example 9.7.6   Find the general solution of

$$(D - 4)(D + 1)y = 15e^{4x}. \tag{9.7.6}$$

Solution   The auxiliary polynomial for the given equation is $P(r) = (r - 4)(r + 1)$, so that the complementary function is

$$y_c = c_1 e^{-x} + c_2 e^{4x}.$$

In this case $F(x) = 15e^{4x}$, which has annihilator $Q(D) = D - 4$. Operating on the given differential equation with $Q(D)$ yields the homogeneous equation

$$(D - 4)^2 (D + 1)y = 0,$$

with general solution

$$y = \underbrace{c_1 e^{-x} + c_2 e^{4x}}_{\text{Complementary function}} + \underbrace{A_0 x e^{4x}}_{\text{Trial solution}}.$$

The first two terms coincide with the complementary function and therefore cannot contribute to a particular solution to (9.7.6). It follows that an appropriate trial solution for (9.7.6) is

$$y_p = A_0 x e^{4x}.$$

To determine A_0 we substitute y_p into (9.7.6). Differentiaing y_p twice yields

$$y_p' = A_0 e^{4x}(4x + 1), \qquad y_p'' = A_0 e^{4x}(16x + 8).$$

Thus, substituting into the given differential equation, it follows that A_0 must satisfy

$$A_0 e^{4x}[(16x + 8) - 3(4x + 1) - 4x] = 15e^{4x},$$

that is,

$$5A_0 = 15,$$

so that

$$A_0 = 3.$$

Consequently, a particular solution of (9.7.6) is

$$y_p = 3xe^{4x},$$

and hence the general solution is

$$y = c_1 e^{-x} + c_2 e^{4x} + 3xe^{4x}.$$

Example 9.7.7   Solve the initial value problem

$$y'' - y' - 2y = 10 \sin x, \qquad\qquad (9.7.7)$$

$$y(0) = 0, \qquad y'(0) = 1.$$

Solution   The auxiliary polynomial is

$$P(r) = r^2 - r - 2 = (r - 2)(r + 1),$$

so that the complementary function is

$$y_c = c_1 e^{2x} + c_2 e^{-x}.$$

The annihilator of $F(x) = 10 \sin x$ is $Q(D) = D^2 + 1$. Thus, operating on (9.7.7) with $Q(D)$ yields

$$(D^2 + 1)(D^2 - D - 2)y = 0,$$

which has general solution

$$y = \underbrace{c_1 e^{2x} + c_2 e^{-x}}_{\text{Complementary function}} + \underbrace{A_0 \sin x + A_1 \cos x.}_{\text{Trial solution}}$$

The first two terms coincide with the complementary function, and hence we choose as our trial solution

$$y_p = A_0 \sin x + A_1 \cos x.$$

Differentiating y_p we obtain

$$y_p' = A_0 \cos x - A_1 \sin x, \qquad y_p'' = -A_0 \sin x - A_1 \cos x.$$

Substituting into (9.7.7) yields

$$(-A_0 \sin x - A_1 \cos x) - (A_0 \cos x - A_1 \sin x) - 2(A_0 \sin x + A_1 \cos x) = 10 \sin x;$$

that is,

$$(-3A_0 + A_1)\sin x - (A_0 + 3A_1)\cos x = 10 \sin x.$$

This equation is satisfied for all x if and only if

$$-3A_0 + A_1 = 10, \qquad A_0 + 3A_1 = 0.$$

The unique solution to this system of equations is

$$A_0 = -3, \qquad A_1 = 1,$$

so that a particular solution to (9.7.7) is

$$y_p = -3 \sin x + \cos x.$$

Consequently (9.7.7) has general solution

$$y = c_1 e^{2x} + c_2 e^{-x} - 3 \sin x + \cos x. \qquad (9.7.8)$$

We now impose the given initial conditions. It is easily shown that $y(0) = 0$ if and only if

$$c_1 + c_2 = -1, \qquad (9.7.9)$$

whereas $y'(0) = 1$ if and only if

$$2c_1 - c_2 = 4. \qquad (9.7.10)$$

Solving (9.7.9) and (9.7.10) yields

$$c_1 = 1, \qquad c_2 = -2,$$

so that, from (9.7.8), the unique solution to the given initial value problem is

$$y = e^{2x} - 2e^{-x} - 3 \sin x + \cos x.$$

Example 9.7.8   Find the general solution of

$$(D^2 - 4D + 5)y = 8xe^{2x}\cos x. \qquad (9.7.11)$$

Solution   The auxiliary equation is $r^2 - 4r + 5 = 0$, with roots $r = 2 \pm i$. Thus

$$y_c = e^{2x}(c_1 \cos x + c_2 \sin x).$$

In this case the annihilator is $Q(D) = (D^2 - 4D + 5)^2$, so that the general solution of (9.7.11) is contained in the general solution to

$$(D^2 - 4D + 5)^3 y = 0.$$

Solving this latter equation yields

$$y = e^{2x}(c_1 \cos x + c_2 \sin x) + xe^{2x}(A_0 \cos x + B_0 \sin x) + x^2 e^{2x}(A_1 \cos x + B_1 \sin x).$$

Neglecting the contribution from the complementary function, we obtain the trial solution

$$y_p = xe^{2x}(A_0 \cos x + B_0 \sin x) + x^2 e^{2x}(A_1 \cos x + B_1 \sin x).$$

Substituting into (9.7.11), a lengthy computation yields

$$(-2A_0 + 2B_1 - 4xA_1)\sin x + (2B_0 + 2A_1 + 4xB_1)\cos x = 8x \cos x,$$

so that A_0, A_1, B_0, B_1 must satisfy

$$-2A_0 \qquad\qquad + 2B_1 = 0,$$
$$2A_1 + 2B_0 \qquad = 0,$$
$$-4A_1 \qquad\qquad = 0,$$
$$4B_1 = 8.$$

We see that $A_1 = 0 = B_0$ and $A_0 = 2 = B_1$, so that a particular solution to (9.7.11) is

$$y_p = 2xe^{2x}(x \sin x + \cos x),$$

and the general solution is

$$y = e^{2x}[c_1\cos x + c_2\sin x + 2x(x \sin x + \cos x)].$$

Example 9.7.9   Determine a trial solution for

$$(D - 1)^2(D - 2)(D^2 + 1)y = e^x + 2 \cos x - 4 \sin x.$$

Solution   The complementary function is

$$y_c = (c_1 + c_2x)e^x + c_3e^{2x} + c_4\cos x + c_5\sin x.$$

The appropriate annihilator is

$$Q(D) = (D - 1)(D^2 + 1),$$

so that a trial solution is obtained from

$$(D - 1)(D^2 + 1)(D - 1)^2(D - 2)(D^2 + 1)y = 0,$$

that is,

$$(D - 1)^3(D - 2)(D^2 + 1)^2y = 0.$$

The general solution of this differential equation is

$$y = y_c + A_0x^2e^x + x(A_1\cos x + A_2\sin x).$$

so that an appropriate trial solution for (9.7.12) is

$$y_p = A_0x^2e^x + x(A_1\cos x + A_2\sin x).$$

As usual, the constants A_0, A_1, A_2 can be obtained by substituting into the given differential equation (although in this example we were *not* asked to find them).

More generally, we can derive appropriate trial solutions to cover any case that may arise. For example, consider

$$P(D)y = cx^ke^{ax}, \qquad\qquad (9.7.13)$$

and let y_c denote the complementary function. The appropriate annihilator for (9.7.13) is $Q(D) = (D - a)^{k+1}$, and so a trial solution for (9.7.13) can be determined from the general solution to

$$Q(D)P(D)y = 0. \tag{9.7.14}$$

Two cases arise.

CASE 1   If $r = a$ is *not* a root of $P(r) = 0$, then the general solution of (9.7.14) will be of the form

$$y = y_c + e^{ax}(A_0 + A_1 x + \cdots + A_k x^k),$$

so that an appropriate trial solution is

$$y_p = e^{ax}(A_0 + A_1 x + \cdots + A_k x^k). \qquad \leftarrow \textit{Usual trial solution}$$

CASE 2   If $r = a$ is a root of multiplicity m of $P(r) = 0$, then the complementary function y_c will contain the terms $e^{ax}[c_0 + c_1 x + \cdots + c_{m-1} x^{m-1}]$. The operator $Q(D)P(D)$ will therefore contain the term $(D - a)^{m+k+1}$, so that the terms in the general solution to (9.7.14) that do not arise in the complementary function are

$$y_p = e^{ax} x^m (A_0 + A_1 x + \cdots + A_k x^k). \qquad \leftarrow \textit{Modified trial solution}$$

The derivation of appropriate trial solutions for $P(D)y = F(x)$ in the case when $F(x) = cx^k e^{ax} \cos bx$ or $F(x) = cx^k e^{ax} \sin bx$ is left as an exercise. We summarize the results in a table.

$F(x)$	Usual trial solution	Modified trial solution
$cx^k e^{ax}$	If $P(a) \neq 0$, choose: $y_p = e^{ax}(A_0 + A_1 x + \cdots + A_k x^k)$.	If a is a root of $P(r) = 0$ of multiplicity m, choose: $y_p = x^m e^{ax}(A_0 + A_1 x + \cdots + A_k x^k)$.
$cx^k e^{ax} \cos bx$ or $cx^k e^{ax} \sin bx$	If $P(a + ib) \neq 0$, choose: $y_p = e^{ax}[(A_0 \cos bx + B_0 \sin bx) + x(A_1 \cos bx + B_1 \sin bx) + \cdots + x^k(A_k \cos bx + B_k \sin bx)]$.	If $a + ib$ is a root of $P(r) = 0$ of multiplicity m, choose: $y_p = x^m e^{ax}[(A_0 \cos bx + B_0 \sin bx) + x(A_1 \cos bx + B_1 \sin bx) + \cdots + x^k(A_k \cos bx + B_k \sin bx)]$.

If $F(x)$ is the sum of functions of the preceding form, then the appropriate trial solution is the corresponding sum.

In the following table we have specialized the trial solutions to the cases that arise most often in applications.

$F(x)$	Usual trial solution	Modified trial solution
ce^{ax}	If $P(a) \neq 0$, choose: $y_p = A_0 e^{ax}$.	If a is a root of $P(r) = 0$ of multiplicity m, choose: $y_p = A_0 x^m e^{ax}$.
$c \cos bx$ or $c \sin bx$	If $P(ib) \neq 0$, choose: $y_p = A_0 \cos bx + A_1 \sin bx$.	If ib is a root of $P(r) = 0$ of multiplicity m, choose: $y_p = x^m (A_0 \cos bx + A_1 \sin bx)$.
cx^k	If $P(0) \neq 0$, choose: $y_p = A_0 + A_1 x + \cdots + A_k x^k$.	If zero is a root of $P(r) = 0$ of multiplicity m, choose: $y_p = x^m (A_0 + A_1 x + \cdots + A_k x^k)$.

EXERCISES 9.7

In problems 1–10 determine the annihilator of the given function.

1. $F(x) = 5e^{-3x}$.

2. $F(x) = 2e^x - 3x$.

3. $F(x) = \sin x + 3xe^{2x}$.

4. $F(x) = x^3e^{7x} + 5\sin 4x$.

5. $F(x) = 4e^{-2x}\sin x$.

6. $F(x) = e^x\sin 2x + 3\cos 2x$.

7. $F(x) = (1 - 3x)e^{4x} + 2x^2$.

8. $F(x) = e^{5x}(2 - x^2)\cos x$.

9. $F(x) = e^{-3x}(2\sin x + 7\cos x)$.

10. $F(x) = e^{4x}(x - 2\sin 5x) + 3x - x^2e^{-2x}\cos x$.

In problems 11–24 determine the general solution of the given differential equation.

11. $y'' - 3y' + 2y = 4e^{3x}$.

12. $(D^2 - 2D - 3)y = 15e^{4x}$.

13. $(D^2 - 2D + 1)y = 3x(x - 4)$.

14. $D(D + 2)y = 49e^x\sin 2x$.

15. $y'' - 2y' - 3y = 8e^{3x}$.

16. $D^2(D^2 + 1)y = 6 - 12x$.

17. $y'' + 3y' + 2y = 6(2e^{2x} + 3e^x)$.

18. $y'' + 16y = 24\cos 4x$.

19. $y'' + 2y' + 2y = 4x^2$.

20. $(D + 2)^2y = 10e^{-2x}$.

21. $(D^2 + 4)y = 16x\cos 2x$.

22. $(D - 1)(D + 3)y = 8e^x - 12e^{3x}$.

23. $y'' - y' - 2y = 40\sin^2 x$.

24. $y'' + 4y' + 5y = 24\sin x$.

In problems 25–29 solve the given initial value problem.

25. $(D^2 + D - 2)y = -10\sin x$, $y(0) = 2$, $y'(0) = 1$.

26. $y'' + 9y = 5\cos 2x$, $y(0) = y(\pi/2) = 2$.

27. $y'' - y = 9xe^{2x}$, $y(0) = 0$, $y'(0) = 7$.

28. $y'' + y' - 2y = 4\cos x - 2\sin x$, $y(0) = -1$, $y'(0) = 4$.

29. $(D - 1)(D - 2)(D - 3)y = 6e^{4x}$, $y(0) = 4$, $y'(0) = 10$, $y''(0) = 30$.

In problems 30–32 determine an appropriate trial solution for the given differential equation. Do *not* solve for the constants that arise in your trial solution.

30. $D^2(D - 1)(D^2 + 4)^2y = e^x - \sin 2x$.

31. $(D^2 - 2D + 2)^3(D - 2)^2(D + 4)y = e^x\cos x - 3e^{2x}$.

32. $D(D^2 - 9)(D^2 - 4D + 5)y = 2e^{3x} + e^{2x}\sin x$.

33. Consider the electric circuit shown in Figure 9.7.1, where R is the resistance, C is the capacitance, L is the inductance, and E is the electromotive force (EMF).

Figure 9.7.1

An application of Kirchhoff's law leads to the following differential equation for the current, i, in the circuit

$$L\frac{di}{dt} + Ri + \frac{1}{C}q = E(t), \qquad \text{(i)}$$

where q is the charge on the capacitor, and we assume that R, C, and L are constants. q and i are related by

$$i = \frac{dq}{dt}, \qquad \text{(ii)}$$

so that (i) can be written as

$$\frac{d^2q}{dt^2} + \frac{R}{L}\frac{dq}{dt} + \frac{1}{LC}q = \frac{1}{L}E(t). \qquad \text{(iii)}$$

(a) Find the general solution of (iii) when $E(t) = 0$ and $R^2 = 4L/C$, and hence find the corresponding current $i(t)$.

(b) Consider the general case when $R = 10\ \Omega$, $C = 1/450$ F, $L = 0.5$ H, and $E(t) = 600\cos 30t$. Find the general solution of (iii) and the corresponding current, $i(t)$. Note that your solution for $i(t)$ consists of an exponential part that dies out quickly (the *transient* current) and an oscillatory part that dominates as $t \to \infty$ (the *steady-state* current).

34. Derive the appropriate trial solution for the differential equation

$$P(D)y = cx^ke^{ax}\cos bx.$$

9.8   COMPLEX-VALUED TRIAL SOLUTIONS

The method of undetermined coefficients for solving

$$P(D)y = F(x)$$

is extremely tedious if $F(x)$ contains terms of the form $x^k e^{ax} \sin bx$ or $x^k e^{ax} \cos bx$. However, the computations can be reduced significantly from the observation that

$$x^k e^{ax} \cos bx = \mathrm{Re}\{x^k e^{(a+ib)x}\} \quad \text{and} \quad x^k e^{ax} \sin bx = \mathrm{Im}\{x^k e^{(a+ib)x}\}.$$

where Re and Im denote the real part and the imaginary part of a complex-valued function, respectively. To see why this observation is useful, we recall from Theorem 9.5.1 that if $y(x) = u(x) + iv(x)$ is a complex-valued solution of $Ly = f(x) + ig(x)$ on an interval I, then $Lu = f$ and $Lv = g$. Consequently, if we solve the complex equation

$$P(D)y = cx^k e^{(a+ib)x}, \tag{9.8.1}$$

then by taking the real and imaginary parts of the resulting complex-valued solution, we can directly determine solutions to

$$P(D)y = cx^k e^{ax} \cos bx \quad \text{and} \quad P(D)y = cx^k e^{ax} \sin bx. \tag{9.8.2}$$

The key point is that (9.8.1) is a simpler equation to solve than its real counterparts given in (9.8.2). We illustrate the technique with some examples.

Example 9.8.1   Solve

$$y'' + y' - 6y = 4 \cos 2x. \tag{9.8.3}$$

Solution   The complementary function for (9.8.3) is

$$y_c = c_1 e^{-3x} + c_2 e^{2x}.$$

In determining a particular solution we consider the complex differential equation

$$z'' + z' - 6z = 4e^{2ix}. \tag{9.8.4}$$

An appropriate complex-valued trial solution for this differential equation is

$$z_p = A_0 e^{2ix}, \tag{9.8.5}$$

where A_0 is a complex constant. The first two derivatives of z_p are

$$z_p' = 2iA_0 e^{2ix}, \qquad z_p'' = -4A_0 e^{2ix},$$

so that z_p is a solution of (9.8.4) if and only if

$$(-4A_0 + 2iA_0 - 6A_0)e^{2ix} = 4e^{2ix},$$

that is, if and only if

$$A_0 = \frac{2}{-5+i} = -\frac{1}{13}(5+i).$$

Substituting into (9.8.5) yields

$$z_p = -\frac{1}{13}(5 + i)e^{2ix} = -\frac{1}{13}(5 + i)(\cos 2x + i \sin 2x).$$

Consequently a particular solution of (9.8.3) is

$$y_p = \text{Re}(z_p) = -\frac{1}{13}(5 \cos 2x - \sin 2x),$$

so that the general solution of (9.8.3) is

$$y = y_c + y_p = c_1 e^{-3x} + c_2 e^{2x} - \frac{1}{13}(5 \cos 2x - \sin 2x).$$

Notice that we can also write down the general solution of the differential equation

$$y'' + y' - 6y = 4 \sin 2x,$$

since a particular solution will just be $\text{Im}(z_p)$.

Example 9.8.2   Find the general solution of the following differential equation:

$$y'' - 2y' + 5y = 8e^x \sin 2x. \tag{9.8.6}$$

Solution   The complementary function is

$$y_c = e^x(c_1 \cos 2x + c_2 \sin 2x).$$

To determine a particular solution to (9.8.6), we consider the complex counterpart namely,

$$z'' - 2z' + 5z = 8e^{(1 + 2i)x}. \tag{9.8.7}$$

Since $1 + 2i$ is a root of the auxiliary equation, an appropriate trial solution for (9.8.7) is

$$z_p = A_0 x e^{(1 + 2i)x}. \tag{9.8.8}$$

Differentiating with respect to x yields

$$z_p' = A_0 e^{(1 + 2i)x}[(1 + 2i)x + 1],$$

$$z_p'' = A_0 e^{(1 + 2i)x}[(1 + 2i)^2 x + 2(1 + 2i)] = A_0 e^{(1 + 2i)x}[(-3 + 4i)x + 2(1 + 2i)].$$

Thus, substituting into (9.8.7), we require A_0 to satisfy

$$A_0[(-3 + 4i)x + 2(1 + 2i) - 2(1 + 2i)x - 2 + 5x] = 8,$$

that is,

$$A_0 = \frac{2}{i} = -2i.$$

It follows from (9.8.8) that a complex-valued solution to (9.8.7) is

$$z_p = -2ixe^{(1 + 2i)x} = -2ixe^x(\cos 2x + i \sin 2x),$$

and so a particular solution to (9.8.6) is

$$y_p = \text{Im}(z_p) = -2xe^x\cos 2x.$$

Consequently, (9.8.6) has general solution

$$y = e^x(c_1\cos 2x + c_2\sin 2x) - 2xe^x\cos 2x.$$

EXERCISES 9.8

In the following problems use a complex-valued trial solution to determine a particular solution to the given differential equation.

1. $y'' + 2y' + y = 50 \sin 3x.$

2. $y'' - y = 10e^{2x}\cos x.$

3. $y'' + 4y' + 4y = 169 \sin 3x.$

4. $y'' - y' - 2y = 40 \sin^2 x.$

5. $y'' + y = 3e^x\cos 2x.$

6. $y'' + 2y' + 2y = 2e^{-x}\sin x.$

7. $y'' - 4y = 100xe^x\sin x.$

8. $y'' + 2y' + 5y = 4e^{-x}\cos 2x.$

9. $y'' - 2y' + 10y = 24e^x\cos 3x.$

10. $y'' + 16y = 34e^x + 16 \cos 4x - 8 \sin 4x.$

11. $\dfrac{d^2x}{dt^2} + \omega_0^2 x = F_0 \cos \omega t,$ where ω_0 and ω are positive constants, and F_0 is an arbitrary constant. You will need to consider the cases $\omega \neq \omega_0$ and $\omega = \omega_0$ separately.

9.9   THE VARIATION-OF-PARAMETERS METHOD

The method of undetermined coefficients has two severe limitations. First, it is only applicable to differential equations with *constant* coefficients; second, it can be applied only to differential equations whose nonhomogeneous terms are of a very special form, namely, $F(x) = x^k e^{ax}\cos bx$, $F(x) = x^k e^{ax}\sin bx$, or sums of such functions.

For example, we could not use the method of undetermined coefficients to find a particular solution to the differential equation

$$y'' + 4y' - 6y = x^2\ln x.$$

In this section we introduce a very powerful method for obtaining particular solutions to nth-order linear *nonhomogeneous* differential equations, assuming that we know the *general solution* of the associated homogeneous equation. Unlike the method of undetermined coefficients, the variation-of-parameters method is not restricted to differential equations with constant coefficients, and, at least in theory, the actual form of the nonhomogeneous term is immaterial. We will begin by considering the *second-order* case, since the generalization to nth-order will then be fairly straightforward.

Consider the linear second-order nonhomogeneous differential equation *written in the standard form*

$$y'' + p(x)y' + q(x)y = F(x), \tag{9.9.1}$$

where we assume that p, q, and F are continuous on an interval I. Suppose that $y = y_1(x)$ and $y = y_2(x)$ are two *linearly independent* solutions of the associated homogeneous equation

$$y'' + p(x)y' + q(x)y = 0, \tag{9.9.2}$$

on I, so that the general solution of (9.9.2) on I is

$$y_c = c_1 y_1(x) + c_2 y_2(x). \tag{9.9.3}$$

The variation-of-parameters method consists of replacing the constants c_1 and c_2 by functions $u_1(x)$ and $u_2(x)$ (that is, we allow the "parameters" c_1, c_2 to "vary") determined in such a way that the resulting function

$$y_p = u_1(x)y_1(x) + u_2(x)y_2(x) \tag{9.9.4}$$

is a particular solution of (9.9.1).

Differentiating (9.9.4) with respect to x yields

$$y_p' = u_1'y_1 + u_1 y_1' + u_2'y_2 + u_2 y_2'.$$

It is tempting to differentiate this expression once more and then to substitute into (9.9.1) to determine u_1 and u_2. However, if we did this, the resulting expression for y_p'' would involve second derivatives of both u_1 and u_2 and hence we would have complicated our problem. Since y_p contains *two* unknown functions, whereas (9.9.1) gives only one condition for determining them, we might expect that we have the freedom to impose a further constraint on u_1 and u_2. In order to eliminate second derivatives of u_1 and u_2 arising in y_p'', we try for solutions of the form (9.9.4) satisfying the constraint

$$u_1'y_1 + u_2'y_2 = 0. \tag{9.9.5}$$

Imposing this condition the expression for y_p' reduces to

$$y_p' = u_1 y_1' + u_2 y_2',$$

so that

$$y_p'' = u_1'y_1' + u_1 y_1'' + u_2'y_2' + u_2 y_2''.$$

Substituting into (9.9.1) it follows that y_p is a solution provided

$$u_1[y_1'' + p(x)y_1' + q(x)y_1] + u_2[y_2'' + p(x)y_2' + q(x)y_2] + [u_1'y_1' + u_2'y_2'] = F(x).$$

The first two terms in parentheses vanish, since y_1 and y_2 satisfy (9.9.2). We therefore require

$$u_1'y_1' + u_2'y_2' = F(x). \tag{9.9.6}$$

Thus we may conclude that $y_p = u_1 y_1 + u_2 y_2$ is a solution of (9.9.1), provided u_1 and u_2 satisfy (9.9.5) and (9.9.6)—that is,

$$y_1 u_1' + y_2 u_2' = 0,$$

$$y_1' u_1' + y_2' u_2' = F(x).$$

This is a linear algebraic system of equations for the unknowns u_1' and u_2'. The matrix of coefficients of this system has determinant

$$\begin{vmatrix} y_1 & y_2 \\ y_1' & y_2' \end{vmatrix},$$

which is just the Wronskian, $W[y_1, y_2](x)$, of y_1 and y_2. Since y_1 and y_2 are linearly independent on I, $W[y_1, y_2](x)$ is *nonzero* on I and hence the system has a unique solution for u_1' and u_2'. Once this solution has been obtained, u_1 and u_2 can be determined by integration. We have therefore proved the following theorem.

Theorem 9.9.1 (Variation-of-Parameters Method):   Consider

$$y'' + p(x)y' + q(x)y = F(x), \tag{9.9.7}$$

where p, q, and F are assumed to be (at least) continuous on the interval I.   Let y_1, y_2 be linearly independent solutions of the associated homogeneous equation

$$y'' + p(x)y' + q(x)y = 0,$$

on I.   Then a particular solution of (9.9.7) is

$$y_p = u_1 y_1 + u_2 y_2, \tag{9.9.8}$$

where u_1 and u_2 satisfy

$$y_1 u_1' + y_2 u_2' = 0, \tag{9.9.9}$$

$$y_1' u_1' + y_2' u_2' = F(x). \tag{9.9.10}$$

REMARK   If we define the determinants W_1 and W_2 by

$$W_1(x) = \begin{vmatrix} 0 & y_2 \\ F & y_2' \end{vmatrix}, \qquad W_2(x) = \begin{vmatrix} y_1 & 0 \\ y_1' & F \end{vmatrix},$$

respectively, then, using Cramer's rule, the solution to (9.9.9), (9.9.10) can be written as

$$u_1' = \frac{W_1(x)}{W(x)}, \qquad u_2' = \frac{W_2(x)}{W(x)},$$

where, for simplicity, we have denoted the Wronskian $W[y_1, y_2](x)$ by $W(x)$.   Consequently, the particular solution (9.9.8) can be written as

$$y_p(x) = y_1(x) \int^x \frac{W_1(t)}{W(t)}\, dt + y_2(x) \int^x \frac{W_2(t)}{W(t)}\, dt. \tag{9.9.11}$$

This formula can easily be generalized to higher-order differential equations (see the following).

Example 9.9.1   Solve $y'' + y = \sec x$.

Solution Two linearly independent solutions to the associated homogeneous equation are $y_1 = \cos x$, $y_2 = \sin x$. Thus a particular solution of the given differential equation is

$$y_p = u_1 y_1 + u_2 y_2, \qquad (9.9.12)$$

where

$$(\cos x)u_1' + (\sin x)u_2' = 0,$$

$$-(\sin x)u_1' + (\cos x)u_2' = \sec x.$$

By Cramer's rule the solution of this system is

$$u_1' = \frac{W_1(x)}{W(x)} = -\frac{\sec x \sin x}{1}, \qquad u_2' = \frac{W_2(x)}{W(x)} = \frac{\sec x \cos x}{1} = 1.$$

Consequently

$$u_1 = -\int \sec x \sin x \, dx = -\int \frac{\sin x}{\cos x} \, dx = \ln|\cos x|,$$

and

$$u_2 = \int \sec x \cos x \, dx = x,$$

where we have set the integration constants to zero, since we require only one particular solution. Substitution into (9.9.12) yields

$$y_p = (\cos x) \ln|\cos x| + x \sin x,$$

so that the general solution of the given differential equation is

$$y = c_1 \cos x + c_2 \sin x + (\cos x) \ln|\cos x| + x \sin x.$$

Example 9.9.2 Find the general solution of $y'' + 4y' + 4y = e^{-2x}\ln x$.

Solution In this case two linearly independent solutions to the associated homogeneous equation are $y_1 = e^{-2x}$, $y_2 = xe^{-2x}$, and hence a particular solution of the given differential equation is

$$y_p = u_1 y_1 + u_2 y_2,$$

where u_1 and u_2 satisfy

$$e^{-2x}u_1' + xe^{-2x}u_2' = 0,$$

$$-2e^{-2x}u_1' + e^{-2x}(1 - 2x)u_2' = e^{-2x}\ln x.$$

The solution to this system is

$$u_1' = \frac{W_1(x)}{W(x)} = \frac{\begin{vmatrix} 0 & xe^{-2x} \\ e^{-2x}\ln x & e^{-2x}(1-2x) \end{vmatrix}}{\begin{vmatrix} e^{-2x} & xe^{-2x} \\ -2e^{-2x} & e^{-2x}(1-2x) \end{vmatrix}} = -\frac{e^{-4x}x\ln x}{e^{-4x}} = -x\ln x,$$

$$u_2' = \frac{W_2(x)}{W(x)} = \frac{\begin{vmatrix} e^{-2x} & 0 \\ -2e^{-2x} & e^{-2x}\ln x \end{vmatrix}}{\begin{vmatrix} e^{-2x} & xe^{-2x} \\ -2e^{-2x} & e^{-2x}(1-2x) \end{vmatrix}} = \frac{e^{-4x}\ln x}{e^{-4x}} = \ln x.$$

Integrating both of these expressions by parts (and setting the integration constants to zero), we obtain

$$u_1 = -\left(\tfrac{1}{2}x^2\ln x - \tfrac{1}{4}x^2\right) = \tfrac{1}{4}x^2(1-2\ln x), \qquad u_2 = x\ln x - x = x(\ln x - 1).$$

Thus

$$y_p = \tfrac{1}{4}x^2 e^{-2x}(1 - 2\ln x) + x^2 e^{-2x}(\ln x - 1),$$

that is,

$$y_p = \tfrac{1}{4}x^2 e^{-2x}(2\ln x - 3).$$

Consequently the general solution of the given differential equation is

$$y = e^{-2x}[c_1 + c_2 x + \tfrac{1}{4}x^2(2\ln x - 3)].$$

GENERALIZATION TO HIGHER ORDER

We now consider the generalization of the variation-of-parameters method to linear nonhomogeneous differential equations of arbitrary order n. In this case the basic equation is

$$y^{(n)} + a_1(x)y^{(n-1)} + \cdots + a_{n-1}(x)y' + a_n(x)y = F(x), \qquad (9.9.13)$$

where we assume that the functions $a_i(x)$ and F are at least continuous on the interval I. Let $\{y_1(x), y_2(x), \ldots, y_n(x)\}$ be a linearly independent set of solutions to the associated homogeneous equation

$$y^{(n)} + a_1(x)y^{(n-1)} + \cdots + a_{n-1}(x)y' + a_n(x)y = 0, \qquad (9.9.14)$$

on I, so that the general solution of (9.9.14) on I is

$$y_c = c_1 y_1(x) + c_2 y_2(x) + \cdots + c_n y_n(x).$$

We now look for a particular solution of (9.9.13) of the form

$$y_p = u_1 y_1(x) + u_2 y_2(x) + \cdots + u_n y_n(x). \qquad (9.9.15)$$

The idea is to substitute for y_p into (9.9.13) and choose the functions $u_1, u_2, \ldots, u_n$ so that the resulting y_p is indeed a solution. However, (9.9.13) will only give one constraint on the functions u_i and their derivatives. Since we have n functions we might expect that we can impose $(n-1)$ further constraints on these functions. Following the steps taken in the second order case we differentiate y_p n times, while imposing the constraint that the sum of the terms involving derivatives of the u_i

that arise at each stage (except the last) should equal zero. For example, at the first stage we obtain

$$y_p' = u_1 y_1' + u_1' y_1 + u_2 y_2' + u_2' y_2 + \cdots + u_n y_n' + u_n' y_n$$

and so we impose the constraint

$$u_1' y_1 + u_2' y_2 + \cdots + u_n' y_n = 0,$$

in which case the preceding expression for y_p' reduces to

$$y_p' = u_1 y_1' + u_2 y_2' + \cdots + u_n y_n'.$$

Continuing in this manner leads to the following expressions for y_p and its derivatives

$$
\begin{aligned}
y_p &= u_1 y_1 \quad + u_2 y_2 \quad + \cdots + u_n y_n, \\
y_p' &= u_1 y_1' \quad + u_2 y_2' \quad + \cdots + u_n y_n', \\
y_p'' &= u_1 y_1'' \quad + u_2 y_2'' \quad + \cdots + u_n y_n'', \\
&\ \vdots \\
y_p^{(n)} &= u_1 y_1^{(n)} + u_2 y_2^{(n)} + \cdots + u_n y_n^{(n)}
\end{aligned}
\qquad (9.9.16)
$$

$$+ \, [u_1' y_1^{(n-1)} + u_2' y_2^{(n-1)} + \cdots + u_n' y_n^{(n-1)}],$$

together with the corresponding constraint conditions

$$
\begin{aligned}
u_1' y_1 \quad + u_2' y_2 \quad &+ \cdots + u_n' y_n \quad = 0, \\
u_1' y_1' \quad + u_2' y_2' \quad &+ \cdots + u_n' y_n' \quad = 0, \\
&\ \vdots \\
u_1' y_1^{(n-2)} + u_2' y_2^{(n-2)} + \cdots &+ u_n' y_n^{(n-2)} = 0.
\end{aligned}
\qquad (9.9.17)
$$

Substitution from (9.9.16) into (9.9.13) yields the following condition in order for y_p to be a solution (simply multiply each equation in (9.9.16) by the appropriate a_i and add the elements in each column):

$$
\begin{aligned}
u_1 [y_1^{(n)} + a_1 y_1^{(n-1)} + \cdots &+ a_{n-1} y_1' + a_n y_1] \\
+ \, u_2 [y_2^{(n)} + a_1 y_2^{(n-1)} + \cdots &+ a_{n-1} y_2' + a_n y_2] \\
+ \cdots + u_n [y_n^{(n)} + a_1 y_n^{(n-1)} + \cdots &+ a_{n-1} y_n' + a_n y_n] \\
+ \, [u_1' y_1^{(n-1)} + u_2' y_2^{(n-1)} + \cdots &+ u_n' y_n^{(n-1)}] = F(x).
\end{aligned}
$$

The terms in each of the brackets except the last vanish since $y_1, y_2, \ldots, y_n$ are solutions of (9.9.14). We are therefore left with the condition

$$u_1' y_1^{(n-1)} + u_2' y_2^{(n-1)} + \cdots + u_n' y_n^{(n-1)} = F(x).$$

Combining this with the constraints given in (9.9.17) leads to the following linear system of equations for determining $u_1', u_2', \ldots, u_n'$

$$y_1 u_1' \quad + y_2 u_2' \quad + \cdots + y_n u_n' \quad = 0,$$
$$y_1' u_1' \quad + y_2' u_2' \quad + \cdots + y_n' u_n' \quad = 0,$$
$$\vdots \qquad\qquad\qquad (9.9.18)$$
$$y_1^{(n-2)} u_1' + y_2^{(n-2)} u_2' + \cdots + y_n^{(n-2)} u_n' = 0,$$
$$y_1^{(n-1)} u_1' + y_2^{(n-1)} u_2' + \cdots + y_n^{(n-1)} u_n' = F(x).$$

The determinant of the matrix of coefficients of this system is the Wronskian of the functions $y_1, y_2, \ldots, y_n$, which is necessarily nonzero on I, since $y_1, y_2, \ldots, y_n$ are linearly independent on I. It follows that the system (9.9.18) has a unique solution for $u_1', u_2', \ldots, u_n'$, from which we can determine $u_1, u_2, \ldots, u_n$ by integration. Having found $u_1, u_2, \ldots, u_n$, we can obtain y_p by substitution into (9.9.15).

Our results are summarized in the following theorem.

Theorem 9.9.2   Consider

$$y^{(n)} + a_1(x)y^{(n-1)} + \cdots + a_{n-1}(x)y' + a_n(x)y = F(x), \qquad (9.9.19)$$

where $a_1, a_2, \ldots, a_n, F$ are assumed to be (at least) continuous on the interval I. Let $y_1, y_2, \ldots, y_n$ be linearly independent solutions of the associated homogeneous equation

$$y^{(n)} + a_1(x)y^{(n-1)} + \cdots + a_{n-1}(x)y' + a_n(x)y = 0,$$

on I. Then a particular solution of (9.9.19) is

$$y_p = u_1 y_1 + u_2 y_2 + \cdots + u_n y_n,$$

where the functions $u_1, u_2, \ldots, u_n$ satisfy (9.9.18).

REMARK   In order to generalize the formula (9.9.11) to the nth-order case, we let $W_k(x)$ denote the determinant that is obtained when the kth column of $W[y_1, y_2, \ldots, y_n](x)$ is replaced by

$$\begin{bmatrix} 0 \\ 0 \\ \vdots \\ F(x) \end{bmatrix}.$$

Then, using Cramer's rule, it follows that the solution of the system (9.9.18) can be written in the form

$$u_k' = \frac{W_k(x)}{W(x)}, \qquad k = 1, 2, \ldots, n,$$

where, for simplicity, we have denoted $W[y_1, y_2, \ldots, y_n](x)$ by $W(x)$. A particular solution to the differential equation (9.9.19) is therefore given by

$$y_p(x) = y_1(x) \int^x \frac{W_1(t)}{W(t)}\, dt + y_2(x) \int^x \frac{W_2(t)}{W(t)}\, dt + \cdots + y_n(x) \int^x \frac{W_n(t)}{W(t)}\, dt.$$

Although this is an elegant mathematical formula, it is usually computationally more efficient to solve the system (9.8.18) using Gaussian elimination rather than Cramer's rule when dealing with differential equations of order higher than two.

Example 9.9.3   Find the general solution of

$$y''' - 3y'' + 3y' - y = 36e^x\ln x. \tag{9.9.20}$$

Solution   In this case the auxiliary polynomial of the associated homogeneous equation is

$$P(r) = r^3 - 3r^2 + 3r - 1 = (r-1)^3.$$

It follows that three linearly independent solutions are

$$y_1 = e^x, \qquad y_2 = xe^x, \qquad y_3 = x^2e^x.$$

The system (9.9.18) therefore assumes the form

$$u_1' + xu_2' + x^2u_3' = 0,$$

$$u_1' + (x+1)u_2' + (x^2+2x)u_3' = 0,$$

$$u_1' + (x+2)u_2' + (x^2+4x+2)u_3' = 36 \ln x,$$

where we have divided each equation by e^x. In order to solve this system, we reduce its augmented matrix to row echelon form.

$$\begin{bmatrix} 1 & x & x^2 \\ 1 & x+1 & x^2+2x \\ 1 & x+2 & x^2+4x+2 \end{bmatrix}\begin{array}{|c} 0 \\ 0 \\ 36 \ln x \end{array}\ \xrightarrow[R_3 \to R_3 - R_1]{R_2 \to R_2 - R_1}\ \begin{bmatrix} 1 & x & x^2 \\ 0 & 1 & 2x \\ 0 & 2 & 4x+2 \end{bmatrix}\begin{array}{|c} 0 \\ 0 \\ 36 \ln x \end{array}$$

$$\xrightarrow{R_3 \to R_3 - 2R_2}\ \begin{bmatrix} 1 & x & x^2 \\ 0 & 1 & 2x \\ 0 & 0 & 2 \end{bmatrix}\begin{array}{|c} 0 \\ 0 \\ 36 \ln x \end{array}\ \xrightarrow{R_3 \to \frac{1}{2}R_3}\ \begin{bmatrix} 1 & x & x^2 \\ 0 & 1 & 2x \\ 0 & 0 & 1 \end{bmatrix}\begin{array}{|c} 0 \\ 0 \\ 18 \ln x \end{array}.$$

Consequently,

$$u_1' = 18x^2 \ln x, \qquad u_2' = -36x \ln x, \qquad u_3' = 18 \ln x.$$

Integrating we obtain

$$u_1(x) = 18 \int x^2\ln x\, dx = 2x^3(3 \ln x - 1),$$

$$u_2(x) = -36 \int x \ln x\, dx = 9x^2(1 - 2 \ln x),$$

$$u_3(x) = 18 \int \ln x\, dx = 18x(\ln x - 1),$$

where we have set the integration constants to zero without loss of generality. It follows that a particular solution to (9.9.20) is

$$y_p = u_1y_1 + u_2y_2 + u_3y_3,$$

that is,

$$y_p = x^3e^x(6 \ln x - 11).$$

The general solution of the given differential equation is therefore

$$y = e^x[c_1 + c_2x + c_3x^2 + x^3(6 \ln x - 11)].$$

EXERCISES 9.9

In problems 1–14 find the general solution to the given differential equation.

1. $y'' + 6y' + 9y = \dfrac{2e^{-3x}}{x^2 + 1}$.

2. $y'' - 4y = \dfrac{8}{e^{2x} + 1}$.

3. $y'' - 4y' + 5y = e^{2x}\tan x$.

4. $(D - 3)^2 y = 4e^{3x}\ln x$.

5. $y'' + 4y' + 4y = \dfrac{e^{-2x}}{x^2}$, $x > 0$.

6. $y'' + 9y = 18\sec^3(3x)$.

7. $y'' - y = 2\tanh x$.

8. $y'' - 2my' + m^2 y = \dfrac{e^{mx}}{1 + x^2}$, m constant.

9. $y'' - 2y' + y = 4e^x x^{-3}\ln x$.

10. $y'' + 2y' + y = \dfrac{e^{-x}}{\sqrt{4 - x^2}}$.

11. $y'' + 2y' + 17y = \dfrac{64e^{-x}}{3 + \sin^2(4x)}$.

12. $y'' + 9y = \dfrac{36}{4 - \cos^2(3x)}$.

13. $y'' - 10y' + 25y = \dfrac{2e^{5x}}{4 + x^2}$.

14. $y'' - 6y' + 13y = 4e^{3x}\sec^2(2x)$, $0 \le x < \pi/4$.

15. Consider the electric circuit given in problem 33 on page 349. Taking $R = 0$ Ω, $L = 10$ H, $C = 0.025$ F, and $E(t) = 10\sec 2t$ volts, find the current, i, in the circuit assuming that the initial charge and current are zero.

16. Consider the differential equation

$$y'' - 2ay' + a^2 y = e^{ax}g(x), \qquad (i)$$

where a is a constant.
(a) Derive the following particular solution to (i):

$$y_p(x) = -e^{ax}\int^x tg(t)\,dt + xe^{ax}\int^x g(t)\,dt.$$

(b) Find the general solution of (i) when
(i) $g(x) = \dfrac{\alpha}{x^2 + \beta^2}$.

(ii) $g(x) = \dfrac{\alpha}{\sqrt{\beta^2 - x^2}}$.

(iii) $g(x) = x^\alpha \ln x$.

In each case α and β are constants.
(c) Show that the solution in (a) can be written in the form

$$y_p = e^{ax}\int_{x_0}^x \int_{t_0}^t g(s)\,ds\,dt,$$

where x_0 and t_0 are constants.

17. Consider the differential equation

$$y'' + p(x)y' + q(x)y = F(x), \qquad (ii)$$

where p, q, and F are continuous on the interval $[a, b]$. Let y_1, y_2 be linearly independent solutions to the associated homogeneous equation on $[a, b]$. Show that for $a \le x \le b$ a particular solution to (ii) is

$$y_p(x) = \int_a^x \left[\dfrac{y_1(t)\,y_2(x) - y_2(t)\,y_1(x)}{W[y_1, y_2](t)} \right] F(t)\,dt.$$

18. Consider the differential equation

$$y'' + y = F(x),$$

where F is continuous on the interval $[a, b]$.
(a) For $a \le x \le b$, derive the following particular solution to this differential equation:

$$y_p = \int_a^x F(t)\,\sin(x - t)\,dt.$$

(b) If $x_0 \in [a, b]$, show that the solution of the initial value problem

$$y'' + y = F(x), \qquad y(x_0) = y_0, \; y'(x_0) = y_1$$

is

$$y = y_0\cos(x - x_0) + y_1\sin(x - x_0)$$

$$+ \int_{x_0}^x F(t)\sin(x - t)\,dt.$$

In problems 19–22 use the variation-of-parameters technique to determine a particular solution of the given differential equation.

19. $y''' - 6y'' + 12y' - 8y = 36e^{2x}\ln x$.

20. $y''' - 3y'' + 3y' - y = \dfrac{2e^x}{x^2}$, $x > 0$.

21. $y''' + 3y'' + 3y' + y = \dfrac{2e^{-x}}{1 + x^2}$.

22. $y''' - 6y'' + 9y' = 12e^{3x}$. Suggest a better method for solving this problem.

23. If F is continuous on the interval $[a, b]$, use the variation-of-parameters technique to show that a particular solution to

$$(D - r)^3 y = F(x), \qquad r \text{ constant,}$$

is

$$y_p(x) = \frac{1}{2} \int_a^x F(t)(x - t)^2 e^{r(x-t)} \, dt.$$

9.10   A DIFFERENTIAL EQUATION WITH NONCONSTANT COEFFICIENTS

In this section we consider a particular type of homogeneous differential equation that has nonconstant coefficients. The solution of this differential equation will be useful in Chapter 13 and also will enable us to give a further illustration of the power of the variation-of-parameters technique introduced in the previous section.

> **Definition 9.10.1:**   A differential equation of the form
>
> $$x^n \frac{d^n y}{dx^n} + a_1 x^{n-1} \frac{d^{n-1} y}{dx^{n-1}} + \cdots + a_{n-1} x \frac{dy}{dx} + a_n y = 0,$$
>
> where $a_1, a_2, \ldots, a_n$ are constants, is called a **Cauchy–Euler** equation.

Cauchy–Euler equations differ from a constant coefficient equation since each derivative term, $d^k y / dx^k$, is multiplied by x^k. Notice that if we replace x by λx, where λ is a constant, then the form of a Cauchy–Euler equation is left unaltered. Such a rescaling of x can be interpreted as a dimensional change (for example, inches $\rightarrow$ centimeters) and so Cauchy–Euler equations are sometimes called *equidimensional* equations.

We begin our analysis by restricting attention to the second-order case and will assume that $x > 0$. Thus consider the differential equation

$$x^2 y'' + a_1 x y' + a_2 y = 0, \qquad x > 0, \tag{9.10.1}$$

where a_1 and a_2 are constants. The solution technique is based on the observation that if we substitute $y = x^r$ into (9.10.1), then each of the resulting terms on the left-hand side will be multiplied by the same power of x, and hence we suspect that there may be solutions of the form

$$y = x^r \tag{9.10.2}$$

for an appropriately chosen constant r. To investigate this possibility we differentiate (9.10.2) twice to obtain

$$y' = r x^{r-1}, \qquad y'' = r(r-1) x^{r-2}.$$

Substituting into (9.10.1) yields the condition

$$x^r [r(r-1) + a_1 r + a_2] = 0,$$

so that (9.10.2) is indeed a solution of (9.10.1) provided r satisfies

$$r(r-1) + a_1 r + a_2 = 0,$$

that is,

$$r^2 + (a_1 - 1)r + a_2 = 0. \tag{9.10.3}$$

This is referred to as the **indicial equation** associated with (9.10.1). The roots of (9.10.3) are

$$r_1 = \frac{-(a_1 - 1) + \sqrt{(a_1 - 1)^2 - 4a_2}}{2}, \qquad r_2 = \frac{-(a_1 - 1) - \sqrt{(a_1 - 1)^2 - 4a_2}}{2},$$

so that there are three cases to consider.

CASE 1 r_1, r_2 real and distinct   In this case two solutions to (9.10.1) are

$$y_1 = x^{r_1}, \qquad y_2 = x^{r_2}.$$

It is easily shown that

$$W[y_1, y_2](x) = (r_2 - r_1)x^{r_1 + r_2 - 1},$$

so that y_1 and y_2 are linearly independent on $(0, \infty)$. Consequently, the general solution to (9.10.1) is

$$y = c_1 x^{r_1} + c_2 x^{r_2}. \tag{9.10.4}$$

CASE 2 $r_1 = r_2 = -\dfrac{(a_1 - 1)}{2}$ In this case we obtain only one solution, namely,

$$y = x^{r_1}.$$

We try for a second solution of the form

$$y_2 = x^{r_1} u(x).$$

Differentiating y_2 twice with respect to x yields

$$y_2' = x^{r_1} u' + r_1 x^{r_1 - 1} u, \qquad y_2'' = x^{r_1} u'' + 2r_1 x^{r_1 - 1} u' + r_1(r_1 - 1)x^{r_1 - 2} u.$$

Substituting into (9.10.1) we obtain the following equation that u must satisfy in order that y_2 is a solution:

$$x^2[x^{r_1} u'' + 2r_1 x^{r_1 - 1} u' + r_1(r_1 - 1)x^{r_1 - 2} u] + a_1 x[x^{r_1} u' + r_1 x^{r_1 - 1} u] + a_2 x^{r_1} u = 0.$$

Equivalently,

$$x^{r_1 + 2} u'' + (2r_1 + a_1)x^{r_1 + 1} u' + x^{r_1}[r_1(r_1 - 1) + a_1 r_1 + a_2] u = 0.$$

The last term on the left-hand side vanishes, since r_1 is a root of the indicial equation (9.10.3). Thus u must satisfy

$$u'' + \frac{1}{x} u' = 0,$$

where we have substituted for $r_1 = -(a_1 - 1)/2$. This separable equation can be written as

$$\frac{u''}{u'} = -\frac{1}{x},$$

which can be integrated directly to yield

$$\ln|u'| = -\ln x + \ln c.$$

We can therefore choose

$$u' = \frac{1}{x}.$$

Integrating once more we obtain

$$u = \ln x,$$

where we have set the integration constant to zero, since we require only one solution. Consequently a second solution to (9.10.1) in this case is

$$y_2 = x^{r_1}\ln x.$$

It is easily shown that

$$W[y_1, y_2](x) = x^{2r_1-1},$$

so that y_1 and y_2 are linearly independent on $(0, \infty)$. The general solution to (9.10.1) is therefore given by

$$y = c_1 x^{r_1} + c_2 x^{r_1}\ln x = x^{r_1}(c_1 + c_2\ln x). \tag{9.10.5}$$

CASE 3 Complex conjugate roots, $r_1 = a + ib$, $r_2 = a - ib$, $b \neq 0$ In this case a complex-valued solution to (9.10.1) is

$$Y = x^{a+ib},$$

that is (see Section 9.5),

$$Y = x^a[\cos(b \ln x) + i \sin(b \ln x)].$$

Taking the real and imaginary parts of this complex-valued solution yields the two real-valued solutions

$$y_1 = x^a\cos(b \ln x), \qquad y_2 = x^a\sin(b \ln x).$$

The Wronskian of these solutions is

$$W[y_1, y_2](x) = bx^{2a-1},$$

which is nonzero because $b \neq 0$. Consequently y_1 and y_2 are linearly independent on $(0, \infty)$, and so the general solution of (9.10.1) in this case is

$$y = x^a[c_1\cos(b \ln x) + c_2\sin(b \ln x)]. \tag{9.10.6}$$

In the above derivation we have assumed that $x > 0$. When $x < 0$, it is easily shown that the substitution $y = (-x)^r$ also yields the indicial equation (9.10.3).[1] Thus, linearly independent solutions to (9.10.1) on $(-\infty, 0)$ can be determined by replacing x with $(-x)$, in (9.10.4), (9.10.5), and (9.10.6). Consequently, if we

---

[1] We cannot use the substitution $y = x^r$ when $x < 0$, since x^r will not be defined for all values of r.

replace x by $|x|$ in these solutions, we will obtain solutions to (9.10.1) that are valid for all $x \neq 0$. These results are summarized in Table 9.10.1.

TABLE 9.10.1

Roots of indicial equation $r^2 + (a_1 - 1)r + a_2 = 0$	Linearly independent solutions of $x^2 y'' + a_1 x y' + a_2 y = 0.$
Real distinct: $r_1 \neq r_2$	$y_1 = |x|^{r_1},$ $\qquad y_2 = |x|^{r_2}.$
Real repeated: $r_1 = r_2$	$y_1 = |x|^{r_1},$ $\qquad y_2 = |x|^{r_1} \ln |x|.$
Complex conjugates: $r_1 = a + ib, r_2 = a - ib$	$y_1 = |x|^a \cos (b \ln |x|),$ $y_2 = |x|^a \sin (b \ln |x|).$

Example 9.10.1 Solve

$$x^2 \frac{d^2 y}{dx^2} - x \frac{dy}{dx} - 8y = 0, \qquad x > 0. \tag{9.10.7}$$

Solution Substituting $y = x^r$ into (9.10.7) yields the indicial equation

$$r(r - 1) - r - 8 = 0,$$

that is,

$$r^2 - 2r - 8 = 0.$$

Factoring the quadratic yields

$$(r - 4)(r + 2) = 0.$$

It follows that two linearly independent solutions to (9.10.7) are

$$y_1 = x^4, \qquad y_2 = x^{-2}.$$

Thus the general solution to (9.10.7) on $(0, \infty)$ is

$$y = c_1 x^4 + c_2 x^{-2}.$$

Example 9.10.2 Find the general solution of

$$x^2 \frac{d^2 y}{dx^2} - 3x \frac{dy}{dx} + 13y = 0, \qquad x > 0. \tag{9.10.8}$$

Solution Substituting $y = x^r$ into (9.10.8) yields the indicial equation

$$r^2 - 4r + 13 = 0,$$

which has the complex conjugate roots

$$r = 2 \pm 3i.$$

It follows that two linearly independent solutions to (9.10.8) on $(0, \infty)$ are

$$y_1 = x^2 \cos(3 \ln x), \qquad y_2 = x^2 \sin(3 \ln x),$$

so that the general solution of (9.10.8) is

$$y = c_1 x^2 \cos(3 \ln x) + c_2 x^2 \sin(3 \ln x).$$

Now consider the nonhomogeneous equation

$$x^2 y'' + a_1 x y' + a_2 y = F(x),$$

where a_1 and a_2 are constants. Since the associated homogeneous equation is a Cauchy–Euler equation, we can determine the complementary function and hence the *variation-of-parameters* method can be used to determine a particular solution. We illustrate with an example.

Example 9.10.3   Find the general solution of

$$x^2 y'' - 3xy' + 4y = x^2 \ln x. \qquad (9.10.9)$$

Solution   The associated homogeneous equation is the Cauchy–Euler equation

$$x^2 y'' - 3xy' + 4y = 0. \qquad (9.10.10)$$

Substituting $y = x^r$ into this equation yields the indicial equation

$$r^2 - 4r + 4 = 0,$$

that is,

$$(r - 2)^2 = 0.$$

It follows that two linearly independent solutions to (9.10.10) on $(0, \infty)$ are

$$y_1 = x^2, \qquad y_2 = x^2 \ln x.$$

In order to apply the variation-of-parameters technique we first rewrite (9.10.9) in standard form:

$$y'' - \frac{3}{x} y' + \frac{4}{x^2} y = \ln x.$$

According to the variation-of-parameters technique, a particular solution to this nonhomogeneous equation is

$$y_p = y_1 u_1 + y_2 u_2 \qquad (9.10.11)$$

where u_1 and u_2 are determined from

$$x^2 u_1' + x^2 (\ln x) u_2' = 0, \qquad (9.10.12)$$

$$2x u_1' + (2x \ln x + x) u_2' = \ln x. \qquad (9.10.13)$$

Therefore,

$$W_1(x) = \begin{vmatrix} 0 & x^2 \ln x \\ \ln x & 2x \ln x + x \end{vmatrix} = -x^2 (\ln x)^2, \qquad W_2(x) = \begin{vmatrix} x^2 & 0 \\ 2x & \ln x \end{vmatrix} = x^2 \ln x,$$

and

$$W[y_1, y_2](x) = \begin{vmatrix} x^2 & x^2 \ln x \\ 2x & 2x \ln x + x \end{vmatrix} = x^3.$$

Consequently, applying Cramer's rule, the solution to (9.10.12) and (9.10.13) is

$$u_1' = -\frac{1}{x}(\ln x)^2, \qquad u_2' = \frac{1}{x}\ln x,$$

which implies that

$$u_1(x) = -\frac{1}{3}(\ln x)^3, \qquad u_2(x) = \frac{1}{2}(\ln x)^2,$$

where we have set the integration constants to zero without loss of generality. Substituting into (9.10.11) yields

$$y_p = -\frac{1}{3}x^2(\ln x)^3 + \frac{1}{2}x^2(\ln x)^3$$

$$= \frac{1}{6}x^2(\ln x)^3.$$

Thus (9.10.9) has general solution

$$y = c_1 x^2 + c_2 x^2 \ln x + \frac{1}{6}x^2(\ln x)^3,$$

that is,

$$y = \frac{1}{6}x^2[c_1 + c_2 \ln x + (\ln x)^3],$$

where we have redefined the constants.

GENERALIZATION TO HIGHER ORDER

Now consider the general Cauchy–Euler equation of order n,

$$x^n y^{(n)} + a_1 x^{n-1} y^{(n-1)} + \cdots + a_{n-1} xy' + a_n y = 0, \qquad (9.10.14)$$

where $a_1, a_2, \ldots, a_n$ are constants, on the interval $x > 0$. Once more we begin by substituting $y = x^r$. The result is the indicial equation

$$r(r-1)(r-2)\cdots(r-n+1) + a_1 r(r-1)\cdots(r-n+2)$$

$$+ \cdots + a_{n-1} r + a_n = 0. \qquad (9.10.15)$$

In the case that (9.10.15) has n distinct roots, say $r_1, r_2, \ldots, r_n$, we directly obtain the n solutions

$$x^{r_1}, x^{r_2}, \ldots, x^{r_n},$$

and it can be shown that these solutions are linearly independent on $(0, \infty)$. Of course if some of the roots are complex, then we must take the real and imaginary parts of the corresponding complex-valued solutions in order to determine real-valued solutions. If, however, $r = r_1$ is a root of (9.10.15) of multiplicity k, then we cannot directly determine the appropriate number of solutions. Based on our

SEC. 9.10: A Differential Equation with Nonconstant Coefficients

experience in the second-order case, we might suspect that there are k solutions of the form

$$x^{r_1}, \; x^{r_1}\ln x, \; x^{r_1}(\ln x)^2, \; \ldots, \; x^{r_1}(\ln x)^{k-1}. \tag{9.10.16}$$

We now show that this is indeed correct. The key idea is that the change of variables $x = e^z$, or equivalently $z = \ln x$, transforms (9.10.14) into a constant-coefficient equation, which can be solved via the technique of Section 9.6. To establish this we require the following lemma.

Lemma 9.10.1: If y is a sufficiently smooth function of x and $x = e^z$, then

$$x^k \frac{d^k y}{dx^k} = D(D-1)(D-2) \cdots (D-k+1)\,y, \qquad k = 1, 2, 3, \ldots, \tag{9.10.17}$$

where $D = d/dz$.

PROOF   The proof of this result requires the chain rule and mathematical induction. Since $x = e^z$, we have $z = \ln x$, so that $dz/dx = 1/x$. Thus, by the chain rule,

$$\frac{dy}{dx} = \frac{dy}{dz}\frac{dz}{dx} = \frac{1}{x}\frac{dy}{dz},$$

which implies that

$$x\frac{dy}{dx} = Dy, \tag{9.10.18}$$

where $D = d/dz$. Thus (9.10.17) is certainly true when $k = 1$. Now suppose that the result is true when $k = m$. That is,

$$x^m \frac{d^m y}{dx^m} = D(D-1)(D-2) \cdots (D-m+1)y. \tag{9.10.19}$$

We must show that this implies its validity when $k = m+1$. We proceed as follows. Using the product rule we have

$$x\frac{d}{dx}\left(x^m \frac{d^m y}{dx^m}\right) = x^{m+1}\frac{d^{m+1}y}{dx^{m+1}} + mx^m \frac{d^m y}{dx^m},$$

so that

$$x^{m+1}\frac{d^{m+1}y}{dx^{m+1}} = x\frac{d}{dx}\left(x^m \frac{d^m y}{dx^m}\right) - mx^m \frac{d^m y}{dx^m}.$$

Substituting from (9.10.18) for $x(d/dx) = D$ and from (9.10.19) for $x^m\,(d^m y/dx^m)$ into this equation yields

$$x^{m+1}\frac{d^{m+1}y}{dx^{m+1}} = D[D(D-1)(D-2) \cdots (D-m+1)y]$$

$$- m[D(D-1)(D-2) \cdots (D-m+1)y];$$

that is, since we can interchange the order of the factors in a polynomial differential operator,

$$x^{m+1}\frac{d^{m+1}y}{dx^{m+1}} = D(D-1)(D-2)\cdots(D-m+1)(D-m)y.$$

We have therefore shown that the validity of (9.10.17) when $k = m$ implies its validity when $k = m + 1$. Since the result is true when $k = 1$, it follows, by induction, that it is valid for all positive integers k. ∎

REMARK   Although the rule for transforming derivatives given in (9.10.17) looks quite formidable, it is quite easy to remember. We write out the first three derivatives in order to elucidate this:

$$x\frac{dy}{dx} = Dy, \qquad x^2\frac{d^2y}{dx^2} = D(D-1)y, \qquad x^3\frac{d^3y}{dx^3} = D(D-1)(D-2)y,$$

where $D \equiv d/dz$.

We can now prove the main result.

Theorem 9.10.1:   The change of variable $x = e^z$ transforms the Cauchy–Euler equation

$$x^n\frac{d^ny}{dx^n} + a_1x^{n-1}\frac{d^{n-1}}{dx^{n-1}} + \cdots + a_{n-1}x\frac{dy}{dx} + a_ny = 0, \qquad x > 0, \qquad (9.10.20)$$

into the constant coefficient equation

$$[D(D-1)(D-2)\cdots(D-n+1) + a_1D(D-1)\cdots(D-n+2)$$
$$+ \cdots + a_{n-1}D + a_n]y = 0. \quad (9.10.21)$$

PROOF   Equation (9.10.21) follows directly by substituting for each of the terms $x^k(d^ky/dx^k)$ in (9.10.20) using the previous lemma. ∎

The auxiliary polynomial of the constant coefficient equation (9.10.21) is

$$r(r-1)(r-2)\cdots(r-n+1) + a_1r(r-1)\cdots(r-n+2) + \cdots + a_{n-1}r + a_n = 0,$$

which coincides with the indicial equation (9.10.15) of the original equation. If $r = r_1$ is a root of multiplicity k, then it follows from our results of Section 9.6 that the corresponding linearly independent solutions to (9.10.21), and hence (9.10.20), are

$$e^{r_1z}, ze^{r_1z}, \ldots, z^{k-1}e^{r_1z};$$

that is, since $z = \ln x$,

$$x^{r_1}, x^{r_1}\ln x, x^{r_1}(\ln x)^2, \ldots, x^{r_1}(\ln x)^{k-1}.$$

If $r_1 = a + ib$ is complex, then we can obtain the appropriate real-valued solutions by extracting the real and imaginary parts of the corresponding complex-valued solutions. Thus, corresponding to complex conjugate roots of the indicial equation of multiplicity k, we obtain the $2k$ real-valued solutions

$$x^a\cos(b \ln x), \ x^a\ln x \ \cos(b \ln x), \ x^a(\ln x)^2\cos(b \ln x), \ \ldots, \ x^a(\ln x)^{k-1}\cos(b \ln x),$$

$$x^a\sin(b \ln x), \ x^a\ln x \ \sin(b \ln x), \ x^a(\ln x)^2\sin(b \ln x), \ \ldots, \ x^a(\ln x)^{k-1}\sin(b \ln x).$$

In summary, to solve a Cauchy–Euler equation we can substitute $y = x^r$ into the differential equation to obtain the indicial equation. Corresponding to each root, r_i, of the indicial equation of multiplicity k there are k solutions

$$x^{r_i}, \ x^{r_i}\ln x, \ x^{r_i}(\ln x)^2, \ \ldots, \ x^{r_i}(\ln x)^{k-1},$$

which are linearly independent on $(0, \infty)$.

Example 9.10.4   Find the general solution of

$$x^3y''' + 2x^2y'' + 4xy' - 4y = 0 \qquad (9.10.22)$$

on $(0, \infty)$.

Solution   Substituting $y = x^r$ into (9.10.22) yields the indicial equation

$$r(r-1)(r-2) + 2r(r-1) + 4r - 4 = 0,$$

that is,

$$r^3 - r^2 + 4r - 4 = 0,$$

which can be factored as

$$(r-1)(r^2 + 4) = 0.$$

The roots of the indicial equation are therefore

$$r = 1, \qquad r = \pm \, 2i,$$

so that three linearly independent solutions to (9.10.22) on $(0, \infty)$ are

$$y_1 = x, \qquad y_2 = \cos(2 \ln x), \qquad y_3 = \sin(2 \ln x).$$

Consequently the general solution is

$$y = c_1x + c_2\cos(2 \ln x) + c_3\sin(2 \ln x).$$

Example 9.10.5   Find the general solution of

$$x^3y''' - 3x^2y'' + 7xy' - 8y = 0, \qquad x > 0.$$

Solution   In this case the substitution $y = x^r$ yields the indicial equation

$$r(r-1)(r-2) - 3r(r-1) + 7r - 8 = 0,$$

that is,

$$r^3 - 6r^2 + 12r - 8 = 0.$$

By inspection we see that $r = 2$ is a root, and the indicial equation can be written as

$$(r - 2)(r^2 - 4r + 4) = 0,$$

that is,

$$(r - 2)^3 = 0.$$

It follows that three linearly independent solutions to the given differential equation are

$$y_1 = x^2, \qquad y_2 = x^2\ln x, \qquad y_3 = x^2(\ln x)^2,$$

so that the general solution is

$$y = x^2[c_1 + c_2\ln x + c_3(\ln x)^2].$$

Finally we mention that on the interval $(-\infty, 0)$, the substitution $y = (-x)^r$ can be used to obtain the indicial equation, and the solutions of the differential equation are the same as those obtained previously, with x replaced by $-x$. Consequently, if we use $|x|$ in place of x, we will obtain solutions of the differential equation that are valid for all $x \neq 0$.

EXERCISES 9.10

In problems 1–17 determine the general solution of the given differential equation on $(0, \infty)$.

1. $x^2y'' - xy' + 5y = 0.$
2. $x^2y'' - 6y = 0.$
3. $x^2y'' - 3xy' + 4y = 0.$
4. $x^2y'' - 4xy' + 4y = 0.$
5. $x^2y'' + 3xy' + y = 0.$
6. $x^2y'' + 5xy' + 13y = 0.$
7. $x^2y'' - xy' - 35y = 0.$
8. $x^2y'' + xy' + 16y = 0.$
9. $x^3y''' + x^2y'' - 2xy' + 2y = 0.$
10. $x^3y''' - x^2y'' + 11xy' - 20y = 0.$
11. $x^3y''' + 5x^2y'' + 4xy' = 0.$
12. $x^3y''' + 3x^2y'' - 2xy' + 2y = 0.$
13. $x^4y''' - 6x^2y' + 12xy = 0.$
14. $x^3y''' + 6x^2y'' + 5xy' + 3y = 0.$
15. $x^3y''' + 6x^2y'' + 20xy' - 20y = 0.$
16. $x^4y^{(4)} + 2x^3y''' + 5x^2y'' - 5xy' + 5y = 0.$
17. $x^4y^{(4)} + 6x^3y''' + 9x^2y'' + 3xy' + y = 0.$

In problems 18–20 solve the given Cauchy–Euler equation on the interval $(0, \infty)$. In each case m and k are positive constants.

18. $x^2y'' + xy' - m^2y = 0.$
19. $x^2y'' - x(2m - 1)y' + m^2y = 0.$
20. $x^2y'' - x(2m - 1)y' + (m^2 + k^2)y = 0.$

21. Consider the second-order Cauchy–Euler equation

$$x^2y'' + xa_1y' + a_2y = 0, \qquad x > 0. \qquad \text{(i)}$$

(a) Show that the change of independent variable defined by $x = e^z$ transforms (i) into the constant-coefficient equation

$$\frac{d^2y}{dz^2} + (a_1 - 1)\frac{dy}{dz} + a_2y = 0. \qquad \text{(ii)}$$

(b) Show that if $y_1(z)$, $y_2(z)$ are linearly independent solutions of (ii), then $y_1(\ln x)$, $y_2(\ln x)$ are linearly independent solutions of (i). [Hint: We already know from (a) that y_1 and y_2 are solutions of (i).   To show that they are linearly independent, show that

$$W[y_1, y_2](x) = \frac{dz}{dx}\, W[y_1, y_2](z).]$$

22. Consider the Cauchy–Euler equation

$$x^2 y'' + a_1 xy' + a_2 y = 0, \ x < 0.$$

Show that the substitution $y = (-x)^r$ yields the indicial equation

$$r^2 + (a_1 - 1)r + a_2 = 0.$$

In problems 23–31 solve the given differential equation on the interval $x > 0$.   Remember to write your equation in standard form before applying the variation-of-parameters method.

23. $x^2 y'' - 4xy' + 6y = x^4 \sin x.$

24. $x^2 y'' + 4xy' + 2y = 4 \ln x.$

25. $x^2 y'' - 3xy' + 4y = \dfrac{x^2}{\ln x}.$

26. $x^2 y'' + 6xy' + 6y = 4e^{2x}.$

27. $x^2 y'' + xy' + 9y = 9 \ln x.$

28. $x^2 y'' + 4xy' + 2y = \cos x.$

29. $x^2 y'' - xy' + 5y = 8x(\ln x)^2.$

30. $x^2 y'' - (2m - 1)xy' + m^2 y = x^m (\ln x)^k,$ where m and k are constants.

31. $x^3 y''' + xy' - y = 24x \ln x.$

9.11   REDUCTION OF ORDER

Finally, in this chapter, we consider a powerful technique for determining the general solution of any second-order linear differential equation, assuming that we already know one solution to the associated homogeneous equation.   The technique is usually referred to as **reduction of order**.

Consider first the general second-order linear homogeneous differential equation written in standard form

$$y'' + p(x)y' + q(x)y = 0, \qquad (9.11.1)$$

where we assume that the functions p and q are continuous on an interval I.   We know that the general solution to (9.11.1) is of the form

$$y = c_1 y_1 + c_2 y_2,$$

where y_1 and y_2 are linearly independent solutions to (9.11.1) on I.   Suppose that we have found one solution, say $y = y_1$.   If $y = y_2$ is a second solution of (9.11.1) then, in order that y_1 and y_2 be linearly independent on I, the ratio y_2/y_1 must be nonconstant.   That is,

$$y_2 = y_1 u(x), \qquad (9.11.2)$$

for some (nonconstant) function u.   We now show that $u(x)$ can, in theory, always be determined from (9.11.1).   Differentiating (9.11.2) twice with respect to x yields

$$y_2' = u'y_1 + uy_1',$$
$$y_2'' = u''y_1 + 2u'y_1' + uy_1''.$$

Substituting into (9.11.1), it follows that y_2 is a solution provided

$$(u''y_1 + 2u'y_1' + uy_1'') + p(u'y_1 + uy_1') + q(uy_1) = 0,$$

that is, provided

$$u(y_1'' + py_1' + qy_1) + u''y_1 + u'(2y_1' + py_1) = 0. \tag{9.11.3}$$

Since $y = y_1$ is a solution of (9.11.1), we have $y_1'' + py_1' + qy_1 = 0$, so that (9.11.3) reduces to

$$u''y_1 + u'(2y_1' + py_1) = 0.$$

This is a first-order differential equation for u' that is separable (and linear).[1] Separating the variables yields

$$\frac{u''}{u'} = -\left(\frac{2y_1'}{y_1} + p\right).$$

The functions appearing on the right-hand side are all known, and so we can formally integrate this equation to obtain

$$\ln|u'| = -2 \ln|y_1| - \int^x p(t)\, dt + c,$$

that is,

$$u' = cy_1^{-2} e^{-\int^x p(s)\, ds},$$

where c is a (nonzero) constant. One more integration yields

$$u = c \int^x y_1^{-2}(t)(e^{-\int^t p(s)\, ds})dt, \tag{9.11.4}$$

where we have set the integration constant to zero without loss of generality. We have therefore shown that

$$y_2 = y_1 u(x)$$

is a solution of (9.11.1) provided u is given by (9.11.4). Further, it is left as an exercise to show that the Wronskian of y_1 and y_2 is

$$W[y_1, y_2](x) = u'y_1^2,$$

so that the solutions are linearly independent on I since $u' \neq 0$. We therefore have the following theorem.

Theorem 9.11.1:   If $y = y_1(x)$ is a solution of

$$y'' + p(x)y' + q(x)y = 0 \tag{9.11.5}$$

on an interval I, then a second linearly independent solution of (9.11.5) on I is

$$y_2 = cy_1 \int^x y_1^{-2}(t)(e^{-\int^t p(s)\, ds})dt,$$

where c is a constant that can be chosen to have any convenient nonzero value.

---

[1] This is why the technique is referred to as reduction of order.

REMARK   Do *not* memorize the formula for y_2. Just remember the idea behind the technique, namely, that if $y = y_1$ is a solution of (9.11.5), then a second linearly independent solution of the form $y_2 = uy_1$ can be determined by substitution into the differential equation.

Example 9.11.1   Find the general solution of

$$xy'' - 2y' + (2 - x)y = 0, \qquad x > 0 \tag{9.11.6}$$

given that one solution is $y_1 = e^x$.

Solution   We know that there is a second linearly independent solution of the form

$$y_2 = y_1 u(x) = e^x u(x). \tag{9.11.7}$$

Differentiating y_2 twice with respect to x yields

$$y_2' = e^x(u' + u),$$

$$y_2'' = e^x(u'' + 2u' + u).$$

Substituting into (9.11.6), it follows that y_2 is a solution if and only if u satisfies

$$x(u'' + 2u' + u) - 2(u' + u) + (2 - x)u = 0,$$

that is, if and only if

$$xu'' + 2u'(x - 1) = 0.$$

Separating the variables yields

$$\frac{u''}{u'} = 2\left(\frac{1}{x} - 1\right).$$

Integrating we obtain

$$\ln|u'| = 2(\ln x - x) + \ln c,$$

which can be written as

$$u' = cx^2 e^{-2x},$$

where c is an arbitrary constant. Integrating once more and setting the resulting integration constant to zero, we obtain

$$u = -\tfrac{1}{4}ce^{-2x}(1 + 2x + 2x^2).$$

Substituting into (9.11.7) yields the second linearly independent solution

$$y_2 = e^{-x}(1 + 2x + 2x^2),$$

where we have set $c = -4$ without loss of generality. It follows that the general solution to (9.11.6) is

$$y = c_1 e^x + c_2 e^{-x}(1 + 2x + 2x^2).$$

   The reduction-of-order technique can also be used to determine the general solution of a second order *nonhomogeneous* linear differential equation, assuming that we know one solution of the associated homogeneous equation. Rather than derive the general result, which we leave as an exercise, we will illustrate with an example.

Example 9.11.2   Find the general solution of the differential equation

$$x^2 y'' + 3xy' + y = 4 \ln x, \tag{9.11.8}$$

given that one solution of the associated homogeneous equation is $y = x^{-1}$.

Solution   We try for a solution of the form

$$y = x^{-1} u(x), \tag{9.11.9}$$

where $u(x)$ is to be determined. Differentiating y twice yields

$$y' = x^{-1} u' - x^{-2} u, \qquad y'' = x^{-1} u'' - 2x^{-2} u' + 2x^{-3} u.$$

Substituting into (9.11.8) and collecting terms we obtain the following differential equation for u:

$$u'' + \frac{1}{x} u' = \frac{4}{x} \ln x. \tag{9.11.10}$$

To facilitate the integration, we let

$$v = u', \tag{9.11.11}$$

in which case (9.11.10) can be written as the first-order linear differential equation

$$v' + \frac{1}{x} v = \frac{4}{x} \ln x. \tag{9.11.12}$$

An integrating factor for this linear equation is (see Section 1.5) $I = e^{\int (1/x)\, dx} = x$, so that (9.11.12) can be written in the equivalent form

$$\frac{d}{dx}(xv) = 4 \ln x.$$

Integrating both sides with respect to x yields

$$xv = 4x(\ln x - 1) + c_1,$$

where c_1 is a constant. Thus

$$v = 4(\ln x - 1) + c_1 x^{-1}.$$

Substituting in (9.11.11) we obtain

$$u' = 4(\ln x - 1) + c_1 x^{-1},$$

which can be integrated directly to yield

$$u = 4x(\ln x - 2) + c_1 \ln x + c_2,$$

where c_2 is another constant. Substituting for u into (9.11.9), we obtain the solution

$$y = 4(\ln x - 2) + c_1 x^{-1}\ln x + c_2 x^{-1}.$$

Is this the general solution of (9.11.8)? The answer is yes. The complementary function is

$$y_c = c_1 x^{-1}\ln x + c_2 x^{-1},$$

whereas

$$y_p = 4(\ln x - 2)$$

is a particular solution.

EXERCISES 9.11

In problems 1–6, y_1 is a solution of the given differential equation. Use the method of reduction of order to determine a second linearly independent solution.

1. $x^2 y'' - 3xy' + 4y = 0$, $x > 0$, $y_1(x) = x^2$.

2. $x^2 y'' - 2xy' + (x^2 + 2)y = 0$, $x > 0$,
$y_1(x) = x \sin x$.

3. $xy'' + (1 - 2x)y' + (x - 1)y = 0$, $x > 0$,
$y_1(x) = e^x$.

4. $y'' - \dfrac{1}{x}y' + 4x^2 y = 0$, $x > 0$, $y_1(x) = \sin x^2$.

5. $(1 - x^2)y'' - 2xy' + 2y = 0$, $y_1(x) = x$,
$-1 < x < 1$.

6. $4x^2 y'' + 4xy' + (4x^2 - 1)y = 0$,
$y_1 = x^{-1/2} \sin x$.

7. Consider the Cauchy–Euler equation

$$x^2 y'' - (2m - 1)xy' + m^2 y = 0, \qquad x > 0, \quad \text{(i)}$$

where m is a constant.
(a) Determine a particular solution to (i) of the form $y_1 = x^r$.
(b) Use your solution from (a) and the method of *reduction of order* to obtain a second linearly independent solution.

8. Determine the values of the constants a_0, a_1, a_2 such that

$$y = a_0 + a_1 x + a_2 x^2$$

is a solution to

$$(4 + x^2)y'' - 2y = 0,$$

and use the reduction-of-order technique to find a second linearly independent solution.

9. Consider the differential equation

$$xy'' - (\alpha x + \beta)y' + \alpha\beta y = 0, \qquad x > 0, \quad \text{(ii)}$$

where α and β are constants.
(a) Show that $y_1 = e^{\alpha x}$ is a solution of (ii).
(b) Use reduction of order to derive the second linearly independent solution

$$y_2 = e^{\alpha x} \int x^\beta e^{-\alpha x}\, dx.$$

(c) In the particular case when $\alpha = 1$ and β is a nonnegative integer, show that a second linearly independent solution to (ii) is

$$y_2 = 1 + x + \frac{1}{2!}x^2 + \cdots + \frac{1}{\beta!}x^\beta.$$

In problems 10–15, y_1 is a solution of the associated homogeneous equation. Use the method of reduction of order to determine the general solution of the given differential equation.

10. $y'' - 6y' + 9y = 15e^{3x}\sqrt{x}$, $y_1 = e^{3x}$.

11. $y'' - 4y' + 4y = 4e^{2x}\ln x$, $y_1 = e^{2x}$.

12. $4x^2 y'' + y = \sqrt{x}\,\ln x$, $y_1 = \sqrt{x}$.

13. $y'' + y = \csc x$, $0 < x < \pi$, $y_1 = \sin x$.

14. $xy'' - (2x + 1)y' + 2y = 8x^2 e^{2x}$, $x > 0$,
$y_1 = e^{2x}$.

15. $x^2 y'' - 3xy' + 4y = 8x^4$, $x > 0$, $y_1 = x^2$.

16. Consider the differential equation

$$y'' + p(x)y' + q(x)y = r(x), \qquad \text{(iii)}$$

where p, q, and r are continuous on an interval I. If $y = y_1$ is a solution of the associated homogeneous equation, show that $y_2 = u(x)y_1$ is a solution of (iii) provided that $v = u'$ is a solution of the linear differential equation

$$v' + \left(2\frac{y_1'}{y_1} + p\right)v = \frac{r}{y_1}.$$

Express the solution to (iii) in terms of integrals. Identify two linearly independent solutions to the associated homogeneous equation and a particular solution to (iii).

9.12   SUMMARY OF RESULTS

In this chapter we have analyzed the nth-order linear differential equation

$$Ly = F$$

on an interval I, where L is the linear differential operator

$$L = D^n + a_1 D^{n-1} + \cdots + a_{n-1}D + a_n.$$

The underlying theory has been developed completely and some special solution techniques have been derived. Below we summarize the key results from the chapter.

UNDERLYING THEORY

1. The set of all solutions to the homogeneous differential equation $Ly = 0$ forms a vector space of dimension n, and hence any set of n linearly independent solutions forms a basis for the solution space of $Ly = 0$.
2. The set of solutions $\{y_1, y_2, \ldots, y_n\}$ to $Ly = 0$ are linearly independent on I if and only if $W[y_1, y_2, \ldots, y_n](x)$ is nonzero on I.
3. The general solution to $Ly = F$ on I is of the form

$$y = y_c + y_p,$$

where $y = y_c$ is the general solution to $Ly = 0$, and $y = y_p$ is a particular solution to $Ly = F$. We refer to y_c as the complementary function to $Ly = F$.

SOLUTION TECHNIQUES

HOMOGENEOUS LINEAR DIFFERENTIAL EQUATIONS

Constant coefficient differential equations: The main type of homogeneous differential equations for which a general solution technique has been derived are those with constant coefficients. Such a differential equation is written as

$$P(D)y = 0,$$

where $P(D)$ is a polynomial differential operator of order n. Corresponding to each root of multiplicity m of the auxiliary equation $P(r) = 0$, we obtain the m solutions

$$e^{rx}, xe^{rx}, \ldots, x^{m-1}e^{rx}.$$

These solutions are linearly independent on any interval. If r is complex, then we can determine linearly independent real-valued solutions by taking the real and imaginary parts of the above complex-valued solutions. In this manner we can obtain n real-valued solutions that are linearly independent on any interval.

Cauchy–Euler equations: The only type of nonconstant coefficient linear differential equation for which a general solution technique has been derived is the Cauchy–Euler equation

$$x^n \frac{d^n y}{dx^n} + a_1 x^{n-1} \frac{d^{n-1} y}{dx^{n-1}} + \cdots + a_{n-1} x \frac{dy}{dx} + a_n y = 0, \qquad x > 0$$

where $a_1, a_2, \ldots, a_n$ are real constants. If we substitute $y = x^r$ into this differential equation, we obtain the indicial equation

$$r(r-1)(r-2) \cdots (r - n + 1) + a_1 r(r-1) \cdots (r - n + 2) + \cdots + a_{n-1} r + a_n = 0.$$

Corresponding to each root of multiplicity k of the indicial equation we obtain the k solutions

$$x^r, x^r \ln x, x^r (\ln x)^2, \ldots, x^r (\ln x)^{k-1}.$$

These solutions are linearly independent on the interval $(0, \infty)$. Once more, if r is complex, we can determine linearly independent real-valued solutions by taking the real and imaginary parts of the preceding complex-valued solutions. In this manner we can obtain n real-valued solutions that are linearly independent on $(0, \infty)$. If $x < 0$, then these solutions are still valid, provided we replace x by $-x$.

NONHOMOGENEOUS DIFFERENTIAL EQUATIONS

The method of undetermined coefficients: This technique is applicable only to differential equations that satisfy:

1. The differential equation is of the form

$$P(D)y = F. \tag{9.12.1}$$

2. $Q(D)F = 0$, for some polynomial differential operator $Q(D)$.

The idea behind the method of undetermined coefficients is to operate on (9.12.1) with $Q(D)$ to obtain the homogeneous differential equation

$$Q(D) P(D)y = 0.$$

From the general solution to this differential equation, we determine a trial solution for (9.12.1) that involves undetermined coefficients. The value of these coefficients is then determined by substitution into (9.12.1).

Variation of parameters: This is more general than the undetermined coefficients method, since it can be applied to any nth-order differential equation of the form

$$Ly = F.$$

The drawback is that we must know the complementary function. If $\{y_1, y_2, \ldots, y_n\}$ is any set of linearly independent solutions to $Ly = 0$, then there is a particular solution to $Ly = F$ of the form

$$y_p = u_1 y_1 + u_2 y_2 + \cdots + u_n y_n.$$

The functions $u_1, u_2, \ldots, u_n$ can be determined by direct substitution into $Ly = F$. It has been shown that they can be chosen to satisfy the linear system

$$y_1 u_1' \quad + y_2 u_2' \quad + \cdots + y_n u_n' \quad = 0,$$
$$y_1' u_1' \quad + y_2' u_2' \quad + \cdots + y_n' u_n' \quad = 0,$$
$$\vdots$$
$$y_1^{(n-2)} u_1' + y_2^{(n-2)} u_2' + \cdots + y_n^{(n-2)} u_n' = 0,$$
$$y_1^{(n-1)} u_1' + y_2^{(n-1)} u_2' + \cdots + y_n^{(n-1)} u_n' = F(x).$$

Having solved this system for $u_1', u_2', \ldots, u_n'$, we can obtain $u_1, u_2, \ldots, u_n$ by integration.

Reduction of order: This is a technique for solving a second-order linear differential equation, assuming that we know one solution of the associated homogeneous equation. If $y = y_1$ is the known solution, then substitution of $y_2 = u y_1$ into the given differential equation reduces it to a first-order linear differential equation for u'. This latter equation can be solved using the technique of Section 1.5.

10

Mechanical Systems:
A Simple Application
of Second-Order
Linear Equations

10.1 MATHEMATICAL FORMULATION
OF THE PROBLEM

We now consider a simple, but important, physical problem that is governed by a linear constant coefficient differential equation. Before solving the problem we carefully construct an appropriate mathematical model of the physical situation under consideration.

Statement of the problem: A mass of m kilograms is attached to the end of a spring whose natural length is l_0 meters.   At $t = 0$ the mass is displaced a distance x_0 meters and released with a velocity v_0 meters/second.   Determine the initial value problem that governs the resulting motion.

Mathematical formulation of the problem: We assume that the motion takes place vertically and adopt the convention that distances are measured positive in the *downward* direction.   To formulate the problem mathematically, we need to determine the forces that are acting on the mass. Consider first the *static equilibrium* position in which the mass hangs freely from the spring with no motion (see Figure 10.1.1). The forces acting on the mass in this equilibrium position are

1. The force due to gravity

$$F_g = mg.$$

2. The *spring force*, F_s, due to the displacement of the spring from its natural length.

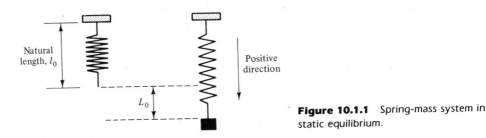

Figure 10.1.1   Spring-mass system in static equilibrium.

This latter force is determined from Hooke's law:

Hooke's Law

When a spring is extended or compressed a (signed) distance L from its natural length, the corresponding spring force is

$$F_s = -kL,$$

where k is a positive constant, called the spring constant.

REMARK   Notice that if the spring is stretched ($L > 0$), then F_s is negative, whereas if the spring is compressed ($L < 0$), then F_s is positive. Thus the spring force always acts in the opposite direction to the extension or compression.

If L_0 denotes the extension of the spring in the static equilibrium position, then

$$F_s = -kL_0.$$

Since the system is in static equilibrium, this force must be exactly balanced by that of gravity, so that $F_s + F_g = 0$, that is,

$$mg = kL_0. \tag{10.1.1}$$

Now consider the situation when the mass has been set in motion (see Figure 10.1.2). We let $x(t)$ denote the position of the mass at time t and take $x = 0$ to coincide with the equilibrium position of the system. The equation of motion of the mass can then be obtained from Newton's second law. The forces that now act on the mass are as follows:

1. The force due to gravity F_g. Once more this is

$$F_g = mg. \tag{10.1.2}$$

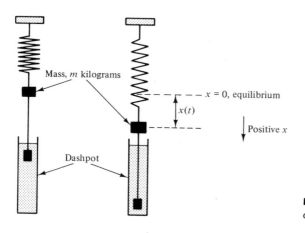

Mass, m kilograms

$x = 0$, equilibrium

$x(t)$

Positive x

Dashpot

Figure 10.1.2 A simple model of a damped spring-mass system.

2. The spring force F_s. At time t the total displacement of the spring from its natural length is $L_0 + x(t)$, so that, according to Hooke's law,

$$F_s = -k[L_0 + x(t)]. \qquad (10.1.3)$$

3. A damping force F_d. In general the motion will be damped due, for example, to air resistance or, as shown in Figure 10.1.2, an external damping system such as a dashpot. We assume that any damping forces that are present are directly proportional to the velocity of the mass. Under this assumption we have

$$F_d = -c \frac{dx}{dt}, \qquad (10.1.4)$$

where c is a positive constant called the **damping constant**. Note that the negative sign is inserted in (10.1.4), since F_d always acts in the opposite direction to that of the motion.

4. Any external driving forces, $F(t)$, that are present. For example, the top of the spring or the mass itself may be subjected to an external force.

The total force acting on the system will be the sum of the preceding forces. Thus, using Newton's second law, the differential equation that governs the motion of the mass is

$$m \frac{d^2x}{dt^2} = F_g + F_s + F_d + F(t).$$

Substituting from (10.1.2)–(10.1.4) yields

$$m \frac{d^2x}{dt^2} = mg - k(L_0 + x) - c \frac{dx}{dt} + F(t);$$

that is, using (10.1.1) and rearranging terms,

$$\frac{d^2x}{dt^2} + \frac{c}{m} \frac{dx}{dt} + \frac{k}{m} x = \frac{1}{m} F(t). \qquad (10.1.5)$$

In addition we also have the initial conditions

$$x(0) = x_0, \qquad \frac{dx}{dt}(0) = v_0.$$

Thus to determine the motion of the system we must solve the initial value problem:

$$\begin{cases} \dfrac{d^2x}{dt^2} + \dfrac{c}{m}\dfrac{dx}{dt} + \dfrac{k}{m}x = \dfrac{1}{m}F(t), \\[2mm] x(0) = x_0, \quad \dfrac{dx}{dt}(0) = v_0. \end{cases} \qquad (10.1.6)$$

In the next section we will consider the case when $F(t) = 0$ and then return to discuss the general case in Section 10.3.

EXERCISES 10.1

1. Consider the double spring-mass system shown in Figure 10.1.3. If there is a damping force proportional to the velocity and an external force $F(t)$ acting on the system, set up the initial value problem governing the motion of the system.

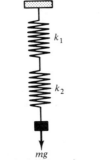

k_1

k_2

mg

Figure 10.1.3   A double spring-mass system.

2. According to Archimedes' principle, when an object is partially or wholly immersed in a fluid, it experiences an upward force equal to the weight of fluid displaced. A cube of side L and density ρ_0 floats, half-submerged, in a fluid whose density is ρ. At $t = 0$ the mass is pushed down a distance x_0 and released from rest.

Determine the initial value problem that governs the resulting motion. (Neglect any damping forces or external force that may act on the object.)

10.2   FREE OSCILLATIONS OF A MECHANICAL SYSTEM

In this section we consider the motion of the spring-mass system under the assumption that there is no external driving force. Thus we are assuming that the system is set in motion and no further external interaction is present. In the general formulation from the previous section, this corresponds to setting $F(t) = 0$, so that the initial value problem (10.1.5) reduces to:

$$\begin{cases} \dfrac{d^2x}{dt^2} + \dfrac{c}{m}\dfrac{dx}{dt} + \dfrac{k}{m}x = 0, \\[2mm] x(0) = x_0, \quad \dfrac{dx}{dt}(0) = v_0. \end{cases} \qquad (10.2.1)$$

For most of the discussion we will concentrate on the differential equation alone, since the initial conditions do not significantly affect the behavior of its solutions. Thus we wish to solve

$$\frac{d^2x}{dt^2} + \frac{c}{m}\frac{dx}{dt} + \frac{k}{m}x = 0. \qquad (10.2.2)$$

This is a homogeneous constant coefficient linear differential equation, which can therefore be solved using the techniques of the previous chapter. We divide the discussion of the solution of (10.2.2) into several subcases.

CASE 1: NO DAMPING   We first consider the case when there is no damping ($c = 0$). This is the simplest case that can arise and is of importance for understanding the more general situation. Our basic differential equation (10.2.2) reduces to

$$\frac{d^2x}{dt^2} + \omega_0^2 x = 0, \qquad (10.2.3)$$

where

$$\omega_0 = \sqrt{\frac{k}{m}}. \qquad (10.2.4)$$

The constant coefficient differential equation (10.2.3) has general solution

$$x(t) = c_1\cos \omega_0 t + c_2\sin \omega_0 t. \qquad (10.2.5)$$

It is instructive to introduce two new constants A_0 and ϕ defined in terms of c_1 and c_2 by (see Figure 10.2.1)

$$A_0\cos \phi = c_1, \qquad A_0\sin \phi = c_2, \qquad (10.2.6)$$

that is,

$$A_0 = \sqrt{c_1^2 + c_2^2}, \qquad \phi = \tan^{-1}\left(\frac{c_2}{c_1}\right). \qquad (10.2.7)$$

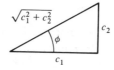

Figure 10.2.1   The definition of the phase angle ϕ.

Substituting from (10.2.6) into (10.2.5) yields

$$x(t) = A_0(\cos \omega_0 t \cos \phi + \sin \omega_0 t \sin \phi).$$

Consequently, we can write

$$x(t) = A_0\cos(\omega_0 t - \phi). \qquad (10.2.8)$$

Clearly, the motion described by (10.2.8) is periodic. We refer to such motion as **simple harmonic motion** (SHM). Figure 10.2.2 depicts this motion for typical

$x(t)$

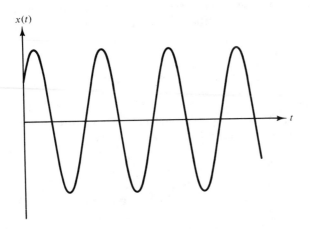

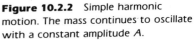

Figure 10.2.2 Simple harmonic
motion. The mass continues to oscillate
with a constant amplitude A.

values of the constants A_0, ω_0, and ϕ. The standard names for the constants arising in the solution are as follows:

A_0:   the **amplitude** of the motion.

ω_0:   the **circular frequency** of the system.

ϕ: the **phase** of the motion.

The fundamental **period** of oscillation (that is, the time for the system to
undergo one complete cycle), T, is

$$T = \frac{2\pi}{\omega_0} = 2\pi \sqrt{\frac{m}{k}}. \tag{10.2.9}$$

Consequently, the frequency of oscillation (number of oscillations per second), f, is
given by

$$f = \frac{1}{T} = \frac{\omega_0}{2\pi} = \frac{1}{2\pi}\sqrt{\frac{k}{m}}.$$

Notice that this is independent of the initial conditions. It is truly a property
of the system.

CASE 2: DAMPING   We now discuss the motion of the spring-mass system when
the damping constant, c, is *nonzero*. In this case the auxiliary polynomial for
(10.2.2) is

$$P(r) = r^2 + \frac{c}{m}r + \frac{k}{m},$$

with roots

$$r = \frac{-c \pm \sqrt{c^2 - 4km}}{2m}.$$

As we might expect, the behavior of the system is dependent on whether the auxiliary polynomial has distinct real roots, coincident real roots, or complex conjugate

roots.   These three situations will arise depending on the magnitude of the (dimensionless) combination of the system variables $c^2/4km$.   For a given spring and mass only the damping can be altered, and so this leads to the following terminology.

We say that the system is

(a) *Underdamped if* $\dfrac{c^2}{4km} < 1$ (complex conjugate roots),

(b) *Critically damped if* $\dfrac{c^2}{4km} = 1$ (repeated real root),

(c) *Overdamped if* $\dfrac{c^2}{4km} > 1$ (two distinct real roots).

The corresponding solutions to (10.2.2) are

(a) $x(t) = e^{-(c/2m)t}(c_1\cos \mu t + c_2\sin \mu t),$ $\mu = \dfrac{\sqrt{4km - c^2}}{2m}.$ (10.2.10)

(b) $x(t) = e^{-(c/2m)t}(c_1 + c_2 t).$ (10.2.11)

(c) $x(t) = e^{-(c/2m)t}(c_1 e^{vt} + c_2 e^{-vt}),$ $v = \dfrac{\sqrt{c^2 - 4km}}{2m}.$ (10.2.12)

In all three cases we have (see the following discussion)

$$\lim_{t \to \infty} x(t) = 0,$$

which implies that the motion dies out for large t.   This is certainly consistent with our everyday experience.   We discuss the different cases separately.

CASE 2a: UNDERDAMPING   In this case, the position of the mass at time t is given in (10.2.10), which reduces to SHM when $c = 0$.   Once more it is convenient to introduce constants A_0 and ϕ defined by

$$A_0\cos \phi = c_1, \qquad A_0\sin \phi = c_2.$$

Then (10.2.10) can be written in the equivalent form

$$x(t) = A_0 e^{-(c/2m)t}\cos(\mu t - \phi). \qquad (10.2.13)$$

We see that the mass oscillates between $\pm A_0 e^{-(c/2m)t}$.   The corresponding motion is depicted in Figure 10.2.3 for the case when $x(0) > 0$ and $(dx/dt)(0) > 0$.

In general the motion *is* oscillatory, but it is *not* periodic.   The amplitude of the motion dies out exponentially with time, although the time interval, T, between successive maxima (or minima) of $x(t)$ has the *constant* value (see problem 13)

$$T = \frac{2\pi}{\mu} = \frac{4\pi m}{\sqrt{4km - c^2}}.$$

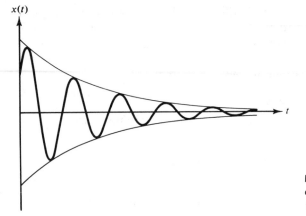

$x(t)$

Figure 10.2.3 Damped harmonic oscillation: The mass oscillates between $\pm A_0 e^{-(c/2m)t}$

CASE 2b: CRITICAL DAMPING   This case arises when $c^2/4km = 1$.   From (10.2.2), the differential equation describing the motion is

$$\frac{d^2x}{dt^2} + \frac{c}{m}\frac{dx}{dt} + \frac{c^2}{4m^2}x = 0,$$

with general solution

$$x(t) = e^{-(c/2m)t}(c_1 + c_2 t). \tag{10.2.14}$$

Now the damping is so severe that the system can pass through the equilibrium position at most once, and so we do not really have an oscillatory behavior.   If we impose the initial conditions

$$x(0) = x_0, \qquad \frac{dx}{dt}(0) = v_0,$$

then it is easily shown (see problem 14) that (10.2.14) can be written in the form

$$x(t) = e^{-(c/2m)t}\left[x_0 + t\left(v_0 + \frac{c}{2m}x_0\right)\right].$$

Consequently, the system will pass through the equilibrium position provided x_0 and $v_0 + (c/2m)x_0$ have opposite signs.   A sketch of the motion described by (10.2.14) is given in Figure 10.2.4.

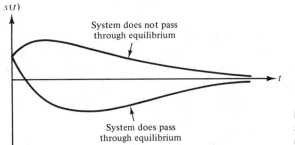

$x(t)$

System does not pass through equilibrium

System does pass through equilibrium

Figure 10.2.4 Critical damping: The system can pass through equilibrium at most once.

CASE 2c: OVERDAMPING   In this case we have $c^2/4km > 1$. The roots of the auxiliary equation corresponding to (10.2.2) are

$$r_1 = \frac{-c + \sqrt{c^2 - 4km}}{2m}, \qquad r_2 = \frac{-c - \sqrt{c^2 - 4km}}{2m}.$$

We let $v = \sqrt{c^2 - 4km}/2m$, so that the general solution of (10.2.2) is

$$x(t) = e^{-(c/2m)t}(c_1 e^{vt} + c_2 e^{-vt}).$$

Since c, k, and m are positive, it follows that both of the roots of the auxiliary equation are negative, which implies that both terms in $x(t)$ decay in time. Once more we do not have oscillatory behavior. The motion is very similar to that of the critically damped case. The system can pass through the equilibrium position at most once. This is illustrated in Figure 10.2.5.

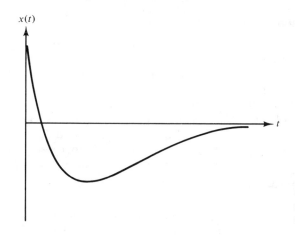

Figure 10.2.5   Overdamping: The system passes through equilibrium once.

EXERCISES 10.2

In problems 1 and 2, consider the spring-mass system whose motion is governed by the given initial value problem. Determine the circular frequency of the system and the amplitude, phase, and period of the motion.

1. $\dfrac{d^2x}{dt^2} + 4x = 0$, $x(0) = 2$, $\dfrac{dx}{dt}(0) = 4$.

2. $\dfrac{d^2x}{dt^2} + \omega_0^2 x = 0$, $x(0) = x_0$, $\dfrac{dx}{dt}(0) = v_0$, where ω_0, x_0, v_0 are constants.

3. A force of 3 N stretches a large spring by 1 m.
(a) Find the spring constant k.

(b) A mass of 4 kg is attached to the spring. At $t = 0$ the mass is pushed up a distance 1 m from equilibrium and released with a downward velocity of $\frac{1}{2}$ m/s. Assuming that damping is negligible, determine an expression for the position of the mass at time t. Find the circular frequency of the system and the amplitude, phase, and period of the motion.

In problems 4–9 determine the motion of the spring-mass system governed by the given initial value problem. In each case state whether the motion is underdamped, critically damped, or overdamped, and make a sketch depicting the motion.

4. $\dfrac{d^2x}{dt^2} + 2\dfrac{dx}{dt} + 5x = 0$, $x(0) = 1$, $\dfrac{dx}{dt}(0) = 3$.

5. $\dfrac{d^2x}{dt^2} + 3\dfrac{dx}{dt} + 2x = 0$, $x(0) = 1$, $\dfrac{dx}{dt}(0) = 0$.

6. $4\dfrac{d^2x}{dt^2} + 12\dfrac{dx}{dt} + 5x = 0$, $x(0) = 1$, $\dfrac{dx}{dt}(0) = -3$.

7. $\dfrac{d^2x}{dt^2} + 2\dfrac{dx}{dt} + x = 0$, $x(0) = -1$, $\dfrac{dx}{dt}(0) = 2$.

8. $4\dfrac{d^2x}{dt^2} + 4\dfrac{dx}{dt} + x = 0$, $x(0) = 4$, $\dfrac{dx}{dt}(0) = -1$.

9. $\dfrac{d^2x}{dt^2} + 4\dfrac{dx}{dt} + 7x = 0$, $x(0) = 2$, $\dfrac{dx}{dt}(0) = 6$.

10. (a) Determine the motion of the spring-mass system governed by
$$\dfrac{d^2x}{dt^2} + 5\dfrac{dx}{dt} + 6x = 0, \ x(0) = -1, \ \dfrac{dx}{dt}(0) = 4.$$

(b) Find the time at which the mass passes through the equilibrium position and determine the maximum positive displacement of the mass from equilibrium.
(c) Make a sketch depicting the motion.

11. Consider the spring-mass system whose motion is governed by the differential equation
$$\dfrac{d^2x}{dt^2} + 2\alpha\dfrac{dx}{dt} + x = 0.$$

Determine all values of the (positive) constant α for which the system is
(a) underdamped,
(b) critically damped, and
(c) overdamped. In the case of overdamping solve the system completely. If the initial velocity of the system is zero, determine whether the mass passes through the equilibrium position.

12. Consider the spring-mass system whose motion is governed by the initial value problem
$$\dfrac{d^2x}{dt^2} + 3\dfrac{dx}{dt} + 2x = 0, \ x(0) = 1, \ \dfrac{dx}{dt}(0) = -3,$$

in the usual notation.
(a) Determine the position of the mass at time t.
(b) Determine the time when the mass passes through the equilibrium position.

(c) Make a sketch depicting the general motion of the system.

13. Consider the general solution for an *underdamped* spring-mass system.
(a) Show that the time between successive maxima (or minima) of $x(t)$ is
$$T = \dfrac{2\pi}{\mu} = \dfrac{4\pi m}{\sqrt{4km - c^2}}.$$

(b) Show that if $c^2/4km \ll 1$, then
$$T \approx 2\pi\sqrt{\dfrac{m}{k}}.$$

Is this result reasonable?

14. Show that the general solution for the motion of a *critically damped* spring-mass system with initial displacement x_0 and initial velocity v_0 can be written in the form
$$x(t) = e^{-(c/2m)t}\left[x_0 + t\left(v_0 + \dfrac{c}{2m}x_0\right)\right],$$

and that the system can pass through the equilibrium position at most once.

15. A cylinder of side L meters lies one quarter submerged and upright in a certain fluid. At $t = 0$ the cylinder is pushed down a distance $L/2$ meters and released from rest. Show that the resulting motion is simple harmonic, and determine the circular frequency and period of the motion.[1]

A simple pendulum consists of a mass, m kilograms, attached to the end of a light rod of length L meters, whose other end is fixed (see Figure 10.2.6). If we let θ radians denote the angle the rod is displaced from the vertical at time t, then the component of the velocity in the direction of motion is $v = L(d\theta/dt)$, so that the component of

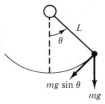

Figure 10.2.6 The simple pendulum.

---

[1] The upward force on the cylinder is determined from Archimedes' principle (see problem 2 on page 382).

the acceleration in this direction is $L(d^2\theta/dt^2)$. Further, the tangential component of the force is $F_T = -mg \sin \theta$, so that, from Newton's second law, the equation of motion of the pendulum is

$$mL\frac{d^2\theta}{dt^2} = -mg \sin \theta,$$

that is,

$$\frac{d^2\theta}{dt^2} + \frac{g}{L} \sin \theta = 0. \qquad \text{(i)}$$

This is a nonlinear differential equation. However, if we recall the Maclaurin expansion for $\sin \theta$, namely,

$$\sin \theta = \theta - \frac{1}{3!}\theta^3 + \frac{1}{5!}\theta^5 - \cdots,$$

it follows that for small oscillations we can approximate $\sin \theta$ by θ. Then (i) can be replaced to reasonable accuracy by the simple *linear* differential equation

$$\frac{d^2\theta}{dt^2} + \frac{g}{L}\theta = 0. \qquad \text{(ii)}$$

Problems 16–19 deal with the simple pendulum whose motion is described by (ii).

16. A pendulum of length 0.5 m is displaced an angle 0.1 rad from the equilibrium position and released from rest. Determine the resulting motion.

17. A pendulum of length L meters is displaced an angle α radians from the vertical and released with an angular velocity of β radians/second. Determine the amplitude, phase, and period of the resulting motion.

18. Show that the period of the simple pendulum is $T = 2\pi\sqrt{L/g}$. Determine the length of a pen-

dulum that takes 1 s to swing from its extreme position on the right to its extreme position on the left. Take $g = 9.8$ m/s^2.

19. A clock has a pendulum of length 90 cm. If the clock ticks each time the pendulum swings from its extreme position on the right to its extreme position on the left, determine the number of times the clock ticks in 1 min. Take $g = 9.8$ m/s^2.

20. An object of mass m is attached to the midpoint of a light elastic string of natural length $6a$. When the ends of the string are fixed at the same level a distance $6a$ apart and the mass is allowed to hang in equilibrium, the length of the stretched string is $10a$ (see Figure 10.2.7). The mass is pulled down a small vertical distance from equilibrium and released. Show that, *for small oscillations*, the period of the resulting motion is

$$T = \frac{20\pi}{7}\sqrt{\frac{a}{g}}.$$

21. Repeat the previous problem if the string has natural length $2L_0$ and in equilibrium the stretched string has length $2L$.

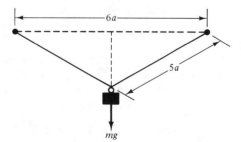

Figure 10.2.7 The static equilibrium position.

10.3   FORCED OSCILLATIONS

We now consider the situation when a spring-mass system is acted upon by an external force. For example, the ceiling to which the spring is attached may itself be vibrating due to somebody walking across the floor above it. As shown in Section 10.1, the appropriate differential equation describing the motion of the system is

$$\frac{d^2x}{dt^2} + \frac{c}{m}\frac{dx}{dt} + \frac{k}{m}x = \frac{F(t)}{m}.$$

The situation of most interest arises when the applied force is periodic in time and so we therefore restrict attention to a driving term of the form

$$F(t) = F_0 \cos \omega t,$$

where F_0 and ω are constants.   Thus the basic differential equation that we must study is

$$\frac{d^2x}{dt^2} + \frac{c}{m}\frac{dx}{dt} + \frac{k}{m}x = \frac{F_0}{m}\cos \omega t. \tag{10.3.1}$$

Once more we will divide our discussion of the resulting motion into several cases.

CASE 1: NO DAMPING   Setting $c = 0$ in (10.3.1) yields

$$\frac{d^2x}{dt^2} + \omega_0^2 x = \frac{F_0}{m} \cos \omega t, \tag{10.3.2}$$

where, as in the previous section, ω_0 denotes the circular frequency of the system and is defined by

$$\omega_0 = \sqrt{\frac{k}{m}}.$$

The complementary function for (10.3.2) is

$$x_c(t) = c_1 \cos \omega_0 t + c_2 \sin \omega_0 t,$$

which can be written in the form

$$x_c = A_0 \cos(\omega_0 t - \phi), \tag{10.3.3}$$

for appropriate constants A_0 and ϕ.   We therefore need to find a particular solution to (10.3.2).   The right-hand side is of an appropriate form to use the method of undetermined coefficients, although the trial solution will depend on whether $\omega \neq \omega_0$ or $\omega = \omega_0$.

$\omega \neq \omega_0$:   In this case the appropriate trial solution is

$$x_p = A \cos \omega t + B \sin \omega t.$$

A straightforward calculation yields the following particular solution for (10.3.2) (see problem 4)

$$x_p = \frac{F_0}{m(\omega_0^2 - \omega^2)} \cos \omega t, \tag{10.3.4}$$

so that the general solution of (10.3.2) is

$$x(t) = A_0 \cos(\omega_0 t - \phi) + \frac{F_0}{m(\omega_0^2 - \omega^2)} \cos \omega t. \tag{10.3.5}$$

Comparing this with (10.2.9) we see that the resulting motion consists of a superposition of two simple harmonic oscillation modes.   One of these modes has the circular frequency, ω_0, of the system, whereas the other mode has the frequency of the driving force. Consequently, the motion is oscillatory and bounded for all time, but it is *not* periodic in general. Indeed, it can be shown (see problem 5) that the motion is periodic only if the ratio ω/ω_0 is a rational number, say

$$\frac{\omega}{\omega_0} = \frac{p}{q}, \tag{10.3.6}$$

where p and q are positive integers. In such a case the fundamental period of the motion is

$$T = \frac{2\pi q}{\omega_0} = \frac{2\pi p}{\omega},$$

where p and q are the smallest integers satisfying (10.3.6). A typical (nonperiodic) motion of the form (10.3.5) is sketched in Figure 10.3.1.

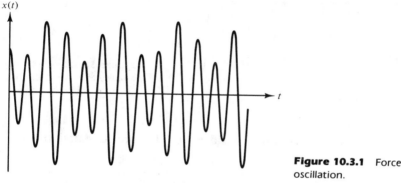

$x(t)$

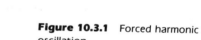

Figure 10.3.1 Forced harmonic oscillation.

$\omega = \omega_0$ *Resonance:* When the frequency of the driving term coincides with the frequency of the system, we must solve

$$\frac{d^2x}{dt^2} + \omega_0^2 x = \frac{F_0}{m}\cos \omega_0 t. \tag{10.3.7}$$

The complementary function can be written as

$$x_c = A_0\cos(\omega_0 t - \phi),$$

and an appropriate trial solution is

$$x_p = t(A \cos \omega_0 t + B \sin \omega_0 t).$$

A straightforward application of the method of undetermined coefficients yields the particular solution (see problem 4)

$$x_p = \frac{F_0}{2m\omega_0} t \sin \omega_0 t, \tag{10.3.8}$$

so that the general solution to (10.3.7) is

$$x(t) = A_0\cos(\omega_0 t - \phi) + \frac{F_0}{2m\omega_0} t \sin \omega_0 t.$$

We see that the motion is oscillatory but that the amplitude increases without bound as $t \to \infty$. This phenomenon, which occurs when the driving and natural frequencies coincide, is called **resonance**. Its physical consequences cannot be overemphasized. For example, the occurrence of resonance in the present situation would eventually cause the spring's elastic limit to be exceeded, and hence the system would be destroyed. This situation is depicted in Figure 10.3.2.

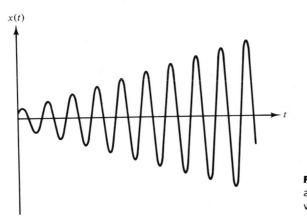

Figure 10.3.2   Resonance: The amplitude of the oscillation increases without bound as $t \to \infty$.

CASE 2: DAMPING   We now consider the full damped equation

$$\frac{d^2x}{dt^2} + \frac{c}{m}\frac{dx}{dt} + \frac{k}{m}x = \frac{F_0}{m}\cos \omega t, \tag{10.3.9}$$

where $c \neq 0$. An appropriate trial solution for this equation is

$$x_p = A \cos \omega t + B \sin \omega t.$$

A fairly lengthy but straightforward computation (best performed using the complex-valued trial solution $x_p = Ae^{i\omega t}$) yields the particular solution (see problem 7)

$$x_p = \frac{F_0}{(k - m\omega^2)^2 + c^2\omega^2}\left[(k - m\omega^2)\cos \omega t + c\omega \sin \omega t\right], \tag{10.3.10}$$

which can be written in the form

$$x_p = \frac{F_0}{H}\cos(\omega t - \eta), \tag{10.3.11}$$

where

$$\cos \eta = \frac{m(\omega_0^2 - \omega^2)}{H}, \qquad \sin \eta = \frac{c\omega}{H}, \qquad H = \sqrt{m^2(\omega_0^2 - \omega^2)^2 + c^2\omega^2},$$

and

$$\omega_0 = \sqrt{\frac{k}{m}}.$$

Consider first the case of underdamping. Using the homogeneous solution from (10.2.13) and the preceding particular solution, it follows that the general solution to (10.3.9) in this case is

$$x(t) = A_0 e^{-(c/2m)t}\cos(\mu t - \phi) + \frac{F_0}{H}\cos(\omega t - \eta). \qquad (10.3.12)$$

For large t we see that x_p is dominant. For this reason we refer to the complementary function as the **transient** part of the solution, and x_p is called the **steady-state** solution. We recognize (10.3.12) as consisting of a superposition of two harmonic oscillations, one damped and the other undamped. The motion is eventually simple harmonic with a frequency coinciding with that of the driving term.

The cases for critical damping and overdamping are similar, since in both cases the complementary function (transient part of the solution) dies out exponentially and the steady-state solution (10.3.11) dominates. A typical motion of a forced mechanical system with damping is shown in Figure 10.3.3.

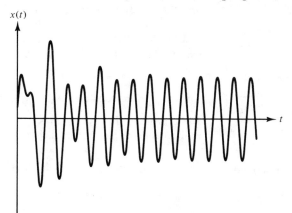

Figure 10.3.3 An example of forced motion with damping.

EXERCISES 10.3

1. Consider the damped spring-mass system whose motion is governed by

$$\frac{d^2x}{dt^2} + 2\frac{dx}{dt} + 5x = 17 \sin 2t,$$

$$x(0) = -2, \qquad \frac{dx}{dt}(0) = 0.$$

(a) Determine whether the motion is underdamped, overdamped, or critically damped.

(b) Determine the solution of the given initial value problem and identify the *transient* and *steady-state* parts.

2. Consider the spring-mass system whose motion is governed by

$$\frac{d^2x}{dt^2} + \omega_0^2 x = F_0 \sin \omega t,$$

$$x(0) = 0, \qquad \frac{dx}{dt}(0) = 0.$$

Determine the solution if the system is resonating.

3. Consider the spring-mass system whose motion is governed by

$$\frac{d^2x}{dt^2} + 3\frac{dx}{dt} + 2x = 10 \sin t.$$

Determine the *steady-state* solution, x_p, and express your answer in the form $x_p = A_0\sin(t - \phi)$, for appropriate constants A_0 and ϕ.

4. Consider the forced undamped spring-mass system whose motion is governed by

$$\frac{d^2x}{dt^2} + \omega_0^2 x = \frac{F_0}{m}\cos \omega t.$$

Derive the particular solutions given in equations (10.3.4) and (10.3.8). (You will need to consider $\omega \neq \omega_0$ and $\omega = \omega_0$ separately.)

5. The general solution of the forced undamped (nonresonating) spring-mass system is

$$x(t) = A_0\cos(\omega_0 t - \phi) + \frac{F_0}{m(\omega_0^2 - \omega^2)}\cos \omega t,$$

If $\omega/\omega_0 = p/q$, where p and q are integers, show that the motion is periodic with period $T = 2\pi q/\omega_0$.

6. Determine the period of the motion for the spring-mass system governed by the differential equation

$$\frac{d^2x}{dt^2} + \frac{9}{16}x = 55 \cos 2t.$$

7. Consider the damped forced motion described by

$$\frac{d^2x}{dt^2} + \frac{c}{m}\frac{dx}{dt} + \frac{k}{m}x = \frac{F_0}{m}\cos \omega t.$$

Derive the steady-state solution (10.3.10) given in the text.

8. Consider the damped forced motion described by

$$\frac{d^2x}{dt^2} + \frac{c}{m}\frac{dx}{dt} + \frac{k}{m}x = \frac{F_0}{m}\cos \omega t.$$

We have shown that the steady-state solution can be written in the form

$$x_p = \frac{F_0}{H}\cos(\omega t - \eta),$$

where

$$\cos \eta = \frac{m(\omega_0^2 - \omega^2)}{H},$$

$$\sin \eta = \frac{c\omega}{H}, \qquad \omega_0 = \sqrt{\frac{k}{m}},$$

and

$$H = \sqrt{m^2(\omega_0^2 - \omega^2)^2 + c^2\omega^2}.$$

Assuming that $c^2/2m^2\omega_0^2 < 1$, show that the amplitude of the steady-state solution is a maximum when

$$\omega = \sqrt{\omega_0^2 - \frac{c^2}{2m^2}}.$$

(*Hint:* The maximum occurs for the value of ω that makes H a minimum. Assume that H is a function of ω, and determine the value of ω that minimizes H.)

9. Consider the damped spring-mass system with $m = 1$, $k = 5$, $c = 2$, and $F(t) = 8\cos \omega t$.
(a) Determine the transient part of the solution and the steady-state solution.
(b) Determine the value of ω that maximizes the amplitude of the steady-state solution and express the corresponding solution in the form

$$x_p = A_0\cos(\omega t - \eta),$$

for appropriate constants A_0, ω, and η.

10. Consider the spring-mass system whose motion is governed by the differential equation

$$\frac{d^2x}{dt^2} + 2\frac{dx}{dt} + 5x = 4e^{-t}\cos 2t.$$

(a) Describe the variation with time of the applied external force.
(b) Determine the motion of the mass. What happens as $t \to \infty$?

11. Consider the spring-mass system whose motion is governed by the differential equation

$$\frac{d^2x}{dt^2} + 16x = 130e^{-t}\cos t.$$

Determine the resulting motion and identify any transient and steady-state parts of your solution.

12. This is concerned with the phenomenon of *beats*, which occur when the frequency of an oscillating system is close—but not equal to—the driving frequency.
(a) Show that the initial value problem

$$\frac{d^2x}{dt^2} + \omega_0^2 x = F_0\cos \omega t,$$

$$x(0) = 0, \qquad \frac{dx}{dt}(0) = 0,$$

has solution

$$x = \frac{F_0}{m(\omega_0^2 - \omega^2)}(\cos \omega t - \cos \omega_0 t).$$

(b) Using the trigonometric identity

$$2 \sin A \sin B = \cos(A - B) - \cos(A + B),$$

show that the solution in (a) can be written in the form

$$x = \frac{2F_0}{m(\omega_0^2 - \omega^2)} \sin \left[\left(\frac{\omega_0 - \omega}{2}\right)t\right] \sin \left[\left(\frac{\omega_0 + \omega}{2}\right)t\right].$$

If $|\omega_0 - \omega| \ll 1$, then $\sin \left[\left(\frac{\omega_0 - \omega}{2}\right)t\right]$ is slowly varying compared to $\sin \left[\left(\frac{\omega_0 + \omega}{2}\right)t\right]$. Thus $x(t)$ behaves like a rapidly oscillating SHM mode whose amplitude is slowly varying in time. See Figure 10.3.4.

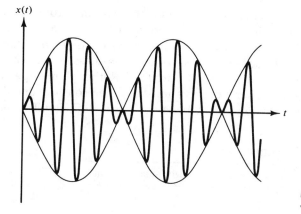

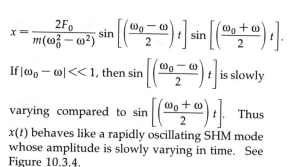

Figure 10.3.4 When $\omega_0 \approx \omega$, the resulting motion can be interpreted as being simple harmonic with a slowly varying amplitude.

10.4   RLC CIRCUITS

In Section 2.6 we used Kirchoff's second law to derive the differential equation

$$\frac{di}{dt} + \frac{R}{L}i + \frac{1}{LC}q = \frac{1}{L}E(t), \tag{10.4.1}$$

which governs the behavior of the RLC circuit shown in Figure 10.4.1. Here q is the charge on the capacitor at time t, the constants R, L, C are the resistance, inductance, and capacitance of the circuit elements, respectively, and $E(t)$ denotes the driving electromotive force (EMF). The current in the circuit is related to the charge on the capacitor via

$$i(t) = \frac{dq}{dt}. \tag{10.4.2}$$

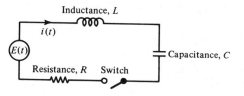

Figure 10.4.1 An RLC circuit.

Consequently (10.4.1) can be written as the second-order constant coefficient differential equation

$$\frac{d^2q}{dt^2} + \frac{R}{L}\frac{dq}{dt} + \frac{1}{LC}q = \frac{1}{L}E(t). \tag{10.4.3}$$

A comparison of (10.4.3) with the basic differential equation governing the motion of a spring-mass system, namely,

$$\frac{d^2x}{dt^2} + \frac{c}{m}\frac{dx}{dt} + \frac{k}{m}x = \frac{F(t)}{m},$$

reveals that although the two problems are distinct physically, from a purely mathematical standpoint they are identical. The correspondence between the variables and parameters in an *RLC* circuit and a spring-mass system is given in Table 10.4.1. It follows that all the results so far derived for a spring-mass system can be translated into corresponding results for *RLC* circuits. Rather than repeating these results, we will make some general observations and then consider one illustrative example. The full investigation of the behavior of an *RLC* circuit is left for the exercises.

TABLE 10.4.1

RLC circuit	Spring-mass system
$q(t)$	$x(t)$
L	m
R	c
$1/C$	k
$E(t)$	$F(t)$

Consider first the homogeneous equation

$$\frac{d^2q}{dt^2} + \frac{R}{L}\frac{dq}{dt} + \frac{1}{LC}q = 0. \tag{10.4.4}$$

This has auxiliary polynomial

$$P(r) = r^2 + \frac{R}{L}r + \frac{1}{LC}$$

with roots

$$r = \frac{-R \pm \sqrt{R^2 - \dfrac{4L}{C}}}{2L}.$$

Three familiar cases arise. The circuit is said to be

(a) *Underdamped if* $R^2 < \dfrac{4L}{C}$.

(b) *Critically damped if* $R^2 = \dfrac{4L}{C}$.

(c) *Overdamped if* $R^2 > \dfrac{4L}{C}$.

The corresponding solutions to (10.4.4) are

(a) $q(t) = e^{-(R/2L)t}[c_1 \cos \mu t + c_2 \sin \mu t]$, $\qquad \mu = \dfrac{\sqrt{\dfrac{4L}{C} - R^2}}{2L}$. $\qquad$ (10.4.5)

(b) $q(t) = e^{-(R/2L)t}(c_1 + c_2 t)$.

(c) $q(t) = e^{-(R/2L)t}(c_1 e^{vt} + c_2 e^{-vt})$, $\qquad v = \dfrac{\sqrt{R^2 - \dfrac{4L}{C}}}{2L}$.

In all cases with $R \neq 0$,

$$\lim_{t \to \infty} q(t) = 0.$$

Equivalently, we can state that the complementary function for (10.4.3), $q = q_c$, satisfies

$$\lim_{t \to \infty} q_c(t) = 0.$$

We refer to q_c as the *transient part* of the solution to (10.4.3), since it decays exponentially with time.

As an example we consider the case of a periodic driving EMF in an underdamped circuit.

Example 10.4.1 Discuss the behavior of the current in the *RLC* circuit

$$\frac{d^2 q}{dt^2} + \frac{R}{L}\frac{dq}{dt} + \frac{1}{LC} q = \frac{E_0}{L} \cos \omega t, \qquad (10.4.6)$$

where E_0 and ω are positive constants and $R^2 < \dfrac{4L}{C}$.

Solution The complementary function that is given in (10.4.5) can be written as

$$q_c = A_0 e^{-(R/2L)t} \cos(\mu t - \phi),$$

where μ is given in (10.4.5), and A_0, ϕ are defined in the usual manner, namely,

$$A_0 \cos \phi = c_1, \qquad A_0 \sin \phi = c_2.$$

A particular solution to (10.4.6) can be obtained by making the appropriate replacements in the solution (10.3.11) for the corresponding spring-mass system. The result is

$$q_p = \frac{E_0}{H} \cos(\omega t - \eta),$$

where

$$H = \sqrt{L^2(\omega_0^2 - \omega^2)^2 + R^2 \omega^2},$$

and

$$\cos \eta = \frac{L(\omega_0^2 - \omega^2)}{H}, \qquad \sin \eta = \frac{R\omega}{H}, \qquad \omega_0 = \sqrt{\frac{1}{LC}}.$$

Consequently, the charge on the capacitor at time t is

$$q(t) = A_0 e^{-(R/2L)t} \cos(\mu t - \phi) + \frac{E_0}{H} \cos(\omega t - \eta),$$

and the corresponding current in the circuit can be determined from

$$i(t) = \frac{dq}{dt}.$$

Rather than compute this derivative, we consider the late-time behavior of q, and the corresponding late-time behavior of the current. Since q_c tends to zero as t tends to $+\infty$, for large t the particular solution q_p will be the dominant part of $q(t)$. For this reason we refer to q_p as the *steady-state* solution. The corresponding *steady-state* current in the circuit, denoted i_s, is given by

$$i_s = \frac{dq_p}{dt} = -\frac{\omega E_0}{H} \sin(\omega t - \eta).$$

We see that this is oscillatory with time and that the frequency of the oscillation coincides with the frequency of the driving EMF. The amplitude of the oscillation is

$$A = \frac{\omega E_0}{H},$$

that is,

$$A = \frac{\omega E_0}{\sqrt{L^2(\omega_0^2 - \omega^2)^2 + R^2\omega^2}}. \tag{10.4.7}$$

It is often required to determine the value of ω that maximizes this amplitude. In order to do so, we rewrite (10.4.7) in the equivalent form

$$A = \frac{E_0}{\sqrt{\omega^{-2}L^2(\omega_0^2 - \omega^2)^2 + R^2}}.$$

This will be a maximum when the term in parentheses vanishes, that is, when

$$\omega^2 = \omega_0^2.$$

Substituting for $\omega_0^2 = \frac{1}{LC}$, it follows that the amplitude of the steady-state current will be a maximum when $\omega = \omega_{\max}$, where

$$\omega_{\max} = \sqrt{\frac{1}{LC}}.$$

The corresponding value of A is

$$A_{\max} = \frac{E_0}{R}.$$

The behavior of A as a function of ω for typical values of E_0, R, L, and C is shown in Figure 10.4.2.

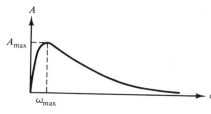

Figure 10.4.2 The behavior of the steady-state current as a function of the driving frequency ω.

EXERCISES 10.4

1. Determine the steady-state current in the RLC circuit which has $R = \frac{3}{2}\ \Omega$, $L = \frac{1}{2}$ H, $C = \frac{2}{3}$ F, and $E = 13 \cos 3t$ volts.

2. Determine the charge on the capacitor at time t in the RLC circuit which has $R = 4\ \Omega$, $L = 4$ H, $C = \frac{1}{17}$ F, and $E = E_0$ V, where E_0 is constant. What happens to the charge on the capacitor as $t \to +\infty$? Describe the behavior of the current in the circuit.

3. Consider the RLC circuit with $E(t) = E_0 \cos \omega t$ volts, were E_0 and ω are constants. If there is no resistor in the circuit, show that the charge on the capacitor satisfies

$$\lim_{t \to \infty} q(t) = +\infty,$$

if and only if $\omega = 1/\sqrt{LC}$. What happens to the current in the circuit as $t \to +\infty$?

4. Consider the RLC circuit with $R = 16\ \Omega$, $L = 8$ H, $C = \frac{1}{40}$ F, and $E(t) = 17 \cos 2t$ volts. Determine the current in the circuit for $t > 0$, given that at $t = 0$ the capacitor is uncharged and there is no current flowing.

5. Consider the RLC circuit with $R = 3\ \Omega$, $L = \frac{1}{2}$ H, $C = \frac{1}{5}$ F, and $E(t) = 2 \cos \omega t$ volts. Determine the current in the circuit at time t, and find the value of ω that maximizes the amplitude of the steady-state current.

6. Show that the differential equation governing the behavior of an RLC circuit can be written

directly in terms of the current $i(t)$ as

$$\frac{d^2 i}{dt^2} + \frac{R}{L}\frac{di}{dt} + \frac{1}{LC}i = \frac{1}{L}\frac{dE}{dt}.$$

7. Determine the current in the general RLC circuit with $R^2 < 4L/C$, if $E(t) = E_0 e^{-at}$, where E_0 and a are constants.

8. Consider the RLC circuit with $R = 2\ \Omega$, $L = \frac{1}{2}$ H, $C = \frac{2}{5}$ F. Initially the capacitor is uncharged and there is no current flowing in the circuit. Determine the current for $t > 0$, if the applied EMF is (see Figure 10.4.3)

$$E(t) = \begin{cases} 50t \text{ volts}, & 0 \le t < \pi \\ 50\pi \text{ volts}, & t \ge \pi. \end{cases}$$

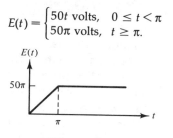

Figure 10.4.3

Hint: You will need to consider both time intervals $[0, \pi)$ and $[\pi, \infty)$ separately. The constants arising on the interval $0 \le t < \pi$ can be determined from the initial conditions. The remaining constants can be determined by ensuring that q and i are continuous at $t = \pi$. We will develop a superior technique for solving this type of problem in Chapter 12.]

11

Linear Systems of Differential Equations

11.1   INTRODUCTION

So far our discussion of differential equations has centered around solving a single differential equation for a single unknown function $y(x)$. However, in practice most applied problems involve more than one unknown function for their formulation and hence require the solution of a system of differential equations. Perhaps the simplest way to see how systems naturally arise is to consider the motion of an object in space. If this object has mass m and is moving under the influence of a force $\mathbf{F} = (F_1, F_2, F_3)$, then, according to Newton's second law of motion, the position of the object at time t, $(x(t), y(t), z(t))$, is obtained by solving the system

$$m \frac{d^2x}{dt^2} = F_1, \qquad m \frac{d^2y}{dt^2} = F_2, \qquad m \frac{d^2z}{dt^2} = F_3.$$

In this chapter we consider the formulation and solution of systems of differential equations. There are some familiar questions to answer:

Question 1: How can we formulate our problems in a way suitable for solution?

Question 2: How many solutions, if any, does our differential system possess?

Question 3: How do we find the solutions that arise in Question 2?

We will restrict our attention to *linear* systems of differential equations, in which case the answer to Question 1 will fall out naturally. In answering Question 2 we will once more require the vector space techniques from Chapters 6 and 7, whereas we will find an elegant answer to Question 3 using eigenvalues and eigenvectors of appropriate matrices.

Before beginning the general development of the theory for systems of differential equations, we consider two physical problems that can be formulated mathematically in terms of such systems.

Consider the coupled spring-mass system which consists of two masses m_1, m_2 connected by two springs whose spring constants are k_1, k_2, respectively (see Figure 11.1.1). Let $x(t)$, $y(t)$ denote the displacement of m_1, m_2 from their positions

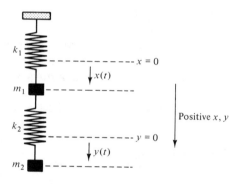

Figure 11.1.1 A coupled spring-mass system.

when the system is in the static equilibrium position. Then, using Hooke's law[1] and Newton's second law, it follows that the motion of the masses is governed by the system of differential equations

$$m_1 \frac{d^2x}{dt^2} = -k_1x + k_2(y - x), \tag{11.1.1}$$

$$m_2 \frac{d^2y}{dt^2} = -k_2(y - x). \tag{11.1.2}$$

We would expect the problem to have a unique solution once the initial positions and velocities of the masses have been specified.

As a second example, consider the mixing problem depicted in Figure 11.1.2. Two tanks contain a solution consisting of a chemical dissolved in water. A solution containing c grams/liter of the chemical flows into tank 1 at a rate of r liters/minute and the solution in tank 2 flows out at the same rate. In addition, a solution

---

[1] See Section 10.1, page 380.

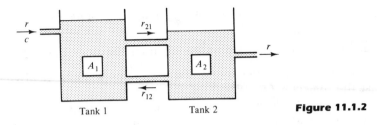

Tank 1   Tank 2   **Figure 11.1.2**

flows into tank 1 from tank 2 at a rate of r_{12} liters/minute and into tank 2 from tank 1 at a rate of r_{21} liters/minute. We wish to determine the amounts of chemical, $A_1(t), A_2(t)$, in tanks 1 and 2 at any time t. A similar analysis to that used in Section 2.5 yields the following system of differential equations governing the behavior of A_1 and A_2:

$$\frac{dA_1}{dt} = -\frac{r_{21}}{V_1}A_1 + \frac{r_{12}}{V_2}A_2 + cr, \tag{11.1.3}$$

$$\frac{dA_2}{dt} = \frac{r_{21}}{V_1}A_1 - \frac{(r_{12}+r)}{V_2}A_2, \tag{11.1.4}$$

where V_1, V_2 denote the volume of solution in each tank at time t.

We will give a full discussion of both of these problems in Section 11.8 once we have developed the theory and solution techniques for linear differential systems.

11.2   FIRST-ORDER LINEAR SYSTEMS

In keeping with our aim of studying linear mathematics, we focus our attention on linear systems of differential equations. The analysis of nonlinear systems is considerably more complicated and is best left for a second course in differential equations. Restricting ourselves to linear systems will enable us to use the power of the vector space methods to derive appropriate results regarding the solution properties of such systems.

Definition 11.2.1: A system of differential equations of the form

$$\frac{dx_1}{dt} = a_{11}(t)x_1(t) + a_{12}(t)x_2(t) + \cdots + a_{1n}(t)x_n(t) + b_1(t),$$

$$\frac{dx_2}{dt} = a_{21}(t)x_1(t) + a_{22}(t)x_2(t) + \cdots + a_{2n}(t)x_n(t) + b_2(t),$$

$$\vdots \tag{11.2.1}$$

$$\frac{dx_n}{dt} = a_{n1}(t)x_1(t) + a_{n2}(t)x_2(t) + \cdots + a_{nn}(t)x_n(t) + b_n(t),$$

where the $a_{ij}(t)$ and $b_i(t)$ are specified functions on some interval I, is called a **first-order linear system**. If $b_1 = b_2 = \cdots = b_n = 0$, then the system is called **homogeneous**; otherwise it is called **nonhomogeneous**.

REMARKS

1. It is important to notice the structure of a first-order linear system. The highest derivative occurring in such a system is a first derivative. Further, there is precisely one equation involving the derivative of each separate unknown function. Finally, the terms that appear on the right-hand side of the equations do *not* involve any derivatives and are linear in the unknown functions $x_1, x_2, \ldots, x_n$.

2. We will usually denote $\dfrac{dx_i}{dt}$ by x_i'.

Example 11.2.1 An example of a nonhomogeneous first-order linear system is

$$x_1' = e^t x_1 + t^2 x_2 + \sin t,$$
$$x_2' = tx_1 + 3x_2 - \cos t,$$

and the associated homogeneous system is

$$x_1' = e^t x_1 + t^2 x_2,$$
$$x_2' = tx_1 + 3x_2.$$

Definition 11.2.2: By a **solution** to the system (11.2.1) on an interval I we mean an n-tuple of functions $x_1(t), x_2(t), \ldots, x_n(t)$, which, when substituted into the left-hand side of the system, yield the right-hand side for all t in I.

Example 11.2.2 Show that

$$x_1 = -2e^{5t} + 4e^{-t},$$
$$x_2 = e^{5t} + e^{-t}, \tag{11.2.2}$$

is a solution to

$$x_1' = x_1 - 8x_2, \tag{11.2.3}$$
$$x_2' = -x_1 + 3x_2, \tag{11.2.4}$$

on $(-\infty, \infty)$.

Solution From (11.2.2) it follows that the left-hand side of (11.2.3) is

$$x_1' = -10e^{5t} - 4e^{-t},$$

whereas the right-hand side is

$$x_1 - 8x_2 = (-2e^{5t} + 4e^{-t}) - 8(e^{5t} + e^{-t}) = -10e^{5t} - 4e^{-t}.$$

Thus we have

$$x_1' = x_1 - 8x_2,$$

so that (11.2.3) is satisfied by the given functions for all $t \in (-\infty, \infty)$. Similarly, it is easily shown that for all $t \in (-\infty, \infty)$,

$$x_2' = -x_1 + 3x_2,$$

so that (11.2.4) is also satisfied. It follows that x_1 and x_2 do define a solution to the given system on $(-\infty, \infty)$.

We now derive a simple technique for solving (11.2.1) that can be used when the coefficients $a_{ij}(t)$ in the system are constants. Although we will develop a superior technique for such systems in the later sections, the method introduced here will be useful in motivating some of the subsequent results. For simplicity we will only consider $n = 2$. Under the assumption that all a_{ij} are constants, the system (11.2.1) reduces to

$$x_1' = a_{11}x_1 + a_{12}x_2 + b_1(t),$$

$$x_2' = a_{21}x_1 + a_{22}x_2 + b_2(t).$$

This system can be written in the equivalent form

$$(D - a_{11})x_1 - a_{12}x_2 = b_1(t), \tag{11.2.5}$$

$$-a_{21}x_1 + (D - a_{22})x_2 = b_2(t), \tag{11.2.6}$$

where D is the differential operator d/dt. The idea behind the technique is that we can now easily eliminate x_2 between these two equations by operating on (11.2.5) with $(D - a_{22})$, multiplying (11.2.6) by a_{12}, and adding the resulting equations. This yields a constant coefficient linear differential equation for x_1 only, which can be solved using the techniques of Chapter 9. Substituting the expression thereby obtained for x_1 into (11.2.5) will then yield x_2.[1] We illustrate the technique with an example.

Example 11.2.3 Solve the system

$$x_1' = x_1 + 2x_2, \tag{11.2.7}$$

$$x_2' = 2x_1 - 2x_2. \tag{11.2.8}$$

Solution We begin by rewriting the system in operator form as

---

[1] If $a_{12} = 0$, we can determine x_1 directly from (11.2.5) and then x_2 can be obtained from (11.2.6).

$$(D-1)x_1 - \quad 2x_2 = 0, \tag{11.2.9}$$

$$-2x_1 + (D+2)x_2 = 0. \tag{11.2.10}$$

In order to eliminate x_2 between these two equations we first operate on (11.2.9) with $(D+2)$ to obtain

$$(D+2)(D-1)x_1 - 2(D+2)x_2 = 0.$$

Adding twice (11.2.10) to this equation yields

$$(D+2)(D-1)x_1 - 4x_1 = 0,$$

that is,

$$(D^2 + D - 6)x_1 = 0.$$

This constant coefficient differential equation has auxiliary polynomial

$$P(r) = r^2 + r - 6 = (r+3)(r-2).$$

Consequently,

$$x_1 = c_1 e^{-3t} + c_2 e^{2t}. \tag{11.2.11}$$

We now determine x_2. From (11.2.9), we have

$$x_2 = \tfrac{1}{2}(D-1)x_1;$$

that is, substituting for x_1 from (11.2.11),

$$x_2 = \tfrac{1}{2}(Dx_1 - x_1) = \tfrac{1}{2}(-4c_1 e^{-3t} + c_2 e^{2t}).$$

Hence the solution of the system (11.2.7), (11.2.8) is

$$x_1 = c_1 e^{-3t} + c_2 e^{2t}, \qquad x_2 = \tfrac{1}{2}(-4c_1 e^{-3t} + c_2 e^{2t}),$$

where c_1, c_2 are arbitrary constants.

In solving an applied problem that is governed by a system of differential equations, we often require only the particular solution of the system that corresponds to the specific problem of interest (rather than finding all solutions of the system). Such a particular solution is obtained by specifying appropriate auxiliary conditions. This leads to the idea of an initial value problem for linear systems.

Definition 11.2.3: Solving the system (11.2.1) subject to n auxiliary conditions imposed at the *same* value of the independent variable is called an **initial value problem**. Thus the general form of the auxiliary conditions for an initial value problem is:

$$x_1(t_0) = \alpha_1, \; x_2(t_0) = \alpha_2, \; \ldots, \; x_n(t_0) = \alpha_n,$$

where $\alpha_1, \alpha_2, \ldots, \alpha_n$ are constants.

Example 11.2.4   Solve the initial value problem

$$x_1' = x_1 + 2x_2, \qquad x_2' = 2x_1 - 2x_2,$$
$$x_1(0) = 1, \, x_2(0) = 0.$$

Solution   We saw in the previous example that the solution of the given system of differential equations is

$$x_1 = c_1 e^{-3t} + c_2 e^{2t}, \qquad x_2 = \tfrac{1}{2}(-4c_1 e^{-3t} + c_2 e^{2t}), \qquad (11.2.12)$$

where c_1 and c_2 are arbitrary constants.   Imposing the two initial conditions yields the following equations for determining c_1 and c_2:

$$c_1 + c_2 = 1,$$
$$-4c_1 + c_2 = 0.$$

Consequently,

$$c_1 = \tfrac{1}{5}, \qquad c_2 = \tfrac{4}{5}.$$

Substituting for c_1, c_2 into (11.2.12) yields the unique solution

$$x_1 = \tfrac{1}{5}(e^{-3t} + 4e^{2t}), \qquad x_2 = \tfrac{2}{5}(e^{2t} - e^{-3t}).$$

It might appear that restricting to *first-order* linear systems means that we are considering only very special types of linear differential systems.   In fact this is incorrect, since most systems of k differential equations that are linear in k unknown functions and their derivatives can be rewritten as equivalent first-order systems by redefining the dependent variables.   We illustrate with an example.

Example 11.2.5   Rewrite the linear system

$$\frac{d^2x}{dt^2} - 4y = e^t, \qquad\qquad (11.2.13)$$

$$\frac{d^2y}{dt^2} + t^2 \frac{dx}{dt} = \sin t \qquad\qquad (11.2.14)$$

as an equivalent first-order system.

Solution   We introduce new dependent variables relative to which (11.2.13) and (11.2.14) reduce to first-order differential equations.   Thus let

$$x_1 = x, \qquad x_2 = \frac{dx}{dt}, \qquad x_3 = y, \qquad x_4 = \frac{dy}{dt}. \qquad (11.2.15)$$

Then (11.2.13) and (11.2.14) become

$$\frac{dx_2}{dt} - 4x_3 = e^t, \qquad \frac{dx_4}{dt} + t^2 x_2 = \sin t.$$

These equations must also be supplemented with equations for x_1 and x_3. From (11.2.15) we see that

$$\frac{dx_1}{dt} = x_2, \qquad \frac{dx_3}{dt} = x_4.$$

Consequently, the given system of differential equations is equivalent to the first-order linear system

$$\frac{dx_1}{dt} = x_2, \qquad \frac{dx_2}{dt} = 4x_3 + e^t, \qquad \frac{dx_3}{dt} = x_4, \qquad \frac{dx_4}{dt} = -t^2x_2 + \sin t.$$

Finally, consider the general nth-order linear differential equation

$$x^{(n)} + a_1(t)x^{(n-1)} + \cdots + a_{n-1}(t)x' + a_n(t)x = F(t). \qquad (11.2.16)$$

If we introduce the new variables $x_1, x_2, \ldots, x_n$ defined by

$$x_1 = x, \; x_2 = x', \ldots, x_n = x^{(n-1)},$$

then (11.2.16) can be replaced by the equivalent first-order linear system

$$x_1' = x_2, \; x_2' = x_3, \ldots, \; x_{n-1}' = x_n$$
$$x_n' = -a_n(t)x_1 - a_{n-1}(t)x_2 - \cdots - a_1(t)x_n + F(t).$$

EXERCISES 11.2

In problems 1–7 solve the given system of differential equations.

1. $x_1' = 2x_1 - 3x_2, \; x_2' = x_1 - 2x_2.$
2. $x_1' = 4x_1 + 2x_2, \; x_2' = -x_1 + x_2.$
3. $x_1' = 2x_1 + 4x_2, \; x_2' = -4x_1 - 6x_2.$
4. $x_1' = 2x_2, \; x_2' = -2x_1.$
5. $x_1' = x_1 - 3x_2, \; x_2' = 3x_1 + x_2.$
6. $x_1' = 2x_1, \; x_2' = x_2 - x_3, \; x_3' = x_2 + x_3.$
7. $x_1' = -2x_1 + x_2 + x_3, \; x_2' = x_1 - x_2 + 3x_3,$ $x_3' = -x_2 - 3x_3.$

In problems 8–10, solve the given initial value problem.

8. $x_1' = 2x_2, \; x_2' = x_1 + x_2, \; x_1(0) = 3, \; x_2(0) = 0.$
9. $x_1' = 2x_1 + 5x_2, \; x_2' = -x_1 - 2x_2, \; x_1(0) = 0,$ $x_2(0) = 1.$
10. $x_1' = 2x_1 + x_2, \; x_2' = -x_1 + 4x_2, \; x_1(0) = 1,$ $x_2(0) = 3.$

In problems 11–13, solve the given nonhomogeneous system.

11. $x_1' = x_1 + 2x_2 + 5e^{4t}, \; x_2' = 2x_1 + x_2.$
12. $x_1' = -2x_1 + x_2 + t, \; x_2' = -2x_1 + x_2 - 1.$
13. $x_1' = x_1 + x_2 + e^{2t}, \; x_2' = 3x_1 - x_2 + 5e^{2t}.$

In problems 14 and 15 convert the given system of differential equations to a first-order linear system.

14. $\dfrac{dx}{dt} - ty = \cos t, \; \dfrac{d^2y}{dt^2} - \dfrac{dx}{dt} + x = e^t.$

15. $\dfrac{d^2x}{dt^2} - 3\dfrac{dy}{dt} + x = \sin t, \; \dfrac{d^2y}{dt^2} - t\dfrac{dx}{dt} - e^ty = t^2.$

In problems 16–18 convert the given linear differential equation into a first-order linear system.

16. $x'' + 2tx' + x = \cos t.$
17. $x'' + ax' + bx = F(t), \; a, \; b$ constants.

18. $x''' + t^2 x' - e^t x = t.$

19. The initial value problem that governs the behavior of a coupled spring-mass system is (see Section 11.1)

$$m_1 \frac{d^2 x}{dt^2} = -k_1 x + k_2(y - x), \quad m_2 \frac{d^2 y}{dt^2} = -k_2(y - x).$$

$$x(0) = \alpha_1, \quad x'(0) = \alpha_2, \quad y(0) = \alpha_3, \quad y'(0) = \alpha_4,$$

where $\alpha_1, \alpha_2, \alpha_3, \alpha_4$ are constants. Convert this problem into an initial value problem for an equivalent first-order linear system. (You must give the appropriate initial conditions in the new variables.)

20. Solve the initial value problem

$$x_1' = -(\tan t)x_1 + 3 \cos^2 t,$$
$$x_2' = x_1 + \tan t\, x_2 + 2 \sin t,$$

$$x_1(0) = 4, \, x_2(0) = 0.$$

11.3   MATRIX FORMULATION OF FIRST-ORDER LINEAR SYSTEMS

In the previous section we introduced a simple technique for solving first-order linear systems with constant coefficients. We now wish to develop the theory underlying the solution of general first-order linear systems. The first step in this development is to formulate the problem of solving a first-order linear system as an appropriate vector space problem. The key to this formulation is the realization that the system

$$x_1' = a_{11}(t)x_1(t) + a_{12}(t)x_2(t) + \cdots + a_{1n}(t)x_n(t) + b_1(t),$$
$$x_2' = a_{21}(t)x_1(t) + a_{22}(t)x_2(t) + \cdots + a_{2n}(t)x_n(t) + b_2(t),$$
$$\vdots$$
$$x_n' = a_{n1}(t)x_1(t) + a_{n2}(t)x_2(t) + \cdots + a_{nn}(t)x_n(t) + b_n(t),$$

can be written in matrix form as

$$x'(t) = A(t)x(t) + b(t), \tag{11.3.1}$$

where

$$x(t) = \begin{bmatrix} x_1 \\ x_2 \\ \vdots \\ x_n \end{bmatrix}, \quad x'(t) = \begin{bmatrix} x_1' \\ x_2' \\ \vdots \\ x_n' \end{bmatrix},$$

and

$$A(t) = \begin{bmatrix} a_{11}(t) & a_{12}(t) & \cdots & a_{1n}(t) \\ a_{21}(t) & a_{22}(t) & \cdots & a_{2n}(t) \\ \vdots & \vdots & & \vdots \\ a_{n1}(t) & a_{n2}(t) & \cdots & a_{nn}(t) \end{bmatrix}, \quad b(t) = \begin{bmatrix} b_1(t) \\ b_2(t) \\ \vdots \\ b_n(t) \end{bmatrix}.$$

Example 11.3.1   The system of equations

$$\frac{dx_1}{dt} = 3x_1 + (\sin t)x_2 + e^t,$$

$$\frac{dx_2}{dt} = 7tx_1 + t^2x_2 - 4e^{-t},$$

can be written as

$$\begin{bmatrix} x_1' \\ x_2' \end{bmatrix} = \begin{bmatrix} 3 & \sin t \\ 7t & t^2 \end{bmatrix}\begin{bmatrix} x_1 \\ x_2 \end{bmatrix} + \begin{bmatrix} e^t \\ -4e^{-t} \end{bmatrix},$$

that is,

$$\mathbf{x}'(t) = A(t)\mathbf{x}(t) + \mathbf{b}(t),$$

where

$$\mathbf{x} = \begin{bmatrix} x_1 \\ x_2 \end{bmatrix}, \qquad A(t) = \begin{bmatrix} 3 & \sin t \\ 7t & t^2 \end{bmatrix}, \qquad \mathbf{b}(t) = \begin{bmatrix} e^t \\ -4e^{-t} \end{bmatrix}.$$

In general the initial value problem for a first-order *linear* system can therefore be formulated as the following matrix problem:

Solve

$$\mathbf{x}'(t) = A(t)\mathbf{x}(t) + \mathbf{b}(t),$$

subject to

$$\mathbf{x}(t_0) = \mathbf{x}_0, \qquad \mathbf{x}_0 = [\alpha_1 \quad \alpha_2 \quad \cdots \quad \alpha_n]^T.$$

We see that matrices provide a natural framework for studying first-order *linear* systems. However, it should be noticed that we are, in fact, dealing with matrix *functions*—that is, matrices whose elements are themselves functions. We briefly list the properties of matrix functions that will be required in the remainder of this chapter.

MATRIX FUNCTIONS

The algebra of matrix functions is the same as that for matrices of constants. The calculus of matrix functions is defined element-wise:

1. A matrix function is continuous on an interval I if each of its elements are continuous on I.

2. The derivative of a matrix function is obtained by differentiating *every* element of the matrix. Thus, if $A(t) = [a_{ij}(t)]$, then $dA/dt = [da_{ij}/dt]$, provided each of the a_{ij} are differentiable.

Of importance to us later will be the product rule for differentiating matrix functions. If A and B are both differentiable and the product AB is defined, then

$$\frac{d}{dt}(AB) = A\frac{dB}{dt} + \frac{dA}{dt}B.$$

The key point to notice is that the order of the multiplication must be preserved.

3. Integration of a matrix function is also defined element-wise. Thus, if $A(t) = [a_{ij}(t)]$, where each $a_{ij}(t)$ is integrable on the interval $[a, b]$, then

$$\int_a^b A(t)\, dt = \left[\int_a^b a_{ij}(t)\, dt\right].$$

Example 11.3.2   If $A(t) = \begin{bmatrix} 2t & 1 \\ 6t^2 & 4e^{2t} \end{bmatrix}$, determine $\dfrac{dA}{dt}$ and $\displaystyle\int_0^1 A(t)\, dt$.

Solution

$$\frac{dA}{dt} = \begin{bmatrix} 2 & 0 \\ 12t & 8e^{2t} \end{bmatrix},$$

whereas

$$\int_0^1 A(t)\, dt = \begin{bmatrix} \int_0^1 2t\, dt & \int_0^1 1\, dt \\ \int_0^1 6t^2\, dt & \int_0^1 4e^{2t}\, dt \end{bmatrix} = \begin{bmatrix} 1 & 1 \\ 2 & 2(e^2 - 1) \end{bmatrix}.$$

An $n \times 1$ matrix function is called a **column vector function**.[1] Thus $\mathbf{x}$, $\mathbf{x}'$, and $\mathbf{b}$ in (11.3.1) are column vector functions. We will let $V_n(I)$ denote the set of all column n-vector functions defined on an interval I. The following results concerning $V_n(I)$ will be needed in the remaining sections.

Theorem 11.3.1: $V_n(I)$ forms a vector space.

PROOF   We leave the verification that $V_n(I)$ satisfies Definition 6.2.1 (page 185) as an exercise. ∎

Since $V_n(I)$ forms a vector space, we can discuss linear dependence and linear independence of column vector functions. We first need the following definition.

---

[1] If we wish to emphasize the number of elements in a column vector function, then we refer to it as a column n-vector function.

> **Definition 11.3.1:**   Let $\mathbf{x}_1(t)$, $\mathbf{x}_2(t)$, . . . , $\mathbf{x}_n(t)$ be vectors in $V_n(I)$. Then the **Wronskian** of these vector functions, denoted by $W[\mathbf{x}_1, \mathbf{x}_2, \ldots, \mathbf{x}_n](t)$, is defined by
>
> $$W[\mathbf{x}_1, \mathbf{x}_2, \ldots, \mathbf{x}_n](t) = \det[\mathbf{x}_1(t), \mathbf{x}_2(t), \ldots, \mathbf{x}_n(t)].$$

REMARK   Notice that the Wronskian introduced in this definition refers to vectors in the vector space $V_n(I)$, whereas the Wronskian defined previously in the text (page 211) refers to functions in $C^n(I)$. The relationship between these two Wronskians is investigated in problem 26.

Example 11.3.3   Determine the Wronskian of the column vector functions

$$\mathbf{x}_1 = \begin{bmatrix} e^t \\ 2e^t \end{bmatrix}, \qquad \mathbf{x}_2 = \begin{bmatrix} 3 \sin t \\ \cos t \end{bmatrix}.$$

Solution   From Definition 11.3.1 we have

$$W[\mathbf{x}_1, \mathbf{x}_2](t) = \begin{vmatrix} e^t & 3 \sin t \\ 2e^t & \cos t \end{vmatrix} = e^t(\cos t - 6 \sin t).$$

Our next theorem indicates that the Wronskian plays a familiar role in determining the linear independence of vectors in $V_n(I)$.

> **Theorem 11.3.2:**   Let $\mathbf{x}_1(t)$, $\mathbf{x}_2(t)$, . . . , $\mathbf{x}_n(t)$ be vectors in $V_n(I)$. If $W[\mathbf{x}_1, \mathbf{x}_2, \ldots, \mathbf{x}_n](t_0)$ is *nonzero* at any point t_0 in I, then $\mathbf{x}_1(t)$, $\mathbf{x}_2(t)$, . . . , $\mathbf{x}_n(t)$ are linearly independent on I.

PROOF   Consider

$$c_1 \mathbf{x}_1(t) + c_2 \mathbf{x}_2(t) + \cdots + c_n \mathbf{x}_n(t) = \mathbf{0},$$

where $c_1, c_2, \ldots, c_n$ are scalars. We can write this as the matrix equation

$$X(t)\mathbf{c} = \mathbf{0},$$

where $\mathbf{c} = [c_1 \quad c_2, \quad \cdots \quad c_n]^T$ and $X(t) = [\mathbf{x}_1(t) \quad \mathbf{x}_2(t) \quad \cdots \quad \mathbf{x}_n(t)]$. Let t_0 be in I. Then if $\det[X(t_0)] \neq 0$, it follows from Corollary 5.6.1 (page 175) that the only solution to this $n \times n$ system of linear equations is $\mathbf{c} = \mathbf{0}$, and therefore $\mathbf{x}_1(t)$, $\mathbf{x}_2(t)$, . . . , $\mathbf{x}_n(t)$ are linearly independent on I. But, $\det[X(t_0)] = W[\mathbf{x}_1, \mathbf{x}_2, \ldots, \mathbf{x}_n](t_0)$, and hence the result follows.   ∎

Example 11.3.4   It follows from the previous example that the vector functions

$$\mathbf{x}_1 = \begin{bmatrix} e^t \\ 2e^t \end{bmatrix}, \qquad \mathbf{x}_2 = \begin{bmatrix} 3 \sin t \\ \cos t \end{bmatrix},$$

are linearly independent on $(-\infty, \infty)$, since, for example, $W[\mathbf{x}_1, \mathbf{x}_2](0) = 1 \neq 0$.

The problem of determining all solutions to the general first-order linear system of differential equations can now be formulated as the vector space problem.

Find all vectors $\mathbf{x} \in V_n(\mathbf{I})$ satisfying $\mathbf{x}' = A(t)\mathbf{x} + \mathbf{b}$.

The vector space $V_n(\mathbf{I})$ is not finite-dimensional since there is no finite set of linearly independent vectors that spans $V_n(\mathbf{I})$. However, the key to solving linear differential systems comes from the realization that the set of all solutions to

$$\mathbf{x}'(t) = A(t)\,\mathbf{x}(t)$$

forms a *finite*-dimensional subspace of $V_n(\mathbf{I})$. We will pursue this in the next section.

EXERCISES 11.3

In problems 1–6 convert the given equation or system to a first-order linear system and write the resulting system in matrix form.

1. $x_1' + 4x_1 - 3x_2 = 4t$, $x_2' - 6x_1 + 4x_2 = t^2$.

2. $x_1' + tx_2 - t^2x_1 = 0$, $x_2' + (\sin t)x_1 - x_2 = 0$.

3. $x_1' + (\sin t)x_2 - x_3 = t$, $x_2' + e^tx_1 - t^2x_3 = t^3$, $x_3' + tx_1 - t^2x_2 = 1$.

4. $x'' + t^2x' - e^tx = \sin t$.

5. $x''' - (\sin t)x'' + (t^2 - a^2)x = e^t$, a constant.

6. $x'' - 4(\sin t)x' + (\cos t)y' - y = \tan t$, $y'' + 5ty' + 4t^2y - x = t$.

In problems 7–10 determine the derivative of the given matrix function.

7. $A(t) = \begin{bmatrix} e^{-2t} \\ \sin t \end{bmatrix}$.

8. $A(t) = \begin{bmatrix} t & \sin t \\ \cos t & 4t \end{bmatrix}$.

9. $A(t) = \begin{bmatrix} e^t & e^{2t} & t^2 \\ 2e^t & 4e^{2t} & 5t^2 \end{bmatrix}$.

10. $A(t) = \begin{bmatrix} \sin t & \cos t & 0 \\ -\cos t & \sin t & t \\ 0 & 3t & 1 \end{bmatrix}$.

11. Let $A = [a_{ij}(t)]$ be an $m \times n$ matrix function and let $B = [b_{ij}(t)]$ be an $n \times p$ matrix function.

Use the definition of matrix multiplication to prove that

$$\frac{d}{dt}(AB) = A\frac{dB}{dt} + \frac{dA}{dt}B.$$

In problems 12 and 13 determine $\displaystyle\int_a^b A(t)\,dt$ for the given matrix function.

12. $A(t) = \begin{bmatrix} \cos t \\ \sin t \end{bmatrix}$, $a = 0$, $b = \dfrac{\pi}{2}$.

13. $A(t) = \begin{bmatrix} e^t & e^{-t} \\ 2e^t & 5e^{-t} \end{bmatrix}$, $a = 0$, $b = 1$.

14. Evaluate $\displaystyle\int A(t)\,dt$ if $A(t) = \begin{bmatrix} 2t \\ 3t^2 \end{bmatrix}$.

In problems 15 and 16 show that the given vector function defines a solution to $\mathbf{x}' = A\mathbf{x} + \mathbf{b}$ for the given A and $\mathbf{b}$.

15. $\mathbf{x} = \begin{bmatrix} e^{4t} \\ -2e^{4t} \end{bmatrix}$, $A = \begin{bmatrix} 2 & -1 \\ -2 & 3 \end{bmatrix}$, $\mathbf{b} = \begin{bmatrix} 0 \\ 0 \end{bmatrix}$.

16. $\mathbf{x} = \begin{bmatrix} 4e^{-2t} + 2\sin t \\ 3e^{-2t} - \cos t \end{bmatrix}$, $A = \begin{bmatrix} 1 & -4 \\ -3 & 2 \end{bmatrix}$,

$\mathbf{b} = \begin{bmatrix} -2(\cos t + \sin t) \\ 7\sin t + 2\cos t \end{bmatrix}$.

In problems 17–20 show that the given vector functions are linearly independent on $(-\infty, \infty)$.

17. $\mathbf{x}_1 = \begin{bmatrix} e^t \\ -e^t \end{bmatrix}$, $\mathbf{x}_2 = \begin{bmatrix} e^t \\ e^t \end{bmatrix}$.

18. $\mathbf{x}_1 = \begin{bmatrix} t \\ t \end{bmatrix}$, $\mathbf{x}_2 = \begin{bmatrix} t \\ t^2 \end{bmatrix}$.

19. $\mathbf{x}_1 = \begin{bmatrix} t+1 \\ t-1 \\ 2t \end{bmatrix}$, $\mathbf{x}_2 = \begin{bmatrix} e^t \\ e^{2t} \\ e^{3t} \end{bmatrix}$, $\mathbf{x}_3 = \begin{bmatrix} 1 \\ \sin t \\ \cos t \end{bmatrix}$.

20. $\mathbf{x}_1 = \begin{bmatrix} \sin t \\ \cos t \\ 1 \end{bmatrix}$, $\mathbf{x}_2 = \begin{bmatrix} t \\ 1-t \\ 1 \end{bmatrix}$, $\mathbf{x}_3 = \begin{bmatrix} \sinh t \\ \cosh t \\ 1 \end{bmatrix}$.

In problems 21 and 22 show that the given vector functions are linearly dependent on $(-\infty, \infty)$.

21. $\mathbf{x}_1 = \begin{bmatrix} e^t \\ 2e^{2t} \end{bmatrix}$, $\mathbf{x}_2 = \begin{bmatrix} 4e^t \\ 8e^{2t} \end{bmatrix}$.

22. $\mathbf{x}_1 = \begin{bmatrix} \sin^2 t \\ \cos^2 t \\ 2 \end{bmatrix}$, $\mathbf{x}_2 = \begin{bmatrix} 2\cos^2 t \\ 2\sin^2 t \\ 1 \end{bmatrix}$, $\mathbf{x}_3 = \begin{bmatrix} 2 \\ 2 \\ 5 \end{bmatrix}$.

23. Prove that $V_n(I)$ forms a vector space.

24. Let $A(t)$ be an $n \times n$ matrix function. Prove that the set of all solutions to $\mathbf{x}' = A(t)\mathbf{x}$ is a subspace of $V_n(I)$. What do you think the dimension of this subspace is?

25. If $A = \begin{bmatrix} 2 & -4 \\ 1 & -3 \end{bmatrix}$, determine two linearly independent solutions to $\mathbf{x}' = A\mathbf{x}$ on $(-\infty, \infty)$. (*Hint:* Use the technique developed in the previous section to obtain the components of $\mathbf{x}_1$ and $\mathbf{x}_2$.)

The next problem investigates the relationship between the Wronskian defined in this section for vectors in $V_n(I)$ and the Wronskian defined previously for functions in $C^n(I)$.

26. Consider the differential equation

$$\frac{d^2 y}{dt^2} + a\,\frac{dy}{dt} + by = 0, \qquad \text{(i)}$$

where a and b are arbitrary functions of t.

(a) Show that (i) can be replaced by the equivalent linear system

$$\mathbf{x}' = A\mathbf{x}, \qquad \text{(ii)}$$

where

$$A = \begin{bmatrix} 0 & 1 \\ -b & -a \end{bmatrix} \quad \text{and} \quad x_1 = y,\ x_2 = y'.$$

(b) If $y_1 = f_1(t)$, and $y_2 = f_2(t)$ are solutions of (i) on an interval I, show that the corresponding solutions to (ii) are

$$\mathbf{x}_1 = \begin{bmatrix} f_1 \\ f_1' \end{bmatrix}, \qquad \mathbf{x}_2 = \begin{bmatrix} f_2 \\ f_2' \end{bmatrix}.$$

(c) Show that

$$W[\mathbf{x}_1, \mathbf{x}_2](t) = W[y_1, y_2](t).$$

11.4 GENERAL RESULTS FOR FIRST-ORDER LINEAR SYSTEMS

In this section we show how the underlying theory for linear differential systems can be obtained as an application of the vector space results from Chapters 6 and 7. Many of the results will parallel what we have already done for nth-order linear differential equations in Chapter 9.

The fundamental theoretical result is the following existence-uniqueness theorem for first-order linear systems.

Theorem 11.4.1: The initial value problem

$$\mathbf{x}'(t) = A(t)\mathbf{x}(t) + \mathbf{b}(t), \qquad \mathbf{x}(t_0) = \mathbf{x}_0,$$

where $A(t)$ and $\mathbf{b}(t)$ are continuous on an interval I, has a unique solution on I.

PROOF   The proof is omitted (see, for example, F. J. Murray and K. S. Miller, *Existence Theorems*, New York University Press, 1954).   ∎

Theorem 11.4.1 guarantees that a given initial value problem will have a solution and, further, that there will be only one solution.

Just as for a single *n*th-order linear equation, the solution of a nonhomogeneous differential system can usually be obtained once we have solved the associated homogeneous differential system. Consequently, we begin by developing the theory for the homogeneous linear differential system

$$\mathbf{x}'(t) = A(t)\mathbf{x}(t), \tag{11.4.1}$$

where A is an $n \times n$ matrix function. This is where the vector space techniques are required. We first show that the set of all solutions to (11.4.1) forms an n-dimensional subspace of the vector space of all column n-vector functions.

Theorem 11.4.2:   The set of all solutions to $\mathbf{x}'(t) = A(t)\mathbf{x}(t)$, where $A(t)$ is an $n \times n$ matrix function that is continuous on an interval I, forms a vector space of dimension n.

PROOF   Let S denote the set of all solutions to $\mathbf{x}' = A\mathbf{x}$. We begin by proving that S is a subspace of $V_n(\mathrm{I})$. Let $\mathbf{u}(t)$ and $\mathbf{v}(t)$ be in S. Then $\mathbf{u}' = A\mathbf{u}$ and $\mathbf{v}' = A\mathbf{v}$. We must show that $\mathbf{u} + \mathbf{v}$ and $c\mathbf{u}$ are also in S. However, this is immediate, since

$$(\mathbf{u} + \mathbf{v})' = \mathbf{u}' + \mathbf{v}' = A\mathbf{u} + A\mathbf{v} = A(\mathbf{u} + \mathbf{v})$$

which implies that $\mathbf{u} + \mathbf{v}$ is in S, and

$$(c\mathbf{u})' = c\mathbf{u}' = cA\mathbf{u} = A(c\mathbf{u})$$

which implies that $c\mathbf{u}$ is also in S. Consequently, since S is certainly nonempty, it is indeed a subspace of $V_n(\mathrm{I})$ and hence is a vector space.

We now prove that the dimension of this vector space is n by constructing a basis for the solution space containing n vectors. We first show that there exist n linearly independent solutions to $\mathbf{x}' = A\mathbf{x}$. Let $\mathbf{e}_i$ denote the ith column vector of the identity matrix I_n. Then, from Theorem 11.4.1, for each i the initial value problem

$$\begin{cases} \mathbf{x}_i'(t) = A(t)\mathbf{x}_i(t), \\ \mathbf{x}_i(t_0) = \mathbf{e}_i, \end{cases} \qquad i = 1, 2, \ldots, n$$

has a unique solution. Further, $W[\mathbf{x}_1, \mathbf{x}_2, \ldots, \mathbf{x}_n](t_0) = \det(I_n) = 1 \neq 0$, so that the solutions $\mathbf{x}_1(t), \mathbf{x}_2(t), \ldots, \mathbf{x}_n(t)$ are linearly independent on I. We now show that these solutions span the solution space. Let $\mathbf{x}(t)$ be any real solution to $\mathbf{x}' = A\mathbf{x}$ on I. Then, since $\mathbf{x}_1(t_0), \mathbf{x}_2(t_0), \ldots, \mathbf{x}_n(t_0)$ are linearly independent in $\mathbf{R}^n$, they form a basis for $\mathbf{R}^n$ and so it is certainly true that

$$\mathbf{x}(t_0) = c_1\mathbf{x}_1(t_0) + c_2\mathbf{x}_2(t_0) + \cdots + c_n\mathbf{x}_n(t_0),$$

for some scalars $c_1, c_2, \ldots, c_n$. It follows that $\mathbf{x}(t)$ is the *unique* (by Theorem 11.4.1) solution to the initial value problem

$$\begin{cases} \mathbf{x}'(t) = A(t)\, \mathbf{x}(t), \\ \mathbf{x}(t_0) = c_1\mathbf{x}_1(t_0) + c_2\mathbf{x}_2(t_0) + \cdots + c_n\mathbf{x}_n(t_0). \end{cases} \tag{11.4.2}$$

But

$$\mathbf{u}(t) = c_1\mathbf{x}_1(t) + c_2\mathbf{x}_2(t) + \cdots + c_n\mathbf{x}_n(t)$$

also satisfies the initial value problem (11.4.2) and so, by uniqueness, we must have

$$\mathbf{x}(t) = \mathbf{u}(t) = c_1\mathbf{x}_1(t) + c_2\mathbf{x}_2(t) + \cdots + c_n\mathbf{x}_n(t).$$

We have therefore shown that any solution of $\mathbf{x}' = A\mathbf{x}$ on I can be written as a linear combination of the n linearly independent solutions $\mathbf{x}_1, \mathbf{x}_2, \ldots, \mathbf{x}_n$, and hence these solutions form a basis for the solution space. Consequently, the dimension of the solution space is n. ∎

It follows from Theorem 11.4.2 that if $\{\mathbf{x}_1(t), \mathbf{x}_2(t), \ldots, \mathbf{x}_n(t)\}$ is *any* set of n linearly independent solutions to (11.4.1), then every solution of the system can be written as

$$\mathbf{x}(t) = c_1\mathbf{x}_1(t) + c_2\mathbf{x}_2(t) + \cdots + c_n\mathbf{x}_n(t), \tag{11.4.3}$$

for appropriate constants $c_1, c_2, \ldots, c_n$. In keeping with the terminology that we have used throughout the text, we will refer to (11.4.3) as the **general solution** of the linear system (11.4.1).

The following definition introduces some important terminology for homogeneous linear differential systems.

Definition 11.4.1:   A set of n linearly independent solutions to $\mathbf{x}' = A\mathbf{x}$ on an interval I, $\{\mathbf{x}_1(t), \mathbf{x}_2(t), \ldots, \mathbf{x}_n(t)\}$, is called a **fundamental solution set** on I, and the corresponding matrix $X(t)$ defined by

$$X(t) = [\mathbf{x}_1(t), \mathbf{x}_2(t), \ldots, \mathbf{x}_n(t)]$$

is called a **fundamental matrix** for the differential system $\mathbf{x}' = A\mathbf{x}$.

REMARK   If $X(t)$ is a fundamental matrix for (11.4.1), then the general solution (11.4.3) can be written in matrix form as $\mathbf{x}(t) = X(t)\mathbf{c}$, where $\mathbf{c} = [c_1 \quad c_2 \quad \cdots \quad c_n]^T$.

Now suppose that $\mathbf{x}_1(t), \mathbf{x}_2(t), \ldots, \mathbf{x}_n(t)$ are solutions to $\mathbf{x}' = A\mathbf{x}$ on an interval I. We showed in the previous section that if $W[\mathbf{x}_1, \mathbf{x}_2, \ldots, \mathbf{x}_n](t) \neq 0$ at any point in I, then the solutions are linearly independent on I. We now prove the converse.

Theorem 11.4.3:   If $\mathbf{x}_1(t), \mathbf{x}_2(t), \ldots, \mathbf{x}_n(t)$ are *linearly independent solutions* to $\mathbf{x}' = A\mathbf{x}$ on an interval I, then $W[\mathbf{x}_1, \mathbf{x}_2, \ldots, \mathbf{x}_n](t) \neq 0$ at every point in I.

PROOF   It is easier to prove the equivalent statement that if $W[\mathbf{x}_1, \mathbf{x}_2, \ldots, \mathbf{x}_n](t_0) = 0$ at some point t_0 in I, then $\mathbf{x}_1, \mathbf{x}_2, \ldots, \mathbf{x}_n$ are linearly dependent. We proceed as follows. If $W[\mathbf{x}_1, \mathbf{x}_2, \ldots, \mathbf{x}_n](t_0) = 0$, then, from Theorem 6.5.2 (page 207), the vectors $\mathbf{x}_1(t_0), \mathbf{x}_2(t_0), \ldots, \mathbf{x}_n(t_0)$ are linearly dependent in $\mathbf{R}^n$. Thus there exist scalars $c_1, c_2, \ldots, c_n$, not all zero, such that

$$c_1 x_1(t_0) + c_2 x_2(t_0) + \cdots + c_n x_n(t_0) = 0. \tag{11.4.4}$$

Now let

$$x(t) = c_1 x_1(t) + c_2 x_2(t) + \cdots + c_n x_n(t). \tag{11.4.5}$$

It follows from (11.4.4), (11.4.5), and Theorem 11.4.1 that $x(t)$ is the unique solution to the initial value problem

$$x'(t) = A(t)x(t), \qquad x(t_0) = 0.$$

However, this initial value problem admits the solution $x(t) = 0$, and so, by uniqueness, we must have:

$$c_1 x_1(t) + c_2 x_2(t) + \cdots + c_n x_n(t) = 0.$$

Since not all of the c_i are zero, it follows that the vector functions $x_1, x_2, \ldots, x_n$ are indeed linearly dependent on I. ∎

Thus to determine whether $\{x_1(t), x_2(t), \ldots, x_n(t)\}$ forms a fundamental solution set for $x' = Ax$ on an interval I we can compute their Wronskian at *any* convenient point, t_0, in I. If $W[x_1, x_2, \ldots, x_n](t_0) \neq 0$, then the solutions are linearly independent on I, whereas if $W[x_1, x_2, \ldots, x_n](t_0) = 0$, then the solutions are linearly dependent on I.

Example 11.4.1   Show that

$$x_1(t) = \begin{bmatrix} 0 \\ 0 \\ e^t \end{bmatrix}, \quad x_2(t) = \begin{bmatrix} e^{2t} \\ e^{2t} \\ 3e^{2t} \end{bmatrix}, \quad x_3(t) = \begin{bmatrix} 3e^{-2t} \\ -3e^{-2t} \\ e^{-2t} \end{bmatrix}$$

are a fundamental set of solutions for the differential system $x' = Ax$ on $(-\infty, \infty)$, where

$$A = \begin{bmatrix} 0 & 2 & 0 \\ 2 & 0 & 0 \\ 1 & 2 & 1 \end{bmatrix}.$$

Use these solutions to find the general solution to the differential system on $(-\infty, \infty)$, and hence find the solution of the initial value problem $x' = Ax$, $x(0) = [5 \ \ -1 \ \ 1]^T$.

Solution   It is easily shown by direct substitution that x_1, x_2, x_3 are solutions of the given differential system for all values of t. To show that they form a fundamental set of solutions on $(-\infty, \infty)$, we must show that they are linearly independent on that interval. The Wronskian of these solutions is

$$W[x_1, x_2, x_3](t) = \begin{vmatrix} 0 & e^{2t} & 3e^{-2t} \\ 0 & e^{2t} & -3e^{-2t} \\ e^t & 3e^{2t} & e^{-2t} \end{vmatrix}.$$

Evaluating at $t = 0$ yields

$$W[\mathbf{x}_1, \mathbf{x}_2, \mathbf{x}_3](0) = \begin{vmatrix} 0 & 1 & 3 \\ 0 & 1 & -3 \\ 1 & 3 & 1 \end{vmatrix} = -6 \neq 0.$$

Since the Wronskian is nonzero, it follows that $\mathbf{x}_1, \mathbf{x}_2, \mathbf{x}_3$ are linearly independent on $(-\infty, \infty)$ and so do form a fundamental set of solutions for the given differential system. Therefore, the general solution of the system on $(-\infty, \infty)$ is

$$\mathbf{x}(t) = c_1\mathbf{x}_1(t) + c_2\mathbf{x}_2(t) + c_3\mathbf{x}_3(t),$$

which can be written as

$$\mathbf{x}(t) = \begin{bmatrix} 0 & e^{2t} & 3e^{-2t} \\ 0 & e^{2t} & -3e^{-2t} \\ e^t & 3e^{2t} & e^{-2t} \end{bmatrix} \begin{bmatrix} c_1 \\ c_2 \\ c_3 \end{bmatrix}.$$

The solution satisfying the given initial condition is obtained by solving

$$\begin{bmatrix} 0 & 1 & 3 \\ 0 & 1 & -3 \\ 1 & 3 & 1 \end{bmatrix} \begin{bmatrix} c_1 \\ c_2 \\ c_3 \end{bmatrix} = \begin{bmatrix} 5 \\ -1 \\ 1 \end{bmatrix}.$$

We find that $c_1 = -6$, $c_2 = 2$, $c_3 = 1$. Consequently,

$$\mathbf{x}(t) = \begin{bmatrix} 0 & e^{2t} & 3e^{-2t} \\ 0 & e^{2t} & -3e^{-2t} \\ e^t & 3e^{2t} & e^{-2t} \end{bmatrix} \begin{bmatrix} -6 \\ 2 \\ 1 \end{bmatrix} = -6 \begin{bmatrix} 0 \\ 0 \\ e^t \end{bmatrix} + 2 \begin{bmatrix} e^{2t} \\ e^{2t} \\ 3e^{2t} \end{bmatrix} + \begin{bmatrix} 3e^{-2t} \\ -3e^{-2t} \\ e^{-2t} \end{bmatrix},$$

which can be written as

$$\mathbf{x}(t) = \begin{bmatrix} 2e^{2t} + 3e^{-2t} \\ 2e^{2t} - 3e^{-2t} \\ -6e^t + 6e^{2t} + e^{-2t} \end{bmatrix}.$$

The preceding results dealt with the case of a homogeneous linear system. We end this section with the main theoretical result that we will need for nonhomogeneous systems. This result should not be too surprising.

Theorem 11.4.4: Let $\mathbf{x}_1, \mathbf{x}_2, \ldots, \mathbf{x}_n$ be linearly independent solutions of the homogeneous linear differential system $\mathbf{x}'(t) = A(t)\mathbf{x}(t)$ on an interval I, and let $\mathbf{x} = \mathbf{x}_p$ be any particular solution of the nonhomogeneous system

$$\mathbf{x}'(t) = A(t)\,\mathbf{x}(t) + \mathbf{b}(t)$$

on I. Then every solution of $\mathbf{x}'(t) = A(t)\mathbf{x}(t) + \mathbf{b}(t)$ on I is of the form

$$\mathbf{x}(t) = c_1\mathbf{x}_1 + c_2\mathbf{x}_2 + \cdots + c_n\mathbf{x}_n + \mathbf{x}_p.$$

PROOF   Since $\mathbf{x} = \mathbf{x}_p$ is a solution of $\mathbf{x}'(t) = A(t)\mathbf{x}(t) + \mathbf{b}(t)$ on I, we have

$$\mathbf{x}'_p(t) = A(t)\mathbf{x}_p(t) + \mathbf{b}(t). \tag{11.4.6}$$

Now let $\mathbf{x} = \mathbf{u}(t)$ be any other solution to $\mathbf{x}'(t) = A(t)\mathbf{x}(t) + \mathbf{b}(t)$ on I. We then also have

$$\mathbf{u}'(t) = A(t)\mathbf{u}(t) + \mathbf{b}(t). \tag{11.4.7}$$

Subtracting (11.4.6) from (11.4.7) yields

$$(\mathbf{u} - \mathbf{x}_p)' = A(\mathbf{u} - \mathbf{x}_p).$$

Thus the vector function $\mathbf{x} = \mathbf{u} - \mathbf{x}_p$ is a solution of the associated homogeneous system $\mathbf{x}' = A\mathbf{x}$ on I. Since $\mathbf{x}_1, \mathbf{x}_2, \ldots, \mathbf{x}_n$ span the solution space of this system it follows that

$$\mathbf{u} - \mathbf{x}_p = c_1\mathbf{x}_1 + c_2\mathbf{x}_2 + \cdots + c_n\mathbf{x}_n$$

for some scalars $c_1, c_2, \ldots, c_n$. Consequently,

$$\mathbf{u} = c_1\mathbf{x}_1 + c_2\mathbf{x}_2 + \cdots + c_n\mathbf{x}_n + \mathbf{x}_p,$$

and the result is proved.   ∎

It follows from Theorem 11.4.4 that in order to solve a nonhomogeneous linear system, we must first find the general solution of the associated homogeneous system. In the next two sections we will concentrate on homogeneous differential systems; then in Section 11.7 we will see how the variation-of-parameters technique can be used to determine a particular solution of a nonhomogeneous linear differential system.

EXERCISES 11.4

In problems 1–3 show that the given functions are solutions of the system $\mathbf{x}' = A\mathbf{x}$ for the given matrix A, and hence find the general solution of the system (remember to check linear independence). If auxiliary conditions are given, find the particular solution that satisfies these conditions.

1. $\mathbf{x}_1 = \begin{bmatrix} e^{4t} \\ 2e^{4t} \end{bmatrix}$, $\mathbf{x}_2(t) = \begin{bmatrix} 3e^{-t} \\ e^{-t} \end{bmatrix}$, $A = \begin{bmatrix} -2 & 3 \\ -2 & 5 \end{bmatrix}$,

$\mathbf{x}(0) = \begin{bmatrix} -2 \\ 1 \end{bmatrix}$.

2. $\mathbf{x}_1 = \begin{bmatrix} e^{2t} \\ -e^{2t} \end{bmatrix}$, $\mathbf{x}_2 = \begin{bmatrix} e^{2t}(1+t) \\ -te^{2t} \end{bmatrix}$, $A = \begin{bmatrix} 3 & 1 \\ -1 & 1 \end{bmatrix}$.

3. $\mathbf{x}_1 = \begin{bmatrix} -3 \\ 9 \\ 5 \end{bmatrix}$, $\mathbf{x}_2 = \begin{bmatrix} e^{2t} \\ 3e^{2t} \\ e^{2t} \end{bmatrix}$, $\mathbf{x}_3 = \begin{bmatrix} e^{4t} \\ e^{4t} \\ e^{4t} \end{bmatrix}$,

$A = \begin{bmatrix} 2 & -1 & 3 \\ 3 & 1 & 0 \\ 2 & -1 & 3 \end{bmatrix}$.

In problems 4–7 determine two linearly independent solutions to the given system.

4. $\mathbf{x}' = A\mathbf{x}$, where $A = \begin{bmatrix} -1 & 2 \\ 2 & 2 \end{bmatrix}$.

5. $x' = Ax$, where $A = \begin{bmatrix} 0 & 3 \\ -3 & 0 \end{bmatrix}$.

6. $x' = Ax$, where $A = \begin{bmatrix} 1 & -2 \\ 2 & 1 \end{bmatrix}$.

7. $x' = Ax$, where $A = \begin{bmatrix} -3 & -1 \\ 4 & 1 \end{bmatrix}$.

8. If $x_1, x_2, \ldots, x_n$ are solutions of $x' = A(t)x$ and $X = [x_1, x_2, \ldots, x_n]$, prove that

$$X' = A(t)X.$$

9. Let $X(t)$ be a fundamental matrix for $x' = A(t)x$ on the interval I.
(a) Show that the general solution to the linear system can be written as

$$x = X(t)c,$$

where c is a vector of constants.
(b) If $t_0 \in I$, show that the solution to the initial value problem

$$x' = Ax, \qquad x(t_0) = x_0,$$

can be written as

$$x = X(t)X^{-1}(t_0)x_0.$$

11.5   HOMOGENEOUS CONSTANT COEFFICIENT LINEAR SYSTEMS: NONDEFECTIVE COEFFICIENT MATRIX

The theory that we have developed in the previous section is valid for any first-order linear system. However, in practice, these are too difficult to solve in general, and so we must make a simplifying assumption in order to develop solution techniques applicable to a broad class of linear systems. The assumption that we will make is that the coefficient matrix is a constant matrix.[1] In the next two sections, we will consider only homogeneous linear systems

$$x' = Ax,$$

where A is an $n \times n$ matrix of real *constants*. For example,

$$\left. \begin{array}{l} x_1' = 2x_1 - 3x_2, \\ x_2' = -x_1 + 4x_2, \end{array} \right\} \Leftrightarrow x' = Ax, \qquad \text{where } A = \begin{bmatrix} 2 & -3 \\ -1 & 4 \end{bmatrix}.$$

To motivate the new solution technique that we wish to develop, consider the system

$$x_1' = \qquad x_2, \tag{11.5.1}$$

$$x_2' = 6x_1 + x_2. \tag{11.5.2}$$

that is, in matrix form,

$$x' = Ax,$$

---

[1] This assumption should not be too surprising in view of the discussion of linear nth-order equations in Chapter 9.

where

$$A = \begin{bmatrix} 0 & 1 \\ 6 & 1 \end{bmatrix}.$$

We can use the method derived in Section 11.2 to solve this system.   Rewriting (11.5.1), (11.5.2) in operator form yields

$$Dx_1 - x_2 = 0, \tag{11.5.3}$$

$$-6x_1 + (D-1)x_2 = 0. \tag{11.5.4}$$

Operating on (11.5.3) with $D-1$ and adding the resulting equation to (11.5.4), we obtain

$$D(D-1)x_1 - 6x_1 = 0,$$

that is,

$$(D^2 - D - 6)x_1 = 0.$$

This constant coefficient differential equation has general solution

$$x_1 = c_1 e^{3t} + c_2 e^{-2t}.$$

Substituting this expression for x_1 into (11.5.2) yields

$$x_2 = 3c_1 e^{3t} - 2c_2 e^{-2t}.$$

Consequently, the solution to the given system is

$$x_1 = c_1 e^{3t} + c_2 e^{-2t},$$

$$x_2 = 3c_1 e^{3t} - 2c_2 e^{-2t},$$

which we can write as

$$\mathbf{x} = c_1 e^{3t} \begin{bmatrix} 1 \\ 3 \end{bmatrix} + c_2 e^{-2t} \begin{bmatrix} 1 \\ -2 \end{bmatrix}.$$

We see that two linearly independent solutions to (11.5.1), (11.5.2) are

$$\mathbf{x}_1 = e^{3t} \begin{bmatrix} 1 \\ 3 \end{bmatrix}, \qquad \mathbf{x}_2 = e^{-2t} \begin{bmatrix} 1 \\ -2 \end{bmatrix}.$$

The key point to notice is that both of these solutions are of the form

$$\mathbf{x} = e^{\lambda t}\mathbf{v}, \tag{11.5.5}$$

where λ is a scalar and $\mathbf{v}$ is a constant vector. This suggests that the general system

$$\mathbf{x}' = A\mathbf{x} \tag{11.5.6}$$

may also have solutions of the form (11.5.5). We now investigate this possibility. Differentiating (11.5.5) with respect to t yields

$$\mathbf{x}' = \lambda e^{\lambda t}\mathbf{v}.$$

Thus $\mathbf{x} = e^{\lambda t}\mathbf{v}$ is a solution of (11.5.6) if and only if

$$\lambda e^{\lambda t}\mathbf{v} = e^{\lambda t}A\mathbf{v},$$

that is, if and only if λ and $\mathbf{v}$ satisfy

$$A\mathbf{v} = \lambda\mathbf{v}.$$

But this is just the statement that λ must be an eigenvalue of A with corresponding eigenvector $\mathbf{v}$. Consequently we have proved the following fundamental result.

Theorem 11.5.1: Let A be an $n \times n$ matrix of real constants and let λ be an eigenvalue of A with corresponding eigenvector $\mathbf{v}$. Then

$$\mathbf{x}(t) = e^{\lambda t}\mathbf{v}$$

is a solution to the constant coefficient linear differential system $\mathbf{x}' = A\mathbf{x}$ on any interval.

REMARK   Notice that we have not assumed that the eigenvalues and eigenvectors of A are real; the preceding result holds in the complex case also.

We now illustrate how Theorem 11.5.1 can be used to find the general solution of linear differential systems.

Example 11.5.1   Find the general solution to

$$\begin{aligned} x_1' &= 2x_1 + x_2, \\ x_2' &= -3x_1 - 2x_2, \end{aligned} \tag{11.5.7}$$

on $(-\infty, \infty)$.

Solution   The given system can be written in matrix form as $\mathbf{x}' = A\mathbf{x}$, where

$$A = \begin{bmatrix} 2 & 1 \\ -3 & -2 \end{bmatrix}.$$

We first find the eigenvalues and eigenvectors of A. A straightforward calculation yields

$$\det(A - \lambda I) = \begin{vmatrix} 2 - \lambda & 1 \\ -3 & -2 - \lambda \end{vmatrix} = \lambda^2 - 1,$$

so that A has eigenvalues $\lambda = \pm 1$.

Eigenvectors
$\lambda = 1$: In this case the system $(A - \lambda I)\mathbf{v} = \mathbf{0}$ is

$$\begin{aligned} v_1 + v_2 &= 0, \\ -3v_1 - 3v_2 &= 0, \end{aligned}$$

with solution $\mathbf{v} = r(1, -1)$. It follows from Theorem 11.5.1 that

$$\mathbf{x}_1(t) = e^t \begin{bmatrix} 1 \\ -1 \end{bmatrix}$$

is a solution of the system (11.5.7) on $(-\infty, \infty)$.

$\lambda = -1$: In this case the system $(A - \lambda I)\mathbf{v} = \mathbf{0}$ is

$$3v_1 + v_2 = 0,$$

$$-3v_1 - v_2 = 0,$$

with solution $\mathbf{v} = s(1, -3)$.   Consequently

$$\mathbf{x}_2(t) = e^{-t} \begin{bmatrix} 1 \\ -3 \end{bmatrix}$$

is also a solution of the system (11.5.7) on $(-\infty, \infty)$.

Further, the Wronskian of these solutions is

$$W[\mathbf{x}_1, \mathbf{x}_2](t) = \begin{vmatrix} e^t & e^{-t} \\ -e^t & -3e^{-t} \end{vmatrix},$$

so that

$$W[\mathbf{x}_1, \mathbf{x}_2](0) = \begin{vmatrix} 1 & 1 \\ -1 & -3 \end{vmatrix} = -2 \neq 0.$$

It follows from Theorem 11.3.2 that $\mathbf{x}_1, \mathbf{x}_2$ are linearly independent on $(-\infty, \infty)$, and hence the general solution to (11.5.7) is

$$\mathbf{x}(t) = c_1\mathbf{x}_1 + c_2\mathbf{x}_2 = c_1 e^t \begin{bmatrix} 1 \\ -1 \end{bmatrix} + c_2 e^{-t} \begin{bmatrix} 1 \\ -3 \end{bmatrix}.$$

To find the general solution of an $n \times n$ linear constant coefficient system of differential equations, we need to find n linearly independent solutions to the system. The preceding example together with our experience with eigenvalues and eigenvectors suggest that we will be able to find n such linearly independent solutions provided the matrix A has n linearly independent eigenvectors (that is, A is nondefective). This is indeed the case, although if the eigenvalues and eigenvectors are complex, we have to do some work in order to obtain real-valued solutions to the system. We first give the result for the case of real eigenvalues.

Theorem 11.5.2:   Let A be an $n \times n$ matrix of real constants. If A has n real linearly independent eigenvectors $\mathbf{v}_1, \mathbf{v}_2, \ldots, \mathbf{v}_n$, with corresponding eigenvalues $\lambda_1, \lambda_2, \ldots, \lambda_n$ (not necessarily distinct), then the vector functions $\{\mathbf{x}_1, \mathbf{x}_2, \ldots, \mathbf{x}_n\}$ defined by

$$\mathbf{x}_k = e^{\lambda_k t}\mathbf{v}_k, \qquad k = 1, 2, \ldots, n,$$

for all t, are linearly independent solutions to $\mathbf{x}' = A\mathbf{x}$ on any interval. The general solution of this differential system is

$$\mathbf{x}(t) = c_1\mathbf{x}_1 + c_2\mathbf{x}_2 + \cdots + c_n\mathbf{x}_n.$$

PROOF   We have already shown (Theorem 11.5.1) that each x_k satisfies $x' = Ax$ for all t. Further, $W[x_1, x_1, \ldots, x_n] = e^{(\lambda_1 + \lambda_2 + \cdots + \lambda_n)t} \det[v_1, v_2, \ldots, v_n] \neq 0$ (since the eigenvectors are linearly independent by assumption), and hence the solutions are linearly independent on any interval.   ■

Example 11.5.2   Find the general solution of $x' = Ax$ if $A = \begin{bmatrix} 0 & 2 & -3 \\ -2 & 4 & -3 \\ -2 & 2 & -1 \end{bmatrix}$.

Solution   We first determine the eigenvalues and eigenvectors of A. For the given matrix we have

$$\det(A - \lambda I) = \begin{vmatrix} -\lambda & 2 & -3 \\ -2 & 4 - \lambda & -3 \\ -2 & 2 & -1 - \lambda \end{vmatrix} = -(\lambda + 1)(\lambda - 2)^2,$$

so that the eigenvalues are $\lambda = -1, 2$.

Eigenvectors
$\lambda = -1$: The corresponding eigenvectors are obtained from

$$v_1 + 2v_2 - 3v_3 = 0,$$
$$-2v_1 + 5v_2 - 3v_3 = 0,$$
$$-2v_1 + 2v_2 \qquad = 0.$$

This system has solution $v = r(1, 1, 1)$, so that we can take

$$v_1 = (1, 1, 1).$$

$\lambda = 2$: The system for the eigenvectors reduces to the single equation

$$2v_1 - 2v_2 + 3v_3 = 0,$$

which has solution $v = r(1, 1, 0) + s(-3, 0, 2)$. Thus two linearly independent eigenvectors corresponding to $\lambda = 2$ are

$$v_2 = (1, 1, 0), \qquad v_3 = (-3, 0, 2).$$

It follows from Theorem 11.5.2 that three linearly independent solutions to the given system of differential equations are

$$x_1 = e^{-t} \begin{bmatrix} 1 \\ 1 \\ 1 \end{bmatrix}, \qquad x_2 = e^{2t} \begin{bmatrix} 1 \\ 1 \\ 0 \end{bmatrix}, \qquad x_3 = e^{2t} \begin{bmatrix} -3 \\ 0 \\ 2 \end{bmatrix},$$

and hence the general solution of the given system is

$$x(t) = c_1 e^{-t} \begin{bmatrix} 1 \\ 1 \\ 1 \end{bmatrix} + c_2 e^{2t} \begin{bmatrix} 1 \\ 1 \\ 0 \end{bmatrix} + c_3 e^{2t} \begin{bmatrix} -3 \\ 0 \\ 2 \end{bmatrix},$$

which can be written in the form

$$\mathbf{x}(t) = \begin{bmatrix} e^{-t} & e^{2t} & -3e^{2t} \\ e^{-t} & e^{2t} & 0 \\ e^{-t} & 0 & 2e^{2t} \end{bmatrix} \begin{bmatrix} c_1 \\ c_2 \\ c_3 \end{bmatrix}.$$

We now consider the case when some (or all) of the eigenvalues are complex. Since we are restricting attention to systems of equations with *real* constant coefficients, it follows that the matrix of the system will have real entries; hence, from Theorem 8.2.1 (page 280), the eigenvalues *and* eigenvectors will occur in conjugate pairs. The corresponding solutions to $\mathbf{x}' = A\mathbf{x}$ guaranteed by Theorem 11.5.1 will also be complex conjugate pairs. However, as we now show, each conjugate pair gives rise to two real-valued solutions in a familiar manner.

Lemma 11.5.1: Let $\mathbf{u}(t)$ and $\mathbf{v}(t)$ be real-valued vector functions. If $\mathbf{x}(t) = \mathbf{u}(t) + i\mathbf{v}(t)$ and $\mathbf{x}(t) = \mathbf{u}(t) - i\mathbf{v}(t)$ are complex conjugate solutions of $\mathbf{x}' = A\mathbf{x}$, then $\mathbf{x}(t) = \mathbf{u}(t)$ and $\mathbf{x}(t) = \mathbf{v}(t)$ are themselves *real-valued* solutions of $\mathbf{x}' = A\mathbf{x}$.

PROOF   Since $\mathbf{x}(t) = \mathbf{u}(t) \pm i\mathbf{v}(t)$ are solutions of $\mathbf{x}' = A\mathbf{x}$, we have

$$[\mathbf{u}(t) \pm i\mathbf{v}(t)]' = A[\mathbf{u} \pm i\mathbf{v}(t)],$$

that is,

$$\mathbf{u}'(t) \pm i\mathbf{v}'(t) = A\mathbf{u}(t) \pm iA\mathbf{v}(t).$$

However, this is satisfied if and only if

$$\mathbf{u}'(t) = A\mathbf{u}(t) \text{ and } \mathbf{v}'(t) = A\mathbf{v}(t),$$

that is, if and only if $\mathbf{x}(t) = \mathbf{u}(t)$ and $\mathbf{x}(t) = \mathbf{v}(t)$ are themselves real-valued solutions of $\mathbf{x}' = A\mathbf{x}$. ∎

We now explicitly derive two appropriate real-valued solutions corresponding to a complex conjugate pair of eigenvalues.

Suppose that $\lambda = a + ib$ ($b \neq 0$) is an eigenvalue of A with corresponding eigenvector $\mathbf{v} = \mathbf{r} + i\mathbf{s}$. It follows from Theorem 11.5.1 that a complex-valued solution to $\mathbf{x}' = A\mathbf{x}$ is

$$\mathbf{u} = e^{(a+ib)t}(\mathbf{r} + i\mathbf{s}) = e^{at}(\cos bt + i \sin bt)(\mathbf{r} + i\mathbf{s}),$$

which can be written as

$$\mathbf{u} = e^{at}[(\cos bt)\mathbf{r} - (\sin bt)\mathbf{s}] + ie^{at}[(\sin bt)\mathbf{r} + (\cos bt)\mathbf{s}].$$

Lemma 11.5.1 implies that two real-valued solutions to $\mathbf{x}' = A\mathbf{x}$ are

$$\mathbf{x}_1 = e^{at}[(\cos bt)\mathbf{r} - (\sin bt)\mathbf{s}], \qquad \mathbf{x}_2 = e^{at}[(\sin bt)\mathbf{r} + (\cos bt)\mathbf{s}].$$

It can further be shown that the set of all real-valued solutions obtained in this manner is linearly independent on any interval.

REMARK   Notice that we do not have to derive the solution corresponding to the conjugate eigenvalue $\lambda = a - ib$, since it does not yield any new linearly independent solutions to $\mathbf{x}' = A\mathbf{x}$.

Example 11.5.3   Find the general solution of $\mathbf{x}' = A\mathbf{x}$ if $A = \begin{bmatrix} 2 & -1 \\ 2 & 4 \end{bmatrix}$.

Solution   The characteristic polynomial of A is

$$\det(A - \lambda I) = \begin{vmatrix} 2 - \lambda & -1 \\ 2 & 4 - \lambda \end{vmatrix} = \lambda^2 - 6\lambda + 10,$$

so that the eigenvalues are $\lambda = 3 \pm i$.   We need only to find the eigenvectors corresponding to one of these conjugate eigenvalues.   When $\lambda = 3 + i$, the eigenvectors are obtained by solving

$$-(1 + i)v_1 - \quad\quad v_2 = 0,$$
$$2v_1 + (1 - i)v_2 = 0,$$

which yield the complex eigenvectors $\mathbf{v} = r(1, -(1 + i))$.   Hence a complex-valued solution of the given system is

$$\mathbf{u} = e^{3t}(\cos t + i \sin t)\begin{bmatrix} 1 \\ -(1 + i) \end{bmatrix} = e^{3t}\begin{bmatrix} \cos t + i \sin t \\ -(1 + i)(\cos t + i \sin t) \end{bmatrix}$$

$$= e^{3t}\begin{bmatrix} \cos t + i \sin t \\ (\sin t - \cos t) - i(\sin t + \cos t) \end{bmatrix}$$

$$= e^{3t}\left\{ \begin{bmatrix} \cos t \\ \sin t - \cos t \end{bmatrix} + i\begin{bmatrix} \sin t \\ -(\sin t + \cos t) \end{bmatrix} \right\}.$$

From Lemma 11.5.1, the real and imaginary parts of this complex-valued solution yield the following two *real-valued* linearly independent solutions

$$\mathbf{x}_1 = e^{3t}\begin{bmatrix} \cos t \\ \sin t - \cos t \end{bmatrix}, \quad \mathbf{x}_2 = e^{3t}\begin{bmatrix} \sin t \\ -(\sin t + \cos t) \end{bmatrix}.$$

The general solution of the given system is thus

$$\mathbf{x} = c_1 e^{3t}\begin{bmatrix} \cos t \\ \sin t - \cos t \end{bmatrix} + c_2 e^{3t}\begin{bmatrix} \sin t \\ -(\sin t + \cos t) \end{bmatrix}$$

$$= e^{3t}\left\{ c_1\begin{bmatrix} \cos t \\ \sin t - \cos t \end{bmatrix} + c_2\begin{bmatrix} \sin t \\ -(\sin t + \cos t) \end{bmatrix} \right\}.$$

The results of this section are summarized in the following theorem.

Theorem 11.5.3: Let A be an $n \times n$ matrix of real constants.

1. Suppose λ is a real eigenvalue of A of multiplicity m with corresponding linearly independent eigenvectors $\mathbf{v}_1, \mathbf{v}_2, \ldots, \mathbf{v}_k$ $(1 \le k \le m)$. Then k linearly independent solutions to $\mathbf{x}' = A\mathbf{x}$ are

$$\mathbf{x}_j(t) = e^{\lambda t}\mathbf{v}_j, \qquad j = 1, 2, \ldots, k.$$

2. Suppose $\lambda = a + ib$ is a complex eigenvalue of multiplicity m with corresponding linearly independent eigenvectors $\mathbf{v}_1, \mathbf{v}_2, \ldots, \mathbf{v}_k$ $(1 \le k \le m)$, where $\mathbf{v}_j = \mathbf{r}_j + i\mathbf{s}_j$. Then k complex-valued solutions to $\mathbf{x}' = A\mathbf{x}$ are

$$\mathbf{u}_j(t) = e^{\lambda t}\mathbf{v}_j, \qquad j = 1, 2, \ldots, k,$$

and $2k$ *real-valued* linearly independent solutions to $\mathbf{x}' = A\mathbf{x}$ are

$$\mathbf{x}_1 = e^{at}[(\cos bt)\mathbf{r}_1 - (\sin bt)\mathbf{s}_1], \ \mathbf{x}_2 = e^{at}[(\cos bt)\mathbf{r}_2 - (\sin bt)\mathbf{s}_2], \ \ldots,$$

$$\mathbf{x}_k = e^{at}[(\cos bt)\mathbf{r}_k - (\sin bt)\mathbf{s}_k], \ \mathbf{x}_{k+1} = e^{at}[(\sin bt)\mathbf{r}_1 + (\cos bt)\mathbf{s}_1], \ \ldots,$$

$$\mathbf{x}_{2k} = e^{at}[(\sin bt)\mathbf{r}_k + (\cos bt)\mathbf{s}_k].$$

Further, the set of all solutions to $\mathbf{x}' = A\mathbf{x}$ obtained in this manner is linearly independent on any interval.

Corollary 11.5.1: If A has a complete set of eigenvectors, then the solutions obtained from (1) and (2) of the previous theorem yield a fundamental set of solutions to $\mathbf{x}' = A\mathbf{x}$, and the general solution of this differential system is

$$\mathbf{x}(t) = c_1\mathbf{x}_1 + c_2\mathbf{x}_2 + \cdots + c_n\mathbf{x}_n.$$

EXERCISES 11.5

In problems 1–15, determine the general solution of the system $\mathbf{x}' = A\mathbf{x}$ for the given matrix A.

1. $\begin{bmatrix} -2 & -7 \\ -1 & 4 \end{bmatrix}$.

2. $\begin{bmatrix} 0 & -4 \\ 4 & 0 \end{bmatrix}$.

3. $\begin{bmatrix} 1 & -2 \\ 5 & -5 \end{bmatrix}$.

4. $\begin{bmatrix} -1 & 2 \\ -2 & -1 \end{bmatrix}$.

5. $\begin{bmatrix} 2 & 0 & 0 \\ 0 & 5 & -7 \\ 0 & 2 & -4 \end{bmatrix}$.

6. $\begin{bmatrix} -1 & 0 & 0 \\ 1 & 5 & -1 \\ 1 & 6 & -2 \end{bmatrix}$.

7. $\begin{bmatrix} 0 & 1 & 0 \\ -1 & 0 & 0 \\ 0 & 0 & 5 \end{bmatrix}$.

8. $\begin{bmatrix} 2 & 0 & 3 \\ 0 & -4 & 0 \\ -3 & 0 & 2 \end{bmatrix}$.

9. $\begin{bmatrix} 3 & 2 & 6 \\ -2 & 1 & -2 \\ -1 & -2 & -4 \end{bmatrix}$.

10. $\begin{bmatrix} 0 & -3 & 1 \\ -2 & -1 & 1 \\ 0 & 0 & 2 \end{bmatrix}$.

11. $\begin{bmatrix} 3 & 0 & -1 \\ 0 & -3 & -1 \\ 0 & 2 & -1 \end{bmatrix}$.

12. $\begin{bmatrix} 1 & 1 & -1 \\ 1 & 1 & 1 \\ -1 & 1 & 1 \end{bmatrix}$.

13. $\begin{bmatrix} 2 & -1 & 3 \\ 2 & -1 & 3 \\ 2 & -1 & 3 \end{bmatrix}$.

14. $\begin{bmatrix} 1 & 2 & 3 & 4 \\ 4 & 3 & 2 & 1 \\ 4 & 5 & 6 & 7 \\ 7 & 6 & 5 & 4 \end{bmatrix}$.

15. $\begin{bmatrix} 0 & 1 & 0 & 0 \\ -1 & 0 & 0 & 0 \\ 0 & 0 & 0 & -1 \\ 0 & 0 & 1 & 0 \end{bmatrix}$.

In problems 16–18, solve the initial value problem $\mathbf{x}' = A\mathbf{x}$, $\mathbf{x}(0) = \mathbf{x}_0$.

16. $A = \begin{bmatrix} -1 & 4 \\ 2 & -3 \end{bmatrix}$, $\mathbf{x}_0 = \begin{bmatrix} 3 \\ 0 \end{bmatrix}$.

17. $A = \begin{bmatrix} -1 & -6 \\ 3 & 5 \end{bmatrix}$, $\mathbf{x}_0 = \begin{bmatrix} 2 \\ 2 \end{bmatrix}$.

18. $A = \begin{bmatrix} 2 & -1 & 3 \\ 3 & 1 & 0 \\ 2 & -1 & 3 \end{bmatrix}$, $\mathbf{x}_0 = \begin{bmatrix} -4 \\ 4 \\ 4 \end{bmatrix}$.

19. Solve the initial value problem $\mathbf{x}' = A\mathbf{x}$, $\mathbf{x}(0) = \begin{bmatrix} 1 \\ 1 \end{bmatrix}$, if $A = \begin{bmatrix} 0 & 2 \\ -2 & 0 \end{bmatrix}$. Sketch the solution in the $x_1 x_2$-plane.

20. Consider the differential equation

$$\frac{d^2 x}{dt^2} + b\frac{dx}{dt} + cx = 0, \qquad \text{(i)}$$

where b and c are constants.
(a) Show that (i) can be replaced by the equivalent first-order linear system

$$\mathbf{x}' = A\mathbf{x},$$

where $A = \begin{bmatrix} 0 & 1 \\ -c & -b \end{bmatrix}$.

(b) Show that the characteristic polynomial of A coincides with the auxiliary polynomial of (i).

21. Let $\lambda = a + ib$, $b \neq 0$, be an eigenvalue of the $n \times n$ (real) matrix A with corresponding eigenvector $\mathbf{v} = \mathbf{r} + i\mathbf{s}$. Then we have shown in the text that two real-valued solutions to $\mathbf{x}' = A\mathbf{x}$ are

$$\mathbf{x}_1 = e^{at}[(\cos bt)\mathbf{r} - (\sin bt)\mathbf{s}],$$

$$\mathbf{x}_2 = e^{at}[(\sin bt)\mathbf{r} + (\cos bt)\mathbf{s}].$$

Prove that $\mathbf{x}_1$, $\mathbf{x}_2$ are linearly independent on any interval. (You may assume that $\mathbf{r}$ and $\mathbf{s}$ are linearly independent in $\mathbf{R}^n$.)

The remaining problems in this section investigate general properties of solutions to $\mathbf{x}' = A\mathbf{x}$, where A is a nondefective matrix.

22. Let A be a 2×2 nondefective matrix. If all eigenvalues of A have negative real part, prove that every solution of $\mathbf{x}' = A\mathbf{x}$ satisfies

$$\lim_{t \to +\infty} \mathbf{x}(t) = \mathbf{0}. \qquad \text{(ii)}$$

23. Let A be a 2×2 nondefective matrix. If *every solution* of $\mathbf{x}' = A\mathbf{x}$ satisfies (ii), prove that all eigenvalues of A have negative real part.

24. Determine the general solution to $\mathbf{x}' = A\mathbf{x}$ if $A = \begin{bmatrix} 0 & b \\ -b & 0 \end{bmatrix}$, $b > 0$. Describe the behavior of the solutions.

25. Describe the behavior of the solutions to $\mathbf{x}' = A\mathbf{x}$ if $A = \begin{bmatrix} a & b \\ -b & a \end{bmatrix}$, $a < 0$, $b > 0$.

26. What conditions on the eigenvalues of an $n \times n$ matrix A would guarantee that the system $\mathbf{x}' = A\mathbf{x}$ has at least one solution satisfying

$$\mathbf{x}(t) = \mathbf{x}_0,$$

for all t, where $\mathbf{x}_0$ is a constant vector?

27. The motion of a certain physical system is described by the system of differential equations

$$x_1' = x_2, \qquad x_2' = -bx_1 - ax_2,$$

where a and b are positive constants and $a \neq 2b$. Show that the motion of the system dies out as $t \to +\infty$.

11.6 HOMOGENEOUS CONSTANT COEFFICIENT LINEAR SYSTEMS: DEFECTIVE COEFFICIENT MATRIX

The results of the previous section enable us to solve any constant coefficient linear differential system $\mathbf{x}' = A\mathbf{x}$, *provided A has a complete set of eigenvectors.* We recall from Chapter 8 that if m denotes the multiplicity of an eigenvalue of A, then the dimension k of the corresponding eigenspace satisfies the inequality

$$1 \le k \le m,$$

and the condition for A to be nondefective is that the dimension of each eigenspace equals the multiplicity of the corresponding eigenvalue (see Theorem 8.3.4, page 287). We now turn our attention to the case when A is defective. That is, for at least one eigenvalue the dimension k of the corresponding eigenspace is strictly less than the multiplicity m of the eigenvalue. In this case, there are only k linearly independent eigenvectors corresponding to λ, and so Theorem 11.5.3 will only yield k linearly independent solutions to the differential system $\mathbf{x}' = A\mathbf{x}$. We must therefore find an additional $m - k$ linearly independent solutions. In order to motivate the main result of this section (which is rather difficult to prove) we consider a particular example.

Example 11.6.1   Find the general solution of

$$x_1' = \qquad x_2, \tag{11.6.1}$$

$$x_2' = -9x_1 + 6x_2. \tag{11.6.2}$$

Solution   We try using the technique from the previous section. The coefficient matrix of the given linear system is $A = \begin{bmatrix} 0 & 1 \\ -9 & 6 \end{bmatrix}$, which has eigenvalue $\lambda = 3$ of multiplicity 2. It is straightforward to show that there is just one corresponding linearly independent eigenvector which we may take to be $\mathbf{v}_0 = (1, 3)$. Consequently we obtain only one linearly independent solution to the differential system, namely,

$$\mathbf{x}_1 = e^{3t} \begin{bmatrix} 1 \\ 3 \end{bmatrix}. \tag{11.6.3}$$

To find the general solution of the system, we must therefore determine a second linearly independent solution; although it is not at all clear where this second solution should come from. In this particular example, however, we can use the technique introduced in Section 11.2 for solving linear systems. Thus we rewrite (11.6.1) and (11.6.2) in the form

$$Dx_1 - x_2 = 0, \tag{11.6.4}$$

$$9x_1 + (D - 6)x_2 = 0. \tag{11.6.5}$$

Operating on (11.6.4) with $D - 6$ and adding the resulting equation to (11.6.5) yields

$$D(D - 6)x_1 + 9x_1 = 0,$$

that is,

$$(D^2 - 6D + 9)x_1 = 0.$$

Hence

$$x_1 = c_1 e^{3t} + c_2 t e^{3t}.$$

From (11.6.4) we therefore obtain

$$x_2 = Dx_1 = 3c_1 e^{3t} + c_2 e^{3t}(3t + 1).$$

Consequently, the solution to (11.6.1), (11.6.2) is

$$\mathbf{x} = \begin{bmatrix} c_1 e^{3t} + c_2 t e^{3t} \\ 3c_1 e^{3t} + c_2 e^{3t}(3t + 1) \end{bmatrix},$$

which can be written in the equivalent form

$$\mathbf{x} = c_1 e^{3t} \begin{bmatrix} 1 \\ 3 \end{bmatrix} + c_2 e^{3t} \left\{ \begin{bmatrix} 0 \\ 1 \end{bmatrix} + t \begin{bmatrix} 1 \\ 3 \end{bmatrix} \right\}.$$

We see that two linearly independent solutions to the given system are

$$\mathbf{x}_1 = e^{3t} \begin{bmatrix} 1 \\ 3 \end{bmatrix}, \qquad \mathbf{x}_2 = e^{3t} \left\{ \begin{bmatrix} 0 \\ 1 \end{bmatrix} + t \begin{bmatrix} 1 \\ 3 \end{bmatrix} \right\}.$$

The first of these solutions coincides with the solution (11.6.3), which was derived using the eigenvalue-eigenvector technique of the previous section. The key point to notice is that in this particular example, there is a second linearly independent solution of the form

$$\mathbf{x}_2 = e^{\lambda t}(\mathbf{v}_1 + t\mathbf{v}_2).$$

It can be shown that the basic *form* of the second linearly independent solution just obtained holds in the general case when the coefficient matrix has an eigenvalue of multiplicity 2 with a corresponding one-dimensional eigenspace. Further, it can be generalized to include the case of arbitrary multiplicity. Indeed, in view of the preceding example we might suspect that if an eigenvalue has multiplicity m but only k corresponding linearly independent eigenvectors, then we should be able to find $m - k$ solutions of the form

$$\mathbf{x}_i = e^{\lambda t}(\mathbf{v}_0 + t\mathbf{v}_1 + \cdots + t^i \mathbf{v}_i), \qquad i = 1, 2, \ldots, m - k.$$

This is indeed the case, although a rigorous proof requires some deeper results from linear algebra (the Jordan canonical form of a matrix) than we have considered in this text. Before dealing with the general case, we first consider eigenvalues with multiplicity either two or three, since these are the only possibilities for 2×2 or 3×3 matrices.

Theorem 11.6.1:   Consider the linear differential system $\mathbf{x}' = A\mathbf{x}$, where A is a constant matrix. Let m denote the multiplicity of the eigenvalue λ and let k denote the number of corresponding linearly independent eigenvectors.

1. $m = 2$, $k = 1$: There exist two linearly independent solutions to $\mathbf{x}' = A\mathbf{x}$ of the form

$$\mathbf{x}_1 = e^{\lambda t}\mathbf{v}_0, \qquad \mathbf{x}_2 = e^{\lambda t}(\mathbf{v}_1 + t\mathbf{v}_2).$$

2. $m = 3$, $k = 1$: There exist three linearly independent solutions to $\mathbf{x}' = A\mathbf{x}$ of the form[1]

$$\mathbf{x}_1 = e^{\lambda t}\mathbf{v}_0, \qquad \mathbf{x}_2 = e^{\lambda t}(\mathbf{v}_1 + t\mathbf{v}_2), \qquad \mathbf{x}_3 = e^{\lambda t}\left(\mathbf{v}_3 + t\mathbf{v}_4 + \frac{t^2}{2!}\mathbf{v}_5\right).$$

3. $m = 3$, $k = 2$: There exist three linearly independent solutions to $\mathbf{x}' = A\mathbf{x}$ of the form

$$\mathbf{x}_1 = e^{\lambda t}\mathbf{v}_0, \qquad \mathbf{x}_2 = e^{\lambda t}\mathbf{v}_1, \qquad \mathbf{x}_3 = e^{\lambda t}(\mathbf{v}_2 + t\mathbf{v}_3),$$

where $\mathbf{v}_0$ and $\mathbf{v}_1$ are linearly independent eigenvectors of A corresponding to the eigenvalue λ.

PROOF   As mentioned before, the proof of this theorem requires some deeper results from linear algebra than we have studied and so is omitted.   ∎

REMARKS

1. Notice that Theorem 11.6.1 tells us only the *form* of the appropriate linearly independent solutions. We can consider them as trial solutions. In order to obtain the solutions themselves, we must substitute the trial solutions into the given differential system and explicitly determine the appropriate $\mathbf{v}_k$. It is important to notice that there could be more than one linearly independent solution corresponding to each consecutive trial solution.

2. If λ is a complex eigenvalue, then the resulting solutions will themselves be complex-valued. In such a situation we can obtain real-valued linearly independent solutions in the same manner as in the previous section.

We illustrate the use of the preceding theorem with several examples.

Example 11.6.2   Solve the initial value problem $\mathbf{x}' = A\mathbf{x}$, $\mathbf{x}(0) = \begin{bmatrix} -1 \\ 1 \end{bmatrix}$, if $A = \begin{bmatrix} 6 & -8 \\ 2 & -2 \end{bmatrix}$.

Solution   We first determine the eigenvalues and eigenvectors of A. We have

$$\det(A - \lambda I) = \begin{vmatrix} 6 - \lambda & -8 \\ 2 & -2 - \lambda \end{vmatrix} = (\lambda - 2)^2, \tag{11.6.6}$$

---

[1] The reason for including the factor $1/2!$ in $\mathbf{x}_3$ is that it simplifies the computation of $\mathbf{v}_5$.

so that there is only one eigenvalue $\lambda = 2$, with multiplicity 2. The eigenvectors corresponding to this eigenvalue are of the form

$$\mathbf{v}_0 = r(2, 1). \tag{11.6.7}$$

We therefore obtain only one linearly independent solution of the given system, namely,

$$\mathbf{x}_1 = e^{2t} \begin{bmatrix} 2 \\ 1 \end{bmatrix}.$$

From Theorem 11.6.1 it follows that there is another solution to the system of the form

$$\mathbf{x}_2 = e^{2t}(\mathbf{v}_1 + t\mathbf{v}_2). \tag{11.6.8}$$

The vectors $\mathbf{v}_1$ and $\mathbf{v}_2$ must be chosen so that $\mathbf{x}_2$ is a solution of the given system, that is, so that $\mathbf{x}_2' = A\mathbf{x}_2$. Differentiating (11.6.8) with respect to t yields

$$\mathbf{x}_2' = e^{2t}[(2\mathbf{v}_1 + \mathbf{v}_2) + 2t\mathbf{v}_2].$$

Consequently, $\mathbf{x}_2' = A\mathbf{x}_2$ if and only if $\mathbf{v}_1$ and $\mathbf{v}_2$ satisfy

$$\underbrace{e^{2t}[(2\mathbf{v}_1 + \mathbf{v}_2) + 2t\mathbf{v}_2]}_{\mathbf{x}_2'} = \underbrace{e^{2t}(A\mathbf{v}_1 + tA\mathbf{v}_2)}_{A\mathbf{x}_2},$$

that is, if and only if

$$(2\mathbf{v}_1 + \mathbf{v}_2) + 2t\mathbf{v}_2 = A\mathbf{v}_1 + tA\mathbf{v}_2,$$

which is satisfied for all t if and only if

$$A\mathbf{v}_1 = 2\mathbf{v}_1 + \mathbf{v}_2, \tag{11.6.9}$$

$$A\mathbf{v}_2 = 2\mathbf{v}_2.$$

This last equation implies that $\mathbf{v}_2$ must be an eigenvector of A corresponding to the eigenvalue $\lambda = 2$ and, hence, from (11.6.7) must be of the form

$$\mathbf{v}_2 = r(2, 1). \tag{11.6.10}$$

We now rewrite (11.6.9) in the equivalent form

$$(A - 2I)\mathbf{v}_1 = \mathbf{v}_2, \tag{11.6.11}$$

which is a nonhomogeneous linear algebraic system for $\mathbf{v}_1$. If we let $\mathbf{v}_1 = (a, b)$, and substitute into (11.6.11) for $A - 2I$ and $\mathbf{v}_2$, we obtain the following system for a and b:[1]

$$4a - 8b = 2r, \tag{11.6.12}$$

$$2a - 4b = r. \tag{11.6.13}$$

These equations are consistent for all values of r, and so we can choose r to be any

---

[1] Note that the elements in the coefficient matrix $A - 2I$ can be obtained directly by setting $\lambda = 2$ in the determinant appearing in (11.6.6).

convenient *nonzero* value. We set $r = 2$, in which case the system (11.6.12), (11.6.13) reduces to the single equation

$$a - 2b = 1.$$

This has solution $a = 1 + 2s$, $b = s$, where s is a free variable. Since we require only one solution, we set $s = 0$, in which case $a = 1$, $b = 0$, so that

$$\mathbf{v}_1 = (1, 0).$$

Substituting $r = 2$ into (11.6.10) yields

$$\mathbf{v}_2 = (4, 2).$$

Equation (11.6.8) therefore implies that a second linearly independent solution to the given system of differential equations is

$$\mathbf{x}_2 = e^{2t} \left\{ \begin{bmatrix} 1 \\ 0 \end{bmatrix} + t \begin{bmatrix} 4 \\ 2 \end{bmatrix} \right\} = e^{2t} \begin{bmatrix} 1 + 4t \\ 2t \end{bmatrix},$$

and hence the general solution of the given system is

$$\mathbf{x}(t) = c_1 e^{2t} \begin{bmatrix} 2 \\ 1 \end{bmatrix} + c_2 e^{2t} \begin{bmatrix} 1 + 4t \\ 2t \end{bmatrix}.$$

We now impose the given initial condition. Setting $t = 0$ in the general solution yields

$$\mathbf{x}(0) = c_1 \begin{bmatrix} 2 \\ 1 \end{bmatrix} + c_2 \begin{bmatrix} 1 \\ 0 \end{bmatrix},$$

so that $\mathbf{x}(0) = \begin{bmatrix} -1 \\ 1 \end{bmatrix}$ if and only if c_1 and c_2 satisfy

$$2c_1 + c_2 = -1,$$

$$c_1 = 1.$$

Thus $c_1 = 1$ and $c_2 = -3$, and so the solution of the given initial value problem is

$$\mathbf{x}(t) = e^{2t} \begin{bmatrix} 2 \\ 1 \end{bmatrix} - 3e^{2t} \begin{bmatrix} 1 + 4t \\ 2t \end{bmatrix},$$

which can be written as

$$\mathbf{x} = e^{2t} \begin{bmatrix} -(1 + 12t) \\ 1 - 6t \end{bmatrix}.$$

In the previous example the matrix A had an eigenvalue of multiplicity 2 but only a one-dimensional eigenspace. If we look carefully at the vectors $\mathbf{v}_1$ and $\mathbf{v}_2$ arising in the solution $\mathbf{x}_2 = e^{2t}(\mathbf{v}_1 + t\mathbf{v}_2)$, we see that they were determined from the two systems

$$(A - \lambda I)\mathbf{v}_2 = \mathbf{0},$$

$$(A - \lambda I)\mathbf{v}_1 = \mathbf{v}_2,$$

where $\lambda = 2$, the eigenvalue of A. The first of these equations tells us that $\mathbf{v}_2$ must be an eigenvector of A corresponding to the eigenvalue λ. Any vector $\mathbf{v}_1$ satisfying the second equation is called a **generalized eigenvector** of A. We now show that these equations hold more generally.

Theorem 11.6.2: If the linear differential system $\mathbf{x}' = A\mathbf{x}$ has a solution of the form

$$\mathbf{x} = e^{\lambda t}(\mathbf{v}_1 + t\mathbf{v}_2),$$

then $\mathbf{v}_1$ and $\mathbf{v}_2$ must satisfy

$$(A - \lambda I)\mathbf{v}_2 = \mathbf{0},$$
$$(A - \lambda I)\mathbf{v}_1 = \mathbf{v}_2.$$

PROOF   If $\mathbf{x} = e^{\lambda t}(\mathbf{v}_1 + t\mathbf{v}_2)$, then, differentiating with respect to t,

$$\mathbf{x}' = e^{\lambda t}[(\lambda\mathbf{v}_1 + \mathbf{v}_2) + \lambda t\mathbf{v}_2].$$

Thus, $\mathbf{x}' = A\mathbf{x}$ if and only if

$$e^{\lambda t}[(\lambda\mathbf{v}_1 + \mathbf{v}_2) + \lambda t\mathbf{v}_2] = e^{\lambda t}(A\mathbf{v}_1 + tA\mathbf{v}_2),$$

that is, if and only if

$$\lambda\mathbf{v}_1 + \mathbf{v}_2 + \lambda t\mathbf{v}_2 = A\mathbf{v}_1 + tA\mathbf{v}_2.$$

This equation can hold for all values of t if and only if

$$A\mathbf{v}_1 = \lambda\mathbf{v}_1 + \mathbf{v}_2,$$
$$A\mathbf{v}_2 = \lambda\mathbf{v}_2.$$

Rearranging these equations yields

$$(A - \lambda I)\mathbf{v}_2 = \mathbf{0},$$
$$(A - \lambda I)\mathbf{v}_1 = \mathbf{v}_2.$$

as required.   ∎

Example 11.6.3   Find the general solution to $\mathbf{x}' = A\mathbf{x}$ if $A = \begin{bmatrix} 6 & 3 & 6 \\ 1 & 4 & 2 \\ -2 & -2 & -1 \end{bmatrix}$.

Solution   In this case,

$$A - \lambda I = \begin{bmatrix} 6-\lambda & 3 & 6 \\ 1 & 4-\lambda & 2 \\ -2 & -2 & -1-\lambda \end{bmatrix}, \tag{11.6.14}$$

and a short computation yields the single eigenvalue $\lambda = 3$ (multiplicity 3). The associated eigenvectors are of the form

$$\mathbf{v} = r(-1, 1, 0) + s(-2, 0, 1). \tag{11.6.15}$$

Thus two linearly independent solutions of our differential system are

$$\mathbf{x}_1 = e^{3t}\begin{bmatrix} -1 \\ 1 \\ 0 \end{bmatrix}, \qquad \mathbf{x}_2 = e^{3t}\begin{bmatrix} -2 \\ 0 \\ 1 \end{bmatrix}.$$

According to Theorem 11.6.1, there exists a third solution to the differential system of the form

$$\mathbf{x}_3 = e^{3t}(\mathbf{v}_1 + t\mathbf{v}_2). \qquad (11.6.16)$$

Substituting from (11.6.16) into the given system, it follows that $\mathbf{x}_3$ is a solution ($\mathbf{x}_3' = A\mathbf{x}_3$) if and only if

$$e^{3t}(3\mathbf{v}_1 + \mathbf{v}_2 + 3t\mathbf{v}_2) = e^{3t}(A\mathbf{v}_1 + tA\mathbf{v}_2).$$

Equating coefficients of e^t and te^t on either side of the equation yields the two conditions

$$A\mathbf{v}_1 = 3\mathbf{v}_1 + \mathbf{v}_2,$$

$$A\mathbf{v}_2 = 3\mathbf{v}_2,$$

which we can write as

$$(A - 3I)\mathbf{v}_2 = \mathbf{0}, \qquad (11.6.17)$$

$$(A - 3I)\mathbf{v}_1 = \mathbf{v}_2. \qquad (11.6.18)$$

(These equations could have been written directly as an application of Theorem 11.6.2.) It follows from (11.6.17) that $\mathbf{v}_2$ must be an eigenvector of A corresponding to the eigenvalue $\lambda = 3$, and hence from (11.6.15) it is of the form

$$\mathbf{v}_2 = r(-1, 1, 0) + s(-2, 0, 1). \qquad (11.6.19)$$

We now solve (11.6.18) for $\mathbf{v}_1$. If we let (a, b, c) denote the components of $\mathbf{v}_1$, then (11.6.18) can be written as [use (11.6.14) with $\lambda = 3$ to get the coefficient matrix]

$$3a + 3b + 6c = -r - 2s,$$

$$a + b + 2c = r, \qquad (11.6.20)$$

$$-2a - 2b - 4c = s.$$

This is a nonhomogeneous linear system and so we must check for consistency. The augmented matrix of the system is

$$\left[\begin{array}{ccc|c} 3 & 3 & 6 & -r - 2s \\ 1 & 1 & 2 & r \\ -2 & -2 & -4 & s \end{array}\right],$$

which is row equivalent to

$$\left[\begin{array}{ccc|c} 1 & 1 & 2 & r \\ 0 & 0 & 0 & -2(2r + s) \\ 0 & 0 & 0 & 2r + s \end{array}\right].$$

It follows that the system (11.6.20) is consistent if and only if $2r + s = 0$. For convenience we take $r = 1$, $s = -2$. It then follows from (11.6.19) that

$$\mathbf{v}_2 = (-1, 1, 0) - 2(-2, 0, 1) = (3, 1, -2). \tag{11.6.21}$$

Setting $r = 1$, $s = -2$ in (11.6.20) yields the single equation

$$a + b + 2c = 1.$$

This system has solution $a = 1 - p - 2q$, $b = p$, $c = q$, where p and q are free variables. Since we require only one solution, we set $p = q = 0$ to obtain

$$\mathbf{v}_1 = (1, 0, 0). \tag{11.6.22}$$

Substituting from (11.6.21) and (11.6.22) into (11.6.16) yields the third linearly independent solution

$$\mathbf{x}_3 = e^{3t} \left\{ \begin{bmatrix} 1 \\ 0 \\ 0 \end{bmatrix} + t \begin{bmatrix} 3 \\ 1 \\ -2 \end{bmatrix} \right\} = e^{3t} \begin{bmatrix} 1 + 3t \\ t \\ -2t \end{bmatrix}.$$

It follows that the general solution of the given system is

$$\mathbf{x}(t) = c_1 e^{3t} \begin{bmatrix} -1 \\ 1 \\ 0 \end{bmatrix} + c_2 e^{3t} \begin{bmatrix} -2 \\ 0 \\ 1 \end{bmatrix} + c_3 e^{3t} \begin{bmatrix} 1 + 3t \\ t \\ -2t \end{bmatrix},$$

or, equivalently,

$$\mathbf{x} = \begin{bmatrix} -e^{3t} & -2e^{3t} & e^{3t}(1 + 3t) \\ e^{3t} & 0 & te^{3t} \\ 0 & e^{3t} & -2te^{3t} \end{bmatrix} \begin{bmatrix} c_1 \\ c_2 \\ c_3 \end{bmatrix}.$$

We give one final example to illustrate the case when the eigenvectors are complex.

Example 11.6.4   Find the general solution of $\mathbf{x}' = A\mathbf{x}$ if

$$A = \begin{bmatrix} 0 & 2 & 1 & 0 \\ -2 & 0 & 0 & 1 \\ 0 & 0 & 0 & 2 \\ 0 & 0 & -2 & 0 \end{bmatrix}.$$

Solution   The characteristic polynomial of A is

$$\det(A - \lambda I) = \begin{vmatrix} -\lambda & 2 & 1 & 0 \\ -2 & -\lambda & 0 & 1 \\ 0 & 0 & -\lambda & 2 \\ 0 & 0 & -2 & -\lambda \end{vmatrix} = (\lambda^2 + 4)^2, \tag{11.6.23}$$

so that we obtain the complex conjugate eigenvalues $\lambda = \pm 2i$, each of multiplicity

2. If we consider the eigenvalue $\lambda = 2i$, then a straightforward computation yields the complex eigenvectors

$$\mathbf{v} = r(-i,\ 1,\ 0,\ 0).$$

This gives the complex-valued solution

$$\mathbf{u}_1 = e^{2it}\begin{bmatrix} -i \\ 1 \\ 0 \\ 0 \end{bmatrix} = (\cos 2t + i \sin 2t)\begin{bmatrix} -i \\ 1 \\ 0 \\ 0 \end{bmatrix}.$$

Taking the real and imaginary parts of $\mathbf{u}_1$ yields the following two linearly independent real-valued solutions:

$$\mathbf{x}_1 = \begin{bmatrix} \sin 2t \\ \cos 2t \\ 0 \\ 0 \end{bmatrix}, \qquad \mathbf{x}_2 = \begin{bmatrix} -\cos 2t \\ \sin 2t \\ 0 \\ 0 \end{bmatrix}.$$

We therefore require two more linearly independent real-valued solutions or, equivalently, one more linearly independent complex-valued solution. According to Theorem 11.6.1, there is another linearly independent complex-valued solution of the form

$$\mathbf{u}_2 = e^{2it}(\mathbf{v}_1 + t\mathbf{v}_2). \tag{11.6.24}$$

Further, from Theorem 11.6.2, $\mathbf{v}_1$ and $\mathbf{v}_2$ satisfy

$$(A - 2iI)\mathbf{v}_2 = \mathbf{0}, \tag{11.6.25}$$

$$(A - 2iI)\mathbf{v}_1 = \mathbf{v}_2. \tag{11.6.26}$$

It follows from (11.6.25) that $\mathbf{v}_2$ must be an eigenvector of A corresponding to the eigenvalue $\lambda = 2i$ and, hence, from (11.6.23) is of the form

$$\mathbf{v}_2 = r(-i,\ 1,\ 0,\ 0). \tag{11.6.27}$$

To determine $\mathbf{v}_1$ we must solve the linear system (11.6.26). It is easily shown that the reduced row echelon form of the augmented matrix of this system is

$$\left[\begin{array}{cccc|c} 1 & i & 0 & 0 & 0 \\ 0 & 0 & 1 & 0 & -ir \\ 0 & 0 & 0 & 1 & r \\ 0 & 0 & 0 & 0 & 0 \end{array}\right],$$

so that (11.6.26) is consistent for all values of r. We choose $r = 1$, in which case

$$\mathbf{v}_1 = (-is,\ s,\ -i,\ 1),$$

where s is a free variable. Since we require only one solution and there are no further consistency requirements to be satisfied, we choose, for simplicity, $s = 0$. It then follows that

$$\mathbf{v}_1 = (0,\ 0,\ -i,\ 1),$$

and, from (11.6.27) with $r = 1$,

$$\mathbf{v}_2 = (-i,\ 1,\ 0,\ 0).$$

Substituting into (11.6.24) yields the second complex-valued solution

$$\mathbf{u}_2(t) = e^{2it} \left\{ \begin{bmatrix} 0 \\ 0 \\ -i \\ 1 \end{bmatrix} + t \begin{bmatrix} -i \\ 1 \\ 0 \\ 0 \end{bmatrix} \right\}$$

which can be written as

$$\mathbf{u}_2(t) = (\cos 2t + i \sin 2t) \begin{bmatrix} -it \\ t \\ -i \\ 1 \end{bmatrix}.$$

Taking the real and imaginary parts of this complex-valued solution, we obtain the two linearly independent real-valued solutions

$$\mathbf{x}_3(t) = \begin{bmatrix} t \sin 2t \\ t \cos 2t \\ \sin 2t \\ \cos 2t \end{bmatrix}, \qquad \mathbf{x}_4(t) = \begin{bmatrix} -t \cos 2t \\ t \sin 2t \\ -\cos 2t \\ \sin 2t \end{bmatrix}.$$

Consequently the general solution of the given differential system is

$$\mathbf{x}(t) = c_1 \begin{bmatrix} \sin 2t \\ \cos 2t \\ 0 \\ 0 \end{bmatrix} + c_2 \begin{bmatrix} -\cos 2t \\ \sin 2t \\ 0 \\ 0 \end{bmatrix} + c_3 \begin{bmatrix} t \sin 2t \\ t \cos 2t \\ \sin 2t \\ \cos 2t \end{bmatrix} + c_4 \begin{bmatrix} -t \cos 2t \\ t \sin 2t \\ -\cos 2t \\ \sin 2t \end{bmatrix}.$$

Finally, we state the generalization of the results of this section to higher-order multiplicities.

Theorem 11.6.3:   Let λ be an eigenvalue of the $n \times n$ matrix A of multiplicity m, and suppose that the dimension of the corresponding eigenspace is k. Then there exist m linearly independent solutions to the linear differential system $\mathbf{x}' = A\mathbf{x}$ of the form

$$\mathbf{x}_{i+1}(t) = e^{\lambda t}\left[\mathbf{v}_{i(i+1)/2} + t\mathbf{v}_{i(i+1)/2+1} + \cdots + \frac{1}{i!}t^i \mathbf{v}_{i(i+3)/2} \right], \qquad i = 0,\ 1,\ \ldots,\ m-k.$$

Further, the set of all solutions to $\mathbf{x}' = A\mathbf{x}$ corresponding to different eigenvalues obtained in this manner are linearly independent on any interval.

PROOF   See, for example, M. W. Hirsch and S. Smale, *Differential Equations, Dynamical Systems, and Linear Algebra*, Academic Press, 1974.   ■

Theorem 11.6.4:   If the differential system $\mathbf{x}' = A\mathbf{x}$ has a solution of the form

$$\mathbf{x} = e^{\lambda t}\left(\mathbf{v}_0 + t\mathbf{v}_1 + \frac{1}{2!}t^2\mathbf{v}_2 + \cdots + \frac{1}{n!}t^n\mathbf{v}_n\right),$$

then $\mathbf{v}_0, \mathbf{v}_1, \ldots, \mathbf{v}_n$ must satisfy

$$(A - \lambda I)\mathbf{v}_n = 0,$$
$$(A - \lambda I)\mathbf{v}_{n-1} = \mathbf{v}_n,$$
$$(A - \lambda I)\mathbf{v}_{n-2} = \mathbf{v}_{n-1},$$
$$\vdots$$
$$(A - \lambda I)\mathbf{v}_0 = \mathbf{v}_1.$$

PROOF   We leave the proof of this theorem as an exercise.   ∎

EXERCISES 11.6:

In problems 1–13 determine the general solution of the system $\mathbf{x}' = A\mathbf{x}$ for the given matrix A.

1. $\begin{bmatrix} 0 & -2 \\ 2 & 4 \end{bmatrix}$.

2. $\begin{bmatrix} -3 & -2 \\ 2 & 1 \end{bmatrix}$.

3. $\begin{bmatrix} 0 & 1 & 0 \\ 0 & 0 & 1 \\ 1 & 1 & -1 \end{bmatrix}$.

4. $\begin{bmatrix} 2 & 2 & -1 \\ 2 & 1 & -1 \\ 2 & 3 & -1 \end{bmatrix}$.

5. $\begin{bmatrix} -2 & 0 & 0 \\ 1 & -3 & -1 \\ -1 & 1 & -1 \end{bmatrix}$.

6. $\begin{bmatrix} 15 & -32 & 12 \\ 8 & -17 & 6 \\ 0 & 0 & -1 \end{bmatrix}$.

7. $\begin{bmatrix} 4 & 0 & 0 \\ 1 & 4 & 0 \\ 0 & 1 & 4 \end{bmatrix}$.

8. $\begin{bmatrix} 1 & 0 & 0 \\ 0 & 3 & 2 \\ 2 & -2 & -1 \end{bmatrix}$.

9. $\begin{bmatrix} 3 & 1 & 0 \\ -1 & 5 & 0 \\ 0 & 0 & 4 \end{bmatrix}$.

10. $\begin{bmatrix} -1 & 1 & 0 \\ -2 & -3 & 1 \\ 1 & 1 & -2 \end{bmatrix}$.

11. $\begin{bmatrix} 0 & -1 & 0 & 0 \\ 1 & 0 & 0 & 0 \\ 1 & 0 & 2 & 1 \\ 0 & 1 & 0 & 2 \end{bmatrix}$.

12. $\begin{bmatrix} -2 & 3 & 0 & 0 \\ 3 & -2 & 0 & 0 \\ 1 & 0 & 1 & -1 \\ 0 & 1 & 0 & 1 \end{bmatrix}$.

13. $\begin{bmatrix} 0 & -1 & 0 & 0 \\ 1 & 0 & 0 & 0 \\ 1 & 0 & 0 & -1 \\ 0 & 1 & 1 & 0 \end{bmatrix}$.

In problems 14 and 15 solve the given initial value problem.

14. $\mathbf{x}' = A\mathbf{x}$, $\mathbf{x}(0) = \mathbf{x}_0$, where $A = \begin{bmatrix} -2 & -1 \\ 1 & -4 \end{bmatrix}$,

$\mathbf{x}_0 = \begin{bmatrix} 0 \\ -1 \end{bmatrix}$.

15. $x' = Ax, x(0) = x_0$, where $A = \begin{bmatrix} -2 & -1 & 4 \\ 0 & -1 & 0 \\ -1 & -3 & 2 \end{bmatrix}$,

$x_0 = \begin{bmatrix} -2 \\ 1 \\ 1 \end{bmatrix}$.

16. (a) Show that if the differential system $x' = Ax$ has a solution of the form

$$x = e^{\lambda t}\left(v_0 + tv_1 + \frac{1}{2!}t^2 v_2\right),$$

then the constant vectors v_0, v_1, v_2 must satisfy

$$(A - \lambda I)v_2 = 0, \qquad (A - \lambda I)v_1 = v_2,$$
$$(A - \lambda I)v_0 = v_1.$$

(b) Prove Theorem 11.6.4.

17. Let A be a 2×2 real matrix. Prove that all solutions to $x' = Ax$ satisfy

$$\lim_{t \to +\infty} x(t) = 0,$$

if and only if all eigenvalues of A have negative real part.

18. Extend the result of the previous exercise to the system $x' = Ax$, where A is an arbitrary (real) $n \times n$ matrix.

11.7   NONHOMOGENEOUS LINEAR SYSTEMS

We now consider solving the nonhomogeneous system of linear differential equations

$$x' = A(t)x + b(t), \tag{11.7.1}$$

where A is an $n \times n$ matrix function and $b(t)$ is a vector function. The homogeneous equation associated with (11.7.1) is

$$x' = A(t)x. \tag{11.7.2}$$

According to Theorem 11.4.4, every solution of the system (11.7.1) is of the form

$$x(t) = c_1 x_1 + c_2 x_2 + \cdots + c_n x_n + x_p,$$

where $x_1, x_2, \ldots, x_n$ are linearly independent solutions of the associated homogeneous system (11.7.2) and x_p is a *particular* solution of (11.7.1). Provided the matrix A has *constant* entries, we can find n linearly independent solutions to the associated homogeneous equation (11.7.2) using our eigenvalue/eigenvector method. Our task in solving (11.7.1) then reduces to obtaining a particular solution to the system. In this section we show how the variation-of-parameters method can be used to fulfill this task.

Theorem 11.7.1 (Variation-of-Parameters Method): If $x_1, x_2, \ldots, x_n$ are linearly independent solutions to $x' = Ax$ and $X = [x_1, x_2, \ldots, x_n]$, then there is a particular solution to $x' = Ax + b$ of the form $x_p = Xu$ where u satisfies $Xu' = b$. Explicitly,

$$x_p(t) = X(t)\int^t X^{-1}(s)b(s)\, ds.$$

PROOF   The general solution of $x' = Ax$ is $x_c = c_1 x_1 + c_2 x_2 + \cdots + c_n x_n$. We try for a particular solution to

$$x' = Ax + b \tag{11.7.3}$$

of the form[1]

$$x_p(t) = u_1(t)\mathbf{x}_1 + u_2(t)\mathbf{x}_2 + \cdots + u_n(t)\mathbf{x}_n,$$

that is,

$$\mathbf{x}_p(t) = X(t)\mathbf{u}(t), \tag{11.7.4}$$

where

$$X(t) = [\mathbf{x}_1, \mathbf{x}_2, \ldots, \mathbf{x}_n] \quad \text{and} \quad \mathbf{u}(t) = [u_1(t) \quad u_2(t) \quad \cdots \quad u_n(t)]^T.$$

Substituting (11.7.4) into (11.7.3), it follows that $\mathbf{x}_p$ is a solution of (11.7.3) provided $\mathbf{u}$ satisfies

$$(X\mathbf{u})' = A(X\mathbf{u}) + \mathbf{b}. \tag{11.7.5}$$

Applying the product rule for differentiation to the left-hand side of this equation we obtain

$$X'\mathbf{u} + X\mathbf{u}' = A(X\mathbf{u}) + \mathbf{b}. \tag{11.7.6}$$

By definition we have

$$X = [\mathbf{x}_1, \mathbf{x}_2, \ldots, \mathbf{x}_n],$$

so that

$$X' = [\mathbf{x}_1', \mathbf{x}_2', \ldots, \mathbf{x}_n']. \tag{11.7.7}$$

Since each of the vector functions $\mathbf{x}_i$ is a solution of the associated homogeneous equation $\mathbf{x}' = A\mathbf{x}$, we can write (11.7.7) in the form

$$X' = [A\mathbf{x}_1, A\mathbf{x}_2, \ldots, A\mathbf{x}_n],$$

that is,

$$X' = AX.$$

Substituting this expression for X' into (11.7.6) yields

$$(AX)\mathbf{u} + X\mathbf{u}' = A(X\mathbf{u}) + \mathbf{b},$$

so that

$$X\mathbf{u}' = \mathbf{b}.$$

This implies that[2]

$$\mathbf{u}' = X^{-1}\mathbf{b}.$$

Consequently

$$\mathbf{u}(t) = \int^t X^{-1}(s)\mathbf{b}(s) \, ds,$$

where we have set the integration constants to zero without loss of generality.

---

[1] That is, we replace the constants in $\mathbf{x}_c$ by arbitrary functions.

[2] Note that X^{-1} exists since $\det(X) \neq 0$. (Why?)

Hence, from (11.7.4), a particular solution to (11.7.3) is

$$\mathbf{x}_p(t) = X(t) \int^t X^{-1}(s)\mathbf{b}(s) \; ds. \qquad \blacksquare$$

REMARK You should not memorize the formula for $\mathbf{x}_p$ given in the previous theorem. Rather you should remember that a particular solution to $\mathbf{x}' = A\mathbf{x} + \mathbf{b}$ is

$$\mathbf{x}_p(t) = X\mathbf{u},$$

where $\mathbf{u}'$ is determined by solving the linear system

$$X\mathbf{u}' = \mathbf{b}. \qquad (11.7.8)$$

Whereas in the proof of the previous theorem we used X^{-1} to determine a simple formula for the solution of this system, it is better in practice to use Gaussian elimination to solve (11.7.8) for $\mathbf{u}'$ and then integrate the resulting vector to determine $\mathbf{u}$.

Example 11.7.1 Solve $\mathbf{x}' = A\mathbf{x} + \mathbf{b}$, $\mathbf{x}(0) = \begin{bmatrix} 3 \\ 0 \end{bmatrix}$, if $A = \begin{bmatrix} 1 & 2 \\ 4 & 3 \end{bmatrix}$ and $\mathbf{b} = \begin{bmatrix} 12e^{3t} \\ 18e^{2t} \end{bmatrix}$.

Solution We first solve the associated homogeneous equation $\mathbf{x}' = A\mathbf{x}$. For the given matrix A we find

$$\det(A - \lambda I) = (\lambda - 5)(\lambda + 1)$$

so that the eigenvalues of A are $\lambda = -1, 5$.

    Eigenvectors:

$$\lambda = -1 \text{ implies that } \mathbf{v}_1 = r(-1, 1),$$
$$\lambda = 5 \text{ implies that } \mathbf{v}_2 = s(1, 2),$$

and so two linearly independent solutions to $\mathbf{x}' = A\mathbf{x}$ are

$$\mathbf{x}_1 = e^{-t} \begin{bmatrix} -1 \\ 1 \end{bmatrix}, \qquad \mathbf{x}_2 = e^{5t} \begin{bmatrix} 1 \\ 2 \end{bmatrix}.$$

If we set

$$X = \begin{bmatrix} -e^{-t} & e^{5t} \\ e^{-t} & 2e^{5t} \end{bmatrix},$$

then it follows from Theorem 11.7.1 that a particular solution to $\mathbf{x}' = A\mathbf{x} + \mathbf{b}$ is

$$\mathbf{x}_p = X\mathbf{u},$$

where

$$X\mathbf{u}' = \mathbf{b}.$$

To find $\mathbf{u}'$ we determine the reduced row echelon form of the augmented matrix of this system. We have

$$\begin{bmatrix} -e^{-t} & e^{5t} \ \Big| \ 12e^{3t} \\ e^{-t} & 2e^{5t} \ \Big| \ 18e^{2t} \end{bmatrix} \xrightarrow[R_1 \to -e^t R_1]{R_2 \to R_2 + R_1} \begin{bmatrix} 1 & -e^{6t} \ \Big| \ -12e^{4t} \\ 0 & 3e^{5t} \ \Big| \ 12e^{3t} + 18e^{2t} \end{bmatrix}$$

$$\xrightarrow{R_2 \to \frac{1}{3}e^{-5t}R_2} \begin{bmatrix} 1 & -e^{6t} \ \Big| \ -12e^{4t} \\ 0 & 1 \ \Big| \ 4e^{-2t} + 6e^{-3t} \end{bmatrix}$$

$$\xrightarrow{R_1 \to R_1 + e^{6t}R_2} \begin{bmatrix} 1 & 0 \ \Big| \ -8e^{4t} + 6e^{3t} \\ 0 & 1 \ \Big| \ 4e^{-2t} + 6e^{-3t} \end{bmatrix}.$$

Thus

$$\mathbf{u}' = \begin{bmatrix} -8e^{4t} + 6e^{3t} \\ 4e^{-2t} + 6e^{-3t} \end{bmatrix},$$

which implies that

$$\mathbf{u} = \begin{bmatrix} -2e^{4t} + 2e^{3t} \\ -2e^{-2t} - 2e^{-3t} \end{bmatrix},$$

where we have set the integration constants equal to zero without loss of generality. It follows that a particular solution to the given differential system is

$$\mathbf{x}_p = X\mathbf{u} = \begin{bmatrix} -e^{-t} & e^{5t} \\ e^{-t} & 2e^{5t} \end{bmatrix} \begin{bmatrix} -2e^{4t} + 2e^{3t} \\ -2e^{-2t} - 2e^{-3t} \end{bmatrix} = \begin{bmatrix} -4e^{2t} \\ -2e^{2t} - 6e^{3t} \end{bmatrix},$$

so that the general solution of the given nonhomogeneous equation is

$$\mathbf{x}(t) = c_1 e^{-t} \begin{bmatrix} -1 \\ 1 \end{bmatrix} + c_2 e^{5t} \begin{bmatrix} 1 \\ 2 \end{bmatrix} - \begin{bmatrix} 4e^{2t} \\ 2e^{2t} + 6e^{3t} \end{bmatrix}.$$

The initial condition $\mathbf{x}(0) = \begin{bmatrix} 3 \\ 0 \end{bmatrix}$ requires

$$c_1 \begin{bmatrix} -1 \\ 1 \end{bmatrix} + c_2 \begin{bmatrix} 1 \\ 2 \end{bmatrix} - \begin{bmatrix} 4 \\ 8 \end{bmatrix} = \begin{bmatrix} 3 \\ 0 \end{bmatrix},$$

that is,

$$\begin{aligned} -c_1 + c_2 &= 7, \\ c_1 + 2c_2 &= 8. \end{aligned}$$

Thus, $c_1 = -2$, $c_2 = 5$, and the solution to the initial value problem is

$$\mathbf{x} = 5e^{5t} \begin{bmatrix} 1 \\ 2 \end{bmatrix} - 2e^{-t} \begin{bmatrix} -1 \\ 1 \end{bmatrix} - \begin{bmatrix} 4e^{2t} \\ 2e^{2t}(1 + 3e^t) \end{bmatrix}.$$

EXERCISES 11.7

In problems 1–8 use the variation-of-parameters technique to find a particular solution, $\mathbf{x}_p$, of $\mathbf{x}' = A\mathbf{x} + \mathbf{b}$ for the given A and $\mathbf{b}$. Also obtain the general solution of the system of differential equations.

1. $A = \begin{bmatrix} 2 & -1 \\ -1 & 2 \end{bmatrix}$, $\mathbf{b} = \begin{bmatrix} 0 \\ 4e^t \end{bmatrix}$.

2. $A = \begin{bmatrix} 4 & -3 \\ 2 & -1 \end{bmatrix}$, $\mathbf{b} = \begin{bmatrix} e^{2t} \\ e^t \end{bmatrix}$.

3. $A = \begin{bmatrix} -1 & 1 \\ 3 & 1 \end{bmatrix}$, $\mathbf{b} = \begin{bmatrix} 20e^{3t} \\ 12e^t \end{bmatrix}$.

4. $A = \begin{bmatrix} -1 & 2 \\ -2 & 4 \end{bmatrix}$, $\mathbf{b} = \begin{bmatrix} 54te^{3t} \\ 9e^{3t} \end{bmatrix}$.

5. $A = \begin{bmatrix} 2 & 4 \\ -2 & -2 \end{bmatrix}$, $\mathbf{b} = \begin{bmatrix} 8 \sin 2t \\ 8 \cos 2t \end{bmatrix}$.

6. $A = \begin{bmatrix} 3 & 2 \\ -2 & -1 \end{bmatrix}$, $\mathbf{b} = \begin{bmatrix} -3e^t \\ 6te^t \end{bmatrix}$.

7. $A = \begin{bmatrix} 1 & 0 & 0 \\ 2 & -3 & 2 \\ 1 & -2 & 2 \end{bmatrix}$, $\mathbf{b} = \begin{bmatrix} -e^t \\ 6e^{-t} \\ e^t \end{bmatrix}$.

8. $A = \begin{bmatrix} -1 & -2 & 2 \\ 2 & 4 & -1 \\ 0 & 0 & 3 \end{bmatrix}$, $\mathbf{b} = \begin{bmatrix} -e^{3t} \\ 4e^{3t} \\ 3e^{3t} \end{bmatrix}$.

9. Let $X(t)$ be a fundamental matrix for the differential system $\mathbf{x}' = A(t)\mathbf{x}$, where $A(t)$ is an $n \times n$ matrix function. Show that the solution of the initial value problem

$$\mathbf{x}' = A(t)\mathbf{x} + \mathbf{b}(t), \qquad \mathbf{x}(t_0) = \mathbf{x}_0,$$

can be written as

$$\mathbf{x}(t) = X(t)X^{-1}(t_0)\mathbf{x}_0 + X(t)\int_{t_0}^t X^{-1}(s)\mathbf{b}(s)\,ds.$$

10. Consider the nonhomogeneous system

$$x_1' = 2x_1 - 3x_2 + 34 \sin t,$$
$$x_2' = -4x_1 - 2x_2 + 17 \cos t.$$

Find the general solution of this system by first solving the associated homogeneous system and then using the *method of undetermined coefficients* to obtain a particular solution. (*Hint:* The form of the nonhomogeneous term suggests a trial solution of the form

$$\mathbf{x}_p = \begin{bmatrix} A_1\cos t + B_1\sin t \\ A_2\cos t + B_2\sin t \end{bmatrix},$$

where the constants A_1, A_2, B_1, B_2 can be determined by substituting into the given system.)

11.8 SOME APPLICATIONS OF LINEAR SYSTEMS OF DIFFERENTIAL EQUATIONS

In this section we analyze the two problems that were briefly introduced in Section 11.1. We begin with the coupled spring-mass system that consists of two masses m_1, m_2 connected by two springs whose spring constants are k_1, k_2, respectively (see Figure 11.8.1). Let $x(t)$, $y(t)$ denote the displacement of m_1, m_2 from their equilibrium positions. When the system is in motion, the extension of spring 1 is

$$L_1(t) = x(t),$$

whereas the *net* extension of spring 2 is

$$L_2(t) = y(t) - x(t).$$

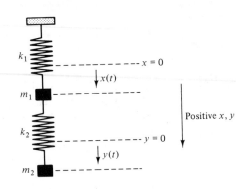

Figure 11.8.1   A coupled spring-mass system.

Consequently, using Hooke's law, the net forces acting on masses m_1 and m_2 at time t are

$$F_1(t) = -k_1 x(t) + k_2[y(t) - x(t)], \qquad F_2(t) = -k_2[y(t) - x(t)],$$

respectively.   Thus, applying Newton's second law to each mass yields the system of differential equations

$$m_1 \frac{d^2 x}{dt^2} = -k_1 x + k_2(y - x), \tag{11.8.1}$$

$$m_2 \frac{d^2 y}{dt^2} = -k_2(y - x). \tag{11.8.2}$$

The motion of the spring-mass system will be fully determined once we have specified appropriate initial conditions of the form

$$x(t_0) = \alpha_1, \qquad \frac{dx}{dt}(t_0) = \alpha_2, \qquad y(t_0) = \alpha_3, \qquad \frac{dy}{dt}(t_0) = \alpha_4, \tag{11.8.3}$$

where $\alpha_1, \alpha_2, \alpha_3, \alpha_4$ are constants.

To apply the techniques that we have developed in this chapter for solving systems of differential equations, we must convert (11.8.1) and (11.8.2) into a first-order system.   We introduce new variables x_1, x_2, x_3, x_4 defined by

$$x_1 = x, \qquad x_2 = x', \qquad x_3 = y, \qquad x_4 = y'. \tag{11.8.4}$$

Then (11.8.1) and (11.8.2) can be replaced by the equivalent system

$$x_1' = x_2, \qquad x_2' = -\frac{k_1}{m_1} x_1 + \frac{k_2}{m_1}(x_3 - x_1), \qquad x_3' = x_4, \qquad x_4' = -\frac{k_2}{m_2}(x_3 - x_1).$$

Rearranging terms yields the first-order linear system

$$x_1' = x_2, \tag{11.8.5}$$

$$x_2' = -\left(\frac{k_1}{m_1} + \frac{k_2}{m_1}\right) x_1 + \frac{k_2}{m_1} x_3, \tag{11.8.6}$$

$$x_3' = x_4, \tag{11.8.7}$$

$$x_4' = \frac{k_2}{m_2} x_1 - \frac{k_2}{m_2} x_3. \tag{11.8.8}$$

In the new variables the initial conditions (11.8.3) are

$$x_1(t_0) = \alpha_1, \qquad x_2(t_0) = \alpha_2, \qquad x_3(t_0) = \alpha_3, \qquad x_4(t_0) = \alpha_4. \tag{11.8.9}$$

This initial value problem for x_1, x_2, x_3, x_4 can be written in matrix form as

$$\mathbf{x}' = A\mathbf{x}, \qquad \mathbf{x}(t_0) = \mathbf{x}_0,$$

where

$$\mathbf{x} = \begin{bmatrix} x_1 \\ x_2 \\ x_3 \\ x_4 \end{bmatrix}, \qquad A = \begin{bmatrix} 0 & 1 & 0 & 0 \\ -\dfrac{1}{m_1}(k_1 + k_2) & 0 & \dfrac{k_2}{m_1} & 0 \\ 0 & 0 & 0 & 1 \\ \dfrac{k_2}{m_2} & 0 & -\dfrac{k_2}{m_2} & 0 \end{bmatrix}, \qquad \mathbf{x}_0 = \begin{bmatrix} \alpha_1 \\ \alpha_2 \\ \alpha_3 \\ \alpha_4 \end{bmatrix}.$$

We leave the analysis of the general system for the exercises and consider a particular example.

Example 11.8.1   Consider the spring-mass system with

$$k_1 = 4 Nm^{-1}, \qquad k_2 = 2 Nm^{-1}, \qquad m_1 = 2 \text{ kg}, \qquad m_2 = 1 \text{ kg}.$$

At $t = 0$ both masses are displaced a distance 1 m from equilibrium and released from rest. Determine the motion of the resulting system.

Solution   The motion of the system is governed by the initial value problem

$$2 \frac{d^2 x}{dt^2} = -4x + 2(y - x), \tag{11.8.10}$$

$$\frac{d^2 y}{dt^2} = -2(y - x), \tag{11.8.11}$$

$$x(0) = 1, \qquad \frac{dx}{dt}(0) = 0, \qquad y(0) = 1, \qquad \frac{dy}{dt}(0) = 0.$$

Introducing new variables $x_1 = x, x_2 = x', x_3 = y, x_4 = y'$, yields the equivalent initial value problem

$$x_1' = x_2,$$

$$x_2' = -3x_1 + x_3,$$

$$x_3' = x_4,$$

$$x_4' = 2x_1 - 2x_3,$$

$$x_1(0) = 1, \qquad x_2(0) = 0, \qquad x_3(0) = 1, \qquad x_4(0) = 0.$$

This can be written as

$$\mathbf{x}' = A\mathbf{x}, \qquad \mathbf{x}(0) = \mathbf{x}_0, \tag{11.8.12}$$

where

$$A = \begin{bmatrix} 0 & 1 & 0 & 0 \\ -3 & 0 & 1 & 0 \\ 0 & 0 & 0 & 1 \\ 2 & 0 & -2 & 0 \end{bmatrix} \quad \text{and} \quad \mathbf{x}_0 = \begin{bmatrix} 1 \\ 0 \\ 1 \\ 0 \end{bmatrix}.$$

The characteristic polynomial of A is[1]

$$
\det(A - \lambda I) = \begin{vmatrix} -\lambda & 1 & 0 & 0 \\ -3 & -\lambda & 1 & 0 \\ 0 & 0 & -\lambda & 1 \\ 2 & 0 & -2 & -\lambda \end{vmatrix}
$$

$$= \lambda^4 + 5\lambda^2 + 4$$

$$= (\lambda^2 + 1)(\lambda^2 + 4).$$

Thus the eigenvalues of A are

$$\lambda = \pm i, \ \pm 2i.$$

We now determine the eigenvectors.
  $\lambda = i$:   The system $(A - \lambda I)\mathbf{v}_1 = \mathbf{0}$ has augmented matrix

$$
\begin{bmatrix} -i & 1 & 0 & 0 & | & 0 \\ -3 & -i & 1 & 0 & | & 0 \\ 0 & 0 & -i & 1 & | & 0 \\ 2 & 0 & -2 & -i & | & 0 \end{bmatrix},
$$

with reduced row echelon form

$$
\begin{bmatrix} 1 & 0 & 0 & \dfrac{i}{2} & | & 0 \\ 0 & 1 & 0 & -\dfrac{1}{2} & | & 0 \\ 0 & 0 & 1 & i & | & 0 \\ 0 & 0 & 0 & 0 & | & 0 \end{bmatrix}.
$$

Consequently the eigenvectors are

$$\mathbf{v}_1 = r(-i, \ 1, \ -2i, \ 2).$$

Thus a complex-valued solution to $\mathbf{x}' = A\mathbf{x}$ is

---

[1] In problem 1 the reader is asked to fill in the missing details of this computation.

$$\mathbf{u}_1 = e^{it}\begin{bmatrix} -i \\ 1 \\ -2i \\ 2 \end{bmatrix} = (\cos t + i \sin t)\begin{bmatrix} -i \\ 1 \\ -2i \\ 2 \end{bmatrix}.$$

Taking the real and imaginary parts of this complex-valued solution yields the two real-valued solutions

$$\mathbf{x}_1 = \begin{bmatrix} \sin t \\ \cos t \\ 2\sin t \\ 2\cos t \end{bmatrix}, \qquad \mathbf{x}_2 = \begin{bmatrix} -\cos t \\ \sin t \\ -2\cos t \\ 2\sin t \end{bmatrix}.$$

$\lambda = 2i$: In this case the augmented matrix of the system $(A - \lambda I)\mathbf{v}_2 = 0$ is

$$\left[\begin{array}{cccc|c} -2i & 1 & 0 & 0 & 0 \\ -3 & -2i & 1 & 0 & 0 \\ 0 & 0 & -2i & 1 & 0 \\ 2 & 0 & -2 & -2i & 0 \end{array}\right],$$

with reduced row echelon form

$$\left[\begin{array}{cccc|c} 1 & 0 & 0 & -\dfrac{i}{2} & 0 \\ 0 & 1 & 0 & 1 & 0 \\ 0 & 0 & 1 & \dfrac{i}{2} & 0 \\ 0 & 0 & 0 & 0 & 0 \end{array}\right].$$

The corresponding eigenvectors are therefore of the form

$$\mathbf{v}_2 = s(i, -2, -i, 2),$$

so that a complex-valued solution to the system $\mathbf{x}' = A\mathbf{x}$ is

$$\mathbf{u}_2 = e^{2it}\begin{bmatrix} i \\ -2 \\ -i \\ 2 \end{bmatrix} = (\cos 2t + i \sin 2t)\begin{bmatrix} i \\ -2 \\ -i \\ 2 \end{bmatrix}.$$

Taking the real and imaginary parts of this complex-valued solution yields the additional real-valued linearly independent solutions

$$\mathbf{x}_3 = \begin{bmatrix} -\sin 2t \\ -2\cos 2t \\ \sin 2t \\ 2\cos 2t \end{bmatrix}, \qquad \mathbf{x}_4 = \begin{bmatrix} \cos 2t \\ -2\sin 2t \\ -\cos 2t \\ 2\sin 2t \end{bmatrix}.$$

Consequently the system in (11.8.12) has general solution

$$\mathbf{x} = c_1 \begin{bmatrix} \sin t \\ \cos t \\ 2 \sin t \\ 2 \cos t \end{bmatrix} + c_2 \begin{bmatrix} -\cos t \\ \sin t \\ -2 \cos t \\ 2 \sin t \end{bmatrix} + c_3 \begin{bmatrix} -\sin 2t \\ -2 \cos 2t \\ \sin 2t \\ 2 \cos 2t \end{bmatrix} + c_4 \begin{bmatrix} \cos 2t \\ -2 \sin 2t \\ -\cos 2t \\ 2 \sin 2t \end{bmatrix}.$$

We now impose the initial condition that

$$\mathbf{x}(0) = \begin{bmatrix} 1 \\ 0 \\ 1 \\ 0 \end{bmatrix}.$$

This requires c_1, c_2, c_3, c_4 to satisfy

$$
\begin{aligned}
- c_2 && + c_4 &= 1, \\
c_1 && - 2c_3 && &= 0, \\
- 2c_2 && - c_4 &= 1, \\
2c_1 && + 2c_3 && &= 0,
\end{aligned}
$$

which has solution $c_1 = 0$, $c_2 = -\frac{2}{3}$, $c_3 = 0$, $c_4 = \frac{1}{3}$. Thus

$$\mathbf{x}(t) = -\frac{2}{3} \begin{bmatrix} -\cos t \\ \sin t \\ -2 \cos t \\ 2 \sin t \end{bmatrix} + \frac{1}{3} \begin{bmatrix} \cos 2t \\ -2 \sin 2t \\ -\cos 2t \\ 2 \sin 2t \end{bmatrix}.$$

Since $x = x_1$ and $y = x_3$, it follows that the motion of the spring-mass system is given by

$$x(t) = \tfrac{1}{3}(2 \cos t + \cos 2t),$$
$$y(t) = \tfrac{1}{3}(4 \cos t - \cos 2t).$$

Now consider the mixing problem depicted in Figure 11.8.2.   Two tanks contain a solution consisting of chemical dissolved in water.   A solution containing c_{in} grams/liter of a chemical flows into tank 1 at a rate of r_{in} liters/minute, and solution of concentration c_{out} grams/liter flows out of tank 2 at a rate of r_{out} liters/minute.   In addition, a solution of concentration c_{12} grams/liter flows into tank 1 from tank 2 at a rate of r_{12} liters/minute, and a solution of concentration c_{21} grams/liter flows into tank 2 from tank 1 at a rate of r_{21} liters/minute.   We wish to determine $A_1(t)$, $A_2(t)$, the amounts of chemical in tank 1 and tank 2, respectively.

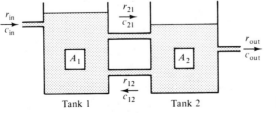

Tank 1   Tank 2   **Figure 11.8.2**

The analysis is similar to that used in Section 2.5. Assuming that the solution in each tank is well mixed, it follows immediately that

$$c_{12} = c_{\text{out}} = \frac{A_2}{V_2},$$

$$c_{21} = \frac{A_1}{V_1},$$

where V_i denotes the volume of solution in tank i at time t.

Consider a short time interval Δt. The total amount of chemical entering tank 1 in this time interval is approximately

$$(c_{\text{in}} r_{\text{in}} + c_{12} r_{12}) \Delta t \text{ grams,}$$

whereas approximately

$$c_{21} r_{21} \Delta t \text{ grams}$$

of chemical leave tank 1 in the same time interval. Consequently, the change in the amount of chemical in tank 1 in the time interval Δt, denoted by ΔA_1, is approximately

$$\Delta A_1 \approx [(c_{\text{in}} r_{\text{in}} + c_{12} r_{12}) - c_{21} r_{21}] \Delta t,$$

that is,

$$\Delta A_1 \approx \left(c_{\text{in}} r_{\text{in}} + r_{12} \frac{A_2}{V_2} - r_{21} \frac{A_1}{V_1} \right) \Delta t. \tag{11.8.13}$$

Similarly, the change in the amount of chemical in tank 2 in the time interval Δt, denoted by ΔA_2, is

$$\Delta A_2 \approx [r_{21} c_{21} - (r_{12} c_{12} + r_{\text{out}} c_{\text{out}})] \Delta t,$$

or, equivalently,

$$\Delta A_2 \approx \left[r_{21} \frac{A_1}{V_1} - (r_{12} + r_{\text{out}}) \frac{A_2}{V_2} \right] \Delta t. \tag{11.8.14}$$

Dividing (11.8.13) and (11.8.14) by Δt and taking the limit as $\Delta t \to 0^+$ yields the following differential equations for A_1 and A_2:

$$\frac{dA_1}{dt} = -r_{21} \frac{A_1}{V_1} + r_{12} \frac{A_2}{V_2} + c_{\text{in}} r_{\text{in}},$$

$$\frac{dA_2}{dt} = r_{21} \frac{A_1}{V_1} - (r_{12} + r_{\text{out}}) \frac{A_2}{V_2}.$$

We will now assume that V_1 and V_2 are constant. This imposes the conditions (see problem 7)

$$r_{\text{in}} + r_{12} - r_{21} = 0,$$

$$r_{21} - r_{12} - r_{\text{out}} = 0.$$

Consequently the preceding system of differential equations reduces to

$$\frac{dA_1}{dt} = -\frac{r_{21}}{V_1} A_1 + \frac{r_{12}}{V_2} A_2 + c_{in} r_{in},$$

$$\frac{dA_2}{dt} = \frac{r_{21}}{V_1} A_1 - \frac{r_{21}}{V_2} A_2.$$

This is a constant coefficient system for A_1 and A_2 and therefore can be solved using the techniques that we have developed in this chapter.

Example 11.8.2   Each of two tanks contains 20 L of a solution consisting of salt dissolved in water. A solution containing 4 g/L of salt flows into tank 1 at a rate of 3 L/min, and the solution in tank 2 flows out at the same rate. In addition, the solution flows into tank 1 from tank 2 at a rate of 1 L/min and into tank 2 from tank 1 at a rate of 4 L/min. Initially tank 1 contained 40 g of salt and tank 2 contained 20 g of salt. Find the amount of salt in each tank at time t.

Solution   The inflow and outflow rates from each tank are indicated in Figure 11.8.3. We notice that the total amount of solution flowing into tank 1 is 4 L/min, and the same volume of solution flows out of tank 1 per minute. Consequently, the volume of solution in tank 1 remains constant at 20 L. The same is true for tank 2. We let $A_1(t)$ and $A_2(t)$ denote the amounts of salt in tanks 1 and 2, respectively, and let c_{ij} denote the concentration of salt in the solution flowing into tank i from tank j. Now consider a short time interval Δt. The overall change in the amount of salt in tank 1 in this time interval is

$$\Delta A_1 \approx (12 + c_{12})\Delta t - 4c_{21}\Delta t,$$

that is,

$$\Delta A_1 \approx \left(12 + \frac{1}{20} A_2 - \frac{1}{5} A_1\right)\Delta t. \tag{11.8.15}$$

A similar analysis of the change in the amount of salt in tank 2 in the time interval Δt yields

$$\Delta A_2 \approx \left(4\frac{A_1}{20} - \frac{A_2}{20} - 3\frac{A_2}{20}\right)\Delta t,$$

that is

$$\Delta A_2 \approx \left(\frac{1}{5} A_1 - \frac{1}{5} A_2\right)\Delta t. \tag{11.8.16}$$

Dividing (11.8.15) and (11.8.16) by Δt and taking the limit as $\Delta t \to 0^+$ yields the system of differential equations

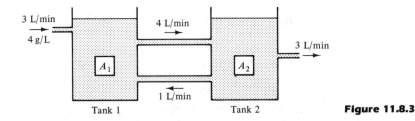

Tank 1   Tank 2   **Figure 11.8.3**

$$\frac{dA_1}{dt} = -\frac{1}{5} A_1 + \frac{1}{20} A_2 + 12,$$

$$\frac{dA_2}{dt} = \frac{1}{5} A_1 - \frac{1}{5} A_2.$$

This system can be written as the matrix equation

$$\mathbf{x}' = A\mathbf{x} + \mathbf{b},$$

where

$$\mathbf{x} = \begin{bmatrix} A_1 \\ A_2 \end{bmatrix}, \qquad A = \begin{bmatrix} -\frac{1}{5} & \frac{1}{20} \\ \frac{1}{5} & -\frac{1}{5} \end{bmatrix}, \qquad \mathbf{b} = \begin{bmatrix} 12 \\ 0 \end{bmatrix}.$$

To solve this nonhomogeneous system we first determine the eigenvalues and eigenvectors of A. We have

$$\det(A - \lambda I) = \begin{vmatrix} -\frac{1}{5} - \lambda & \frac{1}{20} \\ \frac{1}{5} & -\frac{1}{5} - \lambda \end{vmatrix} = (\lambda + \tfrac{1}{5})^2 - \tfrac{1}{100}.$$

Thus the eigenvalues of A are

$$\lambda = -\tfrac{1}{5} \pm \tfrac{1}{10},$$

that is,

$$\lambda_1 = -\tfrac{1}{10}, \qquad \lambda_2 = -\tfrac{3}{10}.$$

The corresponding eigenvectors are scalar multiples of

$$\mathbf{v}_1 = (1, 2), \qquad \mathbf{v}_2 = (1, -2),$$

respectively, so that two linearly independent solutions to $\mathbf{x}' = A\mathbf{x}$ are

$$\mathbf{x}_1 = e^{-t/10} \begin{bmatrix} 1 \\ 2 \end{bmatrix}, \qquad \mathbf{x}_2 = e^{-3t/10} \begin{bmatrix} 1 \\ -2 \end{bmatrix}.$$

Thus, the general solution to $\mathbf{x}' = A\mathbf{x}$ is

$$\mathbf{x}_c = c_1 e^{-t/10} \begin{bmatrix} 1 \\ 2 \end{bmatrix} + c_2 e^{-3t/10} \begin{bmatrix} 1 \\ -2 \end{bmatrix}.$$

We now need a particular solution to $\mathbf{x}' = A\mathbf{x} + \mathbf{b}$. According to the variation-of-parameters technique, a particular solution is

$$\mathbf{x}_p = X\mathbf{u},$$

where

$$X\mathbf{u}' = \mathbf{b}, \qquad\qquad (11.8.17)$$

and

$$X = \begin{bmatrix} e^{-t/10} & e^{-3t/10} \\ 2e^{-t/10} & -2e^{-3t/10} \end{bmatrix}.$$

The system (11.8.17) is

$$e^{-t/10}u_1' + e^{-3t/10}u_2' = 12,$$
$$e^{-t/10}u_1' - e^{-3t/10}u_2' = 0,$$

which has solution

$$u_1' = 6e^{t/10}, \qquad u_2' = 6e^{3t/10}.$$

Integrating, we obtain

$$u_1 = 60e^{t/10}, \qquad u_2 = 20e^{3t/10},$$

where we have set the integration constants to zero without loss of generality. Consequently,

$$\mathbf{x}_p = \begin{bmatrix} e^{-t/10} & e^{-3t/10} \\ 2e^{-t/10} & -2e^{-3t/10} \end{bmatrix} \begin{bmatrix} 60e^{t/10} \\ 20e^{3t/10} \end{bmatrix} = \begin{bmatrix} 80 \\ 80 \end{bmatrix}.$$

Hence the general solution to the system $\mathbf{x}' = A\mathbf{x} + \mathbf{b}$ is

$$\mathbf{x} = c_1 e^{-t/10} \begin{bmatrix} 1 \\ 2 \end{bmatrix} + c_2 e^{-3t/10} \begin{bmatrix} 1 \\ -2 \end{bmatrix} + \begin{bmatrix} 80 \\ 80 \end{bmatrix}.$$

We now impose the initial conditions in order to determine the appropriate values of c_1 and c_2. It is given that $A_1(0) = 40$, $A_2(0) = 20$. Thus

$$\mathbf{x}(0) = \begin{bmatrix} 40 \\ 20 \end{bmatrix}.$$

Imposing this condition requires

$$c_1 \begin{bmatrix} 1 \\ 2 \end{bmatrix} + c_2 \begin{bmatrix} 1 \\ -2 \end{bmatrix} + \begin{bmatrix} 80 \\ 80 \end{bmatrix} = \begin{bmatrix} 40 \\ 20 \end{bmatrix},$$

that is,

$$c_1 + c_2 = -40,$$
$$c_1 - c_2 = -30.$$

Thus $c_1 = -35$ and $c_2 = -5$, and the solution of the initial value problem is

$$\mathbf{x} = -35e^{-t/10} \begin{bmatrix} 1 \\ 2 \end{bmatrix} - 5e^{-3t/10} \begin{bmatrix} 1 \\ -2 \end{bmatrix} + \begin{bmatrix} 80 \\ 80 \end{bmatrix}.$$

Consequently the amounts of salt in tanks 1 and 2 are, respectively,

$$A_1 = 80 - 35e^{-t/10} - 5e^{-3t/10},$$
$$A_2 = 80 - 70e^{-t/10} + 10e^{-3t/10}.$$

EXERCISES 11.8

1. Derive the eigenvalues and eigenvectors given in Example 11.8.1.

2. Determine the motion of the coupled spring-mass system that has

$$k_1 = 3\text{N/m}, \qquad k_2 = \tfrac{1}{2}\text{N/m},$$

$$m_1 = \tfrac{1}{2} \text{ kg}, \qquad m_2 = \tfrac{1}{12} \text{ kg},$$

given that at $t = 0$ both masses are set in motion from their equilibrium positions with a velocity of 1 m/s.

3. Determine the general motion of the coupled spring-mass system that has

$$k_1 = 3 \text{N/m}, \qquad k_2 = 4 \text{N/m},$$

$$m_1 = 1 \text{ kg}, \qquad m_2 = \tfrac{4}{3} \text{ kg}.$$

4. Determine the general motion of the coupled spring-mass system that has

$$k_1 = 2k_2, \qquad m_1 = 2m_2.$$

(*Hint:* Let $\omega^2 = k_2/2m_2$.)

5. Consider the general coupled spring-mass system whose motion is governed by the system (11.8.5)–(11.8.8). Show that the coefficient matrix of the system has characteristic equation

$$\lambda^4 + \left[\frac{k_2}{m_2} + \frac{(k_1 + k_2)}{m_1} \right] \lambda^2 + \frac{k_1 k_2}{m_1 m_2} = 0$$

and that the corresponding eigenvalues are of the form

$$\lambda = \pm i\omega_1, \qquad \pm i\omega_2,$$

where ω_1, ω_2 are positive real numbers.

6. Two masses, m_1, m_2, rest on a horizontal frictionless plane. The masses are attached to fixed walls by springs whose spring constants are k_1 and k_3 (see Figure 11.8.4). The masses are also connected by a spring whose spring constant is k_2. Determine a first-order system of differential equations that governs the motion of the system.

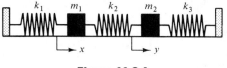

Figure 11.8.4

7. Show that the assumption that V_1 and V_2 are constant in the general mixing problem considered in the text imposes the conditions

$$r_{\text{in}} + r_{12} = r_{21},$$

$$r_{21} - r_{12} = r_{\text{out}}.$$

8. Solve the initial value problem arising in Example 11.8.2 using the technique derived in Section 11.2.

9. Solve the mixing problem depicted in Figure 11.8.5, given that at $t = 0$, the volume of solution in both tanks is 60 L and tank 1 contains 60 g of chemical, whereas tank 2 contains 200 g of chemical.

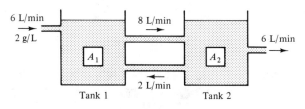

Tank 1   Tank 2

Figure 11.8.5

10. In the mixing problem shown in Figure 11.8.6 there is no inflow from or outflow to the outside. For this reason the system is said to be *closed.* If tank 1 contains 6 L of solution and tank 2 contains 12 L of solution, determine the amount of chemical in each tank at time t, given that initially tank 1 contains 5 g of chemical and tank 2 contains 25 g of chemical.

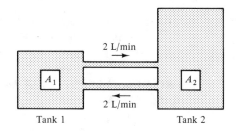

Tank 1   Tank 2

Figure 11.8.6

11. Consider the general closed system depicted in Figure 11.8.7.

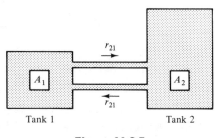

Tank 1   Tank 2

Figure 11.8.7

(a) Derive the system of differential equations that governs the behavior of A_1 and A_2.

(b) Define the constant β by $V_2 = \beta V_1$, when V_1 and V_2 denote the volume of solution in tank 1 and tank 2, respectively. Show that the eigenvalues of the coefficient matrix of the system derived in (a) are

$$\lambda_1 = 0, \qquad \lambda_2 = -\frac{(1 + \beta)}{\beta V_1} r_{21}.$$

(c) Determine A_1 and A_2, given that $A_1(0) = \alpha_1$, $A_2(0) = \alpha_2$, where α_1 and α_2 are positive constants.

(d) Show that

$$\lim_{t \to +\infty} \frac{A_1}{V_1} = \lim_{t \to +\infty} \frac{A_2}{V_2} = \frac{\alpha_1 + \alpha_2}{(1 + \beta)V_1}.$$

Is this result reasonable?

11.9   AN INTRODUCTION TO THE MATRIX EXPONENTIAL FUNCTION

In the final two sections we give a brief introduction to the matrix exponential function and indicate the role that this function plays in the analysis and solution of systems of linear differential equations.

> **Definition 11.9.1:** Let A be an $n \times n$ matrix of constants. We define the **matrix exponential function**, denoted by e^{At}, by
>
> $$e^{At} = I + At + \frac{1}{2!}(At)^2 + \frac{1}{3!}(At)^3 + \cdots + \frac{1}{k!}(At)^k + \cdots. \quad (11.9.1)$$

It can be shown that the infinite series appearing on the right-hand side of (11.9.1) converges for all $n \times n$ matrices A and all values of $t \in (-\infty, \infty)$, so that e^{At} is indeed well-defined.

PROPERTIES OF THE MATRIX EXPONENTIAL FUNCTION

1. If A and B are $n \times n$ matrices satisfying $AB = BA$, then

$$\boxed{e^{(A+B)t} = e^{At}e^{Bt}.}$$

2. For all $n \times n$ matrices A, e^{At} is nonsingular and

$$\boxed{(e^{At})^{-1} = e^{(-A)t} = e^{-At},}$$

that is,

$$e^{At}e^{-At} = I_n.$$

The proof of these results requires a precise definition of convergence of an infinite series of matrices. This would take us too far astray from the main focus of this text and hence the proofs are omitted. (See, for example, M. W. Hirsch and S.

Smale, *Differential Equations, Dynamical Systems, and Linear Algebra*, Academic Press, 1974.)

Before investigating the relationship between the matrix exponential function and systems of differential equations, we discuss some direct methods for finding e^{At}.

Example 11.9.1 Compute e^{At} if $A = \begin{bmatrix} 2 & 0 \\ 0 & -1 \end{bmatrix}$.

Solution In this case we see that

$$At = \begin{bmatrix} 2t & 0 \\ 0 & -t \end{bmatrix}, \quad (At)^2 = \begin{bmatrix} (2t)^2 & 0 \\ 0 & (-t)^2 \end{bmatrix}, \quad (At)^3 = \begin{bmatrix} (2t)^3 & 0 \\ 0 & (-t)^3 \end{bmatrix}, \cdots,$$

$$(At)^k = \begin{bmatrix} (2t)^k & 0 \\ 0 & (-t)^k \end{bmatrix}, \cdots$$

so that

$$e^{At} = \begin{bmatrix} 1 & 0 \\ 0 & 1 \end{bmatrix} + \begin{bmatrix} 2t & 0 \\ 0 & -t \end{bmatrix} + \frac{1}{2!}\begin{bmatrix} (2t)^2 & 0 \\ 0 & (-t)^2 \end{bmatrix} + \cdots + \frac{1}{k!}\begin{bmatrix} (2t)^k & 0 \\ 0 & (-t)^k \end{bmatrix} + \cdots$$

$$= \begin{bmatrix} \sum_{k=0}^{\infty} \frac{1}{k!}(2t)^k & 0 \\ 0 & \sum_{k=0}^{\infty} \frac{1}{k!}(-t)^k \end{bmatrix}$$

Hence,

$$e^{At} = \begin{bmatrix} e^{2t} & 0 \\ 0 & e^{-t} \end{bmatrix}.$$

More generally, it can be shown (see problem 1) that

if $A = \text{diag}(d_1, d_2, \ldots, d_n)$, then $e^{At} = \text{diag}(e^{d_1 t}, e^{d_2 t}, \ldots, e^{d_n t})$.

If A is not a diagonal matrix, then the computation of e^{At} is more involved. The next simplest case that can arise is when A is nondefective. In this case, as we have shown in Section 8.4, A is *similar* to a diagonal matrix, and we might suspect that this would lead to a simplification in the evaluation of e^{At}. We now show that this is indeed the case. Suppose that A has n linearly independent eigenvectors $\mathbf{v}_1, \mathbf{v}_2, \ldots, \mathbf{v}_n$, and define the $n \times n$ matrix S by

$$S = [\mathbf{v}_1, \mathbf{v}_2, \ldots, \mathbf{v}_n].$$

Then, from Theorem 8.4.2 (page 292),

$$S^{-1}AS = \text{diag}(\lambda_1, \lambda_2, \ldots, \lambda_n), \tag{11.9.2}$$

where $\lambda_1, \lambda_2, \ldots, \lambda_n$ are the eigenvalues of A corresponding to the eigenvectors $\mathbf{v}_1$,

$\mathbf{v}_2, \ldots, \mathbf{v}_n$ respectively. Multiplying (11.9.2) on the left by S and on the right by S^{-1} yields

$$A = SDS^{-1},$$

where

$$D = \text{diag}(\lambda_1, \lambda_2, \ldots, \lambda_n).$$

We now compute e^{At}. From Definition 11.9.1,

$$e^{At} = I + At + \frac{1}{2!}(At)^2 + \frac{1}{3!}(At)^3 + \cdots + \frac{1}{k!}(At)^k + \cdots$$

$$= SS^{-1} + (SDS^{-1})t + \frac{1}{2!}(SDS^{-1})^2t^2 + \cdots + \frac{1}{k!}(SDS^{-1})^kt^k + \cdots. \quad (11.9.3)$$

To proceed with this computation we need the following lemma.

Lemma 11.9.1:   If S is an $n \times n$ matrix and D is an $n \times n$ diagonal matrix, then

$$(SDS^{-1})^k = SD^kS^{-1}. \quad (11.9.4)$$

PROOF   The proof is by mathematical induction. The result is trivially true when $k = 1$. Now suppose that for some $m \geq 1$,

$$(SDS^{-1})^m = SD^mS^{-1}. \quad (11.9.5)$$

We must prove that this implies the validity of (11.9.4) when $k = m + 1$. This can be accomplished as follows.

$$(SDS^{-1})^{m+1} = (SDS^{-1})^m(SDS^{-1}),$$

that is, using (11.9.5),

$$(SDS^{-1})^{m+1} = (SD^mS^{-1})(SDS^{-1})$$
$$= SD^m(S^{-1}S)DS^{-1}$$
$$= SD^mIDS^{-1}$$
$$= SD^{m+1}S^{-1},$$

as required. It follows by induction that (11.9.4) is true for all positive integers k.   ∎

We now return to (11.9.3). Using the result of the preceding lemma we can write

$$e^{At} = S\left[I + Dt + \frac{1}{2!}(Dt)^2 + \cdots + \frac{1}{k!}(Dt)^k + \cdots\right]S^{-1},$$

that is,

$$e^{At} = Se^{Dt}S^{-1}.$$

Consequently we have proved the following theorem.

   Theorem 11.9.1:   Let A be a *nondefective* $n \times n$ matrix with linearly independent eigenvectors $\mathbf{v}_1, \mathbf{v}_2, \ldots, \mathbf{v}_n$ and corresponding eigenvalues $\lambda_1, \lambda_2, \ldots, \lambda_n$. Then

$$e^{At} = S e^{Dt} S^{-1},$$

where $S = [\mathbf{v}_1, \mathbf{v}_2, \ldots, \mathbf{v}_n]$ and $D = \text{diag}(\lambda_1, \lambda_2, \ldots, \lambda_n)$.

Example 11.9.2   Determine e^{At} if $A = \begin{bmatrix} 3 & 3 \\ 5 & 1 \end{bmatrix}$.

Solution: It is easily shown that A has eigenvalues $\lambda_1 = 6$, $\lambda_2 = -2$.   Hence A is nondefective.   A straightforward computation yields the following eigenvectors:

   $\lambda_1 = 6$ implies that $\mathbf{v}_1 = (1, 1)$; $\lambda_2 = -2$ implies that $\mathbf{v}_2 = (-3, 5)$.

It follows from Theorem 11.9.1 that if we set

$$S = \begin{bmatrix} 1 & -3 \\ 1 & 5 \end{bmatrix}, \qquad D = \text{diag}(6, -2),$$

then

$$e^{At} = S e^{Dt} S^{-1},$$

that is,

$$e^{At} = S \begin{bmatrix} e^{6t} & 0 \\ 0 & e^{-2t} \end{bmatrix} S^{-1}. \tag{11.9.6}$$

It is easily shown that

$$S^{-1} = \begin{bmatrix} \frac{5}{8} & \frac{3}{8} \\ -\frac{1}{8} & \frac{1}{8} \end{bmatrix},$$

so that, substituting into (11.9.6),

$$e^{At} = \begin{bmatrix} 1 & -3 \\ 1 & 5 \end{bmatrix} \begin{bmatrix} e^{6t} & 0 \\ 0 & e^{-2t} \end{bmatrix} \begin{bmatrix} \frac{5}{8} & \frac{3}{8} \\ -\frac{1}{8} & \frac{1}{8} \end{bmatrix} = \begin{bmatrix} 1 & -3 \\ 1 & 5 \end{bmatrix} \begin{bmatrix} \frac{5}{8}e^{6t} & \frac{3}{8}e^{6t} \\ -\frac{1}{8}e^{-2t} & \frac{1}{8}e^{-2t} \end{bmatrix}.$$

Consequently,

$$e^{At} = \begin{bmatrix} \frac{1}{8}(5e^{6t} + 3e^{-2t}) & \frac{3}{8}(e^{6t} - e^{-2t}) \\ \frac{5}{8}(e^{6t} - e^{-2t}) & \frac{1}{8}(3e^{6t} + 5e^{-2t}) \end{bmatrix}.$$

   The computation of e^{At} when A is a defective matrix is best accomplished by relating e^{At} to the solution of the corresponding system of differential equations $\mathbf{x}' = A\mathbf{x}$.   This is one of the goals of the next section.

EXERCISES 11.9

1. If $A = \text{diag}(d_1, d_2, \ldots, d_n)$, prove that $e^{At} = \text{diag}(e^{d_1 t}, e^{d_2 t}, \ldots, e^{d_n t})$.

2. If $A = \begin{bmatrix} -3 & 0 \\ 0 & 5 \end{bmatrix}$, determine e^{At} and e^{-At}.

3. Prove that for all values of the constant λ,
$$e^{\lambda I_n t} = e^{\lambda t} I_n.$$

4. Consider the matrix $A = \begin{bmatrix} a & b \\ 0 & a \end{bmatrix}$. We can write

$A = B + C$, where $B = \begin{bmatrix} a & 0 \\ 0 & a \end{bmatrix}$ and $C = \begin{bmatrix} 0 & b \\ 0 & 0 \end{bmatrix}$.

(a) Show that $BC = CB$.
(b) Show that $C^2 = 0_2$, and determine e^{Ct}.
(c) Use property 1 of the matrix exponential function to find e^{At}.

5. If $A = \begin{bmatrix} a & b \\ -b & a \end{bmatrix}$, use property 1 of the matrix exponential function and Definition 11.9.1 to show that $e^{At} = e^{at} \begin{bmatrix} \cos bt & \sin bt \\ -\sin bt & \cos bt \end{bmatrix}$.

In problems 6–12 show that A is nondefective and use Theorem 11.9.1 to find e^{At}.

6. $A = \begin{bmatrix} 1 & 2 \\ 0 & 3 \end{bmatrix}$.

7. $A = \begin{bmatrix} 3 & 1 \\ 1 & 3 \end{bmatrix}$.

8. $A = \begin{bmatrix} 0 & 2 \\ -2 & 0 \end{bmatrix}$.

9. $A = \begin{bmatrix} -1 & 3 \\ -3 & -1 \end{bmatrix}$.

10. $A = \begin{bmatrix} a & b \\ -b & a \end{bmatrix}$.

11. $A = \begin{bmatrix} 3 & -2 & -2 \\ 1 & 0 & -2 \\ 0 & 0 & 3 \end{bmatrix}$.

12. $A = \begin{bmatrix} 6 & -2 & -1 \\ 8 & -2 & -2 \\ 4 & -2 & 1 \end{bmatrix}$; you may assume that $p(\lambda) = -(\lambda - 2)^2(\lambda - 1)$.

An $n \times n$ matrix A that satisfies $A^k = 0_n$ for some k is called **nilpotent**. In problems 13–17 show that the given matrix is nilpotent and use Definition 11.9.1 to determine e^{At}.

13. $A = \begin{bmatrix} -3 & 9 \\ -1 & 3 \end{bmatrix}$.

14. $A = \begin{bmatrix} 1 & 1 \\ -1 & -1 \end{bmatrix}$.

15. $A = \begin{bmatrix} 0 & 0 & 0 \\ 1 & 0 & 0 \\ 0 & 1 & 0 \end{bmatrix}$.

16. $A = \begin{bmatrix} -1 & -6 & -5 \\ 0 & -2 & -1 \\ 1 & 2 & 3 \end{bmatrix}$.

17. $A = \begin{bmatrix} 0 & 1 & 0 & 0 \\ 0 & 0 & 1 & 0 \\ 0 & 0 & 0 & 1 \\ 0 & 0 & 0 & 0 \end{bmatrix}$.

18. Let A be the $n \times n$ matrix whose only nonzero elements are

$$a_{i+1\,i} = 1, \qquad i = 1, 2, \ldots, n-1.$$

Determine e^{At}. (See problem 15 for the case $n = 3$.)

11.10 THE MATRIX EXPONENTIAL FUNCTION AND SYSTEMS OF DIFFERENTIAL EQUATIONS

In this section we investigate the relationship between the matrix exponential function, e^{At}, and solutions to the corresponding system of differential equations

$$\mathbf{x}' = A\mathbf{x}.$$

We begin by defining the derivative of e^{At}. It can be shown that the infinite series (11.9.1) defining e^{At} can be differentiated term by term, and the resulting series converges for all $t \in (-\infty, \infty)$. Thus, differentiating (11.9.1), we have

$$\frac{d}{dt}\left(e^{At}\right) = A + A^2 t + \frac{1}{2!} A^3 t^2 + \cdots + \frac{1}{(k-1)!} A^k t^{k-1} + \cdots$$

that is,

$$\frac{d}{dt}\left(e^{At}\right) = A\left[I + At + \frac{1}{2!}(At)^2 + \frac{1}{3!}(At)^3 + \cdots + \frac{1}{k!}(At)^k + \cdots\right].$$

Hence

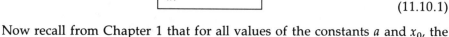

$$\frac{d}{dt}\left(e^{At}\right) = Ae^{At}.$$

$$(11.10.1)$$

Now recall from Chapter 1 that for all values of the constants a and x_0, the unique solution to the initial value problem

$$\frac{dx}{dt} - ax = 0, \qquad x(0) = x_0,$$

is

$$x = x_0 e^{at}.$$

Our next theorem shows that the same formula holds for the linear system

$$\mathbf{x}' = A\mathbf{x},$$

provided we replace e^{at} by e^{At}. This is a very elegant result that has far-reaching consequences in both the computation of e^{At} and the analysis of linear systems of differential equations.

Theorem 11.10.1: Let $\mathbf{x}_0$ be an arbitrary vector in $\mathbf{R}^n$. Then the unique solution to the initial value problem

$$\mathbf{x}' = A\mathbf{x}, \qquad \mathbf{x}(0) = \mathbf{x}_0,$$

is

$$\mathbf{x} = e^{At}\mathbf{x}_0.$$

PROOF   If $\mathbf{x} = e^{At}\mathbf{x}_0$, then $\mathbf{x}' = Ae^{At}\mathbf{x}_0$, that is,

$$\mathbf{x}' = A\mathbf{x}.$$

Further, setting $t = 0$,

$$\mathbf{x}(0) = e^{0A}\mathbf{x}_0$$
$$= I\mathbf{x}_0 = \mathbf{x}_0.$$

Consequently $\mathbf{x} = e^{At}\mathbf{x}_0$ is a solution of the given initial value problem. The uniqueness of the solution follows from Theorem 11.4.1.   ∎

We now investigate how the result of Theorem 11.10.1 combined with our previous techniques for solving $\mathbf{x}' = A\mathbf{x}$ can be used to determine e^{At}. To this end, let $\mathbf{x}_1, \mathbf{x}_2 \ldots, \mathbf{x}_n$ be linearly independent solutions to the system of differential equations

$$\mathbf{x}' = A\mathbf{x}, \tag{11.10.2}$$

where A is an $n \times n$ matrix of constants. We recall from Section 11.4 that the corresponding matrix function

$$X(t) = [\mathbf{x}_1, \mathbf{x}_2, \ldots, \mathbf{x}_n]$$

is called a *fundamental matrix* for (11.10.2) and that the general solution to (11.10.2) can be written as

$$\mathbf{x} = X(t)\mathbf{c},$$

where $\mathbf{c}$ is a column vector of arbitrary constants. If $X(t)$ is any fundamental matrix for (11.10.2) and B is any *nonsingular* matrix of constants, then the matrix function

$$Y(t) = X(t)B,$$

is also a fundamental matrix for (11.10.2), since its column vectors are linear combinations of the column vectors of X and hence are linearly independent solutions of (11.10.2). (The linear independence follows since $Y(t)$ is nonsingular.) We focus our attention on a particular fundamental matrix.

Definition **11.10.1:**   The unique fundamental matrix for $\mathbf{x}' = A\mathbf{x}$ that satisfies

$$X(0) = I_n$$

is called the **transition matrix for $\mathbf{x}' = A\mathbf{x}$ based at $t = 0$** and is denoted by $X_0(t)$.

In terms of the transition matrix, the solution to the initial value problem

$$\mathbf{x}' = A\mathbf{x}, \qquad \mathbf{x}(0) = \mathbf{x}_0,$$

is just

$$\mathbf{x} = X_0(t)\mathbf{x}_0,$$

so that the transition matrix does indeed describe the transition of the system from its state at time $t = 0$ to its state at time t. Further, if $X(t)$ is any fundamental matrix for $\mathbf{x}' = A\mathbf{x}$, then the transition matrix can be determined from (see problem 1)

$$X_0(t) = X(t)X^{-1}(0). \tag{11.10.3}$$

We now prove that $X_0(t)$ is in fact e^{At}. From (11.10.1) we have

$$\frac{d}{dt}(e^{At}) = Ae^{At},$$

so that the column vectors of e^{At} are solutions of $\mathbf{x}' = A\mathbf{x}$. Since e^{At} is also nonsingular [see property 2 on page 454], it follows that the column vectors of e^{At} are linearly independent on any interval. Further, setting $t = 0$ in the definition of e^{At} yields

$$e^{0A} = I_n. \tag{11.10.4}$$

Combining (11.10.4) with the uniqueness of the transition matrix leads to the required conclusion, namely,

$$e^{At} = X_0(t). \tag{11.10.5}$$

Thus, if A is an $n \times n$ matrix and $X(t)$ is *any* fundamental matrix for the corresponding differential system $\mathbf{x}' = A\mathbf{x}$, then (11.10.3) and (11.10.5) imply that

$$\boxed{e^{At} = X(t)X^{-1}(0).} \tag{11.10.6}$$

Consequently, to determine e^{At}, we can use the techniques from Section 11.4 and 11.5 to find a fundamental matrix for $\mathbf{x}' = A\mathbf{x}$, and then e^{At} can be obtained directly from (11.10.6).

Example 11.10.1 Determine e^{At} if $A = \begin{bmatrix} 6 & -8 \\ 2 & -2 \end{bmatrix}$.

Solution We first determine a fundamental matrix for

$$\mathbf{x}' = A\mathbf{x}.$$

This system has been solved in Example 11.6.2, where it was found that two linearly independent solutions are[1]

$$\mathbf{x}_1 = e^{2t}\begin{bmatrix} 2 \\ 1 \end{bmatrix}, \qquad \mathbf{x}_2 = e^{2t}\begin{bmatrix} 1 + 4t \\ 2t \end{bmatrix}.$$

Thus a fundamental matrix for $\mathbf{x}' = A\mathbf{x}$ is

---

[1] Notice that in this example A is defective.

$$X(t) = \begin{bmatrix} 2e^{2t} & e^{2t}(1+4t) \\ e^{2t} & 2te^{2t} \end{bmatrix},$$

which has

$$X^{-1}(0) = \begin{bmatrix} 0 & 1 \\ 1 & -2 \end{bmatrix}.$$

Consequently,

$$e^{At} = X(t)X^{-1}(0) = \begin{bmatrix} 2e^{2t} & e^{2t}(1+4t) \\ e^{2t} & 2te^{2t} \end{bmatrix} \begin{bmatrix} 0 & 1 \\ 1 & -2 \end{bmatrix}$$

that is,

$$e^{At} = \begin{bmatrix} e^{2t}(1+4t) & -8te^{2t} \\ 2te^{2t} & e^{2t}(1-4t) \end{bmatrix},$$

which can be written as

$$e^{At} = e^{2t} \begin{bmatrix} 1+4t & -8t \\ 2t & 1-4t \end{bmatrix}.$$

We now indicate how the matrix exponential function can be used directly to derive linearly independent solutions to $\mathbf{x}' = A\mathbf{x}$. Let $\mathbf{v}_1, \mathbf{v}_2, \ldots, \mathbf{v}_n$ be linearly independent vectors in $\mathbf{R}^n$ (or $\mathbf{C}^n$), and consider the corresponding vector functions

$$\mathbf{x}_1 = e^{At}\mathbf{v}_1, \, \mathbf{x}_2 = e^{At}\mathbf{v}_2, \, \ldots, \, \mathbf{x}_n = e^{At}\mathbf{v}_n.$$

Theorem 11.10.1 implies that each of these vector functions is a solution to $\mathbf{x}' = A\mathbf{x}$; further,

$$\det([\mathbf{x}_1, \mathbf{x}_2, \ldots, \mathbf{x}_n]) = \det(e^{At}[\mathbf{v}_1, \mathbf{v}_2, \ldots, \mathbf{v}_n])$$
$$= \det(e^{At})\det([\mathbf{v}_1, \mathbf{v}_2, \ldots, \mathbf{v}_n]),$$

which is nonzero since $\mathbf{v}_1, \mathbf{v}_2, \ldots, \mathbf{v}_n$ are linearly independent and e^{At} is nonsingular. Consequently

$$X(t) = [\mathbf{x}_1, \mathbf{x}_2, \ldots, \mathbf{x}_n]$$

is a fundamental matrix for $\mathbf{x}' = A\mathbf{x}$. Now, we can certainly write

$$A = \lambda I + (A - \lambda I),$$

and since the matrices $B = \lambda I$ and $C = A - \lambda I$ satisfy $BC = CB$, it follows from property 1 of the matrix exponential function that

$$e^{At}\mathbf{v} = e^{[\lambda It + (A-\lambda I)t]}\mathbf{v} = e^{\lambda It}e^{(A-\lambda I)t}\mathbf{v}.$$

Further,

$$e^{\lambda It} = \text{diag}(e^{\lambda t}, e^{\lambda t}, \ldots, e^{\lambda t}) = e^{\lambda t}I,$$

so that

$$e^{At}\mathbf{v} = e^{\lambda t}\left[\mathbf{v} + t(A - \lambda I)\mathbf{v} + \frac{t^2}{2!}(A - \lambda I)^2\mathbf{v} + \cdots\right]. \tag{11.10.7}$$

In general, the preceding series contains an infinite number of terms. However, if we can find vectors $\mathbf{v}$ such that

$$(A - \lambda I)^k\mathbf{v} = \mathbf{0}$$

for some k, then the series will terminate after a finite number of terms. For example, if $\mathbf{v}$ is an eigenvector of A [that is, $(A - \lambda I)\mathbf{v} = \mathbf{0}$], then the series in (11.10.7) has only one term and we obtain the result of Theorem 11.5.1, namely, that

$$\mathbf{x} = e^{\lambda t}\mathbf{v}$$

is a solution of $\mathbf{x}' = A\mathbf{x}$ whenever λ and $\mathbf{v}$ are an eigenvalue-eigenvector pair for A. Hence, if A is nondefective, (11.10.7) yields n linearly independent solutions to $\mathbf{x}' = A\mathbf{x}$ in the usual manner. Suppose, however, that A is defective. Then there is at least one eigenvalue of A that has an eigenspace whose dimension is less than the multiplicity of the eigenvalue. In this case Theorem 11.6.3 gives us the general form of the solutions. We now derive an equivalent technique for obtaining these solutions based on the matrix exponential function. We need the following theorem from linear algebra.

Theorem 11.10.2: Let A be an $n \times n$ matrix, and suppose that λ is an eigenvalue of multiplicity m. Then the system of equations

$$(A - \lambda I)^m\mathbf{v} = \mathbf{0}, \tag{11.10.8}$$

has m linearly independent solutions. Further, the nontrivial solutions to (11.10.8) corresponding to distinct eigenvalues are linearly independent.

PROOF   The proof of this theorem is best left for a second course in linear algebra. ∎

Suppose that λ is an eigenvalue of multiplicity m. Then according to Theorem 11.10.2, we can obtain m linearly independent vectors, $\mathbf{v}_1, \mathbf{v}_2, \ldots, \mathbf{v}_m$, satisfying (11.10.8). Further, using these vectors, the infinite series (11.10.7) terminates after a finite number of terms, specifically,[1]

$$e^{At}\mathbf{v}_i = e^{\lambda t}\left[\mathbf{v}_i + t(A - \lambda I)\mathbf{v}_i + \frac{t^2}{2!}(A - \lambda I)^2\mathbf{v}_i + \cdots + \frac{t^{m-1}}{(m-1)!}(A - \lambda I)^{m-1}\mathbf{v}_i\right],$$

$i = 1, 2, \ldots, m$. Proceeding in this manner for each eigenvalue, we can obtain n linearly independent solutions to $\mathbf{x}' = A\mathbf{x}$ which determines a fundamental matrix for $\mathbf{x}' = A\mathbf{x}$. We can then obtain e^{At} in the usual manner.

---

[1] Often the series will terminate after fewer than m terms.

Example 11.10.2 If $A = \begin{bmatrix} 6 & 8 & 1 \\ -1 & -3 & 3 \\ -1 & -1 & 1 \end{bmatrix}$, determine a fundamental matrix for

$\mathbf{x}' = A\mathbf{x}$ and use it to find e^{At}.

Solution The characteristic polynomial of A is

$$p(\lambda) = -(\lambda - 3)^2(\lambda + 2).$$

Hence A has eigenvalues $\lambda_1 = 3$ (multiplicity 2), and $\lambda_2 = -2$.

$\lambda_1 = 3$: In this case we must determine two linearly independent solutions to

$$(A - 3I)^2 \mathbf{v} = \mathbf{0}. \tag{11.10.9}$$

The coefficient matrix of this system is

$$(A - 3I)^2 = \begin{bmatrix} 3 & 8 & 1 \\ -1 & -6 & 3 \\ -1 & -1 & -2 \end{bmatrix}\begin{bmatrix} 3 & 8 & 1 \\ -1 & -6 & 3 \\ -1 & -1 & -2 \end{bmatrix} = \begin{bmatrix} 0 & -25 & 25 \\ 0 & 25 & -25 \\ 0 & 0 & 0 \end{bmatrix},$$

so that system (11.10.9) reduces to the single equation

$$v_2 - v_3 = 0,$$

which has two free variables. We set $v_1 = r$, and $v_3 = s$, in which case $v_2 = s$. Hence (11.10.9) has solution

$$\mathbf{v} = r(1, 0, 0) + s(0, 1, 1).$$

Consequently two linearly independent solutions to (11.10.9) are

$$\mathbf{v}_1 = (1, 0, 0), \qquad \mathbf{v}_2 = (0, 1, 1).$$

Thus, since $(A - 3I)^2 \mathbf{v}_1 = \mathbf{0}$, $e^{At}\mathbf{v}_1$ reduces to

$$e^{At}\mathbf{v}_1 = e^{3t}[\mathbf{v}_1 + t(A - 3I)\mathbf{v}_1]$$

$$= e^{3t}\left\{\begin{bmatrix} 1 \\ 0 \\ 0 \end{bmatrix} + t\begin{bmatrix} 3 & 8 & 1 \\ -1 & -6 & 3 \\ -1 & -1 & -2 \end{bmatrix}\begin{bmatrix} 1 \\ 0 \\ 0 \end{bmatrix}\right\}$$

$$= e^{3t}\left\{\begin{bmatrix} 1 \\ 0 \\ 0 \end{bmatrix} + t\begin{bmatrix} 3 \\ -1 \\ -1 \end{bmatrix}\right\}.$$

Hence, one solution to the given system is

$$\mathbf{x}_1 = e^{At}\mathbf{v}_1 = e^{3t}\begin{bmatrix} 1 + 3t \\ -t \\ -t \end{bmatrix}.$$

Similarly,

$$e^{At}\mathbf{v}_2 = e^{3t}[\mathbf{v}_2 + t(A - 3I)\mathbf{v}_2]$$

$$= e^{3t}\left\{\begin{bmatrix}0\\1\\1\end{bmatrix} + t\begin{bmatrix}3 & 8 & 1\\-1 & -6 & 3\\-1 & -1 & -2\end{bmatrix}\begin{bmatrix}0\\1\\1\end{bmatrix}\right\}$$

$$= e^{3t}\left\{\begin{bmatrix}0\\1\\1\end{bmatrix} + t\begin{bmatrix}9\\-3\\-3\end{bmatrix}\right\}.$$

Thus a second linearly independent solution to the given system is

$$\mathbf{x}_2 = e^{At}\mathbf{v}_2 = e^{3t}\begin{bmatrix}9t\\1-3t\\1-3t\end{bmatrix}.$$

$\lambda_2 = -2$: It is easily shown that the eigenvectors corresponding to $\lambda = -2$ are all scalar multiples of

$$\mathbf{v}_3 = (-1, 1, 0).$$

Hence, a third linearly independent solution to the given system is

$$\mathbf{x}_3 = e^{At}\mathbf{v}_3 = e^{-2t}\begin{bmatrix}-1\\1\\0\end{bmatrix}.$$

Consequently a fundamental matrix for $\mathbf{x}' = A\mathbf{x}$ is

$$X(t) = [e^{At}\mathbf{v}_1, e^{At}\mathbf{v}_2, e^{At}\mathbf{v}_3] = \begin{bmatrix}e^{3t}(1+3t) & 9te^{3t} & -e^{-2t}\\-te^{3t} & e^{3t}(1-3t) & e^{-2t}\\-te^{3t} & e^{3t}(1-3t) & 0\end{bmatrix}.$$

We now use $X(t)$ to find e^{At}. Setting $t = 0$ yields

$$X(0) = \begin{bmatrix}1 & 0 & -1\\0 & 1 & 1\\0 & 1 & 0\end{bmatrix},$$

and using Gaussian elimination we find that

$$X^{-1}(0) = \begin{bmatrix}1 & 1 & -1\\0 & 0 & 1\\0 & 1 & -1\end{bmatrix}.$$

Consequently, from (11.10.6),

$$e^{At} = X(t)X^{-1}(0) = \begin{bmatrix}e^{3t}(1+3t) & 9te^{3t} & -e^{-2t}\\-te^{3t} & e^{3t}(1-3t) & e^{-2t}\\-te^{3t} & e^{3t}(1-3t) & 0\end{bmatrix}\begin{bmatrix}1 & 1 & -1\\0 & 0 & 1\\0 & 1 & -1\end{bmatrix},$$

that is,

$$e^{At} = \begin{bmatrix} e^{3t}(1+3t) & e^{3t}(1+3t) - e^{-2t} & e^{3t}(6t-1) + e^{-2t} \\ -te^{3t} & e^{-2t} - te^{3t} & e^{3t}(1-2t) - e^{-2t} \\ -te^{3t} & -te^{3t} & e^{3t}(1-2t) \end{bmatrix}.$$

As the examples in this section indicate, the computation of e^{At} can be quite tedious. The main use of the matrix exponential is theoretical.

We end this chapter by showing that the results we have obtained are a generalization of those from Chapter 1. Consider the initial value problem

$$\frac{dx}{dt} - ax = b(t), \qquad x(0) = x_0,$$

where a is a constant. Using the technique developed in Section 1.5 for solving linear differential equations, it is easily shown that the solution of the initial value problem is

$$x = e^{at}\left[\int_0^t e^{-as} b(s) \, ds + x_0 \right]. \tag{11.10.10}$$

Now consider the corresponding initial value problem for linear differential systems, namely,

$$\mathbf{x}' = A(t)\mathbf{x} + \mathbf{b}(t), \qquad \mathbf{x}(0) = \mathbf{x}_0. \tag{11.10.11}$$

According to the variation-of-parameters method, a particular solution to the system is

$$\mathbf{x}_p = X(t) \int_0^t X^{-1}(s)\mathbf{b}(s) \, ds,$$

where $X(t)$ is any fundamental matrix for $\mathbf{x}' = A(t)\mathbf{x}$. Further, the general solution to $\mathbf{x}' = A(t)\mathbf{x}$ has the form

$$\mathbf{x}_c = X(t)\mathbf{c}.$$

If we use the matrix exponential function e^{At} as the fundamental matrix then, combining $\mathbf{x}_c$ and $\mathbf{x}_p$, the general solution to the system (11.10.11) assumes the form

$$\mathbf{x} = e^{At}\mathbf{c} + e^{At} \int_0^t e^{-As}\mathbf{b}(s) \, ds. \tag{11.10.12}$$

Imposing the initial condition $\mathbf{x}(0) = \mathbf{x}_0$ yields

$$\mathbf{c} = \mathbf{x}_0.$$

Substituting into (11.10.12) and simplifying, we finally obtain

$$\mathbf{x} = e^{At}\left[\mathbf{x}_0 + \int_0^t e^{-As}\mathbf{b}(s) \, ds \right],$$

which is the generalization of (11.10.10) to systems of differential equations.

EXERCISES 11.10

1. If $X(t)$ is any fundamental matrix for $\mathbf{x}' = A\mathbf{x}$, show that the transition matrix based at $t = 0$ is given by

$$X_0 = X(t)X^{-1}(0).$$

In problems 2–4 use the techniques from Sections 11.5 and 11.6 to determine a fundamental matrix for $\mathbf{x}' = A\mathbf{x}$, and hence find e^{At}.

2. $A = \begin{bmatrix} 1 & 2 \\ 0 & -1 \end{bmatrix}$.

3. $A = \begin{bmatrix} 2 & 1 \\ 0 & 2 \end{bmatrix}$.

4. $A = \begin{bmatrix} 3 & 0 & 0 \\ 0 & 3 & -1 \\ 0 & 1 & 1 \end{bmatrix}$.

In problems 5–11 find n linearly independent solutions to $\mathbf{x}' = A\mathbf{x}$ of the form $e^{At}\mathbf{v}$, and hence find e^{At}.

5. $A = \begin{bmatrix} 3 & -1 \\ 4 & -1 \end{bmatrix}$.

6. $A = \begin{bmatrix} -3 & -2 \\ 2 & 1 \end{bmatrix}$.

7. $A = \begin{bmatrix} 2 & 0 & 0 \\ 0 & 1 & -8 \\ 0 & 2 & -7 \end{bmatrix}$.

In problems 8–10, solve $\mathbf{x}' = A\mathbf{x}$ by determining n linearly independent solutions of the form $\mathbf{x} = e^{At}\mathbf{v}$.

8. $A = \begin{bmatrix} -8 & 6 & -3 \\ -12 & 10 & -3 \\ -12 & 12 & -2 \end{bmatrix}$; you may assume that

$$p(\lambda) = -(\lambda + 2)^2(\lambda - 4).$$

9. $A = \begin{bmatrix} 0 & 1 & 3 \\ 2 & 3 & -2 \\ 1 & 1 & 2 \end{bmatrix}$; you may assume that

$$p(\lambda) = -(\lambda + 1)(\lambda - 3)^2.$$

10. $A = \begin{bmatrix} 1 & 0 & 0 & 0 \\ 0 & 6 & -7 & 3 \\ 0 & 0 & 3 & -1 \\ 0 & -4 & 9 & -3 \end{bmatrix}$; you may assume

that $p(\lambda) = (\lambda - 1)(\lambda - 2)^3$.

11. The matrix $A = \begin{bmatrix} 0 & -1 & 0 & 0 \\ 1 & 0 & 0 & 0 \\ 1 & 0 & 0 & -1 \\ 0 & 1 & 1 & 0 \end{bmatrix}$ has characteristic polynomial $p(\lambda) = (\lambda^2 + 1)^2$. Determine two *complex-valued* solutions to $\mathbf{x}' = A\mathbf{x}$ of the form $\mathbf{x} = e^{At}\mathbf{v}$, and hence find four linearly independent *real-valued* solutions to the differential system.

12

The Laplace Transform and Some Elementary Applications

12.1 THE DEFINITION OF THE LAPLACE TRANSFORM

In this chapter we introduce another technique for solving linear, constant coefficient ordinary differential equations. Actually, the technique has a much broader usage than this, for example, in the solution of linear systems of differential equations (see Section 12.4), partial differential equations and also integral equations (see Section 12.9). The reader's immediate reaction is probably to question the need for introducing a new method for solving constant coefficient equations, since our results from Chapter 9 can be applied to any such equation. To answer this question, consider the differential equation

$$y'' + ay' + by = F,$$

where a and b are constants. We have seen how to solve this equation when F is a continuous function on some interval I. However, in many problems that arise in engineering, physics, and applied mathematics, F represents the external force that is acting on the system under investigation, and often this force acts intermittently or even instantaneously.[1]   Although our techniques from Chapter 9 can be extended to cover these cases, the computations involved are tedious.   In contrast, the approach introduced here can handle these types of problems quite easily.

The technique that we develop in this chapter is based on a particular linear transformation called the Laplace transform.   Suppose that $K(s, t)$ is a function that is continuous for all s and t.   Then the transformation $L:C^0[a, b] \rightarrow C^0[0, \infty)$ defined by

$$L[f] = \int_a^b K(s, t)\, f(t)\, dt,$$

is a linear transformation, that is, it satisfies

$$L[f + g] = L[f] + L[g], \qquad L[cf] = cL[f],$$

for all $f, g \in C^0[a, b]$ and all real numbers c.   The Laplace transform arises when we extend this transformation to the interval $[0, \infty)$ and make the special choice

$$K(s, t) = e^{-st}.$$

Definition 12.1.1:   Let f be a function defined on the interval $[0, \infty)$.   The **Laplace transform** of f is the function $F(s)$ defined by

$$F(s) = \int_0^\infty e^{-st} f(t)\, dt, \tag{12.1.1}$$

provided the improper integral converges.   We usually denote the Laplace transform of f by $L[f]$.

Recall that the improper integral appearing in (12.1.1) is defined by

$$\int_0^\infty e^{-st} f(t)\, dt = \lim_{N \to \infty} \int_0^N e^{-st} f(t)\, dt$$

and that this improper integral converges if and only if the limit on the right-hand side exists and is finite.   It follows that *not* all functions defined on $[0, \infty)$ have a Laplace transform.   In the next section we will address some of the theoretical aspects associated with determining the types of functions for which (12.1.1) converges.   In the remainder of this section we focus our attention on gaining familiarity with Definition 12.1.1 and derive some basic Laplace transforms.

Example 12.1.1   Determine the Laplace transform of the following functions:
   (a) $f(t) = 1$.
   (b) $f(t) = t$.

---

[1] For example, the switch in an *RLC* circuit (see Section 10.4) may be turned on and off several times, or the mass in a spring-mass system (see Section 10.1) may be dealt an instantaneous blow at $t = t_0$.

(c) $f(t) = e^{at}$, a constant.
(d) $f(t) = \cos bt$, b constant.

Solution

(a) From the preceding definition we have

$$L[1] = \int_0^\infty e^{-st}\,dt = \lim_{N\to\infty}\left[-\frac{1}{s}e^{-st}\right]_0^N = \lim_{N\to\infty}\left[\frac{1}{s} - \frac{1}{s}e^{-sN}\right] = \frac{1}{s}, \qquad s > 0.$$

Notice that the restriction $s > 0$ is required in order for the improper integral to converge.

(b) In this case we use integration by parts to obtain

$$L[t] = \int_0^\infty e^{-st}t\,dt = \lim_{N\to\infty}\left[-\frac{te^{-st}}{s}\right]_0^N + \int_0^\infty \frac{1}{s}e^{-st}\,dt.$$

But.

$$\lim_{N\to\infty} Ne^{-sN} = 0, \qquad s > 0,$$

so that

$$L[t] = \int_0^\infty \frac{1}{s}e^{-st}\,dt = \lim_{N\to\infty}\left[-\frac{e^{-st}}{s^2}\right]_0^N = \frac{1}{s^2}, \qquad s > 0.$$

It is left as an exercise to show that, more generally, for all positive integers n,

$$\boxed{L[t^n] = \frac{n!}{s^{n+1}}, \qquad s > 0.} \tag{12.1.2}$$

(c) In this case we have

$$L[e^{at}] = \int_0^\infty e^{-st}e^{at}\,dt = \int_0^\infty e^{(a-s)t}\,dt = \lim_{N\to\infty}\left[\frac{1}{a-s}e^{(a-s)t}\right]_0^N = \frac{1}{s-a},$$

provided that $s > a$. Thus

$$\boxed{L[e^{at}] = \frac{1}{s-a}, \qquad s > a.} \tag{12.1.3}$$

(d) From the definition of the Laplace transform:

$$L[\cos bt] = \int_0^\infty e^{-st}\cos bt\,dt.$$

Using the standard integral

$$\int e^{at}\cos bt \ dt = \frac{e^{at}}{a^2+b^2}(a \cos bt + b \sin bt) + c,$$

it follows that

$$L[\cos bt] = \lim_{N\to\infty}\left[\frac{e^{-st}}{s^2+b^2}(b \sin bt - s \cos bt)\right]_0^N = \frac{s}{s^2+b^2},$$

provided $s > 0$. Thus

$$L[\cos bt] = \frac{s}{s^2+b^2}, \qquad s > 0. \tag{12.1.4}$$

Similarly, it can be shown that

$$L[\sin bt] = \frac{b}{s^2+b^2}, \qquad s > 0. \tag{12.1.5}$$

As illustrated by the preceding examples, the range of values that s can assume must often be restricted to ensure the convergence of the improper integral (12.1.1).

LINEARITY OF THE LAPLACE TRANSFORM

Suppose that the Laplace transforms of both f and g exist for $s > \alpha$, where α is a constant. Then, using properties of convergent improper integrals, it follows that for $s > \alpha$,

$$L[f+g] = \int_0^\infty e^{-st}[f(t)+g(t)] \ dt = \int_0^\infty e^{-st}f(t) \ dt + \int_0^\infty e^{-st}g(t) \ dt$$

$$= L[f] + L[g].$$

Further, if c is any real number, then

$$L[cf] = \int_0^\infty e^{-st}cf(t) \ dt = c\int_0^\infty e^{-st}f(t) \ dt = cL[f].$$

Consequently,

1. $L[f+g] = L[f] + L[g],$
2. $L[cf] = cL[f],$

so that the Laplace transform satisfies the basic properties of a linear transformation. This linearity of L enables us to determine the Laplace transform of complicated functions from a knowledge of the Laplace transform of some basic functions, and this will be used continually throughout the chapter.

Example 12.1.2   Determine the Laplace transform of

$$f(t) = 4e^{3t} + 2 \sin 5t - 7t^3.$$

Solution   Since the Laplace transform is linear, it follows that

$$L[4e^{3t} + 2 \sin 5t - 7t^3] = 4L[e^{3t}] + 2L[\sin 5t] - 7L[t^3].$$

Thus, using the results of the previous example we have

$$L[4e^{3t} + 2 \sin 5t - 7t^3] = \frac{4}{s-3} + \frac{10}{s^2 + 25} - \frac{42}{s^4}, \qquad s > 3.$$

PIECEWISE CONTINUOUS FUNCTIONS

The functions that we considered in the preceding examples were continuous on $[0, \infty)$. As we will see in the later sections, the real power of the Laplace transform comes from the fact that piecewise continuous functions can be transformed. Before illustrating this point we recall the definition of a piecewise continuous function.

Definition 12.1.2:   A function f is called **piecewise continuous** on the interval $[a, b]$ if we can divide $[a, b]$ into a finite number of subintervals in such a manner that

1. f is continuous on each subinterval, and
2. f approaches a finite limit as the endpoints of each interval are approached from within.

If f is piecewise continuous on every interval of the form $[0, b]$, where b is a constant, then we say that f is piecewise continuous on $[0, \infty)$.

Example 12.1.3   The function f defined by

$$f(t) = \begin{cases} t^2 + 1, & 0 \le t \le 1, \\ 2 - t, & 1 < t \le 2, \\ 1, & 2 < t \le 3, \end{cases}$$

is piecewise continuous on $[0, 3]$, whereas

$$f(t) = \begin{cases} \dfrac{1}{1-t}, & 0 \le t < 1, \\ t, & 1 \le t \le 3, \end{cases}$$

is not piecewise continuous on $[0, 3]$. The graphs of these functions are shown in Figure 12.1.1.

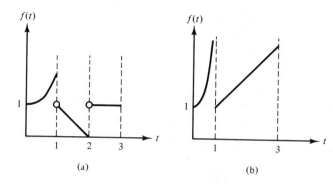

Figure 12.1.1 (a) An example of a piecewise continuous function. (b) An example of a function that is not piecewise continuous on $[0,3]$.

(a)

(b)

Example 12.1.4 Determine the Laplace transform of the piecewise continuous function

$$f(t) = \begin{cases} t, & 0 \le t < 1, \\ -1, & t \ge 1. \end{cases}$$

Solution The function is sketched in Figure 12.1.2. To determine the Laplace transform of f, we use Definition 12.1.1.

$$L[f] = \int_0^\infty e^{-st} f(t)\, dt = \int_0^1 e^{-st} f(t)\, dt + \int_1^\infty e^{-st} f(t)\, dt$$

$$= \int_0^1 e^{-st} t\, dt - \int_1^\infty e^{-st}\, dt = \left[-\frac{1}{s} t e^{-st} - \frac{1}{s^2} e^{-st} \right]_0^1 + \lim_{N \to \infty} \left[\frac{1}{s} e^{-st} \right]_1^N$$

$$= -\frac{1}{s} e^{-s} - \frac{1}{s^2} e^{-s} + \frac{1}{s^2} - \frac{1}{s} e^{-s},$$

provided $s > 0$. Thus,

$$L[f] = \frac{1}{s^2}[1 - e^{-s}(2s + 1)], \qquad s > 0.$$

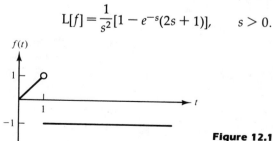

Figure 12.1.2 The piecewise continuous function in Example 12.1.4.

EXERCISES 12.1

In problems 1–12 use (12.1.1) to determine L[f].

1. $f(t) = e^{2t}$.

2. $f(t) = t - 1$.

3. $f(t) = \sin bt$, b constant.

4. $f(t) = te^t$.

5. $f(t) = \cosh bt$, b constant.

6. $f(t) = \sinh bt$, b constant.

7. $f(t) = 2t$.

8. $f(t) = 3e^{2t}$.

9. $f(t) = \begin{cases} 1, & 0 \le t < 2, \\ -1, & t \ge 2. \end{cases}$

10. $f(t) = \begin{cases} t^2, & 0 \le t \le 1, \\ 1, & t > 1. \end{cases}$

11. $f(t) = e^t \sin t$.

12. $f(t) = e^{2t} \cos 3t$.

In problems 13–22 use the linearity of L and the formulas derived in this section to determine L[f].

13. $f(t) = 2t - e^{3t}$.

14. $f(t) = 2 \sin 3t + 4t^3$.

15. $f(t) = \cosh bt$, b constant.

16. $f(t) = \sinh bt$, b constant.

17. $f(t) = 3t^2 - 5 \cos 2t + \sin 3t$.

18. $f(t) = 7e^{-2t} + 1$.

19. $f(t) = 2e^{-3t} + 4e^t - 5 \sin t$.

20. $f(t) = 4 \cos \left(t - \dfrac{\pi}{4} \right)$.

21. $f(t) = 4 \cos^2 bt$, b constant.

22. $f(t) = 2 \sin^2 4t - 3$.

In problems 23–30 sketch the given function and determine whether it is piecewise continuous on $[0, \infty)$.

23. $f(t) = \begin{cases} 3, & 0 \le t < 1, \\ 0, & 1 \le t < 3, \\ -1, & t \ge 3. \end{cases}$

24. $f(t) = \begin{cases} 1, & 0 \le t \le 1, \\ 1 - t, & 1 < t \le 2, \\ 1, & t > 2. \end{cases}$

25. $f(t) = \begin{cases} 1, & 0 \le t \le 1, \\ \dfrac{1}{t-1}, & t > 1. \end{cases}$

26. $f(t) = \begin{cases} t, & 0 \le t \le 1, \\ \dfrac{1}{t^2}, & t > 1. \end{cases}$

27. $f(t) = n$, $n \le t < n + 1$, $n = 0, 1, 2, \ldots$.

28. $f(t) = t$, $0 \le t < 1$, $f(t + 1) = f(t)$.

29. $f(t) = \dfrac{1}{t-2}$.

30. $f(t) = \dfrac{2}{t+1}$.

In problems 31–34 sketch the given function and determine its Laplace transform.

31. $f(t) = \begin{cases} t, & 0 \le t < 1, \\ 0, & t \ge 1. \end{cases}$

32. $f(t) = \begin{cases} 1, & 0 \le t \le 2, \\ -1, & t > 2. \end{cases}$

33. $f(t) = \begin{cases} 0, & 0 \le t \le 1, \\ t, & 1 < t \le 2, \\ 0, & t > 2. \end{cases}$

34. $f(t) = \begin{cases} t, & 0 \le t < 1, \\ 1, & 1 \le t < 3, \\ e^{t-3}, & t \ge 3. \end{cases}$

35. Recall that according to Euler's formula,

$$e^{ibt} = \cos bt + i \sin bt.$$

Since the Laplace transformation is linear, it follows that

$$L[\cos bt] = \text{Re}\{L[e^{ibt}]\}, \qquad L[\sin bt] = \text{Im}\{L[e^{ibt}]\}.$$

Find L[e^{ibt}] and hence derive (12.1.4) and (12.1.5).

36. Use the technique introduced in the previous problem to determine L[$e^{at} \cos bt$] and L[$e^{at} \sin bt$], where a and b are arbitrary constants.

37. Use mathematical induction to prove that for any positive integer n

$$L[t^n] = \frac{n!}{s^{n+1}}.$$

38. (a) By making the change of variables $t = x^2/s$, $s > 0$, in the integral that defines the Laplace transform, show that

$$L[t^{-1/2}] = \frac{2}{s^{1/2}} \int_0^\infty e^{-x^2} \, dx.$$

(b) Use your result in (a) to show that

$$\{L[t^{-1/2}]\}^2 = \frac{4}{s} \int_0^\infty \int_0^\infty e^{-(x^2+y^2)} dx \, dy.$$

(c) By changing to polar coordinates evaluate the double integral in (b) and hence show that

$$L[t^{-1/2}] = \sqrt{\frac{\pi}{s}}, \qquad s > 0.$$

12.2   THE EXISTENCE OF THE LAPLACE TRANSFORM AND THE INVERSE TRANSFORM

In the previous section we derived the Laplace transform of several elementary functions. In this section we address some of the more theoretical aspects of the Laplace transform. The first question that we wish to answer is the following:

What types of functions have a Laplace transform?

We will not be able to answer this question completely, since it requires a deeper mathematical background than we assume of the reader. However we can identify a very large class of functions that are Laplace transformable. By definition, the Laplace transform of a function f is

$$L[f] = \int_0^\infty e^{-st} f(t) \, dt \qquad\qquad (12.2.1)$$

provided the integral converges. If f is piecewise continuous on an interval $[a, b]$, then it is a standard result from calculus that f is also integrable over $[a, b]$. Thus if we restrict attention to functions that are piecewise continuous on $[0, \infty)$, it follows that the integral

$$\int_0^b e^{-st} f(t) \, dt \qquad\qquad (12.2.2)$$

exists for all positive (and finite) b. However, it does not follow that the Laplace transform of f exists, since the improper integral in (12.2.1) may still diverge. In order to guarantee convergence of the integral we must ensure that the integrand in (12.2.1) approaches zero rapidly enough as $t \to \infty$. As we show next, this will be the case provided that, in addition to being piecewise continuous, f also satisfies the following definition.

> *Definition 12.2.1:*   A function f is said to be of **exponential order** if there exist constants M and α such that
>
> $$|f(t)| \le Me^{\alpha t},$$
>
> for all $t > 0$.

Example 12.2.1   The function $f(t) = 10e^{7t}\cos 5t$ is of exponential order, since

$$|f(t)| = 10e^{7t}|\cos 5t| \le 10e^{7t}.$$

Now let $E(0, \infty)$ denote the set of all functions that are both piecewise continuous on $[0, \infty)$ and of exponential order. If we add two functions that are in $E(0, \infty)$, the result is a new function that is also in $E(0, \infty)$. Similarly, if we multiply a function in $E(0, \infty)$ by a constant, the result is once more a function in $E(0, \infty)$. It follows from Theorem 6.3.1 (page 194) that $E(0, \infty)$ forms a subspace of the vector space of all functions defined on $[0, \infty)$. We will show below that the functions in the vector space $E(0, \infty)$ have a Laplace transform. Before doing so, we need to state a basic theorem about the convergence of improper integrals.

Lemma 12.2.1 (The Comparison Test for Improper Integrals):   Suppose that

$$0 \le G(t) \le H(t) \text{ for } 0 \le t < \infty. \quad \text{If } \int_0^\infty H(t)\, dt \text{ converges, then so does } \int_0^\infty G(t)\, dt.$$

PROOF   See any textbook on advanced calculus.   ∎

We also recall that if $\int_0^\infty |G(t)\, dt|$ converges, then so does $\int_0^\infty G(t)\, dt$. We can now prove a key existence theorem for the Laplace transform.

Theorem 12.2.1   If f is in $E(0, \infty)$, then there exists a constant α such that

$$L[f] = \int_0^\infty e^{-st}f(t)\, dt$$

exists for all $s > \alpha$.

PROOF   Since f is piecewise continuous on $[0, \infty)$, $e^{-st}f(t)$ is integrable over any finite interval. Further, since f is in $E(0, \infty)$, there exist constants M and α such that

$$|f(t)| \le Me^{\alpha t},$$

for all $t > 0$. We now use the comparison test for integrals to establish that the improper integral defining the Laplace transform converges. Let

$$G(t) = |e^{-st}f(t)|.$$

Then

$$G(t) = e^{-st}|f(t)| \le Me^{(\alpha - s)t}.$$

But, for $s > \alpha$,

$$\int_0^\infty Me^{(\alpha-s)t} \, dt = \lim_{N\to\infty} \int_0^N Me^{(\alpha-s)t} \, dt = \frac{M}{s-\alpha}.$$

Thus, applying the comparison test for improper integrals with $G(t)$ as before and $H(t) = Me^{(\alpha-s)t}$, it follows that

$$\int_0^\infty |e^{-st}f(t)| \, dt$$

converges for $s > \alpha$ and hence so also does

$$\int_0^\infty e^{-st}f(t) \, dt.$$

Thus we have shown that $L[f]$ exists for $s > \alpha$, as required. ∎

REMARK   The preceding theorem gives only sufficient conditions that guarantee the existence of the Laplace transform. There are functions that are not in $E(0, \infty)$ but do have a Laplace transform. For example $f(t) = t^{-1/2}$ is certainly not in $E(0, \infty)$, but $L[t^{-1/2}] = \sqrt{\pi/s}$ (see problem 38 in the previous section).

THE INVERSE LAPLACE TRANSFORM

Let V denote the subspace of $E(0, \infty)$ consisting of all *continuous* functions of exponential order. We have seen in the previous section that the Laplace transform satisfies

$$L[f + g] = L[f] + L[g], \qquad L[cf] = cL[f].$$

Consequently, L defines a linear transformation of V onto Rng(L). Further, it can be shown that L is also one-to-one; therefore, from the results of Section 7.3, the inverse transformation L^{-1} exists and is defined as follows.

Definition 12.2.2:   The linear transformation L^{-1}: Rng(L) → V defined by

$$L^{-1}[F](t) = f(t) \quad \text{if and only if} \quad L[f](s) = F(s), \qquad (12.2.3)$$

is called the *inverse Laplace transform.*

REMARK   We emphasize the fact that L^{-1} is a *linear transformation*, so that

$$L^{-1}[F + G] = L^{-1}[G] + L^{-1}[H]$$

and

$$L^{-1}[cF] = cL^{-1}[F]$$

for all F, G in Rng[L] and all real numbers c.

In Section 12.1 we derived the transforms

$$L[t^n] = \frac{n!}{s^{n+1}}, \qquad L[e^{at}] = \frac{1}{s-a}, \qquad L[\cos bt] = \frac{s}{s^2+b^2}, \qquad L[\sin bt] = \frac{b}{s^2+b^2},$$

from which we directly obtain the inverse transforms

$$L^{-1}\left[\frac{1}{s^{n+1}}\right] = \frac{1}{n!}t^n, \qquad\qquad L^{-1}\left[\frac{1}{s-a}\right] = e^{at},$$

$$L^{-1}\left[\frac{s}{s^2+b^2}\right] = \cos bt, \qquad\qquad L^{-1}\left[\frac{b}{s^2+b^2}\right] = \sin bt.$$

Example 12.2.2   Find $L^{-1}[F](t)$ in each case.

(a) $F(s) = \dfrac{2}{s^2}$.

(b) $F(s) = \dfrac{3s}{s^2+4}$.

(c) $F(s) = \dfrac{3s+2}{(s-1)(s-2)}$.

Solution

(a) $L^{-1}\left[\dfrac{2}{s^2}\right] = 2L^{-1}\left[\dfrac{1}{s^2}\right] = 2t.$

(b) $L^{-1}\left[\dfrac{3s}{s^2+4}\right] = 3L^{-1}\left[\dfrac{s}{s^2+4}\right] = 3\cos 2t.$

(c) In this case it is not obvious at first sight what the appropriate inverse transform is.   However, decomposing $F(s)$ into partial fractions yields[1]

$$F(s) = \frac{3s+2}{(s-1)(s-2)} = \frac{8}{s-2} - \frac{5}{s-1}.$$

Consequently, using the linearity of L^{-1},

$$L^{-1}\left[\frac{3s+2}{(s-1)(s-2)}\right] = L^{-1}\left[\frac{8}{s-2}\right] - L^{-1}\left[\frac{5}{s-1}\right]$$

$$= 8L^{-1}\left[\frac{1}{s-2}\right] - 5L^{-1}\left[\frac{1}{s-1}\right]$$

$$= 8e^{2t} - 5e^t.$$

---

[1] It is very important in this chapter to be able to perform partial fraction decompositions.   A review of this technique is given in Appendix 2.

If we relax the assumption that V contains only *continuous* functions of exponential order, then it is no longer true that L is one-to-one; in fact, for a given $F(s)$, there will be (infinitely) many piecewise continuous functions f with the property that

$$L[f] = F(s).$$

Thus we lose the uniqueness of $L^{-1}[F]$. However, it can be shown (see, for example, R. V. Churchill, *Modern Operational Mathematics in Engineering*, McGraw-Hill, 1944) that if two functions have the same Laplace transform, then they can differ only in their values at points of discontinuity. This does not affect the solution to our problems, and therefore we will use (12.2.3) to determine the inverse Laplace transform even if f is piecewise continuous.

Example 12.2.3 In the previous section we showed that the Laplace transform of the piecewise continuous function

$$f(t) = \begin{cases} t, & 0 \le t < 1, \\ -1, & t \ge 1, \end{cases}$$

is

$$L[f] = \frac{1}{s^2}[1 - e^{-s}(2s + 1)], \qquad s > 0.$$

Consequently,

$$L^{-1}\left\{\frac{1}{s^2}[1 - e^{-s}(2s + 1)]\right\} = f(t).$$

It is possible to give a general formula for determining the inverse Laplace transform of $F(s)$ in terms of a contour integral in the complex plane. However, this is beyond the scope of the present treatment of the Laplace transform. In practice, as in the previous examples, we determine inverse Laplace transforms by recognizing $F(s)$ as being the Laplace transform of an appropriate function $f(t)$. In order for this approach to work, we need to memorize a few basic transforms and then be able to use these transforms to determine the inverse Laplace transform of more complicated functions. This is similar to the way that we learn how to integrate. The transform pairs that you will need for the remainder of the text are listed in Table 12.2.1 on page 480. Several of the transforms given in this table are derived in the following sections. More generally, very large tables of Laplace transforms have been compiled for use in applications.

EXERCISES 12.2

In problems 1–5 show that the given function is of exponential order.

1. $f(t) = 3 \cos 2t$.

2. $f(t) = e^{2t}$.

3. $f(t) = 5e^{3t}\sin 4t$.

4. $f(t) = te^{-2t}$.

5. $f(t) = t^n e^{at}$, a and n are positive integers.

6. Show that if f and g are in $E(0, \infty)$, then so are $f + g$ and cf for any scalar c.

In problems 7–21 determine the inverse Laplace transform of the given function.

7. $F(s) = \dfrac{2}{s}$.

8. $F(s) = \dfrac{3}{s - 2}$.

9. $F(s) = \dfrac{5}{s + 3}$.

10. $F(s) = \dfrac{1}{s^2 + 4}$.

11. $F(s) = \dfrac{2s}{s^2 + 9}$.

12. $F(s) = \dfrac{4}{s^3}$.

13. $F(s) = \dfrac{s + 6}{s^2 + 1}$.

14. $F(s) = \dfrac{2s + 1}{s^2 + 16}$.

15. $F(s) = \dfrac{2}{s} - \dfrac{3}{s + 1}$.

16. $F(s) = \dfrac{4}{s^2} - \dfrac{s + 2}{s^2 + 9}$.

17. $F(s) = \dfrac{1}{s(s + 1)}$.

18. $F(s) = \dfrac{s - 2}{(s + 1)(s^2 + 4)}$.

19. $F(s) = \dfrac{2s + 3}{(s - 2)(s^2 + 1)}$.

20. $F(s) = \dfrac{s + 4}{(s - 1)(s + 2)(s - 3)}$.

21. $F(s) = \dfrac{2s + 3}{(s^2 + 4)(s^2 + 1)}$.

TABLE 12.2.1

Function $f(t)$	Laplace transform $F(s)$
$f(t) = t^n$, n a nonnegative integer	$F(s) = \dfrac{n!}{s^{n+1}}$, $s > 0$.
$f(t) = e^{at}$, a a constant	$F(s) = \dfrac{1}{s - a}$, $s > a$.
$f(t) = \sin bt$, b constant	$F(s) = \dfrac{b}{s^2 + b^2}$, $s > 0$.
$f(t) = \cos bt$, b constant	$F(s) = \dfrac{s}{s^2 + b^2}$, $s > 0$.
$f(t) = t^{-1/2}$	$F(s) = \sqrt{\dfrac{\pi}{s}}$, $s > 0$.
$f(t) = u_a(t)$ (see Section 12.7)	$F(s) = \dfrac{e^{-as}}{s}$.
$f(t) = \delta(t - a)$ (see Section 12.8)	$F(s) = e^{-as}$.

Transform of Derivatives (see Section 12.4)

f'	$L[f'] = s\,L[f] - f(0)$.
f''	$L[f''] = s^2\,L[f] - sf(0) - f'(0)$.

Shifting Theorems (see Sections 12.5 and 12.7)

$e^{at} f(t)$	$F(s - a)$.
$u_a(t) f(t - a)$	$e^{-as} F(s)$.

**12.3   PERIODIC FUNCTIONS
        AND THE LAPLACE TRANSFORM**

Many of the functions that arise in engineering applications are periodic on some interval. Due to the symmetry associated with a periodic function, we might suspect that the evaluation of the Laplace transform of such a function can be reduced to an integration over one period of the function. We show next that this is indeed the case, but first we recall the definition of a periodic function.

> *Definition 12.3.1:* A function f defined on the interval $[0, \infty)$ is said to be **periodic with period T** if it satisfies
>
> $$f(t + T) = f(t),$$
>
> for all $t \geq 0$.

The most familiar examples of such functions are the trigonometric functions sine and cosine, which have period 2π.

Example 12.3.1   The function f defined by

$$f(t) = \begin{cases} 2, & 0 \leq t \leq 1, \\ \\ 1, & 1 < t < 2, \end{cases} \qquad f(t + 2) = f(t),$$

is periodic on $[0, \infty)$ with period 2 (see Figure 12.3.1).

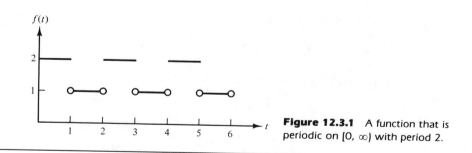

Figure 12.3.1   A function that is periodic on $[0, \infty)$ with period 2.

The following theorem can be used to simplify the evaluation of the Laplace transform of a periodic function.

> *Theorem 12.3.1:*   Let f be in $E(0, \infty)$. If f is periodic on $[0, \infty)$ with period T, then

$$L[f] = \frac{1}{1 - e^{-sT}} \int_0^T e^{-st}f(t)\, dt. \qquad (12.3.1)$$

PROOF   By definition of the Laplace transform we have

$$L[f] = \int_0^\infty e^{-st}f(t)\,dt = \int_0^T e^{-st}f(t)\,dt + \int_T^{2T} e^{-st}f(t)\,dt + \cdots + \int_{nT}^{(n+1)T} e^{-st}f(t)\,dt + \cdots .$$

Now consider the general integral

$$I = \int_{nT}^{(n+1)T} e^{-st}f(t)\,dt.$$

If we let $x = t - nT$, then $dx = dt$. Further, $t = nT$ corresponds to $x = 0$, whereas $t = (n+1)T$ corresponds to $x = T$. Thus I can be written in the equivalent form

$$I = \int_0^T e^{-s(x+nT)}f(x + nT)\,dx = e^{-nsT}\int_0^T e^{-sx}f(x)\,dx,$$

where we have used the fact that f is periodic of period T to replace $f(x + nT)$ by $f(x)$. All the integrals that arise in the expression for $L[f]$ are of the above form for an appropriate value of n. It follows, therefore, that we can write

$$L[f] = (1 + e^{-sT} + e^{-2sT} + \cdots + e^{-nsT} + \cdots)\int_0^T e^{-sx}f(x)\,dx. \qquad (12.3.2)$$

However, the term multiplying the integral in (12.3.2) is just a geometric series[1] with common ratio e^{-sT}. Consequently, the sum of the geometric series is $1/(1 - e^{-sT})$, so that

$$L[f] = \frac{1}{1 - e^{-sT}}\int_0^T e^{-st}f(t)\,dt,$$

where we have replaced the dummy variable x by t in (12.3.2) without loss of generality. ∎

Example 12.3.2   Determine the Laplace transform of

$$f(t) = \begin{cases} \sin t, & 0 \le t < \pi, \\ 0, & \pi \le t < 2\pi, \end{cases} \qquad f(t + 2\pi) = f(t).$$

Solution   The given function is periodic on $[0, \infty)$ with period 2π (see Figure 12.3.2). We can therefore use Theorem 12.3.1 to determine $L[f]$. We have:

$$L[f] = \frac{1}{1 - e^{-2\pi s}}\int_0^{2\pi} e^{-st}f(t)\,dt = \frac{1}{1 - e^{-2\pi s}}\int_0^\pi e^{-st}\sin t\,dt.$$

---

[1] Recall that an infinite series of the form $a + ar + ar^2 + ar^3 + \cdots$ is called a geometric series with common ratio r. If $|r| < 1$, then the sum of such a series is $a/(1 - r)$.

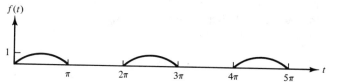

Figure 12.3.2 The periodic function defined in Example 12.3.2.

Using the standard integral:

$$\int e^{at}\sin bt \, dt = \frac{1}{(a^2 + b^2)} e^{at}(a \sin bt - b \cos bt),$$

it follows that

$$L[f] = \frac{1}{1 - e^{-2\pi s}} \left\{ -\frac{1}{s^2 + 1}[e^{-st}(s \sin t + \cos t)]_0^\pi \right\}$$

$$= \frac{1}{1 - e^{-2\pi s}} \left[\frac{e^{-s\pi} + 1}{s^2 + 1} \right].$$

Substituting for

$$1 - e^{-2\pi s} = (1 - e^{-\pi s})(1 + e^{-\pi s}),$$

yields

$$L[f] = \frac{1}{(s^2 + 1)(1 - e^{-\pi s})}.$$

EXERCISES 12.3

In problems 1–8 sketch the given function and determine its Laplace transform.

1. $f(t) = t, \ 0 \le t < 1, \ f(t + 1) = f(t).$

2. $f(t) = t^2, \ 0 \le t < 2, \ f(t + 2) = f(t).$

3. $f(t) = \sin t, \ 0 \le t < \pi, \ f(t + \pi) = f(t).$

4. $f(t) = \cos t, \ 0 \le t < \pi, \ f(t + \pi) = f(t).$

5. $f(t) = e^t, \ 0 \le t < 1, \ f(t + 1) = f(t).$

6. $f(t) = \begin{cases} 1, & 0 \le t < 1, \\ -1, & 1 \le t < 2, \end{cases} \quad f(t + 2) = f(t).$

7. $f(t) = \begin{cases} \dfrac{2}{\pi}t, & 0 \le t < \dfrac{\pi}{2}, \\ \sin t, & \dfrac{\pi}{2} \le t < \pi, \end{cases} \quad f(t + \pi) = f(t).$

8. $f(t) = |\cos t|, \ 0 \le t < \pi, \ f(t + \pi) = f(t).$

9. Determine the Laplace transform of the triangular wave function (see Figure 12.3.3)

$$f(t) = \begin{cases} \dfrac{t}{a}, & 0 \le t < a, \\ \dfrac{2a - t}{a}, & a \le t < 2a, \end{cases} \quad f(t + 2a) = f(t),$$

where a is a positive constant.

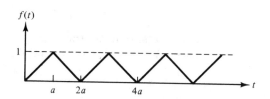

Figure 12.3.3 A triangular wave function.

10. Use Theorem 12.3.1 together with the fact $f(t) = \sin at$ is periodic on the interval $[0, 2\pi/a]$ to determine $L[f]$.

11. Repeat the previous question for the function $f(t) = \cos at$.

12.4   THE TRANSFORM OF DERIVATIVES AND THE SOLUTION OF INITIAL VALUE PROBLEMS

The reason that we have introduced the Laplace transform is that it provides an alternative technique for solving differential equations. To see how this technique arises we must first consider how the derivative of a function transforms.

Theorem 12.4.1:   Suppose that f is of exponential order on $[0, \infty)$ and that f' exists and is piecewise continuous on $[0, \infty)$. Then $L[f']$ exists and is given by

$$L[f'] = sL[f] - f(0).$$

PROOF   For simplicity we consider the case when f' is continuous on $[0, \infty)$. The extension to the case of piecewise continuity is straightforward. Since f is differentiable and of exponential order on $[0, \infty)$, it follows that it belongs to $E(0, \infty)$, and hence its Laplace transform exists. By definition of the Laplace transform we have

$$L[f'] = \int_0^\infty e^{-st} f'(t)\, dt = \left[e^{-st} f(t) \right]_0^\infty + s \int_0^\infty e^{-st} f(t)\, dt;$$

that is, since f is of exponential order on $[0, \infty)$,

$$L[f'] = sL[f] - f(0). \qquad \blacksquare$$

Example 12.4.1   Solve the initial value problem

$$\frac{dy}{dt} = t, \qquad y(0) = 1.$$

Solution   This problem can be solved by a direct integration. However, we will use the Laplace transform. Taking the Laplace transform of both sides of the given differential equation and using the result of the previous theorem, we obtain

$$sY(s) - y(0) = \frac{1}{s^2},$$

where $Y(s)$ denotes the Laplace transform of $y(t)$. This is an *algebraic* equation for $Y(s)$. Substituting in the initial condition and solving algebraically for $Y(s)$ yields:

$$Y(s) = \frac{1}{s^3} + \frac{1}{s}.$$

To determine the solution of the original problem, we now take the inverse Laplace transform of both sides of this equation.   The result is

$$y(t) = L^{-1}\left[\frac{1}{s^3} + \frac{1}{s}\right];$$

that is, since $L^{-1}\left[\dfrac{1}{s^{n+1}}\right] = \dfrac{1}{n!}\, t^n$,

$$y(t) = \frac{1}{2} t^2 + 1.$$

The preceding example illustrates the basic steps in solving an initial value problem using the Laplace transform.   We proceed as follows:

1. Take the Laplace transform of the given differential equation and substitute in the given initial conditions.
2. Solve the resulting equation algebraically for $Y(s)$.
3. Take the inverse Laplace transform of the resulting equation to determine the solution, $y(t)$, of the given initial value problem.

These steps are illustrated in Figure 12.4.1.

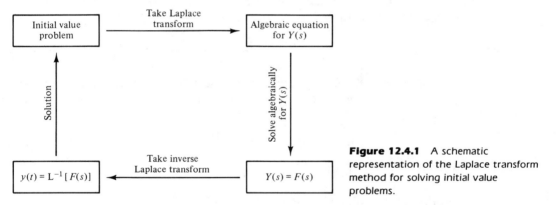

Figure 12.4.1   A schematic representation of the Laplace transform method for solving initial value problems.

To extend the technique introduced in the previous example to higher-order differential equations, we need to determine how the higher-order derivatives transform.   This can be derived quite easily from Theorem 12.4.1.   We illustrate for the case of second-order derivatives and leave the derivation of the general case as an exercise.

Assuming that f'' is sufficiently smooth, it follows from Theorem 12.4.1 that

$$L[f''] = sL[f'] - f'(0).$$

Thus applying Theorem 12.4.1 once more yields

$$\boxed{L[f''] = s^2 L[f] - sf(0) - f'(0).}$$

More generally it is straightforward to show that

$$L[f^{(n)}] = s^n L[f] - s^{n-1}f(0) - s^{n-2}f'(0) - \cdots - sf^{(n-2)}(0) - f^{(n-1)}(0).$$

Example 12.4.2   Use the Laplace transform to solve the initial value problem

$$y'' - y' - 6y = 0, \qquad y(0) = 1, \qquad y'(0) = 2.$$

Solution   We take the Laplace transform of both sides of the differential equation to obtain

$$[s^2 Y(s) - sy(0) - y'(0)] - [sY(s) - y(0)] - 6Y(s) = 0.$$

Substituting in for the given initial values and rearranging terms yields

$$(s^2 - s - 6)\,Y(s) = s + 1,$$

that is,

$$Y(s) = \frac{s+1}{(s-3)(s+2)}.$$

Thus we have solved for the Laplace transform of $y(t)$. In order to find y itself we must take the inverse Laplace transform. We first decompose the right-hand side into partial fractions to obtain

$$Y(s) = \frac{4}{5(s-3)} + \frac{1}{5(s+2)}.$$

We recognize the terms on the right-hand side as being the Laplace transform of appropriate exponential functions. Taking the inverse Laplace transform yields

$$y(t) = \frac{4}{5}e^{-3t} + \frac{1}{5}e^{-2t},$$

and the initial value problem is solved.

Example 12.4.3   Solve the initial value problem

$$y'' + y = e^{2t}, \qquad y(0) = 0, \qquad y'(0) = 1.$$

Solution   Once more we take the Laplace transform of both sides of the differential equation to obtain

$$[s^2 Y(s) - sy(0) - y'(0)] + Y(s) = \frac{1}{s-2},$$

that is, upon substituting for the given initial conditions and simplifying,

$$Y(s) = \frac{s-1}{(s-2)(s^2+1)}.$$

We must now determine the partial fraction decomposition of the right-hand side. We have

$$\frac{s-1}{(s-2)(s^2+1)} = \frac{A}{s-2} + \frac{Bs+C}{s^2+1},$$

for some constants A, B, C. Multiplying both sides of this equality by $(s-2)(s^2+1)$ yields

$$s - 1 = A(s^2+1) + (Bs+C)(s-2).$$

Equating coefficients of s^0, s^1, s^2 yields the three conditions

$$A - 2C = -1, \qquad -2B + C = 1, \qquad A + B = 0.$$

Solving for A, B, and C we obtain

$$A = \tfrac{1}{5}, \qquad B = -\tfrac{1}{5}, \qquad C = \tfrac{3}{5}.$$

Thus,

$$Y(s) = \frac{1}{5(s-2)} - \frac{s-3}{5(s^2+1)},$$

that is,

$$Y(s) = \frac{1}{5(s-2)} - \frac{s}{5(s^2+1)} + \frac{3}{5(s^2+1)}.$$

Taking the inverse Laplace transform of both sides of this equation yields

$$y(t) = \underbrace{\tfrac{1}{5}e^{2t}}_{\text{Particular solution}} \quad \underbrace{-\tfrac{1}{5}\cos t + \tfrac{3}{5}\sin t.}_{\text{Complementary function}}$$

The structure of the solution obtained in the preceding example has a familiar form. The first term represents a particular solution to the differential equation, which could have been obtained by the method of undetermined coefficients, whereas the last two terms represent the complementary function. There are no arbitrary constants in the solution, since we have solved an initial value problem. Notice the difference between solving an initial value problem using the Laplace transform and our previous techniques. In the Laplace transform technique, we impose the initial values at the beginning of the problem and just solve the initial value problem. In our previous techniques we first found the general solution of the differential equation and *then* imposed the initial values to solve the initial value problem. It should be noted, however, that the Laplace transform can also be used to determine the general solution of a differential equation (see problem 27).

It should be apparent from the results of the previous two sections that the main difficulties in applying the Laplace transform technique to the solution of initial value problems is in steps 1 and 3. In order for the technique to be useful, we need to know the transform and inverse transform of a large number of functions.

So far we have determined only the Laplace transform of some very basic functions, namely, t^n, e^{at}, $\sin bt$, $\cos bt$. We will show in the remaining sections how these basic transforms can be used to determine the Laplace transform of almost any function that is likely to arise in the applications. You are once more strongly advised to memorize the basic transforms.

EXERCISES 12.4

In problems 1–26 use the Laplace transform to solve the given initial value problem.

1. $y' + y = 8e^{3t}$, $y(0) = 2$.

2. $y' + 3y = 2e^{-t}$, $y(0) = 3$.

3. $y' + 2y = 4t$, $y(0) = 1$.

4. $y' - y = 6 \cos t$, $y(0) = 2$.

5. $y' - y = 5 \sin 2t$, $y(0) = -1$.

6. $y' + y = 5 \sin t$, $y(0) = 1$.

7. $y'' + y' - 2y = 0$, $y(0) = 1$, $y'(0) = 4$.

8. $y'' + 4y = 0$, $y(0) = 5$, $y'(0) = 1$.

9. $y'' - 3y' + 2y = 4$, $y(0) = 0$, $y'(0) = 1$.

10. $y'' - y' - 12y = 36$, $y(0) = 0$, $y'(0) = 12$.

11. $y'' + y' - 2y = 10e^{-t}$, $y(0) = 0$, $y'(0) = 1$.

12. $y'' - 3y' + 2y = 4e^{3t}$, $y(0) = 0$, $y'(0) = 0$.

13. $y'' - 2y' = 30e^{-3t}$, $y(0) = 1$, $y'(0) = 0$.

14. $y'' - y = 12e^{2t}$, $y(0) = 1$, $y'(0) = 1$.

15. $y'' + 4y = 10e^{-t}$, $y(0) = 4$, $y'(0) = 0$.

16. $y'' - y' - 6y = 6(2 - e^t)$, $y(0) = 5$, $y'(0) = -3$.

17. $y'' - y = 6 \cos t$, $y(0) = 0$, $y'(0) = 4$.

18. $y'' - 9y = 13 \sin 2t$, $y(0) = 3$, $y'(0) = 1$.

19. $y'' - y = 8 \sin t - 6 \cos t$, $y(0) = 2$, $y'(0) = -1$.

20. $y'' - y' - 2y = 10 \cos t$, $y(0) = 0$, $y'(0) = -1$.

21. $y'' + 5y' + 4y = 20 \sin 2t$, $y(0) = -1$, $y'(0) = 2$.

22. $y'' + 5y' + 4y = 20 \sin 2t$, $y(0) = 1$, $y'(0) = -2$.

23. $y'' - 3y' + 2y = 3 \cos t + \sin t$, $y(0) = 1$, $y'(0) = 1$.

24. $y'' + 4y = 9 \sin t$, $y(0) = 1$, $y'(0) = -1$.

25. $y'' + y = 6 \cos 2t$, $y(0) = 0$, $y'(0) = 2$.

26. $y'' + 9y = 7 \sin 4t + 14 \cos 4t$, $y(0) = 1$, $y'(0) = 2$.

27. Use the Laplace transform to find the general solution of $y'' - y = 0$.

28. Use the Laplace transform to solve the initial value problem

$$y'' + \omega^2 y = A \sin \omega_0 t + B \cos \omega_0 t,$$

$$y(0) = y_0, \qquad y'(0) = y_1,$$

where A, B, ω, and ω_0 are positive constants and $\omega \neq \omega_0$.

29. The current, $i(t)$, in the RL circuit shown in Figure 12.4.2 is governed by the differential equation

$$\frac{di}{dt} + \frac{R}{L} i = \frac{1}{L} E(t),$$

where R and L are constants.

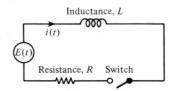

Figure 12.4.2 A simple RL circuit.

(a) Use the Laplace transform to determine $i(t)$ if $E(t) = E_0$, a constant, and there is no current flowing initially.

(b) Repeat (a) in the case when $E(t) = E_0 \sin \omega t$, where ω is a constant.

The Laplace transform can also be used to solve initial value problems for systems of linear differential equations. The remaining problems deal with this.

30. Consider the initial value problem

$$x_1' = a_{11}x_1 + a_{12}x_2 + b_1(t),$$

$$x_2' = a_{21}x_1 + a_{22}x_2 + b_2(t),$$

$$x_1(0) = \alpha_1, \qquad x_2(0) = \alpha_2,$$

where the a_{ij} are constants. Show that the Laplace transforms of $x_1(t)$, $x_2(t)$ must satisfy the linear algebraic system

$$(s - a_{11}) X_1(s) - a_{12} X_2(s) = \alpha_1 + B_1(s),$$

$$-a_{21} X_1(s) + (s - a_{22}) X_2(s) = \alpha_2 + B_2(s).$$

This system can be solved quite easily (for example, by Cramer's rule) to determine $X_1(s)$ and

$X_2(s)$, and then $x_1(t)$, $x_2(t)$ can be obtained by taking the inverse Laplace transform.

In problems 31 and 32 solve the given initial value problem.

31. $x_1' = -4x_1 - 2x_2, \quad x_2' = x_1 - x_2, \quad x_1(0) = 0,$
$x_2(0) = 1.$

32. $x_1' = -3x_1 + 4x_2, x_2' = -x_1 + 2x_2,$
$x_1(0) = 2, x_2(0) = 1.$

12.5   THE FIRST SHIFTING THEOREM

For the Laplace transform to be a useful tool for solving differential equations, we need to be able to find $L[f]$ for a large class of functions. Obviously, trying to apply the definition of the Laplace transform to determine $L[f]$ for every function we encounter is not an appropriate way to proceed. Instead we derive some general theorems that will enable us to derive the Laplace transform of most elementary functions from a knowledge of the transforms of the functions given in Table 12.2.1. For the remainder of the section, we will be assuming that all of the functions that we encounter do have a Laplace transform.

Theorem 12.5.1 (First Shifting Theorem):   If $L[f] = F(s)$, then

$$\boxed{L[e^{at}f(t)] = F(s - a).}$$

Conversely, if $L^{-1}[F(s)] = f(t)$, then

$$\boxed{L^{-1}[F(s - a)] = e^{at}f(t).}$$

PROOF   From the definition of the Laplace transform we have

$$L[e^{at}f(t)] = \int_0^\infty e^{-st}e^{at}f(t)\, dt = \int_0^\infty e^{-(s-a)t}f(t)\, dt. \qquad (12.5.1)$$

But, by assumption,

$$F(s) = \int_0^\infty e^{-st}f(t)\, dt,$$

so that

$$F(s - a) = \int_0^\infty e^{-(s-a)t}f(t)\, dt. \qquad (12.5.2)$$

Comparing (12.5.1) with (12.5.2) we obtain

$$L[e^{at}f(t)] = F(s - a), \qquad (12.5.3)$$

as required.   Taking the inverse Laplace transform of both sides of (12.5.3) yields

$$L^{-1}[F(s-a)] = e^{at}f(t).$$ ∎

We illustrate the use of the preceding theorem with several examples (see also Figure 12.5.1).

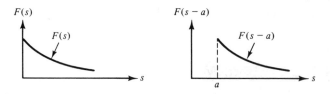

$F(s)$

$F(s)$

s

$F(s-a)$

$F(s-a)$

a

s

Figure 12.5.1   An illustration of the first shifting theorem. Multiplying $f(t)$ by e^{at} has the effect of shifting $F(s)$ by a units to the right in "s-space."

Example 12.5.1   Find $L[f]$ for each $f(t)$.
   (a) $f(t) = e^{5t}\cos 4t$.
   (b) $f(t) = e^{at}\sin bt$, a, b constants.
   (c) $f(t) = e^{at}t^n$, where a is a constant and n is a positive integer.

Solution

   (a) From Table 12.2.1 we have

$$L[\cos 4t] = \frac{s}{s^2 + 16},$$

so that, applying the first shifting theorem with $a = 5$, it follows that

$$L[e^{5t}\cos 4t] = \frac{s - 5}{(s-5)^2 + 16}.$$

   (b) Since $L[\sin bt] = \dfrac{b}{s^2 + b^2}$, it follows that

$$L[e^{at}\sin bt] = \frac{b}{(s-a)^2 + b^2}. \tag{12.5.4}$$

Similarly, it follows from Table 12.2.1 and the first shifting theorem that

$$L[e^{at}\cos bt] = \frac{s - a}{(s-a)^2 + b^2}. \tag{12.5.5}$$

   (c) From Table 12.2.1 we have

$$L[t^n] = \frac{n!}{s^{n+1}},$$

so that

$$L[e^{at}t^n] = \frac{n!}{(s-a)^{n+1}}. \tag{12.5.6}$$

The previous examples deal with the direct use of the first shifting theorem to obtain the Laplace transform of a function. Of equal importance is its use in determining inverse transforms. Once more we illustrate this with several examples.

Example 12.5.2 Determine $L^{-1}[F(s)]$ for the given F.

(a) $F(s) = \dfrac{3}{(s-2)^2 + 9}$.

(b) $F(s) = \dfrac{4}{(s-4)^3}$.

(c) $F(s) = \dfrac{s+4}{s^2 + 6s + 13}$.

(d) $F(s) = \dfrac{s-2}{s^2 + 2s - 1}$.

Solution

(a) It follows from Table 12.2.1 that

$$L^{-1}\left[\frac{3}{s^2 + 9}\right] = \sin 3t,$$

so that, by the first shifting theorem,

$$L^{-1}\left[\frac{3}{(s-2)^2 + 9}\right] = e^{2t}\sin 3t.$$

(b) From Table 12.2.1,

$$L^{-1}\left[\frac{4}{s^3}\right] = 2t^2.$$

Thus, applying the first shifting theorem,

$$L^{-1}\left[\frac{4}{(s-4)^3}\right] = 2t^2 e^{4t}.$$

(c) In this case

$$F(s) = \frac{s+4}{s^2 + 6s + 13},$$

which we do not recognize as being the transform of any of the functions given in Table 12.2.1. The first step in determining the inverse Laplace transform is to complete the square in the denominator of $F(s)$.[1] Thus we can write

$$F(s) = \frac{s+4}{(s+3)^2 + 4}. \tag{12.5.7}$$

_____

[1] Recall that we can always write $x^2 + ax + b = (x + a/2)^2 + b - a^2/4$. This procedure is known as *completing the square*.

We still cannot write down the inverse transform directly; however, by the first shifting theorem, we have

$$L^{-1}\left[\frac{s+3}{(s+3)^2+4}\right] = e^{-3t}\cos 2t, \tag{12.5.8}$$

$$L^{-1}\left[\frac{2}{(s+3)^2+4}\right] = e^{-3t}\sin 2t. \tag{12.5.9}$$

This suggests that we rewrite (12.5.7) in the equivalent form

$$F(s) = \left[\frac{s+3}{(s+3)^2+4} + \frac{1}{(s+3)^2+4}\right],$$

so that, using the linearity of L^{-1},

$$L^{-1}[F(s)] = L^{-1}\left[\frac{s+3}{(s+3)^2+4}\right] + L^{-1}\left[\frac{1}{(s+3)^2+4}\right]$$

$$= e^{-3t}\cos 2t + \frac{1}{2}e^{-3t}\sin 2t,$$

using (12.5.8) and (12.5.9).

   (d) We proceed as in the previous example.   In this case we have

$$F(s) = \frac{s-2}{(s+1)^2-2},$$

which can be written as

$$F(s) = \frac{s+1}{(s+1)^2-2} - \frac{3}{(s+1)^2-2}.$$

Thus, using Table 12.2.1 and the first shifting theorem, it follows that

$$L^{-1}[F(s)] = e^{-t}\cos\sqrt{2}t - \frac{3}{\sqrt{2}}e^{-t}\sin\sqrt{2}t.$$

EXERCISES 12.5

In problems 1–10 determine $f(t-a)$ for the given function f and the given constant a.

1. $f(t) = t$, $a = 1$.
2. $f(t) = 1$, $a = 3$.
3. $f(t) = t^2 - 2t$, $a = -2$.
4. $f(t) = e^{3t}$, $a = 2$.
5. $f(t) = e^{2t}\cos t$, $a = \pi$.

6. $f(t) = te^{2t}$, $a = -1$.
7. $f(t) = e^{-t}\sin 2t$, $a = \dfrac{\pi}{6}$.
8. $f(t) = \dfrac{t}{t^2+4}$, $a = 1$.
9. $f(t) = \dfrac{t+1}{t^2-2t+2}$, $a = 2$.
10. $f(t) = e^{-t}(\sin 2t + \cos 2t)$, $a = \dfrac{\pi}{4}$.

In problems 11–16, determine $f(t)$.

11. $f(t-1) = (t-1)^2$.
12. $f(t-1) = (t-2)^2$.
13. $f(t-2) = (t-2)e^{3(t-2)}$.
14. $f(t-1) = t\sin[3(t-1)]$.
15. $f(t-3) = te^{-(t-3)}$.
16. $f(t-4) = \dfrac{t+1}{(t-1)^2+4}$.

In problems 17–26 determine the Laplace transform of f.

17. $f(t) = e^{3t}\cos 4t$.
18. $f(t) = e^{-4t}\sin 5t$.
19. $f(t) = te^{2t}$.
20. $f(t) = 3te^{-t}$.
21. $f(t) = t^3 e^{-4t}$.
22. $f(t) = e^t - te^{-2t}$.
23. $f(t) = 2e^{3t}\sin t + 4e^{-t}\cos 3t$.
24. $f(t) = e^{2t}(1 - \sin^2 t)$.
25. $f(t) = t^2(e^t - 3)$.
26. $f(t) = e^{-2t}\sin\left(t - \dfrac{\pi}{4}\right)$.

In problems 27–41 determine $L^{-1}[F]$.

27. $F(s) = \dfrac{1}{(s-3)^2}$.
28. $F(s) = \dfrac{4}{(s+2)^3}$.
29. $F(s) = \dfrac{2}{(s+3)^{1/2}}$.
30. $F(s) = \dfrac{2}{(s-1)^2+4}$.
31. $F(s) = \dfrac{s+2}{(s+2)^2+9}$.
32. $F(s) = \dfrac{s}{(s-3)^2+4}$.

33. $F(s) = \dfrac{5}{(s-2)^2+16}$.
34. $F(s) = \dfrac{6}{s^2+2s+2}$.
35. $F(s) = \dfrac{s-2}{s^2+2s+26}$.
36. $F(s) = \dfrac{2s}{s^2-4s+13}$.
37. $F(s) = \dfrac{s}{(s+1)^2+4}$.
38. $F(s) = \dfrac{2s+3}{(s+5)^2+49}$.
39. $F(s) = \dfrac{4}{s(s+2)^2}$.
40. $F(s) = \dfrac{2s+1}{(s-1)^2(s+2)}$.
41. $F(s) = \dfrac{2s+3}{s(s^2-2s+5)}$.

In problems 42–53 solve the given initial value problem.

42. $y'' - y = 8e^t$, $y(0) = 0$, $y'(0) = 0$.
43. $y'' - 4y = 12e^{2t}$, $y(0) = 2$, $y'(0) = 3$.
44. $y'' - y' - 2y = 6e^{-t}$, $y(0) = 0$, $y'(0) = 1$.
45. $y'' + y' - 2y = 3e^{-2t}$, $y(0) = 3$, $y'(0) = -1$.
46. $y'' - 4y' + 4y = 6e^{2t}$, $y(0) = 1$, $y'(0) = 0$.
47. $y'' + 2y' + y = 2e^{-t}$, $y(0) = 2$, $y'(0) = 1$.
48. $y'' - 4y = 2te^t$, $y(0) = 0$, $y'(0) = 0$.
49. $y'' + 3y' + 2y = 12te^{2t}$, $y(0) = 0$, $y'(0) = 1$.
50. $y'' + y = 5te^{-3t}$, $y(0) = 2$, $y'(0) = 0$.
51. $y'' - y = 8e^t\sin 2t$, $y(0) = 2$, $y'(0) = -2$.
52. $y'' + 2y' - 3y = 26e^{2t}\cos t$, $y(0) = 1$, $y'(0) = 0$.
53. Solve the initial value problem

$$x_1' = 2x_1 - x_2, \qquad x_2' = x_1 + 2x_2,$$

$$x_1(0) = 1,\ x_2(0) = 0.$$

12.6 THE UNIT STEP FUNCTION

In the application of differential equations such as

$$y'' + by' + cy = F(t),$$

to engineering problems, it often arises that the forcing term $F(t)$ is either piecewise continuous or even discontinuous. In such a situation the Laplace transform is ideally suited for determining the solution of the differential equation as compared to the techniques developed in Chapter 9. To specify piecewise continuous or discontinuous functions in an appropriate manner, it is useful to introduce the unit step function, defined as follows.

Definition 12.6.1: The **unit step function**, or **Heaviside step function**, $u_a(t)$, is defined by

$$u_a(t) = \begin{cases} 0, & 0 \le t < a, \\ 1, & t \ge a, \end{cases}$$

where a is any positive number (see Figure 12.6.1).

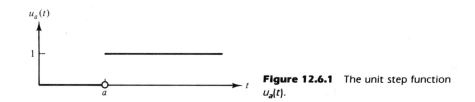

Figure 12.6.1 The unit step function $u_a(t)$.

Example 12.6.1 Sketch the function $f(t) = u_a(t) - u_b(t)$, $b > a$.

Solution By definition of the unit step function we have

$$f(t) = \begin{cases} 0, & 0 \le t < a, \\ 1, & a \le t < b, \\ 0, & t \ge b, \end{cases}$$

so that the graph of f is as given in Figure 12.6.2.

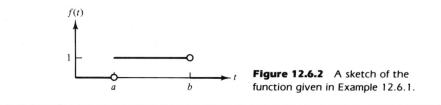

Figure 12.6.2 A sketch of the function given in Example 12.6.1.

The real power of the unit step function is that it enables us to model the situation when a force acts intermittently or in a nonsmooth manner. For example, the function f in Figure 12.6.2 can be interpreted as representing a force of unit magnitude that begins to act at $t = a$ and that stops acting at $t = b$. More generally, it is useful to regard the unit step function $u_a(t)$ as giving a mathematical description of a switch that is turned on at $t = a$.

The remaining examples in this section indicate how $u_a(t)$ can be useful for representing functions that are piecewise continuous, or discontinuous.

Example 12.6.2 Express the following function in terms of the unit step function:

$$f(t) = \begin{cases} 0, & 0 \le t < 1, \\ t - 1, & 1 \le t < 2, \\ 1, & t \ge 2. \end{cases}$$

Solution We view the given function in the following way. The contribution $f_1(t) = t - 1$ is "switched on" at $t = 1$ and is "switched off" again at $t = 2$. Mathematically this can be described by:

$$f_1(t) = \underbrace{u_1(t)\,(t - 1)}_{\substack{\text{Switch on} \\ \text{at } t=1.}} - \underbrace{u_2(t)\,(t - 1)}_{\substack{\text{Switch off} \\ \text{at } t=2.}}.$$

At $t = 2$ the contribution $f_2(t) = 1$ switches on and remains on for all $t \ge 2$. Mathematically, this is described by

$$f_2(t) = u_2(t).$$

Thus we obtain

$$f(t) = f_1(t) + f_2(t) = (t - 1)u_1(t) - (t - 1)u_2(t) + u_2(t),$$

which can be written in the equivalent form

$$f(t) = (t - 1)u_1(t) - (t - 2)u_2(t). \qquad (12.6.1)$$

A sketch of $f(t)$ is given in Figure 12.6.3. Notice that this sketch is more easily determined from the original definition of f rather than from (12.6.1).

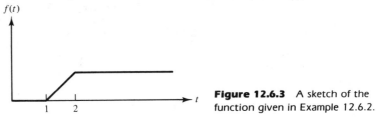

Figure 12.6.3 A sketch of the function given in Example 12.6.2.

Example 12.6.3 Make a sketch of the function $f(t)$ defined by

$$f(t) = \begin{cases} t, & 0 \le t < 2, \\ -1, & 2 \le t < 4, \\ t - 4, & 4 \le t < 5, \\ e^{5-t}, & t \ge 5, \end{cases}$$

and express f in terms of the unit step function.

Solution The function is sketched in Figure 12.6.4. Using the unit step function we see that f consists of the following different parts:

$$f_1(t) = t[1 - u_2(t)], \qquad\qquad f_2(t) = -1[u_2(t) - u_4(t)],$$

$$f_3(t) = (t - 4)[u_4(t) - u_5(t)], \qquad f_4(t) = e^{5-t}u_5(t).$$

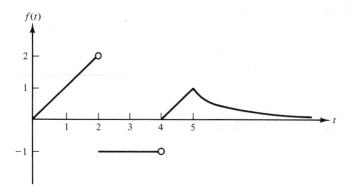

Figure 12.6.4 A sketch of the function defined in Example 12.6.3.

Thus

$$f(t) = t[1 - u_2(t)] - [u_2(t) - u_4(t)] + (t - 4)[u_4(t) - u_5(t)] + e^{5-t}u_5(t).$$

EXERCISES 12.6

In problems 1–6 sketch the given function on the interval $[0, \infty)$.

1. $f(t) = 2u_1(t) - 4u_3(t)$.

2. $f(t) = 1 + (t - 1)u_1(t)$.

3. $f(t) = t[1 - u_1(t)]$.

4. $f(t) = u_1(t) + u_2(t) + u_3(t) + u_4(t)$.

5. $f(t) = u_1(t) + u_2(t) + u_3(t) + \cdots = \sum_{i=1}^{\infty} u_i(t)$.

6. $f(t) = u_1(t) - u_2(t) + u_3(t) - \cdots$

$$= \sum_{i=1}^{\infty} (-1)^{i+1} u_i(t).$$

In problems 7–14 sketch the given function and express it in terms of the unit step function.

7. $f(t) = \begin{cases} 3, & 0 \le t < 1, \\ -1, & t \ge 1. \end{cases}$

8. $f(t) = \begin{cases} t^2, & 0 \le t < 1, \\ 1, & t \ge 1. \end{cases}$

9. $f(t) = \begin{cases} 2, & 0 \le t < 2, \\ 1, & 2 \le t < 4, \\ -1, & t \ge 4. \end{cases}$

10. $f(t) = \begin{cases} 2, & 0 \le t < 1, \\ 2e^{t-1}, & t \ge 1. \end{cases}$

11. $f(t) = \begin{cases} t, & 0 \le t < 3, \\ 6 - t, & 3 \le t < 6, \\ 0, & t \ge 6. \end{cases}$

12. $f(t) = \begin{cases} 0, & 0 \le t < 2, \\ 3 - t, & 2 \le t < 4, \\ -1, & t \ge 4. \end{cases}$

13. $f(t) = \begin{cases} 1, & 0 \le t < \dfrac{\pi}{2}, \\ \sin t, & \dfrac{\pi}{2} \le t < \dfrac{3\pi}{2}, \\ -1 & t \ge \dfrac{3\pi}{2}. \end{cases}$

14. $f(t) = \begin{cases} \sin t, & 2n\pi \le t < (2n + 1)\pi, \\ & \qquad n = 0, 1, 2, \ldots \\ 0, & \text{otherwise.} \end{cases}$

12.7   THE SECOND SHIFTING THEOREM

In the previous section we saw how the unit step function can be used to represent functions that are piecewise continuous or even discontinuous. In this section we show that the Laplace transform provides a straightforward method for solving constant coefficient linear differential equations that have such functions as driving terms. We first need to determine how the unit step function transforms.

Theorem 12.7.1 (Second Shifting Theorem):   Let $L[f(t)] = F(s)$. Then

$$L[u_a(t)f(t-a)] = e^{-as}F(s).$$

(12.7.1)

Conversely,

$$L^{-1}[e^{-as}F(s)] = u_a(t)f(t-a).$$

(12.7.2)

PROOF   Once more we must return to the definition of the Laplace transform. We have

$$L[u_a(t)f(t-a)] = \int_0^\infty e^{-st} u_a(t)f(t-a)\, dt$$

$$= \int_a^\infty e^{-st} f(t-a)\, dt,$$

where we have used the definition of the unit step function. We now make a change of variable in the integral. Let $x = t - a$. Then $dx = dt$, and the lower limit of integration ($t = a$) corresponds to $x = 0$, whereas the upper limit of integration is unchanged. Thus

$$L[u_a(t)f(t-a)] = \int_0^\infty e^{-s(x+a)} f(x)\, dx = e^{-as} \int_0^\infty e^{-sx} f(x)\, dx$$

$$= e^{-as} L[f],$$

as required. Taking the inverse Laplace transform of both sides of (12.7.1) yields (12.7.2).   ∎

This theorem is illustrated in Figure 12.7.1.

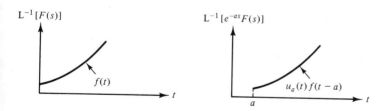

$L^{-1}[F(s)]$

$f(t)$

$L^{-1}[e^{-as}F(s)]$

$u_a(t)f(t-a)$

Figure 12.7.1   An illustration of the second shifting theorem. Multiplying $F(s)$ by e^{-as} has the effect of shifting $f(t)$ by a units to the right in "t-space."

Corollary 12.7.1:   If $L[f] = F(s)$, then

$$L[u_a(t)f(t)] = e^{-as}L[f(t+a)].$$

PROOF   The proof is a direct consequence of the previous theorem.   ∎

Example 12.7.1   Determine L[f] if

$$f(t) = \begin{cases} 0, & 0 \leq t < 1, \\ t-1, & 1 \leq t < 2, \\ 1, & t \geq 2. \end{cases}$$

Solution   We have already shown in Example 12.6.2 that the given function can be expressed in terms of the unit step function as

$$f(t) = (t-1)u_1(t) - (t-2)u_2(t).$$

If we let $g(t) = t$, then

$$f(t) = g(t-1)u_1(t) - g(t-2)u_2(t).$$

Thus using Theorem 12.7.1, it follows that

$$L[f] = e^{-s}L[g] - e^{-2s}L[g] = \frac{1}{s^2}(e^{-s} - e^{-2s}).$$

Example 12.7.2   Find L[f] if

$$f(t) = \begin{cases} 1, & 0 \leq t < 2, \\ e^{-(t-2)}, & t \geq 2. \end{cases}$$

Solution   To determine L[f] we first express f in terms of the unit step function. In this case we have

$$f(t) = [1 - u_2(t)] + e^{-(t-2)}u_2(t);$$

that is,

$$f(t) = 1 + u_2(t)[e^{-(t-2)} - 1].$$

If we let $g(t) = e^{-t} - 1$, then

$$f(t) = 1 + u_2(t)g(t-2),$$

so that, from Theorem 12.7.1,

$$L[f] = \frac{1}{s} + e^{-2s}\left(\frac{1}{s+1} - \frac{1}{s}\right).$$

We can write this in the equivalent form

$$L[f] = \frac{1 - e^{-2s}}{s} + \frac{e^{-2s}}{s+1}.$$

Example 12.7.3   Determine $L^{-1}\left[\dfrac{2e^{-s}}{s^2+4}\right]$.

Solution   From Table 12.2.1 we have

$$L[\sin 2t] = \frac{2}{s^2+4}.$$

Thus we can write

$$L^{-1}\left[\frac{2e^{-s}}{s^2+4}\right] = L^{-1}\{e^{-s}L[\sin 2t]\} = u_1(t)\sin[2(t-1)],$$

using Theorem 12.7.1.

Example 12.7.4   Determine $L^{-1}\left[\dfrac{(s-4)\,e^{-3s}}{s^2-4s+5}\right]$.

Solution   Let

$$G(s) = \frac{(s-4)\,e^{-3s}}{s^2-4s+5}.$$

We first rewrite G in a more suitable form for determining $L^{-1}[G]$. Completing the square in the denominator yields

$$G(s) = \frac{(s-4)\,e^{-3s}}{(s-2)^2+1},$$

which can be written in the equivalent form

$$G(s) = e^{-3s}\left[\frac{(s-2)}{(s-2)^2+1} - \frac{2}{(s-2)^2+1}\right].$$

Thus

$$\begin{aligned}
L^{-1}[G] &= L^{-1}\{e^{-3s}(L[e^{2t}\cos t - 2e^{2t}\sin t])\} \\
&= u_3(t)[e^{2(t-3)}\cos(t-3) - 2e^{2(t-3)}\sin(t-3)] \\
&= e^{2(t-3)}u_3(t)[\cos(t-3) - 2\sin(t-3)].
\end{aligned}$$

We now illustrate how the unit step function can be useful in the solution of initial value problems. For simplicity we will start with a first-order differential equation.

Example 12.7.5   Solve the initial value problem

$$y' - y = 1 - (t-1)u_1(t), \qquad y(0) = 0.$$

Solution   In this case the forcing term on the right-hand side of the differential equation is sketched in Figure 12.7.2. Taking the Laplace transform of both sides of the differential equation yields

$$sY(s) - Y(s) - y(0) = \frac{1}{s} - \frac{e^{-s}}{s^2}.$$

Imposing the given initial condition and simplifying we obtain

$$Y(s) = \frac{1}{s(s-1)} - e^{-s}\left[\frac{1}{s^2(s-1)}\right].$$

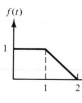

Figure 12.7.2   The forcing function in Example 12.7.5.

Decomposing the terms on the right-hand side into partial fractions yields

$$Y(s) = \frac{1}{s-1} - \frac{1}{s} - e^{-s}\left[\frac{1}{s-1} - \frac{1}{s} - \frac{1}{s^2}\right].$$

Taking the inverse Laplace transform of both sides of this equation, we obtain

$$y(t) = e^t - 1 - u_1(t)[e^{t-1} - 1 - (t-1)],$$

that is,

$$y(t) = e^t - 1 - u_1(t)(e^{t-1} - t).$$

To emphasize the power of the Laplace transform, we now solve the preceding initial value problem using the technique introduced in Section 1.5 for first-order linear equations. In order to use this technique we must solve the differential equation on the intervals $[0, 1)$, and $[1, \infty)$ separately. We can write the differential equation as

$$y' - y = f(t), \qquad \text{where } f(t) = \begin{cases} 1, & 0 \le t < 1, \\ 2 - t, & t \ge 1. \end{cases}$$

Thus on the interval $[0, 1)$ we must solve the initial value problem

$$y' - y = 1, \qquad y(0) = 0.$$

An integrating factor for this equation is $I = e^{-t}$, so that we have

$$\frac{d}{dt}(e^{-t}y) = e^{-t}.$$

Integrating both sides of this equation yields

$$y(t) = -1 + ce^t.$$

Imposing the given initial condition we find that $c = 1$, so that

$$y(t) = e^t - 1, \qquad 0 \le t < 1. \tag{12.7.3}$$

Now consider the interval $t \ge 1$. We must solve the differential equation

$$y' - y = 2 - t.$$

Once more an appropriate integrating factor is $I = e^{-t}$, so that we can write the differential equation in the form

$$\frac{d}{dt}(e^{-t}y) = e^{-t}(2 - t).$$

Integrating both sides of this equation yields

$$y(t) = t - 1 + c_1 e^t, \qquad t \geq 1. \tag{12.7.4}$$

To determine the integration constant c_1, we impose the condition that the solution must be continuous at $t = 1$; that is, we require that

$$\lim_{t \to 1^-} y(t) = y(1).$$

From (12.7.3) it follows that

$$\lim_{t \to 1^-} y(t) = e - 1,$$

whereas from (12.7.4) we have

$$y(1) = c_1 e.$$

Thus we require

$$c_1 e = e - 1,$$

so that

$$c_1 = 1 - e^{-1}.$$

Substituting into (12.7.4) yields

$$y(t) = t - 1 + e^t - e^{(t-1)}, \qquad t \geq 1. \tag{12.7.5}$$

Using the unit step function we can combine the solutions given in (12.7.3) and (12.7.5) into the single expression

$$y(t) = e^t - 1 + u_1(t)(t - e^{t-1}),$$

which is the same as the result obtained using the Laplace transform.   The power of the Laplace transform should be quite clear from this simple example.

Example 12.7.6   Solve the initial value problem

$$y'' - y = f(t), \qquad y(0) = 0, \, y'(0) = 0,$$

if

$$f(t) = \begin{cases} 1, & 0 \leq t < 1, \\ -1, & t \geq 1. \end{cases}$$

Solution   We first express f in terms of the unit step function (see Figure 12.7.3). In this case we have

$$f(t) = 1 - 2u_1(t),$$

so that the differential equation can be written as

$$y'' - y = 1 - 2u_1(t),$$

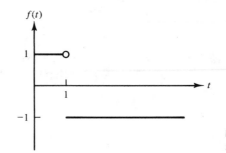

Figure 12.7.3   The forcing function $f(t)$ in Example 12.7.6.

and we can now proceed in the usual manner.   Taking the Laplace transform of both sides of the differential equation yields

$$[s^2Y(s) - sy(0) - y'(0)] - Y(s) = \frac{1}{s}(1 - 2e^{-s}),$$

that is, imposing the given initial conditions and simplifying,

$$Y(s) = \frac{1 - 2e^{-s}}{s(s^2 - 1)}.$$

We must now determine the partial fraction decomposition of the right-hand side. It is easily shown that

$$\frac{1}{s(s^2 - 1)} = \frac{1}{s(s - 1)(s + 1)} = -\frac{1}{s} + \frac{1}{2(s - 1)} + \frac{1}{2(s + 1)},$$

so that

$$Y(s) = (1 - 2e^{-s})\left[-\frac{1}{s} + \frac{1}{2(s - 1)} + \frac{1}{2(s + 1)}\right]$$

$$= -\frac{1}{s} + \frac{1}{2(s - 1)} + \frac{1}{2(s + 1)} - 2e^{-s}\left[-\frac{1}{s} + \frac{1}{2(s - 1)} + \frac{1}{2(s + 1)}\right].$$

Taking the inverse Laplace transform of both sides and using Theorem 12.7.1 yields

$$y(t) = \tfrac{1}{2}(e^t + e^{-t}) - 1 - 2u_1(t)\left[\tfrac{1}{2}(e^{t-1} + e^{-(t-1)}) - 1\right],$$

that is,

$$y(t) = \cosh(t) - 1 - 2u_1(t)[\cosh(t - 1) - 1].$$

We can express this solution in the simple form

$$y(t) = g(t) - 2u_1(t)g(t - 1),$$

where

$$g(t) = \cosh t - 1.$$

REMARK   Notice that y and y' are continuous at $t = 1$ but that y'' is discontinuous at $t = 1$.   This must happen since the driving $f(t)$ has a discontinuity at $t = 1$.

EXERCISES 12.7

In problems 1–10 determine the Laplace transform of the given function f.

1. $f(t) = (t - 1)u_1(t)$.

2. $f(t) = e^{3(t-2)}u_2(t)$.

3. $f(t) = u_{\pi/4}(t)\sin\left(t - \dfrac{\pi}{4}\right)$.

4. $f(t) = u_\pi(t)\cos t$.

5. $f(t) = (t - 2)^2 u_2(t)$.

6. $f(t) = tu_3(t)$.

7. $f(t) = (t - 1)^2 u_2(t)$.

8. $f(t) = e^{t-4}(t - 4)^3 u_4(t)$.

9. $f(t) = u_1(t)e^{-2(t-1)}\sin 3(t - 1)$.

10. $f(t) = u_c(t)e^{a(t-c)}\cos b(t - c)$, a, b, c positive constants.

In problems 11–25 determine the inverse Laplace transform of F.

11. $F(s) = \dfrac{e^{-2s}}{s^2}$.

12. $F(s) = \dfrac{e^{-s}}{s + 1}$.

13. $F(s) = \dfrac{e^{-3s}}{s + 4}$.

14. $F(s) = \dfrac{se^{-s}}{s^2 + 4}$.

15. $F(s) = \dfrac{e^{-3s}}{s^2 + 1}$.

16. $F(s) = \dfrac{e^{-2s}}{s + 2}$.

17. $F(s) = \dfrac{e^{-s}}{(s + 1)(s - 4)}$.

18. $F(s) = \dfrac{e^{-2s}}{s^2 + 2s + 2}$.

19. $F(s) = \dfrac{e^{-s}(s + 6)}{s^2 + 9}$.

20. $F(s) = \dfrac{e^{-5s}}{s^2 + 16}$.

21. $F(s) = \dfrac{e^{-2s}}{(s - 3)^3}$.

22. $F(s) = \dfrac{e^{-4s}(s + 3)}{s^2 - 6s + 13}$.

23. $F(s) = \dfrac{e^{-s}(2s - 1)}{s^2 + 4s + 5}$.

24. $F(s) = \dfrac{2e^{-2s}}{(s - 1)(s^2 + 1)}$.

25. $F(s) = \dfrac{50e^{-3s}}{(s + 1)^2(s^2 + 4)}$.

In problems 26–40, solve the given initial value problem.

26. $y' + 2y = 2u_1(t)$, $y(0) = 1$.

27. $y' - 2y = u_2(t)e^{t-2}$, $y(0) = 2$.

28. $y' - y = 4u_{\pi/4}(t)\cos\left(t - \dfrac{\pi}{4}\right)$, $y(0) = 1$.

29. $y' + 2y = u_\pi(t)\sin 2t$, $y(0) = 3$.

30. $y' + 3y = f(t)$, $y(0) = 1$, where

$$f(t) = \begin{cases} 1, & 0 \le t < 1, \\ 0, & t \ge 1. \end{cases}$$

31. $y' - 3y = f(t)$, $y(0) = 2$, where

$$f(t) = \begin{cases} \sin t, & 0 \le t < \dfrac{\pi}{2}, \\ 1, & t \ge \dfrac{\pi}{2}. \end{cases}$$

32. $y' - 3y = 10e^{-(t-a)}[\sin 2(t - a)]u_a(t)$, $y(0) = 5$, where a is a positive constant.

33. $y'' - y = u_1(t)$, $y(0) = 2$, $y'(0) = 0$.

34. $y'' - y' - 2y = 1 - 3u_2(t)$, $y(0) = 1$, $y'(0) = -2$.

35. $y'' - 4y = u_1(t) - u_2(t)$, $y(0) = 0$, $y'(0) = 4$.

36. $y'' + y = t - u_1(t)(t - 1)$, $y(0) = 2$, $y'(0) = 1$.

37. $y'' + 3y' + 2y = 10u_{\pi/4}(t)\sin\left(t - \dfrac{\pi}{4}\right)$, $y(0) = 1$, $y'(0) = 0$.

38. $y'' + y' - 6y = 30u_1(t)e^{-(t-1)}$, $y(0) = 3$, $y'(0) = -4$.

39. $y'' + 4y' + 5y = 5u_3(t)$, $y(0) = 2$, $y'(0) = 1$.

40. $y'' - 2y' + 5y = 2\sin t +$
$u_{\pi/2}(t)\left[1 - \sin\left(t - \dfrac{\pi}{2}\right)\right]$, $y(0) = 0$, $y'(0) = 0$.

In problems 41–44, solve the given initial value problem.

41. $y' + y = f(t)$, $y(0) = 2$, where $f(t)$ is given in Figure 12.7.4.

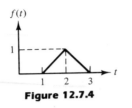

Figure 12.7.4

42. $y' + 2y = f(t)$, $y(0) = 0$, where $f(t)$ is given in Figure 12.7.5.

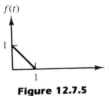

Figure 12.7.5

43. $y' - y = f(t)$, $y(0) = 2$, where $f(t)$ is given in Figure 12.7.6.

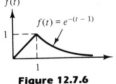

$f(t) = e^{-(t-1)}$

Figure 12.7.6

44. $y' - 2y = f(t)$, $y(0) = 0$, where $f(t)$ is given in Figure 12.7.7.

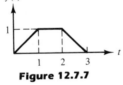

Figure 12.7.7

45. Solve the initial value problem

$$y' - y = f(t), \quad y(0) = 1,$$

where

$$f(t) = \begin{cases} 2, & 0 \le t < 1, \\ -1, & t \ge 1, \end{cases}$$

in the following two ways:
(a) Directly using the Laplace transform.
(b) Using the technique for solving first-order linear equations developed in Section 1.5.

46. The current, $i(t)$, in a simple RL circuit is governed by the differential equation

$$\frac{di}{dt} + \frac{R}{L} i = \frac{1}{L} E(t),$$

where R and L are constants and $E(t)$ represents the applied EMF. At $t = 0$ the switch in the circuit is closed, and the applied EMF increases linearly from 0 to 10 V in a time interval of 5 s. The EMF then remains constant for $t \ge 5$. Determine the current in the circuit for $t \ge 0$.

47. The differential equation governing the charge, $q(t)$, on the capacitor in a simple RC circuit is

$$\frac{dq}{dt} + \frac{1}{RC} q = \frac{1}{R} E(t),$$

where R and C are constants and $E(t)$ represents the applied EMF. Over a time interval of 10 s, the applied EMF has the constant value 20 V. Thereafter the EMF decays exponentially according to $E(t) = 20 \, e^{-(t-10)}$. If the capacitor is initially uncharged, and assuming $RC \ne 1$, determine the current in the circuit for $t > 0$. [*Recall:* The current $i(t)$ is related to the charge on the capacitor by $i(t) = dq/dt$.]

12.8   IMPULSIVE DRIVING TERMS: THE DIRAC DELTA FUNCTION

Consider the differential equation

$$y'' + by' + cy = f(t).$$

We saw in the previous sections that the Laplace transform is useful in the case when the forcing term, $f(t)$, is either piecewise continuous or even discontinuous. We now consider another type of forcing term, namely, that describing an impulsive force. Such a force arises when an object is dealt an instantaneous blow—for example, when an object is hit by a hammer (see Figure 12.8.1). The aim of this section is to develop a way of representing impulsive forces mathematically and then to show how the Laplace transform can be used to solve differential equations when the driving term is due to an impulsive force.

Impulsive force

M

Figure 12.8.1 An example of an impulsive force.

Suppose that a force of magnitude F acts on an object over the time interval $[t_1, t_2]$. The *impulse* of this force, I, is defined by[1]

$$I = \int_{t_1}^{t_2} F(t) \, dt.$$

Since $F(t)$ is zero for t outside the interval $[t_1, t_2]$, we can write

$$I = \int_{-\infty}^{\infty} F(t) \, dt.$$

Mathematically, I gives the area under the curve $y = F(t)$ lying over the t-axis (see Figure 12.8.2). We now introduce a mathematical description of a force that instantaneously imparts an impulse of unit magnitude to an object at $t = a$. Thus the two properties that we wish to characterize are the following:

1. The force acts instantaneously.
2. The force has unit impulse.

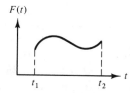

$F(t)$

t_1 t_2 t

Figure 12.8.2 When a force of magnitude F newtons acts on an object over a time interval $[t_1, t_2]$ seconds, the impulse of the force is given by the area under the curve.

In order to do so, we proceed in the following manner. Define the function $d_\varepsilon(t - a)$ by

$$d_\varepsilon(t - a) = \frac{u_a(t) - u_{a+\varepsilon}(t)}{\varepsilon}, \qquad (12.8.1)$$

where u_a is the unit step function (see Figure 12.8.3). We can interpret $d_\varepsilon(t - a)$ as representing a force of magnitude $1/\varepsilon$ that acts for a time interval of ε starting at

---

[1] This represents the change in momentum of the object due to the applied force.

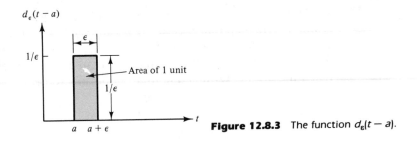

Figure 12.8.3   The function $d_\epsilon(t-a)$.

$t = a$. Notice that this force does have unit impulse, since

$$I = \int_{-\infty}^{\infty} d_\epsilon(t-a)\,dt = \int_{-\infty}^{\infty} \frac{u_a(t) - u_{a+\epsilon}(t)}{\epsilon}\,dt = \int_{a}^{a+\epsilon} \frac{1}{\epsilon}\,dt = 1.$$

To capture the idea of an instantaneous force, we take the limit as $\epsilon \to 0^+$. It follows from (12.8.1) that

$$\lim_{\epsilon \to 0^+} d_\epsilon(t-a) = 0 \qquad \text{whenever } t \neq a.$$

Also, since $I = 1$ for all t,

$$\lim_{\epsilon \to 0^+} I = 1.$$

These properties characterize mathematically the idea of a force of unit impulse acting instantaneously at $t = a$. We use them to define the *unit impulse function*.

Definition 12.8.1: The **unit impulse function**, or **Dirac delta function**, $\delta(t - a)$, is the (generalized) function that satisfies

1. $\delta(t - a) = 0, \qquad t \neq a,$

2. $\displaystyle\int_{-\infty}^{\infty} \delta(t - a)\,dt = 1.$

REMARK   The unit impulse function is not a function in the usual sense. It is an example of what is called a *generalized function*. The detailed study of such functions is beyond the scope of the present text. However, all that we will require are properties 1 and 2 of Definition 12.8.1.

Thus, to summarize:

$\delta(t - a)$ describes a force that instantaneously imparts a unit impulse to a system at $t = a$.

We now consider the possibility of determining the Laplace transform of $\delta(t - a)$. The natural way to do this is to return to the function $d_\epsilon(t - a)$ and define the Laplace transform of $\delta(t - a)$ in the following manner:

$$L[\delta(t-a)] = \lim_{\varepsilon \to 0^+} L[d_\varepsilon(t-a)] = \lim_{\varepsilon \to 0^+} \int_0^\infty e^{-st}\left[\frac{u_a(t) - u_{a+\varepsilon}(t)}{\varepsilon}\right] dt$$

$$= \lim_{\varepsilon \to 0^+} \frac{1}{\varepsilon} \int_a^{a+\varepsilon} e^{-st}\, dt$$

$$= \lim_{\varepsilon \to 0^+} \frac{1}{\varepsilon}\left[-\frac{1}{s}\left(e^{-s(a+\varepsilon)} - e^{-sa}\right)\right]$$

$$= \frac{e^{-sa}}{s} \lim_{\varepsilon \to 0^+}\left[\frac{1 - e^{-\varepsilon s}}{\varepsilon}\right].$$

Using l'Hôpital's rule to evaluate the preceding limit yields

$$\boxed{L[\delta(t-a)] = e^{-sa}.}$$

$$(12.8.2)$$

In particular,

$$\boxed{L[\delta(t)] = 1.}$$

$$(12.8.3)$$

More generally, it can be shown that if g is a continuous function on $(-\infty, \infty)$, then

$$\int_{-\infty}^\infty g(t)\, \delta(t-a)\, dt = g(a).$$

$$(12.8.4)$$

Example 12.8.1   Solve the initial value problem

$$y'' + 4y' + 13y = \delta(t-\pi), \qquad y(0) = 2, \qquad y'(0) = 1.$$

Solution   We proceed in the usual manner. Taking the Laplace transform of both sides of the given differential equation and imposing the initial conditions yields

$$[s^2 Y - 2s - 1] + 4[sY - 2] + 13Y = e^{-\pi s},$$

which implies that

$$Y = \frac{e^{-\pi s} + 2s + 9}{s^2 + 4s + 13}$$

$$= \frac{e^{-\pi s} + 2s + 9}{(s+2)^2 + 9}$$

$$= \frac{e^{-\pi s}}{(s+2)^2 + 9} + \frac{2(s+2)}{(s+2)^2 + 9} + \frac{5}{(s+2)^2 + 9}.$$

Taking the inverse Laplace transform of both sides, we obtain

$$y = L^{-1}\left\{\frac{e^{-\pi s}}{3} L[e^{-2t}\sin 3t]\right\} + 2e^{-2t}\cos 3t + \frac{5}{3} e^{-2t}\sin 3t$$

$$= \frac{1}{3} u_\pi(t)e^{-2(t-\pi)}\sin 3(t-\pi) + 2e^{-2t}\cos 3t + \frac{5}{3} e^{-2t}\sin 3t.$$

Since $\sin 3(t - \pi) = -\sin 3t$, we finally obtain

$$y(t) = -\tfrac{1}{3}\, u_\pi(t) e^{-2(t-\pi)} \sin 3t + e^{-2t}(2 \cos 3t + \tfrac{5}{3} \sin 3t).$$

Example 12.8.2   Consider the spring-mass system depicted in Figure 12.8.4. At $t = 0$, the mass is pulled down a distance 1 unit from equilibrium and released from rest. After 1 s the mass is dealt an instantaneous blow of unit impulse in the upward direction. The initial value problem governing the motion of the mass is

$$\frac{d^2x}{dt^2} + 4x = -\delta(t - 1), \qquad x(0) = 1, \qquad \frac{dx}{dt}(0) = 0,$$

where $\delta(t - 1)$ describes the impulsive force that acts on the mass at $t = 1$. Determine the motion of the mass for all $t > 0$.

Impulsive force acts after 1 s

Figure 12.8.4   A spring-mass system in which friction is neglected and the only external force acting on the system is an impulsive force of unit impulse that acts at $t = 1$.

Solution   Taking the Laplace transform of the differential equation and imposing the initial conditions yields

$$s^2 X(s) - s + 4X(s) = -e^{-s},$$

so that

$$X(s) = \frac{-e^{-s} + s}{s^2 + 4}$$

$$= \frac{-e^{-s}}{s^2 + 4} + \frac{s}{s^2 + 4}.$$

Taking the inverse Laplace transform of both sides of this equation yields

$$x(t) = L^{-1}\{-\tfrac{1}{2}\, e^{-s} L[\sin 2t]\} + \cos 2t.$$

Thus

$$x(t) = -\tfrac{1}{2}\, u_1(t) \sin 2(t - 1) + \cos 2t.$$

The first term on the right-hand side represents the contribution from the impulsive force. Obviously, this does not affect the motion of the mass until $t = 1$ but then contributes for all $t \geq 1$. More explicitly, we can write the solution as

$$x(t) = \begin{cases} \cos 2t, & 0 \leq t < 1, \\ \cos 2t - \tfrac{1}{2} \sin 2(t - 1), & t \geq 1. \end{cases}$$

EXERCISES 12.8

In problems 1–12 solve the given initial value problem.

1. $y' - 2y = \delta(t - 2)$, $y(0) = 1$.

2. $y' + 4y = 3\delta(t - 1)$, $y(0) = 2$.

3. $y' - 5y = 2e^{-t} + \delta(t - 3)$, $y(0) = 0$.

4. $y'' - 3y' + 2y = \delta(t - 1)$, $y(0) = 1$, $y'(0) = 0$.

5. $y'' - 4y = \delta(t - 3)$, $y(0) = 0$, $y'(0) = 1$.

6. $y'' + 2y' + 5y = \delta\left(t - \dfrac{\pi}{2}\right)$, $y(0) = 0$, $y'(0) = 2$.

7. $y'' - 4y' + 13y = \delta\left(t - \dfrac{\pi}{4}\right)$, $y(0) = 3$, $y'(0) = 0$.

8. $y'' + 4y' + 3y = \delta(t - 2)$, $y(0) = 1$, $y'(0) = -1$.

9. $y'' + 6y' + 13y = \delta\left(t - \dfrac{\pi}{4}\right)$, $y(0) = 5$, $y'(0) = 5$.

10. $y'' + 9y = 15\sin 2t + \delta\left(t - \dfrac{\pi}{6}\right)$, $y(0) = 0$, $y'(0) = 0$.

11. $y'' + 16y = 4\cos 3t + \delta\left(t - \dfrac{\pi}{3}\right)$, $y(0) = 0$, $y'(0) = 0$.

12. $y'' + 2y' + 5y = 4\sin t + \delta\left(t - \dfrac{\pi}{6}\right)$, $y(0) = 0$, $y'(0) = 1$.

13. The motion of a spring-mass system is governed by the initial value problem

$$\frac{d^2 x}{dt^2} + 4x = F_0 \cos 3t, \quad x(0) = 0, \quad \frac{dx}{dt}(0) = 0,$$

where F_0 is a constant. At $t = 1$ s, the mass is dealt a blow in the upward direction that instantaneously imparts 5 units of impulse to the system. Determine the resulting motion of the mass.

14. The motion of a spring-mass system is governed by the initial value problem

$$\frac{d^2 x}{dt^2} + 4\frac{dx}{dt} + 13x = 10\sin 5t,$$

$$x(0) = 0, \quad \frac{dx}{dt}(0) = 0.$$

At $t = 10$ s, the mass is dealt a blow in the downward direction that instantaneously imparts 2 units of impulse to the system. Determine the resulting motion of the mass.

15. Consider the spring-mass system whose motion is governed by the initial value problem

$$\frac{d^2 x}{dt^2} + \omega_0^2 x = F_0 \sin \omega t + A\delta(t - t_0),$$

$$x(0) = 0, \quad \frac{dx}{dt}(0) = 0,$$

where ω_0, ω, F_0, A, and t_0 are positive constants, and $\omega \neq \omega_0$. Solve the initial value problem to determine the position of the mass at time t.

12.9   THE CONVOLUTION INTEGRAL

Very often in solving a differential equation using the Laplace transform method, we require the inverse Laplace transform of an expression of the form

$$H(s) = F(s)G(s),$$

where $F(s)$ and $G(s)$ are functions whose inverse Laplace transform is known. In such a situation it is tempting to suspect that

$$L^{-1}[H(s)] = L^{-1}[F(s)] \, L^{-1}[G(s)].$$

It is important to realize that this is *not* true. For example,

$$L^{-1}\left[\frac{1}{(s-1)(s^2+1)}\right] = L^{-1}\left[\frac{1}{2(s-1)} - \frac{s+1}{2(s^2+1)}\right]$$

$$= \frac{1}{2} e^t - \frac{1}{2} \cos t - \frac{1}{2} \sin t,$$

whereas,

$$L^{-1} \left[\frac{1}{s-1} \right] L^{-1} \left[\frac{1}{s^2+1} \right] = e^t \sin t,$$

so that

$$L^{-1} \left[\frac{1}{(s-1)(s^2+1)} \right] \neq L^{-1} \left[\frac{1}{s-1} \right] L^{-1} \left[\frac{1}{s^2+1} \right].$$

However, it is possible, at least in theory, to determine $L^{-1}[F(s)G(s)]$ directly in terms of an integral involving $f(t)$ and $g(t)$. Before showing this, we require a definition.

Definition 12.9.1: Suppose that f and g are continuous on the interval $[0, b]$. Then for t in $(0, b]$, the **convolution product**, $f * g$, of f and g is defined by

$$(f * g)(t) = \int_0^t f(t - \tau) g(\tau) \, d\tau.$$

Notice that $f * g$ is indeed a function of t. The integral

$$\int_0^t f(t - \tau) g(\tau) \, d\tau \tag{12.9.1}$$

is called a **convolution integral**.

Example 12.9.1   If $f(t) = t$, and $g(t) = \sin t$, determine $f * g$.

Solution   From Definition 12.9.1 we have

$$(f * g)(t) = \int_0^t (t - \tau) \sin \tau \, d\tau = t \int_0^t \sin \tau \, d\tau - \int_0^t \tau \sin \tau \, d\tau$$

$$= t(1 - \cos t) - [\sin t - t \cos t]$$

$$= t - \sin t.$$

The convolution product satisfies the three basic properties of the ordinary multiplicative product, namely,

1. $f * g = g * f$ (Commutative).
2. $f * (g * h) = (f * g) * h$ (Associative).
3. $f * (g + h) = f * g + f * h$ (Distributive over addition).

The proofs of these three properties are left as exercises.
    We now show how the convolution product can be useful in evaluating inverse Laplace transforms.

Theorem 12.9.1 (The Convolution Theorem): If f and g are in $E(0, \infty)$, then

$$L[(f * g)(t)] = L[f] \, L[g]. \qquad (12.9.2)$$

Conversely,

$$L^{-1}[F(s)G(s)] = (f * g)(t). \qquad (12.9.3)$$

PROOF   We must use the definition of the Laplace transform and the convolution product.

$$L[(f * g)(t)] = \int_0^\infty e^{-st} \left\{ \int_0^t f(t - \tau)g(\tau) \, d\tau \right\} dt$$

$$= \int_0^\infty \int_0^t e^{-st} f(t - \tau)g(\tau) \, d\tau \, dt.$$

It is not clear how to proceed at this point.   However, since we are dealing with an iterated double integral, it is usually worth changing the order of integration to see if any simplification arises.   In this case the limits of integration are $0 \le \tau \le t$, $0 \le t < \infty$, so that the region of integration is that part of the $t\tau$-plane that lies above the t-axis and below the line $\tau = t$.   This region is shown in Figure 12.9.1.

Reversing the order of integration, the new limits are $\tau \le t < \infty, 0 \le \tau < \infty$. Thus we can write

$$L[(f * g)(t)] = \int_0^\infty \int_\tau^\infty e^{-st} f(t - \tau)g(\tau) \, dt \, d\tau.$$

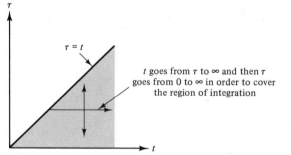

t goes from τ to ∞ and then τ goes from 0 to ∞ in order to cover the region of integration

Figure 12.9.1   Changing the order of integration in Theorem 12.9.1.

We now make the change of variable $u = t - \tau$ in the first iterated integral.   Then $du = d\tau$ (remember that τ is treated as a constant when performing the first integration) and the new u-limits are $0 \le u < \infty$.   Thus

$$L[(f * g)(t)] = \int_0^\infty \int_0^\infty e^{-s(u+\tau)} g(\tau) f(u) \, du \, d\tau = \int_0^\infty e^{-s\tau} g(\tau) \left[\int_0^\infty e^{-su} f(u) \, du \right] d\tau$$

$$= \left[\int_0^\infty e^{-su} f(u) \, du \right] \left[\int_0^\infty e^{-s\tau} g(\tau) \, d\tau \right]$$

$$= L[f]L[g],$$

as required.   The converse, (12.9.3), is obtained in the usual manner by taking the inverse Laplace transform of (12.9.2).   ■

REMARK   More generally, it can be shown that

$$L^{-1}[F_1(s)F_2(s) \cdots F_n(s)] = f_1(t) * f_2(t) * \cdots * f_n(t).$$

Example 12.9.2   Determine L[f] if

$$f(t) = \int_0^t \sin(t - \tau)e^{-\tau} \, d\tau.$$

Solution   In this case we recognize that

$$f(t) = \sin t * e^{-t},$$

so that, by the convolution theorem,

$$L[f] = L[\sin t]L[e^{-t}] = \frac{1}{(s^2 + 1)(s + 1)}.$$

Example 12.9.3   Find $L^{-1}\left[\dfrac{1}{s^2(s - 1)}\right]$.

Solution   We could determine the inverse Laplace transform in the usual manner by first using a partial fraction decomposition.   However, we will use the convolution theorem.

$$L^{-1}\left[\frac{1}{s^2(s - 1)}\right] = L^{-1}\left[\frac{1}{s^2}\right] * L^{-1}\left[\frac{1}{s - 1}\right] = \int_0^t (t - \tau)e^\tau \, dt.$$

Integrating by parts we obtain

$$L^{-1}\left[\frac{1}{s^2(s - 1)}\right] = \left[te^\tau - (\tau e^\tau - e^\tau) \right]_0^t$$

$$= e^t - t - 1.$$

Example 12.9.4   Find $L^{-1}\left[\dfrac{G(s)}{(s - 1)^2 + 1}\right]$.

Solution   Using the convolution theorem,

$$L^{-1}\left[\frac{G(s)}{(s - 1)^2 + 1}\right] = L^{-1}\left[\frac{1}{(s - 1)^2 + 1}\right] * L^{-1}[G(s)]$$

$$= e^t \sin t * g(t),$$

that is,

$$L^{-1}\left[\frac{G(s)}{(s - 1)^2 + 1}\right] = \int_0^t e^{t-\tau}\sin(t - \tau)g(\tau) \, dt.$$

Example 12.9.5   Solve the initial value problem

$$y'' + \omega^2 y = f(t), \qquad y(0) = \alpha, \qquad y'(0) = \beta,$$

where α, β, and ω ($\neq 0$) are constants and $f(t)$ is Laplace transformable.

Solution   Taking the Laplace transform of the differential equation and imposing the given initial conditions yields:

$$[s^2 Y(s) - \alpha s - \beta] + \omega^2 Y(s) = F(s),$$

where $F(s)$ denotes the Laplace transform of f.   Thus, simplifying, we obtain

$$Y(s) = \frac{F(s)}{s^2 + \omega^2} + \frac{\alpha s}{s^2 + \omega^2} + \frac{\beta}{s^2 + \omega^2}.$$

Taking the inverse Laplace transform of both sides of this equation and using the convolution theorem yields

$$y(t) = \frac{1}{\omega} \int_0^t \sin \omega(t - \tau) f(\tau) \, d\tau + \alpha \cos \omega t + \frac{\beta}{\omega} \sin \omega t.$$

VOLTERRA INTEGRAL EQUATIONS

All the applications of the Laplace transform that we have so far considered have been to solve differential equations.   We now briefly discuss another type of equation whose solution can often be obtained using the Laplace transform.
    An equation of the form

$$x(t) = f(t) + \int_0^t k(t - \tau) x(\tau) \, d\tau, \tag{12.9.4}$$

is called a **Volterra integral equation**.   In this equation the unknown function is $x(t)$.   The functions f and k are specified, and k is called the **kernel** of the equation. For example,

$$x(t) = 2 \sin t + \int_0^t \cos (t - \tau) x(\tau) \, d\tau$$

is a Volterra integral equation.   We now show how the convolution theorem for Laplace transforms can be used to determine, up to the evaluation of an inverse transform, the function $x(t)$ that satisfies (12.9.4).   The key to solving (12.9.4) is to notice that the integral that appears in this equation is, in fact, a convolution integral.   Thus, taking the Laplace transform of both sides of (12.9.4) we obtain

$$X(s) = F(s) + L[k(t) * x(t)],$$

that is, using the convolution theorem,

$$X(s) = F(s) + K(s)X(s).$$

Solving algebraically for $X(s)$ yields

$$X(s) = \frac{F(s)}{1 - K(s)}.$$

Consequently,

$$x(t) = L^{-1}\left[\frac{F(s)}{1 - K(s)}\right].$$

This technique can be used to solve a wide variety of Volterra integral equations.

Example 12.9.6   Solve the Volterra integral equation

$$x(t) = 3 \cos t + 5 \int_0^t \sin(t - \tau)x(\tau)\, d\tau.$$

Solution   Taking the Laplace transform of the given integral equation and using the convolution theorem yields

$$X(s) = \frac{3s}{s^2 + 1} + \frac{5}{s^2 + 1}\, X(s),$$

that is,

$$X(s)\left(\frac{s^2 - 4}{s^2 + 1}\right) = \frac{3s}{s^2 + 1},$$

so that

$$X(s) = \frac{3s}{s^2 - 4}.$$

Decomposing the right-hand side into partial fractions we obtain

$$X(s) = \frac{3}{2}\left(\frac{1}{s - 2} + \frac{1}{s + 2}\right).$$

Taking the inverse Laplace transform yields

$$x(t) = \frac{3}{2}\left(e^{2t} + e^{-2t}\right) = 3 \cosh 2t.$$

EXERCISES 12.9

In problems 1–5 determine $f * g$.

1. $f(t) = t$, $g(t) = 1$.

2. $f(t) = \cos t$, $g(t) = t$.

3. $f(t) = e^t$, $g(t) = t$.

4. $f(t) = t^2$, $g(t) = e^t$.
5. $f(t) = e^t$, $g(t) = e^t \sin t$.
6. Prove that $f * g = g * f$.
7. Prove that $f * (g * h) = (f * g) * h$.
8. Prove that $f * (g + h) = f * g + f * h$.

In problems 9–13 determine L[$f * g$].

9. $f(t) = t$, $g(t) = \sin t$.

10. $f(t) = e^{2t}$, $g(t) = 1$.

11. $f(t) = \sin t$, $g(t) = \cos 2t$.

12. $f(t) = e^t$, $g(t) = te^{2t}$.

13. $f(t) = t^2$, $g(t) = e^{3t}\sin 2t$.

In problems 14–19 determine $L^{-1}[F(s)G(s)]$ in the following two ways: **(a)** using the convolution theorem, **(b)** using partial fractions.

14. $F(s) = \dfrac{1}{s}$, $G(s) = \dfrac{1}{s-2}$.

15. $F(s) = \dfrac{1}{s+1}$, $G(s) = \dfrac{1}{s}$.

16. $F(s) = \dfrac{s}{s^2+4}$, $G(s) = \dfrac{2}{s}$.

17. $F(s) = \dfrac{1}{s+2}$, $G(s) = \dfrac{s+2}{s^2+4s+13}$.

18. $F(s) = \dfrac{1}{s^2+9}$, $G(s) = \dfrac{2}{s^3}$.

19. $F(s) = \dfrac{1}{s^2}$, $G(s) = \dfrac{e^{-\pi s}}{s^2+1}$.

In problems 20–24 express $L^{-1}[F(s)G(s)]$ in terms of a convolution integral.

20. $F(s) = \dfrac{4}{s^3}$, $G(s) = \dfrac{s-1}{s^2-2s+5}$.

21. $F(s) = \dfrac{s+1}{s^2+2s+2}$, $G(s) = \dfrac{1}{(s+3)^2}$.

22. $F(s) = \dfrac{2}{s^2+6s+10}$, $G(s) = \dfrac{2}{s-4}$.

23. $F(s) = \dfrac{s+4}{s^2+8s+25}$, $G(s) = \dfrac{se^{-\pi s/2}}{s^2+16}$.

24. $F(s) = \dfrac{1}{s-4}$.

In problems 25–31 solve the given initial value problem up to the evaluation of a convolution integral.

25. $y'' + y = e^{-t}$, $y(0) = 0$, $y'(0) = 1$.

26. $y'' - 2y' + 10y = \cos 2t$, $y(0) = 0$, $y'(0) = 1$.

27. $y'' + 16y = f(t)$, $y(0) = \alpha$, $y'(0) = \beta$, where α and β are constants.

28. $y' - ay = f(t)$, $y(0) = \alpha$, where a and α are constants.

29. $y'' - a^2 y = f(t)$, $y(0) = \alpha$, $y'(0) = \beta$, where a, α, and β are constants, and $a \neq 0$.

30. $y'' - (a+b)y' + aby = f(t)$, $y(0) = \alpha$, $y'(0) = \beta$, where a, b, α, and β are constants, and $a \neq b$.

31. $y'' - 2ay' + (a^2 + b^2)y = f(t)$, $y(0) = \alpha$, $y'(0) = \beta$, where a, b, α, and β are constants, and $b \neq 0$.

In problems 32–37 solve the given Volterra integral equation.

32. $x(t) = e^{-t} + 4 \displaystyle\int_0^t (t-\tau)x(\tau)\,d\tau$.

33. $x(t) = 2e^{3t} - \displaystyle\int_0^t e^{2(t-\tau)}x(\tau)\,d\tau$.

34. $x(t) = 4e^t + 3 \displaystyle\int_0^t e^{-(t-\tau)}x(\tau)\,d\tau$.

35. $x(t) = 1 + 2 \displaystyle\int_0^t \sin(t-\tau)x(\tau)\,d\tau$.

36. $x(t) = e^{2t} + 5 \displaystyle\int_0^t [\cos 2(t-\tau)]x(\tau)\,d\tau$.

37. $x(t) = 2(1 + \displaystyle\int_0^t [\cos 2(t-\tau)]x(\tau)\,d\tau)$.

38. Show that the initial value problem
$$y'' + y = f(t), \quad y(0) = 0, \quad y'(0) = 0,$$
can be reformulated as the integral equation
$$x(t) = f(t) - \int_0^t (t-\tau)x(\tau)\,d\tau,$$
where $y''(t) = x(t)$.

13

Series Solutions of Linear Differential Equations

13.1   INTRODUCTION

So far the techniques that we have developed for solving differential equations have involved determining a closed form solution for a given equation (or system) in terms of familiar elementary functions. Essentially the only equations that we can derive such solutions for are

1. Constant coefficient equations;
2. Cauchy–Euler equations.

For example, we cannot at the present time determine the solution to the seemingly simple differential equation

$$y'' + e^x y = 0.$$

In this chapter we consider the possibility of determining solutions to differential equations that can be represented as an infinite series of some sort. We begin in

Section 13.3 with the simplest case, namely, differential equations whose solutions can be represented as a convergent power series,

$$y(x) = \sum_{n=0}^{\infty} a_n x^n,$$

where a_n are constants.   This can be considered as a generalization of the method of undetermined coefficients to the case when we have an infinite number of constants.   We will determine the appropriate values of these constants by substitution into the differential equation.

Not all differential equations have solutions that can be represented by a convergent power series.   We will find that the next simplest type of series solution that is applicable to a broad class of differential equations is one of the form

$$y(x) = x^r \sum_{n=0}^{\infty} a_n x^n,$$

called a Frobenius series.   Here, in addition to the coefficients a_n, we must also determine the value of the constant r (which in general will *not* be a positive integer).   The analysis of this problem is quite involved, and the computations can be extremely tedious.   However, the technique is an important and useful addition to the applied mathematician's tools for solving differential equations.

Before beginning the development of the theory, we note that for simplicity we will restrict our attention in this chapter to second-order linear homogeneous differential equations whose standard form is

$$y'' + p(x)y' + q(x)y = 0,$$

where p and q are functions that are specified on some interval I.   The techniques can be extended easily to higher-order linear differential equations and also to systems of linear differential equations.

13.2   A REVIEW OF POWER SERIES

We begin with a very brief review of the main facts about power series, which should be familiar from a previous calculus course.   They will be required throughout the remainder of the chapter.

BASIC DEFINITION   An infinite series of the form

$$\sum_{n=0}^{\infty} a_n(x - x_0)^n, \tag{13.2.1}$$

where a_n and x_0 are constants is called a **power series** centered at $x = x_0$.   The substitution $u = x - x_0$ has the effect of transforming (13.2.1) to

$$\sum_{n=0}^{\infty} a_n u^n,$$

so that we can, without loss of generality, restrict attention to power series of the form

$$\sum_{n=0}^{\infty} a_n x^n, \tag{13.2.2}$$

whose center is $x = 0$. The series (13.2.2) is said to **converge** at $x = x_1$ if

$$\lim_{k \to \infty} \sum_{n=0}^{k} a_n x_1^n$$

exists and is finite. The set of all x for which (13.2.2) converges is called the **interval of convergence.**

BASIC CONVERGENCE THEOREM For the power series (13.2.2) precisely one of the following is true:

1. $\displaystyle\sum_{n=0}^{\infty} a_n x^n$ converges only at $x = 0$.

2. $\displaystyle\sum_{n=0}^{\infty} a_n x^n$ converges for all real x.

3. There is a positive number R such that $\displaystyle\sum_{n=0}^{\infty} a_n x^n$ converges (absolutely) for $|x| < R$ and diverges for $|x| > R$.

REMARK The number R occurring in (3) is called the **radius of convergence** (see Figure 13.2.1). The convergence or divergence of the series at the endpoints $x = \pm R$ must be treated separately. In (2) we define the radius of convergence to be $R = \infty$.

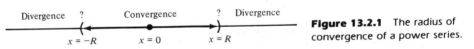

Divergence ? Convergence ? Divergence

$x = -R$ $x = 0$ $x = R$

Figure 13.2.1 The radius of convergence of a power series.

RATIO TEST The ratio test for the convergence of a power series can be stated as follows.

Ratio Test: Consider the power series $\displaystyle\sum_{n=0}^{\infty} a_n x^n$. If $\displaystyle\lim_{n \to \infty} |a_{n+1}/a_n| = L$, then the radius of convergence of the power series is $R = 1/L$. If $L = 0$, the series converges for all x, whereas if $L = \infty$, the power series converges only at $x = 0$.

Example 13.2.1 Determine the radius of convergence of $\displaystyle\sum_{n=0}^{\infty} \frac{n^2 x^n}{3^n}$.

Solution In this case we have

$$\lim_{n \to \infty} \left| \frac{a_{n+1}}{a_n} \right| = \lim_{n \to \infty} \frac{(n+1)^2}{3^{n+1}} \cdot \frac{3^n}{n^2} = \lim_{n \to \infty} \frac{(n+1)^2}{3n^2} = \frac{1}{3}.$$

Thus $L = \frac{1}{3}$, so that the radius of convergence is $R = 3$. It is easy to see that the series diverges at the endpoints[1] $x = \pm 3$, so that the interval of convergence is $(-3, 3)$.

THE ALGEBRA OF POWER SERIES Two power series $\sum\limits_{n=0}^{\infty} a_n x^n$ and $\sum\limits_{n=0}^{\infty} b_n x^n$ are equal if and only if each corresponding coefficient is equal, that is, $a_n = b_n$ for all n. In particular,

$$\sum_{n=0}^{\infty} a_n x^n = 0, \text{ if and only if } a_n = 0 \text{ for every } n.$$

We will use this result repeatedly throughout the chapter.

Now let $\sum\limits_{n=0}^{\infty} a_n x^n$ and $\sum\limits_{n=0}^{\infty} b_n x^n$ be power series with radii of convergence R_1 and R_2, respectively, and let $R = \min(R_1, R_2)$. For $|x| < R$, define the functions f and g by

$$f(x) = \sum_{n=0}^{\infty} a_n x^n, \qquad g(x) = \sum_{n=0}^{\infty} b_n x^n.$$

Then, for $|x| < R$,

1. $f(x) + g(x) = \sum\limits_{n=0}^{\infty} (a_n + b_n) x^n$ (addition of power series).

2. $cf(x) = \sum\limits_{n=0}^{\infty} (ca_n) x^n$ (multiplication of a power series by a real number c).

3. $f(x)g(x) = \sum\limits_{n=0}^{\infty} c_n x^n$, where $c_n = \sum\limits_{k=0}^{n} a_{n-k} b_k$ (multiplication of power series). The coefficients c_n appearing in this formula can be written in the equivalent form
$c_n = \sum\limits_{k=0}^{n} a_k b_{n-k}$.

Example 13.2.2 It is known that the coefficients in the expansion

$$f(x) = \sum_{n=0}^{\infty} a_n x^n$$

---

[1] Recall that a necessary (but not sufficient) condition for the convergence of the infinite series $\sum\limits_{n=0}^{\infty} a_n x^n$ is that $\lim\limits_{n \to \infty} a_n = 0$.

satisfy

$$\sum_{n=1}^{\infty} na_n x^n - \sum_{n=0}^{\infty} a_n x^{n+1} = 0. \tag{13.2.3}$$

Express all a_n in terms of a_0.

Solution   We replace n by $n-1$ in the second summation in (13.2.3) (and alter the range of n appropriately) to obtain a common power of x^n in both sums. The result is

$$\sum_{n=1}^{\infty} na_n x^n - \sum_{n=1}^{\infty} a_{n-1} x^n = 0,$$

which can be written as

$$\sum_{n=1}^{\infty} (na_n - a_{n-1})x^n = 0.$$

It follows that the a_n must satisfy the *recurrence relation*

$$na_n - a_{n-1} = 0, \qquad n = 1, 2, 3, \ldots,$$

that is,

$$a_n = \frac{1}{n} a_{n-1}, \qquad n = 1, 2, 3, \ldots.$$

Substituting for successive values of n, we obtain

$$n = 1: \quad a_1 = a_0, \qquad\qquad n = 2: \quad a_2 = \frac{1}{2} a_1 = \frac{1}{2} a_0,$$

$$n = 3: \quad a_3 = \frac{1}{3} a_2 = \frac{1}{3 \cdot 2} a_0, \qquad n = 4: \quad a_4 = \frac{1}{4} a_3 = \frac{1}{4 \cdot 3 \cdot 2} a_0.$$

Continuing in this manner we see that

$$a_n = \frac{1}{n!} a_0.$$

Consequently we can write[1]

$$f(x) = a_0 \sum_{n=0}^{\infty} \frac{1}{n!} x^n.$$

---

[1] The power series is just the Maclaurin expansion of e^x (see following material), so that we can write $f(x) = a_0 e^x$.

THE CALCULUS OF POWER SERIES Suppose that $\sum\limits_{n=0}^{\infty} a_n x^n$ has radius of convergence R, and let

$$f(x) = \sum_{n=0}^{\infty} a_n x^n, \qquad |x| < R.$$

Then $f(x)$ can be differentiated an arbitrary number of times on the interval $|x| < R$; furthermore, the derivatives can be obtained by termwise differentiation. Thus,

$$f'(x) = \sum_{n=0}^{\infty} n a_n x^{n-1} = \sum_{n=1}^{\infty} n a_n x^{n-1},$$

$$f''(x) = \sum_{n=1}^{\infty} n(n-1) a_n x^{n-2} = \sum_{n=2}^{\infty} n(n-1) a_n x^{n-2},$$

and so on for higher-order derivatives.

ANALYTIC FUNCTIONS AND TAYLOR SERIES We now introduce one of the main definitions of the section.

Definition 13.2.1:   A function is said to be **analytic at $x = x_0$** if it can be represented by a convergent power series centered at $x = x_0$ with nonzero radius of convergence.

In a previous calculus course you should have seen that if a function is analytic at $x = x_0$, then the power series representation of that function is unique and is given by

$$f(x) = \sum_{n=0}^{\infty} \frac{f^{(n)}(x_0)}{n!} (x - x_0)^n. \tag{13.2.4}$$

This is the **Taylor series** expansion of $f(x)$ about $x = x_0$. If $x_0 = 0$, then (13.2.4) reduces to

$$f(x) = \sum_{n=0}^{\infty} \frac{f^{(n)}(0)}{n!} x^n, \tag{13.2.5}$$

which is called the **Maclaurin series** expansion of $f(x)$. Many of the familiar elementary functions are analytic at all points. In particular the Maclaurin expansions of e^x, $\sin x$, and $\cos x$ are, respectively,

$$e^x = 1 + x + \frac{1}{2!} x^2 + \frac{1}{3!} x^3 + \cdots + \frac{1}{n!} x^n + \cdots = \sum_{n=0}^{\infty} \frac{1}{n!} x^n,$$

$$\tag{13.2.6}$$

$$\sin x = x - \frac{1}{3!} x^3 + \frac{1}{5!} x^5 + \cdots + \frac{(-1)^n}{(2n+1)!} x^{2n+1} + \cdots = \sum_{n=0}^{\infty} \frac{(-1)^n}{(2n+1)!} x^{2n+1},$$

$$\tag{13.2.7}$$

$$\cos x = 1 - \frac{1}{2!} x^2 + \frac{1}{4!} x^4 + \cdots + \frac{(-1)^n}{(2n)!} x^{2n} + \cdots = \sum_{n=0}^{\infty} \frac{(-1)^n}{(2n)!} x^{2n},$$

$$\tag{13.2.8}$$

and each of the series on the right of these equations converges to the function on the left for all real values of x.

We can determine many other analytic functions using the following theorem.

Theorem 13.2.1: If $f(x)$ and $g(x)$ are analytic at x_0, then so also are $f(x) \pm g(x)$, $f(x)g(x)$, and $f(x)/g(x)$ [provided $g(x_0) \neq 0$].

Of particular importance to us throughout this chapter will be polynomial functions—that is, functions of the form

$$p(x) = a_0 + a_1 x + a_2 x^2 + \cdots + a_n x^n, \tag{13.2.9}$$

where $a_0, a_1, \ldots, a_n$ are real numbers. Such a function is analytic at all points. In particular, (13.2.9) can be considered as the Maclaurin series expansion of p about $x = 0$. Since the series has only a finite number of terms, it converges for all real x. Now suppose that $p(x)$ and $q(x)$ are polynomials and hence are analytic at all points.   According to Theorem 13.2.1, the rational function r defined by $r(x) = p(x)/q(x)$ is analytic at all points $x = x_0$ such that $q(x) \neq 0$.   However, Theorem 13.2.1 does not give us any indication of the radius of convergence of the series representation of $r(x)$. The following theorem deals with this point.

Theorem 13.2.2: If $p(x)$ and $q(x)$ are polynomials and $q(x_0) \neq 0$, then the power series representation of p/q has radius of convergence R, where R is the distance in the complex plane from x_0 to the nearest zero of q.

PROOF   The proof of this theorem requires results from complex analysis with which we do not assume the reader is familiar. For this reason the proof is omitted. ■

REMARK   If $z = a + ib$ is a zero of q, then the distance from the center, $x = x_0$, of the power series to z is (see Figure 13.2.2)

$$|z - x_0| = \sqrt{(a - x_0)^2 + b^2}.$$

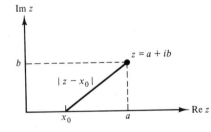

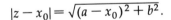

Figure 13.2.2   Determining the radius of convergence of the power series representation of a rational function centered at $x = x_0$.

Example 13.2.3   Determine the radius of convergence of the power series representation of the function

$$f(x) = \frac{1 - x}{x^2 - 4},$$

centered at (a) $x = 0$, (b) $x = 1$.

Solution Taking $p(x) = 1 - x$, and $q(x) = x^2 - 4$, we have

$$f(x) = \frac{p(x)}{q(x)},$$

and the zeros of q are $x = \pm 2$.

(a) In this case the center of the power series is $x = 0$, so that the distance to the nearest zero of q is 2 (see Figure 13.2.3). Consequently, the radius of convergence of the power series representation of f centered at $x = 0$ is $R = 2$.

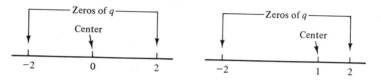

Figure 13.2.3 Determining the radius of convergence of the power series representation of the rational function given in Example 13.2.3.

(b) If the center of the power series is $x = 1$, then the nearest zero of q is at $x = 2$ (see Figure 13.2.3), and hence the radius of convergence of the power series representation of f is $R = 1$.

Example 13.2.4 Determine the radius of convergence of the power series expansion of

$$f(x) = \frac{1 - x}{(x^2 + 2x + 2)(x - 2)},$$

centered at $x = 0$.

Solution We take $p(x) = 1 - x$ and $q(x) = (x^2 + 2x + 2)(x - 2)$. According to Theorem 13.2.2, the radius of convergence of the required power series will be given by the distance from $x = 0$ to the nearest zero of q. It is easily seen that the zeros of q are $x_1 = -1 + i$, $x_2 = -1 - i$, $x_3 = 2$. The corresponding distances from $x = 0$ are

$$d_1 = \sqrt{(1)^2 + (-1)^2} = \sqrt{2}, \qquad d_2 = \sqrt{(1)^2 + (1)^2} = \sqrt{2}, \qquad d_3 = 2.$$

Consequently, the radius of convergence is $R = \sqrt{2}$ (see Figure 13.2.4).

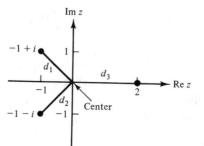

Figure 13.2.4 Determining the radius of convergence of the power series representation of the rational function given in Example 13.2.4.

EXERCISES 13.2

In problems 1–5 determine the radius of convergence of the given power series.

1. $\displaystyle\sum_{n=0}^{\infty} \frac{x^n}{2^{2n}}.$

2. $\displaystyle\sum_{n=0}^{\infty} \frac{x^n}{n^2}.$

3. $\displaystyle\sum_{n=0}^{\infty} \frac{2^n x^n}{n}.$

4. $\displaystyle\sum_{n=0}^{\infty} n! x^n.$

5. $\displaystyle\sum_{n=0}^{\infty} \frac{5^n x^n}{n!}.$

In problems 6–10 determine the radius of convergence of the power series representation of the given function with center x_0.

6. $f(x) = \dfrac{x^2 - 1}{x + 2}, \ x_0 = 0.$

7. $f(x) = \dfrac{x}{x^2 + 1}, \ x_0 = 0.$

8. $f(x) = \dfrac{2x}{x^2 + 16}, \ x_0 = 1.$

9. $f(x) = \dfrac{x^2 - 3}{x^2 - 2x + 5}, \ x_0 = 0.$

10. $f(x) = \dfrac{x}{(x^2 + 4x + 13)(x - 3)}, \ x_0 = -1.$

11. (a) Determine all values of x at which the function

$$f(x) = \frac{1}{x^2 - 1} \qquad \text{(i)}$$

is analytic.
(b) Determine the radius of convergence of a power series representation of the function (i) centered at $x = x_0$. (You will need to consider the cases $-1 < x_0 < 1$ and $|x_0| > 1$ separately.)

12. By redefining the ranges of the summations appearing on the left-hand side, show that

$$\sum_{n=2}^{\infty} n(n-1)a_{n-1}x^{n-2} + \sum_{n=1}^{\infty} na_n x^{n-1}$$

$$= \sum_{n=0}^{\infty} (n+1)(n+3)a_{n+1}x^n.$$

13. If $f(x) = \displaystyle\sum_{n=0}^{\infty} a_n x^n$, where the coefficients in the expansion satisfy

$$\sum_{n=0}^{\infty} n(n+2)a_n x^n + \sum_{n=1}^{\infty} (n-3)a_{n-1}x^n = 0,$$

determine $f(x)$.

14. Suppose it is known that the coefficients in the expansion

$$f(x) = \sum_{n=0}^{\infty} a_n x^n$$

satisfy

$$\sum_{n=0}^{\infty} (n+2)a_{n+1}x^n - \sum_{n=0}^{\infty} a_n x^n = 0.$$

Show that

$$f(x) = \frac{a_0}{x} \sum_{n=0}^{\infty} \frac{1}{(n+1)!} x^{n+1},$$

and express this in terms of familiar elementary functions.

15. If

$$\sum_{n=1}^{\infty} (n+1)(n+2)a_{n+1}x^n - \sum_{n=1}^{\infty} na_{n-1}x^n = 0,$$

show that

$$a_{2k} = \frac{1 \cdot 3 \cdot 5 \cdot \cdots \cdot (2k-1)}{(2k+1)!} a_0,$$

$$a_{2k+1} = \frac{2^{k+1} k!}{(2k+2)!} a_1, \qquad k = 1, 2, \ldots.$$

**13.3   SERIES SOLUTIONS
ABOUT AN ORDINARY POINT**

We now consider the second-order linear homogeneous differential equation written in standard form,

$$y'' + p(x)y' + q(x)y = 0.$$

Our aim is to determine a series representation of the general solution to this differential equation centered at $x = x_0$. We will see that the existence and form of the solution is dependent on the behavior of the functions p and q at $x = x_0$.

Definition 13.3.1:   The point $x = x_0$ is called an **ordinary point** of the differential equation

$$y'' + p(x)y' + q(x)y = 0 \qquad\qquad (13.3.1)$$

if p and q are *both* analytic at $x = x_0$. Any point that is not an ordinary point of (13.3.1) is called a **singular point** of the differential equation.

Example 13.3.1   The differential equation

$$y'' + \frac{1}{x^2 - 4}\, y' + \frac{1}{x + 1}\, y = 0,$$

has

$$p(x) = \frac{1}{x^2 - 4} \quad \text{and} \quad q(x) = \frac{1}{x + 1}.$$

We see by inspection that the only points at which p fails to be analytic are $x = \pm\, 2$, whereas q is analytic at all points except $x = -1$. Consequently, the only singular points of the differential equation are $x = \pm 2, -1$. All other points are ordinary points.

In this section we restrict our attention to ordinary points. Since the functions p and q are analytic at an ordinary point $x = x_0$, we might suspect that any solution to (13.3.1) valid at $x = x_0$ is also analytic there and hence can be represented as a convergent power series

$$y = \sum_{n=0}^{\infty} a_n(x - x_0)^n$$

for appropriate constants a_n. This is indeed the case, and the power series representation of the solution will, in general, have a nonzero radius of convergence. Before stating the general result, we consider a familiar differential equation.

Example 13.3.2 Determine two linearly independent power series solutions to the differential equation

$$y'' + y = 0 \tag{13.3.2}$$

centered at $x = 0$. Identify the solutions in terms of familiar elementary functions.

Solution Since $x = 0$ is an ordinary point of the differential equation, we try for a power series solution of the form

$$y = \sum_{n=0}^{\infty} a_n x^n. \tag{13.3.3}$$

We must determine the coefficients a_n such that (13.3.3) does indeed define a solution to (13.3.2). We proceed in a similar manner to the method of undetermined coefficients by substituting (13.3.3) into (13.3.2) and determining the values of the a_n such that (13.3.3) is indeed a solution. Differentiating (13.3.3) twice with respect to x yields

$$y' = \sum_{n=1}^{\infty} na_n x^{n-1}, \qquad y'' = \sum_{n=2}^{\infty} n(n-1)a_n x^{n-2},$$

where we have shifted the starting point on the summations without loss of generality. Substituting into (13.3.2), it follows that (13.3.3) does define a solution provided

$$\sum_{n=2}^{\infty} n(n-1)a_n x^{n-2} + \sum_{n=0}^{\infty} a_n x^n = 0. \tag{13.3.4}$$

If we replace n by $k+2$ in the first summation and replace n by k in the second summation, the result is

$$\sum_{k=0}^{\infty} (k+2)(k+1)a_{k+2} x^k + \sum_{k=0}^{\infty} a_k x^k = 0.$$

Combining the summations yields

$$\sum_{k=0}^{\infty} [(k+2)(k+1)a_{k+2} + a_k] x^k = 0.$$

This implies that the coefficients of x^k must vanish for $k = 0, 1, 2, \ldots$. Consequently we obtain the *recurrence relation*

$$(k+2)(k+1)a_{k+2} + a_k = 0, \qquad k = 0, 1, 2, \ldots.$$

Since $(k+2)(k+1)$ is never zero, we can write this recurrence relation in the equivalent form

$$a_{k+2} = -\frac{1}{(k+2)(k+1)} a_k, \qquad k = 0, 1, 2, \ldots. \tag{13.3.5}$$

We now use this relation to determine the appropriate values of the coefficients. It is convenient to consider the two cases k even and k odd separately.

Even k: Consider the case when k is an even integer. Substituting successively into (13.3.5) we obtain the following.

$k = 0$:

$$a_2 = -\frac{1}{2} a_0.$$

$k = 2$:

$$a_4 = -\frac{1}{4 \cdot 3} a_2 = \frac{1}{4 \cdot 3 \cdot 2} a_0,$$

that is,

$$a_4 = \frac{1}{4!} a_0.$$

$k = 4$:

$$a_6 = -\frac{1}{6 \cdot 5} a_4,$$

so that

$$a_6 = -\frac{1}{6!} a_0.$$

Continuing in this manner, we soon recognize the pattern that is emerging, namely,

$$a_{2n} = \frac{(-1)^n}{(2n)!} a_0. \tag{13.3.6}$$

Thus all the even coefficients are determined in terms of a_0, but a_0 itself is arbitrary.

Odd k: Now consider the recurrence relation (13.3.5) when k is an odd positive integer.

$k = 1$:

$$a_3 = -\frac{1}{3 \cdot 2} a_1 = -\frac{1}{3!} a_1.$$

$k = 3$:

$$a_5 = -\frac{1}{5 \cdot 4} a_3 = \frac{1}{5!} a_1.$$

$k = 5$:

$$a_7 = -\frac{1}{7 \cdot 6} a_5 = -\frac{1}{7!} a_1.$$

Continuing in this manner, we see that the general odd coefficient is

$$a_{2n+1} = \frac{(-1)^n}{(2n+1)!} a_1. \tag{13.3.7}$$

Thus, we have shown that for all values of a_0, a_1, a solution to the given differential equation is

$$y = a_0 \left(1 - \frac{1}{2!} x^2 + \frac{1}{4!} x^4 - \cdots \right) + a_1 \left(x - \frac{1}{3!} x^3 + \frac{1}{5!} x^5 - \cdots \right),$$

that is,

$$y = a_0 \sum_{n=0}^{\infty} \frac{(-1)^n}{(2n)!} x^{2n} + a_1 \sum_{n=0}^{\infty} \frac{(-1)^n}{(2n+1)!} x^{2n+1}. \tag{13.3.8}$$

Setting $a_1 = 0$, and $a_0 = 1$ yields the solution

$$y_1 = \sum_{n=0}^{\infty} \frac{(-1)^n}{(2n)!} x^{2n},$$

whereas if we set $a_0 = 0$ and $a_1 = 1$, we obtain the solution

$$y_2 = \sum_{n=0}^{\infty} \frac{(-1)^n}{(2n+1)!} x^{2n+1}.$$

Applying the ratio test it is straightforward to show that both series converge for all real x. Finally,

$$W[y_1, y_2](0) = \begin{vmatrix} 1 & 0 \\ 0 & 1 \end{vmatrix} = 1 \neq 0,$$

so that the solutions are linearly independent on $(-\infty, \infty)$. It follows that (13.3.8) is the general solution of the given differential equation. Indeed, the power series representing y_1 is just the Maclaurin series expansion of $\cos x$, whereas the series defining y_2 is the Maclaurin series expansion of $\sin x$. Thus we can write (13.3.8) in the more familiar form

$$y = a_0 \cos x + a_1 \sin x.$$

The solution of the preceding example consisted of the following four steps.

1. Assume a power series solution of the form

$$y = \sum_{n=0}^{\infty} a_n x^n$$

exists.
2. Determine the values of the coefficients, a_n, such that y is a formal solution of the differential equation. This led to two distinct solutions, one determined in terms of the constant a_0 and the other in terms of the constant a_1.
3. Use the ratio test to determine the radius of convergence of the solutions and hence the interval over which the solutions are valid.
4. Check that the solutions are linearly independent on the interval of existence.

The following theorem justifies these steps and shows that the technique can be applied about any ordinary point of a differential equation.

Theorem 13.3.1: Let p and q be analytic at $x = x_0$, and suppose that their power series expansions are valid for $|x - x_0| < R$. Then the general solution of the differential equation

$$y'' + p(x)y' + q(x)y = 0 \qquad (13.3.9)$$

can be represented as a power series centered at $x = x_0$, with radius of convergence *at least R*. The coefficients in this series solution can be determined in terms of a_0 and a_1 by directly substituting $\sum_{n=0}^{\infty} a_n(x - x_0)^n$ into (13.3.9). The resulting solution is of the form

$$y = a_0 y_1 + a_1 y_2,$$

where y_1 and y_2 are linearly independent solutions to (13.3.9) on the interval of existence. If the initial conditions $y(x_0) = \alpha$, $y'(x_0) = \beta$ are imposed, then $a_0 = \alpha$, $a_1 = \beta$.

IDEA BEHIND PROOF   We outline the steps required to prove this theorem but do not give details. The first step is to expand p and q in a power series centered at $x = x_0$. Then we assume a solution exists of the form $y = \sum_{n=0}^{\infty} a_n(x - x_0)^n$ and substitute this into the differential equation. Upon collecting the coefficients of like powers of $(x - x_0)$, a recurrence relation is obtained, and it can be shown that this relation determines all of the coefficients in terms of a_0 and a_1. These steps are tedious computationally but quite straightforward. The hard part is to show that the power series solution that has been obtained has a radius of convergence at least equal to R. This requires some ideas from advanced calculus. Having determined a power series solution, y_1 and y_2 arise as the special cases $a_0 = 1$, $a_1 = 0$, and $a_0 = 0$, $a_1 = 1$, respectively. The Wronskian of these functions satisfies $W[y_1, y_2](x_0) = 1$, so that they are linearly independent on their interval of existence. Finally, it is easy to show that $y(x_0) = a_0$ and that $y'(x_0) = a_1$. ■

We now illustrate the use of the preceding theorem with some examples.

Example 13.3.3   Show that

$$(1 + x^2)y'' + 3xy' + y = 0, \qquad (13.3.10)$$

has two linearly independent series solutions centered at $x = 0$, and determine a lower bound on the radius of convergence of these solutions.

Solution   We first rewrite (13.3.10) in the standard form

$$y'' + \frac{3x}{1 + x^2}\, y' + \frac{1}{1 + x^2}\, y = 0,$$

from which we can conclude that $x = 0$ is an ordinary point and hence (13.3.10) does indeed have two linearly independent series solutions centered at $x = 0$. In this case

$$p(x) = \frac{3x}{1 + x^2} \quad \text{and} \quad q(x) = \frac{1}{1 + x^2}.$$

According to Theorem 13.3.1 the radius of convergence of the power series solutions will be *at least* equal to the smaller of the radii of convergence of the power series representations of p and q. Using Theorem 13.2.2 we see directly that the series expansions of both p and q about $x = 0$ have radius of convergence $R = 1$, so that a lower bound on the radius of convergence of the power series solutions to (13.3.10) is also $R = 1$.

Example 13.3.4   Determine two linearly independent series solutions in powers of x to

$$y'' - 2xy' - 4y = 0, \tag{13.3.11}$$

and find the radius of convergence of these solutions.

Solution   The point $x = 0$ is an ordinary point of the differential equation, and therefore Theorem 13.3.1 can be applied with $x_0 = 0$. In this case we have

$$p(x) = -2x, \qquad q(x) = -4.$$

Since these are both polynomials, their power series expansions about $x = 0$ are valid for all x; hence, from the previous theorem, the general solution to (13.3.11) can be represented in the form

$$y = \sum_{n=0}^{\infty} a_n x^n, \tag{13.3.12}$$

and this power series solution will converge for all real x. Differentiating (13.3.12) we obtain

$$y' = \sum_{n=1}^{\infty} n a_n x^{n-1}, \qquad y'' = \sum_{n=2}^{\infty} n(n - 1) a_n x^{n-2}.$$

Substitution into (13.3.11) yields

$$\sum_{n=2}^{\infty} n(n - 1) a_n x^{n-2} - 2 \sum_{n=1}^{\infty} n a_n x^n - 4 \sum_{n=0}^{\infty} a_n x^n = 0.$$

We now redefine the ranges in the summations in order to obtain a common x^k in all terms. This is accomplished by replacing n by $k + 2$ in the first summation and, for consistency in notation, we replace n by k in the other summations. The result is

$$\sum_{k=0}^{\infty} [(k + 2)(k + 1) a_{k+2} - 2k a_k - 4 a_k] x^k = 0.$$

This equation requires that the coefficient of x^k vanish, and hence we obtain the recurrence relation

$$(k + 2)(k + 1) a_{k+2} - 2k a_k - 4 a_k = 0, \qquad k = 0, 1, 2, \ldots,$$

which can be written in the equivalent form

$$a_{k+2} = \frac{2(k+2)}{(k+1)(k+2)} a_k, \qquad k = 0, 1, 2, \ldots,$$

that is,

$$a_{k+2} = \frac{2}{(k+1)} a_k, \qquad k = 0, 1, 2, \ldots . \qquad (13.3.13)$$

We see from this relation that, as in the previous example, all the even coefficients can be expressed in terms of a_0, whereas all the odd coefficients can be expressed in terms of a_1. We now determine the exact form of these coefficients.

Consider first the case when k is an even integer. From (13.3.13) we have

$$k = 0: \quad a_2 = 2a_0.$$

$$k = 2: \quad a_4 = \frac{2}{3} a_2 = \frac{2^2}{3} a_0.$$

$$k = 4: \quad a_6 = \frac{2}{5} a_4 = \frac{2^3}{1 \cdot 3 \cdot 5} a_0.$$

$$k = 6: \quad a_8 = \frac{2}{7} a_6 = \frac{2^4}{1 \cdot 3 \cdot 5 \cdot 7} a_0.$$

The general even term is thus

$$a_{2n} = \frac{2^n}{1 \cdot 3 \cdot 5 \cdot \cdots \cdot (2n-1)} a_0, \qquad n = 1, 2, \ldots .$$

Now consider the case when k is an odd integer. Substituting successively into (13.3.13) yields

$$k = 1: \quad a_3 = a_1.$$

$$k = 3: \quad a_5 = \frac{2}{4} a_3 = \frac{1}{1 \cdot 2} a_1.$$

$$k = 5: \quad a_7 = \frac{2}{6} a_5 = \frac{1}{1 \cdot 2 \cdot 3} a_1.$$

$$k = 7: \quad a_9 = \frac{2}{8} a_7 = \frac{1}{1 \cdot 2 \cdot 3 \cdot 4} a_1.$$

The general odd term can therefore be written as

$$a_{2n+1} = \frac{1}{n!} a_1, \qquad n = 1, 2, \ldots .$$

Substituting back into (13.3.12) we obtain the solution

$$y = a_0 \left[1 + 2x^2 + \frac{2^2}{1 \cdot 3} x^4 + \frac{2^3}{1 \cdot 3 \cdot 5} x^6 + \cdots + \frac{2^n}{1 \cdot 3 \cdot 5 \cdots (2n-1)} x^{2n} + \cdots \right]$$

$$+ a_1 \left[x + x^3 + \frac{1}{2!} x^5 + \frac{1}{3!} x^7 + \cdots + \frac{1}{n!} x^{2n+1} + \cdots \right],$$

that is,

$$y = a_0 \left[1 + \sum_{n=1}^{\infty} \frac{2^n}{1 \cdot 3 \cdot 5 \cdots (2n-1)} x^{2n} \right] + a_1 \left[\sum_{n=0}^{\infty} \frac{1}{n!} x^{2n+1} \right].$$

Consequently, from Theorem 13.3.1, two linearly independent solutions to (13.3.11) on $(-\infty, \infty)$ are

$$y_1 = 1 + \sum_{n=1}^{\infty} \frac{2^n}{1 \cdot 3 \cdot 5 \cdots (2n-1)} x^{2n}, \qquad y_2 = \sum_{n=0}^{\infty} \frac{1}{n!} x^{2n+1}.$$

In Examples 13.3.2 and 13.3.4 we have been able to solve the recurrence relation that has arisen from the power series technique. In general this is not possible, and hence we must be satisfied with obtaining just a finite number of terms in each power series solution.

Example 13.3.5   Determine the terms up to x^5 in each of two linearly independent power series solutions to

$$y'' + (2 - 4x^2)y' - 8xy = 0,$$

centered at $x = 0$. Also find the radius of convergence of these solutions.

Solution   The functions $p(x) = 2 - 4x^2$, and $q(x) = -8x$ are polynomials, and hence, from Theorem 13.3.1, the power series solutions will converge for all real x. We now determine the solutions. If

$$y = \sum_{n=0}^{\infty} a_n x^n, \tag{13.3.14}$$

then

$$y' = \sum_{n=1}^{\infty} n a_n x^{n-1}, \qquad y'' = \sum_{n=2}^{\infty} n(n-1) a_n x^{n-2}.$$

Substitution into the given differential equation yields

$$\sum_{n=2}^{\infty} n(n-1) a_n x^{n-2} + 2 \sum_{n=1}^{\infty} n a_n x^{n-1} - 4 \sum_{n=1}^{\infty} n a_n x^{n+1} - 8 \sum_{n=0}^{\infty} a_n x^{n+1} = 0.$$

Replacing n by $k + 2$ in the first summation, $k + 1$ in the second summation, and $k - 1$ in the third and fourth summations, we obtain

$$\sum_{k=0}^{\infty}(k + 2)(k + 1)a_{k+2}x^k + 2\sum_{k=0}^{\infty}(k + 1)a_{k+1}x^k$$

$$- 4\sum_{k=2}^{\infty}(k - 1)a_{k-1}x^k - 8\sum_{k=1}^{\infty}a_{k-1}x^k = 0.$$

Separating out the terms corresponding to $k = 0$ and $k = 1$, it follows that this can be written as

$$\underbrace{2a_2 + 2a_1}_{k=0} + \underbrace{(6a_3 + 4a_2 - 8a_0)x}_{k=1} + \sum_{k=2}^{\infty}\{(k + 2)(k + 1)a_{k+2}$$

$$+ 2(k + 1)a_{k+1} - [4(k - 1) + 8]a_{k-1}\}x^k = 0.$$

Setting the coefficients of consecutive powers of x to zero yields

$$k = 0: \quad 2a_2 + 2a_1 = 0, \tag{13.3.15}$$
$$k = 1: \quad 6a_3 + 4a_2 - 8a_0 = 0, \tag{13.3.16}$$
$$k \geq 2: \quad (k + 2)(k + 1)a_{k+2} + 2(k + 1)a_{k+1} - [4(k - 1) + 8]a_{k-1} = 0. \tag{13.3.17}$$

It follows from (13.3.15) and (13.3.16) that

$$a_2 = -a_1, \qquad a_3 = \frac{2}{3}(2a_0 - a_2) = \frac{2}{3}(2a_0 + a_1), \tag{13.3.18}$$

and (13.3.17) yields the general recurrence relation

$$a_{k+2} = \frac{4a_{k-1} - 2a_{k+1}}{k + 2}, \qquad k = 2, 3, 4, \ldots . \tag{13.3.19}$$

In this case the recurrence relation is quite difficult to solve. However we were asked to determine terms only up to x^5 in the series solutions, and so we proceed to do so. We already have a_2 and a_3 expressed in terms of a_0 and a_1. Setting $k = 2$ in (13.3.19) yields

$$a_4 = \tfrac{1}{4}(4a_1 - 2a_3) = \tfrac{1}{4}[4a_1 - \tfrac{4}{3}(2a_0 + a_1)],$$

where we have substituted from (13.3.18) for a_3. Simplifying this expression we obtain

$$a_4 = \tfrac{2}{3}(a_1 - a_0).$$

We still require one more term. Setting $k = 3$ in (13.3.19) yields

$$a_5 = \tfrac{1}{5}(4a_2 - 2a_4) = \tfrac{1}{5}[-4a_1 - \tfrac{4}{3}(a_1 - a_0)],$$

so that

$$a_5 = \tfrac{4}{15}(a_0 - 4a_1).$$

Substituting for the coefficients a_2, a_3, a_4, a_5 into (13.3.14) we obtain

$$y = a_0 + a_1x - a_1x^2 + \tfrac{2}{3}(2a_0 + a_1)x^3 + \tfrac{2}{3}(a_1 - a_0)x^4 + \tfrac{4}{15}(a_0 - 4a_1)x^5 + \cdots,$$

that is,

$$y = a_0[1 + \tfrac{4}{3}x^3 - \tfrac{2}{3}x^4 + \tfrac{4}{15}x^5 + \cdots] + a_1[x - x^2 + \tfrac{2}{3}x^3 + \tfrac{2}{3}x^4 - \tfrac{16}{15}x^5 + \cdots].$$

Thus two linearly independent solutions to the given differential equation on $(-\infty, \infty)$ are

$$y_1 = 1 + \tfrac{4}{3}x^3 - \tfrac{2}{3}x^4 + \tfrac{4}{15}x^5 + \cdots,$$

$$y_2 = x - x^2 + \tfrac{2}{3}x^3 + \tfrac{2}{3}x^4 - \tfrac{16}{15}x^5 + \cdots.$$

EXERCISES 13.3

In problems 1–8 determine two linearly independent power series solutions to the given differential equation centered at $x = 0$. Also determine the radius of convergence of the series solutions.

1. $y'' - y = 0$.

2. $y'' - 2xy' - 2y = 0$.

3. $y'' + 2xy' + 4y = 0$.

4. $y'' + xy = 0$.

5. $y'' - x^2y' - 2xy = 0$.

6. $y'' - x^2y' - 3xy = 0$.

7. $y'' + xy' + 3y = 0$.

8. $y'' + 2x^2y' + 2xy = 0$.

In problems 9–12 determine two linearly independent power series solutions to the given differential equation centered at $x = 0$. Give a lower bound on the radius of convergence of the series solutions obtained.

9. $(1 + x^2)y'' + 4xy' + 2y = 0$.

10. $(x^2 - 3)y'' - 3xy' - 5y = 0$.

11. $(x^2 - 1)y'' - 6xy' + 12y = 0$.

12. $(1 - 4x^2)y'' - 20xy' - 16y = 0$.

In problems 13–16 determine terms up to and including x^5 in two linearly independent power series solutions of the given differential equation. State the radius of convergence of the series solutions.

13. $y'' + xy' + (2 + x)y = 0$.

14. $y'' + 2y' + 4xy = 0$.

15. $y'' - e^xy = 0$. $\left(\text{Hint:}\right.$

$$e^x = 1 + x + \frac{1}{2!}x^2 + \frac{1}{3!}x^3 + \cdots.\left.\right)$$

16. $y'' + (\sin x)y' + y = 0$.

17. Consider the differential equation

$$xy'' - (x - 1)y' - xy = 0. \qquad \text{(i)}$$

(a) Is $x = 0$ an ordinary point?

(b) Determine the first three nonzero terms in each of two linearly independent series solutions to (i) centered at $x = 1$. (*Hint:* Make the change of variables $z = x - 1$ and obtain a series solution in powers of z.) Give a lower bound on the radius of convergence of each of your solutions.

18. Determine a series solution to the initial value problem

$$(1 + 2x^2)y'' + 7xy' + 2y = 0,$$

$$y(0) = 0, \; y'(0) = 1.$$

19. (a) Determine a series solution to the initial value problem

$$4y'' + xy' + 4y = 0,$$

$$y(0) = 1, \; y'(0) = 0. \qquad \text{(ii)}$$

(b) Find a polynomial that approximates the solution to (ii) with an error less than 10^{-5} on the interval $[-1, 1]$. (*Hint:* The series obtained is a convergent alternating series.)

20. Consider the differential equation

$$(x^2 - 1)y'' + [1 - (a + b)]xy' + aby = 0, \quad \text{(iii)}$$

where a and b are constants.

(a) Show that the coefficients in a series solution to (iii) centered at $x = 0$ must satisfy the recurrence relation

$$a_{n+2} = \frac{(n - a)(n - b)}{(n + 2)(n + 1)} a_n, \quad n = 0, 1, \ldots,$$

and determine two linearly independent series solutions.

(b) Show that if either a or b is a nonnegative integer, then one of the solutions obtained in (a) is a polynomial.

(c) Show that if a is an odd positive integer and b is an even positive integer, then *both* of the solutions obtained in (a) are polynomials.

(d) If $a = 5$ and $b = 4$, determine two linearly independent polynomial solutions to (iii).

Notice that in this case the radius of convergence of the solutions obtained is $R = \infty$, whereas Theorem 13.3.1 guarantees a radius of convergence of only $R \geq 1$.

The power series technique can also be used to solve nonhomogeneous differential equations of the form

$$y'' + p(x)y' + q(x)y = r(x),$$

provided p, q, and r are analytic at the point about which we are expanding. In problems 21 and 22, determine terms up to x^6 in the power series representation of the general solution to the given differential equation centered at $x = 0$. Identify those terms in your solution that correspond to the complementary function and those that correspond to a particular solution of the differential equation.

21. $y'' + xy' - 4y = 6e^x$.

22. $y'' + 2x^2y' + xy = 2 \cos x$.

13.4   THE LEGENDRE EQUATION

There are several linear differential equations that tend to arise frequently in applied mathematics and whose solutions can only be obtained using a power series technique. Among the most important of these are the following:

$$(1 - x^2)y'' - 2xy' + \alpha(\alpha + 1)y = 0 \qquad \text{(Legendre equation)},$$

$$y'' - 2xy' + \alpha y = 0 \qquad \text{(Hermite equation)},$$

$$(1 - x^2)y'' - xy' + \alpha^2 y = 0 \qquad \text{(Chebyshev equation)},$$

where α is an arbitrary constant. Since $x = 0$ is an ordinary point of these equations, we can obtain a series solution in powers of x. We will consider only the Legendre equation and leave the analysis of the remaining equations for the exercises.

Example 13.4.1   Obtain two linearly independent series solutions to Legendre's equation in powers of x. Give a lower bound for the radii of convergence of the series solutions.

Solution   The Legendre equation is

$$(1 - x^2)y'' - 2xy' + \alpha(\alpha + 1)y = 0, \qquad (13.4.1)$$

where α is an arbitrary constant.   In order to determine a lower bound on the radius of convergence of the series solutions to this equation, we divide by $1 - x^2$ to obtain

$$y'' - \frac{2x}{(1 - x^2)} y' + \frac{\alpha(\alpha + 1)}{(1 - x^2)} y = 0.$$

Since the power series expansion of $1/(1 - x^2)$ about $x = 0$ is valid for $|x| < 1$, it follows that a lower bound on the radius of convergence of the power series solutions of (13.4.1) about $x = 0$ is 1.   We now determine the series solutions.   Let

$$y = \sum_{n=0}^{\infty} a_n x^n.$$

Then

$$y' = \sum_{n=1}^{\infty} n a_n x^{n-1}, \; y'' = \sum_{n=2}^{\infty} n(n - 1) a_n x^{n-2}.$$

Substituting into (13.4.1) yields

$$\sum_{n=2}^{\infty} n(n - 1) a_n x^{n-2} - \sum_{n=2}^{\infty} n(n - 1) a_n x^n - 2 \sum_{n=1}^{\infty} n a_n x^n + \sum_{n=0}^{\infty} \alpha(\alpha + 1) a_n x^n = 0.$$

That is, upon redefining the ranges of the summations,

$$\sum_{n=0}^{\infty} [(n + 2)(n + 1) a_{n+2} - n(n - 1) a_n - 2 n a_n + \alpha(\alpha + 1) a_n] x^n = 0.$$

Thus we obtain the recurrence relation

$$a_{n+2} = \frac{[n(n+1) - \alpha(\alpha + 1)]}{(n + 1)(n + 2)} a_n, \qquad n = 0, 1, 2, \ldots, \qquad (13.4.2)$$

which can be written as

$$a_{n+2} = -\frac{(\alpha - n)(\alpha + n + 1)}{(n + 1)(n + 2)} a_n, \qquad n = 0, 1, 2, \ldots.$$

Even n:

$$n = 0 \Rightarrow a_2 = -\frac{\alpha(\alpha + 1)}{2} a_0;$$

$$n = 2 \Rightarrow a_4 = -\frac{(\alpha - 2)(\alpha + 3)}{3 \cdot 4} a_2 = \frac{(\alpha - 2)\alpha(\alpha + 1)(\alpha + 3)}{4!} a_0;$$

$$n = 4 \Rightarrow a_6 = -\frac{(\alpha - 4)(\alpha + 5)}{5 \cdot 6} a_4 = -\frac{(\alpha - 4)(\alpha - 2)\alpha(\alpha + 1)(\alpha + 3)(\alpha + 5)}{6!} a_0.$$

In general,

$$
a_{2k} =
$$
$$
(-1)^k \frac{(\alpha - 2k + 2)(\alpha - 2k + 4) \cdots \cdots (\alpha - 2)\alpha(\alpha + 1)(\alpha + 3) \cdots \cdots (\alpha + 2k - 1)}{(2k)!} a_0,
$$
$$
k = 1, 2, 3, \ldots.
$$

Odd n:

$$
n = 1 \Rightarrow a_3 = -\frac{(\alpha - 1)(\alpha + 2)}{2 \cdot 3} a_1;
$$

$$
n = 3 \Rightarrow a_5 = -\frac{(\alpha - 3)(\alpha + 4)}{4 \cdot 5} a_3 = \frac{(\alpha - 3)(\alpha - 1)(\alpha + 2)(\alpha + 4)}{5!} a_1;
$$

$$
n = 5 \Rightarrow a_7 = -\frac{(\alpha - 5)(\alpha + 6)}{6 \cdot 7} a_5 = -\frac{(\alpha - 5)(\alpha - 3)(\alpha - 1)(\alpha + 2)(\alpha + 4)(\alpha + 6)}{7!} a_1.
$$

In general,

$$
a_{2k+1} = (-1)^k \frac{(\alpha - 2k + 1) \cdots \cdots (\alpha - 3)(\alpha - 1)(\alpha + 2)(\alpha + 4) \cdots \cdots (\alpha + 2k)}{(2k + 1)!} a_1,
$$
$$
k = 1, 2, 3, \ldots.
$$

Consequently, for $a_0 \neq 0$ and $a_1 \neq 0$, two linearly independent solutions to the Legendre equation are

$$
y_1 = a_0 \left[1 - \frac{\alpha(\alpha + 1)}{2} x^2 + \frac{(\alpha - 2)\alpha(\alpha + 1)(\alpha + 3)}{4!} x^4 \right.
$$
$$
\left. - \frac{(\alpha - 4)(\alpha - 2)\alpha(\alpha + 1)(\alpha + 3)(\alpha + 5)}{6!} x^6 + \cdots \right], \quad (13.4.3)
$$

and

$$
y_2 = a_1 \left[x - \frac{(\alpha - 1)(\alpha + 2)}{3!} x^3 + \frac{(\alpha - 3)(\alpha - 1)(\alpha + 2)(\alpha + 4)}{5!} x^5 + \cdots \right], \quad (13.4.4)
$$

and both of these solutions are valid for $|x| < 1$.

THE LEGENDRE POLYNOMIALS

Consider the Legendre equation

$$
(1 - x^2)y'' - 2xy' + N(N + 1)y = 0,
$$

where N is a nonnegative *integer*. The recurrence relation (13.4.2) is

$$
a_{n+2} = \frac{[n(n + 1) - N(N + 1)]}{(n + 1)(n + 2)} a_n, \quad n = 0, 1, 2, \ldots,
$$

which implies that

$$a_{N+2} = a_{N+4} = \cdots = 0.$$

Consequently, one of the solutions to Legendre's equation in this case is a polynomial of degree N. (Notice that such a solution converges for all x, and hence we have a radius of convergence greater than is guaranteed by Theorem 13.3.1.)

Definition 13.4.1: Let N be a nonnegative integer. The **Legendre polynomial of degree** N, denoted $P_N(x)$, is defined to be the polynomial solution of

$$(1 - x^2)y'' - 2xy' + N(N+1)y = 0$$

which has been normalized so that $P_N(1) = 1$.

Example 13.4.2 Determine P_0, P_1, P_2.

Solution Substituting $\alpha = N$ into (13.4.3), (13.4.4) yields

$N = 0$: $y_1(x) = a_0$, which implies that $P_0(x) = 1$;
$N = 1$: $y_2(x) = a_1 x$, which implies that $P_1(x) = x$;
$N = 2$: $y_1(x) = a_0(1 - 3x^2)$. Imposing the normalizing condition that $y_1(1) = 1$ we require $a_0 = -\frac{1}{2}$, so that $P_2(x) = \frac{1}{2}(3x^2 - 1)$.

In general it is tedious to determine $P_N(x)$ directly from (13.4.3) and (13.4.4), and various other methods have been derived. Among the most useful are the following.

Rodrigues's formula

$$P_N(x) = \frac{1}{2^N N!} \frac{d^N}{dx^N}[(x^2 - 1)^N], \qquad N = 0, 1, 2, \ldots.$$

Recurrence relation

$$P_{N+1}(x) = \frac{(2N+1)xP_N(x) - NP_{N-1}(x)}{(N+1)}, \qquad N = 1, 2, \ldots.$$

We can use Rodrigues's formula to obtain P_N directly. Alternatively, starting with P_0 and P_1, we can use the recurrence relation to generate all P_N.

Example 13.4.3 According to Rodrigues's formula,

$$P_2(x) = \frac{1}{8}\frac{d^2}{dx^2}[(x^2 - 1)^2] = \frac{1}{8}\frac{d}{dx}[4x(x^2 - 1)] = \frac{1}{2}(3x^2 - 1),$$

which does indeed coincide with that given in Example 13.4.2.

The first five Legendre polynomials are tabulated next.

TABLE 13.4.1 THE FIRST FIVE
LEGENDRE POLYNOMIALS

N	Legendre polynomial of degree N
0	$P_0(x) = 1.$
1	$P_1(x) = x.$
2	$P_2(x) = \frac{1}{2}(3x^2 - 1).$
3	$P_3(x) = \frac{1}{2}x(5x^2 - 3).$
4	$P_4(x) = \frac{1}{8}(35x^4 - 30x^2 + 3).$

ORTHOGONALITY OF THE LEGENDRE POLYNOMIALS

In Section 6.8 we defined an inner product on the vector space $C^0[a, b]$ by

$$<f, g> = \int_a^b f(x)g(x)\ dx,$$

for all f, g in $C^0(a, b)$. We now show that the Legendre polynomials are orthogonal relative to the above inner product on the interval $[-1, 1]$.

Theorem 13.4.1: The Legendre polynomials $\{P_0, P_1, \ldots\}$ are orthogonal on the interval $[-1, 1]$, that is,

$$\int_{-1}^1 P_M(x)P_N(x)\ dx = 0 \qquad \text{whenever } M \neq N.$$

PROOF It is easily seen that Legendre's equation

$$(1 - x^2)y'' - 2xy' + \alpha(\alpha + 1)y = 0$$

can be written in the form

$$[(1 - x^2)y']' + \alpha(\alpha + 1)y = 0.$$

It follows that the Legendre polynomials $P_N(x)$ and $P_M(x)$ satisfy

$$[(1 - x^2)P_N']' + N(N + 1)P_N = 0, \tag{13.4.5}$$

$$[(1 - x^2)P_M']' + M(M + 1)P_M = 0, \tag{13.4.6}$$

respectively. Multiplying (13.4.5) by P_M and (13.4.6) by P_N and subtracting yields

$$[(1 - x^2)P_N']'P_M - [(1 - x^2)P_M']'P_N + [N(N + 1) - M(M + 1)]P_M P_N = 0,$$

which can be written as

$$\{[(1-x^2)P_N'P_M]' - (1-x^2)P_N'P_M'\} - \{[(1-x^2)P_M'P_N]' - (1-x^2)P_N'P_M'\}$$
$$+ [N(N+1) - M(M+1)]P_MP_N = 0,$$

that is,

$$[(1-x^2)(P_N'P_M - P_M'P_N)]' + [N(N+1) - M(M+1)]P_MP_N = 0.$$

Integrating over the interval $[-1, 1]$, we obtain

$$\left[(1-x^2)(P_N'P_M - P_M'P_N)\right]_{-1}^1 + [N(N+1) - M(M+1)]\int_{-1}^1 P_M(x)P_N(x)\, dx = 0.$$

The first term vanishes at $x = \pm 1$, and the term multiplying the integral can be factored to yield

$$(N-M)(N+M+1)\int_{-1}^1 P_M(x)P_N(x)\, dx = 0.$$

Consequently, since M and N are nonnegative integers,

$$\int_{-1}^1 P_M(x)P_N(x)\, dx = 0 \qquad \text{whenever } M \neq N. \qquad \blacksquare$$

It can also be shown, although it is more difficult (see N. N. Lebedev, *Special Functions and their Applications*, Dover, 1972), that

$$\int_{-1}^1 P_N^2(x)\, dx = \frac{2}{2N+1}. \tag{13.4.7}$$

Consequently, $\{\sqrt{(2N+1)/2}\, P_N(x)\}$ is an orthonormal set of polynomials on $[-1, 1]$.

Since the Legendre polynomials $P_0, P_1, \ldots, P_N$ are linearly independent on any interval, they form a basis for the vector space of all polynomials of degree less than or equal to N. Thus if $p(x)$ is any polynomial of degree less than or equal to N, there exist scalars $a_0, a_1, \ldots, a_N$ such that

$$p(x) = \sum_{k=0}^N a_k P_k(x). \tag{13.4.8}$$

We can use the orthogonality of the Legendre polynomials to determine the coefficients a_k in this expansion as follows. Multiplying (13.4.8) by $P_j(x)$, $0 \leq j \leq N$, and integrating over the interval $[-1, 1]$ yields

$$\int_{-1}^1 p(x)P_j(x)\, dx = \int_{-1}^1 \sum_{k=0}^N a_k P_k(x)P_j(x)\, dx = \sum_{k=0}^N a_k \int_{-1}^1 P_k(x)P_j(x)\, dx.$$

However, due to the orthogonality of the Legendre polynomials, all the terms in the summation with $k \neq j$ vanish, so that

$$\int_{-1}^1 p(x)P_j(x)\, dx = a_j \int_{-1}^1 P_j(x)P_j(x)\, dx.$$

Consequently, using (13.4.7), we obtain

$$\int_{-1}^{1} p(x)P_j(x)\, dx = \frac{2}{2j+1}\, a_j,$$

which implies that

$$a_j = \frac{2j+1}{2} \int_{-1}^{1} p(x)P_j(x)\, dx. \tag{13.4.9}$$

Example 13.4.4 Express $f(x) = x^2 - x + 2$ as a linear combination of Legendre polynomials.

Solution Since $f(x)$ has degree 2, we can write

$$x^2 - x + 2 = a_0 P_0 + a_1 P_1 + a_2 P_2,$$

where, from (13.4.9), the coefficients are given by

$$a_j = \frac{2j+1}{2} \int_{-1}^{1} (x^2 - x + 2)P_j(x)\, dx.$$

From Table 13.4.1,

$$P_0 = 1, \qquad P_1 = x, \qquad P_2 = \tfrac{1}{2}(3x^2 - 1),$$

so that

$$a_0 = \tfrac{1}{2} \int_{-1}^{1} (x^2 - x + 2)\, dx = \tfrac{7}{3},$$

$$a_1 = \tfrac{3}{2} \int_{-1}^{1} (x^2 - x + 2)x\, dx = -1,$$

$$a_2 = \tfrac{5}{2} \int_{-1}^{1} \tfrac{1}{2}(x^2 - x + 2)(3x^2 - 1)\, dx = \tfrac{2}{3}.$$

Consequently,

$$x^2 - x + 2 = \tfrac{7}{3} P_0 - P_1 + \tfrac{2}{3} P_2.$$

More generally, it can be shown that an arbitrary (sufficiently smooth) function $f(x)$ can be expanded in a series of Legendre polynomials and, further, that the coefficients in the expansion are given by (13.4.9). It should be noted, however, that the series representation will, in general, involve an infinite number of terms.

EXERCISES 13.4

1. Use (13.4.3) and (13.4.4) to determine polynomial solutions to Legendre's equation when $\alpha = 3$ and $\alpha = 4$. Hence determine the Legendre polynomials $P_3(x)$ and $P_4(x)$.

2. Starting with $P_0(x) = 1$, $P_1(x) = x$, use the recurrence relation

$$(n + 1)P_{n+1} + nP_{n-1} = (2n + 1)xP_n,$$

$$n = 1, 2, 3, \ldots,$$

to determine P_2, P_3, P_4.

3. Use Rodrigues's formula to determine the Legendre polynomial of degree 3.

4. Determine the values of the constants a_0, a_1, a_2, a_3 such that

$$x^3 + 2x = a_0 P_0 + a_1 P_1 + a_2 P_2 + a_3 P_3.$$

5. Express $p(x) = 2x^3 + x^2 + 5$ as a linear combination of Legendre polynomials.

6. Let $Q(x)$ be a polynomial of degree less than N. Prove that

$$\int_{-1}^{1} Q(x)P_N(x)\, dx = 0.$$

7. Show that

$$\frac{d^2 Y}{d\phi^2} + \cot\phi\, \frac{dY}{d\phi} + \alpha(\alpha + 1)Y = 0, \qquad 0 < \phi < \pi,$$

is transformed into Legendre's equation by the change of variables $x = \cos\phi$.

Problems 8–10 deal with Hermite's equation,

$$y'' - 2xy' + 2\alpha y = 0, \qquad -\infty < x < \infty. \quad \text{(i)}$$

8. Determine two linearly independent series solutions to (i) centered at $x = 0$.

9. Show that if $\alpha = N$, a positive integer, then (i) has a polynomial solution. Determine the polynomial solutions when $\alpha = 0, 1, 2, 3$.

10. When suitably normalized, the polynomial solutions to (i) are called the **Hermite polynomials**, and are denoted by $H_N(x)$.
(a) Use (i) to show that $H_N(x)$ satisfies

$$(e^{-x^2}H_N')' + 2Ne^{-x^2}H_N = 0. \quad \text{(ii)}$$

[*Hint:* Replace α by N in (i) and multiply the resulting equation by e^{-x^2}.]
(b) Use (ii) to prove that the Hermite polynomials satisfy

$$\int_{-\infty}^{\infty} e^{-x^2}H_N(x)H_M(x)\, dx = 0, \qquad M \neq N. \quad \text{(iii)}$$

[*Hint:* Follow the steps taken in proving orthogonality of the Legendre polynomials. You will need to recall that $\lim_{x \to \pm\infty} e^{-x^2}p(x) = 0$ for any polynomial p.]
(c) Let $p(x)$ be a polynomial of degree N. Then we can write

$$p(x) = \sum_{k=1}^{N} a_k H_k(x). \quad \text{(iv)}$$

Given that

$$\int_{-\infty}^{\infty} e^{-x^2}H_N^2(x)\, dx = 2^n n! \sqrt{\pi},$$

use (iii) to prove that the constants in (iv) are given by

$$a_j = \frac{1}{2^j j! \sqrt{\pi}} \int_{-\infty}^{\infty} e^{-x^2}H_j(x)p(x)\, dx.$$

11. Consider the Chebyshev equation

$$(1 - x^2)y'' - xy' + \alpha^2 y = 0, \quad \text{(v)}$$

where α is a constant.
(a) Show that if $\alpha = N$, a nonnegative integer, then (v) has a polynomial solution of degree N. When suitably normalized, these polynomials are called the **Chebyshev polynomials** and are denoted by $T_N(x)$.
(b) Use (v) to show that $T_N(x)$ satisfies

$$(\sqrt{1 - x^2}\,T_N')' + \frac{N^2}{\sqrt{1 - x^2}}\,T_N = 0.$$

(c) Use the result from (b) to prove that

$$\int_{-1}^{1} \frac{T_N(x)T_M(x)}{\sqrt{1 - x^2}}\, dx = 0, \qquad M \neq N.$$

13.5   SERIES SOLUTIONS ABOUT A REGULAR SINGULAR POINT

The power series technique for solving

$$y'' + P(x)y' + Q(x)y = 0 \qquad (13.5.1)$$

developed in Section 13.3 is only directly applicable at ordinary points, that is, points where P and Q are both analytic. According to Definition 13.3.1, any points at which P or Q fail to be analytic are called *singular points* of (13.5.1), and the general analysis of the behavior of solutions to (13.5.1) in the neighborhood of a singular point is quite complicated. However, singular points often turn out to be the points of major interest in an applied problem, and so it is of some importance that we pursue this analysis. In the next two sections we will show that, provided the functions P and Q are not too badly behaved at a singular point, the power series technique can be extended to obtain solutions of the corresponding differential equation that are valid in the neighborhood of the singular point. We will restrict our attention to differential equations whose singular points satisfy the following definition.

Definition 13.5.1: The point $x = x_0$ is called a **regular singular point** of the differential equation (13.5.1) if and only if the following two conditions are satisfied:

1. x_0 is a singular point of (13.5.1).
2. *Both* of the functions

$$p(x) = (x - x_0)P(x) \quad \text{and} \quad q(x) = (x - x_0)^2 Q(x)$$

are analytic at $x = x_0$.

A singular point of (13.5.1) that does not satisfy (2) is called an **irregular singular point**.

Example 13.5.1 Determine whether $x = 0$, $x = 1$, and $x = 2$ are ordinary points, regular singular points, or irregular singular points of the differential equation

$$y'' + \frac{1}{x(x-1)^2} y' + \frac{(x+1)}{x(x-1)^3} y = 0. \tag{13.5.2}$$

Solution In this case we have

$$P(x) = \frac{1}{x(x-1)^2}, \qquad Q(x) = \frac{(x+1)}{x(x-1)^3},$$

and, by inspection, P and Q are analytic at all points except $x = 0, 1$. Hence the only singular points of (13.5.2) are $x = 0, 1$. Consequently $x = 2$ is an ordinary point. We now determine whether the singular points are regular or irregular.
  (a) Consider the singular point $x = 0$. The functions

$$p(x) = xP(x) = \frac{1}{(x-1)^2}, \qquad q(x) = x^2 Q(x) = \frac{x(x+1)}{(x-1)^3},$$

are both analytic at $x = 0$, so that $x = 0$ is a *regular singular point* of (13.5.2).
  (b) Now consider the singular point $x = 1$. Since

$$p(x) = (x-1)P(x) = \frac{1}{x(x-1)},$$

is nonanalytic at $x = 1$, it follows that $x = 1$ is an *irregular singular point* of (13.5.2).

Now suppose that $x = x_0$ is a regular singular point of the differential equation

$$y'' + P(x)y' + Q(x)y = 0.$$

Multiplying this equation by $(x - x_0)^2$ yields

$$(x-x_0)^2 y'' + (x-x_0)[(x-x_0)P(x)]y' + (x-x_0)^2 Q(x)y = 0,$$

which we can write as

$$(x-x_0)^2 y'' + (x-x_0)p(x)y' + q(x)y = 0,$$

where

$$p(x) = (x-x_0)P(x), \qquad q(x) = (x-x_0)^2 Q(x).$$

Since, by assumption, $x = x_0$ is a regular singular point, it follows that the functions p and q are analytic at $x = x_0$. By the change of variables $z = x - x_0$ we can always transform a regular singular point to $x = 0$, and so we will restrict attention to differential equations that can be written in the form

$$\boxed{x^2 y'' + xp(x)y' + q(x)y = 0,} \tag{13.5.3}$$

where p and q are analytic at $x = 0$. *This is the standard form of any equation that has a regular singular point at $x = 0$.* The simplest type of equation that falls into this category is the second-order Cauchy–Euler equation,

$$x^2 y'' + p_0 xy' + q_0 y = 0, \tag{13.5.4}$$

where p_0 and q_0 are constants. The solution techniques that we will develop for solving (13.5.3) will be motivated by the solutions of (13.5.4). Recall from Section 9.10 that (13.5.4) has solutions on the interval $(0, \infty)$ of the form $y = x^r$, where r is a root of the indicial equation

$$r(r-1) + p_0 r + q_0 = 0. \tag{13.5.5}$$

Now consider (13.5.3). Since p and q are analytic at $x = 0$ by assumption, we can write

$$p(x) = p_0 + p_1 x + p_2 x^2 + \cdots, \qquad q(x) = q_0 + q_1 x + q_2 x^2 + \cdots, \tag{13.5.6}$$

for x in some interval of the form $(-R, R)$. It follows that (13.5.3) can be written as

$$x^2 y'' + x[p_0 + p_1 x + p_2 x^2 + \cdots]y' + [q_0 + q_1 x + q_2 x^2 + \cdots]y = 0.$$

For $|x| \ll 1$, this is approximately the Cauchy–Euler equation (13.5.4) and so it is reasonable to *expect* that for x in the interval $(0, R)$, (13.5.3) has solutions of the form

Power series

$$\downarrow$$

$$y = x^r \sum_{n=0}^{\infty} a_n x^n, \qquad a_0 \neq 0, \tag{13.5.7}$$

$$\uparrow$$

Cauchy–Euler solution

where r is a root of the *indicial equation*

$$r(r-1) + p_0 r + q_0 = 0. \tag{13.5.8}$$

A series of the form (13.5.7) is called a **Frobenius series**. We can assume without loss of generality that $a_0 \neq 0$ since if this were not the case we could always factor the leading power of x out of the series and combine it into x^r.

The following theorem confirms our expectations.

Theorem 13.5.1: Consider the differential equation

$$x^2 y'' + x p(x) y' + q(x) y = 0, \qquad x > 0, \tag{13.5.9}$$

where p and q are analytic at $x = 0$. Suppose that

$$p(x) = \sum_{n=0}^{\infty} p_n x^n, \qquad q(x) = \sum_{n=0}^{\infty} q_n x^n,$$

for $|x| < R$. Let r_1, r_2 denote the roots of the indicial equation

$$r(r-1) + p_0 r + q_0 = 0,$$

and assume that $r_1 \geq r_2$ if these roots are real. Then (13.5.9) has a solution of the form

$$y_1 = x^{r_1} \sum_{n=0}^{\infty} a_n x^n, \qquad a_0 \neq 0.$$

This solution is valid (at least) for $0 < x < R$. Further, *provided r_1 and r_2 are distinct and do *not* differ by an integer*, there exists a second solution to (13.5.9), valid (at least) for $0 < x < R$, of the form

$$y_2 = x^{r_2} \sum_{n=0}^{\infty} b_n x^n, \qquad b_0 \neq 0.$$

The solutions y_1 and y_2 are linearly independent on their interval of existence.

PROOF   The proof of this theorem, and its extension to the case when the roots of the indicial equation do differ by an integer, will be discussed fully in the next section.   ∎

REMARK   Using the formula for the Maclaurin expansion of p and q, it follows that the constants p_0 and q_0 appearing in (13.5.6) are given by

$$p_0 = p(0), \qquad q_0 = q(0).$$

Consequently the indicial equation (13.5.8) for

$$x^2y'' + xp(x)y' + q(x)y = 0$$

can be written directly as

$$r(r-1) + p(0)r + q(0) = 0.$$

We conclude this section with some examples that illustrate the implementation of the preceding theorem.

Example 13.5.2   Show that the differential equation

$$x^2y'' + xe^{2x}y' - 2(\cos x)y = 0, \qquad x > 0,$$

has two linearly independent Frobenius series solutions and determine the interval on which these solutions are valid.

Solution   Comparing the given differential equation with the standard form (13.5.3), we see that

$$p(x) = e^{2x}, \qquad q(x) = -2 \cos x.$$

Consequently,

$$p(0) = 1, \qquad q(0) = -2,$$

and so the indicial equation is

$$r(r-1) + r - 2 = 0,$$

that is,

$$r^2 - 2 = 0.$$

Thus the roots of the indicial equation are $r_1 = \sqrt{2}$, $r_2 = -\sqrt{2}$. Since r_1 and r_2 are distinct and do not differ by an integer, it follows from Theorem 13.5.1 that the given differential equation has two Frobenius series solutions of the *form*

$$y_1 = x^{\sqrt{2}} \sum_{n=0}^{\infty} a_n x^n, \qquad y_1 = x^{-\sqrt{2}} \sum_{n=0}^{\infty} b_n x^n.$$

Further, since the power series expansions of p and q about $x = 0$ are valid for all x, it follows that the preceding solutions will be defined and linearly independent on $(0, \infty)$.

Example 13.5.3   Find the general solution of

$$2x^2y'' + xy' - (1 + x)y = 0, \qquad x > 0. \tag{13.5.10}$$

Solution   In this case $p(x) = \frac{1}{2}$ and $q(x) = -(1 + x)/2$, both of which are analytic at $x = 0$. Thus $x = 0$ is a regular singular point of (13.5.10), and so there is at least one Frobenius series solution. Further, since p and q are polynomials, their power series expansions about $x = 0$ converge for all real x. Consequently, from Theorem

13.5.1, any Frobenius series solution will be valid for $0 < x < \infty$. In order to determine the solutions, we let

$$y = x^r \sum_{n=0}^{\infty} a_n x^n = \sum_{n=0}^{\infty} a_n x^{r+n}, \qquad a_0 \neq 0,$$

so that

$$y' = \sum_{n=0}^{\infty} (r+n) a_n x^{r+n-1}, \qquad y'' = \sum_{n=0}^{\infty} (r+n)(r+n-1) a_n x^{r+n-2}.$$

Substituting into (13.5.10) yields

$$\sum_{n=0}^{\infty} 2(r+n)(r+n-1) a_n x^{r+n} + \sum_{n=0}^{\infty} (r+n) a_n x^{r+n} - \sum_{n=0}^{\infty} a_n x^{r+n} - \sum_{n=0}^{\infty} a_n x^{r+n+1} = 0;$$

that is, combining the first three terms, replacing n with $n-1$ in the fourth sum, and dividing by x^r,

$$\sum_{n=0}^{\infty} [2(r+n)(r+n-1) + (r+n) - 1] a_n x^n - \sum_{n=1}^{\infty} a_{n-1} x^n = 0. \qquad (13.5.11)$$

Thus the coefficients of x^n must vanish for $n = 0, 1, 2, \ldots$. When $n = 0$, we obtain

$$[2r(r-1) + r - 1] a_0 = 0.$$

Since $a_0 \neq 0$ by assumption, we must have

$$2r(r-1) + r - 1 = 0,$$

which is just the indicial equation for (13.5.10). This can be written as

$$(2r+1)(r-1) = 0,$$

so that the roots of the indicial equation are

$$r_1 = 1, \qquad r_2 = -\tfrac{1}{2}. \qquad (13.5.12)$$

Since these roots are distinct and do not differ by an integer, there exist two linearly independent Frobenius series solutions. From (13.5.11) when $n = 1, 2, \ldots$, we obtain the *recurrence relation*

$$(r+n-1)(2r+2n+1) a_n - a_{n-1} = 0. \qquad (13.5.13)$$

We now substitute the values of r obtained in (13.5.12) into this relation in order to determine the corresponding Frobenius series solutions.

$r = 1$ Substitution into the recurrence relation (13.5.13) yields

$$a_n = \frac{1}{n(2n+3)} a_{n-1}, \qquad n = 1, 2, 3, \ldots. \qquad (13.5.14)$$

Thus,

$$n = 1: \quad a_1 = \frac{1}{1 \cdot 5} a_0;$$

$$n = 2: \quad a_2 = \frac{1}{2 \cdot 7} a_1 = \frac{1}{(2!)(5 \cdot 7)} a_0;$$

$$n = 3: \quad a_3 = \frac{1}{3 \cdot 9} a_2 = \frac{1}{(3!)(5 \cdot 7 \cdot 9)} a_0;$$

$$n = 4: \quad a_4 = \frac{1}{4 \cdot 11} a_3 = \frac{1}{(4!)(5 \cdot 7 \cdot 9 \cdot 11)} a_0.$$

It follows that in general,

$$a_n = \frac{1}{(n!)[5 \cdot 7 \cdot 9 \cdot \dots \cdot (2n + 3)]} a_0, \qquad n = 1, 2, 3 \dots,$$

so that the corresponding Frobenius series solution is

$$y_1 = x\left[1 + \frac{1}{5} x + \frac{1}{(2!)(5 \cdot 7)} x^2 + \frac{1}{(3!)(5 \cdot 7 \cdot 9)} x^3 + \dots \right.$$

$$\left. + \frac{1}{(n!)[5 \cdot 7 \cdot 9 \cdot \dots \cdot (2n + 3)]} x^n + \dots \right],$$

where we have set $a_0 = 1$. We can write this solution as

$$y_1 = x\left[1 + \sum_{n=1}^{\infty} \frac{1}{(n!)[5 \cdot 7 \cdot 9 \cdot \dots \cdot (2n + 3)]} x^n \right], \qquad x > 0.$$

$r = -\frac{1}{2}$ In this case the recurrence relation (13.5.13) reduces to

$$a_n = \frac{1}{n(2n - 3)} a_{n-1}, \qquad n = 1, 2, \dots .$$

We therefore obtain

$$n = 1: \quad a_1 = -a_0;$$

$$n = 2: \quad a_2 = \frac{1}{2 \cdot 1} a_1 = -\frac{1}{2!} a_0;$$

$$n = 3: \quad a_3 = \frac{1}{3 \cdot 3} a_2 = -\frac{1}{(3!)(1 \cdot 3)} a_0;$$

$$n = 4: \quad a_4 = \frac{1}{4 \cdot 5} a_3 = -\frac{1}{(4!)(1 \cdot 3 \cdot 5)} a_0.$$

In general we have

$$a_n = -\frac{1}{n![1 \cdot 3 \cdot 5 \cdot \dots \cdot (2n - 3)]} a_0, \qquad n = 1, 2, 3, \dots .$$

It follows that a second linearly independent Frobenius series solution to the differential equation (13.5.10) on $(0, \infty)$ is

$$y_2 = x^{-1/2}\left[1 - x - \frac{1}{2!}x^2 - \frac{1}{(3!)(1 \cdot 3)}x^3 - \frac{1}{(4!)(1 \cdot 3 \cdot 5)}x^4 - \cdots\right.$$

$$\left. - \frac{1}{(n!)[1 \cdot 3 \cdots \cdots (2n-3)]}x^n - \cdots\right],$$

where we have once more set $a_0 = 1$. This can be written as

$$y_2 = x^{-1/2}\left[1 - \sum_{n=1}^{\infty}\frac{1}{(n!)[1 \cdot 3 \cdot 5 \cdots \cdots (2n-3)]}x^n\right], \qquad x > 0.$$

Consequently, the general solution to (13.5.10) on $(0, \infty)$ is

$$y = c_1 y_1 + c_2 y_2.$$

The differential equation in the previous example had two linearly independent Frobenius series solutions, and we were therefore able to determine its general solution. In the following example there is only one linearly independent *Frobenius series* solution.

Example 13.5.4 Determine a Frobenius series solution to

$$x^2 y'' + x(3 + x)y' + (1 + 3x)y = 0, \qquad x > 0. \tag{13.5.15}$$

Solution By inspection we see that $x = 0$ is a regular singular point of the differential equation (13.5.15), and so from Theorem 13.5.1 the differential equation has a Frobenius series solution. Further, since $p(x) = 3 + x$ and $q(x) = 1 + 3x$ are both polynomials, their power series expansions about $x = 0$ are valid for all x. It follows that any Frobenius series solution will be valid for $0 < x < \infty$. In order to determine a solution, we let

$$y = x^r \sum_{n=0}^{\infty} a_n x^n .$$

Differentiating twice with respect to x yields

$$y' = \sum_{n=0}^{\infty}(r + n)a_n x^{r+n-1}, \qquad y'' = \sum_{n=0}^{\infty}(r + n)(r + n - 1)a_n x^{r+n-2},$$

so that y is indeed a solution of (13.5.15) provided a_n and r satisfy

$$\sum_{n=0}^{\infty}(r + n)(r + n - 1)a_n x^{r+n} + 3\sum_{n=0}^{\infty}(r + n)a_n x^{r+n} + \sum_{n=0}^{\infty}(r + n)a_n x^{r+n+1}$$

$$+ \sum_{n=0}^{\infty}a_n x^{r+n} + 3\sum_{n=0}^{\infty}a_n x^{r+n+1} = 0.$$

Dividing by x^r and replacing n by $n - 1$ in the third and fifth sums yields

$$\sum_{n=0}^{\infty}[(r+n)(r+n-1)+3(r+n)+1]a_nx^n + \sum_{n=1}^{\infty}(r+n+2)a_{n-1}x^n = 0. \quad (13.5.16)$$

Once more this implies that the coefficients of x^n must vanish for $n = 0, 1, 2, \ldots$. When $n = 0$, we obtain the indicial equation

$$r(r-1)+3r+1=0,$$

that is,

$$(r+1)^2=0.$$

Thus the only value of r for which a Frobenius series solution exists is

$$r=-1.$$

For $n \geq 1$, (13.5.16) yields the recurrence relation

$$[(r+n)(r+n-1)+3(r+n)+1]a_n + (r+n+2)a_{n-1} = 0.$$

Setting $r = -1$, we obtain

$$[(n-1)(n-2)+3(n-1)+1]a_n + (n+1)a_{n-1} = 0,$$

which, upon simplifying, yields

$$a_n = -\frac{(n+1)}{n^2}a_{n-1}, \qquad n = 1, 2, \ldots.$$

Solving this recurrence relation we have

$$n = 1: \quad a_1 = -2a_0. \qquad n = 2: \quad a_2 = -\frac{3}{4}a_1 = \frac{3\cdot 2}{4}a_0.$$

$$n = 3: \quad a_3 = -\frac{4}{9}a_2 = -\frac{4!}{4\cdot 9}a_0.$$

The general term is

$$a_n = (-1)^n \frac{(n+1)!}{2^2\cdot 3^2\cdot\ldots\cdot n^2}a_0, \qquad n = 1, 2, 3, \ldots,$$

which can be written as

$$a_n = (-1)^n \frac{(n+1)!}{(n!)^2}a_0,$$

that is,

$$a_n = (-1)^n \frac{(n+1)}{n!}a_0, \qquad n = 1, 2, 3, \ldots.$$

Consequently the corresponding Frobenius series solution is

$$y = x^{-1}\left[1 + \sum_{n=1}^{\infty}(-1)^n\frac{(n+1)}{n!}x^n\right], \qquad x > 0,$$

where we have set $a_0 = 1$.

Notice that in this problem there is only *one* linearly independent Frobenius series solution to the given differential equation. In order to determine a second linearly independent solution, we could, for example, use the reduction-of-order method introduced in Section 9.11. We will have more to say about this in the next section.

EXERCISES 13.5

In problems 1–4, determine all singular points of the given differential equation and classify them as regular or irregular singular points.

1. $y'' + \dfrac{1}{1-x} y' + xy = 0.$

2. $x^2 y'' + \dfrac{x}{(1-x^2)^2} y' + y = 0.$

3. $(x-2)^2 y'' + (x-2)e^x y' + \dfrac{4}{x} y = 0.$

4. $y'' + \dfrac{2}{x(x-3)} y' - \dfrac{1}{x^3(x+3)} y = 0.$

In problems 5–8, determine the roots of the indicial equation of the given differential equation.

5. $x^2 y'' + x(1-x)y' - 7y = 0.$
6. $4x^2 y'' + xe^x y' - y = 0.$
7. $xy'' - xy' - 2y = 0.$
8. $x^2 y'' - x(\cos x)y' + 5e^{2x} y = 0.$

In problems 9–16, show that the indicial equation of the given differential equation has distinct roots that do *not* differ by an integer, and find two linearly independent Frobenius series solutions on $(0, \infty)$.

9. $4x^2 y'' + 3xy' + xy = 0.$
10. $6x^2 y'' + x(1+18x)y' + (1+12x)y = 0.$
11. $x^2 y'' + xy' - (2+x)y = 0.$
12. $2xy'' + y' - 2xy = 0.$
13. $3x^2 y'' - x(x+8)y' + 6y = 0.$
14. $2x^2 y'' - x(1+2x)y' + 2(4x-1)y = 0.$
15. $x^2 y'' + x(1-x)y' - (5+x)y = 0.$
16. $3x^2 y'' + x(7+3x)y' + (1+6x)y = 0.$

17. Consider the differential equation

$$x^2 y'' + xy' + (1-x)y = 0, \qquad x > 0. \qquad \text{(i)}$$

(a) Find the indicial equation and show that the roots are $r = \pm i$.
(b) Determine the first three terms in a complex-valued Frobenius series solution to (i).
(c) Use the solution in (b) to determine two linearly independent real-valued solutions to (i).

18. Determine the first five nonzero terms in each of two linearly independent Frobenius series solutions to

$$3x^2 y'' + x(1+3x^2)y' - 2xy = 0, \qquad x > 0.$$

19. Consider the differential equation

$$4x^2 y'' - 4x^2 y' + (1+2x)y = 0.$$

(a) Show that the indicial equation has a repeated root and find the corresponding Frobenius series solution.
(b) Use the *reduction-of-order* technique to find a second linearly independent solution on $(0, \infty)$.

(*Hint:* To evaluate $\displaystyle\int x^{-1}e^x \, dx$, expand e^x in a Maclaurin series.)

20. Find two linearly independent solutions to

$$x^2 y'' + x(3-2x)y' + (1-2x)y = 0,$$

on $(0, \infty)$.

21. Consider the differential equation

$$x^2 y'' + x(1+2N-x)y' + N^2 y = 0, \qquad x > 0. \text{(ii)}$$

(a) Find the indicial equation and show that it has the repeated root, $r = -N$.

(b) If N is a nonnegative integer, show that the Frobenius series solution to (ii) terminates after N terms.

(c) Determine the Frobenius series solutions when $N = 0, 1, 2, 3$.

(d) Show that if N is a positive integer then the Frobenius series solution to (ii) can be written as

$$y =$$

$$x^{-N}\left[1 + \sum_{k=1}^{N} (-1)^k \frac{N(N-1)\cdots\cdots(N+1-k)}{1^2 \cdot 2^2 \cdots\cdots k^2} x^k\right]$$

$$= x^{-N}\left[1 + \sum_{k=1}^{N} (-1)^k \prod_{i=1}^{k} \frac{(N+1-i)}{i^2} x^k\right].$$

22. Consider the general "perturbed" Cauchy–Euler equation

$$x^2 y'' + x[1 - (a+b) + \beta x]y' + (ab + \gamma x)y$$
$$= 0, \qquad x > 0, \qquad \text{(iii)}$$

where a, b, β, γ are constants. Assuming that a and b are distinct and do not differ by an integer, determine two linearly independent Frobenius series solutions to (iii). (*Hint:* Use symmetry to get the second solution.)

13.6   FROBENIUS THEORY

In the previous section we saw how Frobenius series solutions can be obtained to the differential equation

$$x^2 y'' + x p(x) y' + q(x) y = 0, \qquad x > 0. \qquad (13.6.1)$$

In this section we give some justification for Theorem 13.5.1 and extend this theorem to the case when the roots of the indicial equation for (13.6.1) differ by an integer. We will first assume $x > 0$, since our results can easily be extended to $x < 0$. We begin by establishing the existence of at least one Frobenius series solution.

Assuming that $x = 0$ is a regular singular point of (13.6.1), it follows that p and q are analytic at $x = 0$, and hence we can write

$$p(x) = \sum_{n=0}^{\infty} p_n x^n, \qquad q(x) = \sum_{n=0}^{\infty} q_n x^n,$$

for $|x| < R$. Consequently (13.6.1) can be written as

$$x^2 y'' + x \sum_{n=0}^{\infty} p_n x^n y' + \sum_{n=0}^{\infty} q_n x^n y = 0. \qquad (13.6.2)$$

We try for a Frobenius series solution and therefore let

$$y = x^r \sum_{n=0}^{\infty} a_n x^n, \qquad a_0 \neq 0,$$

where r and a_n are constants to be determined. Differentiating y twice yields

$$y' = \sum_{n=0}^{\infty} (r+n) a_n x^{r+n-1}, \qquad y'' = \sum_{n=0}^{\infty} (r+n)(r+n+1) a_n x^{r+n-2}.$$

We now substitute into (13.6.2) to obtain

$$\sum_{n=0}^{\infty}(r+n)(r+n-1)a_nx^n + \left[\sum_{n=0}^{\infty}p_nx^n\right]\left[\sum_{n=0}^{\infty}(r+n)a_nx^n\right]$$

$$+ \left[\sum_{n=0}^{\infty}q_nx^n\right]\left[\sum_{n=0}^{\infty}a_nx^n\right] = 0.$$

Using the formula given on page 519 for the product of two infinite series gives

$$\sum_{n=0}^{\infty}(r+n)(r+n-1)a_nx^n + \sum_{n=0}^{\infty}\left[\sum_{k=0}^{n}p_{n-k}(k+r)a_k\right]x^n + \sum_{n=0}^{\infty}\left[\sum_{k=0}^{n}q_{n-k}a_k\right]x^n = 0,$$

which can be written as

$$\sum_{n=0}^{\infty}\left\{(r+n)(r+n-1)a_n + \sum_{k=0}^{n}[p_{n-k}(k+r)+q_{n-k}]a_k\right\}x^n = 0.$$

Thus the a_n must satisfy the *recurrence relation*

$$(r+n)(r+n-1)a_n + \sum_{k=0}^{n}[p_{n-k}(k+r)+q_{n-k}]a_k = 0, \qquad n = 0, 1, 2, \ldots. \quad (13.6.3)$$

Evaluating (13.6.3) when $n = 0$ yields

$$[r(r-1)+p_0r+q_0]a_0 = 0,$$

so that, since $a_0 \neq 0$ by assumption, r must satisfy

$$r(r-1)+p_0r+q_0 = 0, \qquad (13.6.4)$$

which we recognize as being the indicial equation for (13.6.1). When $n \geq 1$, we combine the coefficients of a_n in (13.6.3) to obtain

$$[(r+n)(r+n-1)+p_0(r+n)+q_0]a_n + \sum_{k=0}^{n-1}[p_{n-k}(k+r)+q_{n-k}]a_k = 0,$$

that is,

$$[(r+n)(r+n-1)+p_0(r+n)+q_0]a_n = -\sum_{k=0}^{n-1}[p_{n-k}(k+r)+q_{n-k}]a_k,$$

$$n = 1, 2, \ldots. \quad (13.6.5)$$

If we define $F(r)$ by

$$F(r) = r(r-1)+p_0r+q_0,$$

then

$$F(r+n) = (r+n)(r+n-1)+p_0(r+n)+q_0,$$

so that the indicial equation (13.6.4) is

$$F(r) = 0,$$

whereas the recurrence relation (13.6.5) can be written as

$$F(r+n)a_n = -\sum_{k=0}^{n-1} [p_{n-k}(k+r) + q_{n-k}]a_k, \qquad n = 1, 2, 3, \ldots. \qquad (13.6.6)$$

It is tempting to divide (13.6.6) by $F(r+n)$, thereby determining a_n in terms of a_1, $a_2, \ldots, a_{n-1}$. However, we can do this only if $F(r+n) \neq 0$. Let r_1, r_2 denote the roots of (13.6.4). Three familiar cases arise:

1. r_1, r_2 real and distinct.
2. r_1, r_2 real and coincident.
3. r_1, r_2 complex conjugate.

If r_1 and r_2 are real, we assume without loss of generality that $r_1 \geq r_2$. Consider (13.6.6) when $r = r_1$. We have

$$F(r_1+n)a_n = -\sum_{k=0}^{n-1} [p_{n-k}(k+r_1) + q_{n-k}]a_k, \qquad n = 1, 2, 3, \ldots. \qquad (13.6.7)$$

In (1) and (2) it follows, since $r = r_1$ is the largest root of $F(r) = 0$, that $F(r_1+n) \neq 0$ for any n. Also in (3) $F(r_1+n) \neq 0$, and so in all three cases we can write (13.6.7) as

$$a_n = -\frac{1}{F(r_1+n)} \sum_{k=0}^{n-1} [p_{n-k}(k+r_1) + q_{n-k}]a_k, \qquad n = 1, 2, 3, \ldots. \qquad (13.6.8)$$

Starting from $n = 1$ we can therefore determine all the a_n in terms of a_0, and so we formally obtain the Frobenius series solution

$$y_1 = x^{r_1}\left[1 + \sum_{n=1}^{\infty} a_n(r_1)x^n\right],$$

where $a_n(r_1)$ denotes the coefficients obtained from (13.6.8) upon setting $a_0 = 1$. A fairly delicate analysis shows that this series solution converges for (at least) $0 < x < R$. This justifies the steps in the preceding derivation and establishes the first part of Theorem 13.5.1, stated in the last section.

We now consider the problem of determining a second linearly independent solution to (13.6.1). We must consider the three cases separately.

CASE 1: r_1, r_2 real and distinct. Setting $r = r_2$ in (13.6.6) yields

$$F(r_2+n)a_n = -\sum_{k=0}^{n-1} [p_{n-k}(k+r_2) + q_{n-k}]a_k. \qquad (13.6.9)$$

Thus provided $F(r_2+n) \neq 0$ for any positive integer n, the same procedure as we used when $r = r_1$ will yield a second Frobenius series solution. But, since r_1 and r_2 are the only zeros of F, it follows that $F(r_2+n) = 0$ if and only if there exists a positive integer n such that $r_2 + n = r_1$. Consequently, *provided*

$$r_1 - r_2 \neq \text{positive integer},$$

there exists a second Frobenius series solution of the form

$$y_2 = x^{r_2}\left[1 + \sum_{n=1}^{\infty} a_n(r_2)x^n\right],$$

where $a_n(r_2)$ denotes the values of the coefficients obtained from (13.6.6) when $r = r_2$ and, once more, we have set $a_0 = 1$. Since $r_1 \neq r_2$, it follows that the Frobenius series solutions y_1 and y_2 are linearly independent on (at least) $0 < x < R$.

Now suppose that $r_1 - r_2 = N$, where N is a positive integer. Then substituting for $r_2 = r_1 - N$ in (13.6.9), we obtain

$$F(r_1 + (n - N))a_n = -\sum_{k=0}^{n-1}[p_{n-k}(k + r_2) + q_{n-k}]a_k,$$

which, when $n = N$, leads to the consistency condition

$$0 \cdot a_N = -\sum_{k=0}^{N-1}[P_{N-k}(k + r_2) + q_{N-k}]a_k. \tag{13.6.10}$$

Since all of the coefficients $a_1, a_2, \ldots, a_{N-1}$ will already have been determined in terms of a_0, (13.6.10) will be of the form

$$0 \cdot a_N = \alpha a_0, \tag{13.6.11}$$

where α is a constant. By assumption a_0 is nonzero, and therefore two possibilities arise.[1]

(a) First it may happen that $\alpha = 0$. If this occurs, then, from (13.6.11), a_N can be specified arbitrarily, and (13.6.9) determines all of the remaining Frobenius coefficients. We therefore do obtain a second linearly independent Frobenius series solution.

(b) In the more general case α will be nonzero. Then, since $a_0 \neq 0$ by assumption, (13.6.11) *cannot* be satisfied, and so we cannot compute the Frobenius coefficients. Hence there *does not exist* a Frobenius series solution corresponding to $r = r_2$. The reduction-of-order technique can be used, however, to prove that there exists a second linearly independent solution to (13.6.1) on $(0, R)$, of the form

$$y_2 = Ay_1\ln x + x^{r_2}\sum_{n=0}^{\infty} b_nx^n,$$

where the constants A and b_n can be determined by direct substitution into the differential equation (13.6.1). The derivation is straightforward but quite long-winded, and so the details have been relegated to Appendix 5. Notice that the above form includes case (a) which arises when $A = 0$.

CASE 2: $r_1 = r_2$. In this case there certainly cannot exist a second linearly independent *Frobenius series* solution. However, once more we could use the reduction-of-order technique to obtain a second linearly independent solution. Based on our experience with Cauchy–Euler equations, it should not be too surprising to

---

[1] Notice that (13.6.11) is *not* an equation for determining α; rather, it is a consistency condition for the validity of the recurrence relation.

learn that the resulting solution involves a term of the form $y_1 \ln x$. Indeed it can be shown (see Appendix 5) that when $r_1 = r_2$, there exists a second linearly independent solution of the form:

$$y_2 = y_1 \ln x + x^{r_1} \sum_{n=1}^{\infty} b_n x^n,$$

valid for (at least) $0 < x < R$. The coefficients b_n can be obtained by substituting this expression for y_2 into the differential equation (13.6.1).

CASE 3: r_1, r_2 complex conjugate.   In this case the solution obtained when $r = r_1$ will be a complex-valued solution.   It follows from the general theory that we have developed for linear differential equations in Chapter 9 that the real and imaginary parts of this complex-valued solution will themselves be real-valued solutions.   It can be shown that these real-valued solutions are linearly independent on their interval of existence.   Thus we can always, in theory, obtain the general solution in this case.

Finally we mention that the validity of the preceding solutions can be extended to $-R < x < 0$ by the replacement

$$x^{r_1} \rightarrow |x|^{r_1}, \qquad x^{r_2} \rightarrow |x|^{r_2}, \qquad \ln x \rightarrow \ln|x|.$$

The preceding discussion is summarized in the following theorem.

Theorem 13.6.1:   Consider the differential equation

$$x^2 y'' + xp(x)y' + q(x)y = 0, \tag{13.6.12}$$

where p and q are analytic at $x = 0$.   Suppose that

$$p(x) = \sum_{n=0}^{\infty} p_n x^n, \qquad q(x) = \sum_{n=0}^{\infty} q_n x^n,$$

for $|x| < R$.   Let r_1, r_2 denote the roots of the indicial equation and assume that $r_1 \geq r_2$ if these roots are real.   Then (13.6.12) has two linearly independent solutions valid (at least) on the interval $(0, R)$.   The form of the solution is determined as follows:

1. $r_1 - r_2 \neq$ integer:

$$y_1 = x^{r_1} \sum_{n=0}^{\infty} a_n x^n, \qquad a_0 \neq 0, \tag{13.6.13}$$

$$y_2 = x^{r_2} \sum_{n=0}^{\infty} b_n x^n, \qquad b_0 \neq 0. \tag{13.6.14}$$

2. $r_1 = r_2 = r$:

$$y_1 = x^r \sum_{n=0}^{\infty} a_n x^n, \qquad a_0 \neq 0, \tag{13.6.15}$$

$$y_2 = y_1 \ln x + x^r \sum_{n=1}^{\infty} b_n x^n.$$

3. $r_1 - r_2 = $ positive integer:

$$y_1 = x^{r_1} \sum_{n=0}^{\infty} a_n x^n, \qquad a_0 \neq 0, \tag{13.6.16}$$

$$y_2 = A y_1 \ln x + x^{r_2} \sum_{n=0}^{\infty} b_n x^n, \qquad b_0 \neq 0. \tag{13.6.17}$$

The coefficients in each of these solutions can be determined by direct substitution into (13.6.12). Finally, if x^{r_1} and x^{r_2} are replaced by $|x|^{r_1}$ and $|x|^{r_2}$, and $\ln x$ is replaced by $\ln|x|$, we obtain linearly independent solutions that are valid for (at least) $0 < |x| < R$.

REMARK Since a solution of a homogeneous linear differential equation is defined only up to a multiplicative constant, we can use this freedom to set $a_0 = 1$ in (13.6.13), (13.6.15), and (13.6.16) and to set $b_0 = 1$ in (13.6.14) and (13.6.17). It is often convenient to make these choices in solving our problems.

We now consider several examples to illustrate the use of the preceding theorem.

Example 13.6.1 Consider the differential equation

$$x^2 y'' - x(3 + x)y' + (4 - x)y = 0. \tag{13.6.18}$$

Determine the general form of two linearly independent series solutions in the neighborhood of the regular singular point $x = 0$.

Solution By inspection we see that $x = 0$ is a regular singular point of (13.6.18) and that the indicial equation is

$$r(r - 1) - 3r + 4 = r^2 - 4r + 4 = (r - 2)^2 = 0.$$

Since $r = 2$ is a repeated root, it follows from Theorem 13.6.1 that there exist two linearly independent solutions to (13.6.18) of the form

$$y_1 = x^2 \sum_{n=0}^{\infty} a_n x^n, \qquad y_2 = y_1 \ln|x| + x^2 \sum_{n=1}^{\infty} b_n x^n.$$

The coefficients in each of these series solutions could be obtained by direct substitution into (13.6.18). In this problem we have

$$p(x) = -(3 + x), \qquad q(x) = 4 - x.$$

Since these are polynomials, their power series expansions about $x = 0$ converge for all real x. It follows from Theorem 13.6.1 that the series solutions to (13.6.18) will be valid for all $x \neq 0$.

Example 13.6.2 Consider the differential equation

$$x^2 y'' + x(1 + 2x)y' - \tfrac{1}{4}(1 - 4\gamma x)y = 0, \qquad x > 0. \tag{13.6.19}$$

where γ is a constant. Determine the form of two linearly independent series solutions of this differential equation about the regular singular point $x = 0$.

Solution   In this case we have $p_0 = 1$ and $q_0 = -\frac{1}{4}$, so that

$$F(r) = r(r-1) + r - \tfrac{1}{4} = r^2 - \tfrac{1}{4}.$$

Thus the roots of the indicial equation are $r = \pm\frac{1}{2}$. Setting $r_1 = \frac{1}{2}$, $r_2 = -\frac{1}{2}$, it follows that $r_1 - r_2 = 1$, so that we are in case 3 of Theorem 13.6.1. Thus there exist two linearly independent solutions to (13.6.19) on $(0, \infty)$ of the form

$$y_1 = x^{1/2} \sum_{n=0}^{\infty} a_n x^n, \qquad y_2 = A y_1 \ln x + x^{-1/2} \sum_{n=0}^{\infty} b_n x^n. \tag{13.6.20}$$

In order to determine whether the constant A is zero or nonzero, we need the general recurrence relation. Substituting

$$y = x^r \sum_{n=0}^{\infty} a_n x^n, \qquad a_0 \neq 0,$$

into (13.6.19) yields

$$\left[\frac{4(r+n)^2 - 1}{4}\right] a_n = -[2(r+n-1) + \gamma]a_{n-1}, \qquad n = 1, 2, \ldots .$$

When $r = -\frac{1}{2}$, this reduces to

$$n(n-1)a_n = -(2n - 3 + \gamma)a_{n-1}. \tag{13.6.21}$$

As predicted from our general theory, the coefficient of a_n is zero when $n = r_1 - r_2 = 1$. Thus a second Frobenius series solution exists [$A = 0$ in (13.6.20)] if and only if the term on the right-hand side also vanishes when $n = 1$. However, setting $n = 1$ in the right-hand side of (13.6.21) yields

$$0 \cdot a_1 = (1 - \gamma)a_0. \tag{13.6.22}$$

Thus, since $a_0 \neq 0$, it follows that when $\gamma \neq 1$, there does not exist a second linearly independent Frobenius series solution and so the constant A in (13.6.20) is necessarily nonzero. However, if $\gamma = 1$, then (13.6.22) is identically satisfied independently of the value of a_1. In this case we can specify a_1 arbitrarily and then the recurrence relation (13.6.21) will determine the remaining coefficients in a second linearly independent Frobenius series solution. Thus we have the following in summary:

1. If $\gamma \neq 1$, there exist two linearly independent solutions of the form (13.6.20) with A necessarily nonzero.

2. If $\gamma = 1$, then there exist two linearly independent Frobenius series solutions:

$$y_1 = x^{1/2} \sum_{n=0}^{\infty} a_n x^n, \qquad y_2 = x^{-1/2} \sum_{n=0}^{\infty} b_n x^n.$$

In both cases the series solutions will be valid for $0 < x < \infty$.

Example 13.6.3   Determine two linearly independent series solutions to

$$x^2 y'' + x(3 - x)y' + y = 0, \qquad x > 0. \tag{13.6.23}$$

Solution We see by inspection that $x = 0$ is a regular singular point, and since the functions

$$p(x) = 3 - x, \qquad q(x) = 1,$$

are polynomials, Theorem 13.6.1 provides us with the form of two linearly independent solutions on $(0, \infty)$.

We begin by obtaining a Frobenius series solution. Substituting

$$y = x^r \sum_{n=0}^{\infty} a_n x^n, \qquad a_0 \neq 0,$$

into (13.6.23) yields

$$\sum_{n=0}^{\infty} (r+n)(r+n-1) a_n x^{r+n} + 3 \sum_{n=0}^{\infty} (r+n) a_n x^{r+n}$$

$$- \sum_{n=0}^{\infty} (r+n) a_n x^{r+n+1} + \sum_{n=0}^{\infty} a_n x^{r+n} = 0,$$

which, upon collecting coefficients of x^{r+n} and replacing n by $(n-1)$ in the third sum, can be written as

$$\sum_{n=0}^{\infty} [(r+n)(r+n+2)+1] a_n x^{r+n} - \sum_{n=1}^{\infty} (r+n-1) a_{n-1} x^{r+n} = 0.$$

Dividing by x^r yields

$$\sum_{n=0}^{\infty} [(r+n)(r+n+2)+1] a_n x^n - \sum_{n=1}^{\infty} (r+n-1) a_{n-1} x^n = 0. \qquad (13.6.24)$$

We determine the a_n in the usual manner. When $n = 0$, we obtain

$$[r(r+2)+1] a_0 = 0,$$

so that the indicial equation is

$$r^2 + 2r + 1 = 0,$$

that is

$$(r+1)^2 = 0.$$

Hence there is only one distinct root, namely,

$$r = -1.$$

From (13.6.24) the remaining coefficients must satisfy the recurrence relation

$$[(r+n)(r+n+2)+1] a_n - (r+n-1) a_{n-1} = 0, \qquad n = 1, 2, \ldots ,$$

that is, upon setting $r = -1$ and simplifying,

$$a_n = \frac{(n-2)}{n^2} a_{n-1}, \qquad n = 1, 2, 3, \ldots .$$

Consequently,

$$a_1 = -a_0, \qquad a_2 = a_3 = a_4 = \cdots = 0.$$

Thus, a Frobenius series solution to (13.6.23) is

$$y_1 = x^{-1}(1 - x), \tag{13.6.25}$$

where we have set $a_0 = 1$.

Since the indicial equation for (13.6.23) has only one root, there does not exist a second linearly independent Frobenius series solution. However, according to Theorem 13.6.1, there is a second linearly independent solution of the form

$$y_2 = y_1 \ln x + x^{-1} \sum_{n=1}^{\infty} b_n x^n, \tag{13.6.26}$$

where the coefficients b_n can be determined by substitution into (13.6.23). We now determine such a solution. The computations are quite straightforward, but they are long and tedious. The student is encouraged to pay full attention and not to be overwhelmed by the formidable look of the equations. Differentiating (13.6.26) with respect to x yields

$$y_2' = y_1' \ln x + x^{-1} y_1 + \sum_{n=1}^{\infty} (n-1) b_n x^{n-2},$$

$$y_2'' = y_1'' \ln x + 2x^{-1} y_1' - x^{-2} y_1 + \sum_{n=1}^{\infty} (n-2)(n-1) b_n x^{n-3}.$$

Substituting into (13.6.23) we obtain the following equation for the b_n:

$$x^2 \left[y_1'' \ln x + 2x^{-1} y_1' - x^{-2} y_1 + \sum_{n=1}^{\infty} (n-2)(n-1) b_n x^{n-3} \right]$$

$$+ x(3 - x) \left[y_1' \ln x + x^{-1} y_1 + \sum_{n=1}^{\infty} (n-1) b_n x^{n-2} \right] + y_1 \ln x + x^{-1} \sum_{n=1}^{\infty} b_n x^n = 0.$$

Collecting the terms that multiply $\ln x$ yields

$$[x^2 y_1'' + x(3 - x) y_1' + y_1] \ln x + 2xy_1' + 2y_1 - xy_1 + \sum_{n=1}^{\infty} (n-1)(n-2) b_n x^{n-1}$$

$$+ 3 \sum_{n=1}^{\infty} (n-1) b_n x^{n-1} - \sum_{n=1}^{\infty} (n-1) b_n x^n + \sum_{n=1}^{\infty} b_n x^{n-1} = 0.$$

Since y_1 is a solution of (13.6.23), the terms in the first parentheses vanish.[1] Combining the coefficients of x^{n-1}, we obtain

$$2xy_1' + 2y_1 - xy_1 + \sum_{n=1}^{\infty} [(n-1)(n-2) + 3(n-1) + 1] b_n x^{n-1}$$

$$- \sum_{n=1}^{\infty} (n-1) b_n x^n = 0.$$

---

[1] It should be noted that this is not just a fortuitous result for this particular example. In general the terms multiplying $\ln x$ will vanish at this stage of the computation.

Simplifying the terms in the first sum and replacing n by $n-1$ in the second sum yields

$$2xy_1' + 2y_1 - xy_1 + \sum_{n=1}^{\infty} n^2 b_n x^{n-1} - \sum_{n=2}^{\infty} (n-2) b_{n-1} x^{n-1} = 0. \qquad (13.6.27)$$

We now substitute for y_1 and y_1'. From (13.6.25) we have

$$y_1 = x^{-1}(1-x), \qquad y_1' = -x^{-2}.$$

Substituting into (13.6.27) we obtain

$$-2x^{-1} + 2x^{-1}(1-x) - (1-x) + \sum_{n=1}^{\infty} n^2 b_n x^{n-1} - \sum_{n=2}^{\infty} (n-2) b_{n-1} x^{n-1} = 0,$$

that is,

$$-3 + x + \sum_{n=1}^{\infty} n^2 b_n x^{n-1} - \sum_{n=2}^{\infty} (n-2) b_{n-1} x^{n-1} = 0.$$

Equating the corresponding coefficients of x^{n-1} to zero for $n = 1, 2, \ldots$, we obtain the b_n in a familiar manner.

$n = 1$: $-3 + b_1 = 0$, which implies that $b_1 = 3$.

$n = 2$: $1 + 4b_2 = 0$, which implies that $b_2 = -\frac{1}{4}$.

$n \geq 3$: $n^2 b_n - (n-2) b_{n-1} = 0$,

that is,

$$b_n = \frac{(n-2)}{n^2} b_{n-1}, \qquad n = 3, 4, \ldots. \qquad (13.6.28)$$

When $n = 3$, we have

$$b_3 = \frac{1}{3^2} b_2.$$

Substituting for $b_2 = -1/4 = -1/2^2$ yields

$$b_3 = \frac{-1}{2^2 \cdot 3^2}.$$

When $n = 4$, (13.6.28) implies that

$$b_4 = \frac{2}{4^2} b_3 = -\frac{1 \cdot 2}{2^2 \cdot 3^2 \cdot 4^2}.$$

It is now straightforward to see that in general, for $n \geq 3$,

$$b_n = -\frac{(n-2)!}{2^2 \cdot 3^2 \cdot \ldots \cdot n^2},$$

which can be written as

$$b_n = \frac{(n-2)!}{(n!)^2}, \qquad n = 3, 4, \ldots .$$

Finally, substituting for the b_n into (13.6.26) yields the following solution to (13.6.23):

$$y_2 = x^{-1}(1-x)\ln x + x^{-1}\left[3x - \frac{1}{4}x^2 - \sum_{n=3}^{\infty} \frac{(n-2)!}{(n!)^2} x^n \right].$$

Theorem 13.6.1 implies that y_1 and y_2 are linearly independent on $(0, \infty)$.

___

We give one final example to illustrate the case when the roots of the indicial equation differ by an integer.

Example 13.6.4   Determine two linearly independent solutions to

$$x^2 y'' + xy' - (4+x)y = 0, \qquad x > 0. \tag{13.6.29}$$

Solution   Since $x = 0$ is a regular singular point, we try for Frobenius series solutions. Substituting

$$y = \sum_{n=0}^{\infty} a_n x^{r+n}$$

into (13.6.29) and simplifying yields the indicial equation

$$r^2 - 4 = 0$$

and the recurrence relation

$$[(r+n)^2 - 4]a_n = a_{n-1}, \qquad n = 1, 2, \ldots . \tag{13.6.30}$$

It follows that the roots of the indicial equation are

$$r_1 = 2, \qquad r_2 = -2,$$

which differ by an integer. Substituting $r = 2$ into (13.6.30) yields

$$a_n = \frac{1}{n(n+4)} a_{n-1}, \qquad n = 1, 2, \ldots .$$

This is easily solved to obtain

$$a_n = \frac{4!}{n!(n+4)!} a_0.$$

Consequently, one Frobenius series solution to (13.6.29) is

$$y_1 = a_0 \sum_{n=0}^{\infty} \frac{4!}{n!(n+4)!} x^{n+2}.$$

Choosing $a_0 = 1/4!$, this solution reduces to

$$y_1 = \sum_{n=0}^{\infty} \frac{1}{n!(n+4)!} x^{n+2}. \tag{13.6.31}$$

We now determine whether there exists a second linearly independent *Frobenius series* solution. Substituting $r = -2$ into (13.6.30) yields

$$n(n-4)a_n = a_{n-1}, \qquad n = 1, 2, \ldots . \tag{13.6.32}$$

Thus when $n = 1, 2, 3$, we obtain

$$a_1 = -\frac{1}{3} a_0, \qquad a_2 = \frac{1}{12} a_0, \qquad a_3 = -\frac{1}{36} a_0.$$

However, when $n = 4$, (13.6.32) requires

$$0 \cdot a_4 = a_3 = -\frac{1}{36} a_0,$$

which is clearly impossible, since $a_0 \neq 0$. It follows that a second linearly independent *Frobenius series* solution does not exist. However, according to Theorem 13.6.1, there is a second linearly independent solution of the form

$$y_2 = Ay_1 \ln x + x^{-2} \sum_{n=0}^{\infty} b_n x^n, \tag{13.6.33}$$

where the constants A and b_n can be determined by substitution into (13.6.29). Differentiating y_2 yields

$$y_2' = Ay_1' \ln x + Ax^{-1}y_1 + \sum_{n=0}^{\infty} (n-2)b_n x^{n-3}.$$

$$y_2'' = Ay_1'' \ln x + 2Ax^{-1}y_1' - Ax^{-2}y_1 + \sum_{n=0}^{\infty} (n-3)(n-2)b_n x^{n-4}.$$

Substituting into (13.6.29) and simplifying we obtain

$$A[x^2 y_1'' + xy_1' - (4+x)y]\ln x + 2Axy_1' + \sum_{n=0}^{\infty} (n-3)(n-2)b_n x^{n-2}$$

$$+ \sum_{n=0}^{\infty} (n-2)b_n x^{n-2} - 4 \sum_{n=0}^{\infty} b_n x^{n-2} - \sum_{n=0}^{\infty} b_n x^{n-1} = 0.$$

The terms in the first brackets vanish, since y_1 is a solution of (13.6.29).[1] Combining the coefficients of x^{n-2} and simplifying yields

$$2Axy_1' + \sum_{n=1}^{\infty} n(n-4)b_n x^{n-2} - \sum_{n=0}^{\infty} b_n x^{n-1} = 0. \tag{13.6.34}$$

We must now determine y_1'. Differentiating (13.6.31) we obtain

$$y_1' = \sum_{n=0}^{\infty} \frac{(n+2)}{n!(n+4)!} x^{n+1}.$$

---

[1] Once more we note that this will always happen at this stage of the computation.

Substituting into (13.6.34) yields

$$2A \sum_{n=0}^{\infty} \frac{(n+2)}{n!(n+4)!} x^{n+2} + \sum_{n=1}^{\infty} n(n-4)b_n x^{n-2} - \sum_{n=0}^{\infty} b_n x^{n-1} = 0.$$

We now replace n by $n - 4$ in the first sum and replace n by $n - 1$ in the third sum to obtain a common power of x in all sums.   The result is

$$2A \sum_{n=4}^{\infty} \frac{(n-2)}{(n-4)!n!} x^{n-2} + \sum_{n=1}^{\infty} n(n-4)b_n x^{n-2} - \sum_{n=1}^{\infty} b_{n-1} x^{n-2} = 0;$$

that is, upon multiplying through by x^2,

$$2A \sum_{n=4}^{\infty} \frac{(n-2)}{(n-4)!n!} x^{n} + \sum_{n=1}^{\infty} n(n-4)b_n x^{n} - \sum_{n=1}^{\infty} b_{n-1} x^{n} = 0.$$

We can now determine the appropriate values of the constants by setting successive coefficients of x^n to zero in the usual manner.

$n = 1$: $-3b_1 - b_0 = 0$; hence

$$b_1 = -\frac{1}{3} b_0.$$

$n = 2$: $-4b_2 - b_1 = 0$; hence

$$b_2 = -\frac{1}{4} b_1 = \frac{1}{12} b_0.$$

$n = 3$: $-3b_3 - b_2 = 0$; hence

$$b_3 = -\frac{1}{3} b_2 = -\frac{1}{36} b_0.$$

$n = 4$: $\frac{1}{6} A - b_3 = 0$; hence

$$A = 6 b_3 = -\frac{1}{6} b_0.$$

$n \geq 5$:

$$b_n = \frac{1}{n(n-4)} \left[b_{n-1} - \frac{2A(n-2)}{(n-4)!n!} \right].$$

Using this recurrence relation we could continue to determine values of the b_n. Notice that b_4 is unconstrained in this problem and so can be set equal to any convenient value; for example, we will set $b_4 = 0$.   Substituting the preceding values of the coefficients into (13.6.33), we obtain

$$y_2 = b_0 \left[-\tfrac{1}{6} y_1 \ln x + x^{-2} (1 - \tfrac{1}{3} x + \tfrac{1}{12} x^2 - \tfrac{1}{36} x^3 + \cdots) \right].$$

Setting $b_0 = -6$ yields

$$y_2 = y_1 \ln x - 6x^{-2}(1 - \tfrac{1}{3}x + \tfrac{1}{12}x^2 - \tfrac{1}{36}x^3 + \cdots).$$

REMARK   In the previous example we had to be content with determining only a few terms in the second linearly independent solution, since we could not solve the recurrence relation that arose for the b_n. This is usually the case in these types of problems.

EXERCISES 13.6

In problems 1–7, determine the roots of the indicial equation of the given differential equation. Also obtain the general *form* of two linearly independent solutions to the differential equation on an interval $(0, R)$. Finally, if $r_1 - r_2$ equals a positive integer obtain the recurrence relation and determine whether the constant A in

$$y_2 = Ay_1 \ln x + x^{r_2} \sum_{n=0}^{\infty} b_n x^n,$$

is zero or nonzero.

1. $4x^2 y'' + 2x^2 y' + y = 0$.
2. $x^2 y'' + x(\cos x)y' - 2e^x y = 0$.
3. $x^2 y'' + x^2 y' - (2 + x)y = 0$.
4. $x^2 y'' + 2x^2 y' + \left(x - \dfrac{3}{4}\right)y = 0$.
5. $x^2 y'' + xy' + (2x - 1)y = 0$.
6. $x^2 y'' + x^3 y' - (2 + x)y = 0$.
7. $x^2(x^2 + 1)y'' + 7xe^x y' + 9(1 + \tan x)y = 0$.
8. Determine all values of the constant α for which

$$x^2 y'' + x(1 - 2x)y' + [2(\alpha - 1)x - \alpha^2]y = 0$$

has two linearly independent *Frobenius series* solutions on $(0, \infty)$.

9. The indicial equation and recurrence relation for the differential equation

$$x^2 y'' + x[(2 - b) + x]y' - (b - \gamma x)y = 0, \quad \text{(i)}$$

are, respectively,

$$(r + 1)(r - b) = 0,$$

$$(r + n + 1)(r + n - b)a_n$$

$$= -[(r + n - 1) + \gamma]a_{n-1}, \quad n = 1, 2, 3, \ldots,$$

in the usual notation, where b and γ are con-

stants.   Determine the *form* of two linearly independent series solutions to (i) on $(0, \infty)$ in the following cases:
(a) $b \neq$ integer;
(b) $b = -1$;
(c) $b = N$, a nonnegative integer. (For solutions containing a term of the form $Ay_1 \ln x$, you must determine whether A is zero or nonzero.)

10. Show that

$$x^2(1 + x)y'' + x^2 y' - 2y = 0$$

has two linearly independent Frobenius series solutions on $(-1, 1)$ and find them.

11. Consider the differential equation

$$x^2 y'' + 3xy' + (1 - x)y = 0, \quad x > 0. \quad \text{(ii)}$$

(a) Determine the indicial equation and show that it has the repeated root $r = -1$.
(b) Obtain the corresponding Frobenius series solution.
(c) It follows from Theorem 13.6.1 that (ii) has a second linearly independent solution of the form

$$y_2 = y_1 \ln x + x^{-1} \sum_{n=1}^{\infty} b_n x^n.$$

Show that $b_1 = -2$ and that in general:

$$b_n = \frac{1}{n^2}\left[b_{n-1} - \frac{2}{n!(n-1)!}\right], \quad n = 2, 3, 4, \ldots.$$

Use this to find the first three terms of y_2.

12. Consider the differential equation

$$xy'' - y = 0, \quad x > 0. \quad \text{(iii)}$$

(a) Show that the roots of the indicial equation are $r_1 = 1$, $r_2 = 0$, and determine the Frobenius series solution corresponding to $r_1 = 1$.
(b) Show that there does not exist a second lin-

early independent Frobenius series solution.

(c) According to Theorem 13.6.1, (iii) has a second linearly independent solution of the form

$$y_2 = Ay_1 \ln x + \sum_{n=0}^{\infty} b_n x^n.$$

Show that $A = b_0$, and determine the first three terms in y_2.

In problems 13–26, determine two linearly independent solutions to the given differential equation on $(0, \infty)$.

13. $x^2 y'' + x(1 - x)y' - y = 0.$

14. $x^2 y'' + x(6 + x^2)y' + 6y = 0.$

15. $xy'' + y' - 2y = 0.$

16. $4x^2 y'' + (1 - 4x)y = 0.$

17. $x^2 y'' - x(x + 3)y' + 4y = 0.$

18. $x^2 y'' + xy' - (1 + x)y = 0.$

19. $x^2 y'' - x^2 y' - (3x + 2)y = 0.$

20. $x^2 y'' - x^2 y' - 2y = 0.$

21. $4x^2 y'' + 4x(1 - x)y' + (2x - 9)y = 0.$

22. $x^2 y'' + x(5 - x)y' + 4y = 0.$

23. $x^2 y'' - x(1 - x)y' + (1 - x)y = 0.$

24. $x^2 y'' + 2x(2 + x)y' + 2(1 + x)y = 0.$

25. $4x^2 y'' - (3 + 4x)y = 0.$

26. $4x^2 y'' + 4x(1 + 2x)y' + (4x - 1)y = 0.$

In problems 27 and 28, determine a Frobenius series solution to the given differential equation and use the *reduction-of-order* technique to find a second linearly independent solution on $(0, \infty)$.

27. $xy'' - xy' + y = 0.$

28. $x^2 y'' + x(4 + x)y' + (2 + x)y = 0.$

29. Consider the *Laguerre* differential equation

$$x^2 y'' + x(1 - x)y' + Nxy = 0, \qquad \text{(iv)}$$

where N is a constant.   Show that in the case

when N is a positive integer, (iv) has a solution that is a polynomial of degree N, and find it. When properly normalized, these solutions are called the **Laguerre polynomials**.

30. Consider the differential equation

$$x^2 y'' + x(1 + 2N - x)y' + N^2 y = 0, \qquad \text{(v)}$$

where N is a positive integer.

(a) Show that there is only one Frobenius series solution and that it terminates after $N + 1$ terms.   Find this solution.

(b) Show that the change of variables $Y = x^N y$ transforms (v) into the Laguerre equation (iv).

31. Consider the differential equation

$$x^2 y'' + x(1 + \beta x)y' + [\beta(1 - N)x - N^2]y = 0,$$
$$x > 0, \qquad \text{(vi)}$$

where N is a positive integer, and β is a constant.

(a) Show that the roots of the indicial equation are $r = \pm N$.

(b) Show that the Frobenius series solution corresponding to $r = N$ is

$$y_1 = a_0 x^N \sum_{n=0}^{\infty} \frac{(2N)!(-\beta)^n}{(2N + n)!} x^n,$$

and that by an appropriate choice of a_0, one solution to (vi) is

$$y_1 = x^{-N} \left(e^{-\beta x} - \sum_{n=0}^{2N-1} \frac{(-\beta x)^n}{n!} \right).$$

(c) Show that (vi) has a second linearly independent Frobenius series solution, which can be taken as

$$y_2 = x^{-N} \sum_{n=0}^{2N-1} \frac{(-\beta x)^n}{n!}.$$

Hence conclude that (vi) has linearly independent solutions

$$y_1 = x^{-N} e^{-\beta x}, \qquad y_2 = x^{-N} \sum_{n=0}^{2N-1} \frac{(-\beta x)^n}{n!}.$$

13.7   BESSEL'S EQUATION OF ORDER p

One of the most important differential equations in applied mathematics and mathematical physics is *Bessel's equation of order p*, defined by

$$x^2 y'' + xy' + (x^2 - p^2)y = 0, \qquad (13.7.1)$$

where p is a *nonnegative* constant.   In general it is not possible to obtain closed form solutions to this equation.   However, since $x = 0$ is clearly a regular singular point, we can apply our Frobenius series technique to obtain series solutions.   We will assume that $x > 0$.   The indicial equation for (13.7.1) is

$$r(r-1) + r - p^2 = 0,$$

with roots

$$r = \pm p.$$

It follows that, provided $2p$ is not an integer, there will exist two linearly independent Frobenius series solutions.   In order to obtain these solutions we let

$$y = x^r \sum_{n=0}^{\infty} a_n x^n, \tag{13.7.2}$$

so that

$$y' = \sum_{n=0}^{\infty} (r+n)a_n x^{r+n-1},$$

$$y'' = \sum_{n=0}^{\infty} (r+n)(r+n-1)a_n x^{r+n-2}.$$

Substituting into (13.7.1) and rearranging yields

$$\sum_{n=0}^{\infty} [(r+n)^2 - p^2]a_n x^{r+n} + \sum_{n=2}^{\infty} a_{n-2}x^{r+n} = 0. \tag{13.7.3}$$

When $n = 0$, we obtain the indicial equation whose roots are, as we have just seen,

$$r = \pm p. \tag{13.7.4}$$

When $n = 1$, (13.7.3) implies that

$$[(r+1)^2 - p^2]a_1 = 0, \tag{13.7.5}$$

and for $n \geq 2$, we obtain the general recurrence relation

$$[(r+n)^2 - p^2]a_n = -a_{n-2}, \qquad n = 2, 3, \ldots. \tag{13.7.6}$$

Consider the first root $r = p$.   In this case (13.7.5) implies that

$$(2p+1)a_1 = 0,$$

so that, since $p \geq 0$, we must have

$$a_1 = 0. \tag{13.7.7}$$

Setting $r = p$ in (13.7.6) yields

$$a_n = -\frac{1}{n(2p+n)} a_{n-2}, \qquad n = 2, 3, \ldots. \tag{13.7.8}$$

It follows from (13.7.7) and (13.7.8) that all the odd coefficients are zero, that is,

$$a_{2k+1} = 0, \qquad k = 0, 1, 2, \ldots. \tag{13.7.9}$$

Now consider the even coefficients.   From (13.7.8) we obtain

$$a_2 = -\frac{1}{2(2p+2)} a_0, \quad a_4 = -\frac{1}{4(2p+4)} a_2 = \frac{1}{2 \cdot 4(2p+2)(2p+4)} a_0,$$

and so on.   The general even coefficient is

$$a_{2k} = \frac{(-1)^k}{2 \cdot 4 \cdot \cdots \cdot (2k)(2p+2)(2p+4) \cdots (2p+2k)} a_0, \quad k = 1, 2, \ldots,$$

which can be written as

$$a_{2k} = \frac{(-1)^k}{2^{2k}k!(p+1)(p+2) \cdots (p+k)} a_0, \quad k = 1, 2, 3, \ldots.$$

The corresponding Frobenius series solution of Bessel's equation is thus

$$y_1 = a_0 x^p \left[1 + \sum_{k=1}^{\infty} \frac{(-1)^k}{2^{2k}k!(p+1)(p+2) \cdots (p+k)} x^{2k} \right]. \qquad (13.7.10)$$

This solution is valid for all $x > 0$.

BESSEL FUNCTIONS OF THE FIRST KIND[1]

In order to study the solutions of Bessel's equation obtained above it is convenient to first rewrite (13.7.10) in a different, but equivalent, form.   The analysis splits into two cases.

$p = N$, **A POSITIVE INTEGER**   In this case the solution (13.7.10) can be written as

$$y_1 = a_0 x^N \left[1 + \sum_{k=1}^{\infty} \frac{(-1)^k}{2^{2k}k!(N+1)(N+2) \cdots (N+k)} x^{2k} \right], \qquad (13.7.11)$$

where the constant a_0 can be chosen arbitrarily.   It is convenient to make the choice

$$a_0 = \frac{1}{N!2^N}.$$

The corresponding solution of Bessel's equation is denoted by $J_N(x)$ and is called the **Bessel function of the first kind of integer order** N.   Thus, substituting for a_0 into (13.7.11), we obtain

$$J_N(x) = \sum_{k=0}^{\infty} \frac{(-1)^k}{k!(N+k)!} \left(\frac{x}{2}\right)^{2k+N} \qquad (13.7.12)$$

The most important Bessel functions of integer order are $J_0(x)$ and $J_1(x)$, since, as we shall see presently, all other integer order Bessel functions of the first kind can be expressed in terms of these two.   Writing out the first few terms in (13.7.12) when $N = 0, 1$ yields, respectively,

---

[1] This section includes only a brief introduction to Bessel functions.   For more details and the proofs of the results stated in this section, the reader is referred to N. N. Lebedev, *Special Functions and Their Applications*, Dover, 1972.

$$J_0(x) = 1 - \frac{x^2}{4} + \frac{x^4}{64} - \cdots,$$

$$J_1(x) = \frac{x}{2}\left(1 - \frac{x^2}{8} + \frac{x^4}{192} - \cdots\right).$$

An analysis of these functions shows that they both oscillate with decaying amplitude. Further, each has an infinite number of nonnegative zeros. A sketch of J_0 and J_1 on the interval $(0, 10]$ is given in Figure 13.7.1.

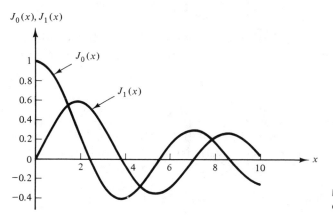

Figure 13.7.1   The Bessel functions of the first kind $J_0(x)$, $J_1(x)$.

$p > 0$, NONINTEGER   In order to obtain a formula for the Frobenius series solution (13.7.10) when p is nonintegral that is analogous to (13.7.12), we need to introduce the *gamma function*. This function can be considered as the generalization of the factorial function to the case of noninteger real numbers.

Definition 13.7.1:   The **gamma function** $\Gamma(p)$ is defined by

$$\Gamma(p) = \int_0^\infty t^{p-1} e^{-t}\, dt, \qquad p > 0.$$

It can be shown that this improper integral converges for all $p > 0$, so that the gamma function is well-defined for all such p. In order to show that the gamma function is a generalization of the factorial function, we first require the following result.

Lemma 13.7.1:   For all $p > 0$,

$$\Gamma(p + 1) = p\Gamma(p). \tag{13.7.13}$$

PROOF   The proof consists of integrating the expression for $\Gamma(p+1)$ by parts:

$$\Gamma(p+1) = \int_0^\infty t^p e^{-t}\, dt = \left[-t^p e^{-t}\right]_0^\infty + p\int_0^\infty t^{p-1} e^{-t}\, dt$$

$$= p\Gamma(p). \qquad \blacksquare$$

We also require

$$\Gamma(1) = \int_0^\infty e^{-t}\, dt = 1. \tag{13.7.14}$$

Equations (13.7.13) and (13.7.14) imply that

$$\Gamma(2) = 1 \cdot \Gamma(1) = 1, \qquad \Gamma(3) = 2 \cdot \Gamma(2) = 2!, \qquad \Gamma(4) = 3 \cdot \Gamma(3) = 3!,$$

and in general, for all nonnegative integers N,

$$\Gamma(N + 1) = N!.$$

This justifies the claim that the gamma function generalizes the factorial function. We now extend the definition of the gamma function to $p < 0$. From (13.7.13)

$$\Gamma(p) = \frac{\Gamma(p + 1)}{p} \tag{13.7.15}$$

for $p > 0$. We use this expression to *define* $\Gamma(p)$ for $p < 0$ as follows. If p is in the interval $(-1, 0)$, then $p + 1$ is in the interval $(0, 1)$, and so (13.7.15) is well-defined. We continue in this manner to successively define $\Gamma(p)$ in the intervals $(-2, -1)$, $(-3, -2), \ldots$. From (13.7.13) and (13.7.15) it follows that

$$\lim_{p \to 0^+} \Gamma(p) = +\infty, \qquad \lim_{p \to 0^-} \Gamma(p) = -\infty,$$

so that the graph of the gamma function has the general form given in Figure 13.7.2.

We note that the gamma function is continuous and, indeed, infinitely differentiable at all points of its domain. Finally, before returning to our discussion of Bessel's equation, we require the following formula:

$$\Gamma(p + 1)[(p + 1)(p + 2) \cdots (p + k)] = \Gamma(p + k + 1). \tag{13.7.16}$$

The proof of this follows by repeated application of (13.7.15), in the form

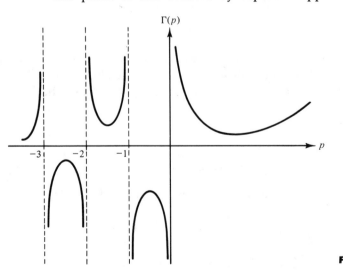

Figure 13.7.2   The gamma function.

$p\Gamma(p) = \Gamma(p + 1)$, to the left-hand side of the preceding equality and is left as an exercise.

Now let us return to the solution (13.7.10) of Bessel's equation. Once more we make a specific choice for a_0. We set

$$a_0 = \frac{1}{2^p \Gamma(p + 1)}.$$

Substituting this value for a_0 into (13.7.10) and using (13.7.16) yields the **Bessel function of the first kind of order** p, $J_p(x)$, defined by

$$J_p(x) = \sum_{k=0}^{\infty} \frac{(-1)^k}{\Gamma(k + 1)\Gamma(p + k + 1)} \left(\frac{x}{2}\right)^{2k+p} \tag{13.7.17}$$

Notice that this does reduce to $J_N(x)$ when N is a nonnegative integer.

BESSEL FUNCTIONS OF THE SECOND KIND

Now consider determining the *general* solution of Bessel's equation. For all $p \geq 0$ we have shown that one solution of Bessel's equation on $(0, \infty)$ is given in (13.7.17). We therefore require a second linearly independent solution. Since the roots of the indicial equation are $r = \pm p$, it follows from our general Frobenius theory that provided $2p \neq$ integer, there will exist a second, linearly independent Frobenius series solution corresponding to the root $r = -p$. It is not too difficult to see that this solution can be obtained by replacing p by $-p$ in (13.7.17). Thus we obtain

$$J_{-p}(x) = \sum_{k=0}^{\infty} \frac{(-1)^k}{\Gamma(k + 1)\Gamma(k - p + 1)} \left(\frac{x}{2}\right)^{2k-p}. \tag{13.7.18}$$

Consequently, the general solution to Bessel's equation of order p when $2p$ is not an integer is

$$y = c_1 J_p(x) + c_2 J_{-p}(x). \tag{13.7.19}$$

When $2p$ is an integer, two subcases arise, depending on whether p is itself an integer or a half-integer (that is, $\frac{1}{2}, \frac{3}{2}, \ldots$). In the latter case a straightforward analysis of the recurrence relation (13.7.6) with $r = -p$, $p = (2j + 1)/2$ (j a nonnegative integer), shows that a second Frobenius series also exists in this case, and it is, in fact, given by (13.7.18). Thus, (13.7.19) represents the general solution of Bessel's equation provided $p \neq$ integer. In practice, rather than using J_{-p} as the second linearly independent solution of Bessel's equation, it is usual to use the following linear combination of these two solutions:

$$Y_p(x) = \frac{J_p(x)\cos p\pi - J_{-p}(x)}{\sin p\pi}. \tag{13.7.20}$$

The function defined in (13.7.20) is called the **Bessel function of the second kind of order** p. Using Y_p we can therefore write the general solution of Bessel's equation, when $p \neq$ positive integer, in the form

$$y = c_1 J_p(x) + c_2 Y_p(x). \tag{13.7.21}$$

The determination of a second linearly independent solution to Bessel's equation when p is a positive integer, n, is quite a bit more complicated. From our general Frobenius theory we certainly know the form of the second linearly independent solution, namely,

$$y_2 = AJ_n(x)\ln x + x^{-n} \sum_{k=0}^{\infty} b_k x^k,$$

and the coefficients A, b_n could be determined by direct substitution into (13.7.1). However, if we extend the definition of the Bessel function of the second kind to the case of positive integers by

$$Y_n(x) = \lim_{p \to n} \left\{ \frac{J_p(x)\cos p\pi - J_{-p}(x)}{\sin p\pi} \right\}, \qquad (13.7.22)$$

it can be shown that the preceding limit exists and that the resulting function is indeed a second linearly independent solution of Bessel's equation of order n. We could derive the series representation of (13.7.22) by evaluating the limit explicitly. The calculations are lengthy and tedious and so we omit them. The result, for those who are interested, is

$$Y_n(x) = \frac{2}{\pi} J_n(x)\left[\ln\left(\frac{x}{2}\right) + \gamma \right] - \frac{1}{\pi}\left[\sum_{k=0}^{n-1} \frac{(n-k-1)!}{k!}\left(\frac{x}{2}\right)^{2k-n} \right.$$
$$\left. + \frac{s_n}{n!}\left(\frac{x}{2}\right)^n + \sum_{k=1}^{\infty}(-1)^k \frac{(s_k + s_{n+k})}{k!(n+k)!}\left(\frac{x}{2}\right)^{2k+n} \right],$$

where

$$s_k = 1 + \frac{1}{2} + \frac{1}{3} + \cdots + \frac{1}{k}$$

and

$$\gamma = \lim_{k \to \infty}(s_k - \ln k) \approx 0.577215664 \ldots$$

is Euler's constant.

Thus (13.7.21) represents the general solution to Bessel's equation of arbitrary order p.

PROPERTIES OF BESSEL FUNCTIONS OF THE FIRST KIND

In practice we are usually interested in solutions of Bessel's equation that are bounded at $x = 0$. However, the Bessel functions of the second kind are always unbounded at $x = 0$. (When $p \neq$ integer, this follows from the fact that x^{-p} is negative, and when $p =$ integer, n, it is due to the second term in the expansion of Y_n given previously.) Thus we usually require only the Bessel functions of the first kind. We now list various properties of the Bessel functions of the first kind that help either in tabulating values of the functions or in working with the functions themselves.

We first recall from (13.7.17) the definition of the Bessel functions of the first kind of order p:

$$J_p(x) = \sum_{k=0}^{\infty} \frac{(-1)^k}{\Gamma(k+1)\Gamma(p+k+1)} \left(\frac{x}{2}\right)^{2k+p} \tag{13.7.23}$$

Property 1: $\dfrac{d}{dx}[x^p J_p(x)] = x^p J_{p-1}(x).$ $\qquad\qquad\qquad\qquad\qquad$ (13.7.24)

Property 2: $\dfrac{d}{dx}[x^{-p} J_p(x)] = -x^{-p} J_{p+1}(x).$ $\qquad\qquad\qquad\qquad$ (13.7.25)

PROOF OF PROPERTY 1 Multiplying (13.7.23) by x^p and differentiating the result with respect to x yields

$$\frac{d}{dx}[x^p J_p(x)] = \sum_{k=0}^{\infty} \frac{(-1)^k}{\Gamma(k+1)\Gamma(p+k+1)} (2k+2p)2^{p-1} \left(\frac{x}{2}\right)^{2k+2p-1}$$

$$= x^p \sum_{k=0}^{\infty} \frac{(-1)^k}{\Gamma(k+1)\Gamma(p+k+1)} (k+p) \left(\frac{x}{2}\right)^{2k+p-1}$$

But, from (13.7.15), $\Gamma(p+k+1) = (p+k)\Gamma(p+k)$, so that

$$\frac{d}{dx}[x^p J_p(x)] = x^p \sum_{k=0}^{\infty} \frac{(-1)^k}{\Gamma(k+1)\Gamma(p+k)} \left(\frac{x}{2}\right)^{2k+p-1}$$

$$= x^p J_{p-1}(x).$$

Property 2 is proved similarly. $\qquad\qquad\qquad\qquad\qquad\qquad\qquad\qquad$ ■

We now derive two identities satisfied by the derivatives of J_p. Expanding the derivatives on the left-hand sides of (13.7.24) and (13.7.25) and dividing the resulting equations by x^p and x^{-p}, respectively, yields

$$J_p'(x) + \frac{p}{x} J_p(x) = J_{p-1}(x), \tag{13.7.26}$$

$$J_p'(x) - \frac{p}{x} J_p(x) = -J_{p+1}(x). \tag{13.7.27}$$

Subtracting (13.7.27) from (13.7.26) and rearranging terms we obtain

Property 3: $J_{p+1}(x) = \dfrac{2p}{x} J_p(x) - J_{p-1}(x).$ $\qquad\qquad\qquad\qquad$ (13.7.28)

Similarly, adding (13.7.26) and (13.7.27) and rearranging yields

Property 4: $J_p'(x) = \dfrac{1}{2}[J_{p-1}(x) - J_{p+1}(x)].$ $\qquad\qquad\qquad\qquad$ (13.7.29)

These formulas allow us to express high-order Bessel functions and their first derivatives in terms of lower-order functions. For example all integer-order Bessel functions can be expressed in terms of $J_0(x)$ and $J_1(x)$.

Example 13.7.1   Express $J_2(x)$ and $J_3(x)$ in terms of $J_0(x)$ and $J_1(x)$.

Solution   Applying (13.7.28) with $p = 1$, we obtain

$$J_2(x) = \frac{2}{x} J_1(x) - J_0(x).$$

Similarly, when $p = 2$,

$$J_3(x) = \frac{4}{x} J_2(x) - J_1(x) = \left(\frac{8}{x^2} - 1\right) J_1(x) - \frac{4}{x} J_0(x).$$

A BESSEL FUNCTION EXPANSION THEOREM

It can be shown that every Bessel function of the first kind has an infinite number of positive zeros. We now show how this can be used to obtain a Bessel function expansion of an arbitrary function.

Let $\lambda_1, \lambda_2, \ldots$ denote the positive zeros of the Bessel function $J_p(x)$, where $p \geq 0$ is fixed, and consider the corresponding functions

$$u_m(x) = J_p(\lambda_m x), \qquad u_n(x) = J_p(\lambda_n x)$$

for x in the interval $[0, 1]$. We begin by deriving an orthogonality relation for the functions u_m and u_n. Since $J_p(x)$ solves Bessel's equation

$$y'' + \frac{1}{x} y' + \left(1 - \frac{p^2}{x^2}\right) y = 0,$$

it is not too difficult to show that u_m and u_n satisfy

$$u_m'' + \frac{1}{x} u_m' + \left(\lambda_m^2 - \frac{p^2}{x^2}\right) u_m = 0, \tag{13.7.30}$$

$$u_n'' + \frac{1}{x} u_n' + \left(\lambda_n^2 - \frac{p^2}{x^2}\right) u_n = 0. \tag{13.7.31}$$

Multiplying (13.7.30) by u_n, (13.7.31) by u_m, and subtracting yields

$$(u_m'' u_n - u_n'' u_m) + \frac{1}{x} (u_m' u_n - u_n' u_m) + (\lambda_m^2 - \lambda_n^2) u_m u_n = 0,$$

which can be written as

$$(u_m' u_n - u_n' u_m)' + \frac{1}{x} (u_m' u_n - u_n' u_m) + (\lambda_m^2 - \lambda_n^2) u_m u_n = 0.$$

Multiplying this equation by x and combining the first two terms of the resulting equation, we obtain

$$\frac{d}{dx} [x(u_m' u_n - u_n' u_m)] + x(\lambda_m^2 - \lambda_n^2) u_m u_n = 0.$$

Integrating from 0 to 1 and using the fact that $u_m(1) = u_n(1) = 0$ [since λ_m and λ_n are zeros of $J_p(x)$] yields

$$(\lambda_m^2 - \lambda_n^2) \int_0^1 x u_m(x) u_n(x) \, dx = 0,$$

that is, since λ_m and λ_n are distinct and positive,

$$\int_0^1 x u_m(x) u_n(x) \, dx = 0.$$

Substituting for $u_m(x)$ and $u_n(x)$, we finally obtain

$$\int_0^1 x J_p(\lambda_m x) J_p(\lambda_n x) \, dx = 0, \qquad \text{whenever } m \neq n. \tag{13.7.32}$$

It can further be shown (although this is more difficult) that when $m = n$,

$$\int_0^1 x [J_p(\lambda_n x)]^2 \, dx = \tfrac{1}{2} [J_{p+1}(\lambda_n)]^2. \tag{13.7.33}$$

We can now state a Bessel function expansion theorem.

Theorem 13.7.1:   If f and f' are continuous on the interval $[0, 1]$, then for $0 < x < 1$,

$$f(x) = a_1 J_p(\lambda_1 x) + a_2 J_p(\lambda_2 x) + \cdots + a_n J_p(\lambda_n x) + \cdots = \sum_{n=1}^{\infty} a_n J_p(\lambda_n x), \tag{13.7.34}$$

where the coefficients can be determined from

$$a_n = \frac{2}{[J_{p+1}(\lambda_n)]^2} \int_0^1 x f(x) J_p(\lambda_n x) \, dx. \tag{13.7.35}$$

PROOF   We show only that if a series of the form (13.7.34) exists, then the coefficients are given by (13.7.35). The proof of convergence is omitted. Multiplying (13.7.34) by $x J_p(\lambda_m x)$ and integrating the resulting equation with respect to x from 0 to 1, we obtain

$$\int_0^1 x f(x) J_p(\lambda_m x) \, dx = a_m \int_0^1 x [J_p(\lambda_m x)]^2 \, dx,$$

where we have used (13.7.32). Using (13.7.33) to substitute for the integral on the right-hand side and rearranging yields

$$a_m = \frac{2}{[J_{p+1}(\lambda_m)]^2} \int_0^1 x f(x) J_p(\lambda_m x) \, dx,$$

which is what we wished to show.   ■

EXERCISES 13.7

1. Use relations (13.7.5) and (13.7.6) to show that if p is a half integer, then Bessel's equation of order p has two linearly independent Frobenius series solutions.

2. Find two linearly independent solutions to

$$x^2 y'' + xy' + (x^2 - \tfrac{9}{4})y = 0,$$

on the interval $(0, \infty)$.

3. Let $\Gamma(p)$ denote the gamma function. Show that

$$\Gamma(p+1)[(p+1)\,(p+2)\cdots(p+k)]$$
$$= \Gamma(p+k+1).$$

4. (a) By making the change of variables $t = x^2$ in the integral that defines the gamma function show that

$$\Gamma(\tfrac{1}{2}) = 2 \int_0^\infty e^{-x^2}\, dx.$$

(b) Use your result in (a) to show that

$$[\Gamma(\tfrac{1}{2})]^2 = 4 \int_0^\infty \int_0^\infty e^{-(x^2+y^2)}dx\,dy.$$

(c) By changing to polar coordinates evaluate the double integral in (b) and hence show that

$$\Gamma(\tfrac{1}{2}) = \sqrt{\pi}.$$

(d) Use your result in (c) to find $\Gamma(\tfrac{3}{2})$ and $\Gamma(-\tfrac{1}{2})$.

5. Let $J_p(x)$ denote the Bessel function of the first kind of order p. Prove that

$$\frac{d}{dx}\,[x^{-p}J_p(x)] = -x^{-p}J_{p+1}(x).$$

6. By manipulating the general expression for

$J_{1/2}(x)$, show that it can be written in closed form as:

$$J_{1/2}(x) = \sqrt{\frac{2}{\pi x}}\,\sin x.$$

7. Given that

$$J_{1/2}(x) = \sqrt{\frac{2}{\pi x}}\,\sin x, \qquad J_{-1/2}(x) = \sqrt{\frac{2}{\pi x}}\,\cos x,$$

express $J_{3/2}(x)$ and $J_{-3/2}(x)$ in closed form. Convince yourself that all half integer–order Bessel functions of the first kind can be expressed as a finite sum of terms involving products of $\sin x$, $\cos x$, and powers of x.

8. By integrating the recurrence relation for derivatives of the Bessel functions of the first kind, show that

(a) $\displaystyle \int x^p J_{p-1}(x)\, dx = x^p J_p(x) + C.$

(b) $\displaystyle \int x^{-p} J_{p+1}(x)\, dx = -x^{-p}J_p(x) + C.$

9. Determine the Bessel series expansion in the functions $J_p(\lambda_n x)$ for $f(x) = x^p$ on the interval $(0, 1)$. [Here λ_n denote the positive zeros of $J_p(x)$. Hint: You will need to use one of the results from the previous question.]

10. Let $J_p(x)$ denote the Bessel function of the first kind of order p, and let λ be a positive constant. If $u(x) = J_p(\lambda x)$, show that u satisfies the differential equation

$$\frac{d^2u}{dx^2} + \frac{1}{x}\frac{du}{dx} + \left(\lambda^2 - \frac{p^2}{x^2}\right)u = 0.$$

A Review
of Complex Numbers

Any number, z, of the form $z = a + ib$, where a and b are real numbers and $i = \sqrt{-1}$, is called a **complex number**. If $z = a + ib$, then we refer to a as the **real part** of z, denoted Re(z), and we refer to b as the **imaginary part** of z, denoted Im(z). Thus

$$\boxed{\text{If } z = a + ib, \text{ then } \mathrm{Re}(z) = a \text{ and } \mathrm{Im}(z) = b.}$$

Example A1.1   If $z = 2 - 3i$, then Re(z) = 2 and Im(z) = −3.

Complex numbers can be added, subtracted, and multiplied in the usual manner, and the result is once more a complex number. Further, these operations satisfy all of the basic properties satisfied by the real numbers. All that we must remember is that whenever we encounter the term i^2, it must be replaced by −1.

Example A1.2   If $z_1 = 3 + 4i$ and $z_2 = -1 + 2i$, find $z_1 - 3z_2$ and $z_1 z_2$.

Solution

$$z_1 - 3z_2 = (3 + 4i) - 3(-1 + 2i) = 6 - 2i = 2(3 - i).$$

$$z_1 z_2 = (3 + 4i)(-1 + 2i) = (-3 + 6i - 4i + 8i^2) = -11 + 2i.$$

Example A1.3   If $z_1 = 4 + 3i$ and $z_2 = 4 - 3i$, determine $z_1 z_2$.

Solution   In this case

$$z_1 z_2 = (4 + 3i)(4 - 3i) = (16 - 12i + 12i - 9i^2) = 16 + 9 = 25.$$

Notice that in the previous example the product $z_1 z_2$ turned out to be a real number. This was not an accident. If we look at the definition of z_2 we see that it can be obtained from z_1 by replacing the imaginary part of z_1 by its negative. Complex numbers that are related in this manner are called conjugates of one another.

Definition A1.1:   If $z = a + ib$, then the complex number, $\bar{z}$, defined by

$$\bar{z} = a - ib,$$

is called the **conjugate** of z.

Example A1.4   If $z = 2 + 5i$, then $\bar{z} = 2 - 5i$, whereas if $z = 3 - 4i$, then $\bar{z} = 3 + 4i$.

Properties of the Conjugate

1. $\bar{\bar{z}} = z$.

2. $z\bar{z} = \bar{z}z = a^2 + b^2$.

PROOF

1. If $z = a + ib$, then $\bar{z} = a - ib$, so that $\bar{\bar{z}} = a + ib = z$.
2. $z\bar{z} = (a + ib)(a - ib) = a^2 - i\,ab + i\,ab - (ib)^2 = a^2 + b^2$.   ∎

If $z = a + ib$, then the real number $\sqrt{a^2 + b^2}$ is often called the **modulus** of z, or **absolute value** of z, and is denoted by $|z|$. It follows from property 2 that

$$|z|^2 = z\bar{z}.$$

Example A1.5   Determine $|z|$ if $z = 2 - 3i$.

Solution   By definition,

$$|z| = \sqrt{(2)^2 + (-3)^2} = \sqrt{13}.$$

We now recall from elementary algebra that an expression of the form $1/(a + \sqrt{b})$ can always be written with the radical in the numerator. To accomplish this we multiply by

$$\frac{a - \sqrt{b}}{a - \sqrt{b}},$$

the result being

$$\frac{a - \sqrt{b}}{a^2 - b}.$$

The reason that this works is because

$$(a + \sqrt{b})(a - \sqrt{b}) = a^2 - b.$$

This is similar to $(a + ib)(a - ib) = a^2 + b^2$. Now consider an expression of the form

$$\frac{1}{a + ib}.$$

As this is written, we cannot say that it is a complex number, since it is not of the form $a + ib$. However, if we multiply by

$$\frac{a - ib}{a - ib}$$

and use property 2 of the conjugate, we obtain

$$\frac{1}{a + ib} = \frac{1}{(a + ib)} \cdot \frac{(a - ib)}{(a - ib)} = \frac{a}{a^2 + b^2} - i\left(\frac{b}{a^2 + b^2}\right)$$

which is a complex number.

Example A1.6 Express $z = \dfrac{1}{2 + 5i}$ in the form $a + ib$.

Solution

$$z = \frac{1}{2 + 5i} = \frac{1}{(2 + 5i)} \frac{(2 - 5i)}{(2 - 5i)} = \frac{2}{29} - \frac{5}{29} i.$$

More generally, if $z_1 = a + ib$ and $z_2 = x + iy$, then

$$\frac{z_1}{z_2} = \frac{a + ib}{x + iy} = \frac{(a + ib)}{(x + iy)} \frac{(x - iy)}{(x - iy)}$$

$$= \frac{1}{x^2 + y^2} [(ax + by) + i(-ay + bx)].$$

Thus we can divide two complex numbers and the result is once more a complex number.

Example A1.7 If $z_1 = 2 + 3i$ and $z_2 = 3 + 4i$, determine $\dfrac{z_1}{z_2}$.

Solution   In this case we have

$$\frac{z_1}{z_2} = \frac{2+3i}{3+4i} = \frac{(2+3i)}{(3+4i)}\frac{(3-4i)}{(3-4i)} = \frac{1}{25}(18+i).$$

EXERCISES A1

In problems 1–5 determine $\bar{z}$ and $|z|$ for the given complex number.

1. $z = 2 + 5i.$

2. $z = 3 - 4i.$

3. $z = 5 - 2i.$

4. $z = 7 + i.$

5. $z = 1 + 2i.$

In problems 6–10 express $z_1 z_2$ and z_1/z_2 in the form $a + ib.$

6. $z_1 = 1 + i, z_2 = 3 + 2i.$

7. $z_1 = -1 + 3i, z_2 = 2 - i.$

8. $z_1 = 2 + 3i, z_2 = 1 - i.$

9. $z_1 = 4 - i, z_2 = 1 + 3i.$

10. $z_1 = 1 - 2i, z_2 = 3 + 4i.$

11. Prove that if z_1 and z_2 are complex numbers, then

$$\overline{z_1 + z_2} = \bar{z}_1 + \bar{z}_2.$$

12. Generalize the previous example to the case when $z_1, z_2, \ldots, z_n$ are all complex numbers.

13. Prove that if z_1 and z_2 are complex numbers, then

$$\overline{(z_1 z_2)} = \bar{z}_1 \, \bar{z}_2.$$

14. Prove that if z_1 and z_2 are complex numbers, then

$$\overline{\left(\frac{z_1}{z_2}\right)} = \frac{\bar{z}_1}{\bar{z}_2}.$$

A Review
of Partial Fractions

In this appendix we review the partial fraction decomposition of rational functions. No proofs are given, since the reader is assumed to have seen the results in a previous calculus course.

Recall that a function of the form

$$p(x) = a_n x^n + a_{n-1} x^{n-1} + \cdots + a_1 x + a_0, \tag{A2.1}$$

with $a_n \neq 0$, is called a **polynomial of degree** n. According to the fundamental theorem of algebra, the equation $p(x) = 0$ has precisely n roots (not all necessarily distinct). If we let $x_1, x_2, \ldots, x_n$ denote these roots, then $p(x)$ can be factored as

$$p(x) = K(x - x_1)(x - x_2) \cdots (x - x_n), \tag{A2.2}$$

where $K = a_n$. Some of the roots may be complex. We will assume that the coefficients in (A2.1) are real numbers, in which case any complex roots must occur in conjugate pairs.

A quadratic factor of the form

$$ax^2 + bx + c,$$

which has no *real* linear factors, is said to be *irreducible*.

Theorem A2.1: Any real polynomial can be factored into linear and irreducible quadratic terms with real coefficients.[1]

PROOF   Let $p(x)$ be a real polynomial and suppose that $x = \alpha$ is a complex root of $p(x) = 0$. Then $x = \overline{\alpha}$ is also a root. Thus (A2.2) will contain the terms $(x - \alpha)(x - \overline{\alpha})$. These linear terms have complex coefficients. However, if we expand the product the result is

$$(x - \alpha)(x - \overline{\alpha}) = [x^2 - (\alpha + \overline{\alpha})x + \alpha\overline{\alpha}].$$

But

$$\alpha + \overline{\alpha} = 2\,\mathrm{Re}(\alpha), \qquad \alpha\overline{\alpha} = |\alpha|^2,$$

which are both real, so that the irreducible quadratic term does indeed have real coefficients. ∎

If $p(x)$ and $q(x)$ are two polynomials (not necessarily of the same degree), then a function of the form

$$R(x) = \frac{p(x)}{q(x)}$$

is called a rational function. Suppose that $q(x)$ has been factored into linear and irreducible quadratic terms. Then $q(x)$ will consist of a product of terms of the form

$$(ax - b)^k \quad \text{or} \quad (ax^2 + bx + c)^k, \tag{A2.3}$$

where a, b, c, and k are constants. An example is

$$\frac{x^2 - 1}{(x + 2)(x^2 + 3)}.$$

The idea behind a partial fraction decomposition is to express a rational function as a sum of terms whose denominators are of the form (A2.3). The following rules tell us the form that such a decomposition must take.

1. Each factor of the form $(ax - b)^k$ in $q(x)$ contributes the following terms to the partial fraction decomposition of $p(x)/q(x)$:

$$\frac{A_1}{ax - b} + \frac{A_2}{(ax - b)^2} + \cdots + \frac{A_k}{(ax - b)^k},$$

where $A_1, A_2, \ldots, A_k$ are constants.

2. Each irreducible quadratic factor of the form $(ax^2 + bx + c)^k$ contributes the

---

[1] By a "real polynomial," we mean a polynomial with real coefficients.

following terms to the partial fraction decomposition of $p(x)/q(x)$:

$$\frac{A_1 x + B_1}{ax^2 + bx + c} + \frac{A_2 x + B_2}{(ax^2 + bx + c)^2} + \cdots + \frac{A_k x + B_k}{(ax^2 + bx + c)^k}.$$

Thus, for example,

$$\frac{x^2 + 1}{x(x-1)(x^2+4)} = \frac{A}{x} + \frac{B}{x-1} + \frac{Cx+D}{x^2+4},$$

for appropriate values of the constants A, B, C, D. Similarly,

$$\frac{x-2}{(x+2)^2(x^2+2x+2)} = \frac{A}{x+2} + \frac{B}{(x+2)^2} + \frac{Cx+D}{(x^2+2x+2)},$$

for appropriate A, B, C, D.

The preceding rules give only the form of a partial fraction decomposition. The next question that needs answering is the following: How do we determine the constants that arise in the partial fraction decomposition? A standard way to proceed is as follows:

---

1. Determine the general form of the partial fraction decomposition of $p(x)/q(x)$.
2. Multiply both sides of the resulting decomposition by $q(x)$.
3. Equate the coefficients of like powers of x on both sides of the resulting equation in order to determine the constants in the partial fraction decomposition.

We illustrate the procedure with several examples.

Example A2.1 Determine the partial fraction decomposition of

$$\frac{2x}{(x-1)(x+3)}.$$

Solution The general form of the partial fraction decomposition is

$$\frac{2x}{(x-1)(x+3)} = \frac{A}{x-1} + \frac{B}{x+3}.$$

Multiplying both sides of this equation by $(x-1)(x+3)$ yields

$$2x = A(x+3) + B(x-1).$$

Equating the coefficients of like powers of x on both sides of this equation yields

$$A + B = 2, \qquad 3A - B = 0.$$

Solving for A and B, we obtain

$$A = \frac{1}{2}, \qquad B = \frac{3}{2},$$

so that

$$\frac{2x}{(x-1)(x+3)} = \frac{1}{2(x-1)} + \frac{3}{2(x+3)}.$$

Example A2.2   Determine the partial fraction decomposition of

$$\frac{x^2 + 1}{(x+1)(x^2+4)}.$$

Solution   In this case the general form of the partial fraction decomposition is

$$\frac{x^2 + 1}{(x+1)(x^2+4)} = \frac{A}{x+1} + \frac{Bx+C}{x^2+4}.$$

Multiplying both sides by $(x+1)(x^2+4)$, we obtain

$$x^2 + 1 = A(x^2+4) + (Bx+C)(x+1).$$

Equating coefficients of like powers of x on both sides of this equality yields

$$A + B = 1, \qquad B + C = 0, \qquad 4A + C = 1.$$

Solving this system of equations we obtain

$$A = \frac{2}{5}, \qquad B = \frac{3}{5}, \qquad C = -\frac{3}{5},$$

so that

$$\frac{x^2 + 1}{(x+1)(x^2+4)} = \frac{2}{5(x+1)} + \frac{3(x-1)}{5(x^2+4)}.$$

Example A2.3   Determine the partial fraction decomposition of

$$\frac{2x - 1}{(x+2)^2(x^2+2x+2)}.$$

Solution   The term $x^2 + 2x + 2$ is irreducible. Thus the partial fraction decomposition has the general form

$$\frac{2x - 1}{(x+2)^2(x^2+2x+2)} = \frac{A}{x+2} + \frac{B}{(x+2)^2} + \frac{Cx+D}{x^2+2x+2}.$$

Clearing the fractions yields

$$2x - 1 = A(x+2)(x^2+2x+2) + B(x^2+2x+2) + (Cx+D)(x+2)^2.$$

Equating the coefficients of like powers of x, we obtain

$$A + \qquad C \qquad = 0,$$

$$4A + B + 4C + D = 0,$$

$$6A + 2B + 4C + 4D = 2,$$

$$4A + 2B \qquad + 4D = -1.$$

Solving this system yields

$$A = -\frac{3}{2}, \qquad B = -\frac{5}{2}, \qquad C = \frac{3}{2}, \qquad D = \frac{5}{2},$$

so that

$$\frac{2x - 1}{(x + 2)^2(x^2 + 2x + 2)} = \frac{3x + 5}{2(x^2 + 2x + 2)} - \frac{3}{2(x + 2)} - \frac{5}{2(x + 2)^2}.$$

SOME SHORTCUTS

The preceding technique for determining the constants that arise in the partial fraction decomposition of a rational function will always work. However, in practice it is often tedious to apply. We now present, without justification, some shortcuts that can circumvent many of the computations.

LINEAR FACTORS: THE COVER-UP RULE   If $q(x)$ contains a linear factor of the form $x - a$, then this factor contributes a term of the form

$$\frac{A}{x - a}$$

to the partial fraction decomposition of $p(x)/q(x)$. Let $P(x)$ denote the expression obtained by omitting the $x - a$ term in $p(x)/q(x)$. Then the constant A is given by

$$\boxed{A = P(a).}$$

Example A2.4   Determine the partial fraction decomposition of $\dfrac{3x - 1}{(x - 3)(x + 2)}$.

Solution   The general form of the decomposition is

$$\frac{3x - 1}{(x - 3)(x + 2)} = \frac{A}{x - 3} + \frac{B}{x + 2}.$$

To determine A, we neglect the $x - 3$ term in the given rational function and set

$$P(x) = \frac{3x - 1}{x + 2}.$$

Then, according to the preceding rule,

$$A = P(3) = \frac{8}{5}.$$

Similarly, to determine B we neglect the $x + 2$ term in the given function and set

$$P(x) = \frac{3x - 1}{x - 3}.$$

Using the cover-up rule it then follows that

$$B = P(-2) = \frac{-7}{-5} = \frac{7}{5}.$$

Thus,

$$\frac{3x - 1}{(x - 3)(x + 2)} = \frac{8}{5(x - 3)} + \frac{7}{5(x + 2)}.$$

The idea behind the technique is to cover up the linear factor $x - a$ in the given rational function and set $x = a$ in the remaining part of the function. The result will be the constant A in the contribution $A/(x - a)$ to the partial fraction decomposition of the rational function.

REPEATED LINEAR FACTORS   The cover-up rule can also be extended to the case of repeated linear factors. Suppose that $q(x)$ contains a factor of the form $(x - a)^k$. Then this contributes the terms

$$\frac{A_1}{x - a} + \frac{A_2}{(x - a)^2} + \cdots + \frac{A_k}{(x - a)^k}$$

to the partial fraction decomposition of $p(x)/q(x)$. Let $P(x)$ be the expression obtained when the $(x - a)^k$ term is neglected in $p(x)/q(x)$. Then the constants A_1, $A_2, \ldots, A_k$ are given by

$$A_k = P(a), \; A_{k-1} = P'(a), \; A_{k-2} = \frac{1}{2!} P''(a), \; \ldots, \; A_1 = \frac{1}{(k - 1)!} P^{(k-1)}(a),$$

where a prime denotes differentiation with respect to x.

REMARKS

1. The preceding formulas look rather formidable to begin with. However, they are often easy to apply in practice.
2. Notice that in the case $k = 1$, we are back to the cover-up rule.

Example A2.5   Determine the partial fraction decomposition of $\dfrac{x}{(x - 1)(x + 2)^2}$.

Solution   The general form of the partial fraction decomposition is

$$\frac{x}{(x - 1)(x + 2)^2} = \frac{A_1}{x + 2} + \frac{A_2}{(x + 2)^2} + \frac{A_3}{x - 1}.$$

To determine A_1 and A_2 we omit the term $(x + 2)^2$ in the given function to obtain

$$P(x) = \frac{x}{x - 1}.$$

Applying the preceding rule with $k = 2$ yields

$$A_2 = P(-2) = \frac{2}{3}, \quad A_1 = P'(-2) = \left.\frac{-1}{(x - 1)^2}\right|_{x=-2} = -\frac{1}{9}.$$

We now use the cover-up rule to determine A_3.   Neglecting the $x - 1$ term in the given function and setting $x = 1$ in the resulting expression yields

$$A_3 = \frac{1}{9}.$$

Thus

$$\frac{x}{(x - 1)(x + 2)^2} = -\frac{1}{9(x + 2)} + \frac{2}{3(x + 2)^2} + \frac{1}{9(x - 1)}.$$

IRREDUCIBLE QUADRATIC FACTORS OF THE FORM $x^2 + a^2$ The final case that we will consider is when $q(x)$ contains a factor of the form $x^2 + a^2$.   This will contribute a term of the form

$$\frac{Ax + B}{x^2 + a^2}$$

to the partial fraction decomposition of $p(x)/q(x)$.   Let $P(x)$ be the expression obtained by deleting the term $x^2 + a^2$ in $p(x)/q(x)$.   Then the constants A and B are given by

$$\boxed{A = \frac{1}{a} \text{Im}[P(ia)], \quad B = \text{Re}[P(ia)].}$$

Example A2.6   Determine the partial fraction decomposition of $\dfrac{x - 1}{(x + 2)(x^2 + 4)}$.

Solution   In this case the general form of the partial fraction decomposition is

$$\frac{x - 1}{(x + 2)(x^2 + 4)} = \frac{Ax + B}{x^2 + 4} + \frac{C}{x + 2}.$$

In order to determine A and B we delete the $x^2 + 4$ term from the given function to obtain

$$P(x) = \frac{x - 1}{x + 2}.$$

Since in this case $a = 2i$, we first compute

$$P(2i) = \frac{2i - 1}{2i + 2} = \frac{1}{4}(1 + 3i).$$

Thus

$$A = \frac{1}{2}\operatorname{Im}\left[\frac{1}{4}(1 + 3i)\right] = \frac{3}{8}, \qquad B = \operatorname{Re}\left[\frac{1}{4}(1 + 3i)\right] = \frac{1}{4}.$$

In order to determine C we use the cover-up rule. Neglecting the $x + 2$ factor in the given function and setting $x = -2$ in the result yields

$$C = -\frac{3}{8}.$$

Thus

$$\frac{x - 1}{(x + 2)(x^2 + 4)} = \frac{3x + 2}{8(x^2 + 4)} - \frac{3}{8(x + 2)}.$$

REMARK   These techniques can be extended to the case of irreducible factors of the form $(ax^2 + bx + c)^k$.

EXERCISES A2

In problems 1–18 determine the partial fraction decomposition of the given rational function.

1. $\dfrac{2x - 1}{(x + 1)(x + 2)}$.

2. $\dfrac{x - 2}{(x - 1)(x + 4)}$.

3. $\dfrac{x + 1}{(x - 3)(x + 2)}$.

4. $\dfrac{x^2 - x + 4}{(x + 3)(x - 1)(x + 2)}$.

5. $\dfrac{2x - 1}{(x + 4)(x - 2)(x + 1)}$.

6. $\dfrac{3x^2 - 2x + 14}{(2x - 1)(x + 5)(x + 2)}$.

7. $\dfrac{2x + 1}{(x + 2)(x + 1)^2}$.

8. $\dfrac{5x^2 + 3}{(x + 1)(x - 1)^2}$.

9. $\dfrac{3x + 4}{x^2(x^2 + 4)}$.

10. $\dfrac{3x - 2}{(x - 5)(x^2 + 1)}$.

11. $\dfrac{x^2 + 6}{(x - 2)(x^2 + 16)}$.

12. $\dfrac{10}{(x - 1)(x^2 + 9)}$.

13. $\dfrac{7x + 2}{(x - 2)(x + 2)^2}$.

14. $\dfrac{7x^2 - 20}{(x - 2)(x^2 + 4)}$.

15. $\dfrac{7x + 4}{(x + 1)^3(x - 2)}$.

16. $\dfrac{x(2x + 3)}{(x + 1)(x^2 + 2x + 2)}$.

17. $\dfrac{3x + 4}{(x - 3)(x^2 + 4x + 5)}$.

18. $\dfrac{7 - 2x^2}{(x - 1)(x^2 + 4)}$.

A Review
of Integration
Techniques

In this appendix we review some of the basic integration techniques that are required throughout the text. This is a very brief refresher and should *not* be considered as a substitute for a calculus text.

INTEGRATION BY PARTS   The basic formula for integration by parts can be written in the form

$$\int u \, dv = uv - \int v \, du.$$

To derive this, we start with the product rule for differentiation, namely,

$$\frac{d}{dx}[u(x)v(x)] = u\frac{dv}{dx} + v\frac{du}{dx}.$$

Integrating both sides of this equation with respect to x yields

$$u(x)v(x) = \int \left(u\frac{dv}{dx} + v\frac{du}{dx} \right) dx;$$

that is, upon rearranging terms,

$$\int u\frac{dv}{dx}\, dx = uv - \int v\frac{du}{dx}\, dx.$$

Consequently,

$$\int u\, dv = uv - \int v\, du.$$

Example A3.1   Evaluate $\int xe^{2x}\, dx$.

Solution   Choosing

$$u = x, \qquad dv = e^{2x}dx,$$

it follows that

$$\frac{du}{dx} = 1, \qquad v = \tfrac{1}{2}\, e^{2x},$$

so that

$$\int xe^{2x}\, dx = \tfrac{1}{2}\, xe^{2x} - \tfrac{1}{2}\int e^{2x}\, dx$$

$$= \tfrac{1}{2}\, xe^{2x} - \tfrac{1}{4}\, e^{2x} + C,$$

where C is an integration constant.

Example A3.2   Evaluate $\int x^2\sin x\, dx$.

Solution   In this case we take

$$u = x^2, \qquad dv = \sin x\, dx,$$

so that

$$\frac{du}{dx} = 2x, \qquad v = -\cos x.$$

Thus

$$\int x^2\sin x\, dx = -x^2\cos x + 2\int x\cos x\, dx.$$

We must now evaluate the second integral.   Once more, integration by parts is appropriate.   This time we take

$$u = x, \qquad dv = \cos x \, dx.$$

Then

$$\frac{du}{dx} = 1, \qquad v = \sin x,$$

so that

$$\int x^2 \sin x \, dx = -x^2 \cos x + 2 \left(x \sin x - \int \sin x \, dx \right)$$

$$= -x^2 \cos x + 2 \left(x \sin x + \cos x \right) + C.$$

As the previous two examples illustrate, the integration-by-parts technique is extremely useful for evaluating integrals of the form

$$\int x^k f(x) \, dx$$

when k is a positive integer.   Such an integral can often be evaluated by successively applying the integration-by-parts formula until the power of x is reduced to zero.   However, this will not always work.

Example A3.3   Evaluate $\int x \ln x \, dx$.

Solution   In this case if we set $u = x$ and $dv = \ln x \, dx$, then we require the integral of $\ln x$.   Instead we choose

$$u = \ln x, \qquad dv = x \, dx,$$

then

$$\frac{du}{dx} = \frac{1}{x}, \qquad v = \frac{1}{2} x^2.$$

Thus, applying the integration-by-parts formula we obtain

$$\int x \ln x \, dx = \tfrac{1}{2} x^2 \ln x - \tfrac{1}{2} \int x \, dx$$

$$= \tfrac{1}{2} x^2 \ln x - \tfrac{1}{4} x^2 + C$$

$$= \tfrac{1}{4} x^2 (2 \ln x - 1) + C.$$

INTEGRATION BY SUBSTITUTION   This is one of the most important integration techniques.   Many of the standard integrals can be derived using a substitution. We illustrate with some examples.

Example A3.4   Evaluate the following integrals:

(a) $\int x e^{3x^2} \, dx$.

(b) $\int \dfrac{1}{\sqrt{1 - x^2}} \, dx$.

(c) $\int \dfrac{1}{x} (\ln x)^2 \, dx$.

Solution

(a) If we let $u = x^2$, then $du = 2x \, dx$, so that

$$\int x e^{3x^2} \, dx = \tfrac{1}{2} \int e^{3u} \, du = \tfrac{1}{6} e^u + C = \tfrac{1}{6} e^{3x^2} + C.$$

(b) Recalling that $1 - \sin^2\theta = \cos^2\theta$, the form of the integrand suggests that we let $x = \sin\theta$. Then $dx = \cos\theta \, d\theta$ and the given integral can be written as

$$\int \dfrac{1}{\sqrt{1 - x^2}} \, dx = \int \dfrac{\cos\theta}{\sqrt{1 - \sin^2\theta}} \, d\theta = \int d\theta = \theta + C.$$

Substituting back for $\theta = \sin^{-1} x$, we obtain

$$\int \dfrac{1}{\sqrt{1 - x^2}} \, dx = \sin^{-1} x + C.$$

(c) In this case we recognize that the derivative of $\ln x$ is $1/x$. This suggests that we make the substitution

$$u = \ln x,$$

so that

$$du = \dfrac{1}{x} \, dx.$$

Then the given integral can be written in the form

$$\int \dfrac{1}{x} (\ln x)^2 \, dx = \int u^2 \, du = \dfrac{1}{3} u^3 + C.$$

Substituting back for $u = \ln x$ yields

$$\int \dfrac{1}{x} (\ln x)^2 \, dx = \dfrac{1}{3} (\ln x)^3 + C.$$

Now consider the general integral

$$\int \dfrac{f'(x)}{f(x)} \, dx.$$

If we let $u = f(x)$, then $du = f'(x)\,dx$, so that

$$\int \frac{f'(x)}{f(x)}\,dx = \int \frac{1}{u}\,du = \ln|u| + C.$$

Substituting back for $u = f(x)$ yields the important formula

$$\boxed{\int \frac{f'(x)}{f(x)}\,dx = \ln|f(x)| + C.}$$

Example A3.5   Evaluate $\displaystyle\int \frac{x-2}{x^2 - 4x + 3}\,dx$.

Solution   If we rewrite the integral in the equivalent form

$$\int \frac{x-2}{x^2 - 4x + 3}\,dx = \frac{1}{2}\int \frac{2(x-2)}{x^2 - 4x + 3}\,dx,$$

then we see that the numerator in the second integral is the derivative of the denominator.   Thus

$$\int \frac{x-2}{x^2 - 4x + 3}\,dx = \frac{1}{2}\ln|x^2 - 4x + 3| + C.$$

INTEGRATION BY PARTIAL FRACTIONS   We can always evaluate an integral of the form

$$\int \frac{p(x)}{q(x)}\,dx \qquad\qquad\qquad\qquad (A3.1)$$

when $p(x)$ and $q(x)$ are polynomials in x.   Consider first the case when the degree of $p(x)$ is *less than* the degree of $q(x)$.   To evaluate (A3.1) we first determine the partial fraction decomposition of the integrand (a review of partial fractions is given in Appendix 2).   The result will always be integrable, although we might need a substitution to carry out this integration.

Example A3.6   Evaluate $\displaystyle\int \frac{3x+2}{(x+1)(x+2)}\,dx$.

Solution   In this case we require the partial fraction decomposition of the integrand.   Using the rules for partial fractions it follows that

$$\frac{3x+2}{(x+1)(x+2)} = \frac{A}{x+1} + \frac{B}{x+2}.$$

Multiplying both sides of this equality by $(x+1)(x+2)$ yields

$$3x + 2 = A(x+2) + B(x+1).$$

Equating coefficients of like powers of x on both sides of this equality, we obtain

$$A + B = 3, \qquad 2A + B = 2.$$

Solving for A and B yields

$$A = -1, \qquad B = 4.$$

Thus,

$$\frac{3x + 2}{(x + 1)(x + 2)} = -\frac{1}{x + 1} + \frac{4}{x + 2},$$

so that

$$\int \frac{3x + 2}{(x + 1)(x + 2)} \, dx = -\ln|x + 1| + 4 \ln|x + 2| + C.$$

Example A3.7   Evaluate $\displaystyle\int \frac{2x - 3}{(x - 2)(x^2 + 1)} \, dx.$

Solution   We first determine the partial fraction decomposition of the integrand. From the general rules of partial fractions it follows that there are constants A, B, C such that

$$\frac{2x - 3}{(x - 2)(x^2 + 1)} = \frac{A}{x - 2} + \frac{Bx + C}{x^2 + 1}. \qquad \text{(A3.2)}$$

In order to determine the values of A, B, and C, we multiply both sides of (A3.2) by $(x - 2)(x^2 + 1)$. This yields

$$2x - 3 = A(x^2 + 1) + (Bx + C)(x - 2).$$

Equating coefficients of like powers of x on both sides of this equality we obtain

$$A + B = 0, \qquad -2B + C = 2, \qquad A - 2C = -3.$$

Solving for A, B, and C yields

$$A = \frac{1}{5}, \qquad B = -\frac{1}{5}, \qquad C = \frac{8}{5},$$

so that

$$\frac{(2x - 3)}{(x - 2)(x^2 + 1)} = \frac{1}{5(x - 2)} + \frac{8 - x}{5(x^2 + 1)}.$$

Thus,

$$\int \frac{(2x - 3)}{(x - 2)(x^2 + 1)} \, dx = \frac{1}{5} \int \frac{1}{x - 2} \, dx + \frac{8}{5} \int \frac{1}{x^2 + 1} \, dx - \frac{1}{5} \int \frac{x}{x^2 + 1} \, dx$$

$$= \frac{1}{5} \ln|x - 2| + \frac{8}{5} \tan^{-1} x - \frac{1}{10} \ln(x^2 + 1) + C.$$

Now return to the integral (A3.1). If the degree of $p(x)$ is greater than or equal to the degree of $q(x)$, then we first divide $q(x)$ into $p(x)$. The resulting expression will be integrable, although in general we will need to perform a further partial fraction decomposition.

Example A3.8 Evaluate $\int \dfrac{2x-1}{x+3}\, dx.$

Solution We first divide the denominator into the numerator to obtain

$$\frac{2x-1}{x+3} = 2 - \frac{7}{x+3}.$$

Thus

$$\int \frac{2x-1}{x+3}\, dx = \int 2\, dx - 7\int \frac{1}{x+3}\, dx$$

$$= 2x - 7\ln|x+3| + C.$$

Example A3.9 Evaluate $\int \dfrac{3x^2+5}{x^2-3x+2}\, dx.$

Solution Once more we must first divide the denominator into the numerator. It is easily shown that

$$\frac{3x^2+5}{x^2-3x+2} = 3 + \frac{9x-1}{x^2-3x+2}. \tag{A3.3}$$

The next step is to determine the partial fraction decomposition of the second term on the right-hand side. We first notice that the denominator can be factored as $(x-2)(x-1)$. Using the rules of partial fractions it follows that

$$\frac{9x-1}{(x-2)(x-1)} = \frac{A}{x-2} + \frac{B}{x-1}.$$

Clearing the fractions yields

$$9x-1 = A(x-1) + B(x-2).$$

Equating coefficients of like powers of x on both sides of this equation, we obtain

$$A + B = 9, \qquad A + 2B = 1.$$

Solving for A and B yields

$$A = 17, \qquad B = -8.$$

Substituting into (A3.3) we have

$$\frac{3x^2+5}{x^2-3x+2} = 3 + \frac{17}{x-2} - \frac{8}{x-1},$$

so that

$$\int \frac{3x^2 + 5}{x^2 - 3x + 2}\, dx = 3x + 17\, \ln|x - 2| - 8\, \ln|x - 1| + C.$$

Table A3.1 lists some of the more important integrals.   Notice that we have omitted the integration constant.

TABLE A3.1   SOME BASIC INTEGRALS

Function $F(x)$	Integral $\int F(x)dx$
$x^n,\ n \neq -1$	$\dfrac{1}{n+1}x^{n+1}$
$\dfrac{1}{x}$	$\ln|x|$
$e^{ax},\ a \neq 0$	$\dfrac{1}{a}e^{ax}$
$\sin x$	$-\cos x$
$\cos x$	$\sin x$
$\tan x$	$\ln|\sec x|$
$\sec x$	$\ln|\sec x + \tan x|$
$\csc x$	$\ln|\csc x - \cot x|$
$e^{ax}\sin bx$	$\dfrac{1}{a^2 + b^2}e^{ax}(a\,\sin bx - b\,\cos bx)$
$e^{ax}\cos bx$	$\dfrac{1}{a^2 + b^2}e^{ax}(a\,\cos bx + b\,\sin bx)$
$\ln x$	$x\,\ln x - x$
$\dfrac{1}{a^2 + x^2}$	$\dfrac{1}{a}\tan^{-1}\left(\dfrac{x}{a}\right)$
$\dfrac{1}{\sqrt{a^2 - x^2}},\ a > 0$	$\sin^{-1}\left(\dfrac{x}{a}\right)$
$\dfrac{1}{\sqrt{a^2 + x^2}}$	$\ln(x + \sqrt{a^2 + x^2})$
$\dfrac{f'(x)}{f(x)}$	$\ln|f(x)|$
$e^{u(x)}\dfrac{du}{dx}$	$e^{u(x)}$

EXERCISES A3

Evaluate the given integral.

1. $\displaystyle\int x\,\cos x\, dx.$

2. $\displaystyle\int x^2 e^{-x}\, dx.$

3. $\displaystyle\int \ln x\, dx.$

4. $\displaystyle\int \tan^{-1} x \, dx.$

5. $\displaystyle\int x^3 e^{x^2} \, dx.$

6. $\displaystyle\int \frac{x}{x^2+1} \, dx.$

7. $\displaystyle\int \frac{x-1}{x+2} \, dx.$

8. $\displaystyle\int \frac{x+2}{(x-1)(x+3)} \, dx.$

9. $\displaystyle\int \frac{2x+1}{x(x^2+4)} \, dx.$

10. $\displaystyle\int \frac{x^2+5}{(x-1)(x+4)} \, dx.$

11. $\displaystyle\int \frac{x+3}{2x-1} \, dx.$

12. $\displaystyle\int \frac{2x+3}{x^2+3x+4} \, dx.$

13. $\displaystyle\int \frac{3x+2}{x(x+1)^2} \, dx.$

14. $\displaystyle\int \frac{1}{\sqrt{4-x^2}} \, dx.$

15. $\displaystyle\int \frac{1}{x^2+2x+2} \, dx.$

16. $\displaystyle\int \frac{1}{x \ln x} \, dx.$

17. $\displaystyle\int \tan x \, dx.$

18. $\displaystyle\int \frac{x+1}{x^2-x-6} \, dx.$

19. $\displaystyle\int \cos^2 x \, dx.$

20. $\displaystyle\int \sqrt{1-x^2} \, dx.$

21. $\displaystyle\int e^{3x} \sin 2x \, dx.$

22. $\displaystyle\int e^x \sin^2 x \, dx.$

4

An Existence-Uniqueness Theorem for First-Order Ordinary Differential Equations

In this section we discuss an existence-uniqueness theorem for first-order ordinary differential equations. This discussion will require the introduction of an important theoretical process known as Picard iteration. First we state the basic theorem.

Theorem A4.1: Let $f(x, y)$ be a function that is defined and continuous on the rectangle

$$R = \{(x, y) : |x - x_0| \leq a, |y - y_0| \leq b\},$$

where a and b are constants. Suppose further that $\partial f / \partial y$ is also continuous in R. Then there exists an interval I containing x_0 such that the initial value problem

$$\begin{cases} \dfrac{dy}{dx} = f(x, y), \\ y(x_0) = y_0, \end{cases}$$

has a unique solution for all x in I. (See Figure A4.1.)

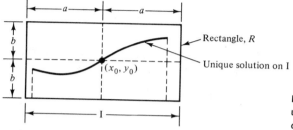

Figure A4.1 The existence-uniqueness theorem for first-order differential equations.

The idea behind the proof of Theorem A4.1 is to construct a sequence of functions that converges to a solution of the initial value problem and then show that there can be no other solution. The proof is quite involved and will be obtained in several steps.

REFORMULATION OF AN INITIAL VALUE PROBLEM AS AN EQUIVALENT INTEGRAL EQUATION

The first step in the proof is to show that an initial value problem can be reformulated in terms of an equivalent integral equation. This is proved in the following lemma.

Lemma A4.1: Consider the initial value problem

$$\begin{cases} \dfrac{dy}{dx} = f(x, y), & \text{(A4.1)} \\[2mm] y(x_0) = y_0, & \text{(A4.2)} \end{cases}$$

where we assume that f is continuous on the rectangle

$$R = \{(x, y) : |x - x_0| \le a, |y - y_0| \le b\}.$$

Then $y(x)$ is a solution of (A4.1) and (A4.2) on an interval I if and only if it is a *continuous* solution of the integral equation

$$y(x) = y_0 + \int_{x_0}^{x} f(t, y(t)) \, dt, \tag{A4.3}$$

on I.

PROOF If y is a solution of (A4.1) and (A4.2) on an interval I, then $y(x)$ and $f(x, y(x))$ are continuous on I, so that we can formally integrate both sides of (A4.1) to obtain

$$y(x) - y(x_0) = \int_{x_0}^{x} f(t, y(t)) \, dt,$$

that is,

$$y(x) = y_0 + \int_{x_0}^{x} f(t, y(t))\, dt.$$

Conversely, suppose that $y(x)$ is a continuous solution of the integral equation (A4.3). Then $f(t, y(t))$ is continuous on I, which implies that

$$\int_{x_0}^{x} f(t, y(t))\, dt$$

is differentiable on I. We can thus differentiate (A4.3) to obtain

$$\frac{dy}{dx} = f(x, y).$$

Further, setting $x = x_0$ in (A4.3) yields

$$y(x_0) = y_0.$$ ■

Example A4.1   Reformulate the following initial value problem as an equivalent integral equation:

$$y' = yx^2 - \sin x, \qquad y(\pi) = 1.$$

Solution   In this case we have

$$f(x, y) = yx^2 - \sin x, \qquad x_0 = \pi, \qquad y_0 = 1.$$

Thus the equivalent integral equation is

$$y(x) = 1 + \int_{\pi}^{x} (yt^2 - \sin t)\, dt.$$

It follows from the previous lemma that we can replace Theorem A4.1 by the following equivalent existence-uniqueness theorem.

Theorem A4.2:   Let $f(x, y)$ be a function that is defined and continuous on the rectangle

$$R = \{(x, y) : |x - x_0| \le a, |y - y_0| \le b\},$$

where a and b are constants. Suppose further that $\partial f/\partial y$ is also continuous in R. Then there exists an interval I containing x_0 such that the integral equation

$$y(x) = y_0 + \int_{x_0}^{x} f(t, y(t))\, dt$$

has a unique *continuous* solution for all x in I.

We will concentrate on this formulation of the existence-uniqueness theorem.

THE PICARD ITERATES   As mentioned before, the idea behind proving the existence part of Theorem A4.1 (or, equivalently, Theorem A4.2) is to determine a sequence of functions that converges to a solution of the initial value problem. Consider the integral equation (A4.3), that is,

$$y(x) = y_0 + \int_{x_0}^x f(t, y(t)) \, dt.$$

We associate with this integral equation the sequence of functions $\{y_1, y_2, \ldots\}$ defined by

$$y_1(x) = y_0 + \int_{x_0}^x f(t, y_0) \, dt,$$

$$y_2(x) = y_0 + \int_{x_0}^x f(t, y_1(t)) \, dt,$$

$$\vdots$$

$$y_n(x) = y_0 + \int_{x_0}^x f(t, y_{n-1}(t)) \, dt,$$

$$\vdots$$

The functions in this sequence are called the **Picard iterates** for the integral equation (A4.3) or, equivalently, for the initial value problem (A4.1) and (A4.2). Notice that each of the Picard iterates satisfies the initial condition $y_i(x_0) = y_0$.

Example A4.2   Determine the Picard iterates for the initial value problem

$$\frac{dy}{dx} = x - y, \qquad y(0) = 0. \tag{A4.4}$$

Solution   In this case we have

$$f(x, y) = x - y, \qquad x_0 = 0, \qquad y_0 = 0,$$

so that

$$y_n(x) = \int_0^x f(t, y_{n-1}(t)) \, dt = \int_0^x [t - y_{n-1}(t)] \, dt.$$

Thus

$$y_1(x) = \int_0^x (t - 0) \, dt = \frac{1}{2} x^2,$$

$$y_2(x) = \int_0^x \left(t - \frac{1}{2} t^2 \right) dt = \frac{1}{2} x^2 - \frac{1}{3 \cdot 2} x^3,$$

$$y_3(x) = \int_0^x \left(t - \frac{1}{2} t^2 + \frac{1}{3 \cdot 2} t^3 \right) dt = \frac{1}{2} x^2 - \frac{1}{3 \cdot 2} x^3 + \frac{1}{4 \cdot 3 \cdot 2} x^4,$$

and so on. We fairly soon recognize the pattern that is emerging, namely,

$$y_n(x) = \frac{1}{2} x^2 - \frac{1}{3!} x^3 + \frac{1}{4!} x^4 - \cdots + \frac{(-1)^{n+1}}{(n+1)!} x^{n+1}.$$

In order to motivate the proof of the existence-uniqueness theorem, we write this expression for y_n in the following form:

$$y_n(x) = (x - 1) + 1 - x + \frac{1}{2!} x^2 - \frac{1}{3!} x^3 + \cdots + \frac{(-1)^{n+1}}{(n+1)!} x^{n+1},$$

that is,

$$y_n(x) = (x - 1) + \sum_{k=0}^{n+1} \frac{(-1)^k}{k!} x^k.$$

Taking the limit as $n \to \infty$ and recalling the Maclaurin expansion for e^{-x}, namely,

$$e^{-x} = \sum_{k=0}^{\infty} \frac{(-1)^k}{k!} x^k,$$

we obtain

$$\lim_{n \to \infty} y_n(x) = (x - 1) + e^{-x}.$$

Thus if we define $y(x)$ by

$$y(x) = (x - 1) + e^{-x},$$

we have shown that the Picard iterates for the initial value problem (A4.4) converge to $y(x)$. The important point that we wish to make is the following: A direct substitution into the differential equation (A4.4) shows that $y(x)$ is, in fact, a solution (for all x); further, it satisfies the given initial condition. We have thus shown that, *in this particular case*, the Picard iterates converge to a solution of the initial value problem or, equivalently, to a continuous solution of the corresponding integral equation. We now show that this is true in general.

PROOF OF THE EXISTENCE PART OF THEOREM A4.2   We wish to prove that the Picard iterates for the integral equation (A4.3) converge to a continuous solution of the integral equation for x in some interval I. It might be thought that the appropriate interval would be $[x_0 - a, x_0 + a]$. However this is not necessarily the case. The reason is that we must ensure that the Picard iterates are well-defined, that is, that each of the y_i are themselves contained within the rectangle R. (If one of the iterates failed to remain in R, then the succeeding iterates would not necessarily be defined or continuous, since they would require the evaluation of f at points outside of R.) In Lemma A4.2 we will derive an appropriate interval. However, before proceeding with the proof we need to recall some basic results from calculus:

1. The triangle inequality for integrals is

$$\left| \int_{x_0}^{x} f(t,\, y(t))\, dt \right| \le \left| \int_{x_0}^{x} |f(t,\, y(t))|\, dt \right|.$$

2. The mean value theorem which, when applied to f, states that for fixed t,

$$f(t,\, y_2(t)) - f(t,\, y_1(t)) = f_y(t,\, \xi)[y_2(t) - y_1(t)],$$

where $\xi \in (y_1(t),\, y_2(t))$.

3. A function that is continuous on a closed region in the xy-plane is necessarily bounded in that region. Thus, since f and f_y in Theorem A4.1 are both continuous on the closed rectangle R, it follows that there are positive constants M and N such that

$$|f(x,\, y)| \le M, \qquad |f_y(x,\, y)| \le N, \qquad \text{for all } x,\, y \text{ in } R.$$

We will use these results repeatedly throughout the proof of Theorem A4.2.

Lemma A4.2 Let f be continuous on the rectangle

$$R = \{(x,\, y) : |x - x_0| \le a,\, |y - y_0| \le b\},$$

and let M be a bound for f on R. If $I = [x_0 - \alpha,\, x_0 + \alpha]$, where $\alpha = \min(a,\, b/M)$, then the Picard iterates for the integral equation (A4.3) are well-defined for all $x \in I$.

PROOF The proof is by induction. It is certainly true that y_0 is well-defined for $x \in I$. Now suppose that $y_0, y_1, \ldots, y_k$ lie in R for all $x \in I$. We must show that this implies that y_{k+1} also lies in R for all $x \in I$. Since y_k lies in R for all $x \in I$, it follows, from the continuity of f in R, that $|f(t,\, y_k(t))| \le M$ for all $t \in I$. Further, by definition of the Picard iterates, we have, for $x \in I$,

$$|y_{k+1}(x) - y_0| = \left| \int_{x_0}^{x} f(t,\, y_k(t))\, dt \right| \le \left| \int_{x_0}^{x} |f(t,\, y_k(t))|\, dt \right|$$

$$\le M\, |x - x_0| \le M\alpha \le b,$$

so that $y_{k+1}(x)$ also lies in R for $x \in I$. Hence, by induction, all of the Picard iterates are well-defined for $x \in I$. ∎

From now on we will restrict our attention to the interval $I = [x_0 - \alpha,\, x_0 + \alpha]$. The next step in the proof of Theorem A4.2 is to show that the Picard iterates converge to a function $y(x)$ for all $x \in I$. This is the content of the next lemma:

Lemma A4.3: The Picard iterates for the integral equation (A4.3) converge to a function $y(x)$ for all $x \in I$.

PROOF The key behind the proof is to recognize that we can write $y_n(x)$ in the form

$$y_n(x) = y_0(x) + [y_1(x) - y_0(x)] + [y_2(x) - y_1(x)] + \cdots + [y_n(x) - y_{n-1}(x)],$$

that is,

$$y_n(x) = y_0 + \sum_{k=0}^{n-1} [y_{k+1}(x) - y_k(x)]. \tag{A4.5}$$

It follows that convergence of the sequence of Picard iterates is equivalent to the convergence of the infinite series

$$y_0(x) + \sum_{k=0}^{\infty} [y_{k+1}(x) - y_k(x)].$$

However, the convergence of this series can be established by the comparison test. Consider the kth term in the series. By definition of the Picard iterates

$$|y_{k+1}(x) - y_k(x)| = \left| \int_{x_0}^x [f(t, y_k(t)) - f(t, y_{k-1}(t))] \, dt \right|$$

$$\leq \left| \int_{x_0}^x |f(t, y_k(t)) - f(t, y_{k-1}(t))| \, dt \right|.$$

But, by the mean value theorem, we have

$$|f(t, y_k(t)) - f(t, y_{k-1}(t))| \leq |f_y(t, \xi)||y_k(t) - y_{k-1}(t)| \leq N|y_k(t) - y_{k-1}(t)|,$$

where N is a bound for f_y on R. Thus

$$|y_{k+1}(x) - y_k(x)| \leq N \left| \int_{x_0}^x |y_k(t) - y_{k-1}(t)| \, dt \right|. \tag{A4.6}$$

Also,

$$|y_1(x) - y_0| = \left| \int_{x_0}^x f(t, y_0) \, dt \right| \leq M|x - x_0|. \tag{A4.7}$$

It follows from (A4.6) and (A4.7) that

$$|y_2(x) - y_1(x)| \leq N \left| \int_{x_0}^x M|t - x_0| \, dt \right| \leq NM \frac{\alpha^2}{2!},$$

and it is easily shown, by induction, that

$$|y_{k+1}(x) - y_k(x)| \leq \frac{NM^k \alpha^{k+1}}{(k+1)!}, \tag{A4.8}$$

for all x in I. But

$$\sum_{k=0}^{\infty} \frac{(M\alpha)^{k+1}}{(k+1)!} = e^{M\alpha} - 1,$$

so that, by the comparison test,

$$\sum_{k=0}^{\infty} [y_{k+1}(x) - y_k(x)]$$

converges (absolutely and uniformly) for all x in I, and hence

$$y_0 + \sum_{k=0}^{\infty} [y_{k+1}(x) - y_k(x)]$$

does also. It follows from (A4.5) that the sequence of Picard iterates $\{y_1, y_2, \ldots\}$ also converges for all x in I. We let $y(x)$ denote the limit function, so that we have proved that

$$\lim_{n \to \infty} y_n(x) = y(x),$$

for all x in I.   ■

It is now possible to prove that the limit function $y(x)$ is continuous on I and also satisfies the integral equation on I. These results can be elegantly derived from a closer analysis of the manner in which the sequence of Picard iterates approach the limit function $y(x)$.

POINTWISE AND UNIFORM CONVERGENCE   When dealing with convergence of functions on an interval, two different types of convergence can be distinguished:

1. *Pointwise convergence:* Given $\varepsilon > 0$, for each x in I there exists a positive integer $N_0(\varepsilon, x)$ such that

$$|y(x) - y_n(x)| < \varepsilon \quad \text{whenever} \quad n > N_0(\varepsilon, x).$$

2. *Uniform convergence:* Given $\varepsilon > 0$, there exists a positive integer $N_0(\varepsilon)$, such that *for all x in I,*

$$|y(x) - y_n(x)| < \varepsilon \quad \text{whenever} \quad n > N_0(\varepsilon).$$

The difference between these two types of convergence is that in the case of uniform convergence for a given ε, there is one integer $N_0(\varepsilon)$ that works for *all* x in I, whereas in the case of pointwise convergence, we, in general, require different integers $N_0(\varepsilon, x)$ for different points in I. Clearly uniform convergence is a stronger condition than pointwise convergence.

We now state two important results for uniformly convergent sequences of functions:

UC1. If the sequence of functions $\{y_1, y_2, \ldots\}$ converges *uniformly* to the function $y(x)$ on the interval I, and if each $y_k(x)$ is continuous on I, then the limit function $y(x)$ is also continuous on I.

UC2. If the sequence of functions $\{y_1, y_2, \ldots\}$ converges *uniformly* to the function

$y(x)$ on the interval I, then for all x and x_0 in I,

$$\lim_{n \to \infty} \int_{x_0}^{x} y_n(x) \, dx = \int_{x_0}^{x} \lim_{n \to \infty} y_n(x) \, dx.$$

A proof of both of these results can be found, for example, in T. M. Apostol, *Mathematical Analysis*, Addison-Wesley, 1974.

We now prove that the convergence of the Picard iterates is uniform.

Lemma A4.4: The Picard iterates converge uniformly to a limit function $y(x)$ for all x in I.

PROOF   We have already established in the previous lemma that the Picard iterates converge to the function

$$y(x) = y_0 + \sum_{k=0}^{\infty} [y_{k+1}(x) - y_k(x)],$$

for all x in I.   We must show that the convergence is uniform.   Since

$$y_n(x) = y_0 + \sum_{k=0}^{n-1} [y_{k+1}(x) - y_k(x)],$$

it follows that

$$|y(x) - y_n(x)| \le \left| \sum_{k=n}^{\infty} [y_{k+1}(x) - y_k(x)] \right| \le \sum_{k=n}^{\infty} |y_{k+1}(x) - y_k(x)|.$$

Substituting from (A4.8), we thus have

$$|y(x) - y_n(x)| \le N \sum_{k=n}^{\infty} \frac{M^k \alpha^{k+1}}{(k+1)!} = \frac{N}{M} (M\alpha)^{n+1} \sum_{j=0}^{\infty} \frac{(M\alpha)^j}{(n+j+1)!}$$

$$\le \frac{N}{M} \frac{(M\alpha)^{n+1}}{n!} \sum_{j=0}^{\infty} \frac{(M\alpha)^j}{(n+1) \cdots (n+1+j)} \le \frac{N}{M} \frac{(M\alpha)^{n+1}}{n!} \sum_{j=0}^{\infty} \frac{(M\alpha)^j}{(j+1)!}$$

$$\le \frac{N}{M} \frac{(M\alpha)^{n+1}}{n!} \sum_{j=0}^{\infty} \frac{(M\alpha)^j}{j!};$$

that is,

$$|y(x) - y_n(x)| \le \frac{N}{M} \frac{(M\alpha)^{n+1}}{n!} e^{M\alpha}. \tag{A4.9}$$

But

$$\lim_{n \to \infty} \frac{(M\alpha)^{n+1}}{n!} = 0,$$

so that, for any $\varepsilon > 0$, there exists an $N_0(\varepsilon)$ such that

$$\frac{N}{M} \frac{(M\alpha)^{n+1}}{n!} e^{M\alpha} < \varepsilon \qquad \text{for all } n > N_0(\varepsilon).$$

Thus, from (A4.9), for any $\varepsilon > 0$ and all $x \in I$,

$$|y(x) - y_n(x)| < \varepsilon \qquad \text{for all } n > N_0(\varepsilon),$$

so that the convergence of the Picard iterates is indeed uniform on I. ∎

We now use the result of this lemma together with UC1 and UC2 to establish that the limit function $y(x)$ is a continuous solution of the integral equation (A4.3).

Lemma A4.5:   The limit function $y(x)$ is a continuous solution of the integral equation (A4.3).

PROOF   Since each of the Picard iterates are continuous on I, it follows directly from UC1 that the limit function $y(x)$ is also continuous on I. Now consider the nth Picard iterate, namely,

$$y_n(x) = y_0 + \int_{x_0}^{x} f(t, y_{n-1}(t))\, dt.$$

Taking the limit as $n \to \infty$ yields

$$y(x) = \lim_{n \to \infty} y_n(x) = y_0 + \lim_{n \to \infty} \int_{x_0}^{x} f(t, y_{n-1}(t))\, dt.$$

From UC2, the uniform convergence of $\{y_n\}$ allows us to take the limit inside the integral to obtain

$$y(x) = y_0 + \int_{x_0}^{x} \lim_{n \to \infty} f(t, y_{n-1}(t))\, dt.$$

Further, we can use the assumed continuity of f on I to take the limit inside the function, thereby obtaining

$$y(x) = y_0 + \int_{x_0}^{x} f(t, \lim_{n \to \infty} y_{n-1}(t))\, dt,$$

that is, since $\{y_{n-1}\}$ converges to y,

$$y(x) = y_0 + \int_{x_0}^{x} f(t, y(t))\, dt.$$

Thus $y(x)$ *is a continuous* solution of the given integral equation. ∎

This completes the proof of the existence part of Theorem A4.2.

PROOF OF UNIQUENESS   The previous lemmas have established the existence of a continuous solution to the integral equation (A4.3), and hence to the initial value problem (A4.1) and (A4.2), on the interval I.   We now establish uniqueness.   Once more it is convenient to consider the integral equation formulation.

Lemma A4.6:   The integral equation

$$y(x) = y_0 + \int_{x_0}^{x} f(t, y(t))\, dt, \tag{A4.10}$$

where f and f_y are continuous on the rectangle

$$R = \{(x, y) : |x - x_0| \le a, |y - y_0| \le b\},$$

has a unique continuous solution on $I = [x_0 - \alpha, x_0 + \alpha]$.

PROOF   We have already shown in the previous lemmas that the integral equation admits a continuous solution on I.   We must now show that there is only one such solution.   In order to do so we assume the existence of two solutions to the integral equation and show that they are the same.

Let Y_1 and Y_2 be two continuous solutions to (A4.10) on I.   Then

$$Y_2(x) - Y_1(x) = \int_{x_0}^{x} [f(t, Y_2(t)) - f(t, Y_1(t))]\, dt,$$

so that

$$|Y_2(x) - Y_1(x)| \le \left| \int_{x_0}^{x} \left| f(t, Y_2(t)) - f(t, Y_1(t)) \right| dt \right|. \tag{A4.11}$$

But, using the mean value theorem,

$$f(t, Y_2(t)) - f(t, Y_1(t)) = f_y(t, \xi)[Y_2(t) - Y_1(t)],$$

so that, since f_y is continuous on R,

$$|f(t, Y_2(t)) - f(t, Y_1(t))| \le N|Y_2(t) - Y_1(t)|,$$

where N is a bound for f_y on I.   Substituting into (A4.11) yields

$$|Y_2(x) - Y_1(x)| \le N \left| \int_{x_0}^{x} |Y_2(t) - Y_2(t)|\, dt \right|. \tag{A4.12}$$

Now define the function $F(x)$ by

$$F(x) = \int_{x_0}^{x} |Y_2(t) - Y_1(t)|\, dt. \tag{A4.13}$$

The remainder of the proof depends on whether $x \ge x_0$, or $x < x_0$.   Suppose that $x \ge x_0$.   Then, (A4.12) can be written as

$$F'(x) - NF(x) \le 0, \tag{A4.14}$$

for $x \geq x_0$, whereas (A4.13) implies that $F(x)$ satisfies the following conditions:

$$F(x) \geq 0 \qquad \text{for } x \geq x_0, \tag{A4.15}$$

$$F(x_0) = 0. \tag{A4.16}$$

Multiplying (A4.14) by e^{-Nx} yields

$$(e^{-Nx}F)' \leq 0$$

for $x \geq x_0$, which, upon integrating from x_0 to x and imposing the condition (A4.16), yields

$$e^{-Nx}F(x) \leq 0$$

for $x \geq x_0$; that is,

$$F(x) \leq 0 \tag{A4.17}$$

for $x \geq x_0$. It follows from (A4.15) and (A4.17) that we must have

$$F(x) = 0$$

for $x \geq x_0$. We leave it as an exercise to show that $F(x) = 0$ for $x < x_0$ also, so that $F(x)$ is identically zero on I. It now follows from (A4.13) that

$$Y_1(x) = Y_2(x)$$

for all x in I, and hence we have uniqueness. ∎

In summary we have the following result.

Theorem A4.3: Let $f(x, y)$ be a function that is defined and continuous on the rectangle

$$R = \{(x, y) : |x - x_0| \leq a, |y - y_0| \leq b\},$$

where a and b are constants. Suppose further that $\partial f / \partial y$ is also continuous in R. Then the Picard iterates converge to the unique solution of the initial value problem

$$\begin{cases} \dfrac{dy}{dx} = f(x, y), \\ y(x_0) = y_0, \end{cases}$$

on the interval $[x_0 - \alpha, x_0 + \alpha]$ where $\alpha = \min(a, b/M)$, and M is a bound for f on R.

The existence-uniqueness theorem can be generalized to include differential equations of arbitrary order and also systems of differential equations (see, for example, F. J. Murray and K. S. Miller, *Existence Theorems*, New York University Press, 1954). It forms the theoretical basis for differential equation theory.

EXERCISES A4

1. Convert the given initial value problem into an equivalent integral equation and determine the first two Picard iterates y_1, y_2:

$$\frac{dy}{dx} = x(y - x), \qquad y(0) = 1.$$

2. Consider the initial value problem

$$\frac{dy}{dx} = x(x + y), \qquad y(0) = 0. \qquad \text{(i)}$$

(a) Show that the nth Picard iterate is

$$y_n(x) = \sum_{k=1}^{n} \frac{1}{1 \cdot 3 \cdot \cdots \cdot (2k + 1)} x^{2k+1}.$$

(b) Prove that

$$\lim_{n \to \infty} y_n(x) = \sum_{k=1}^{\infty} \frac{1}{1 \cdot 3 \cdot \cdots \cdot (2k + 1)} x^{2k+1},$$

by showing that the infinite series on the right-hand side converges for all real x.

(c) Show by direct substitution that

$$y(x) = \sum_{k=1}^{\infty} \frac{1}{1 \cdot 3 \cdot \cdots \cdot (2k + 1)} x^{2k+1}$$

is a solution of the initial value problem (i).

3. Convert the given initial value problem into an equivalent integral equation and find the solution by Picard iteration:

$$\frac{dy}{dx} = 2xy, \qquad y(0) = 1.$$

4. Use the existence-uniqueness theorem to prove that $y(x) = 3$ is the only solution to the initial value problem

$$\frac{dy}{dx} = \frac{x}{x^2 + 1}(y^2 - 9), \qquad y(0) = 3.$$

5. One solution to the initial value problem

$$\frac{dy}{dx} = \frac{2}{3}(y - 1)^{1/2}, \qquad y(1) = 1,$$

is $y(x) = 1$. Determine another solution to this initial value problem. Does this contradict the existence-uniqueness theorem? Explain.

6. Consider the initial value problem

$$\frac{dy}{dx} = y^2 + 4 \sin^2 x, \qquad y(0) = 0. \qquad \text{(ii)}$$

Since f and f_y are continuous for all (x, y), it follows that the existence-uniqueness theorem can be applied on any rectangle

$$R = \{(x, y) : |x| \le a, |y| \le b\},$$

where a and b are constants.

(a) Show that the unique solution guaranteed by the existence-uniqueness theorem is valid at least on the interval $I = [-\alpha, \alpha]$, where

$$\alpha = \min\left(a, \frac{b}{b^2 + 4}\right).$$

(b) Determine the maximum interval of existence of a solution to (ii) guaranteed by the existence-uniqueness theorem.

7. Fill in the missing details in the proof of Lemma A4.6 when $x < x_0$.

Linearly Independent Solutions to $x^2 y'' + xp(x)y' + q(x)y = 0$

Consider the differential equation

$$x^2 y'' + xp(x)y' + q(x)y = 0, \tag{A5.1}$$

where p and q are analytic at $x = 0$. Writing

$$p(x) = \sum_{n=0}^{\infty} p_n x^n, \qquad q(x) = \sum_{n=0}^{\infty} q_n x^n, \tag{A5.2}$$

on $(0, R)$, it follows that the indicial equation for (A5.1) is

$$r(r-1) + p_0 r + q_0 = 0. \tag{A5.3}$$

If the roots of the indicial equation are distinct and do not differ by an integer, then there exist two linearly independent Frobenius series solutions in the neighborhood of $x = 0$. In this appendix we derive the general form for two linearly independent solutions to (A5.1) in the case that the roots of the indicial equation differ by an integer. The analysis of the case when the roots of the indicial equation coincide is left as an exercise. Let r_1 and r_2 denote the roots of the indicial equation and suppose that

$$r_1 - r_2 = N, \tag{A5.4}$$

where N is a positive integer. Then, we know that one solution to (A5.1) on $(0, R)$ is given by the Frobenius series

$$y_1 = x^{r_1}(1 + a_1x + a_2x^2 + \cdots). \tag{A5.5}$$

According to the reduction-of-order technique, a second linearly independent solution to (A5.1) on $(0, R)$ is given by

$$y_2 = uy_1, \tag{A5.6}$$

where the function u can be determined by substitution into (A5.1). Differentiating (A5.6) twice and substituting into (A5.1) yields

$$x^2(y_1u'' + 2y_1'u' + y_1''u) + xp(x)(y_1u' + y_1'u) + q(x)uy_1 = 0;$$

that is, since y_1 is a solution to (A5.1),

$$x^2(y_1u'' + 2y_1'u') + xp(x)(y_1u') = 0.$$

Consequently u can be determined by solving

$$\frac{u''}{u'} = -\left(\frac{p(x)}{x} + 2\frac{y_1'}{y_1}\right),$$

which, upon integrating yields

$$u' = y_1^{-2}e^{-\int(p(x)/x)dx}. \tag{A5.7}$$

We now determine the two terms that appear on the right-hand side. From (A5.2) we can write

$$\frac{p(x)}{x} = \frac{p_0}{x} + p_1 + p_2x + \cdots,$$

so that

$$\int\frac{p(x)}{x}\,dx = p_0\ln x + P(x),$$

where

$$P(x) = p_1x + \frac{1}{2}p_2x^2 + \frac{1}{3}p_3x^3 + \cdots.$$

Consequently,

$$e^{-\int(p(x)/x)\,dx} = x^{-p_0}e^{P(x)}. \tag{A5.8}$$

Since $P(x)$ is analytic at $x = 0$, it follows that

$$e^{P(x)} = \alpha_0 + \alpha_1x + \alpha_2x^2 + \cdots$$

for appropriate constants $\alpha_0, \alpha_1, \alpha_2, \ldots$. Hence, from (A5.8),

$$e^{-\int(p(x)/x)\,dx} = x^{-p_0}(\alpha_0 + \alpha_1x + \alpha_2x^2 + \cdots). \tag{A5.9}$$

Now consider y_1^{-2}. From (A5.5),

$$y_1^{-2} = \frac{x^{-2r_1}}{(1 + a_1x + a_2x^2 + \cdots)^2}. \tag{A5.10}$$

Further, since the series

$$1 + a_1x + a_2x^2 + \cdots$$

converges for $0 \leq x < R$ and is nonzero at $x = 0$, it follows that $(1 + a_1x + a_2x^2 + \cdots)^{-2}$ is analytic at $x = 0$ and hence there exist constants $\beta_1, \beta_2,$ $\ldots$, such that

$$(1 + a_1x + a_2x^2 + \cdots)^{-2} = 1 + \beta_1x + \beta_2x^2 + \cdots.$$

We can therefore write (A5.10) in the form

$$y_1^{-2} = x^{-2r_1}(1 + \beta_1x + \beta_2x^2 + \cdots). \tag{A5.11}$$

Substituting from (A5.9) and (A5.11) into (A5.7) yields

$$u' = x^{-(p_0 + 2r_1)}(\alpha_0 + \alpha_1x + \alpha_2x^2 + \cdots)(1 + \beta_1x + \beta_2x^2 + \cdots),$$

which can be written as

$$u' = x^{-(p_0 + 2r_1)}(A_0 + A_1x + A_2x^2 + \cdots), \tag{A5.12}$$

for appropriate constants $A_0, A_1, A_2, \ldots$. Now, since the roots of the indicial equation are r_1 and $r_2 = r_1 - N$, it follows that the indicial equation can be written as

$$(r - r_1)(r - r_1 + N) = 0,$$

that is,

$$r^2 - (2r_1 - N)r + r_1(r_1 - N) = 0.$$

Comparison with (A5.3) reveals that

$$p_0 - 1 = -(2r_1 - N),$$

so that

$$p_0 + 2r_1 = N + 1.$$

Consequently, (A5.12) can be written as

$$u' = x^{-(N+1)}(A_0 + A_1x + A_2x^2 + \cdots),$$

that is,

$$u' = A_0x^{-(N+1)} + A_1x^{-N} + \cdots + A_Nx^{-1} + A_{N+1} + A_{N+2}x + \cdots,$$

which can be integrated directly to yield

$$u = -\frac{A_0}{N}x^{-N} + \frac{A_1}{1 - N}x^{1-N} + \cdots - A_{N-1}x^{-1} + A_N\ln x + A_{N+1}x + \cdots.$$

Rearranging terms we can write this as

$$u = A \ln x + x^{-N}(B_0 + B_1x + B_2x^2 + \cdots),$$

where we have redefined the coefficients. Substituting this expression for u into (A5.6) we obtain

$$y_2 = [A \ln x + x^{-N}(B_0 + B_1x + B_2x^2 + \cdots)]y_1,$$

that is,

$$y_2 = Ay_1 \ln x + x^{-N}y_1(B_0 + B_1x + B_2x^2 + \cdots).$$

Substituting for y_1 from (A5.5) into the second term on the right-hand side yields

$$y_2 = Ay_1 \ln x + x^{r_1-N}(1 + a_1x + a_2x^2 + \cdots)(B_0 + B_1x + B_2x^2 + \cdots).$$

Finally, multiplying the two power series together and substituting $r_2 = r_1 - N$ from (A5.4), we obtain

$$y_2 = Ay_1 \ln x + x^{r_2}(b_0 + b_1x + b_2x^2 + \cdots),$$

for appropriate constants $b_0, b_1, b_2, \ldots$, that is,

$$y_2 = Ay_1 \ln x + x^{r_2} \sum_{n=0}^{\infty} b_n x^n.$$

The derivation of a second solution to the differential equation (A5.1) in the case when the indicial equation has two equal roots follows exactly the same lines as the preceding computation and is left as an exercise.

Answers to Odd-Numbered Problems

SECTION 1.1 (page 9)

1. 2, nonlinear. **3.** 2, nonlinear. **5.** 4, linear. **7.** $(-\infty, -4)$ or $(-4, \infty)$. **9.** $(-\infty, \infty)$. **11.** $(-\infty, 0)$ or $(0, \infty)$. **13.** $(-\infty, \infty)$. **15.** $(-\infty, \infty)$. **19.** $y = \dfrac{\ln x}{x}, x > 0$. **21.** $y = \dfrac{\sqrt{1 + \sin x}}{x}, x > 0$.

23. $y = 2x^{1/2} + c, x > 0$. **25.** $y = \dfrac{1}{(n + 1)(n + 2)} x^{n+2} + c_1 x + c_2$, if $n \ne -2, -1$; $y = c_1 x + c_2 - \ln|x|$, if $n = -2$; $y = x \ln|x| + c_1 x + c_2$, if $n = -1$. Interval: $(-\infty, \infty)$ if $n \ge 0$; $(-\infty, 0)$ or $(0, \infty)$ if $n < 0$.

27. $y = 3 + x - \cos x$. **29.** $y = e^{-x} - \dfrac{1}{e} x$.

SECTION 1.2 (page 16)

1. $x^2(e^y + y^2) dx - dy = 0$. **3.** $y = 2x^2 + c$. **5.** $y = \dfrac{1}{x} + c$. **15.** $y' = -\dfrac{y}{x}$. **17.** $\dfrac{dy}{dx} = \dfrac{y^2 - x^2}{2xy}$.

19. $y' = \dfrac{x}{-y \pm \sqrt{x^2 + y^2}}$. **21.** Intervals: $(-\infty, \infty)$ if $c > 0$; $(-\infty, 0)$ and $(0, \infty)$ if $c = 0$; $(-\infty, -\alpha)$, $(-\alpha, \alpha)$ and (α, ∞) if $c = -\alpha^2$.

SECTION 1.3 (page 21)

1. $y = ce^{x^2}$. **3.** $y = \ln(c - e^{-x})$. **5.** $y = c(x - 2)$. **7.** $y = \dfrac{cx - 3}{2x - 1}$. **9.** $y = \dfrac{(x - 1) + c(x - 2)^2}{(x - 1) - c(x - 2)^2}$ and

$y = -1$. **11.** $y = c + c_1 \left(\dfrac{x - a}{x - b}\right)^{1/(a-b)}$ **13.** $y = a(1 + \sqrt{1 - x^2})$. **15.** $y(x) = 0$.

17. (a) $v(t) = a \tanh \dfrac{gt}{a}$, where $a = \sqrt{\dfrac{mg}{k}}$. (b) No. (c) $y = \dfrac{a^2}{g} \ln\left(\cosh \dfrac{gt}{a}\right)$.

SECTION 1.4 (page 27)

1. $F(V) = \dfrac{1 - V^2}{V}$. **3.** $F(V) = \dfrac{\sin \dfrac{1}{V} - V\cos V}{V}$. **5.** Not homogeneous. **7.** $F(V) = -\sqrt{1 + V^2}$.

9. $y^2 = ce^{-3x/y}$. **11.** $y = x \cos^{-1} \dfrac{c}{x}$. **13.** $4 \tan^{-1} \dfrac{y}{x} - \ln(x^2 + y^2) = c$. **15.** $y = xe^{(cx + 1)}$.

17. $y^2 = x^2 \ln(\ln cx)$. **19.** $y^2 = c^2 - 2cx$. **21.** $y = x \sin^{-1} cx$. **23.** $2y^2 + xy - x^2 = 2$.

25. $y + \sqrt{x^2 + y^2} = 9$. **29.** $y = 3(3x - \tanh 3x)$.

31. (b) $2 \tan^{-1}\left(\dfrac{y - 1}{x + 1}\right) - \dfrac{1}{2} \ln[(x + 1)^2 + (y - 1)^2] = c$.

SECTION 1.5 (page 32)

1. $y = e^x(e^x + c)$. **3.** $y = x^2 - 1 + ce^{-x^2}$. **5.** $y = \dfrac{1}{1 + x^2}(4 \tan^{-1}x + c)$. **7.** $y = \dfrac{x^3(3 \ln x - 1) + c}{\ln x}$.

9. $x = \dfrac{4e^t(t - 1) + c}{t^2}$. **11.** $y = (\tan x + c)\cos x$.

13. $y = \begin{cases} \dfrac{1}{\alpha + \beta} e^{\beta x} + ce^{-\alpha x}, & \alpha + \beta \neq 0, \\ e^{-\alpha x}(x + c), & \alpha + \beta = 0. \end{cases}$ **15.** $y = x^{-2}(x^4 + 1)$.

17. $x = (4 - t)(1 + t)$. **19.** $y = x^3 + c_1 \ln x + c_2$. **21.** $y = e^{-x}(c - e^{-x})$. **23.** $y = x(x \ln x - x + c)$.

25. (a) $i = \dfrac{E_0}{L(a^2 + \omega^2)}(a \cos \omega t + \omega \sin \omega t) + Ae^{-at}$, where $a = \dfrac{R}{L}$ and A is a constant.

SECTION 1.6 (page 36)

1. $y^2 = x^2(8 \sin x + c)$. **3.** $y^{2/3} = x[x^2(2 \ln x - 1) + c]$. **5.** $y = \dfrac{1}{x^2(c - 2x^3)}$.

7. $y = \dfrac{1}{4}\left(\dfrac{x - b}{x - a}\right)^2 [x + (b - a)\ln |x - b| + c]^2$. **9.** $y = [(x^2 - 1) + ce^{-x^2}]^2$. **11.** $y = \left(\dfrac{x^3 + c}{x}\right)^{1/(1 - \pi)}$.

13. $y = \left(1 + \dfrac{c}{\sec x + \tan x}\right)^{1/(1 - \sqrt{3})}$. **15.** $y^2 = \dfrac{1}{\sin^2 x(2 \cos x + 1)}$. **17.** $y = xe^{x^2}$.

19. $y = \tan^{-1}(1 + ce^{-\sqrt{1 + x}})$.

SECTION 1.7 (page 43)

1. Exact. **3.** Not exact. **5.** $xy^2 - \cos y + \sin x = c$. **7.** $6e^{2x} + 3x^2y - 3xy^2 + y^3 = c$.

9. $\tan^{-1}\frac{y}{x} + \ln x = c$. **11.** $\sin xy + \cos x = c$. **13.** $y = \dfrac{x^3 \ln x + 5}{x}$. **15.** $y = \dfrac{\ln(2 - \sin x)}{x}$.

17. $(2xy - e^x)\,dx + x^2\,dy = 0$.

SECTION 1.8 (page 48)

1. Yes. **3.** Yes. **5.** $2x - y^4 = cy^2$. **7.** $y = \dfrac{c + 2x^{5/2}}{10\sqrt{x}}$. **9.** $y = \dfrac{c + \tan^{-1}x}{1 + x^2}$. **11.** $r = 2,\ s = 4$.

13. $r = 1,\ s = 2$.

SECTION 1.9 (page 50)

1. $y^2 = 2[(\ln x)^2 + c]$. **3.** $y^2 + x^2y + c = 0$. **5.** $y = \dfrac{2\cos x}{\sin^2 x + c}$. **7.** $y = x \sin(\ln cx)$.

9. $y^2 = \dfrac{1}{x^2}[x^5(5\ln x - 1) + c]$. **11.** $y = \dfrac{\ln(c \sec x)}{\sin x}$. **13.** $y = xe^{cx}$. **15.** $y = 1 + ce^{\cos x}$.

17. $y^2 = \dfrac{\ln x}{c + x^3(3\ln x - 1)}$. **19.** $y = \left(\dfrac{x+1}{x-1}\right)(x - 2\ln|x+1| + c)$.

SECTION 1.10 (page 57)

1. $y = c_1x^3 + x^4 + c_2$. **3.** $y = (c_1 + c_2e^x)^{1/3}$. **5.** $y = c_2 - \ln|c_1 - \sin x|$. **7.** $y = \frac{1}{3}x^6 + c_1x^3 + c_2$.

9. $y = -\dfrac{1}{\alpha}\ln|c_1 + c_2e^{\beta x}|$. **11.** $y = c_1\tan^{-1}x + c_2$. **13.** $y = \ln(\sec x) + c_1\ln(\sec x + \tan x) + c_2$.

15. $y = a \cosh \omega x$. **17.** $y = \frac{1}{2}x^2 + c_1x^3 + c_2x + c_3$.

SECTION 2.1 (page 63)

1. $y = cx^4$. **3.** $x^2 + 2y^2 = c$. **5.** $2x^2 + y^2 = c$. **7.** $y^2 = -2x + c$. **9.** $y = -\dfrac{1}{m}x + c$. **11.** $y^m = cx$.

13. (b) $(x + cm)^2 + (y - c)^2 = c^2(1 + m^2)$. **15.** $y = cx^2$. **17.** $y = -\frac{1}{3}x + c$.

19. $\ln(x^2 + y^2) - 2\tan^{-1}(y/x) = c$. **21.** $(x - c)^2 + (y - c)^2 = 2c^2$.

SECTION 2.2 (page 68)

1. 2560. **3.** $t \approx 35.86$ h. **5.** $P(t) = P_0e^{(\alpha - \beta)t}$. $\displaystyle\lim_{t \to \infty} P(t) = \begin{cases} 0, & \text{if } \alpha - \beta < 0, \\ P_0, & \text{if } \alpha = \beta, \\ +\infty, & \text{if } \alpha - \beta > 0. \end{cases}$

Doubling time is $t_d = \dfrac{1}{\alpha - \beta}\ln 2$. **7.** $t_{1/2} \approx 38.87$ y. **9.** $t_{1/2} = 60$ d. **11.** ≈ 851 years ago

15. $\dfrac{dx}{dt} = kt$, where x denotes the amount of A unconverted at time t, and k is a negative constant.

SECTION 2.3 (page 72)

1. 2:00 P.M. **3. (a)** $500°$F; **(b)** $\approx 6:07$ P.M. **5.** ≈ 5.3 lb.

SECTION 2.4 (page 75)

1. See text.   **3.** $x(20) = \dfrac{150}{7}$ g.   **5.** 18.75 g.

7. $\left(1 - \dfrac{x}{\alpha}\right)^{\beta-\gamma}\left(1 - \dfrac{x}{\beta}\right)^{\gamma-\alpha}\left(1 - \dfrac{x}{\gamma}\right)^{\alpha-\beta} = e^{(\alpha-\beta)(\beta-\gamma)(\gamma-\alpha)t}.$

SECTION 2.5 (page 79)

1. 196 g.   **3.** 300 g.   **5. (a)** 6.75 g.   **(b)** $15(2)^{1/3}$L.

SECTION 2.6 (page 85)

1. $i = 5(1 - e^{-40t})$.   **3.** $i = \frac{3}{5}(3 \sin 4t - 4 \cos 4t + 4e^{-3t})$.   **7.** $i(t) = \dfrac{E_0}{R^2 + L^2\omega^2}(R \sin \omega t - \omega L \cos \omega t) +$

$Ae^{-Rt/L}$, $i_s = \dfrac{E_0}{R^2 + L^2\omega^2}(R \sin \omega t - \omega L \cos \omega t)$, $i_T = Ae^{-Rt/L}$.   **9.** $i = \dfrac{q_0}{\sqrt{LC}} \sin \dfrac{1}{\sqrt{LC}} t$.

SECTION 3.1 (page 90)

1. $a_{31} = 0$, $a_{24} = -1$, $a_{14} = 2$, $a_{32} = 2$, $a_{21} = 7$, $a_{34} = -4$.   **3.** $A = \begin{bmatrix} 2 & 1 & -1 \\ 0 & 4 & -2 \end{bmatrix}$.

5. $\begin{bmatrix} 1 & -3 & -2 \\ 3 & 6 & 0 \\ 2 & 7 & 4 \\ -4 & -1 & 5 \end{bmatrix}$.   **7.** 4.   **9.** -1.   **11.** Column vectors: $\begin{bmatrix} 1 \\ -1 \\ 2 \end{bmatrix}$, $\begin{bmatrix} 3 \\ -2 \\ 6 \end{bmatrix}$, $\begin{bmatrix} -4 \\ 5 \\ 7 \end{bmatrix}$.

Row vectors: $\begin{bmatrix} 1 & 3 & -4 \end{bmatrix}$, $\begin{bmatrix} -1 & -2 & 5 \end{bmatrix}$, $\begin{bmatrix} 2 & 6 & 7 \end{bmatrix}$.   **13.** $A = \begin{bmatrix} 1 & 2 \\ 3 & 4 \\ 5 & 1 \end{bmatrix}$.   **15.** $q \times p$.

SECTION 3.2 (page 98)

1. $2A = \begin{bmatrix} 2 & 4 & -2 \\ 6 & 10 & 4 \end{bmatrix}$, $-3B = \begin{bmatrix} -6 & 3 & -9 \\ -3 & -12 & -15 \end{bmatrix}$, $A - 2B = \begin{bmatrix} -3 & 4 & -7 \\ 1 & -3 & -8 \end{bmatrix}$.

3. $AB = \begin{bmatrix} 5 & 10 & -3 \\ 27 & 22 & 3 \end{bmatrix}$, CA not possible, $DB = \begin{bmatrix} 6 & 14 & -4 \end{bmatrix}$, $CD = \begin{bmatrix} 2 & -2 & 3 \\ -2 & 2 & -3 \\ 4 & -4 & 6 \end{bmatrix}$.

5. $AB = \begin{bmatrix} -9 - 21i & 11 + 10i \\ -1 + 19i & 9 + 15i \end{bmatrix}$.   **7.** $ABC = \begin{bmatrix} -12 & -22 \\ 14 & -126 \end{bmatrix}$.

9. B must have dimensions $n \times r$. $(ABC)_{ij} = \sum\limits_{k=1}^{n} \sum\limits_{l=1}^{r} a_{ik}b_{kl}c_{lj}$.   **13.** $x = 2$, $y = -1$, or $x = -1$, $y = 2$.

SECTION 3.3 (page 104)

1. $\begin{bmatrix} 1 & 0 & 0 \\ 0 & 2 & 0 \\ 0 & 0 & 3 \end{bmatrix}$.   **3.** $\begin{bmatrix} 0 & 1 & 2 & -5 \\ -1 & 0 & -3 & 4 \\ -2 & 3 & 0 & 6 \\ 5 & -4 & -6 & 0 \end{bmatrix}$.   **7.** diag $(2, 2, 2, 2)$.   **9.** $B^T = \begin{bmatrix} 0 & -1 & 1 & 2 \\ 1 & 2 & 1 & 1 \end{bmatrix}$.

13. $S = \begin{bmatrix} 1 & -1 & 5 \\ -1 & 2 & 1 \\ 5 & 1 & 6 \end{bmatrix}$, $T = \begin{bmatrix} 0 & -4 & -2 \\ 4 & 0 & 3 \\ 2 & -3 & 0 \end{bmatrix}$.

SECTION 4.1 (page 112)

7. $A = \begin{bmatrix} 1 & 1 & 1 & -1 \\ 2 & 4 & -3 & 7 \end{bmatrix}$, $b = \begin{bmatrix} 3 \\ 2 \end{bmatrix}$, $A^{\#} = \begin{bmatrix} 1 & 1 & 1 & -1 & | & 3 \\ 2 & 4 & -3 & 7 & | & 2 \end{bmatrix}$. **9.** $\begin{aligned} x_1 - x_2 + 2x_3 + 3x_4 &= 1, \\ x_1 + x_2 - 2x_3 + 6x_4 &= -1, \\ 3x_1 + x_2 + 4x_3 + 2x_4 &= 2. \end{aligned}$

SECTION 4.2 (page 122)

1. Row echelon form. **3.** Reduced row echelon form. **5.** Reduced row echelon form.

7. Reduced row echelon form. **9.** $\begin{bmatrix} 1 & -3 \\ 0 & 1 \end{bmatrix}$, rank$(A) = 2$. **11.** $\begin{bmatrix} 1 & 1 & 2 \\ 0 & 1 & 0 \\ 0 & 0 & 0 \end{bmatrix}$, rank$(A) = 2$.

13. $\begin{bmatrix} 1 & 3 \\ 0 & 1 \\ 0 & 0 \end{bmatrix}$, rank$(A) = 2$. **15.** $\begin{bmatrix} 1 & -2 & 1 & 3 \\ 0 & 1 & \frac{1}{3} & -\frac{2}{3} \\ 0 & 0 & 0 & 0 \end{bmatrix}$, rank$(A) = 2$. **17.** $\begin{bmatrix} 1 & 2 & 1 & 2 \\ 0 & 1 & 0 & 1 \\ 0 & 0 & 0 & 0 \\ 0 & 0 & 0 & 0 \end{bmatrix}$, rank$(A) = 2$.

19. I_2, rank$(A) = 2$. **21.** $\begin{bmatrix} 1 & -1 & 2 \\ 0 & 0 & 0 \\ 0 & 0 & 0 \end{bmatrix}$, rank$(A) = 1$. **23.** I_4, rank$(A) = 4$.

25. $\begin{bmatrix} 0 & 1 & 0 & 0 \\ 0 & 0 & 1 & 0 \\ 0 & 0 & 0 & 1 \end{bmatrix}$, rank$(A) = 3$. **27.** $M_2(\frac{1}{5})$. **29.** $M_2(-\frac{1}{8})M_{21}(-4)M_1(\frac{1}{2})$.

31. $P_{12} = \begin{bmatrix} 0 & 1 & 0 \\ 1 & 0 & 0 \\ 0 & 0 & 1 \end{bmatrix}$, $P_{13} = \begin{bmatrix} 0 & 0 & 1 \\ 0 & 1 & 0 \\ 1 & 0 & 0 \end{bmatrix}$, $P_{23} = \begin{bmatrix} 1 & 0 & 0 \\ 0 & 0 & 1 \\ 0 & 1 & 0 \end{bmatrix}$.

$M_1(k) = \begin{bmatrix} k & 0 & 0 \\ 0 & 1 & 0 \\ 0 & 0 & 1 \end{bmatrix}$, $M_2(k) = \begin{bmatrix} 1 & 0 & 0 \\ 0 & k & 0 \\ 0 & 0 & 1 \end{bmatrix}$, $M_3(k) = \begin{bmatrix} 1 & 0 & 0 \\ 0 & 1 & 0 \\ 0 & 0 & k \end{bmatrix}$.

$M_{12}(k) = \begin{bmatrix} 1 & k & 0 \\ 0 & 1 & 0 \\ 0 & 0 & 1 \end{bmatrix}$, $M_{13}(k) = \begin{bmatrix} 1 & 0 & k \\ 0 & 1 & 0 \\ 0 & 0 & 1 \end{bmatrix}$, $M_{23}(k) = \begin{bmatrix} 1 & 0 & 0 \\ 0 & 1 & k \\ 0 & 0 & 1 \end{bmatrix}$,

$M_{21}(k) = \begin{bmatrix} 1 & 0 & 0 \\ k & 1 & 0 \\ 0 & 0 & 1 \end{bmatrix}$, $M_{31}(k) = \begin{bmatrix} 1 & 0 & 0 \\ 0 & 1 & 0 \\ k & 0 & 1 \end{bmatrix}$, $M_{32}(k) = \begin{bmatrix} 1 & 0 & 0 \\ 0 & 1 & 0 \\ 0 & k & 1 \end{bmatrix}$.

33. $M_{32}(1)P_{23}M_{31}(-2)M_{21}(-3)M_1(\frac{1}{5})$.

SECTION 4.3 (page 131)

1. $(-2, 2, -1)$.   **3.** No solution.   **5.** $(2, -1, 3)$.   **7.** $\{(1 - 2r + s - t, r, s, t) : r, s, t \in \mathbf{R}\}$
9. $\{(1 - 2r + 3s - t, r, 2 - 4s + 3t, s, t) : r, s, t \in \mathbf{R}\}$.   **11.** No solution.   **13.** $\{(-s + t, -3 + s + t,$
$s, t) : s, t \in \mathbf{R}\}$.   **15.** $(1, -3, 4, -4, 2)$.   **17.** $\{(-5t, -2t - 1, t) : t \in \mathbf{R}\}$.   **19.** No solution.
21. (a) $k = 2$.   (b) $k = -2$.   (c) $k \neq \pm 2$.   **25.** $(1, 1, 2)$ and $(1, -1, 2)$.

SECTION 4.4 (page 136)

1. Trivial solution.   **3.** $\{(-t, -3t, t) : t \in \mathbf{R}\}$.   **5.** $\{(t, 0, -3t) : t \in \mathbf{R}\}$.   **7.** $\{(2t(i - 1), t, (2 - i)t : t \in$
$\mathbf{C}\}$.   **9.** $\{(2r - 3s, r, s) : r, s \in \mathbf{R}\}$.   **11.** Trivial solution.   **13.** $\{(t(1 - i), t) : t \in \mathbf{C}\}$.   **15.** Trivial solution.
17. $\{(r(1 - 5i) + s(5 + i), 13r, 13s) : r, s \in \mathbf{C}\}$.   **19.** $\{(-3t, -2t, t): t \in \mathbf{R}\}$.   **21.** $\{(3s, r, s, t): r, s, t \in$
$\mathbf{R}\}$.   **23.** (a) $k \neq -1$.

SECTION 4.5 (page 144)

5. $A^{-1} = \begin{bmatrix} -1 & 1+i \\ 1-i & -1 \end{bmatrix}$.   **7.** A is singular.   **9.** $\begin{bmatrix} 8 & -29 & 3 \\ -5 & 19 & -2 \\ 2 & -8 & 1 \end{bmatrix}$.

11. $A^{-1} = \begin{bmatrix} 18 & -34 & -1 \\ -29 & 55 & 2 \\ 1 & -2 & 0 \end{bmatrix}$.   **13.** $A^{-1} = \begin{bmatrix} -i & 1 & 0 \\ 1-5i & i & 2i \\ -2 & 0 & 1 \end{bmatrix}$.

15. $A^{-1} = \begin{bmatrix} 27 & 10 & -27 & 35 \\ 7 & 3 & -8 & 11 \\ -14 & -5 & 14 & -18 \\ 3 & 1 & -3 & 4 \end{bmatrix}$.   **17.** $(4, -1)$.   **19.** $(-2, 2, 1)$.

21. $(-6, 1, 3)$.   **31.** $\mathbf{x}_1 = (0, -1, 0)$, $\mathbf{x}_2 = (9, 8, -2)$, $\mathbf{x}_3 = (-5, -5, 2)$.

33. (b) All right inverses are of the form $\begin{bmatrix} 7 + 5r & -3 + 5s \\ -2 - 2r & 1 - 2s \\ r & s \end{bmatrix}$, where r and s are arbitrary real

numbers.

SECTION 5.1 (page 150)

1. $N(1, 2, 3) = 0$, $N(1, 3, 2) = 1$, $N(2, 1, 3) = 1$, $N(2, 3, 1) = 2$, $N(3, 1, 2) = 2$, $N(3, 2, 1) = 3$.

SECTION 5.2 (page 154)

3. 0.   **5.** 7.   **9.** (a) $p = 4$, $q = 1$, $+$.   (b) $p = 1$, $q = 4$, $+$.   (c) $p = 2$, $q = 1$, $+$.   **13.** 19.

SECTION 5.3 (page 162)

1. -36.   **3.** 45.   **5.** -103.   **7.** 624.   **9.** 21.   **11.** 84.   **13.** $\det(A) = 14 = \det(A^T)$.   **15.** $\det(AB) = 1$.
25. (b) -26.

SECTION 5.4 (page 169)

1. $M_{11} = 4$, $M_{21} = -3$, $M_{12} = 2$, $M_{22} = 1$, $A_{11} = 4$, $A_{21} = 3$, $A_{12} = -2$, $A_{22} = 1$. **3.** Cofactors: $A_{11} = -5$, $A_{12} = 0$, $A_{13} = 4$, $A_{21} = -47$, $A_{22} = -2$, $A_{23} = 38$, $A_{31} = 3$, $A_{32} = 0$, $A_{33} = -2$. **5.** 5.
7. -153. **9.** 0. **11.** 9. **13.** 3. **15.** -4. **17.** 11997. **19.** -170.

SECTION 5.5 (page 174)

1. Nonsingular. **3.** Nonsingular. **5.** Nonsingular. **7.** Singular.

9. $A^{-1} = \begin{bmatrix} \frac{1}{7} & \frac{2}{7} \\ -\frac{4}{7} & -\frac{1}{7} \end{bmatrix}$. **11.** $A^{-1} = \begin{bmatrix} \frac{7}{26} & \frac{3}{13} & -\frac{15}{26} \\ -\frac{2}{13} & \frac{2}{13} & -\frac{5}{13} \\ -\frac{1}{13} & \frac{1}{13} & \frac{4}{13} \end{bmatrix}$. **13.** $A^{-1} = \begin{bmatrix} -\frac{11}{6} & \frac{3}{2} & -\frac{1}{3} \\ -\frac{1}{6} & -\frac{1}{2} & \frac{1}{3} \\ \frac{4}{3} & -1 & \frac{1}{3} \end{bmatrix}$.

15. $A^{-1} = \begin{bmatrix} -\frac{9}{14} & \frac{16}{7} & -\frac{13}{14} \\ \frac{1}{14} & -\frac{1}{7} & \frac{3}{14} \\ \frac{1}{2} & -1 & \frac{1}{2} \end{bmatrix}$. **17.** $A^{-1} = \begin{bmatrix} \frac{14}{67} & -\frac{27}{67} & \frac{3}{67} & -\frac{5}{67} \\ \frac{23}{201} & \frac{10}{67} & \frac{19}{201} & \frac{13}{201} \\ \frac{29}{402} & \frac{33}{134} & -\frac{11}{402} & \frac{65}{201} \\ \frac{27}{134} & -\frac{9}{134} & \frac{1}{134} & -\frac{12}{67} \end{bmatrix}$.

19. $\frac{9}{16}$. **21.** $\begin{bmatrix} e^{-t}\sin 2t & e^{-t}\cos 2t \\ -e^{t}\cos 2t & e^{t}\sin 2t \end{bmatrix}$.

SECTION 5.6 (page 178)

1. $(\frac{16}{7}, \frac{6}{7})$. **3.** $(0, 0, 0)$. **5.** $(\frac{2}{3} e^{-t}(3 \sin t + 2 \cos t), \frac{1}{3} e^{2t}(3 \sin t - 4 \cos t)$. **7.** $\lambda = \pm 1, 4$. When $\lambda = 1$, the solution set is $S = \{(r, r, -2r) : r \in \mathbf{R}\}$.

SECTION 6.2 (page 192)

1. Vector space. **3.** Vector space. **5.** Vector space. **7.** Not a vector space. **9.** Vector space.
11. Vector space. **13.** Not a vector space. **15.** Not a vector space. **17.** Not a vector space. **19.** Not a vector space. **21.** Vector space. **23.** Vector Space. **25.** A2 and A3. **27.** A2.

SECTION 6.3 (page 198)

1. Subspace. **3.** Subspace. **5.** Not a subspace. **7.** Not a subspace. **9.** Not a subspace.
11. Subspace, the line through the origin in $\mathbf{R}^3$ whose direction is determined by $\mathbf{x}$.
13. Not a subspace. **15.** Subspace. **17.** Not a subspace. **19.** No.

SECTION 6.4 (page 203)

1. Yes. $(x_1, x_2) = x_1 \mathbf{v}_1 + (2x_1 - x_2)\mathbf{v}_2$. **3.** $\mathbf{v} = \frac{31}{7} \mathbf{v}_1 - \frac{9}{7} \mathbf{v}_2$. **5.** $\mathbf{v}_1$ and $\mathbf{v}_2$ do span $\mathbf{R}^2$.

7. $\mathbf{v} = x_1(1, 0, -1, 1) + x_2(0, 1, 1, -2)$. $\mathbf{v}_1 = (1, 0, -1, 1)$ and $\mathbf{v}_2 = (0, 1, 1, -2)$ span S.

9. $S = \{\mathbf{x} \in \mathbf{R}^3 : \mathbf{x} = t(-1, -1, 1), t \in \mathbf{R}\}$. $\mathbf{v} = (-1, -1, 1)$ spans S. **11.** $A = \begin{bmatrix} 0 & 1 \\ -1 & 0 \end{bmatrix}$ spans S.

13. $S = \{\mathbf{x} \in \mathbf{R}^3 : \mathbf{x} = c_1(1, -1, 2), c_1 \in \mathbf{R}\}$. **15.** $\text{span}\{\mathbf{v}_1, \mathbf{v}_2\} = \{\mathbf{x} \in \mathbf{R}^3 : \mathbf{x} = c_1(1, 2, -1)\}$, the line

through the origin determined by the vector $\mathbf{v}_1$. **17.** Yes, $\mathbf{v} = 2\mathbf{v}_2 - \mathbf{v}_1$. **19.** No.

21. $\text{span}\{A_1, A_2, A_3\} = \left\{ A \in M_2(\mathbf{R}) : A = \begin{bmatrix} c_1 + 3c_3 & -c_1 + c_2 \\ 2c_1 - 2c_2 + c_3 & c_2 + 2c_3 \end{bmatrix} \right\}$.

23. (a) $k(x) = c_1 \cosh x + c_2 \sinh x$.

SECTION 6.5 (page 209)

1. Linearly independent. **3.** Linearly independent. **5.** Linearly dependent. **7.** Linearly dependent.
9. Linearly independent. **11.** $k \neq -1, 2$. **13.** Linearly dependent. **15.** Linearly dependent.
19. $\alpha \neq \pm 1$.

SECTION 6.6 (page 215)

1. Linearly independent. **3.** Linearly dependent. **5.** Linearly independent. **7.** Linearly
independent. **9.** Linearly independent. **11.** Linearly independent. **13.** Linearly dependent.
17. Linearly independent.

SECTION 6.7 (page 224)

1. Basis: $\{(3, 1, 0), (-1, 0, 1)\}$, $\dim[S] = 2$. **3.** Basis: $\{(-2, 2, 1)\}$, $\dim[S] = 1$.
5. Basis: $\{(1, 0, 1), (0, 1, 1)\}$, $\dim[S] = 2$. **7.** Basis: $\{e^x, e^{-x}\}$, $\dim[S] = 2$. **9.** Basis. **11.** Not a basis.
13. $k \neq 1$. **15.** $\mathbf{e}_1 = -\mathbf{v}_1 + \mathbf{v}_2$, $\mathbf{e}_2 = 3\mathbf{v}_1 - 2\mathbf{v}_2$.
17. $p(x) = (2a_0 - a_1 - a_2)p_1(x) + (2a_0 - 2a_1 - a_2)p_2(x) + (a_1 + a_2 - a_0)p_3(x)$.
19. Basis: $\{1, x, x^2, x^3\}$. **23.** $\dim[\text{Sym}_n(\mathbf{R})] = \dfrac{n(n+1)}{2}$, $\dim[\text{Skew}_n(\mathbf{R})] = \dfrac{n(n-1)}{2}$.

SECTION 6.8 (page 233)

1. Orthonormal set: $\left\{ \dfrac{1}{\sqrt{6}}(2, -1, 1), \dfrac{1}{\sqrt{3}}(1, 1, -1), \dfrac{1}{\sqrt{2}}(0, 1, 1) \right\}$. **3.** Not orthogonal.

5. Orthonormal set: $\left\{ \dfrac{1}{\sqrt{14}}(1, 2, 3), \dfrac{1}{\sqrt{3}}(1, 1, -1), \dfrac{1}{\sqrt{42}}(5, -4, 1) \right\}$.

7. Orthonormal set: $\left\{ \dfrac{1}{\sqrt{5}}(1 - i, 1 + i, i), \dfrac{1}{\sqrt{3}}(0, i, 1 - i), \dfrac{1}{\sqrt{30}}(-3 + 3i, 2 + 2i, 2i) \right\}$.

9. $\langle f_1, f_2 \rangle = \dfrac{2}{3}$, $\|f_1\| = \sqrt{2}$, $\|f_2\| = \sqrt{\dfrac{2}{5}}$, not orthogonal.

11. $\langle f_1, f_2 \rangle = 1$, $\|f_1\| = \sqrt{\dfrac{e^2 - 1}{2}}$, $\|f_2\| = \sqrt{\dfrac{e^2 - 1}{2e^2}}$, not orthogonal.

13. Orthonormal set: $\left\{ \dfrac{1}{\sqrt{2}}f_1, \sqrt{\dfrac{3}{2}}f_2, \sqrt{\dfrac{5}{2}}f_3 \right\}$. **17.** $\langle A, B \rangle = 12$, $\|A\| = \sqrt{39}$, $\|B\| = \sqrt{15}$. **21. (a)** Yes.

(b) No. **23. (a)** No. **(b)** Yes. **25.** $\mathbf{x} = r(1, 1)$ or $\mathbf{x} = s(1, -1)$, $r, s \in \mathbf{R}$.

27. All vectors of the form (x_1, x_2), with $x_1^2 - x_2^2 > 0$.

SECTION 6.9 (page 240)

1. $\left\{\dfrac{1}{\sqrt{3}}(1, -1, -1), \dfrac{1}{\sqrt{42}}(4, 5, -1)\right\}$. 3. $\left\{\dfrac{1}{2}(-1, 1, 1, 1), \dfrac{1}{\sqrt{6}}(2, 1, 0, 1)\right\}$.

5. $\left\{\dfrac{1}{\sqrt{6}}(1, 2, 0, 1), \dfrac{1}{\sqrt{30}}(4, -1, 3, -2), \dfrac{1}{2\sqrt{5}}(-1, -1, 3, 3)\right\}$.

7. $\left\{\dfrac{1}{\sqrt{3}}(1 - i, 0, i), \dfrac{1}{\sqrt{21}}(1, 3 + 3i, 1 - i)\right\}$. 9. $\left\{1, \dfrac{1}{2}(2x - 1), \dfrac{1}{6}(6x^2 - 6x + 1)\right\}$.

11. $\left\{1, \sin x, \dfrac{1}{\pi}(\pi \cos x - 2)\right\}$. 13. $\left\{\begin{bmatrix} 0 & 1 \\ 1 & 0 \end{bmatrix}, \begin{bmatrix} 1 & 0 \\ 0 & 0 \end{bmatrix}, \begin{bmatrix} 0 & 0 \\ 0 & 1 \end{bmatrix}\right\}$, the subspace of all symmetric

matrices in $M_2(\mathbf{R})$. 15. $\{1 + x^2, 1 - x - x^2 + x^3, -3 - 5x + 3x^2 + x^3\}$.

SECTION 7.1 (page 250)

1. Linear. 3. Nonlinear. 5. Nonlinear. 7. $\begin{bmatrix} 3 & -2 \\ 1 & 5 \end{bmatrix}$. 9. $\begin{bmatrix} 1 & -1 & 1 \\ -1 & 0 & 1 \end{bmatrix}$.

19. $T(ax^2 + bx + c) = bx^2 + (3a + c)x + (2a - b + c)$.
21. $T(1) = 3 - 2x, T(x) = 2x, T(x^2) = x^2 - x$.
23. $(T_1 + T_2)(x_1, x_2) = (3x_1 + x_2, 5x_1 + x_2), (5T_1)(x_1, x_2) = (10x_1 - 5x_2, 15x_1 + 10x_2)$,
$(2T_1 - 3T_2)(x_1, x_2) = (x_1 - 8x_2, 7x_2)$.
25. If $\mathbf{v} = c_1\mathbf{v}_1 + c_2\mathbf{v}_2$, then $(T_1 + T_2)(\mathbf{v}) = (3c_1 + 2c_2)\mathbf{v}_1 + 2c_2\mathbf{v}_2$.

SECTION 7.2 (page 260)

3. $\mathrm{Ker}(T) = \{\mathbf{0}\}, \dim[\mathrm{Ker}(T)] = 0; \mathrm{Rng}(T) = \mathbf{R}^3, \dim[\mathrm{Rng}(T)] = 3$. 5. $\mathrm{Ker}(T) = \{\mathbf{x} \in \mathbf{R}^3 : \mathbf{x} = r(-2, 0, 1) + s(1, 1, 0), r, s \in \mathbf{R}\}, \dim[\mathrm{Ker}(T)] = 2; \mathrm{Rng}(T) = \{\mathbf{y} \in \mathbf{R}^2 : \mathbf{y} = t(1, -3), t \in \mathbf{R}\}$,
$\dim[\mathrm{Rng}(T)] = 1$.

7. Basis: $\{1, x\}, \dim[\mathrm{Ker}(T)] = 2$. 9. Basis: $\left\{\begin{bmatrix} 1 & 0 \\ 0 & 0 \end{bmatrix}, \begin{bmatrix} 0 & 1 \\ 1 & 0 \end{bmatrix}, \begin{bmatrix} 0 & 0 \\ 0 & 1 \end{bmatrix}\right\}, \dim[\mathrm{Ker}(T)] = 3$.

11. (b) $\mathrm{Rng}(T) = \{\alpha(x^2 + 4x) + \beta(x^2 + 4) : \alpha, \beta \in \mathbf{R}\}, \dim[\mathrm{Rng}(T)] = 2$.
13. $\mathrm{Ker}(T) = \{\mathbf{0}\}, \dim[\mathrm{Ker}(T)] = 0$. $\mathrm{Rng}(T) = \{\alpha(3x^2 + 1) + \beta(x - x^2) : \alpha, \beta \in \mathbf{R}\}, \dim[\mathrm{Rng}(T)] = 2$.

SECTION 7.3 (page 269)

1. $(T_1T_2)(x_1, x_2) = (-5(x_1 + x_2), x_1 + 15x_2), (T_2T_1)(x_1, x_2) = (7(2x_1 + x_2), 2(x_1 - 2x_2))$.
3. $\mathrm{Ker}(T_1) = \{\mathbf{x} \in \mathbf{R}^2 : \mathbf{x} = r(1, 1), r \in \mathbf{R}\}, \mathrm{Ker}(T_2) = \{\mathbf{0}\}, \mathrm{Ker}(T_1T_2) = \{\mathbf{x} \in \mathbf{R}^2 : \mathbf{x} = s(2, 1), s \in \mathbf{R}\}$,
$\mathrm{Ker}(T_2T_1) = \mathrm{Ker}(T_1)$. 5. $[T_1(f)](x) = \cos(x - a), [T_2(f)](x) = 1 - \cos(x - a)$. 9. One-to-one and onto.

11. Not one-to-one, onto. 13. $T^{-1}(Ax + B) = \dfrac{1}{3}(2B - A)x + \dfrac{1}{3}(A + B)$.

15. $T^{-1}(A\mathbf{v}_1 + B\mathbf{v}_2) = \dfrac{1}{7}(3A + 2B)\mathbf{v}_1 + \dfrac{1}{7}(2A - B)\mathbf{v}_2$.

SECTION 8.2 (page 281)

5. $\lambda_1 = 4, \mathbf{v}_1 = r(1, -1); \lambda_2 = -2, \mathbf{v}_2 = s(1, 5)$. 7. $\lambda = 5, \mathbf{v}_1 = r(-2, 1)$.
9. $\lambda_1 = 1 + 2i, \mathbf{v}_1 = r(1, 1 - i); \lambda_2 = 1 - 2i, \mathbf{v}_2 = s(1, 1 + i)$. 11. $\lambda = 2, \mathbf{v} = r(3, 2, 0) + s(-1, 0, 1)$.
13. $\lambda = 1, \mathbf{v} = r(0, -1, 1)$. 15. $\lambda = -1, \mathbf{v} = r(-3, 0, 4) + s(1, 1, 0)$.
17. $\lambda_1 = 1, \mathbf{v}_1 = r(1, 0, 0); \lambda_2 = i, \mathbf{v}_2 = s(0, 1, i); \lambda_3 = -i, \mathbf{v}_3 = t(0, 1, -i)$.

19. $\lambda_1 = 0$, $\mathbf{v}_1 = r(-3, 9, 5)$; $\lambda_2 = 2$, $\mathbf{v}_2 = s(1, 3, 1)$; $\lambda_3 = 4$, $\mathbf{v}_3 = t(1, 1, 1)$.
21. $\lambda_1 = -2$, $\mathbf{v}_1 = r(-1, 0, 1) + s(-1, 1, 0)$; $\lambda_2 = 4$, $\mathbf{v}_2 = t(1, 1, 1)$.
23. $\lambda_1 = i$, $\mathbf{v}_1 = a(0, 0, i, 1) + b(-i, 1, 0, 0,)$; $\lambda_2 = -i$, $\mathbf{v}_2 = r(0, 0, -i, 1) + s(i, 1, 0, 0)$.

25. (c) $A^{-1} = \begin{bmatrix} \frac{2}{3} & \frac{1}{6} \\ -\frac{1}{3} & \frac{1}{6} \end{bmatrix}$.   **27.** $A(3\mathbf{v}_1 - \mathbf{v}_2) = (12, -3)$.

SECTION 8.3 (page 287)

1. $\lambda_1 = 5$, $m_1 = 1$, basis for $E_1 : \{(1, 1)\}$, $n_1 = 1$; $\lambda_2 = -1$, $m_2 = 1$ basis for $E_2 : \{(-2, 1)\}$, $n_2 = 1$. Nondefective. **3.** $\lambda_1 = 3$, $m_1 = 2$, basis for $E_1 : \{(1, 1)\}$, $n_1 = 1$. Defective. **5.** $\lambda_1 = -2$, $m_1 = 1$, basis for $E_1 : \{(1, 1, 1)\}$, $n_1 = 1$; $\lambda_2 = 3$, $m_2 = 2$, basis for $E_2 : \{(1, 0, 0), (0, -1, 4)\}$, $n_2 = 2$. Nondefective. **7.** $\lambda_1 = 4$, $m_1 = 3$, basis for $E_1 : \{(0, 0, 1), (1, 1, 0)\}$, $n_1 = 2$. Defective. **9.** $\lambda_1 = 2$, $m_1 = 2$, basis for $E_1 : \{(-1, 2, 0)\}$, $n_1 = 1$; $\lambda_2 = -3$, $m_2 = 1$, basis for $E_2 : \{(-1, 1, 1)\}$, $n_2 = 1$. Defective. **11.** $\lambda_1 = -1$, $m_1 = 3$, basis for $E_1 : \{(-3, 0, 4), (1, 1, 0)\}$, $n_1 = 2$; $\lambda_2 = 2$, $m_2 = 1$, basis for $E_2 : \{(1, 1, 1)\}$, $n_2 = 1$. **13.** $\lambda_1 = 0$, $m_1 = 2$, basis for $E_1 : \{(-2, 0, 1), (1, 1, 0)\}$, $n_1 = 2$; $\lambda_2 = 2$, $m_2 = 1$, basis for $E_2 : \{(1, 1, 1)\}$, $n_2 = 1$. Nondefective. **15.** $\lambda_1 = 1$, $m_1 = 2$, basis for $E_1 : \{(-1, 0, 1), (-1, 1, 0)\}$, $n_1 = 2$; $\lambda_2 = -2$, $m_2 = 1$, basis for $E_2 : \{(1, 1, 1)\}$, $n_2 = 1$. Nondefective. **17.** Defective. **19.** Nondefective. **21.** $\lambda_1 = 1$, basis for $E_1 : \{(-1, 1)\}$; $\lambda_2 = 5$, basis for $E_2 : \{(1, 3)\}$. **23.** $\lambda_1 = 5$, basis for $E_1 : \{(1, 0), (0, 1)\}$. **25.** $\lambda_1 = -2$, basis for $E_1 : \{(1, 1, 0)\}$. **27.** $\lambda_1 = 2$, orthogonal basis for $E_1 : \{(1, 0, 1), (-1, 2, 1)\}$; $\lambda_2 = -1$, basis for $E_2 : \{(-1, -1, 1)\}$. Vectors in E_2 are orthogonal to the vectors in E_1.
29. (d) Sum of eigenvalues $= 24$. Product of eigenvalues $= -607$.

SECTION 8.4 (page 294)

1. $S = \begin{bmatrix} 1 & 2 \\ -2 & 1 \end{bmatrix}$, $S^{-1}AS = \text{diag}(3, -2)$. **3.** Not diagonalizable.

5. $S = \begin{bmatrix} 15 & 0 & 0 \\ -7 & 7 & 1 \\ 2 & 1 & -1 \end{bmatrix}$, $S^{-1}AS = \text{diag}(1, 4, -4)$. **7.** $S = \begin{bmatrix} 1 & -1 & -1 \\ 1 & 0 & 1 \\ 1 & 1 & 0 \end{bmatrix}$, $S^{-1}AS = \text{diag}(-4, 2, 2)$.

9. Not diagonalizable. **11.** $S = \begin{bmatrix} -2 & 4+3i & 4-3i \\ 1 & -2+6i & -2-6i \\ 2 & 5 & 5 \end{bmatrix}$, $S^{-1}AS = \text{diag}(0, 3i, -3i)$.

SECTION 8.5 (page 301)

1. $S = \begin{bmatrix} -\frac{1}{\sqrt{5}} & \frac{2}{\sqrt{5}} \\ \frac{2}{\sqrt{5}} & \frac{1}{\sqrt{5}} \end{bmatrix}$, $S^TAS = \text{diag}(-2, 3)$. **3.** $S = \begin{bmatrix} \frac{1}{\sqrt{2}} & -\frac{1}{\sqrt{2}} \\ \frac{1}{\sqrt{2}} & \frac{1}{\sqrt{2}} \end{bmatrix}$, $S^TAS = \text{diag}(3, -1)$.

5. $S = \begin{bmatrix} -\frac{1}{\sqrt{2}} & -\frac{1}{\sqrt{3}} & \frac{1}{\sqrt{6}} \\ 0 & \frac{1}{\sqrt{3}} & \frac{2}{\sqrt{6}} \\ \frac{1}{\sqrt{2}} & -\frac{1}{\sqrt{3}} & \frac{1}{\sqrt{6}} \end{bmatrix}$, $S^TAS = \text{diag}(0, 0, 6)$. **7.** $S = \begin{bmatrix} -\frac{1}{\sqrt{2}} & 0 & \frac{1}{\sqrt{2}} \\ \frac{1}{\sqrt{2}} & 0 & \frac{1}{\sqrt{2}} \\ 0 & 1 & 0 \end{bmatrix}$, $S^TAS = \text{diag}(-1, 1, 1)$.

9. $S = \begin{bmatrix} \dfrac{1}{\sqrt{6}} & \dfrac{1}{\sqrt{2}} & -\dfrac{1}{\sqrt{3}} \\ -\dfrac{1}{\sqrt{6}} & \dfrac{1}{\sqrt{2}} & \dfrac{1}{\sqrt{3}} \\ \dfrac{2}{\sqrt{6}} & 0 & \dfrac{1}{\sqrt{3}} \end{bmatrix}$, $S^T A S = \text{diag}(-1, 1, 2)$.

11. $S = \begin{bmatrix} \dfrac{1}{\sqrt{3}} & -\dfrac{1}{\sqrt{2}} & -\dfrac{1}{\sqrt{6}} \\ \dfrac{1}{\sqrt{3}} & \dfrac{1}{\sqrt{2}} & -\dfrac{1}{\sqrt{6}} \\ \dfrac{1}{\sqrt{3}} & 0 & \dfrac{2}{\sqrt{6}} \end{bmatrix}$, $S^T A S = \text{diag}(1, -5, -5)$. **15.** $\mathbf{v}_2 = (-2, 1)$.

17. $S = \begin{bmatrix} \dfrac{1}{\sqrt{3}} & \dfrac{1}{\sqrt{2}} & -\dfrac{1}{\sqrt{6}} \\ -\dfrac{1}{\sqrt{3}} & \dfrac{1}{\sqrt{2}} & \dfrac{1}{\sqrt{6}} \\ \dfrac{1}{\sqrt{3}} & 0 & \dfrac{2}{\sqrt{6}} \end{bmatrix}$, $A = \begin{bmatrix} a & -b & b \\ -b & a & -b \\ b & -b & a \end{bmatrix}$, where $a = \frac{1}{3}(\lambda_1 + 2\lambda_2)$, $b = \frac{1}{3}(\lambda_1 - \lambda_2)$.

SECTION 9.2 (page 310)

1. $Lf = 2(1 - x^2) + 3e^{2x}(x - 2)$. **3.** $Lf = 24e^{2x}(x - 1)$.
7. $L_1 L_2 = xD^2 + (1 - 2x^3)D - 6x^2$, $L_2 L_1 = xD^2 + 2(1 - x^3)D - 2x^2$.
9. $L_1 L_2 = x^2 D^3 + 2xD^2 - x^3 D$, $L_2 L_1 = x^2 D^3 + xD^2 - x^3 D - 2x^2$.

SECTION 9.3 (page 318)

1. $(D^3 + x^2 D^2 - (\sin x)D + e^x)y = x^3$, $y''' + x^2 y'' - (\sin x)y' + e^x y = 0$.
5. Basis: $\{e^{3x}, e^{-2x}\}$; general solution: $y = c_1 e^{3x} + c_2 e^{-2x}$.
7. Basis: $\{e^{-x}, xe^{-x}\}$; general solution: $y = e^{-x}(c_1 + c_2 x)$.
9. $r = 2, -4$; general solution: $y = c_1 x^2 + c_2 x^{-4}$.
11. (a) $y_c = c_1 e^{-2x} + c_2 e^x$. **(b)** $y_p = -(3 + 2x + 2x^2)$. **(c)** $y = c_1 e^{-2x} + c_2 e^x - (3 + 2x + 2x^2)$.

SECTION 9.4 (page 324)

1. $P(D) = D^2 - 3D + 6$. **3.** $P(r) = r^2 + r - 12$, $P(D) = (D + 4)(D - 3)$. **5.** $P(D) = (D - 1)^3$.
7. $P(D) = D^2 + D - 2$. **9.** $P(D) = D^3 - 5D^2 + 9D - 5$. **11.** $y = c_1 e^{-2x} + c_2 e^{3x}$. **13.** $y = c_1 e^{2x} + c_2 e^{3x}$.
15. $y = c_1 + c_2 e^x$. **17.** $y = c_1 e^{-4x} + c_2 e^{2x} + c_3 e^{3x} + c_4 e^{-3x}$. **19.** $y'' - y = 0$. **21.** $y'' + y' - 12y = 0$.
23. (a) r_1, r_2 must both be negative.

SECTION 9.5 (page 329)

1. $\cos 2x + i \sin 2x$. **3.** $\cos 5x - i \sin 5x$. **5.** $x^2[\cos(\ln x) - i \sin(\ln x)]$.

7. $x^{-1}[\cos(2 \ln x) + i \sin(2 \ln x)]$. **11.** $\dfrac{1}{2i}(e^{4ix} - e^{-4ix})$. **13.** $\dfrac{1}{i}\left(\dfrac{e^{ix} - e^{-ix}}{e^{ix} + e^{-ix}}\right)$.

15. $y_1 = e^x \cos 2x$, $y_2 = e^x \sin 2x$; general solution: $y = c_1 e^x \cos 2x + c_2 e^x \sin 2x$.
17. $r = 2 \pm i$, $y_1 = e^{2x} \cos x$, $y_2 = e^{2x} \sin x$.
19. $r = 1, \pm 2i$, $y_1 = e^x$, $y_2 = \cos 2x$, $y_3 = \sin 2x$.

21. $r = \pm 5i$, $y_1 = \cos(5 \ln x)$, $y_2 = \sin(5 \ln x)$.
23. $r = -2 \pm 2i$, $y_1 = x^{-2}\cos(2 \ln x)$, $y_2 = x^{-2}\sin(2 \ln x)$.

SECTION 9.6 (page 337)

1. $y = c_1 e^{2x} + c_2 e^{-x}$. **3.** $y = e^{-3x}(c_1 \cos 4x + c_2 \sin 4x)$. **5.** $y = (c_1 + c_2 x)e^{-2x}$.
7. $y = e^{3x}(c_1 \cos 5x + c_2 \sin 5x)$. **9.** $y = (c_1 + c_2 x)e^{-5x}$.
11. $y = c_1 e^{2x} + c_2 e^{4x} + c_3 e^{-4x}$. **13.** $y = e^{-4x}(c_1 \cos 2x + c_2 \sin 2x)$.
15. $y = c_1 e^{-2x} + (c_2 + c_3 x)e^{2x}$. **17.** $y = c_1 + c_2 x + c_3 e^x$.
19. $y = e^x(c_1 \cos x + c_2 \sin x) + x e^x(c_3 \cos x + c_4 \sin x) + c_5 e^x + c_6 e^{-x}$.
21. $y = 2e^{2x} + e^{-3x}$. **23.** $y = e^{2x}(3 \cos x - \sin x)$.
27. (b) $u = e^{x/\alpha}[e^{-p\xi}(A \sin q\xi + B \cos q\xi)]$. **29.** $y'' + 4y = 0$.
31. $(D^2 - 2D + 5)(D - 3)y = 0$. **33.** $D(D^2 + 4)y = 0$. **35.** $(D^2 - 2D + 2)^2 y = 0$.

SECTION 9.7 (page 349)

1. $D + 3$. **3.** $(D^2 + 1)(D - 2)^2$. **5.** $D^2 + 4D + 5$. **7.** $D^3(D - 4)^2$. **9.** $D^2 + 6D + 10$.
11. $y = c_1 e^{2x} + c_2 e^x + 2e^{3x}$. **13.** $y = e^x(c_1 + c_2 x) + 3x^2 - 6$.
15. $y = c_1 e^{-x} + e^{3x}(c_2 + 2x)$. **17.** $y = c_1 e^{-2x} + c_2 e^{-x} + e^{2x} + 3e^x$.
19. $y = e^{-x}(c_1 \cos x + c_2 \sin x) + 2 - 4x + 2x^2$.
21. $y = (c_1 + 2x^2)\sin 2x + (c_2 + x)\cos 2x$.
23. $y = c_1 e^{-x} + c_2 e^{2x} - 10 + 3 \cos 2x + \sin 2x$. **25.** $y = e^{-2x} + \cos x + 3 \sin x$.
27. $y = 8e^x - 4e^{-x} + e^{2x}(3x - 4)$. **29.** $y = e^x + e^{2x} + e^{3x} + e^{4x}$.
31. $y_p = x^3 e^x(A_0 \cos x + B_0 \sin x) + A_1 x^2 e^{2x}$.
33. $q(t) = e^{-10t}(A_0 \cos 20\sqrt{2}t + B_0 \sin 20\sqrt{2}t) + 2 \sin 30t$.
$i(t) = e^{-10t}[(-10A_0 + 20\sqrt{2}B_0)\cos 20\sqrt{2}t - (20\sqrt{2}A_0 + 10B_0)\sin 20\sqrt{2}t] + 60 \cos 30t$.

SECTION 9.8 (page 352)

1. $y_p = -(3 \cos 3x + 4 \sin 3x)$. **3.** $y_p = -(12 \cos 3x + 5 \sin 3x)$.

5. $y_p = \frac{3}{10}e^x(2 \sin 2x - \cos 2x)$. **7.** $y_p = e^x[2 \sin x - 14 \cos x - 10x(2 \sin x + \cos x)]$.

9. $y_p = 4x e^x \sin 3x$. **11.** $x_p = \dfrac{F_0}{\omega_0^2 - \omega^2} \cos \omega t$, $\omega_0 \neq \omega$; $x_p = \dfrac{F_0}{2\omega_0} t \sin \omega_0 t$, if $\omega = \omega_0$.

SECTION 9.9 (page 360)

1. $y = e^{-3x}[c_1 + c_2 x + 2x \tan^{-1}x - \ln(x^2 + 1)]$.
3. $y = e^{2x}[\cos x(c_1 - \ln|\sec x + \tan x|) + c_2 \sin x]$. **5.** $y = e^{-2x}(c_1 + c_2 x - \ln x)$.
7. $y = c_1 e^x + c_2 e^{-x} + 4 \tan^{-1}(e^x)\cosh x$. **9.** $y = e^x[c_1 + c_2 x + x^{-1}(2 \ln x + 3)]$.

11. $y = e^{-x}\left[c_1 \cos 4x + c_2 \sin 4x + \cos 4x \ln \left(\dfrac{\cos 4x + 2}{\cos 4x - 2} \right) + \dfrac{4}{\sqrt{3}} \sin 4x \tan^{-1}\left(\dfrac{\sin 4x}{\sqrt{3}} \right) \right]$.

13. $y = e^{5x}[c_1 + c_2 x - \ln(4 + x^2) + x \tan^{-1}(x/2)]$. **15.** $i(t) = t \cos 2t - \frac{1}{2} \sin 2t \ln|\cos 2t|$.

19. $y_p = x^3 e^{2x}(6 \ln x - 11)$. **21.** $y = e^{-x}[(x^2 - 1)\tan^{-1}x - x \ln(1 + x^2)]$.

SECTION 9.10 (page 370)

1. $y = x[c_1 \sin(2 \ln x) + c_2 \cos(2 \ln x)]$. **3.** $y = x^2(c_1 + c_2 \ln x)$. **5.** $y = x^{-1}(c_1 + c_2 \ln x)$.
7. $y = c_1 x^7 + c_2 x^{-5}$. **9.** $y = c_1 x^{-1} + c_2 x + c_3 x^2$. **11.** $y = c_1 + x^{-1}(c_2 + c_3 \ln x)$.
13. $y = c_1 x^{-2} + c_2 x^2 + c_3 x^3$. **15.** $y = c_1 x + x^{-2}[c_2 \cos(4 \ln x) + c_3 \sin(4 \ln x)]$.

17. $y = (c_1 + c_2 \ln x)\cos(\ln x) + (c_3 + c_4 \ln x)\sin(\ln x)$ 19. $y = x^m(c_1 + c_2 \ln x)$.
23. $y = c_1 x^3 + x^2(c_2 - \sin x)$ 25. $y = x^2[c_1 + c_2 \ln x + \ln x(\ln|\ln x| - 1)]$.
27. $y = c_1\cos(3 \ln x) + c_2\sin(3 \ln x) + \ln x$.   29. $y = x[c_1\cos(2 \ln x) + c_2\sin(2 \ln x) + 2(\ln x)^2 - 1]$.
31. $y = x[c_1 + c_2 \ln x + c_3(\ln x)^2 + (\ln x)^4]$.

SECTION 9.11 (page 375)

1. $y_2 = x^2 \ln x$.   3. $y_2 = e^x \ln x$.   5. $y_2 = \dfrac{1}{2}x \ln\left(\dfrac{1+x}{1-x}\right) - 1$.   7. (a) $y_1 = x^m$.   (b) $y_2 = x^m \ln x$.

11. $y = e^{2x}[c_1 + c_2 x + x^2(2 \ln x - 3)]$.   13. $y = c_1\cos x + c_2\sin x + \sin x[\ln(\sin x) - x \cot x]$.
15. $y = x^2(c_1 + c_2 \ln x + 2x^2)$.

SECTION 10.1 (page 382)

1. $\dfrac{d^2x}{dt^2} + \dfrac{c}{m}\dfrac{dx}{dt} + \dfrac{1}{m}(k_1 + k_2)x = \dfrac{F(t)}{m}$, $x(0) = x_0$, $\dfrac{dx}{dt}(0) = v_0$.

SECTION 10.2 (page 387)

1. $\omega_0 = 2$, $A_0 = 2\sqrt{2}$, $\phi = \dfrac{\pi}{4}$, $T = \pi$.   3. (a) $k = 3$.   (b) $\omega_0 = \dfrac{\sqrt{3}}{2}$, $A_0 = \dfrac{2\sqrt{3}}{3}$, $\phi = \dfrac{5\pi}{6}$, $T = 4\pi\dfrac{\sqrt{3}}{3}$.

5. Overdamped, $x = e^{-2t}(2e^t - 1)$.   7. Critically damped, $x = e^{-t}(t - 1)$.

9. Underdamped, $x = \dfrac{2}{3}e^{-2t}(3 \cos\sqrt{3}t + 5\sqrt{3}\sin\sqrt{3}t)$.

11. (a) $0 < \alpha < 1$.   (b) $\alpha = 1$.   (c) $\alpha > 1$.   In (c) $x = e^{-\alpha t}(c_1 e^{\mu t} + c_2 e^{-\mu t})$, $\mu = \sqrt{\alpha^2 - 1}$.   The system does pass through the equilibrium.

15. The initial value problem governing the motion is $\dfrac{d^2x}{dt^2} + \dfrac{4g}{L}x = 0$, $x(0) = \dfrac{L}{2}$, $\dfrac{dx}{dt}(0) = 0$.   Circular

frequency: $\omega_0 = 2\sqrt{\dfrac{g}{L}}$.   Period: $T = \pi\sqrt{\dfrac{L}{g}}$.   17. $A_0 = \sqrt{\dfrac{\alpha^2 g + \beta^2 L}{g}}$; the phase, ϕ, is defined by

$\cos\phi = \dfrac{\alpha}{A_0}$, $\sin\phi = \dfrac{\beta}{A_0}\sqrt{\dfrac{L}{g}}$.   Period: $T = 2\pi\sqrt{\dfrac{L}{g}}$.   19. ≈ 63.   21. $T = \dfrac{2\pi}{\omega}$, where
$\omega^2 = \dfrac{g(L_0^2 + L_0 L + L^2)}{L^2\sqrt{L^2 - L_0^2}}$.

SECTION 10.3 (page 393)

1. (a) Underdamped.   (b) $x = 2e^{-t}\cos 2t - 4 \cos 2t + \sin 2t$.   Transient part of solution: $x_T = 2e^{-t}\cos 2t$.   Steady-state solution: $x_s = \sin 2t - 4 \cos 2t$.   3. $x_p = \sqrt{10}\sin(t - \phi)$, $\phi = \tan^{-1}3$.
9. (a) Transient part of solution: $x_T = A_0 e^{-t}\cos(2t - \phi)$.   Steady-state solution:

$x_s = \dfrac{8}{\sqrt{(5 - \omega^2)^2 + 4\omega^2}}\cos(\omega t - \eta)$, $\tan\eta = \dfrac{2\omega}{5 - \omega^2}$.   (b) $\omega = \sqrt{3}$, $x_T = 2 \cos(\sqrt{3}t - \dfrac{\pi}{3})$.
11. $x = A_0\cos(4t - \phi) + e^{-t}(8 \cos t - \sin t)$.   Transient solution: $x_T = e^{-t}(8 \cos t - \sin t)$.   Steady-state part of solution: $x_s = A_0\cos(4t - \phi)$.

SECTION 10.4 (page 399)

1. $i_s = 2(3 \cos 3t + 2 \sin 3t)$. **5.** $i = -A_0 e^{-3t}[3 \cos(t - \phi) + \sin(t - \phi)] - \dfrac{2\omega}{H} \sin(\omega t - \eta)$, where

$H = \dfrac{1}{2} \sqrt{(10 - \omega^2)^2 + 36\omega^2}$, $\cos \eta = \dfrac{(10 - \omega^2)}{2H}$, $\sin \eta = \dfrac{3\omega}{H}$. The maximum value of the amplitude

occurs when $\omega = \sqrt{10}$. **7.** $i = A_0 e^{-(R/2L)t} \cos(\mu t - \phi) - \dfrac{aCE_0}{(a^2 LC - acR + 1)} e^{-at}$, where $\mu = \dfrac{\sqrt{\dfrac{4L}{C} - R^2}}{2L}$.

SECTION 11.2 (page 407)

1. $x_1 = c_1 e^t + c_2 e^{-t}$, $x_2 = \frac{1}{3}(c_1 e^t + 3 c_2 e^{-t})$. **3.** $x_1 = e^{-2t}(c_1 + c_2 t)$, $x_2 = \frac{1}{4} e^{-2t}[-4c_1 + c_2(1 - 4t)]$.

5. $x_1 = e^t(c_1 \cos 3t + c_2 \sin 3t)$, $x_2 = e^t(c_1 \sin 3t - c_2 \cos 3t)$. **7.** $x_1 = -e^{-2t}(2c_1 + c_2 \cos t + c_3 \sin t)$,
$x_2 = -e^{-2t}[c_1 + c_2(\cos t - \sin t) + c_3(\sin t + \cos t)]$, $x_3 = e^{-2t}(c_1 + c_2 \cos t + c_3 \sin t)$.
9. $x_1 = 5 \sin t$, $x_2 = \cos t - 2 \sin t$. **11.** $x_1 = c_1 e^{-t} + c_2 e^{3t} + 3e^{4t}$, $x_2 = -c_1 e^{-t} + c_2 e^{3t} + 2e^{4t}$.
13. $x_1 = c_1 e^{2t} + c_2 e^{-2t} + 2te^{2t}$, $x_2 = c_1 e^{2t} - 3c_2 e^{-2t} + e^{2t}(1 + 2t)$.
15. $x_1' = x_2$, $x_2' = -x_1 + 3x_4 + \sin t$, $x_3' = x_4$, $x_4' = tx_2 + e^t x_3 + t^2$.
17. $x_1' = x_2$, $x_2' = -bx_1 - ax_2 + F(t)$.

19. $x_1' = x_2$, $x_2' = -\dfrac{(k_1 + k_2)}{m_1} x_1 + \dfrac{k_2}{m_1} x_3$, $x_3' = x_4$, $x_4' = \dfrac{k_2}{m_2} x_1 - \dfrac{k_2}{m_2} x_3$, $x_1(0) = \alpha_1$, $x_2(0) = \alpha_2$,

$x_3(0) = \alpha_3$, $x_4(0) = \alpha_4$.

SECTION 11.3 (page 412)

1. $x_1' = -4x_1 + 3x_2 + 4t$, $x_2' = 6x_1 - 4x_2 + t^2$. Matrix form: $\mathbf{x}' = A\mathbf{x} + \mathbf{b}$, where $A = \begin{bmatrix} -4 & 3 \\ 6 & -4 \end{bmatrix}$ and

$\mathbf{b} = \begin{bmatrix} 4t \\ t^2 \end{bmatrix}$. **3.** $\mathbf{x}' = A\mathbf{x} + \mathbf{b}$, where $A = \begin{bmatrix} 0 & -\sin t & 1 \\ -e^t & 0 & t^2 \\ -t & t^2 & 0 \end{bmatrix}$ and $\mathbf{b} = \begin{bmatrix} t \\ t^3 \\ 1 \end{bmatrix}$.

5. New variables: $x_1 = x$, $x_2 = x'$, $x_3 = x''$. Linear system: $\mathbf{x}' = A\mathbf{x} + \mathbf{b}$, where

$A = \begin{bmatrix} 0 & 1 & 0 \\ 0 & 0 & 1 \\ a^2 - t^2 & 0 & \sin t \end{bmatrix}$ and $\mathbf{b} = \begin{bmatrix} 0 \\ 0 \\ e^t \end{bmatrix}$. **7.** $\dfrac{dA}{dt} = \begin{bmatrix} -2e^{-2t} \\ \cos t \end{bmatrix}$. **9.** $\dfrac{dA}{dt} = \begin{bmatrix} e^t & 2e^{2t} & 2t \\ 2e^t & 8e^{2t} & 10t \end{bmatrix}$.

13. $\begin{bmatrix} e - 1 & 1 - e^{-1} \\ 2(e - 1) & 5(1 - e^{-1}) \end{bmatrix}$. **17.** $W[\mathbf{x}_1, \mathbf{x}_2](0) = 2 \neq 0$. **21.** $4x_1 - x_2 = 0$.

25. $\mathbf{x}_1 = \begin{bmatrix} 4e^t \\ e^t \end{bmatrix}$, $\mathbf{x}_2 = \begin{bmatrix} e^{-2t} \\ e^{-2t} \end{bmatrix}$.

SECTION 11.4 (page 418)

1. General solution: $\mathbf{x} = c_1 \begin{bmatrix} e^{4t} \\ 2e^{4t} \end{bmatrix} + c_2 \begin{bmatrix} 3e^{-t} \\ e^{-t} \end{bmatrix}$. Particular solution: $\mathbf{x} = \begin{bmatrix} e^{4t} & 3e^{-t} \\ 2e^{4t} & e^{-t} \end{bmatrix} \begin{bmatrix} 1 \\ -1 \end{bmatrix}$.

3. General solution: $\mathbf{x} = \begin{bmatrix} -3 & e^{2t} & e^{4t} \\ 9 & 3e^{2t} & e^{4t} \\ 5 & e^{2t} & e^{4t} \end{bmatrix} \begin{bmatrix} c_1 \\ c_2 \\ c_3 \end{bmatrix}$. **5.** $\mathbf{x}_1 = \begin{bmatrix} \cos 3t \\ -\sin 3t \end{bmatrix}$, $\mathbf{x}_2 = \begin{bmatrix} \sin 3t \\ \cos 3t \end{bmatrix}$.

7. $\mathbf{x}_1 = \begin{bmatrix} e^{-t} \\ -2e^{-t} \end{bmatrix}$, $\mathbf{x}_2 = \begin{bmatrix} te^{-t} \\ -e^{-t}(1+2t) \end{bmatrix}$.

SECTION 11.5 (page 426)

1. $\mathbf{x} = c_1 e^{-3t} \begin{bmatrix} 7 \\ 1 \end{bmatrix} + c_2 e^{5t} \begin{bmatrix} -1 \\ 1 \end{bmatrix}$. **3.** $\mathbf{x} = e^{-2t} \left\{ c_1 \begin{bmatrix} 3\cos t - \sin t \\ 5\cos t \end{bmatrix} + c_2 \begin{bmatrix} \cos t + 3\sin t \\ 5\sin t \end{bmatrix} \right\}$.

5. $\mathbf{x} = c_1 e^{-2t} \begin{bmatrix} 0 \\ 1 \\ 1 \end{bmatrix} + c_2 e^{2t} \begin{bmatrix} 1 \\ 0 \\ 0 \end{bmatrix} + c_3 e^{3t} \begin{bmatrix} 0 \\ 7 \\ 2 \end{bmatrix}$. **7.** $\mathbf{x} = c_1 e^{5t} \begin{bmatrix} 0 \\ 0 \\ 1 \end{bmatrix} + c_2 \begin{bmatrix} \sin t \\ \cos t \\ 0 \end{bmatrix} + c_3 \begin{bmatrix} -\cos t \\ \sin t \\ 0 \end{bmatrix}$.

9. $\mathbf{x} = c_1 e^{-3t} \begin{bmatrix} -1 \\ 0 \\ 1 \end{bmatrix} + c_2 e^{t} \begin{bmatrix} -1 \\ -2 \\ 1 \end{bmatrix} + c_3 e^{2t} \begin{bmatrix} 2 \\ -10 \\ 3 \end{bmatrix}$.

11. $\mathbf{x} = c_1 e^{3t} \begin{bmatrix} 1 \\ 0 \\ 0 \end{bmatrix} + e^{-2t} \left\{ c_2 \begin{bmatrix} 5\cos t - \sin t \\ -13(\cos t + \sin t) \\ 26\cos t \end{bmatrix} + c_3 \begin{bmatrix} \cos t + 5\sin t \\ 13(\cos t - \sin t) \\ 26\sin t \end{bmatrix} \right\}$.

13. $\mathbf{x} = c_1 \begin{bmatrix} -3 \\ 0 \\ 2 \end{bmatrix} + c_2 \begin{bmatrix} 1 \\ 2 \\ 0 \end{bmatrix} + c_3 e^{4t} \begin{bmatrix} 1 \\ 1 \\ 1 \end{bmatrix}$. **15.** $\mathbf{x} = c_1 \begin{bmatrix} 0 \\ 0 \\ -\sin t \\ \cos t \end{bmatrix} + c_2 \begin{bmatrix} 0 \\ 0 \\ \cos t \\ \sin t \end{bmatrix} + c_3 \begin{bmatrix} \sin t \\ \cos t \\ 0 \\ 0 \end{bmatrix} + c_4 \begin{bmatrix} -\cos t \\ \sin t \\ 0 \\ 0 \end{bmatrix}$.

17. $\mathbf{x} = e^{2t} \begin{bmatrix} 2\cos 3t - 6\sin 3t \\ 2\cos 3t + 4\sin 3t \end{bmatrix}$. **19.** $\mathbf{x} = \begin{bmatrix} \cos 2t + \sin 2t \\ -\sin 2t + \cos 2t \end{bmatrix}$.

SECTION 11.6 (page 438)

1. $\mathbf{x} = e^{2t} \left\{ c_1 \begin{bmatrix} -1 \\ 1 \end{bmatrix} + c_2 \begin{bmatrix} 1-2t \\ 2t \end{bmatrix} \right\}$. **3.** $\mathbf{x} = c_1 e^{t} \begin{bmatrix} 1 \\ 1 \\ 1 \end{bmatrix} + e^{-t} \left\{ c_2 \begin{bmatrix} 1 \\ -1 \\ 1 \end{bmatrix} + c_2 \begin{bmatrix} 1+t \\ -t \\ t-1 \end{bmatrix} \right\}$.

5. $\mathbf{x} = e^{-2t} \left\{ c_1 \begin{bmatrix} 1 \\ 0 \\ 1 \end{bmatrix} + c_2 \begin{bmatrix} 1 \\ 1 \\ 0 \end{bmatrix} + c_3 \begin{bmatrix} 1 \\ t \\ -t \end{bmatrix} \right\}$. **7.** $\mathbf{x} = e^{4t} \left\{ c_1 \begin{bmatrix} 0 \\ 0 \\ 1 \end{bmatrix} + c_2 \begin{bmatrix} 0 \\ 1 \\ t \end{bmatrix} + c_3 \begin{bmatrix} 2 \\ 2t \\ t^2 \end{bmatrix} \right\}$.

9. $\mathbf{x} = e^{4t} \left\{ c_1 \begin{bmatrix} 0 \\ 0 \\ 1 \end{bmatrix} + c_2 \begin{bmatrix} 1 \\ 1 \\ 0 \end{bmatrix} + c_3 \begin{bmatrix} t \\ 1+t \\ 0 \end{bmatrix} \right\}$.

11. $\mathbf{x} = c_1 \begin{bmatrix} 5(2\sin t - \cos t) \\ -5(2\cos t + \sin t) \\ -2(\cos t + 2\sin t) \\ -\sin t \end{bmatrix} + c_2 \begin{bmatrix} -5(2\cos t + \sin t) \\ 5(\cos t - 2\sin t) \\ 2(\cos t - \sin t) \\ \cos t \end{bmatrix} + e^{2t} \left\{ c_3 \begin{bmatrix} 0 \\ 0 \\ 1 \\ 0 \end{bmatrix} + c_4 \begin{bmatrix} 0 \\ 0 \\ t \\ 1 \end{bmatrix} \right\}$.

13. $\mathbf{x} = c_1 \begin{bmatrix} 0 \\ 0 \\ -\sin t \\ \cos t \end{bmatrix} + c_2 \begin{bmatrix} 0 \\ 0 \\ \cos t \\ \sin t \end{bmatrix} + c_3 \begin{bmatrix} 0 \\ \cos t \\ \cos t - t\sin t \\ \sin t + t\cos t \end{bmatrix} + c_4 \begin{bmatrix} \cos t \\ \sin t \\ \sin t + t\cos t \\ t\sin t - \cos t \end{bmatrix}$. **15.** $\mathbf{x} = \begin{bmatrix} 7 - 2t - 9e^{-t} \\ e^{-t} \\ 3 - t - 2e^{-t} \end{bmatrix}$.

SECTION 11.7 (page 443)

1. $\mathbf{x}_p = \begin{bmatrix} e^t(2t+1) \\ e^t(2t-1) \end{bmatrix}$. 3. $\mathbf{x}_p = \begin{bmatrix} 4e^t(2e^{2t}-1) \\ 4e^t(3e^{2t}-2) \end{bmatrix}$. 5. $\mathbf{x}_p = \begin{bmatrix} (12t+1)\sin 2t + (1-4t)\cos 2t \\ (8t-1)\cos 2t - 4t\sin 2t \end{bmatrix}$.

7. $\mathbf{x}_p = \begin{bmatrix} -te^t \\ 9e^{-t} \\ te^t + 6e^{-t} \end{bmatrix}$.

SECTION 11.8 (page 452)

3. $x = 2(c_1\sin t - c_2\cos t - c_3\sin 3t + c_4\cos 3t),\ y = 3c_1\sin t - 3c_2\cos t + c_3\sin 3t - c_4\cos 3t$.
9. $A_1(t) = 120 - 50e^{-t/5} - 10e^{-t/15},\ A_2(t) = 120 + 100e^{-t/5} - 20e^{-t/15}$.

SECTION 11.9 (page 458)

7. $e^{At} = \begin{bmatrix} \frac{1}{2}(e^{4t}+e^{2t}) & \frac{1}{2}(e^{4t}-e^{2t}) \\ \frac{1}{2}(e^{4t}-e^{2t}) & \frac{1}{2}(e^{4t}+e^{2t}) \end{bmatrix}$. 9. $e^{At} = \begin{bmatrix} e^{-t}\cos 3t & e^{-t}\sin 3t \\ -e^{-t}\sin 3t & e^{-t}\cos 3t \end{bmatrix}$.

11. $e^{At} = \begin{bmatrix} 2e^{2t}-e^t & 2(e^t-e^{2t}) & e^t-e^{3t} \\ e^{2t}-e^t & 2e^t-e^{2t} & e^t-e^{3t} \\ 0 & 0 & e^{3t} \end{bmatrix}$. 13. $e^{At} = \begin{bmatrix} 1-3t & 9t \\ -t & 1+3t \end{bmatrix}$. 15. $e^{At} = \begin{bmatrix} 1 & 0 & 0 \\ t & 1 & 0 \\ \frac{1}{2}t^2 & t & 1 \end{bmatrix}$.

SECTION 11.10 (page 467)

3. $e^{At} = e^{2t}\begin{bmatrix} 1 & t \\ 0 & 1 \end{bmatrix}$. 5. $e^{At} = e^t\begin{bmatrix} 1+2t & -t \\ 4t & 1-2t \end{bmatrix}$.

7. $\mathbf{x}_1 = e^{2t}\begin{bmatrix} 1 \\ 0 \\ 0 \end{bmatrix},\ \mathbf{x}_2 = e^{-3t}\begin{bmatrix} 1+4t \\ 2t \end{bmatrix},\ \mathbf{x}_3 = e^{-3t}\begin{bmatrix} 0 \\ -8t \\ 1-4t \end{bmatrix},\ e^{At} = \begin{bmatrix} e^{2t} & 0 & 0 \\ 0 & e^{-3t}(1+4t) & -8te^{-3t} \\ 0 & 2te^{-3t} & e^{-3t}(1-4t) \end{bmatrix}$.

9. $\mathbf{x} = c_1 e^{-t}\begin{bmatrix} -7 \\ 4 \\ 1 \end{bmatrix} + e^{3t}\left\{ c_2\begin{bmatrix} 1 \\ 0 \\ 1 \end{bmatrix} + c_3\begin{bmatrix} t \\ 1 \\ t \end{bmatrix} \right\}$.

11. $\mathbf{x}_1 = \begin{bmatrix} -\sin t \\ \cos t \\ -t\sin t \\ t\cos t \end{bmatrix},\ \mathbf{x}_2 = \begin{bmatrix} \cos t \\ \sin t \\ t\cos t \\ t\sin t \end{bmatrix},\ \mathbf{x}_3 = \begin{bmatrix} 0 \\ 0 \\ -\sin t \\ \cos t \end{bmatrix},\ \mathbf{x}_4 = \begin{bmatrix} 0 \\ 0 \\ \cos t \\ \sin t \end{bmatrix}$.

SECTION 12.1 (page 474)

1. $\dfrac{1}{s-2}$. 3. $\dfrac{b}{s^2+b^2}$. 5. $\dfrac{s}{s^2-b^2}$. 7. $\dfrac{2}{s^2}$. 9. $\dfrac{1}{s}(1-2e^{-2s})$. 11. $\dfrac{1}{(s-1)^2+1}$. 13. $\dfrac{2}{s^2}-\dfrac{1}{s-3}$.

15. $\dfrac{s}{s^2-b^2}$. 17. $\dfrac{6}{s^3}-\dfrac{5s}{s^2+4}+\dfrac{3}{s^2+9}$. 19. $\dfrac{2}{s+3}+\dfrac{4}{s-1}-\dfrac{5}{s^2+1}$. 21. $\dfrac{4(s^2+2b^2)}{s(s^2+4b^2)}$.

23. Piecewise continuous. **25.** Not piecewise continuous. **27.** Piecewise continuous. **29.** Not piecewise continuous. **31.** $\dfrac{1}{s^2}[1 - e^{-s}(s+1)]$. **33.** $\dfrac{1}{s^2}e^{-2s}[e^s(s+1) - (2s+1)]$.

SECTION 12.2 (page 479)

7. 2. **9.** $5e^{-3t}$. **11.** $2\cos 3t$. **13.** $\cos t + 6\sin t$. **15.** $2 - 3e^{-t}$. **17.** $1 - e^{-t}$.

19. $\frac{1}{5}(7e^{2t} - 7\cos t - 4\sin t)$. **21.** $\frac{1}{6}(4\cos t + 6\sin t - 4\cos 2t - 3\sin 2t)$.

SECTION 12.3 (page 483)

1. $F(s) = \dfrac{1}{s^2(1 - e^{-s})}[1 - e^{-s}(s+1)]$. **3.** $F(s) = \dfrac{1 + e^{-\pi s}}{(1 - e^{-\pi s})(s^2 + 1)}$. **5.** $F(s) = \dfrac{e^{1-s} - 1}{(1 - s)(1 - e^{-s})}$.

7. $F(s) = \dfrac{1}{1 - e^{-\pi s}}\left[\dfrac{2 - e^{-\pi s/2}(\pi s + 2)}{\pi s^2} + \dfrac{se^{-\pi s/2} + e^{-\pi s}}{s^2 + 1}\right]$. **9.** $F(s) = \dfrac{1}{as^2}\tanh\dfrac{as}{2}$.

SECTION 12.4 (page 488)

1. $y = 2e^{3t}$. **3.** $y = 2t - 1 + 2e^{-2t}$. **5.** $y = e^t - \sin 2t - 2\cos 2t$. **7.** $y = 2e^t - e^{-2t}$.
9. $y = 2 + 3e^{2t} - 5e^t$. **11.** $y = 2e^t + 3e^{-2t} - 5e^{-t}$. **13.** $y = 3e^{2t} + 2e^{-3t} - 4$.

15. $y = 2\cos 2t + \sin 2t + 2e^{-t}$. **17.** $y = \frac{7}{2}e^t - \frac{1}{2}e^{-t} - 3\cos t$.

19. $y = e^t - 2e^{-t} - 4\sin t + 3\cos t$. **21.** $y = 2e^{-t} - e^{-4t} - 2\cos 2t$.

23. $y = \frac{1}{5}(7e^{2t} - 5e^t + 3\cos t - 4\sin t)$. **25.** $y = 2(\cos t + \sin t - \cos 2t)$.

27. $y = y(0)\cosh t + y'(0)\sinh t$. **29.** $i(t) = \dfrac{E_0}{R}(1 - e^{-(R/L)t})$.

31. $x_1 = 2(e^{-3t} - e^{-2t})$, $x_2 = 2e^{-2t} - e^{-3t}$.

SECTION 12.5 (page 492)

1. $(t - 1)$. **3.** $t(t + 2)$. **5.** $-e^{2(t-\pi)}\cos t$. **7.** $\frac{1}{2}e^{-(t-\pi/6)}(\sin 2t - \sqrt{3}\cos 2t)$. **9.** $\dfrac{t - 1}{t^2 - 6t + 10}$.

11. t^2. **13.** te^{3t}. **15.** $(t + 3)e^{-t}$. **17.** $\dfrac{s - 3}{(s - 3)^2 + 16}$. **19.** $\dfrac{1}{(s - 2)^2}$. **21.** $\dfrac{6}{(s + 4)^4}$.

23. $\dfrac{2(2s^3 - 9s^2 + 10s + 30)}{[(s - 3)^2 + 1][(s + 1)^2 + 9]}$. **25.** $\dfrac{2}{(s - 1)^3} - \dfrac{6}{s^3}$. **27.** te^{3t}. **29.** $\dfrac{2}{\sqrt{\pi}}e^{-3t}t^{-1/2}$. **31.** $e^{-2t}\cos 3t$.

33. $\frac{5}{4}e^{2t}\sin 4t$. **35.** $\frac{1}{5}e^{-t}(5\cos 5t - 3\sin 5t)$. **37.** $\frac{1}{2}e^{-t}(2\cos 2t - \sin 2t)$. **39.** $1 - e^{-2t}(1 + 2t)$.

41. $\frac{1}{10}[6 + e^t(13\sin 2t - 6\cos 2t)]$. **43.** $y = e^{2t}(1 + 3t) + e^{-2t}$. **45.** $y = 2e^t + e^{-2t}(1 - t)$.

47. $y = e^{-t}(2 + 3t + t^2)$. **49.** $y = \frac{7}{3}e^{-t} - \frac{7}{4}e^{-2t} + \frac{1}{12}e^{2t}(12t - 7)$.

51. $y = 2e^t + e^{-t} - e^t(\sin 2t + \cos 2t)$. **53.** $x_1 = e^{2t}\cos t$, $x_2 = e^{2t}\sin t$.

SECTION 12.6 (page 496)

7. $f(t) = 3 - 4u_1(t)$. 9. $f(t) = 2 - u_2(t) - 2u_4(t)$. 11. $f(t) = t + 2(3 - t)u_3(t) - (6 - t)u_6(t)$.
13. $f(t) = 1 + (\sin t - 1)u_{\pi/2}(t) - (\sin t + 1)u_{3\pi/2}(t)$.

SECTION 12.7 (page 503)

1. $F(s) = \dfrac{1}{s^2} e^{-s}$. 3. $F(s) = \dfrac{e^{-\pi s/4}}{s^2 + 1}$. 5. $F(s) = \dfrac{2}{s^3} e^{-2s}$. 7. $F(s) = \left(\dfrac{s^2 + s + 2}{s^3}\right) e^{-2s}$.

9. $F(s) = \dfrac{3}{(s+2)^2 + 9} e^{-s}$. 11. $f(t) = (t - 2)u_2(t)$. 13. $f(t) = e^{-4(t-3)}u_3(t)$.

15. $f(t) = u_3(t)\sin(t - 3)$. 17. $f(t) = \frac{1}{5}u_1(t)[e^{4(t-1)} - e^{-(t-1)}]$.

19. $f(t) = [\cos 3(t - 1) + 2 \sin 3(t - 1)]u_1(t)$. 21. $f(t) = \frac{1}{2} u_2(t)(t - 2)^2 e^{3(t-2)}$.

23. $f(t) = e^{-2(t-1)}[2 \cos(t - 1) - 5 \sin(t - 1)]u_1(t)$.

25. $f(t) = [2e^{-(t-3)}(5t - 13) - 4 \cos 2(t - 3) - 3 \sin 2t - 3]u_3(t)$.

27. $y = 2e^{2t} - u_2(t)[e^{t-2} - e^{2(t-2)}]$. 29. $y = 3e^{-2t} + \frac{1}{4}u_\pi(t)[e^{-2(t-\pi)} - \cos 2t + \sin 2t]$.

31. $y(t) = \frac{1}{10} \{21e^{3t} - \cos t - 3 \sin t + \frac{1}{3} u_{\pi/2}(t)[e^{3(t-\pi/2)} - 10 + 9 \sin t + 3 \cos t]\}$.

33. $y = 2 \cosh t + u_1(t)[\cosh(t - 1) - 1]$.

35. $y = 2 \sinh 2t + \frac{1}{4}u_1(t)[\cosh 2(t - 1) - 1] - \frac{1}{4}u_2(t)[\cosh 2(t - 2) - 1]$.

37. $y = 2e^{-t} - e^{-2t} + u_{\pi/4}(t)[5e^{-(t-\pi/4)} - 2e^{-2(t-\pi/4)} - 3 \cos(t - \pi/4) + \sin(t - \pi/4)]$.

39. $y = e^{-2t}(2 \cos t + 5 \sin t) + u_3(t)\{1 - e^{-2(t-3)}[\cos(t - 3) - 2 \sin(t - 3)]\}$.

41. $y = 2e^{-t} + u_1(t)[e^{-(t-1)} + t - 2] - 2u_2(t)[e^{-(t-2)} + t - 3] + u_3(t)[e^{-(t-3)} + t - 4]$.

43. $y = 3e^t - 1 - t - \frac{1}{2}u_1(t)[3e^{(t-1)} + e^{-(t-1)} - 2 - 2t]$. 45. $y = 3e^t - 2 + 3u_1(t)(1 - e^{t-1})$.

47. $i(t) = \dfrac{20}{R} \left\{ e^{-at} + u_{10}(t) \left[\dfrac{1}{a - 1} (ae^{-a(t-10)} - e^{-(t-10)}) - e^{-at} \right] \right\}$ where $a = 1/RC$.

SECTION 12.8 (page 509)

1. $y = e^{2t} + u_2(t)e^{2(t-2)}$. 3. $y = \frac{1}{3}(e^{5t} - e^{-t}) + u_3(t)e^{5(t-3)}$.

5. $y = \frac{1}{2}[\sinh 2t + u_3(t)\sinh 2(t - 3)]$.

7. $y = e^{2t}(3 \cos 3t - 2 \sin 3t) - \dfrac{\sqrt{2}}{6} e^{2(t-\pi/4)}(\sin 3t + \cos 3t)u_{\pi/4}(t)$.

9. $y = 5e^{-3t}(\cos 2t + 2 \sin 2t) - \frac{1}{2} e^{-3(t-\pi/4)} u_{\pi/4}(t)\cos 2t$.

11. $y = \frac{4}{7} (\cos 3t - \cos 4t) + \frac{1}{4} u_{\pi/3}(t)\sin 4(t - \pi/3)$.

13. $x = \dfrac{F_0}{5}(\cos 2t - \cos 3t) - \dfrac{5}{2}u_1(t)\sin 2(t-1).$

15. $x = \dfrac{F_0}{\omega_0(\omega^2 - \omega_0^2)}\,(\omega \sin \omega_0 t - \omega_0 \sin \omega t) + \dfrac{A}{\omega_0}u_{t_0}(t)\sin \omega_0(t - t_0).$

SECTION 12.9 (page 514)

1. $f * g = \frac{1}{2}t^2.$ **3.** $f * g = e^t - 1 - t.$ **5.** $f * g = e^t(1 - \cos t).$ **9.** $L[f * g] = \dfrac{1}{s^2(s^2 + 1)}.$

11. $L[f * g] = \dfrac{s}{(s^2 + 1)(s^2 + 4)}.$ **13.** $L[f * g] = \dfrac{4}{s^3[(s - 3)^2 + 4]}.$ **15.** $L^{-1}[FG] = 1 - e^{-t}.$ **17.** $L^{-1}[FG] =$

$\dfrac{1}{3}e^{-2t}\sin 3t.$ **19.** $L^{-1}[FG] = \begin{cases} 0, & \text{if } t < \pi, \\ t - \pi + \sin t, & \text{if } t \geq \pi. \end{cases}$ **21.** $L^{-1}[FG] = \displaystyle\int_0^t e^{-(t+2\tau)}\tau \cos(t - \tau)\, d\tau.$

23. $L^{-1}[FG] = \begin{cases} 0, & 0 \leq t < \dfrac{\pi}{2}, \\ \displaystyle\int_{\pi/2}^t e^{-4(t-\tau)}\cos[3(t - \tau)]\cos 4\tau\, d\tau, & t \geq \dfrac{\pi}{2}. \end{cases}$ **25.** $y = \displaystyle\int_0^t e^{-\tau}\sin(t - \tau)\, d\tau + \sin t.$

27. $y = \dfrac{1}{4}\displaystyle\int_0^t f(t - \tau)\sin 4\tau\, d\tau + \dfrac{1}{4}(4\alpha \cos 4t + \beta \sin 4t).$

29. $y = \dfrac{1}{a}\displaystyle\int_0^t f(t - \tau)\sinh a\tau\, d\tau + \dfrac{(\alpha a + \beta)}{2a}e^{at} + \dfrac{(\alpha a - \beta)}{2a}e^{-at}.$

31. $y = \dfrac{1}{b}\displaystyle\int_0^t f(t - \tau)e^{a\tau}\sin b\tau\, d\tau + e^{at}\left[\alpha \cos bt - \dfrac{(2\alpha a - \beta)}{b}\sin bt\right].$ **33.** $x = e^{3t} + e^t.$

35. $x = 2 \cosh t - 1.$ **37.** $x = 2 + \dfrac{4}{\sqrt{3}}\sin \sqrt{3}t.$

SECTION 13.2 (page 524)

1. $R = 4.$ **3.** $R = \frac{1}{2}.$ **5.** $R = \infty.$ **7.** $R = 1.$ **9.** $R = \sqrt{5}.$ **11. (a)** Analytic for $x \neq \pm 1.$

(b) $R = \begin{cases} 1 - |x_0|, & \text{if } |x_0| < 1, \\ |x_0| - 1, & \text{if } |x_0| > 1. \end{cases}$ **13.** $f(x) = a_0(1 + \frac{2}{3}x + \frac{1}{12}x^2).$

SECTION 13.3 (page 534)

1. $y_1 = \displaystyle\sum_{n=0}^{\infty} \dfrac{1}{(2n)!}x^{2n},\ y_2 = \displaystyle\sum_{n=0}^{\infty}\dfrac{1}{(2n + 1)!}x^{2n+1},\ R = \infty.$

3. $y_1 = 1 + \displaystyle\sum_{n=1}^{\infty}\dfrac{(-2)^n}{1\cdot 3\cdots\cdot(2n - 1)}x^{2n},\ y_2 = \displaystyle\sum_{n=0}^{\infty}\dfrac{(-1)^n}{n!}x^{2n+1},\ R = \infty.$

5. $y_1 = \displaystyle\sum_{n=0}^{\infty}\dfrac{1}{3^n n!}x^{3n},\ y_2 = \displaystyle\sum_{n=0}^{\infty}\dfrac{1}{1\cdot 4\cdots\cdot(3n + 1)}x^{3n+1},\ R = \infty.$

7. $y_1 = \displaystyle\sum_{n=0}^{\infty}(-1)^n\dfrac{1\cdot 3\cdots\cdot(2n + 1)}{(2n)!}x^{2n},\ y_2 = \displaystyle\sum_{n=0}^{\infty}(-2)^n\dfrac{(n + 1)!}{(2n + 1)!}x^{2n+1},\ R = \infty.$

9. $y_1 = \sum\limits_{n=0}^{\infty} (-1)^n x^{2n}$, $y_2 = \sum\limits_{n=0}^{\infty} (-1)^n x^{2n+1}$, $R = 1$. **11.** $y_1 = x(1 + x^2)$, $y_2 = 1 + 6x^2 + x^4$, $R = \infty$.

13. $y_1 = 1 - x^2 - \frac{1}{6} x^3 + \frac{1}{3} x^4 + \frac{11}{120} x^5 + \cdots$, $y_2 = 1 - \frac{1}{2} x^3 - \frac{1}{12} x^4 + \frac{1}{8} x^5 + \cdots$, $R = \infty$.

15. $y_1 = 1 + \frac{1}{2} x^2 + \frac{1}{6} x^3 + \frac{1}{12} x^4 + \frac{1}{24} x^5 + \cdots$, $y_2 = x + \frac{1}{6} x^3 + \frac{1}{12} x^4 + \frac{1}{30} x^5 + \cdots$, $R = \infty$.

17. (a) No. **(b)** $y_1 = 1 + \frac{1}{2}(x-1)^2 + \frac{1}{8}(x-1)^4 + \cdots$, $y_2 = (x-1) + \frac{1}{3}(x-1)^3 - \frac{1}{12}(x-1)^4 + \cdots$,

R is at least 1. **19. (a)** $y = \sum\limits_{n=0}^{\infty} (-1)^n \dfrac{(n+1)!}{2^n (2n)!} x^{2n}$. **(b)** Polynomial approximation: $y_8 = 1 - \frac{1}{2} x^2 +$

$\frac{1}{16} x^4 - \frac{1}{240} x^6 + \frac{1}{5376} x^8$, error is less than 6.3×10^{-6} on $[-1, 1]$. **21.** $y = a_0(1 + 2x^2 + \frac{1}{3} x^4 + \cdots) +$

$a_1(x + \frac{1}{2} x^3 + \frac{1}{40} x^5 + \cdots) + (3x^2 + x^3 + \frac{13}{24} x^4 + \frac{1}{10} x^5 + \frac{1}{120} x^6 + \cdots)$.

SECTION 13.4 (page 541)

1. $\alpha = 3$: $y_2 = a_1 x(1 - \frac{5}{3} x^2)$; $\alpha = 4$: $y_1 = a_0(1 - 10x^2 + \frac{35}{3} x^4)$. $P_3(x)$ and $P_4(x)$ are given in Table 13.4.1.

5. $2x^3 + x^2 + 5 = \frac{16}{3} P_0 + \frac{6}{5} P_1 + \frac{2}{3} P_2 + \frac{4}{5} P_3$.

9. $\alpha = 0$: $y_1 = 1$; $\alpha = 1$: $y_2 = x$; $\alpha = 2$: $y_1 = 1 - 2x^2$; $\alpha = 3$: $y_2 = x(1 - \frac{2}{3} x^3)$.

SECTION 13.5 (page 551)

1. $x = 1$ is a regular singular point. All other points are ordinary points.
3. $x = 0, 2$ are regular singular points. All other points are ordinary points. **5.** $r = \pm \sqrt{7}$.

7. $r = 0, 1$. **9.** $y_1 = x^{1/4}\left\{ 1 + \sum\limits_{n=1}^{\infty} \dfrac{(-1)^n}{n![5 \cdot 9 \cdot \cdots \cdot (4n+1)]} x^n \right\}$, $y_2 = 1 + \sum\limits_{n=1}^{\infty} \dfrac{(-1)^n}{n![3 \cdot 7 \cdot \cdots \cdot (4n-1)]} x^n$.

11. $y_1 = x^{\sqrt{2}}\left[1 + \sum\limits_{n=1}^{\infty} \dfrac{1}{n!(1 + 2\sqrt{2})(2 + 2\sqrt{2}) \cdots (n + 2\sqrt{2})} x^n \right]$,

$y_2 = x^{-\sqrt{2}}\left[1 + \sum\limits_{n=1}^{\infty} \dfrac{1}{n!(1 - 2\sqrt{2})(2 - 2\sqrt{2}) \cdots \cdot (n - 2\sqrt{2})} x^n \right]$.

13. $y_1 = x^3\left[1 + \sum\limits_{n=1}^{\infty} \dfrac{(n+1)(n+2)}{10 \cdot 13 \cdot \cdots \cdot (3n+7)} x^n \right]$, $y_2 = x^{2/3}\left[1 + \sum\limits_{n=1}^{\infty} \dfrac{(3n-4)(3n-1)}{n! 3^n} x^n \right]$.

15. $y_1 = x^{\sqrt{5}}\left[1 + \sum\limits_{n=1}^{\infty} \dfrac{(1 + \sqrt{5})(2 + \sqrt{5}) \cdot \cdots \cdot (n + \sqrt{5})}{n!(1 + 2\sqrt{5})(2 + 2\sqrt{5}) \cdot \cdots \cdot (n + 2\sqrt{5})} x^n \right]$,

$y_2 = x^{-\sqrt{5}}\left[1 + \sum\limits_{n=1}^{\infty} \dfrac{(1 - \sqrt{5})(2 - \sqrt{5}) \cdot \cdots \cdot (n - \sqrt{5})}{n!(1 - 2\sqrt{5})(2 - 2\sqrt{5}) \cdot \cdots \cdot (n - 2\sqrt{5})} x^n \right]$.

17. $y_1 = (1 + \frac{1}{5} x - \frac{3}{100} x^2 + \cdots)\cos(\ln x) + \frac{x}{25}(10 + x + \cdots)\sin(\ln x)$,

$y_2 = \left(1 + \frac{1}{5} x - \frac{3}{100} x^2 + \cdots\right)\sin(\ln x) - \frac{x}{25}(10 + x + \cdots)\cos(\ln x)$.

19. $y_1 = \sqrt{x}$, $y_2 = \sqrt{x}\left(\ln x + \sum\limits_{n=1}^{\infty} \dfrac{1}{n \cdot n!} x^n \right)$.

21. $N = 0$: $y = 1$; $N = 1$: $y = x^{-1}(1 - x)$; $N = 2$: $y = x^{-2}(1 - 2x + \frac{1}{2}x^2)$; $N = 3$: $y = x^{-3}(1 - 3x + \frac{3}{2}x^2 - \frac{1}{6}x^3)$.

SECTION 13.6 (page 565)

1. $y_1 = x^{1/2} \sum_{n=0}^{\infty} a_n x^n$, $y_2 = y_1 \ln x + x^{1/2} \sum_{n=1}^{\infty} b_n x^n$. **3.** $y_1 = x^2 \sum_{n=0}^{\infty} a_n x^n$, $y_2 = x^{-1} \sum_{n=0}^{\infty} b_n x^n$.

5. $y_1 = x \sum_{n=0}^{\infty} a_n x^n$, $y_2 = A y_1 \ln x + x^{-1} \sum_{n=0}^{\infty} b_n x^n$, $A \neq 0$.

7. $y_1 = x^{-3} \sum_{n=0}^{\infty} a_n x^n$, $y_2 = y_1 \ln x + x^{-3} \sum_{n=1}^{\infty} b_n x^n$.

9. (b) $y_1 = x^{-1} \sum_{n=0}^{\infty} a_n x^n$, $y_2 = y_1 \ln x + x^{-1} \sum_{n=1}^{\infty} b_n x^n$.

(c) $y_1 = x^N \sum_{n=0}^{\infty} a_n x^n$, $y_2 = A y_1 \ln x + x^{-1} \sum_{n=0}^{\infty} b_n x^n$, where $A = 0$ if and only if $\gamma = 1, 0, \ldots, (1 - N)$.

11. $y_1 = x^{-1} \sum_{n=0}^{\infty} \frac{1}{(n!)^2} x^n$, $y_2 = y_1 \ln x - (2 + \frac{3}{4}x + \frac{11}{108}x^2 + \cdots)$.

13. $y_1 = x \sum_{n=0}^{\infty} \frac{1}{(n + 2)!} x^n = x^{-1}(e^x - x - 1)$, $y_2 = x^{-1}(1 + x)$.

15. $y_1 = \sum_{n=0}^{\infty} \frac{(2x)^n}{(n!)^2}$, $y_2 = y_1 \ln x - (4x + 3x^2 + \frac{22}{27}x^3 + \cdots)$.

17. $y_1 = x^2 \sum_{n=0}^{\infty} \frac{(n + 1)}{n!} x^n$, $y_2 = y_1 \ln x - x^2 (3x + \frac{13}{4}x^2 + \frac{31}{18}x^3 + \cdots)$.

19. $y_1 = x^2 \left(1 + \sum_{n=1}^{\infty} \frac{(n + 4)}{n!} x^n \right)$, $y_2 = 2 y_1 \ln x + x^{-1}(1 - x + \frac{3}{2}x^2 - 12 x^4 + \cdots)$.

21. $y_1 = x^{3/2} \sum_{n=0}^{\infty} \frac{1}{(n + 3)!} x^n = x^{-3/2}(e^x - 1 - x - \frac{1}{2}x^2)$, $y_2 = x^{-3/2}(1 + x + \frac{1}{2}x^2)$.

23. $y_1 = x$, $y_2 = y_1 \ln x + \sum_{n=1}^{\infty} \frac{(-1)^n}{n \cdot n!} x^{n+1}$.

25. $y_1 = x^{3/2} \sum_{n=0}^{\infty} \frac{1}{n!(n + 2)!} x^n$, $y_2 = 2 y_1 \ln x - 5 x^{-1/2}(1 - x + \frac{1}{5}x^3 + \frac{37}{960}x^4 + \cdots)$.

27. $y_1 = x$, $y_2 = x \ln x - 1 + \sum_{n=2}^{\infty} \frac{1}{n!(n - 1)} x^n$. **29.** $y = 1 + \sum_{k=1}^{N} \frac{(-1)^k N(N - 1) \cdots \cdots (N - k + 1)}{(k!)^2} x^k$.

SECTION 13.7 (page 575)

7. $J_{3/2}(x) = \sqrt{\frac{2}{\pi}} x^{-3/2}(\sin x - x \cos x)$, $J_{-3/2}(x) = -\sqrt{\frac{2}{\pi}} x^{-3/2}(x \sin x + \cos x)$.

9. $x^p = \sum_{n=1}^{\infty} \frac{2}{\lambda_n J_{p+1}(\lambda_n)} J_p(\lambda_n x)$.

APPENDIX 1 (page 580)

1. $\bar{z} = 2 - 5i$, $|z| = \sqrt{29}$. **3.** $\bar{z} = 5 + 2i$, $|z| = \sqrt{29}$. **5.** $\bar{z} = 1 - 2i$, $|z| = \sqrt{5}$.

7. $z_1 z_2 = 1 + 7i$, $\dfrac{z_1}{z_2} = -1 + i$. **9.** $z_1 z_2 = 7 + 11i$, $\dfrac{z_1}{z_2} = \dfrac{1}{10}(1 - 13i)$.

APPENDIX 2 (page 588)

1. $\dfrac{5}{x+2} - \dfrac{3}{x+1}$. **3.** $\dfrac{4}{5(x-3)} + \dfrac{1}{5(x+2)}$. **5.** $\dfrac{1}{3(x+1)} + \dfrac{1}{6(x-2)} - \dfrac{1}{2(x+4)}$.

7. $\dfrac{3}{x+1} - \dfrac{1}{(x+1)^2} - \dfrac{3}{(x+2)}$. **9.** $\dfrac{3}{4x} + \dfrac{1}{x^2} - \dfrac{3x+4}{4(x^2+4)}$. **11.** $\dfrac{1}{2(x-2)} + \dfrac{x+2}{2(x^2+16)}$.

13. $\dfrac{1}{x-2} - \dfrac{1}{x+2} + \dfrac{3}{(x+2)^2}$. **15.** $\dfrac{2}{3(x-2)} - \dfrac{2}{3(x+1)} - \dfrac{2}{(x+1)^2} + \dfrac{1}{(x+1)^3}$.

17. $\dfrac{1}{2(x-3)} - \dfrac{x+1}{2(x^2+4x+5)}$.

APPENDIX 3 (page 596)

Note that we have omitted the integration constants.

1. $\cos x + x \sin x$. **3.** $x \ln x - x$. **5.** $\frac{1}{2} e^{x^2}(x^2 - 1)$. **7.** $x - 3 \ln|x + 2|$.

9. $\dfrac{1}{4} \ln|x| - \dfrac{1}{8} \ln(x^2 + 4) + \tan^{-1}\dfrac{x}{2}$. **11.** $\frac{1}{2}x + \frac{7}{4} \ln|2x - 1|$. **13.** $2 \ln|x| - \dfrac{1}{x+1} - 2 \ln|x + 1|$.

15. $\tan^{-1}(x + 1)$. **17.** $-\ln|\cos x|$. **19.** $\frac{1}{4}(2x + \sin 2x)$. **21.** $\frac{1}{13} e^{3x}(3 \sin 2x - 2 \cos 2x)$.

APPENDIX 4 (page 610)

1. $y_1 = 1 + \dfrac{x^2}{2} - \dfrac{x^3}{3}$; $y_2 = 1 + \dfrac{x^2}{2} - \dfrac{x^3}{3} + \dfrac{x^4}{8} - \dfrac{x^5}{15}$.

3. $y(x) = \sum_{k=0}^{\infty} \dfrac{1}{k!} x^{2k} = e^{x^2}$.

Index